THEORY AND APPLICATION OF INFINITE SERIES

BY

DR. KONRAD KNOPP

LATE PROFESSOR OF MATHEMATICS AT THE
UNIVERSITY OF TÜBINGEN

Translated
from the Second German Edition
and revised
in accordance with the Fourth by
Miss R. C. H. Young, Ph.D., L. ès Sc.

DOVER PUBLICATIONS, INC.
New York

Published in Canada by General Publishing Company, Ltd., 30 Lesmill Road, Don Mills, Toronto, Ontario.

Published in the United Kingdom by Constable and Company, Ltd.

This Dover edition, first published in 1990 is an unabridged and unaltered republication of the second English edition published by Blackie & Son, Ltd., in 1951. It corresponds to the fourth German edition of 1947. The work was first published in German in 1921 and in English in 1928. This edition is published by special arrangement with Blackie & Son, Ltd., Bishopbriggs, Glasgow G64 2NZ, Scotland.

Manufactured in the United States of America
Dover Publications, Inc., 31 East 2nd Street, Mineola, N.Y. 11501

Library of Congress Cataloging-in-Publication Data

Knopp, Konrad, 1882–1957.
[Theorie und Anwendung der unendlichen Reihen. English]
Theory and application of infinite series / by Konrad Knopp ; translated from the second German edition and revised in accordance with the fourth by R.C.H. Young.
p. cm.
Translation of: Theorie und Anwendung der unendlichen Reihen.
Reprint. Originally published: London : Blackie, [1951].
Includes bibliographical references.
ISBN 0-486-66165-2
1. Series, Infinite. I. Title.
QA295.K74 1990
515′.243—dc20 89-71388
CIP

From the preface to the first (German) edition.

There is no general agreement as to where an account of the theory of infinite series should begin, what its main outlines should be, or what it should include. On the one hand, the whole of higher analysis may be regarded as a field for the application of this theory, for all limiting processes — including differentiation and integration — are based on the investigation of infinite sequences or of infinite series. On the other hand, in the strictest (and therefore narrowest) sense, the only matters that are in place in a textbook on infinite series are their definition, the manipulation of the symbolism connected with them, and the theory of convergence.

In his "Vorlesungen über Zahlen- und Funktionenlehre", Vol. 1, Part 2, *A. Pringsheim* has treated the subject with these limitations. There was no question of offering anything similar in the present book.

My aim was quite different: namely, to give a comprehensive account of all the investigations of higher analysis in which infinite series are the chief object of interest, the treatment to be as free from assumptions as possible and to start at the very beginning and lead on to the extensive frontiers of present-day research. To set all this forth in as interesting and intelligible a way as possible, but of course without in the least abandoning exactness, with the object of providing the student with a convenient introduction to the subject and of giving him an idea of its rich and fascinating variety — such was my vision.

The material grew in my hands, however, and resisted my efforts to put it into shape. In order to make a convenient and useful book, the field had to be restricted. But I was guided throughout by the experience I have gained in teaching — I have covered the whole of the ground several times in the general course of my work and in lectures at the universities of Berlin and Königsberg — and also by the aim of the book. *It was to give a thorough and reliable treatment which would be of assistance to the student attending lectures and which would at the same time be adapted for private study.*

The latter aim was particularly dear to me, and this accounts for the form in which I have presented the subject-matter. Since it is generally easier — especially for beginners — to prove a deduction in pure mathematics than to recognize the restrictions to which the train of reasoning is subject, I have always dwelt on *theoretical difficulties*, and have tried to remove them by means of repeated illustrations; and although I have thereby deprived myself of a good deal of space for important matter, I hope to win the gratitude of the student.

I considered that an introduction to the theory of real numbers was indispensable as a beginning, in order that the first facts relating to convergence might have a firm foundation. To this introduction I have added a fairly extensive account of the theory of sequences, and, finally, the actual theory of infinite series. The latter is then constructed in two storeys, so to speak: a ground-floor, in which the classical part of the theory (up to about the stage of *Cauchy*'s Analyse algébrique) is expounded, though with the help of very limited resources, and a superstructure, in which I have attempted to give an account of the later developments of the 19th century.

For the reasons mentioned above, I have had to omit many parts of the subject to which I would gladly have given a place for their own sake. Semi-convergent series, *Euler*'s summation formula, a detailed treatment of the Gamma-function, problems arising from the hypergeometric series, the theory of double series, the newer work on power series, and, in particular, a more thorough development of the last chapter, that on divergent series — all these I was reluctantly obliged to set aside. On the other hand, I considered that it was essential to deal with sequences and series of complex terms. As the theory runs almost parallel with that for real variables, however, I have, from the beginning, formulated all the definitions and proved all the theorems concerned in such a way that they remain valid without alteration, whether the "arbitrary" numbers involved are real or complex. These definitions and theorems are further distinguished by the sign °.

In choosing the examples — in this respect, however, I lay no claim to originality; on the contrary, in collecting them I have made extensive use of the literature — I have taken pains to put practical applications in the fore-front and to leave mere playing with theoretical niceties alone. Hence there are e. g. a particularly large number of exercises on Chapter VIII and only very few on Chapter IX. Unfortunately there was no room for solutions or even for hints for the solution of the examples.

A list of the most important papers, comprehensive accounts, and textbooks on infinite series is given at the end of the book, immediately in front of the index.

Königsberg, September 1921.

From the preface to the second (German) edition.

The fact that a second edition was called for after such a remarkably short time could be taken to mean that the first had on the whole been on the right lines. Hence the general plan has not been altered, but it has been improved in the details of expression and demonstration on almost every page.

The last chapter, that dealing with *divergent series*, has been wholly rewritten, with important extensions, so that it now in some measure provides an introduction to the theory and gives an idea of modern work on the subject.

Königsberg, December 1923.

Preface to the third (German) edition.

The main difference between the third and second editions is that it has become possible to add a new chapter on Euler's summation formula and asymptotic expansions, which I had reluctantly omitted from the first two editions. This important chapter had meanwhile appeared in a similar form in the English translation published by *Blackie & Son Limited*, London and Glasgow, in 1928.

In addition, the whole of the book has again been carefully revised, and the proofs have been improved or simplified in accordance with the progress of mathematical knowledge or teaching experience. This applies especially to theorems **269** and **287**.

Dr. *W. Schöbe* and Herr *P. Securius* have given me valuable assistance in correcting the proofs, for which I thank them heartily.

Tübingen, March 1931.

Preface to the fourth (German) edition.

In view of present difficulties no large changes have been made for the fourth edition, but the book has again been revised and numerous details have been improved, discrepancies removed, and several proofs simplified. The references to the literature have been brought up to date.

Tübingen, July 1947.

Preface to the first English edition.

This translation of the second German edition has been very skilfully prepared by Miss *R. C. H. Young*, L. ès Sc. (Lausanne), Research Student, Girton College, Cambridge. The publishers, Messrs. *Blackie and Son, Ltd.*, Glasgow, have carefully superintended the printing.

In addition, the publishers were kind enough to ask me to add a chapter on *Euler's summation formula and asymptotic expansions*. I agreed to do so all the more gladly because, as I mentioned in the original preface, it was only with great reluctance that I omitted this part of the subject in the German edition. This chapter has been translated by Miss *W. M. Deans*, B.Sc. (Aberdeen), M.A. (Cantab.), with equal skill.

I wish to take this opportunity of thanking the translators and the publishers for the trouble and care they have taken. If — as I hope — my book meets with a favourable reception and is found useful by English-speaking students of Mathematics, the credit will largely be theirs.

Tübingen, February 1928.

Konrad Knopp.

Preface to the second English edition.

The second English edition has been produced to correspond to the fourth German edition (1947).

Although most of the changes are individually small, they have nonetheless involved a considerable number of alterations, about half of the work having been re-set.

The translation has been carried out by Dr. *R. C. H. Young* who was responsible for the original work.

Contents.

Part I.

Real numbers and sequences.

Chapter I.

Principles of the theory of real numbers.

Chapter II.

Sequences of real numbers.

Part II.

Foundations of the theory of infinite series.

Chapter III.

Series of positive terms.

Chapter IV.

Page

Series of arbitrary terms.

Chapter V.

Power series.

Chapter VI.

The expansions of the so-called elementary functions.

Chapter VII.

Infinite products.

Chapter VIII.

Closed and numerical expressions for the sums of series.

Page.

Part III.

Development of the theory.

Chapter IX.

Series of positive terms.

Chapter X.

Series of arbitrary terms.

Chapter XI.

Series of variable terms (Sequences of functions).

Chapter XII.

Series of complex terms.

Chapter XIII.

Divergent series.

Chapter XIV.

Euler's summation formula and asymptotic expansions.

Introduction.

The foundation on which the structure of higher analysis rests is the *theory of real numbers*. Any strict treatment of the foundations of the differential and integral calculus and of related subjects must inevitably start from there; and the same is true even for e. g. the calculation of roots and logarithms. The theory of real numbers first creates the material on which Arithmetic and Analysis can subsequently build, and with which they deal almost exclusively.

The necessity for this has not always been realized. The great creators of the infinitesimal calculus — *Leibniz* and *Newton* [1] — and the no less famous men who developed it, of whom *Euler* [2] is the chief, were too intoxicated by the mighty stream of learning springing from the newly-discovered sources to feel obliged to criticize fundamentals. To them the results of the new methods were sufficient evidence for the security of their foundations. It was only when the stream began to ebb that critical analysis ventured to examine the fundamental conceptions. About the end of the 18th century such efforts became stronger and stronger, chiefly owing to the powerful influence of *Gauss* [3]. Nearly a century had to pass, however, before the most essential matters could be considered thoroughly cleared up.

Nowadays rigour in connection with the underlying number concept is the most important requirement in the treatment of any mathematical subject. Ever since the later decades of the past century the last word on the matter has been uttered, so to speak, — by *Weierstrass* [4] in the sixties, and by *Cantor* [5] and *Dedekind* [6] in 1872. No lecture or treatise

---

[1] *Gottfried Wilhelm Leibniz*, born in Leipzig in 1646, died in Hanover in 1716. *Isaac Newton*, born at Woolsthorpe in 1642, died in London in 1727. Each discovered the foundations of the infinitesimal calculus independently of the other.

[2] *Leonhard Euler*, born in Basle in 1707, died in St. Petersburg in 1783.

[3] *Karl Friedrich Gauss*, born at Brunswick in 1777, died at Göttingen in 1855.

[4] *Karl Weierstrass*, born at Ostenfelde in 1815, died in Berlin in 1897. The first rigorous account of the theory of real numbers which *Weierstrass* had expounded in his lectures since 1860 was given by *G. Mittag-Leffler*, one of his pupils, in his essay: Die Zahl, Einleitung zur Theorie der analytischen Funktionen, The Tôhoku Mathematical Journal, Vol. 17, pp. 157—209. 1920.

[5] *Georg Cantor*, born in St. Petersburg in 1845, died at Halle in 1918: cf. Mathem. Annalen, Vol. 5, p. 123. 1872.

[6] *Richard Dedekind*, born at Brunswick in 1831, died there in 1916: cf. his book: Stetigkeit und irrationale Zahlen, Brunswick 1872.

dealing with the fundamental parts of higher analysis can claim validity unless it takes the refined concept of the real number as its starting-point.

Hence the theory of real numbers has been stated so often and in so many different ways since that time that it might seem superfluous to give another very detailed exposition [7]: for in this book (at least in the later chapters) we wish to address ourselves only to those already acquainted with the elements of the differential and integral calculus. Yet it would scarcely suffice merely to point to accounts given elsewhere. For a theory of infinite series, as will be sufficiently clear from later developments, would be up in the clouds throughout, if it were not firmly based on the system of real numbers, the only possible foundation. On account of this, and in order to leave not the slightest uncertainty as to the hypotheses on which we shall build, we shall discuss in the following pages those ideas and data from the theory of real numbers which we shall need further on. We have no intention, however, of constructing a statement of the theory compressed into smaller space but otherwise complete. We merely wish to make the main ideas, the most important questions, and the answers to them, as clear and prominent as possible. So far as the latter are concerned, our treatment throughout will certainly be detailed and without omissions; it is only in the cases of details of subsidiary importance, and of questions as to the completeness and uniqueness of the system of real numbers which lie outside the plan of this book, that we shall content ourselves with shorter indications.

---

[7] An account which is easy to follow and which includes all the essentials is given by *H. v. Mangoldt*, Einführung in die höhere Mathematik, Vol. I, 8th edition (by K. Knopp), Leipzig 1944. — The treatment of *G. Kowalewski*, Grundzüge der Differential- und Integralrechnung, 6th edition, Leipzig 1929, is accurate and concise. — A rigorous construction of the system of real numbers, which goes into the minutest details, is to be found in *A. Loewy*, Lehrbuch der Algebra, Part I, Leipzig 1915, in *A. Pringsheim*, Vorlesungen über Zahlen- und Funktionenlehre, Vol. I, Part I, 2nd edition, Leipzig 1923 (cf. also the review of the latter work by *H. Hahn*, Gött. gel. Anzeigen 1919, pp. 321—47), and in a book by *E. Landau* exclusively devoted to this purpose, Grundlagen der Analysis (Das Rechnen mit ganzen, rationalen, irrationalen, komplexen Zahlen), Leipzig 1930. A critical account of the whole problem is to be found in the article by *F. Bachmann*, Aufbau des Zahlensystems, in the Enzyklopädie d. math. Wissensch., Vol. I, 2nd edition, Part I, article 3, Leipzig and Berlin 1938.

Part I.

Real numbers and sequences.

Chapter I.

Principles of the theory of real numbers.

§ 1. The system of rational numbers and its gaps.

What do we mean by saying that a particular number is "known" or "given" or may be "calculated"? What does one mean by saying that he *knows* the value of $\sqrt{2}$ or π, or that he *can calculate* $\sqrt{5}$? A question like this is easier to ask than to answer. Were I to say that $\sqrt{2} = 1{\cdot}414$, I should obviously be wrong, since, on multiplying out, $1{\cdot}414 \times 1{\cdot}414$ does *not* give 2. If I assert, with greater caution, that $\sqrt{2} = 1{\cdot}4142135$ *and so on*, even that is no tenable answer, and indeed in the first instance it is entirely meaningless. The question is, after all, *how* we are to go on, and this, without further indication, we cannot tell. Nor is the position improved by carrying the decimal further, even to hundreds of places. In this sense it may well be said that no one has ever beheld the whole of $\sqrt{2}$, — not held it completely in his own hands, so to speak—whilst a statement that $\sqrt{9} = 3$ or that $35 \div 7 = 5$ has a finished and thoroughly satisfactory appearance. The position is no better as regards the number π, or a logarithm or sine or cosine from the tables. Yet we feel certain that $\sqrt{2}$ and π and log 5 really *do have* quite definite values, and even that we actually know these values. But a clear notion of what these impressions exactly amount to or imply we do not as yet possess. Let us endeavour to form such an idea.

Having raised doubts as to the justification for such statements as "I know $\sqrt{2}$", we must, to be consistent, proceed to examine how far one is justified even in asserting that he *knows* the number $-\frac{23}{7}$ or is *given* (for some specific calculation) the number $\frac{9}{4}$. Nay more, the significance of such statements as "I know the number 97" or "for such and such a calculation I am *given* $a = 2$ and $b = 5$" would require scrutiny. We should have to enquire into the whole significance or *concept of the natural numbers* 1, 2, 3, . . .

This last question, however, strikes us at once as distinctly transgressing the bounds of Mathematics and as belonging to an order of ideas quite apart from that which we propose to develop here.

No science rests entirely within itself: each borrows the strength of its ultimate foundations from strata above or below it, such as experience, or theory of knowledge, or logic, or metaphysics, . . . Every science must accept *something* as simply given, and on that it may proceed to build. In this sense neither mathematics nor any other science starts without assumptions. The only question which has to be settled by a criticism of the foundation and logical structure of any science is what shall be assumed as in this sense "given"; or better, what minimum of initial assumptions will suffice, to serve as a basis for the subsequent development of all the rest.

For the problem we are dealing with, that of constructing the system of real numbers, these preliminary investigations are tedious and troublesome, and have actually, it must be confessed, not yet reached any entirely satisfactory conclusion at all. A discussion adequate to the present position of the subject would consequently take us far beyond the limits of the work we are contemplating. Instead, therefore, of shouldering an obligation to assume as basis only a minimum of hypotheses, we propose to regard at once as known (or "given", or "secured") a group of data whose deducibility from a smaller body of assumptions is familiar to everyone — namely, the *system of rational numbers*, i. e. of numbers integral and fractional, positive and negative, including zero. Speaking broadly, it is a matter of common knowledge how this system may be constructed, if — as a smaller body of assumptions — only the ordered sequence of natural numbers 1, 2, 3, . . . , and their combinations by addition and multiplication, are regarded as "given". For everyone knows — and we merely indicate it in passing — how fractional numbers arise from the need of inverting the process of multiplication, — negative numbers and zero from that of inverting the process of addition [1].

The totality, or aggregate, of numbers thus obtained is *called the system* (or *set*) *of rational numbers*. Each of these can be completely and literally "*given*" or "*written down*" or "*made known*" with the help of at most two natural numbers, a dividing bar and possibly a minus sign. For brevity, we represent them by small italic characters; a, b, . . . , x, y, . . . The following are the essential properties of this system:

---

[1] See the works of *Loewy*, *Pringsheim*, and *Landau* mentioned in the Introduction; also *O. Hölder*, Die Arithmetik in strenger Begründung, 2nd edition, Berlin 1929; and *O. Stolz and J. A. Gmeiner*, Theoretische Arithmetik, 3rd edition, Leipzig 1911.

1. Rational numbers form an *ordered* aggregate; meaning that between any two, say a and b, one and only one of the three relations

$$a < b. \qquad a = b, \qquad a > b$$

necessarily holds [2]; and these relations of "order" between rational numbers are subject to a set of quite simple laws, which we assume known, the only essential ones for our purposes being the

Fundamental Laws of Order. **1.**

1. Invariably [3] $a = a$.
2. $a = b$ always implies $b = a$.
3. $a = b$, $b = c$ implies $a = c$.
4. $a \leqq b$, $b < c$, — or $a < b$, $b \leqq c$, — implies [4] $a < c$.

2. Any two rational numbers may be combined in four distinct ways, referred to respectively as the four processes (or basic operations) of Addition, Subtraction, Multiplication, and Division. These operations can always be carried out to one definite result, with the single exception of division by 0, which is undefined and should be regarded as an entirely impossible or meaningless process; the four processes also obey a number of simple laws, the so-called *Fundamental Laws of Arithmetic*, and further rules deducible therefrom.

These too we shall regard as known, and state, concisely, those **Fundamental Laws** or **Axioms of Arithmetic** from which all the others may **2.** be inferred, by purely formal rules (i. e. by the laws of pure logic).

I. Addition. 1. *Every* pair of numbers a and b has invariably associated with it a third, c, called their *sum* and denoted by $a + b$.

2. $a = a'$, $b = b'$ always imply $a + b = a' + b'$.
3. Invariably, $a + b = b + a$ (Commutative Law).
4. Invariably, $(a + b) + c = a + (b + c)$ (Associative Law).
5. $a < b$ always implies $a + c < b + c$ (Law of Monotony).

II. Subtraction.

To *every* pair of numbers a and b there corresponds a third number c, such that $a + c = b$.

---

[2] $a > b$ and $b < a$ are merely two different expressions of the same relation. Strictly speaking, the one symbol "$<$" would therefore suffice.

[3] With regard to this seemingly trivial "law" cf. footnote 11, p. 9, remark 1, p. 28, and footnote 24, p. 29.

[4] To express that one of the relations of order: $a < b$, $a = b$, or $a > b$, does not hold, we write, respectively, $a \geqq b$ ("greater than or equal to", "at least equal to", "not less than"), $a \neq b$ ("unequal to", "different from") or $a \leqq b$. Each of these statements (negations) definitely excludes one of the three relations and leaves undecided which of the other two holds good.

III. Multiplication.

1. To *every* pair of numbers a and b there corresponds a third number c, called their *product* and denoted by $a\,b$.
2. $a = a'$, $b = b'$ always implies $a\,b = a'\,b'$.
3. In all cases $a\,b = b\,a$ (Commutative Law).
4. In all cases $(a\,b)\,c = a\,(b\,c)$ (Associative Law).
5. In all cases $(a + b)\,c = a\,c + b\,c$ (Distributive Law).
6. $a < b$ implies, *provided c is positive*, $a\,c < b\,c$ (Law of Monotony).

IV. Division.

To *every* pair of numbers a and b of which the first is not 0 there corresponds a third number c, such that $a\,c = b$.

As already remarked, all the known rules of arithmetic, — and hence ultimately all mathematical results, — are deduced from these few laws, with the help of the laws of pure logic alone. Among these laws, one is distinguished by its primarily mathematical character, namely the

V. Law of Induction, which may be reckoned among the fundamental laws of arithmetic and is normally stated as follows:

If a set $\mathfrak{M}$ of natural numbers includes the number 1, and if, every time a certain natural number n and all those less than n can be taken to belong to the aggregate, the number $(n + 1)$ may be inferred also to belong to it, then $\mathfrak{M}$ includes *all* the natural numbers.

This law of induction itself follows quite easily from the following theorem, which appears even more obvious and is therefore normally called the fundamental law of the natural numbers:

Law of the Natural Numbers. In every set of natural numbers that is not "empty" there is always a number less than all the rest.

For if, according to the hypotheses of the Induction Law, we consider the set $\mathfrak{N}$ of natural numbers not belonging to $\mathfrak{M}$, this set $\mathfrak{N}$ must be "empty", that is, $\mathfrak{M}$ must contain all the natural numbers. For otherwise, by the law of the natural numbers, $\mathfrak{N}$ would include a number less than all the rest. This least number would exceed 1, for it was assumed that 1 belongs to $\mathfrak{M}$; hence it could be denoted by $n + 1$. Then n would belong to $\mathfrak{M}$, but $(n + 1)$ would not, which contradicts the hypotheses in the law of induction.[5]

In applications it is usually an advantage to be able to make statements not merely about the natural numbers but about *any* whole numbers.

---

[5] The following rather more general form of the law of induction can be deduced in exactly the same way from the fundamental law of the natural numbers. If set $\mathfrak{M}$ of natural numbers includes the number 1, and if the number $(n + 1)$ can be proved to belong to the aggregate provided the number n does, then $\mathfrak{M}$ contains all the natural numbers.

The laws then take the following forms, obviously equivalent to those above:

Law of Induction. If a statement involves a natural number n (e. g. "if $n \geqq 10$, then $2^n > n^3$", or the like) and if

a) this statement is correct for $n = p$,

and

b) its correctness for $n = p, p + 1, \ldots, k$ (where k is any natural number $\geqq p$) always implies its correctness for $n = k + 1$, then the statement is correct for every natural number $\geqq p$.

Law of Integers. In every set of integers all $\geqq p$ that is not "empty", there is always a number less than all the rest.[6]

We will lastly mention a theorem susceptible, in the domain of rational numbers, of immediate proof, although it becomes axiomatic in character very soon after this domain is left; namely the

VI. Theorem of Eudoxus.

If a and b are any two *positive* rational numbers, then a natural number n always exists [7] such that $n b > a$.

The four ways of combining two rational numbers give in every case as the result another rational number. In this sense the system of rational numbers forms a *closed aggregate* (*natürlicher Rationalitätsbereich* or *number corpus*). This property of forming a closed system with respect to the four rules is obviously not possessed by the aggregate of all natural numbers, or of all positive and negative integers. These are, so to speak, too sparsely sown to meet all the demands which the four rules make upon them.

This closed aggregate of all rational numbers and the laws which hold in it, are then all that we regard as given, known, secured.

As that type of argument which makes use of *inequalities* and *absolute values* may be a little unfamiliar to some, its most important rules may be set down here, briefly and without proof: **3.**

I. Inequalities. Here all follows from the laws of order and monotony. In particular

1. The statements in the laws of monotony are reversible; e. g. $a + c < b + c$ always implies $a < b$; and so does $a c < b c$, *provided* $c > 0$.
2. $a < b$, $c < d$ always implies $a + c < b + d$.
3. $a < b$, $c < d$ implies, provided b and c are positive, $a c < b d$.
4. $a < b$ always implies $-b < -a$,

   and also, provided a is positive, $\frac{1}{b} < \frac{1}{a}$.

---

[6] To reduce these forms of the laws to the previous ones, we need only consider the natural numbers m such that, in the one case, the statement in question is correct for $n = (p - 1) + m$, or, in the other, that $(p - 1) + m$ belongs to the non-"empty" set under consideration.

[7] This theorem is usually, but incorrectly, ascribed to *Archimedes*; it is already to be found in *Euclid*, Elements, Book V, Def. 4.

Also these theorems, as well as the laws of order and monotony, hold (with appropriate modifications) when the signs "$\leqq$", "$>$", "$\geqq$" and "$\neq$" are substituted for "$<$", provided we maintain the assumptions that c, b and a are positive, in 1, 3, and 4 respectively.

II. Absolute values. *Definition*: By $|a|$, the *absolute value* (or *modulus*) of a, is meant that one of the two numbers $+a$ and $-a$ which is positive, supposing $a \neq 0$; and the number 0, if $a = 0$. (Hence $|0| = 0$ and if $a \neq 0$, $|a| > 0$.)
The following theorems hold, amongst others:

1. $|a| = |-a|$. 2. $|ab| = |a| \cdot |b|$.

3. $\left|\dfrac{1}{a}\right| = \dfrac{1}{|a|}$; $\left|\dfrac{b}{a}\right| = \dfrac{|b|}{|a|}$, provided $a \neq 0$.

4. $|a+b| \leqq |a| + |b|$; $|a+b| \geqq |a| - |b|$, and indeed $|a+b| \geqq \big||a| - |b|\big|$.

5. The two relations $|a| < r$ and $-r < a < r$ are exactly equivalent; similarly for $|x - a| < r$ and $a - r < x < a + r$.

6. $|a - b|$ is the *distance* between the *points* a and b, with the representation of numbers on a straight line described immediately below.

Proof of the first relation in 4: $\pm a \leqq |a|$, $\pm b \leqq |b|$, so that by **3**, I, 2, $\pm(a+b) \leqq |a| + |b|$, and hence $|a+b| \leqq |a| + |b|$.

We also assume it to be known how the relations of magnitude between rational numbers may be illustrated graphically by relations of positions between points on a straight line. On a straight line or *number-axis*, any two distinct points are marked, one O, the origin (0) and one U, the unit point (1). The point P which is to represent a number $a = \dfrac{p}{q}$ ($q > 0$, $p \lesseqgtr 0$, both integers) is obtained by marking off on the axis, $|p|$ times in succession, beginning at O, the qth part of the distance OU (immediately constructed by elementary geometry) either in the direction OU, if $p > 0$, or if p is negative, in the opposite direction. This point [8] we call for brevity *the point* a, and the totality of points corresponding in this way to all rational numbers we shall refer to as *the rational points* of the axis. — The straight line is usually thought of as drawn from *left to right* and U chosen to the right of O. In this case, the words *positive* and *negative* obviously become equivalents of the phrases: *to the right of O* and *to the left of O*, respectively; and, more generally, $a < b$ signifies that a lies to the left of b, b to the right of a. This mode of expression may often assist us in illustrating abstract relations between numbers.

---

[8] The position of this point is independent of the particular representation of the number a, i. e. if $a = p'/q'$ is another representation with $q' > 0$ and $p' \gtreqless 0$ both integers, and if the construction is performed with q', p' in place of q, p, the *same* point P is obtained.

This completes the sketch of what we propose to take as the previously secured foundation of our subject. We shall now regard the description of these foundations as characterizing the ***concept of number;*** in other words, we shall call any system of conceptually well-distinguished objects (elements, symbols) a ***number system,*** and its elements ***numbers,*** if — to put it quite briefly for the moment — we can operate with them in essentially the same ways as we do with rational numbers.

We proceed to give this somewhat inaccurate statement a precise formulation.

We consider a *system* S of well-distinguished objects, which we **4.** denote by $\alpha, \beta, \ldots$. S will be called a *number system* and its elements $\alpha, \beta, \ldots$ will be called *numbers* if, besides being capable of definition exclusively by means of rational numbers (i. e. ultimately by means of natural numbers alone) [9], these symbols $\alpha, \beta, \ldots$ satisfy the following four conditions:

1. Between any two elements α and β of S one and only one of the three relations [10]

$$\alpha < \beta, \quad \alpha = \beta, \quad \alpha > \beta$$

necessarily holds (this is expressed briefly by saying that *S is an ordered system*) and these *relations of order* between the elements of S are subject to the same fundamental laws **1** as their analogues in the system of rational numbers [11].

2. Four distinct methods of combining any two elements of S are defined, called Addition, Subtraction, Multiplication and Division. With a single exception, to be mentioned immediately (3.), these processes can always be carried out to one definite result, and obey the same Fundamental Laws **2**, I—IV, as their analogues in the system of the rational

---

[9] We shall come across actual examples in § 3 and § 5; for the moment, we may think of decimal fractions, or similar symbols constructed from rational numbers. See also footnote 16, p. 12.

[10] Cf. also footnotes 2 and 4.

[11] As to what we may call the *practical* meaning of these relations, nothing is implied; "<" may as usual stand for "less than", but it may equally well mean "before", "to the left of", "higher than", "lower than", "subsequent to", in fact may express any relation of order (including "greater than"). This meaning merely has to be defined without ambiguity and kept consistent. Similarly, "equality" need not imply identity. Thus, for example, within the system of symbols of the form p/q, where p, q are integers and $q \neq 0$, the symbols $3/4$, $6/8$, $-9/-12$ are generally said to be "equal"; that is, for certain purposes (calculating, measuring, and so on) we *define* equality within our system of symbols in such a way that $3/4 = 6/8 = -9/-12$, although $3/4$, $6/8$, $-9/-12$ are in the first instance different elements of that system (see also **14**, note 1).

numbers [12]. (The "zero" of the system, which must be known in order that the elements can be divided into positive and negative, is to be defined as explained in footnote 14 below.)

3. With every rational number we can associate an element of S (and all others "equal" to it) in such a manner that, if a and b denote rational numbers, α, β their associates from S:

a) the relation 1. holding between α and β is of the same form as that holding between a and b.

b) the element resulting from a combination of α and β (i. e. $\alpha + \beta$, $\alpha - \beta$, $\alpha \cdot \beta$, or $\alpha \div \beta$) has for its associated rational number the result of the similar combination of a and b (i. e. $a + b$, $a - b$, $a \cdot b$, or $a \div b$ respectively).

[This is also expressed, more shortly, by saying that the system S contains a sub-system S' ***similar*** and ***isomorphous*** to the system of rational numbers. Such a sub-system is in fact constituted by those elements of S which we have associated with rational numbers [13].]

In such a correspondence, an element of S associated with the rational number zero, and all elements equal to it, may be shortly referred to as the "zero" of the system of elements. The exception mentioned in 2. then relates to division by zero [14].

---

[12] With reference to these four processes it should be noted, as in the case of the symbols $<$ and $=$, that no practical interpretation is implied. — We also draw attention to the fact that subtraction is already completely defined in terms of addition, and division in terms of multiplication, so that, properly speaking, only *two* modes of combining elements need be assumed *known*.

[13] Two ordered systems are ***similar*** if it is possible to associate each element of the one with an element of the other in such a way that the same one of the relations **4**, 1 as holds between two elements of the one system also holds between the two associated elements of the other: they are ***isomorphous*** relatively to the possible modes of combining their elements, if the element resulting from a combination of two elements of the one system is associated with that resulting from the similar combination of the two associated elements of the other system.

[14] The third of the stipulations by means of which we here characterise the concept of number is fulfilled, moreover, as a consequence of the first and second. For our purposes, this fact is not essential; but as it is significant from a systematic point of view, we briefly indicate its proof as follows: By **4**, 2, there is an element ζ for which $\alpha + \zeta = \alpha$. From the fundamental laws **2**, 1, it then quite easily follows that *one and the same* element ζ of S satisfies $\alpha + \zeta = \alpha$, for *every* α. This element ζ, with all elements equal to it, is called the *neutral* element relatively to the process of addition, or for brevity the "zero" in S. If α is different from this "zero", there is, further, an element ϵ for which $\alpha\,\epsilon = \alpha$; and it again appears that this element is the same as that satisfying $\alpha\,\epsilon = \alpha$ for any other α in S. This ϵ, with all elements equal to it, is called the *neutral* element relatively to the process of multiplication, or, briefly, the "unit" in S. The elements of S produced by repeated addition or subtraction of this "unit", and any others equal to them, are then called "integers" of S. All further elements of S (and all equal to them) which result from these by the process of division then form the sub-system S' of S in question; that it is *similar* and *isomorphous* to the system of all rational numbers is in fact easily deduced from **4**, 1 and **4**, 2. — Thus, as asserted, our concept of number is already determined by the requirements of **4**, 1, 2 and **4**.

4. For any two elements α and β of S both standing in the relation "$>$" to the "zero" of the system, there exists a natural number n for which $n\beta > \alpha$. Here $n\beta$ denotes the sum $\beta + \beta + \ldots + \beta$ containing the element β n times. (***Postulate of Eudoxus;*** cf. **2**, VI.)

To this abstract characterisation of the concept of number we will append the following remark [15]: If the system S contains no other elements than those corresponding to rational numbers as specified in 3, then our system does not differ in any essential feature from the system of rational numbers, but only in the (purely external) *designation* of the elements by symbols, or in the (purely practical) *interpretation* which we give to these symbols; differences almost as irrelevant, at bottom, as those which occur when we write figures at one time in Arabic characters, at another, in Roman or Chinese, or take them to denote now temperature, now velocity or electric charge. Disregarding external characteristics of notation and practical interpretation, we should thus be perfectly justified in considering the system S as *identical* with the system of rational numbers and in this sense we may put $a = \alpha$, $b = \beta, \ldots$.

If, however, the system S contains other elements besides the above mentioned, then we shall say that S *includes* the system of rational numbers, and is an *extension* of it. Whether a system of this more comprehensive kind exists at all, remains for the moment an open question;

---

[15] We have defined the concept of number by a set of properties characterising it. A critical construction of the foundations of arithmetic, which is quite out of the question within the limits of this volume, would have to comprise a strict investigation as to the extent to which these properties are independent of one another, i. e. whether any one of them can or cannot be deduced from the rest as a *provable fact*. Further, it would have to be shewn that none of these fundamental stipulations is in contradiction with any other — and other matters too would require consideration. These investigations are tedious and have not yet reached a final conclusion.

In the treatment by *E. Landau* mentioned on p. 2, footnote 7, it is proved with absolute rigour that the fundamental laws of arithmetic which we have set up can all be deduced from the following 5 axioms relating to the *natural numbers*:

Axiom 1: 1 is a natural number.

Axiom 2: For every natural number n there is just one other number that is called the successor of n. (Let it be denoted by n'.)

Axiom 3: We have always $n' \neq 1$.

Axiom 4: From $m' = n'$, it follows that $m = n$.

Axiom 5: The induction law V is valid (in its first form).

These 5 axioms, first formulated as here by *G. Peano*, but in substance set up by *R. Dedekind*, assume that the natural numbers as a whole are regarded as given, that a relation of equality (and hence also inequality) is defined between them, and that this equality satisfies the relations **1**, **1**, **2**, **3** (which belong to pure logic).

but an example will come before our notice presently in the system of real numbers [16].

Having thus agreed as to the amount of preliminary assumption we require, we may now drop all argument on the subject, and again raise the question: *What do we mean by saying that we know the number* $\sqrt{2}$ *or* π*?*

It must in the first instance be termed altogether paradoxical that a number having its square equal to 2 does not exist in the system so far constructed [17], — or, in geometrical language, that the point A of the number-axis, whose distance from O equals the diagonal of the square of side $O\,U$, coincides with none of the "rational points". For the rational numbers are *dense*, i. e. between any two of them (which are distinct) we can point out as many more as we please (since, if $a < b$, the n rational numbers given by $a + \nu \frac{b-a}{n+1}$, for $\nu = 1, 2, \ldots, n$, evidently all lie between a and b and are distinct from these and from one another); but they are not, as we might say, dense enough to symbolise all conceivable points. Rather, as the aggregate of all integers proved too scanty to meet the requirements of the four processes of arithmetic,

---

[16] The mode of defining the number-concept given in **4** is of course not the only possible one. Frequently the designation of number is still ascribed to objects which fail to satisfy some one or other of the requirements there laid down. Thus for instance we may relinquish the condition that the objects under consideration should be constructively developed from rational numbers, regarding *any* entities (for instance points, or distances, or such like) as numbers, provided only they satisfy the conditions **4**, 1—4, or, in short, are similar and isomorphous to the system we have just set up. — This conception of the notion of number, in accordance with which all isomorphous systems must be regarded as in the abstract sense identical, is perfectly justified from a mathematical point of view, but objections necessarily arise in connection with the theory of knowledge. — We shall encounter another modification of the number-concept when we come to deal with complex numbers.

[17] *Proof*: There is certainly no *natural* number of square equal to 2, as $1^2 = 1$ and all other integers have their squares $\geqq 4$. Thus $\sqrt{2}$ could only be a (positive) fraction $\frac{p}{q}$, where q may be taken $\geqq 2$ and prime to p (i. e. the fraction is in its lowest terms). But if $\frac{p}{q}$ is in its lowest terms, so is $\left(\frac{p}{q}\right)^2 = \frac{p \cdot p}{q \cdot q}$, which therefore cannot reduce to the whole number 2. In a slightly different form: For any two natural numbers p and q without common factor, we have necessarily $p^2 \neq 2\,q^2$. For since two integers without common factors cannot both be even, either p is odd, or else p is even and q odd. In the first case p^2 is again odd, hence cannot equal an even integer $2\,q^2$. In the second case $p^2 = (2\,p')^2$ is divisible by 4, but $2\,q^2$ is not, since it is double an odd number. So $p^2 \neq 2\,q^2$ again. This *Pythagoras* is said to have already known (cf. *M. Cantor*, Gesch. d. Mathem., Vol. 1, 2nd ed., pp. 142 and 169. 1894).

so also the aggregate of all rational numbers contains too many gaps [18] to satisfy the more exacting demands of root extraction. One feels, nevertheless, that a perfectly definite numerical value belongs to the point A and therefore to the symbol $\sqrt{2}$. What are the tangible facts which underlie this feeling?

Obviously, in the first instance, this: We do, it is true, know perfectly well that the values 1·4 or 1·41 or 1·414 etc. for $\sqrt{2}$ are inaccurate, in fact that these (rational) numbers have squares < 2, i. e. are too small. But we also know that the values 1·5 or 1·42 or 1·415 etc. are in the same sense too large; that the value which we are attempting to reach would have therefore to lie between the corresponding too large and too small values. We thus reach the definite conviction that the value of $\sqrt{2}$ is within our grasp, although the given values are all incorrect. The root of this conviction can only lie in the fact that we have at our command a *process*, by which the above values may be continued *as far as we please*; we can, that is, form pairs of decimal fractions, with 1, 2, 3, . . . places of decimals, one fraction of each pair being too large, and the other too small, and the two differing only by one unit in the last decimal place, i. e. by $(\frac{1}{10})^n$, if n is the number of decimal places. As this difference may be made *as small as we please*, by sufficiently increasing the number n of given decimal places, we are taught through the above process to enclose the value which we are in search of between two numbers as near as we please to one another. By a metaphor, somewhat bold at the present stage, we say that through this process $\sqrt{2}$ itself is "given", — in virtue of it, $\sqrt{2}$ is "known", — by it, $\sqrt{2}$ may be "calculated", and so on.

We have precisely the same situation with regard to any other value which cannot actually be denoted by a rational number, as for instance π, $\log 2$, $\sin 10°$ etc. If we say, these numbers are *known*, nothing more is implied than that we know some process (in most cases an *extremely* laborious one) by which, as detailed in the case of $\sqrt{2}$, the desired value may be imprisoned, hemmed in, within a narrower and narrower space between rational numbers, — and this space ultimately narrowed down as much as we please.

For the purpose of a somewhat more general and more accurate

---

[18] This is the paradox, scarcely capable of any direct illustration, that a set of points, *dense* in the sense just explained, may already be marked on the number axis, and yet not comprise *all* the points of the straight line. The situation may be described thus: Integers form a first rough partition into compartments; rational numbers fill these compartments as with a fine sand, which on minute inspection inevitably still discloses gaps. To fill these will be our next problem.

statement of these matters, we insert a discussion of sequences of rational numbers, provisional in character, but nevertheless of fundamental importance for all that comes after.

§ 2. Sequences of rational numbers[1].

In the process indicated above for calculating $\sqrt{2}$, successive well-defined rational numbers were constructed; their expression in decimal form was material in the description; from this form we now propose to free it, and start with the following

5. **Definition.** *If, by means of any suitable process of construction, we can form successively a first, a second, a third, . . . (rational) number and if to every positive integer n one and only one well-defined (rational) number x_n thus corresponds, then the numbers*

$$x_1, \quad x_2, \quad x_3, \ldots, x_n, \ldots$$

(in this order, corresponding to the natural order of the integers 1, 2, 3, . . . , *n, . . .) are said to form a* **sequence.** *We denote it for brevity by* (x_n) *or* $(x_1, x_2, \ldots)$.

6. Examples.

1. $x_n = \frac{1}{n}$; i. e. the sequence $\left(\frac{1}{n}\right)$, or $1, \frac{1}{2}, \frac{1}{3}, \ldots, \frac{1}{n}, \ldots$
2. $x_n = 2^n$; i. e. the sequence 2, 4, 8, 16, . . .
3. $x_n = a^n$; i. e. the sequence $a, a^2, a^3, \ldots$, where a is a given number.
4. $x_n = \frac{1}{2}\{1 - (-1)^n\}$; i. e. the sequence 1, 0, 1, 0, 1, 0, . . .
5. $x_n =$ the decimal fraction for $\sqrt{2}$, terminated at the nth digit.
6. $x_n = \frac{(-1)^{n-1}}{n}$; i. e. the sequence $1, -\frac{1}{2}, +\frac{1}{3}, -\frac{1}{4}, \ldots$
7. Let $x_1 = 1$, $x_2 = 1$, $x_3 = x_1 + x_2 = 2$ and, generally, for $n \geqq 3$, let $x_n = x_{n-1} + x_{n-2}$. We thus obtain the sequence 1, 1, 2, 3, 5, 8, 13, 21, . . . , usually called *Fibonacci's* sequence.
8. $1, 2, \frac{1}{2}, -2, -\frac{1}{2}, 3, \frac{1}{3}, -3, -\frac{1}{3}, \ldots$
9. $2, \frac{3}{2}, \frac{4}{3}, \frac{5}{4}, \ldots, \frac{n+1}{n}, \ldots$
10. $0, \frac{1}{2}, \frac{2}{3}, \frac{3}{4}, \frac{4}{5}, \ldots, \frac{n-1}{n}, \ldots$
11. $x_n =$ the nth prime number[2]; i. e. the sequence 2, 3, 5, 7, 11, 13, . . .
12. The sequence $1, \frac{3}{2}, \frac{11}{6}, \frac{25}{12}, \frac{137}{60}, \ldots$, in which $x_n = \left(1 + \frac{1}{2} + \ldots + \frac{1}{n}\right)$.

---

[1] In this section all literal symbols will continue to stand for *rational numbers* only.

[2] Euclid proved that there is an infinity of primes. If $p_1, p_2, \ldots, p_k$ are any prime numbers, then the integer $m = (p_1 p_2 \ldots p_k) + 1$ is either a prime different from $p_1, p_2, \ldots, p_k$, or else a product of such primes. Hence no finite set of prime numbers can include all primes.

Remarks. 7.

1. The law of formation may be quite arbitrary; it need not, in particular, be embodied in any explicit formula enabling us to obtain x_n, for a given n, by direct calculation. In examples **6**, 5, 7 and 11, clearly no such formula can be immediately written down. If the terms of the sequence are individually given, neither the law of formation (cf. **6**, 5 and 12) nor any other kind of regularity (cf. **6**, 11) among the successive numbers is necessarily apparent.

2. It is sometimes advantageous to start the sequence with a "0th" term x_0, or even with a $(-1)^{\text{th}}$ or $(-2)^{\text{th}}$ term, x_{-1}, x_{-2}. Occasionally, it pays better to start indexing with 2 or 3. The only essential is that there should be an integer $m \gtreqless 0$ such that x_n is defined for every $n \geqq m$. The term x_m is then called the *initial term* of the sequence. We will however, even then, continue to designate as the nth term that which bears the index n. In § **6**, 2, 3 and 4, for instance, we can without further difficulties take a 0th term or even $(-1)^{\text{th}}$ or $(-2)^{\text{th}}$ to head the sequence. The "first term" of a sequence is then not necessarily the term with which the sequence begins. The notation will be preferably $(x_0, x_1, \ldots)$ or $(x_{-1}, x_0, \ldots)$, etc., as the case may be, unless it is either quite clear or irrelevant where our enumeration begins, and the abbreviated notation (x_n) can be adopted.

3. A sequence is frequently characterised as *infinite*. The epithet is then merely intended to emphasize the fact that *every* term is succeeded by other terms. It is also said that there is an *infinite number* of terms. More generally, there is said to be a *finite number* or an *infinite number* of things under consideration according as the number of these things can be indicated by a definite integral number or not. And we may remark here that the word *infinite*, when otherwise used in the sequel, will have a symbolic significance only, intended as a concise expression of some perfectly definite (and usually quite simple) circumstance.

4. If *all* the terms of a sequence have one and the same value c, the sequence is said to be *identically equal to* c, and in symbols $(x_n) \equiv c$. More generally, we shall write $(x_n) \equiv (x_n')$ if the two sequences (x_n) and (x_n') agree term for term, i. e. for every index in question $x_n = x_n'$.

5. It is often helpful and convenient to represent a sequence graphically by marking off its terms on the number-axis, or to think of them as so marked. We thus obtain a *sequence of points*. But in doing this it should be borne in mind that, in a sequence, one and the same number may occur repeatedly, even "infinitely often" (cf. **6**, 4); the corresponding point has then to be counted (i. e. considered as a term of the sequence of points) repeatedly, or infinitely often, as the case may be.

6. A graphical representation of a different kind is obtained by marking, with respect to a pair of rectangular coordinate axes, the points whose coordinates are (n, x_n) for $n = 1, 2, 3, \ldots$ and joining consecutive points by straight segments. The broken line so constructed gives a picture (diagram, or graph) of the sequence.

To consider from the most diverse points of view the *sequences* hereby introduced, and the *real sequences* that will shortly be defined, will be the main object of the following chapters. We shall be interested more particularly in properties which hold, or are stipulated to hold, for *all* the terms of the sequence, or at least *for all terms beyond* (*or following*) *some definite term*[3]. With reference to this last restriction, it may sometimes

---

[3] E. g. all the terms of the sequence **6**, 9 are > 1. Or, all the terms of the sequence **6**, 2 after the 6th are > 100 (or more shortly: for $n > 6$, $x_n > 100$).

be said that particular considerations in hand are valid "a finite number of terms being disregarded", or only concern the *ultimate* behaviour of the sequence. Our first examples of considerations of the kind referred to are afforded by the following definitions:

8. **Definitions.** I. *A sequence is said to be* **bounded**[4], *if there is a positive number K such that each term x_n of the sequence satisfies the inequality*

$$|x_n| \leqq K \quad \text{or} \quad -K \leqq x_n \leqq K.$$

The number K is then called a **bound** *of the sequence.*

Remarks and Examples.

1. In definition 8, it is a matter of practical indifference whether we write "$\leqq K$" or "$< K$". For if $|x_n| \leqq K$ holds always (i. e. for every n in question), then we can also find a constant K' such that $|x_n| < K'$ holds always; indeed, clearly *any* $K' > K$ will serve the purpose. Conversely, if $|x_n| < K$ always, then *a fortiori* $|x_n| \leqq K$. When the exact magnitude of the bound comes in of course the distinction may be essential.

2. If K is a bound of (x_n), then so is any *larger* number K'.

3. The sequences 6, 1, 4, 5, 6, 9, 10 are evidently bounded; so is 6, 3, provided $|a| \leqq 1$. The sequences 6, 2, 7, 8, 11 are certainly not so. Whether 6, 3 for every $|a| > 1$, or 6, 12, is bounded or not, is not immediately obvious.

4. If all we know is the existence of a constant K_1, such that $x_n < K_1$, for every n, then the sequence is said to be *bounded on the right* (or *above*) and K_1 is called *a bound above* (or *a right hand bound*) of the sequence.

If there is a constant K_2 such that $x_n > K_2$ always, then (x_n) is said to be *bounded on the left* (or *below*) and K_2 is called *a bound below* (or *a left hand bound*) of the sequence.

Here K_1 and K_2 need not be positive.

5. Supposing a given sequence is bounded on the right, it may still happen that among its numbers none is the greatest. For instance, 6, 10 is bounded on the right, yet *every* term of this sequence is exceeded by *all* that follow it, and *none* can be the greatest[5]. Similarly, a sequence bounded on the left need contain no least term; cf. 6, 1 and 9.—(With this fact, which will appear at first sight paradoxical, the beginner should make himself thoroughly familiar.)

Among a *finite number* of values there is of course always both a greatest and a least, i. e. a value not exceeded by any of the others, and one which none of the others falls below. (There may, however, be several *equal* to this greatest or least value.)

6. The property of boundedness of a sequence x_n (though not the actual value of one of the bounds) is a property of the *tail-end* of the sequence; it is unaffected by any alteration to an isolated term of the sequence. (Proof?)

---

[4] This nomenclature appears to have been introduced by *C. Jordan*, Cours d'analyse, Vol. 1, p. 22. Paris 1893.

[5] The beginner should guard against modes of expression such as these, which may often be heard: "for n infinitely large, $x_n = 1$"; "1 is the greatest number of the sequence". Anything of this sort is sheer nonsense (cf. on this point 7, 3). For the terms of the sequence are $0, \frac{1}{2}, \frac{2}{3}, \frac{3}{4}, \ldots$ and *none* of these is $= 1$, on the contrary *all* of them are < 1. And there is no such thing as an "infinitely large n".

II. *A sequence is said to be* **monotone ascending** *or* **increasing** *if, for every value of n,* **9.**

$$x_n \leqq x_{n+1};$$

it is said to be **monotone descending** *or* **decreasing** *if, for every n,*

$$x_n \geqq x_{n+1}.$$

Both kinds will also be referred to as **monotone** *sequences.*

Remarks and Examples.

1. A sequence need not of course be either monotone increasing, or monotone decreasing; cf. **6**, 4, 6, 8. Monotone sequences are, however, extremely common, and usually easier to deal with than those which are not monotone. That is why it is convenient to give them a distinguishing name.

2. Instead of "ascending" we should more strictly say "non-descending", and instead of "descending", "non-ascending". This, however, is not customary. If in any special instance the sign of equality is excluded, so that $x_n < x_{n+1}$ or $x_n > x_{n+1}$, as the case may be, for every n, then the sequence is said to be *strictly monotone* (increasing or decreasing).

3. The sequences **6**, 2, 5, 7, 10, 11, 12 and **6**, 1, 9 are monotone; the first-named ascending, the others descending. **6**, 3 is monotone descending, if $0 \leqq a \leqq 1$, but monotone ascending if $a \geqq 1$; for $a < 0$, it is not monotone.

4. The designation of "monotone" is due to *C. Neumann* (Über die nach Kreis-, Kugel- und Zylinderfunktionen fortschreitenden Entwickelungen, pp. 26, 27. Leipzig 1881).

We now come to a definition to which the reader should pay the greatest attention, sparing no effort to make himself master of its meaning and all that it implies.

III. *A sequence will be called a* **null sequence** *if it possesses the following property: given any arbitrary positive (rational) number* ϵ, *the inequality* **10.**

$$|x_n| < \epsilon$$

is satisfied by all the terms, with at most a finite number[6] *of exceptions.* In other words: *an arbitrary positive number* ϵ *being chosen, it is always possible to designate a term* x_m *of the sequence, beyond which the terms are less than* ϵ *in absolute value. Or a number* n_0 *can always be found, such that*

$$|x_n| < \epsilon \quad \textit{for every} \quad n > n_0.$$

Remarks and Examples.

1. If, in a given sequence, these conditions are fulfilled for a *particular* ϵ, they will certainly be fulfilled for every greater ϵ (cf. **8**, 1), but not necessarily for any smaller ϵ. (In **6**, 10, for instance, the conditions are fulfilled for $\epsilon = 1$ and therefore for every larger ϵ, if we put $n_0 = 0$; for $\epsilon = \frac{1}{2}$ it is not possible to satisfy them.) In the case of a null sequence, the conditions have to be fulfilled for *every* positive

---

[6] Cf. **7**, 3.

ϵ, and in particular, therefore, for every very small $\epsilon > 0$. On this account, it is usual to formulate the definition somewhat more emphatically as follows: (x_n) is a null sequence if, *to every* $\epsilon > 0$, *however small*, there corresponds a number n_0 such that

$$|x_n| < \epsilon \text{ for every } n > n_0.$$

Here n_0 need not be an integer.

2. The sequence **6**, 1 is clearly a null sequence; for

$$|x_n| < \epsilon, \text{ provided } n > \frac{1}{\epsilon},$$

whatever be the value of ϵ. It is thus sufficient to put $n_0 = \frac{1}{\epsilon}$.

3. The place in a given sequence beyond which the terms remain numerically $< \epsilon$, will naturally depend in general on the magnitude of ϵ; speaking broadly, it will lie further and further to the right (i. e. n_0 will be larger and larger), the smaller the given ϵ is (cf. 2). This dependence of the number n_0 on ϵ is often emphasised by saying explicitly: "To each given ϵ corresponds a number $n_0 = n_0(\epsilon)$ such that . . ."

4. The positive number below which $|x_n|$ is to lie from some stage onwards need not always be denoted by ϵ. Any positive number, however designated, may serve. In the sequel, where ϵ, α, $K, \ldots$, denoting any given positive numbers, we may often use instead $\frac{\epsilon}{2}, \frac{\epsilon}{3}, \frac{\epsilon}{K}, \epsilon^2, \alpha\,\epsilon, \epsilon^{\alpha}$, etc.

5. The sign of x_n plays no part here, since $|-x_n| = |x_n|$. Accordingly **6**, 6 is also a null sequence.

6. In a null sequence, *no* term need be *equal to* zero. But all terms, whose index is *very large*, must be *very small*. For if I choose $\epsilon = 10^{-6}$, say, then for every $n >$ a certain n_0, $|x_n|$ must be $< 10^{-6}$. Similarly for $\epsilon = 10^{-10}$ and for any other ϵ.

7. *The sequence* (a^n) *specified in* **6**, 3 *is also a null sequence provided* $|a| < 1$.

Proof. If $a = 0$, the assertion is trivial, since then, for *every* $\epsilon > 0$, $|x_n| < \epsilon$ for every n. If $0 < |a| < 1$, then (by **3**, 1, 4) $\frac{1}{|a|} > 1$. If therefore we put

$$\frac{1}{|a|} = 1 + p, \quad \text{then} \quad p > 0.$$

But in that case, for every $n \geqq 2$, we have

$$\text{(a)} \qquad (1 + p)^n > 1 + n\,p.$$

For when $n = 2$, we have $(1 + p)^2 = 1 + 2p + p^2 > 1 + 2p$; the stated relation therefore holds in that case. If, for $n = k \geqq 2$,

$$(1 + p)^k > 1 + k\,p,$$

then by **2**, III, 6

$$(1 + p)^{k+1} > (1 + k\,p)(1 + p) = 1 + (k + 1)\,p + k\,p^2 > 1 + (k + 1)\,p,$$

therefore our relation, assumed true for $n = k$, is true for $n = k + 1$. By **2**, V it therefore holds [7] for *every* $n \geqq 2$.

---

[7] The proof shows moreover that (a) is valid for $n \geqq 2$ provided only $1 + p > 0$, i. e. $p > -1$, but $\neq 0$. For $p = 0$ and for $n = 1$, (a) becomes an equality. For $p > 0$, the validity of (a) follows immediately from the expansion of the left-hand side by the binomial theorem. — The relation (a) is called *Bernoulli's Inequality* (*James Bernoulli*, Propositiones arithmeticae de seriebus, 1689, Prop. 4).

Accordingly, we now have

$$|x_n| = |a^n| = |a|^n = \frac{1}{(1+p)^n} < \frac{1}{1+np} < \frac{1}{np},$$

so that, however small $\epsilon > 0$ may be, we have

$$|x_n| = |a^n| < \epsilon \quad \text{for every} \quad n > \frac{1}{p\,\epsilon}.$$

8. In particular, besides the sequence $\left(\frac{1}{n}\right)$ mentioned in 2., $\left(\frac{1}{2^n}\right)$, $\left(\frac{1}{3^n}\right)$, $\frac{1}{10^n}$, $\left(\left(\frac{4}{5}\right)^n\right)$, etc., are also null sequences.

9. A similar remark to that of **8**, 1 may be appended to Definition **10**: no essential modification is produced by reading "$\leqq \epsilon$" for "$< \epsilon$" there. In fact, if, for every $n > n_0$, $|x_n| < \epsilon$, then *a fortiori* $|x_n| \leqq \epsilon$; conversely, if, given any ϵ, n_0 can be so determined that $|x_n| \leqq \epsilon$ for every $n > n_3$, then choosing any positive number $\epsilon_1 < \epsilon$ there is certainly an n_1 such that $|x_n| \leqq \epsilon_1$, for every $n > n_1$, and consequently

$$|x_n| < \epsilon \quad \text{for every} \quad n > n_1;$$

the conditions in their original form are thus also fulfilled. — Precisely analogous considerations show that in Definition **10** "$> n_0$" and "$\geqq n_0$" are practically interchangeable alternatives.

In any individual case, however, the distinction must of course be taken into account.

10. Although in a sequence every term stands entirely by itself, with a definite fixed value, and is not necessarily in any particular relation with the preceding or following terms, yet it is quite customary to ascribe "to the terms x_n", or "to the general term" any peculiarities in the sequence which may be observed on running through it. We might say, for instance, in **6**, 1 the terms diminish; in **6**, 2 the terms increase; in **6**, 4 or **6**, 6 the terms oscillate; in **6**, 11 the general term cannot be expressed by a formula; and so on. — In this sense, the characteristic behaviour of a null sequence may be described by saying that *the terms become arbitrarily small*, or *infinitely small* [8]; by which neither more nor less is meant than is contained in Definition [9] **10**, viz. that *for every* $\epsilon > 0$ *however small* the terms are *ultimately* (i. e. for all indices $n >$ a suitable n_0; or from and after, or *beyond, a certain* n_0) numerically less than ϵ.

11. A null sequence is ipso facto *bounded*. For if we choose $\epsilon = 1$, then there must be an integer n, such that, for every $n > n_1$, $|x_n| < 1$. Among the finite number of values $|x_1|, |x_2|, \ldots, |x_{n1}|$, however, one (cf. **8**, 5) is greatest, $= M$ say. Then for $K = M + 1$, obviously $|x_n|$ is *always* $< K$.

12. To prove that a given sequence is a null sequence, it is indispensable to show that for a prescribed $\epsilon > 0$, the corresponding n_0 can actually be proved to exist (for instance, as in the examples that follow, by actually designating such a number). Conversely, if a sequence (x_n) is assumed to be a null sequence, it is thereby *assumed* that, for every ϵ, the corresponding n_0 may really be regarded as existent. On the other hand, the student should make sure that he understands clearly what is meant by a sequence *not* being a null sequence. The meaning is this: it is not true that, for *every* positive number ϵ, beyond a certain point $|x_n|$

---

[8] This mode of expression is due to *A. L. Cauchy* (Analyse algébrique, pp. 4 and 26).

[9] There need of course be no question here of the sequence being monotone. Also, in any case, some $|x_n|$'s of index $\leqq n_0$ may already be $< \epsilon$.

is always $< \epsilon$; there exists a *special* positive number ϵ_0, such that $|x_n|$ is not, beyond any n_0, *always* $< \epsilon_0$; after every n_0 there is a larger index n (and therefore an infinite number of such indices) for which $|x_n| \geqq \epsilon_0$.

13. Finally we may indicate a means of interpreting geometrically the special character of a null sequence.

Using the graphical representation **7**, 5, the sequence is a null sequence if its terms ultimately (for $n > n_0$) all belong to the *interval*[10] $-\epsilon \ldots +\epsilon$. Let us call such an interval for brevity an **ϵ-neighbourhood** of the origin; then we may state: (x_n) is a null sequence if every ϵ-neighbourhood of the origin (however small) contains *all but a finite number, at most,* of the terms of the sequence.

Similarly, using the graphical representation **7**, 6, we can state: (x_n) is a null sequence if every *ϵ-strip* (however narrow) *about the axis of abscissae* contains the entire graph, with the exception, at most, of a finite initial portion, the ϵ-strip being limited by parallels to the axis of abscissae through the two points $(0, \pm\epsilon)$.

14. The concept of a null sequence, the "arbitrarily small given positive number ϵ", to which we shall from now on have continually and indispensably to appeal, and which may thus be said to form a main support for the whole superstructure of analysis, appears to have been first used in 1655 by *J. Wallis* (v. Opera I., p. 382/3). Substantially, however, it is already to be found in *Euclid*, Elements V.

We are already in a better position to comprehend what is involved in the idea, discussed above, of a meaning for $\sqrt{2}$ or π or $\log 5$. — In forming on the one hand (we keep to the instance of $\sqrt{2}$) the numbers

$$x_1 = 1{\cdot}4; \quad x_2 = 1{\cdot}41; \quad x_3 = 1{\cdot}414; \quad x_4 = 1{\cdot}4142; \ldots$$

on the other, the numbers

$$y_1 = 1{\cdot}5; \quad y_2 = 1{\cdot}42; \quad y_3 = 1{\cdot}415; \quad y_4 = 1{\cdot}4143; \ldots$$

we are obviously constructing two sequences of (rational) numbers (x_n) and (y_n) according to a perfectly definite (though possibly very laborious) method of procedure. These two sequences are *both monotone*, (x_n) increasing, (y_n) decreasing. Furthermore x_n is $< y_n$ for every n, but the differences, i. e. the numbers

$$y_n - x_n = d_n$$

form, by **10**, 8, a null sequence, since $d_n = \frac{1}{10^n}$. These are clearly the facts which convince us that we "know" $\sqrt{2}$, and can "calculate" it, and so on, although — as we said before — no one has yet had the value $\sqrt{2}$ completely within his view, so to speak. — If we refer again to the more suggestive representation on the number-axis, then, obviously (cf. fig. 1, p. 25): the points x_1 and y_1 determine an interval

---

[10] The word *interval* denotes a portion of the number-axis between a definite pair of its points. According as we reckon these points themselves as belonging to the interval or not, this is termed *closed* or *open*. Unless otherwise stated, the interval will always in the sequel be regarded as *closed*. (For **10**, 13 this is immaterial, by **10**, 9.) Supposing a to be the left end point, b the right end point, of an interval, we call this for brevity the interval $a \ldots b$.

J_1 of length d_1; the points x_2 and y_2, similarly, an interval J_2 of length d_2. Since

$$x_1 \leqq x_2 < y_2 \leqq y_1,$$

the second interval lies wholly within the first. Similarly, the points x_3 and y_3 determine an interval of length d_3, completely within J_2, and generally, the points x_n and y_n determine an interval J_n completely inside J_{n-1}. The lengths of these intervals form a null sequence; the intervals themselves shrink up, — one surmises, — about a definite number, — contract to a quite definite point.

It only remains to examine how near this surmise is to truth. With this purpose in view, we state, more generally, the following:

Definition. *To express the fact that a monotone ascending sequence* **11.**
(x_n) *and a monotone descending sequence* (y_n) *are given, whose terms for every* n *satisfy the condition*

$$x_n \leqq y_n$$

and for which the differences

$$d_n = y_n - x_n$$

form a null sequence, — we say for brevity that we are given a ***nest of intervals*** *(**Intervallschachtelung**)*. The* n^{th} *interval stretches from* x_n *to* y_n *and has length* d_n. *The nest itself will be denoted by* (J_n) *or by* $(x_n \mid y_n)$.

The conjecture which we made above now finds its first confirmation in the following:

Theorem †. *There is at most* ***one*** *(rational) point* s *belonging to all* **12.**
the intervals of a given nest, that is to say satisfying, for every n, *the inequality*

$$x_n \leqq s \leqq y_n.$$

Proof: If there were, besides s, another number s' differing from it, and also satisfying the inequality

$$x_n \leqq s' \leqq y_n$$

for every n, then, for every n, besides

$$x_n \leqq s \leqq y_n,$$

---

* A set or series of similar objects is said to form a *nest* or to be *nested* (ineinander geschachtelt) when each smaller one is enclosed or fits into that which is next in size to it. The word *nest* is here used with the additional (ideal) characteristic implied, that *the sizes diminish to zero*. When this is *not* implied, we shall use the more explicit phrase that each is contained in the preceding (or we might say that they are *nested*).

† We note here for future reference that this theorem continues to hold unaltered when the numbers which occur are arbitrary real numbers.

we should also have (v. **3**, I, 4)

$$-y_n \leqq -s' \leqq -x_n;$$

by **3**, I, 2 and **3**, II, 5, the inequalities

$$-d_n \leqq s - s' \leqq d_n, \quad \text{or} \quad |s - s'| \leqq d_n,$$

would therefore hold for every n. Choosing $\epsilon = |s - s'|$, d_n would *never* (*a fortiori* not for every n beyond a certain n_0) be $< \epsilon$. This contradicts the hypothesis that (d_n) is a null sequence. The assumption that two distinct points belong to *all* the intervals is therefore inadmissible[11]. Q. E. D.

Remarks and Examples.

1. Let $x_n = \frac{n-1}{n}$, $y_n = \frac{n+1}{n}$; that is to say, $J_n = \frac{n-1}{n} \ldots \frac{n+1}{n}$, $d_n = \frac{2}{n}$. We can at once verify that we actually have a nest of intervals here, since $x_n < x_{n+1} < y_{n+1} < y_n$ for every n, and since, for every $n > \frac{2}{\epsilon}$, we have $d_n < \epsilon$, however $\epsilon > 0$ be chosen.

The number $s = 1$ here belongs to all the J_n's, since $\frac{n-1}{n} < 1 < \frac{n+1}{n}$ for every n. No number other than 1 can belong therefore to *all* the intervals.

2. Let J_n be defined as follows[12]: J_0 is the interval $0 \ldots 1$; J_1 the *left* half of J_0; J_2 the *right* half of J_1; J_3 the *left* half of J_2; and so on. These intervals are obviously each contained in the preceding; and since J_n has length $d_n = \frac{1}{2^n}$, and these numbers form a null sequence, we have a nest of intervals. A little consideration shows that the sequence of the x_n's consists of the numbers

$$0,\ \frac{1}{4},\ \frac{1}{4} + \frac{1}{16} = \frac{5}{16},\ \frac{1}{4} + \frac{1}{16} + \frac{1}{64} = \frac{21}{64}, \ldots$$

each taken *twice* running; and that the sequence of y_n's begins with 1 and continues with

$$1 - \frac{1}{2} = \frac{1}{2},\ 1 - \frac{1}{2} - \frac{1}{8} = \frac{3}{8},\ 1 - \frac{1}{2} - \frac{1}{8} - \frac{1}{32} = \frac{11}{32}, \ldots$$

each taken *twice* running. Now

$$\frac{1}{4} + \frac{1}{16} + \frac{1}{64} + \ldots + \frac{1}{4^k} = \frac{1}{3}\left(1 - \frac{1}{4^k}\right) < \frac{1}{3}$$

and[13]

$$1 - \frac{1}{2} - \frac{1}{8} - \ldots - \frac{1}{2 \cdot 4^{k-1}} = \frac{1}{3}\left(1 + \frac{2}{4^k}\right) > \frac{1}{3}.$$

---

[11] From a graphical point of view, what the proof indicates is that if s and s' belong to *all* the intervals, then *each* interval has a length at least equal to the distance $|s - s'|$ between s and s' (v. **3**, II, 6); these lengths cannot, therefore, form a null sequence.

[12] Here we let the index start from 0 (cf. **7**, 2).

[13] For *any* two numbers a and b, and every positive integer k, the formula

$$a^k - b^k = (a - b)(a^{k-1} + a^{k-2}\, b + \ldots + a\, b^{k-2} + b^{k-1})$$

is known to hold. Whence, more particularly, for $a \neq 1$, the formulae

$$1 + a + \ldots + a^{k-1} = \frac{1 - a^k}{1 - a} \text{ and } a + a^2 + \ldots + a^k = \frac{1 - a^k}{1 - a} \cdot a.$$

Hence, for every n, $x_n < \frac{1}{3} < y_n$; thus $s = \frac{1}{3}$ is the *single* number which belongs to all the intervals. Here, therefore, (J_n) "defines" or "determines" the number $\frac{1}{3}$, or (J_n) shrinks up to the number $\frac{1}{3}$.

3. If we are given a nest of intervals (J_n), and a number s has been recognised as belonging to all the J_n's, then by our theorem, s is *quite uniquely* determined by (J_n). We therefore say, more pointedly, that the nest (J_n) *"defines" or "encloses" the number s.* We also say that s is the *innermost point* of all the intervals.

4. If s is any given rational number and we put, for $n = 1, 2, \ldots, x_n = s - \frac{1}{n}$ and $y_n = s + \frac{1}{n}$, then $(x_n \mid y_n)$ is evidently a nest of intervals determining the number s itself. But this is also the case if we put, *for every n, $x_n = s$ and $y_n = s$.* — Manifestly, we can, in the most various ways, form nests of intervals defining a *given* number.

This theorem, however, only confirms what we may regard as one half of our previously described impression; namely, that if a number s belongs to all the intervals of a nest, then there is none other besides with this property, — s is *uniquely* determined by the nest.

The other half of our impression, namely, that there must also always be a (rational) number belonging to all the intervals of a nest, is *erroneous*, and it is precisely this fact which will become our inducement for extending the system of rational numbers.

This the following example shows. As on p. 20, let $x_1 = 1{\cdot}4$; $x_2 = 1{\cdot}41$; . . .; $y_1 = 1{\cdot}5$; $y_2 = 1{\cdot}42$; . . . Then there is *no rational number s*, for which $x_n \leqq s \leqq y_n$ for every n. In fact, if we put

$$x_n' = x_n^2, \qquad y_n' = y_n^2$$

then the intervals $J_n' = x_n' \ldots y_n'$ also form a nest[14]. But $x_n' = x_n^2 < 2$ for all n, and $y_n' = y_n^2 > 2$ for all n (because this was how x_n and y_n were chosen), i. e. $x_n' < 2 < y_n'$. On the other hand, if $x_n \leqq s \leqq y_n$ we should have, by squaring (as we may, by **3**, 1, 3), $x_n' \leqq s^2 \leqq y_n'$ for all n. By our theorem **12** this would involve $s^2 = 2$, which is however impossible, by the proof given in footnote 17 on p. 12. Here, therefore, there is certainly no (rational) number belonging to all the intervals.

In the following paragraphs, we will investigate what, in a case such as this, should be done.

§ 3. Irrational numbers.

We must come to terms with the fact that there is no rational number whose square is 2, that the system of rational numbers is too defective, too incomplete, too full of gaps, to furnish a solution for the

---

[14] For it follows from $x_n \leqq x_{n+1} < y_{n+1} \leqq y_n$ — since all the numbers are positive, so that squaring (cf. **3**, I, 3) is allowed — that $x_n' \leqq x'_{n+1} < y'_{n+1} \leqq y_n'$; further $y_n' - x_n' = (y_n + x_n)(y_n - x_n)$; therefore, since x_n and y_n are certainly < 2 for every n, $y_n' - x_n' < \frac{4}{10^n}$, i. e. $< \varepsilon$, provided $\frac{1}{10^n} < \frac{\epsilon}{4}$; and this, by **10**, 8, is certainly the case for every $n >$ a certain n_0.

equation $x^2 = 2$. Indeed, this is only one of many equations for whose solution the material of the system of rational numbers proves insufficient. Almost all the numerical values which we are in the habit of denoting by $\sqrt[p]{n}$, $\log n$, $\sin \alpha$, $\tan \alpha$ and so on, are non-existent in the system of rational numbers and can no more be immediately "obtained", or "determined", or be "stated in figures", than can $\sqrt{2}$. The material is too coarse for such finer purposes.

The considerations brought forward in the preceding paragraphs point to means for providing ourselves with more suitable material. We saw, on the one hand, that, behind the conviction that we do know $\sqrt{2}$, there lay no more, substantially, than the fact that we possess a method by which a perfectly definite nest of intervals may be obtained; for its construction, the solution of the equation $x^2 = 2$ of course gave the occasion[15]. We saw, on the other hand, that if a nest (J_n) encloses any number s capable of specification at all (this still implying that it is a *rational* number) then this number s is quite uniquely defined by the nest (J_n), — so unambiguously, indeed, that it is entirely indifferent, whether I give (write down, indicate) the number directly, or give, instead, the nest (J_n) — with the tacit addition that, by the latter, I mean precisely the number s which it uniquely encloses or defines. In this sense, the two data (the two symbols) are equivalent, and may to a certain extent be considered equal[16], so that we may write indeed:

$$(J_n) = s \quad \text{or} \quad (x_n \mid y_n) = s.$$

---

[15] The kernel of this procedure is in fact as follows: We ascertain that $1^2 < 2$, $2^2 > 2$, and accordingly put $x_0 = 1$, $y_0 = 2$. We then divide the interval $J_0 = x_0 \ldots y_0$ into 10 equal parts, and taking the points of division, $1 + \frac{k}{10}$, for $k = 0, 1, 2, \ldots, 9, 10$, determine by trial whether their squares are > 2 or < 2. We find that the squares corresponding to $k = 0, 1, 2, 3, 4$ are too small, those corresponding to $k = 5, 6, \ldots, 10$ too large, and accordingly we put $x_1 = 1{\cdot}4$ and $y_1 = 1{\cdot}5$. Next, we divide the interval $J_1 = x_1 \ldots y_1$ into 10 equal parts, and go through a similar test with regard to the new points of division — and so on. The known process for extracting the square root of 2 is intended mainly to make the successive trials as mechanical as possible. — The corresponding treatment of, for instance, the equation $10^x = 2$ (i. e. determination of the common logarithm of 2) involves the following nest of intervals: Since $10^0 < 2$, $10^1 > 2$, we here put $x_0 = 0$, $y_0 = 1$ and divide $J_0 = x_0 \ldots y_0$ into 10 equal parts. For the points of division, $\frac{k}{10}$, we next test whether $10^{k/10} < 2$ or > 2, that is to say, whether $10^k < 2^{10}$ or $> 2^{10}$. As a result of this trial, we shall have to put $x_1 = 0{\cdot}3$, $y_1 = 0{\cdot}4$. The interval $J_1 = x_1 \ldots y_1$ is again divided into 10 equal parts, the same procedure instituted for the points of division $\frac{3}{10} + \frac{k}{100}$ and, in consequence, x_2 put equal to $0{\cdot}30$ and y_2 to $0{\cdot}31$ — and so on. — This obvious procedure is of course much too laborious for practical calculations.

[16] The justification for this is provided by Theorems **14** to **19**.

Consequently, we will not say merely: "the nest (J_n) defines the number s" but rather "(J_n) is only *another symbol* for the number s", or in fine, "(J_n) *is* the number s" — exactly as we are used to look upon the decimal fraction 0·333 . . . as merely another symbol for the number $\frac{1}{3}$, or as being precisely the number $\frac{1}{3}$ itself.

It now becomes extremely natural to *introduce* **tentatively** *an analogous mode of expression* with regard to those nests of intervals which contain *no* rational number. Thus if x_n, y_n denote the numbers constructed previously in connection with the equation $x^2 = 2$, one might — seeing that in the system of rational numbers there is not a single one whose square $= 2$ — decide to say that this nest $(x_n \mid y_n)$ determines the "true" "value of $\sqrt{2}$" though one incapable of being symbolised by means of rational numbers, — that it encloses this

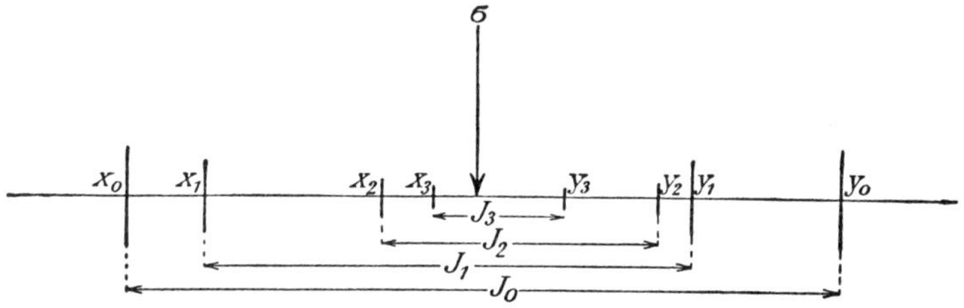

Fig. 1.

value unambiguously — in fine, "it is a newly created symbol for this number", or, for brevity, "it is the number itself". And similarly in every other case. If $(J_n) = (x_n \mid y_n)$ is any nest of intervals and no rational number s belongs to all its intervals, we might finally resolve to say that this nest encloses a perfectly definite *value*, — though one incapable of being directly symbolised by means of rational numbers, — it determines a perfectly definite *number*, — though one unfortunately non-existent in the system of rational numbers, — it is a newly created *symbol* for this number, or briefly: *is the number* itself; and this number, in contradistinction to the rational numbers, would then have to be called an *irrational number*.

Here certainly the question arises: *Can this be done without further justification? Is it allowable?* May we, without more ado, designate these new symbols, the nests $(x_n \mid y_n)$, as *numbers?* The following considerations are intended to show that to this course there is no obstacle whatever.

In the first instance, a simple graphical illustration of these facts on the number-axis (see fig. 1) gives every appearance of justification to our resolution. If, by any construction, we have marked a point P on the number-axis (e. g. by marking off to the right of O the length of the diagonal of a square of side $O\,U$) then we can in any number of ways define a nest of intervals enclosing the point P. We may do so in this way, for instance. First of all we imagine all integers $\gtreqless 0$ marked on the axis. Of these, there will be exactly one, say p, such that our point P lies in the stretch from p *inclusive* to $(p+1)$ *exclusive*. Accordingly we put $x_0 = p$, $y_0 = p + 1$, and divide the interval $J_0 = x_0 \ldots y_0$ into 10 equal parts[17]. The points of division are $p + \frac{k}{10}$ (with $k = 0, 1, 2, \ldots, 10$), and among them, there will again be exactly one, say $p + \frac{k_1}{10}$, such that P lies between $x_1 = p + \frac{k_1}{10}$ inclusive and $y_1 = p + \frac{k_1 + 1}{10}$ exclusive. The interval $J_1 = x_1 \ldots y_1$ is again divided into 10 equal parts, and so on. If we imagine this process continued indefinitely, we obtain a perfectly definite nest (J_n) all of whose intervals J_n contain the point P. *No other point P' besides P can lie in all the intervals J_n.* For, if that were so, all the intervals would have to contain the whole stretch $P\,P'$, which is impossible, as the lengths of the intervals $\left(J_n \text{ has length } \frac{1}{10^n}\right)$ form a null sequence.

For every arbitrarily given point P on the number-axis (rational or not) there are thus nests of intervals — obviously, indeed, any number of such nests — which contain that point and no other. And in the present instance, — i. e. in the graphical representation on the number-axis — the converse appears most plausible; if we consider *any* nest of intervals, there seems to be *always* one point (and by the reasoning above, only this one) belonging to all its intervals, which is thus determined by it. We believe, at any rate, that we may infer this directly from our *conception of the continuity, or gaplessness, of the straight line*[18].

Thus in this geometrical representation we should have complete reciprocity: every point can be enclosed in a suitable nest of intervals and every such nest invariably encloses one and only one point.

This gives us a high degree of confidence in the adequacy of our resolve to consider nests of intervals as numbers, — which we now formulate more precisely as follows:

13. Definition. *We will say of every nest of intervals (J_n) or $(x_n \mid y_n)$, that it* **defines** *or, for brevity,* **it is**, *a determinate number. To represent*

---

[17] Instead of 10 we may of course take any other integer $\geqq 2$. For further detail, see § 5.

[18] The proposition, by which the "continuity of the straight line" is expressly *postulated* — for a *proof* cannot be here expected, since it is essentially a *description of the form of our concept* of the straight line which is involved — is called the ***Cantor-Dedekind axiom.***

it, we use the symbol denoting the nest of intervals itself, and only as an abbreviation replace this by a small Greek letter, writing in this sense[19], *e. g.*

$$(J_n) \quad or \quad (x_n | y_n) = \sigma.$$

Now, in spite of all we have said, this cannot but seem a very arbitrary step, — the question has to be repeated most insistently: *will it pass without further justification?* These purely ideal objects which we have just defined — these nests of intervals (or else that still extremely questionable 'something' which such a nest encloses or determines) — can we speak of these as numbers? *Are* they after all numbers in the same sense as the rational numbers, — more precisely, in the sense in which the number concept was defined by our conditions **4**?

The answer can only consist in deciding, whether the totality or aggregate of all conceivable nests of intervals, or of the symbols (J_n) or $(x_n \mid y_n)$ or σ introduced to denote them, forms *a system of objects* satisfying these conditions **4**[20]; a system therefore — to recapitulate these conditions briefly — whose elements are derived from the rational numbers, and 1. are capable of being ordered; 2. are capable of being combined by the four processes (rules), obeying at the same time the fundamental laws **1** and **2**, I—IV; 3. contain a sub-system similar and isomorphous to the system of rational numbers; and 4. satisfy the Postulate of Eudoxus.

If and only if the decision turns out to be favourable, all will be well; our new symbols will then have vindicated their numerical character, and we shall have established that ***they are numbers,*** whose totality we shall then designate as the *system or set of real numbers.*

Now the decision in question does not present the slightest difficulty, and we may accordingly be brief in expounding the details:

Nests of intervals — or our new symbols $(x_n \mid y_n)$ — are certainly constructed by means of rational number-symbols alone; we have therefore only to settle the points **4**, 1—4. For this, we shall go to work in the following way: Certain of the nests of intervals define a rational number[21], something, therefore, for which both meaning and mode of combination have been previously established. We consider two such rational-valued nests, say $(x_n \mid y_n) = s$ and $(x_n' \mid y_n') = s'$. With the two rational number-symbols s and s', we can immediately distinguish whether the first s is $<$, $=$ or $>$ the second s'; and we can combine the two by the four processes of arithmetic. Essentially, what we have to do is to endeavour directly to recognise the former fact, and to carry out the latter processes, on the two nests of intervals themselves by which s and s' were

---

[19] σ is an abbreviated notation for the nest of intervals (J_n) or $(x_n \mid y_n)$.

[20] The reader should here read these conditions through again.

[21] We will describe such nests for brevity as *rational-valued.*

given, and finally to extend the result to the *aggregate of all nests of intervals.* Each *provable proposition* (*A*) relating to rational-valued nests will accordingly give rise to a corresponding *definition* (*B*). We begin by setting down concisely side by side these pairs of propositions (*A*) and definitions (*B*) [22].

14. Equality: A. Theorem. *If* $(x_n \mid y_n) = s$ *and* $(x_n' \mid y_n') = s'$ *are two rational-valued nests of intervals, then* $s = s'$ *holds if, and only if, besides*

$$x_n \leqq y_n \text{ and } x_n' \leqq y_n',$$

we have [23]

$$x_n' \leqq y_n \quad \text{and} \quad x_n \leqq y_n'$$

for every n.

On this theorem we now base the following:

B. Definition. *Two arbitrary nests of intervals* $\sigma = (x_n \mid y_n)$ *and* $\sigma' = (x_n' \mid y_n')$ *are said to be equal if and only if*

$$x_n \leqq y_n', \qquad x_n' \leqq y_n$$

for every n.

Remarks and Examples.

1. The numbers x_n and x_n' on the one hand, y_n and y_n' on the other, need of course have nothing whatever to do with one another. This is no more surprising than that rational numbers so entirely different in appearance as $\frac{3}{8}$, $\frac{21}{56}$, and 0·375 should be referred to as "equal". *Equality* is indeed something which

---

[22] The import of proposition and definition should in each case be interpreted in relation to the number-axis.

[23] Into the very simple proofs of the propositions **14** to **19** we do not propose to enter, for the general reasons explained on p. 2. They will not present the slightest difficulty to the reader, once he has mastered the contents of Chapter II, whereas at this stage they would appear to him strange; moreover they will serve as exercises in that chapter. Merely as a specimen and example for the solution of those problems, we will here prove Theorem **14**:

a) If $s = s'$, then we have both $x_n \leqq s \leqq y_n$ and $x_n' \leqq s \leqq y_n'$, whence at once, $x_n \leqq y_n'$ and $x_n' \leqq y_n$ for every n.

b) If conversely $x_n \leqq y_n'$ for every n, then $s \leqq s'$ must hold. For if we had $s > s'$, i. e. $s - s' > 0$, then, since $(y_n - x_n)$ is a null sequence, we could so choose the index p, that

$$y_p - x_p < s - s' \quad \text{or} \quad x_p - s' > y_p - s.$$

As however s is certainly $\leqq y_p$, this would imply $x_p - s' > 0$. We could therefore choose a further index r for which

$$y_r' - x_r' < x_p - s'.$$

Since $x_r' \leqq s'$, this would imply $y_r' < x_p$. Choosing an integer m exceeding both p and r, we could deduce, in view of the respective ascending and descending monotony of our sequences of numbers, that *a fortiori* $y_m' < x_m$, — which contradicts the hypothesis that $x_n \leqq y_n'$ *for every n*. Thus $s \leqq s'$ is ensured.

By interchanging throughout the above proof the accented and non-accented letters, we deduce in the same manner that if $x_n' \leqq y_n$ for every n, then $s' \leqq s$ — If then we have both $x_n' \leqq y_n$ and $x_n \leqq y_n'$ holding for every n, then $s = s$ necessarily follows. Q. E. D.

is not fixed *a priori*, but needs to be established by some form of definition, and it is perfectly compatible with marked dissimilarity in a purely external aspect.

2. The two nests $\left(\frac{n-1}{3n} \,\middle|\, \frac{n+1}{3n}\right)$ and **12**, 2 are equal in accordance with our present definition

3. By **14**, we may write e. g. $\left(s - \frac{1}{n} \,\middle|\, s + \frac{1}{n}\right) = s = (s \mid s)$, the latter symbol denoting a nest *all* of whose intervals have both their left and their right endpoints $= s$. In particular, $\left(-\frac{1}{n} \,\middle|\, +\frac{1}{n}\right) = (0 \mid 0) = 0$.

4. It still remains to establish — but the proof is so simple that we will not go into it further — that (cf. Footnote 23), in consequence of our definition, we have a) $\sigma = \sigma$ (Footnote 24), b) $\sigma = \sigma'$ always implies $\sigma' = \sigma$, and c) $\sigma = \sigma'$, $\sigma' = \sigma''$ involve $\sigma = \sigma''$.

Inequality: A. Theorem. *If $(x_n \mid y_n) = s$ and $(x_n' \mid y_n') = s'$ are two rational-valued nests, then we have $s < s'$, if and only if* **15.**

$$x_n \leqq y_n' \textit{ for every } n, \quad \textit{but not} \quad x_n' \leqq y_n \textit{ for every } n,$$

i. e. $y_m < x_m'$ for at least one m.

B. Definition. *Given any two nests of intervals $\sigma = (x_n \mid y_n)$ and $\sigma' = (x_n' \mid y_n')$, then we shall say $\sigma < \sigma'$, if*

$$x_n \leqq y_n' \textit{ for every } n, \quad \textit{but not} \quad x_n' \leqq y_n \textit{ for every } n,$$

i. e. for at least one m, $y_m < x_m'$.

Remarks and Examples.

1. It is clear that by **14** and **15** the totality of all conceivable nests is *ordered.* For if σ and σ' are any two of them, either there is equality, $\sigma = \sigma'$, or, for at least one p, we have $y_p < x_p'$, implying $\sigma < \sigma'$, or finally, for at least one r, $y_r' < x_r$, implying $\sigma' < \sigma$. The last two cases cannot occur simultaneously, since, for m greater than r and p, we should then have, *a fortiori*, $y_m' < x_m'$, which is impossible. Thus between σ and σ' one and only one of the three relations

$$\sigma < \sigma', \quad \sigma = \sigma', \quad \sigma' < \sigma$$

always holds, and the totality of these new symbols is thus *ordered* by **14** and **15**.

2. Here again it would have to be established in all detail that the laws of order **1** continue to hold good with the adopted definitions of equality and inequality. Taking as model the proof in the footnote to Theorem **14**, this presents so few essential difficulties that we will not enter into it further: *The laws of order do, effectually, all remain valid.*

3. In consequence of **14** and **15** we now have, therefore, for every n

$$x_n \leqq \sigma \leqq y_n.$$

What does this mean? It means that each of the rational numbers x_n is, in accordance with **14** and **15**, not greater than the nest $\sigma = (x_n \mid y_n)$. Or: if we con-

[24] Here it may be clearly recognised that this "law" is by no means trivial: it has indeed to be proved that with the given *definition* of equality every nest of intervals is effectually "equal" to itself, that is to say that the conditions of that definition are fulfilled, when the *same* nest is taken for both of the nests of intervals which we are comparing.

sider any particular one of the numbers x_n, say x_p, and denote it for brevity by x, then we may write (see **14**, Rem. 3)

$$(x_p =) \quad x = \left(x - \frac{1}{n} \,\middle|\, x + \frac{1}{n}\right) \quad \text{or} \quad = (x \mid x)$$

and our statement takes the form

$$(x \mid x) \leqq (x_n \mid y_n).$$

We may prove it as follows. If it were not true, then for at least one r,

$$y_r < x, \quad \text{i. e.} \quad y_r < x_p,$$

and so *a fortiori*, if m is greater than r and p,

$$y_m < x_m,$$

which certainly cannot be the case. In the same way we see that $\sigma \leqq y_n$. *Accordingly, σ is to be regarded as lying between x_n and y_n for each n, in other words, as contained within the interval J_n.*

The fact that no other number σ', besides σ, can possess the same property is now easily proved. If in fact there were a second nest of intervals $\sigma' = (x_n' \mid y_n')$ such that for every definite index p we also had $x_p \leqq \sigma' \leqq y_p$, then the left hand inequality means, more precisely (cf. 3), that $(x_p \mid x_p) \leqq (x_n' \mid y_n')$ and so, by **14** and **15**, $x_p \leqq y_n'$ for every n. Since this must hold in particular for $n = p$, we deduce $x_p \leqq y_p'$ for every p, which signifies, by **14** and **15**, that $\sigma \leqq \sigma'$. In the same manner the right hand inequality is seen to imply that $\sigma' \leqq \sigma$. Thus necessarily $\sigma = \sigma'$, which was what we set out to prove.

4. By **15**, σ is > 0, i. e. "*positive*", if and only if $(x_n \mid y_n) > (0 \mid 0)$, that is to say, if for some suitable index p, $x_p > 0$. But in this case, as the x_n's increase with n, we have *a fortiori* $x_n > 0$ for every $n > p$. We may therefore say: $\sigma = (x_n \mid y_n)$ is *positive* if, and only if, *all* the endpoints x_n, y_n are positive from and after a definite index. — The exact analogue holds of course for $\sigma < 0$.

5. If $\sigma > 0$, and, for every $n \geqq p$, $x_n > 0$, let us form a new nest $(x_n' \mid y_n') = \sigma'$ by putting $x_1' = x'_2 = \ldots = x'_{p-1}$ all equal to x_p, but every other x_n' and y_n' equal to the corresponding x_n and y_n. By **14**, obviously $\sigma = \sigma'$; and we may say: If σ is positive, then there are always nests of intervals equal to it, for which *all* the endpoints of intervals are positive. The exact analogue holds for $\sigma < 0$.

So far then, in respect of the possibility of ordering them, our nests of intervals may be said to vindicate their character as numbers completely. It is no more difficult to establish a similar conclusion with regard to the possibilities of combining them.

16. Addition: A. Theorem[25]. *If $(x_n \mid y_n)$ and $(x_n' \mid y_n')$ are any two nests of intervals, then $(x_n + x_n', y_n + y_n')$ is also one, and if the former are both rational-valued and respectively $= s$ and $= s'$, then the latter is also rational-valued, and determines the number $s + s'$.*

B. Definition. *If $(x_n \mid y_n) = \sigma$ and $(x_n' \mid y_n') = \sigma'$ are any two nests of intervals and σ'' denotes the nest $(x_n + x_n', y_n + y_n')$ deduced from them, then we write*

$$\sigma'' = \sigma + \sigma'$$

and σ'' is called the **sum** *of σ and σ'.*

[25] With regard to the proof, cf. footnote 23.

Subtraction: A. Theorem. *If* $(x_n \mid y_n)$ *is a nest of intervals, then so is* $(-y_n \mid -x_n)$; *and if the former is rational-valued* $= s$, *then the latter is also rational-valued, and determines the number* $-s$. **17.**

B. Definition. *If* $\sigma = (x_n \mid y_n)$ *is any nest of intervals and* σ' *denote the nest of intervals* $(-y_n \mid -x_n)$, *we write*

$$\sigma' = -\sigma$$

and say σ' *is the* ***opposite*** *of* σ. — *By the* ***difference*** *of two nests of intervals we then mean the sum of the first and of the opposite of the second.*

Multiplication: A. Theorem. *If* $(x_n \mid y_n)$ *and* $(x_n' \mid y_n')$ *are any two positive nests of intervals, — replaced, if necessary, (in accordance with* **15**, 5) *by two nests of intervals equal to them, for which* ***all*** *the endpoints of intervals are positive (or at least non-negative), — then* $(x_n x_n' \mid y_n y_n')$ *is also a nest of intervals; and if the former are rational-valued and respectively* $= s$ *and* $= s'$, *then the latter is also rational-valued, and determines the number* $s\,s'$. **18.**

B. Definition. *If* $(x_n \mid y_n) = \sigma$ *and* $(x_n' \mid y_n') = \sigma'$ *are any two positive nests of intervals for which* ***all*** *the endpoints of intervals are positive — which is no restriction, by* **15**, 5 — *and* σ'' *denote the nest* $(x_n x_n' \mid y_n y_n')$ *derived from them, then we write*

$$\sigma'' = \sigma \cdot \sigma'$$

and call σ'' *the* ***product*** *of* σ *and* σ'.

The slight modifications which have to be made in this definition if one or both of σ and σ' are negative or zero, we leave to the reader, and henceforth consider the product of any two nests of intervals as defined.

Division: A. Theorem. *If* $(x_n \mid y_n)$ *is any positive nest of intervals for which all endpoints of intervals are positive, (cf.* **15**, 5) *then so is* $\left(\frac{1}{y_n} \middle| \frac{1}{x_n}\right)$; *and if the former is rational-valued, and* $= s$, *the latter is also rational-valued, and determines the number* $\frac{1}{s}$. **19.**

B. Definition. *If* $(x_n \mid y_n) = \sigma$ *is any positive nest of intervals for which all endpoints are positive, and* σ' *denote the nest* $\left(\frac{1}{y_n} \middle| \frac{1}{x_n}\right)$, *then we write*

$$\sigma' = \frac{1}{\sigma}$$

and say σ' *is the* ***reciprocal*** *of* σ. — *By the* ***quotient*** *of a first by a second* ***positive*** *nest of intervals we then mean the product of the first by the reciprocal of the second.*

The slight modifications necessary in this definition, if σ (in the one case) or the second of the two nests of intervals (in the other) is negative, we may again leave to the reader, and henceforth consider the quotient of any two nests of intervals of which the second is different from 0, as defined. — If $(x_n \mid y_n) = \sigma = 0$, then the above method fails to produce a "reciprocal" nest: *division by 0 is here also impossible.*

The result of the preceding considerations is thus as follows: By *definitions* **14** to **19**, the system of all nests of intervals is ordered in the sense of **4**, 1, and admits of having its elements combined by the four processes in the sense of **4**, 2. In consequence of the *theorems* **14** to **19**, as stated in each case, this system possesses further, *in the aggregate of all rational-valued nests*, a sub-system, similar and isomorphous to the system of rational numbers, in the sense of **4**, 3. It remains to show that the system also fulfils the Postulate of Eudoxus. But if $(x_n \mid y_n) = \sigma$ and $(x_n' \mid y_n') = \sigma'$ are any two positive nests for which all endpoints of intervals are positive (cf. **15**, 5), let x_m and y_m' be a definite pair of these endpoints; the theorem of Eudoxus ensures the existence of an integer p, for which $p\,x_m > y_m'$, and the nest $p\,\sigma$, or $(p\,x_n \mid p\,y_n)$, in accordance with **15**, is then effectually $> \sigma'$.

The next step should be to establish in all detail (cf. **14**, 4 and **15**, 2) that the four processes defined in **16** to **19** for nests of intervals obey the fundamental laws **2**. This again offers not the slightest difficulty and we will accordingly spare ourselves the trouble of setting it forth [26]. *The Fundamental Laws of Arithmetic, and thereby the entire body of rules valid in calculations with rational numbers, effectually retain their validity in the new system.*

By this, our nests of intervals have finally proved themselves in every respect to be *numbers* in the sense of **4**: The system of all nests of intervals is *a number-system*, the nests themselves *are numbers* [27].

---

[26] As regards addition, for instance, it should be shown that:

a) Addition can always be carried out. (This follows at once from the definition.)

b) The result is unique; i. e. $\sigma = \sigma'$, $\tau = \tau'$ (in the sense of **14**) imply $\sigma + \tau = \sigma' + \tau'$, — if the sums are formed in accordance with **16** and the test for equality carried out in accordance with **14**. In the corresponding sense, it should be shown further that

c) $\sigma + \tau = \tau + \sigma$ always.

d) $(\varrho + \sigma) + \tau = \varrho + (\sigma + \tau)$ always.

e) $\sigma < \sigma'$ implies $\sigma + \tau < \sigma' + \tau$ always. —

And similarly for the other three processes of combination.

[27] Whether, as above, we regard nests of intervals as themselves numbers, or imagine some hypothetical entity introduced, which belongs to all the intervals J_n (cf. **15**, 3) and thus appears to be in a special sense the number enclosed by the nest of intervals and, consequently, the *common* element in all equal nests — this at bottom is a pure matter of taste and makes no essential difference. — The equality $\sigma = (x_n \mid y_n)$ we may, at any rate, from now on, (cf. **13**, footnote 19) read indifferently either as "σ is an abbreviated notation for the nest of intervals $(x_n \mid y_n)$", or as "σ is the number defined by the nest of intervals $(x_n \mid y_n)$".

This system we shall henceforth designate as *the system of real numbers*. It is an *extension* of the system of rational numbers, — in the sense in which the expression was used on p. 11, — since there are not only rational-valued nests but also others besides.

This system of real numbers is in one-one correspondence with the whole aggregate of points of the number-axis. For, on the strength of the considerations set forth on pp. 24, 25, we can immediately assert that to every nest of intervals σ corresponds one and only one point, namely that common to all the intervals J_n, which on account of the *Cantor-Dedekind* axiom is considered in each case as existing. Also two nests of intervals σ and σ' have, corresponding to them, one and the same point, if and only if they are equal, in the sense of **14**. *To each number* σ (that is to say, to all nests of intervals equal to each other) *corresponds exactly one point, and to each point exactly one number.* The point corresponding in this manner to a particular number is called its *image* (or *representative*) point, and we may now assert that *the system of real numbers can be* ***uniquely and reversibly*** *represented by the points of a straight line.*

§ 4. Completeness and uniqueness of the system of real numbers.

Two last doubts remain to be dispelled [28]: Our starting point in § 3 was the fact that the system of rational numbers, by reason of its "gaps", could not satisfy all demands which would appear in the course of the elementary processes of calculation. Our newly created number-system — the system Z as we will call it for brevity — is in this respect certainly more efficient. E. g. it contains [29] a number σ for which $\sigma^2 = 2$. Yet the possibility is not excluded that the new system may still show gaps like the old, or that in some other way it may be susceptible of still further extension.

Accordingly, we raise the following question: Is it conceivable that a system $\bar{Z}$, recognizable as a number-system in the sense of **4**, and containing all the elements of the system Z, *should also contain additional elements distinct from these?* [30]

---

[28] Cf. the closing words of the Introduction (p. 2).

[29] For if $\sigma = (x_n \mid y_n)$ denote the nest of intervals constructed on p. 20 in connection with the equation $x^2 = 2$, then by **18** we have $\sigma^2 = (x_n^2 \mid y_n^2)$. Since, however, $x_n^2 < 2$ and $y_n^2 > 2$, it follows that $\sigma^2 = 2$. Q. E. D.

[30] I. e. $\bar{Z}$ would have to represent an extension of Z in the same sense as Z itself represents an extension of the system of rational numbers.

It is not difficult to see that this cannot be so, so that we have in fact the following theorem:

20. Theorem of completeness. *The system Z of all real numbers is incapable of further extension compatible with the conditions* **4**.

Proof: Let $\bar{Z}$ be a system which satisfies the conditions **4** and contains all the elements of Z. If α denote an arbitrary element of $\bar{Z}$, then **4**, 4 — in which we choose for β the number 1, contained in Z, and also, therefore, in $\bar{Z}$ — shows that there exists an integer $p > \alpha$, and similarly another $p' > -\alpha$. For these [31] we have $-p' < \alpha < p$. Considering successively the (finite number of) integers between $-p'$ and p, starting with $-p'$, we know that we must come to a last one which is still $\leqq \alpha$. If this be called g, then

$$g \leqq \alpha < g + 1.$$

By applying to this interval $g \ldots g+1$ the method, already repeatedly used, of subdivision into ten parts, a perfectly definite nest of intervals $(x_n \mid y_n)$ is obtained. And a repetition word for word of the proof in **15**, 3 shows that the number thus defined can neither be $>$ nor $< \alpha$. Every element of $\bar{Z}$ is therefore equal to a real number, so that $\bar{Z}$ can contain no elements other than real numbers.

A final objection might be this: We have succeeded in forming the system Z in a comparatively natural, but after all an arbitrary, manner. Other measures, obviously, might be adopted for filling up the gaps in the system of rational numbers. (In the very next section we shall come across other, equally ready means to this end.) It is conceivable that a *different method* would lead to *other numbers*, i. e. to number-systems differing, in more or less essential particulars, from the one constructed by us. — The question thus indicated may be given a precise formulation as follows:

Let us suppose that we have somehow, starting with the system of rational numbers, succeeded in constructing a system $\mathfrak{Z}$ of elements which, besides still satisfying the conditions **4**, — as is the case with our system Z, — and therefore deserving the name of a number-system, also fulfils a further requirement, usually referred to as the Postulate of completeness, on account of the theorem proved above. — On the strength of **4**, 3, $\mathfrak{Z}$ contains elements, corresponding to the rational numbers. Let $(x_n \mid y_n)$ be any nest and let $\mathfrak{x}_n$ and $\mathfrak{y}_n$ be the elements of $\mathfrak{Z}$ associated with x_n, y_n in accordance with **4**, 3; the stipulation then runs thus: $\mathfrak{Z}$ *shall always contain at least one element* $\mathfrak{z}$ *satisfying, for every* n, *the conditions* $\mathfrak{x}_n \leqq \mathfrak{z} \leqq \mathfrak{y}_n$.

In exact form, our problem is now: Can such a system $\mathfrak{Z}$ differ in

---

[31] At this point, the Postulate of Eudoxus gains its axiomatic significance.

any essential particulars from the system Z of real numbers, or must the two systems be regarded as substantially identical, in the perfectly definite sense that they can be brought into relation as similar and isomorphous to one another?

The theorem stated below, by solving this problem in the sense which we should anticipate, closes the construction of the system of real numbers.

Theorem of Uniqueness. *Every such system $\mathfrak{Z}$ is necessarily similar and isomorphous to the system Z of real numbers as constructed by us. Essentially, only* **one** *such system therefore exists.* **21.**

Proof. By **4**, 3, $\mathfrak{Z}$ contains a sub-system $\mathfrak{Z}'$, which is similar and isomorphous to the system of rational numbers contained in Z, and whose elements may therefore be called, for short, the rational elements of $\mathfrak{Z}$. If $\sigma = (x_n \mid y_n)$ is any real number, $\mathfrak{Z}$ must, according to our new stipulation, contain an element $\mathfrak{s}$, which for every n satisfies the conditions $\mathfrak{x}_n \leqq \mathfrak{s} \leqq \mathfrak{y}_n$, if $\mathfrak{x}_n$ and $\mathfrak{y}_n$ are the elements of $\mathfrak{Z}$ corresponding to the rational numbers x_n and y_n.

Also, these conditions define $\mathfrak{s}$ *uniquely*. For if a second element $\mathfrak{s}'$, simultaneously with $\mathfrak{s}$, satisfied the conditions $\mathfrak{x}_n \leqq \mathfrak{s} \leqq \mathfrak{y}_n$ for every n, then it would follow, word for word as in the proof of **12**, that for every n

$$\mathfrak{y}_n - \mathfrak{x}_n \geqq | \mathfrak{s} - \mathfrak{s}' | ,$$

i. e. $\geqq$ the non-negative one of the two elements $\mathfrak{s} - \mathfrak{s}'$ and $\mathfrak{s}' - \mathfrak{s}$. Let r stand for an arbitrary positive rational number, and $\mathfrak{r}$ for the corresponding element in $\mathfrak{Z}$ (therefore in $\mathfrak{Z}'$); then, on account of the similarity and isomorphism of $\mathfrak{Z}'$ with the system of rational numbers, we must have, simultaneously with $y_p - x_p < r$, the relation $\mathfrak{y}_p - \mathfrak{x}_p < \mathfrak{r}$ holding for a suitable index p. For *every* such r therefore

$$| \mathfrak{s} - \mathfrak{s}' | < \mathfrak{r}.$$

If therefore $\mathfrak{r}_1$ denotes one particular such $\mathfrak{r}$ and if $\mathfrak{r}_n$, $n = 1, 2, \ldots$, denotes the element (certainly present in $\mathfrak{Z}'$, by **4**, 2) which, when repeated n times, yields the sum $\mathfrak{r}_1$, we see, after writing down the above inequality for $\mathfrak{r} = \mathfrak{r}_n$ and adding it to itself n times, that for *every* $n = 1, 2, \ldots$,

$$n \cdot | \mathfrak{s} - \mathfrak{s}' | \leqq \mathfrak{r}_1$$

must also hold. Since, however, $\mathfrak{Z}$ satisfies the postulate **4**, 4, it follows that $\mathfrak{s} = \mathfrak{s}'$.

If we proceed to associate this uniquely defined element $\mathfrak{s}$ and the real number σ, it becomes clear that $\mathfrak{Z}$ contains a sub-system $\mathfrak{Z}^*$, similar and isomorphous to the system Z of all real numbers. That such a system $\mathfrak{Z}^*$ is not susceptible of further extension compatible with the conditions **4**, but must be identical with $\mathfrak{Z}$, was the import of the previously established theorem of completeness. Thereby, it is proved that $\mathfrak{Z}$ and Z are similar and isomorphous to one another, and therefore may be regarded, in all essentials, as identical: *Our system Z of all real numbers is in all essentials the only one possible satisfying both the conditions* **4** *and the postulate of completeness.*

After these somewhat abstract considerations, the main result of our whole investigation may be summarised as follows:

Besides the *rational* numbers with which we are familiar, there exist others, the so-called *irrational numbers.* Each of them may be enclosed (determined, given, . . .) by a suitable nest of intervals and this indeed in many ways. These irrational numbers fit in consistently with the rational numbers, in such a manner that the conditions stated in **4** are fulfilled by the joint system of all rational and irrational numbers, with which, to be brief, all calculations may be effected, *formally, exactly as* with the rational numbers alone, *but with greater success.*

This wider system is moreover incapable of any further extension compatible with conditions **4**, and is in all essentials the only system of symbols which satisfies these conditions **4** and also the postulate of completeness.

We call it *the system of real numbers.*

It is with the elements of this system, with *the real numbers*, that we work (at first exclusively) in the sequel. We consider a particular real number as given (known, determined, defined, calculable, . . .) if either it is a rational number and so can be literally written down with the help of integers — inserting if need be a fractional bar or a minus sign — or (and this holds in any case) we are given [32] a nest of intervals defining the number.

We shall very soon see, however, that many other ways and means, besides the nests of intervals, exist, for defining a real number. In proportion as such ways become known to us, we shall widen the above-mentioned conditions, under which we consider a number as *given.*

---

[32] I. e. by the complete explicit specification of the (rational) endpoints in the manner just described.

§ 5. Radix fractions and the *Dedekind* section.

A few of the methods for defining real numbers may be mentioned at once, as particularly important from the points of view of both theory and practice.

In the first place, a nest of intervals need not always be given in the form $(x_n \mid y_n)$ considered by us; it may often be written in a more convenient form. Thus, as we have already seen, a decimal fraction, e.g. 1·41421 . . . , may be immediately interpreted as a nest of intervals, with the assumptions

$$x_1 = 1{\cdot}4; \quad x_2 = 1{\cdot}41; \quad x_3 = 1{\cdot}414; \quad \ldots,$$

and, generally, x_n equal to the decimal fraction broken off after the nth digit; y_n being derived from x_n by raising the last digit by one, i. e. $y_n = x_n + \frac{1}{10^n}$. Practically, we may thus say that decimal fractions represent a peculiarly clear and convenient specification of nests of intervals [33].

It is obviously quite an unessential part that the base or radix 10 of the ordinary scale of notation plays in this connection. If g is any integer $\geqq 2$, we have the exact analogue for *fractions in a scale of radix* g or *radix fractions with base* g. To begin with, given a real number σ, an integer p ($>$, $=$, or < 0) is uniquely defined by the condition

$$p \leqq \sigma < p + 1.$$

The interval J_0 between p and $p + 1$ is next divided into g equal parts, and each of these parts considered — both here and similarly in the following steps — as including its left endpoint, but *not* its right one. Then σ belongs to one, and to one only, of these parts, i. e. among the numbers $0, 1, 2, \ldots, g-1$ there is one and only one — which we shall call for brevity a "digit" and denote by z_1 — for which

$$p + \frac{z_1}{g} \leqq \sigma < p + \frac{z_1 + 1}{g}.$$

---

[33] The drawback to it is that we can seldom perceive the law of succession of the digits, — i. e. the *law of formation* of the x_n's and y_n's.

The interval J_1 thus defined we proceed to divide again into g equal parts, and σ will, as before, belong to one, and to one only, of these parts, i. e. a definite "digit" z_2 will be found for which

$$p + \frac{z_1}{g} + \frac{z_2}{g^2} \leqq \sigma < p + \frac{z_1}{g} + \frac{z_2 + 1}{g^2}.$$

The interval J_2 thus defined we proceed to divide again into g equal parts, and so on. The nest of intervals $(J_n) = (x_n \mid y_n)$ determined by this process, for which

$$\left.\begin{aligned} x_n &= p + \frac{z_1}{g} + \frac{z_2}{g^2} + \dots + \frac{z_{n-1}}{g^{n-1}} + \frac{z_n}{g^n} \\ y_n &= p + \frac{z_1}{g} + \frac{z_2}{g^2} + \dots + \frac{z_{n-1}}{g^{n-1}} + \frac{z_n + 1}{g^n} \end{aligned}\right\} (n = 1, 2, 3, \dots)$$

clearly defines the number σ, so that [34] $\sigma = (x_n \mid y_n)$. But on the analogy of decimal fractions we may now write

$$\sigma = p + 0{\cdot}z_1 \dots$$

— where of course the base g of the radix fraction must be known from the context.

We have therefore the

22. **Theorem 1.** *Every real number can be represented in one and essentially only one* [35] *way by a radix fraction in the scale of base g.*

We mention the following theorem relating further to this representation, but shall make no use of it in the sequel:

Theorem 2. *The radix fraction for a real number σ — whatever be*

---

[34] That we have a nest of intervals is immediately obvious, since $x_{n-1} \leqq x_n < y_n \leqq y_{n-1}$ throughout, and $y_n - x_n = \frac{1}{g^n}$ forms a null sequence, by **10**, 7.

[35] The slight alteration in our method, required if all the intervals are considered as including their right and *not* their left endpoints, the reader will doubtless be able to carry out for himself. The two results differ if, and only if, the given number σ is rational, and can be written as a fraction having, as denominator, a power of g, so that the point σ is an endpoint of one of our intervals. — Actually the two nests of intervals

$$p + 0{\cdot}z_1 z_2 \dots z_{r-1} (z_r - 1) (g - 1) (g - 1) \dots \text{ and } p + 0{\cdot}z_1 z_2 \dots z_{r-1} z_r 0 0 \dots,$$

where the digit z_r is supposed $\geqq 1$, are equal by **14**. In every other case, two radix fractions which are not identical are unequal, by **14**. — The reader will easily prove for himself that, except in this case, the representation of any real number σ as a radix fraction with base g is absolutely unique.

the chosen radix $g \geqq 2$ — *will prove periodic (or recurring) if and only if* σ *is rational*[36].

A particularly advantageous choice to make is often $g = 2$; the process for expressing the number σ is then called briefly the **method of bisection** and the resulting radix fraction, whose digits can in that case only be 0 or 1, is called a *binary* fraction. The method, in a somewhat more general light, is this: we start from a definite interval J_0 and, in accordance with some particular rule or point of view, definitely select one of its two halves, calling it J_1; we then again make a definite choice of one of the two halves of J_1, calling it J_2; and so on. By so doing, we specify, in every case, a well-defined real number, determined with absolute uniqueness by the method which regulates at each stage the choice between the two half-intervals[37].

In radix fractions, just as in decimal fractions, we accordingly see a peculiarly clear and convenient mode of specifying nests of intervals. They shall accordingly in future be admitted for the definition of real numbers on the same footing as decimal fractions.

The distinction lies somewhat deeper between nests of intervals and the following method of definition of real numbers.

We suppose given, in any particular way[38], two classes of numbers A and B, subject to the following three conditions:

1) Each of the two classes contains *at least one* number.

2) *Every* number of the class A is $\leqq$ *every* number of the class B.

3) If an arbitrary positive (small) number ϵ is prescribed, then two numbers can be so chosen from the two classes, — a', say, from A and b', say, from B, — that[39]

$$b' - a' < \epsilon.$$

— Then the following theorem holds:

---

[36] Here for simplicity we regard *terminating* radix fractions as periodic with period 0. — That every rational number can be represented by a recurring decimal fraction was proved by *J. Wallis*, De Algebra tractatus, p. 364, 1693. That conversely every irrational number can always, and in one way only, be represented as a non-recurring decimal fraction was first proved generally by *O. Stolz* (Allgemeine Arithmetik I, p. 119, 1885).

[37] An example was given in **12**, 2.

[38] E. g. A contains all rational numbers whose cube is < 5, B all rational numbers whose cube is > 5.

[39] We say for short: the numbers of the two classes *approach arbitrarily near to one another*. In the example of the preceding footnote, we see at once that conditions 1) and 2) are satisfied; that 3) is also satisfied we recognise from the possibility of calculating (by the method of partition into tenth parts, for instance) two decimal fractions x_n and y_n with n places of decimals, differing only by a unit in the last place, and such that $x_n^3 < 5$, $y_n^3 > 5$; n being so chosen that $\frac{1}{10^n} < \epsilon$.

Theorem 3. *There exists one and only one real number σ such that for every number a in A and every number b in B the relation*

$$a \leqq \sigma \leqq b$$

is always true.

Proof. It is again obvious that no two different numbers σ, σ' with this property can exist. For putting $|\sigma - \sigma'| = \epsilon$, we should have $\epsilon > 0$, yet $b - a \geqq \epsilon$ for *every* pair of elements a and b from A and B respectively, contrary to condition 3.

There exists then at most one such number σ. We find it in the following way: By hypothesis, there is at least *one* number a_1 in A and *one* number b_1 in B. If $a_1 = b_1$, then the common value is manifestly the number σ which we are in search of. If $a_1 \neq b_1$, and therefore by 2), $a_1 < b_1$, then we choose two rational numbers $x_1 \leqq a_1$, and $y_1 \geqq b_1$ and apply the method of bisection to the interval J_1 which they determine; we denote the left or right half by J_2, *according as the left half (endpoints included) does or does not still contain a point of the class B.* By the *same* rule we next select one of the halves of J_2, calling it J_3, and so on.

The intervals $J_1, J_2, \ldots, J_n, \ldots$, being obtained by the method of bisection, necessarily form a nest

$$(J_n) = (x_n \mid y_n) = \sigma.$$

From their mode of formation, they possess moreover the property that no number of B can lie to the left of any of their left endpoints, and no number of A to the right of their right endpoints.

But from this it follows at once that the number σ enclosed by them is the number required by theorem 3. In fact, if, contrary to the assertion in that theorem, a particular number $\bar{a}$ of A were $> \sigma$, so that $\bar{a} - \sigma > 0$, then we could choose from the succession of intervals J_n a particular one, say $J_p = x_p \ldots y_p$, with length $< \bar{a} - \sigma$. Since $x_p \leqq \sigma \leqq y_p$, this would imply

$$y_p - \sigma \leqq y_p - x_p < \bar{a} - \sigma, \quad \text{i.e.} \quad y_p < \bar{a},$$

whereas, actually, no point of A lies to the right of the right endpoint y_p of J_p. If on the other hand, in any instance, $\bar{b} < \sigma$, it would similarly follow that for a suitable index q, $\bar{b} < x_q$, whereas actually no point of B lies to the left of the left endpoint of an interval J_q. *Hence we must invariably have $a \leqq \sigma \leqq b$.* Q. E. D.

As a special corollary, we have the following theorem, which supplements Theorem **12**, forming an extension of it to the case when the numbers there occurring are arbitrary real numbers. In the formulation, we anticipate the obvious definitions **23**—**25** of next paragraph.

Theorem 4. *If (x_n) is a monotone ascending, and (y_n) a monotone descending, sequence of (any) real numbers; if, further, $x_n \leqq y_n$ for every n, and the differences $y_n - x_n = d_n$ form a null sequence; then there is invariably one and only one real number σ, such that for every n*

$$x_n \leqq \sigma \leqq y_n.$$

We then say, as before (cf. Definition **11**), *that the two given sequences define a nest of intervals $(x_n \mid y_n)$ and that σ is the number which it (uniquely) determines.*

Proof. If with all the left endpoints x_n we constitute a class A, and with all the right endpoints y_n a class B, of real numbers, these clearly satisfy conditions 1) to 3) of Theorem 3, from which the correctness of the above statement at once follows.

Remarks and Examples.

1. Instead of 3), it is often more convenient to stipulate that e. g. *every rational number* should belong either to A or to B (as was the case in the example of last footnote). In fact, in that case, since rational numbers are *dense* on the number axis, the requirement 3) is fulfilled *of itself*. To see this, we have only to imagine the whole number-axis subdivided into equal portions of length $< \epsilon/2$. Now consider any one of the portions containing an element from A, and, to the right of it, take another portion containing an element from B; together with these two portions, take the finite number of portions, if any, between them. One of these considered portions must be the *first* of them to contain an element b from B. Either this particular portion, or the preceding one, will contain an element a from A, and we have $b - a < \epsilon$.

2. It is often still more convenient to divide *all real* numbers into two classes A and B. In that case of course 3) is, *a fortiori*, also satisfied of itself.

3. If the two classes A and B are given in one of the last-mentioned ways, then we say that a ***Dedekind section*** is made in the domain of either rational or real numbers, as the case may be [40]. The somewhat more general specification of two classes [41] involved in our theorem 3 will also for brevity be termed a ***section*** and denoted by $(A \mid B)$. Our theorem 3 can then be stated briefly in the form: *A section $(A \mid B)$ invariably defines a determinate real number.* And its proof consists simply in pointing out that the specification of a section carries with it the specification of a nest of intervals, which furnishes a number σ with the properties required.

4. Seeing then that every section immediately provides a definite nest of intervals, we shall henceforth regard sections as permissible means of defining (determining, specifying, . . .) real numbers; also, we now write, if the section $(A \mid B)$ defines the number σ,

$$(A \mid B) = \sigma.$$

---

[40] Cf. p. 1, footnote 6.

[41] This was given in the above form by *A. Capelli*, Giornale di Matematica, Vol. 35, p. 209, 1897.

5. The converse is of course equally true and even more easily proved. Given a nest $(x_n \mid y_n) = \sigma$, we can consider all left endpoints x_n as forming a class A, and right endpoints a class B, and these two classes evidently furnish a section, which defines the same number σ as the nest itself. — A nest can accordingly be regarded as a particular kind of section.

6. By our last remark, the method of sections (for the definition of real numbers) is superior in generality to that of nests. It is also quite as convenient from the intuitional point of view. For if we take, say, the section $(A \mid B)$ in the somewhat more special form, mentioned in 2, of a section in the domain of real numbers, then what our theorem implies is this: If we imagine all points of the number-axis separated into two classes A and B, thinking e. g. of points of the one class as marked black and those of the other as white; and if, when this is done, (1) there is at least one point of each kind, (2) every black point lies to the left of every white point, and (3) *every* point on the number-axis is effectually coloured either black or white; then the two classes must come into contact at a perfectly definite place, and to the left of this place all is black, to the right of it all is white.

7. We must take care, however, not to accept the illustration just given as a *proof*. Had we not already with the help of nests of intervals *invented* the class of real numbers, our theorem could not be proved at all — any more than it could be *proved that* every nest defines a number. We simply agreed — and were amply justified by the result — to regard every nest as a number. In exactly the same way we can agree — and this is actually the course followed by *R. Dedekind*[42] in his construction of the system of real numbers — to regard every section in the domain of *rational* numbers as a "*real number*"; and we should then, exactly as in our investigations in § 3, only have to examine whether this is permissible; i. e. we should have to make sure whether the totality of all such sections $(A \mid B)$ forms a number system in the sense of conditions **4** — which is not more difficult than the analogous investigations carried out in § 3.

Henceforward — and for the present exclusively — *real numbers* form our working material. We may even, if we please, drop the word "real": For the present, "*number*" shall invariably mean *a real number*.

Exercises on Chapter I.

1. From the fundamental laws **1** and **2** deduce the most important of the further arithmetical rules, e. g. (a) the product of two negative numbers is positive; (b) $a + c < b + c$ invariably implies $a < b$; (c) for every a we have $a \cdot 0 = 0$; etc.

2. When in **3**, II, 4 are the signs of equality correct?

3. Express the following numbers as binary and as ternary fractions (i. e. in scales of notation of which the bases are respectively 2 and 3):

$$\frac{1}{2}, \frac{3}{8}, \frac{1}{3}, \frac{1}{7}, \frac{10}{17};$$

find the first few figures of the binary and ternary fractions for $\sqrt{2}$, $\sqrt{3}$, π and e.

[42] Stetigkeit und irrationale Zahlen, Brunswick 1872.

4. In the sequence **6**, 7 prove $x_n = \frac{\alpha^n - \beta^n}{\alpha - \beta}$, where α and β are the roots of the quadratic equation $x^2 = x + 1$. (Hint: the sequences (α^n) and (β^n) have the same law of formation as the sequence **6**, 7.)

5. Form the sequence (x_n) of numbers given, for $n \geqq 1$, by the formula

$$x_{n+1} = a\,x_n + b\,x_{n-1},$$

where a and b are given positive numbers and the initial terms $x_0, x_1 = 0, 1; = 1, 0; = 1, \alpha; = 1, \beta$; or are arbitrary. (Here α and β denote respectively the positive and the negative root of the equation $x^2 = a\,x + b$.) In each of the four cases give an explicit formula for x_n.

6. If $J_0, J_1, J_2, \ldots$ is a sequence of nested intervals (i. e. each contained in the preceding) about whose lengths nothing further is known, then there is *at least* one point which belongs to all the J_n's.

7. A real number σ is irrational, if we can find an ascending sequence of integers (q_n), such that $q_n\sigma$ is not an integer for any n, but if, when p_n stands for the integer *nearest* to $q_n\sigma$, $(q_n\sigma - p_n)$ is a null sequence.

8. Prove that $(x_n \mid y_n)$ is a nest in each of the following examples:

a) $x_n = \frac{1^2 + 2^2 + \ldots + (n-1)^2}{n^3}$, $y_n = \frac{1^2 + 2^2 + \ldots + n^2}{n^3}$, $(n = 1, 2, \ldots)$;

b) $0 < x_1 < y_1$ and for every $n \geqq 1$, $x_{n+1} = \sqrt{x_n y_n}$, $y_{n+1} = \frac{1}{2}(x_n + y_n)$;

c) $0 < x_1 < y_1$ „ „ „ „ , $x_{n+1} = \frac{1}{2}(x_n + y_n)$, $y_{n+1} = \sqrt{x_{n+1} \cdot y_n}$;

d) $0 < x_1 < y_1$ „ „ „ „ , $y_{n+1} = \frac{1}{2}(x_n + y_n)$, $x_{n+1} = \sqrt{x_n \cdot y_{n+1}}$;

e) $0 < x_1 < y_1$ „ „ „ „ , $x_{n+1} = \sqrt{x_n y_n}$, $y_{n+1} = \frac{1}{2}(x_{n+1} + y_n)$;

f) $0 < x_1 < y_1$ „ „ „ „ , $y_{n+1} = \sqrt{x_n y_n}$, $x_{n+1} = \frac{1}{2}(x_n + y_{n+1})$;

g) $0 < x_1 < y_1$ „ „ „ „ , $y_{n+1} = \frac{1}{2}(x_n + y_n)$, $x_{n+1} = \frac{x_n \cdot y_n}{y_{n+1}}$.

Evaluate the numbers defined in examples (a) and (g). (Cf. problems 91 and 92.)

Chapter II.

Sequences of real numbers.

§ 6. Arbitrary sequences and arbitrary null sequences.

We now resume our considerations of § 2, — and generalise them by allowing all the numbers which there occur to be arbitrary real numbers. Since, with these, we may operate precisely as with rational numbers, both the definitions and the theorems of § 2 will, in all essentials, remain unchanged. We may accordingly be brief.

23. °**Definition**[1]. *If to each positive integer* 1, 2, 3, ..., *corresponds a definite real number* x_n, *then the numbers*

$$x_1, \quad x_2, \quad x_3, \ldots, \quad x_n, \ldots$$

are said to form a sequence.

Examples **6**, 1—12, may, of course, also serve here. Similarly, the Remarks **7**, 1—6 retain full validity. We give a few more examples, in which it is not immediately apparent whether the numbers in question are rational or not.

Examples.

1. Let $a = 0{\cdot}3010\ldots$, i. e. equal to the decimal fraction whose first few digits were obtained in a footnote (p. 24) from the equation $10^x = 2$; and put

$$x_n = a^n \quad \text{for} \quad n = 1, 2, 3, \ldots$$

2. With the same meaning for a, let $x_n = \dfrac{1}{a+n}$.

3. Apply the method of successive bisection to the interval $J_0 = 0 \ldots 1$, taking first the left half, then twice running the right half, then for the next three steps again the left half, then four times running the right half, and so on. Denote the number[2] so defined by b (what is its value, approximately?), and put for x_n, successively,

$$+b, \; -b, \; +\frac{1}{b}, \; -\frac{1}{b}, \; +b^2, \; -b^2, \; +\frac{1}{b^2}, \; -\frac{1}{b^2}, \; +b^3, \ldots$$

4. With the same meaning for b, put for x_n, successively,

$$1-b, \; 1+b, \; 1-b^2, \; 1+b^2, \; 1-b^3, \; 1+b^3, \ldots$$

5. With the same meaning for a and b, let x_1 be the middle point of the stretch between them, i. e. $x_1 = \frac{1}{2}(a+b)$; x_2 the middle point between x_1 and b; x_3, that between x_2 and a; x_4, that between x_3 and b; — i. e. generally, x_{n+1}, the middle point between x_n and either a, or b, according as n is even or odd.

24. **Definitions:** °1. *A sequence* (x_n) *is said to be bounded if a constant* K *exists, such that the inequality*

$$|x_n| \leqq K$$

is satisfied for every n.

2. *A sequence* (x_n) *is said to be monotone increasing if* $x_n \leqq x_{n+1}$ *for every* n; *monotone decreasing, if* $x_n \geqq x_{n+1}$ *for every* n.

All remarks made in **8** and **9** retain their full validity.

---

[1] For the meaning of the mark ° cf. the preface, as also later the beginning of § 52.

[2] Written as a *binary* fraction, $b = 0{\cdot}01100011110\ldots$

Examples.

1. The sequences **23**, 1, 2, 4 and 5 are evidently bounded. Sequence 3 is not bounded, and in fact neither on the left nor on the right; for we certainly have $0 < b < \frac{1}{2}$ and therefore $\frac{1}{b^m} > 2^m > m$, and accordingly $-\frac{1}{b^m} < -m$. Terms of the sequence may therefore always be found, which are $> K$ or $< -K$, however large the constant K is chosen. — For 5, the boundedness follows from the fact that all the terms lie between a and b.

2. The sequences **23**, 1 and 2 are monotone decreasing: the others are not monotone.

The definition **10** of a null sequence and the appended remarks — which the student should read through again carefully — also remain unchanged.

°Definition. *A sequence (x_n) shall be termed a* **null sequence** *if, subsequently to the choice of an arbitrary positive number ε, a number $n_0 = n_0(\varepsilon)$ may always be assigned, such that the inequality*[3] **25.**

$$|x_n| < \varepsilon$$

is fulfilled for every $n > n_0$.

Examples.

1. The sequence **23**, 1 is a null sequence, for the proof **10**, 7 is valid for any real a, for which $|a| < 1$.

2. **23**, 2 is also a null sequence, for here $|x_n| < \frac{1}{n}$, therefore $< \varepsilon$, provided $n > \frac{1}{\varepsilon}$.

For null sequences — these will later on play a dominating part — a number of quite simple theorems, which will be continually applied in the sequel, will also be proved here. The following two, in the first place, are obvious enough:

°Theorem 1. *If (x_n) is a null sequence and the terms of the sequence (x_n'), for every n beyond a certain value m, satisfy the condition $|x_n'| \leqq |x_n|$, or, more generally, the condition* **26.**

$$|x_n'| \leqq K \cdot |x_n|,$$

in which K is an arbitrary (fixed) positive number, — then x_n' is also a null sequence. (Comparison test.)

---

[3] Given any positive *real* number ε, a positive *rational* number $\varepsilon' < \varepsilon$ can be designated; in fact, by the fundamental law **2**, VI, we can find a natural number $n > \frac{1}{\varepsilon}$, and $\varepsilon' = \frac{1}{n}$ satisfies the requirements. From this it follows that, for rational sequences, the above definition is equivalent to the definition **10**, in spite of the fact that only rational ε were allowed there.

Proof. If the condition $|x_n'| \leqq K \cdot |x_n|$ is satisfied for $n > m$ and $\varepsilon > 0$ is given, then by the assumptions we can assign $n_0 > m$, so that for every $n > n_0$, $|x_n| < \frac{\varepsilon}{K}$. Since for these values of n we then also have $|x_n'| < \varepsilon$, (x_n') is therefore a null sequence.

The following theorem is only a special case of the preceding:

°**Theorem 2.** *If (x_n) is a null sequence, and (a_n) any bounded sequence, then the numbers*

$$x_n' = a_n x_n$$

also form a null sequence.

On account of this theorem we say for short: A null sequence "may" be multiplied by a bounded factor.

Examples.

1. If (x_n) is a null sequence,

$$10\,x_1,\ \frac{x_2}{10},\ 10\,x_3,\ \frac{x_4}{10},\ 10\,x_5 \ldots$$

is also a null sequence.

2. If (x_n) is a null sequence, so is $(|x_n|)$.

3. A sequence, all of whose terms have the same value, say c, is certainly bounded. If (x_n) is a null sequence, $(c\,x_n)$ is therefore also a null sequence. In particular, $\left(\frac{c}{n}\right)$, $(c\,a^n)$ for $|a| < 1$, etc. are null sequences.

The next propositions are less obvious:

27. °**Theorem 1.** *If (x_n) is a null sequence, then every sub-sequence (x_n') of (x_n) is a null sequence*[4].

Proof. If, for every $n > n_0$, $|x_n| < \varepsilon$, then we have, *ipso facto*, for any such n,

$$|x_n'| = |x_{k_n}| < \varepsilon,$$

since k_n is certainly $> n_0$, when n is.

°**Theorem 2.** *Let an arbitrary sequence (x_n) be separated into two sub-sequences (x_n') and (x_n''), — so that, therefore, every term of (x_n) belongs to one and only one of these sub-sequences. If (x_n') and (x_n'') are both null sequences, then so is (x_n) itself.*

---

[4] If $k_1 < k_2 < k_3 < \ldots < k_n < \ldots$ is any sequence of positive integers, then the numbers

$$x_n' = x_{k_n} \qquad (n = 1, 2, 3, \ldots)$$

are said to form a ***sub-sequence*** of the given sequence.

Proof. If a number $\varepsilon > 0$ be chosen, then by hypothesis a number n' exists, such that for every $n > n'$, $|x_n'| < \varepsilon$, and also a number n'', such that for every $n > n''$, $|x_n''| < \varepsilon$. The terms x_n' with index $\leqq n'$ and the terms x_n'' with index $\leqq n''$, have definite places, i. e. definite indices, in the original sequence (x_n). If n_0 is the higher of these indices, then for every $n > n_0$, obviously $|x_n| < \varepsilon$, q. e. d.

°**Theorem 3.** *If (x_n) is a null sequence and (x_n') an arbitrary rearrangement*[5] *of it, then (x_n') is also a null sequence.*

Proof. For every $n > n_0$, $|x_n| < \varepsilon$. Among the indices belonging to the finite number of places which the terms $x_1, x_2, \ldots, x_{n_0}$ occupy in the sequence (x_n'), let n' be the largest. Then obviously, for every $n > n'$, $|x_n'| < \varepsilon$; hence (x_n') is also a null sequence.

°**Theorem 4.** *If (x_n) is a null sequence and (x_n') is obtained from it by any finite number of alterations*[6], *then (x_n') is also a null sequence*[7].

The proof follows immediately from the fact, that for a suitable integer $p \gtreqless 0$, from some n onwards we must have $x_n' = x_{n+p}$. For if every x_n for $n \geqq n_1$ has remained unchanged, and x_{n_1} has received the index n' in the sequence (x_n'), then in point of fact for every $n > n'$,

$$x_n' = x_{n+p},$$

if we put $p = n_1 - n'$.

Theorem 5. *If (x_n') and (x_n'') are two null sequences and if the sequence (x_n) is so related to them that from a certain m onwards*

$$x_n' \leqq x_n \leqq x_n'' \qquad (n > m)$$

then (x_n) is also a null sequence.

Proof. Having chosen $\varepsilon > 0$, we can chose $n_0 > m$ so that, for every $n > n_0$, $-\varepsilon < x_n'$ and $x_n'' < +\varepsilon$. For these n's we then have, *ipso facto*, $-\varepsilon < x_n < +\varepsilon$, that is $|x_n| < \varepsilon$; q. e. d.

---

[5] If $k_1, k_2, \ldots, k_n, \ldots$ is a sequence of positive integers such that *every* integer occurs once and only once in the sequence, then the sequence formed by

$$x_n' = x_{k_n}$$

is said to be a ***rearrangement*** of the given sequence.

[6] We will describe this concept as follows: If we alter any sequence, by omitting, or inserting, or changing, a finite number of terms (or by doing all three things at once), and then renumber the altered sequence, without changing the order of the terms left untouched, so as to exhibit it as a sequence (x_n'), then we shall say, (x_n') is obtained or has resulted from (x_n) *by a finite number of alterations.*

[7] It is precisely because of this theorem that one may say of a sequence that the property of being a *null sequence* concerns only the *ultimate behaviour* of its terms (cf. p. 16).

Calculations with null sequences, finally, are founded on the following theorems:

28. °**Theorem 1.** *If* (x_n) *and* (x_n') *are two null sequences, then*

$$(y_n) \equiv (x_n + x_n'),$$

i. e. the sequence whose terms are the numbers $y_n = x_n + x_n'$, *is also a null sequence.* — Briefly: *Two null sequences "may" be added term by term.*

Proof. If $\varepsilon > 0$ has been chosen arbitrarily, then by hypothesis (cf. **10**, 4 and 12) a number n_1 and a number n_2 exist such that for every $n > n_1$, $|x_n| < \frac{\varepsilon}{2}$, and for every $n > n_2$, $|x_n'| < \frac{\varepsilon}{2}$. If n_0 is a number $\geqq n_1$ and $\geqq n_2$, then for $n > n_0$

$$|y_n| = |x_n + x_n'| \leqq |x_n| + |x_n'| < \frac{\varepsilon}{2} + \frac{\varepsilon}{2} = \varepsilon.$$

(y_n) is therefore a null sequence [8].

Since, by **26**, 3 (or **10**, 5), $(-x_n')$ is a null sequence if (x_n') is, $(y_n') = (x_n - x_n')$ is then by the above also a null sequence, i. e. we have the theorem:

°**Theorem 2.** *If* (x_n) *and* (x_n') *are null sequences, then so is* $(y_n') \equiv (x_n - x_n')$. Or briefly: *null sequences "may" be subtracted term by term*

Remarks.

1. Since we may add *two* null sequences term by term, we may also do so with *three* or any *definite number* of null sequences. For supposing this proved for $(p-1)$ null sequences (x_n'), (x_n''), ..., $(x_n^{(p-1)})$, i. e. supposing the sequence

$$(x_n' + x_n'' + \cdots + x_n^{(p-1)})$$

to be already recognised as a null sequence, Theorem 1 ensures that the sequence (x_n), for which

$$x_n = (x_n' + \cdots + x_n^{(p-1)}) + x_n^{(p)},$$

is also a null sequence. The theorem thus holds for *every fixed number* of null sequences.

2. That two null sequences "may" also be *multiplied* term by term, is immediately clear from **26**, 1, since null sequences, by **10**, 11, are necessarily bounded.

3. Term by term *division*, on the contrary, is in general not allowed, as is already obvious, for instance, from the fact that when $x_n \neq 0$, $\frac{x_n}{x_n}$ is constantly $= 1$. If we take $x_n = \frac{1}{n}$, $x_n' = \frac{1}{n^2}$, then the ratios $\frac{x_n}{x_n'}$ do not even provide a bounded sequence.

---

[8] For the last inequality **3**, II, 4 is used.

4. In the case of other sequences (x_n) also, little can be said in the first instance about the sequence $\left(\frac{1}{x_n}\right)$ of the reciprocal values. The following is an obvious, but often useful theorem:

○**Theorem 3.** *If the sequence* $(|x_n|)$ *of absolute values of the terms of* (x_n) *have a positive lower bound, — if, therefore, a number* $\gamma > 0$ *exists, such that for every* n,

$$|x_n| \geqq \gamma > 0,$$

then the sequence $\left(\frac{1}{x_n}\right)$ *of reciprocal values is bounded.*

In fact, from $|x_n| \geqq \gamma > 0$ it at once follows that for $K = \frac{1}{\gamma}$ we have

$$\left|\frac{1}{x_n}\right| \leqq K$$

for every n.

In order to increase the scope both of the application of our concepts and of the construction and solution of examples, we insert a paragraph on powers, roots, logarithms and circular functions.

§ 7. Powers, roots and logarithms. Special null sequences.

As, in the discussion of the system of real numbers, it was not our intention to give an exhaustive treatment of all details, but rather to put fundamental ideas alone in a clear light, assuming as known, thereafter, the body of arithmetical rules and concepts, with which after all everyone is thoroughly conversant, so here, in the discussion of powers, roots and logarithms, we will restrict ourselves to an exact elucidation of the basic facts, and then assume known the details of their application.

I. Powers with integral exponents.

If x is an arbitrary number, we know that the symbol x^k for positive integral exponents $k \geqq 2$ is defined as the product of k factors, all equal to x. Here we have therefore only another *notation* for something we know already. By x^1 we mean the number x itself, and if $x \neq 0$, it is convenient to agree, besides, that

$$x^0 \text{ represents the number } 1, \quad x^{-k} \text{ the number } \frac{1}{x^k} \quad (k = 1, 2, 3, \ldots),$$

so that x^p is defined for every integral $p \gtreqless 0$. For these powers* with integral exponents, we merely emphasize the following facts:

1. For arbitrary integral exponents p and q $\left(\gtreqless 0\right)$ the three **29.**
fundamental rules hold:

$$x^p \cdot x^q = x^{p+q}; \quad x^p \cdot y^p = (x\,y)^p; \quad (x^p)^q = x^{p\,q},$$

---

* x^p is a *power* of *base* x and *exponent* p. This *continental* use of the word *power* cannot be here dispensed with, in spite of the slight ambiguity resulting from by far the most frequent use of the word in English to designate the exponent. This sense should be entirely discarded from the reader's mind, notably for § 35, 2a and others. (Tr.)

from which all further rules may be deduced, which regulate calculations with powers[9].

2. Since, in a power with integral exponent, merely a repeated multiplication or division is involved, its calculation has of course to be effected by **18** and **19**. If therefore x is positive and defined for instance by the nest $(x_n \mid y_n)$, with all its endpoints $\geqq 0$ (cf. **15**, 5), then we have simultaneously with

$$= (x_n \mid y_n), \quad x^k = (x_n^k \mid y_n^k) \quad \text{at once,}$$

for all positive integral exponents; and similarly — with appropriate restrictions — for $x \leqq 0$ or $k \leqq 0$.

3. For a *positive* x we have furthermore

$$x^{k+1} \gtreqless x^k \quad \text{according as} \quad x \gtreqless 1$$

as we at once deduce from $x \gtreqless 1$, if we multiply (v. **3**, I, 3) by x^k. — And quite as simply we find:

If x_1, x_2 and the integral exponent k are *positive, then*

$$x_1^k \lesseqgtr x_2^k \quad \text{according as} \quad x_1 \lesseqgtr x_2 .$$

4. For positive integral exponents n and arbitrary a and b we have the formula

$$(a+b)^n = a^n + \binom{n}{1} a^{n-1} b + \binom{n}{2} a^{n-2} b^2 + \cdots$$
$$+ \binom{n}{k} a^{n-k} b^k + \cdots + \binom{n}{n} b^n,$$

where $\binom{n}{k}$, for $1 \leqq k \leqq n$, has the meaning

$$\binom{n}{k} = \frac{n(n-1)(n-2)\ldots(n-k+1)}{1 \cdot 2 \cdot 3 \ldots k}$$

and $\binom{n}{0}$ will be put $= 1$. (Binomial Theorem.)

II. Roots.

If a be any *positive* real number, and k a positive integer, then

$$\sqrt[k]{a}$$

shall denote a number whose k^{th} power $= a$. What interests us here is solely the existence question: Is there such a number, and to what extent is it determined by the problem thus set?

This is dealt with in the

[9] In this, the value 0 for the base x or y is only admissible if the corresponding exponent is positive.

Theorem. *There is, invariably,* one and only one *positive number* ξ *satisfying the equation* **30.**

$$\xi^k = a \qquad (a > 0).$$

We write $\xi = \sqrt[k]{a}$ *and call* ξ *the* k^{th} *root of* a.

Proof. *One* such number may immediately be determined by a nest of intervals, and its existence thereby established: We use the decimal-section method. Since $0^k = 0 < a$, but, p denoting any positive integer $> a$, $p^k \geqq p > a$, — there is one and only one integer $g \geqq 0$ for which[10]

$$g^k \leqq a < (g+1)^k.$$

The interval J_0 determined by g and $(g+1)$ we divide into 10 equal parts and obtain, in the manner now repeatedly worked out, a definite one of the digits 0, 1, 2, ..., 9, — which we may denote, say, by z_1, — and for which

$$\left(g + \frac{z_1}{10}\right)^k \leqq a < \left(g + \frac{z_1+1}{10}\right)^k$$

and so on, and so on. We therefore obtain a nest of intervals $(J_n) = (x_n \mid y_n)$ whose endpoints have definite values of the form

$$x_n = g + \frac{z_1}{10} + \frac{z_2}{10^2} + \cdots + \frac{z_{n-1}}{10^{n-1}} + \frac{z_n}{10^n} \qquad (n = 1, 2, 3, \ldots)$$

and

$$y_n = g + \frac{z_1}{10} + \frac{z_2}{10^2} + \cdots + \frac{z_{n-1}}{10^{n-1}} + \frac{z_n + 1}{10^n}.$$

If $\xi = (x_n \mid y_n)$ be the number thereby determined, then since here all endpoints of intervals are $\geqq 0$, it at once follows by **29**, 2 that

$$\xi^k = (x_n^k \mid y_n^k).$$

But, by construction, $x_n^k \leqq a \leqq y_n^k$ for every n; hence, by § 5, Theorem 4, we must have

$$\xi^k = a.$$

That this number ξ is, moreover, the *only* positive solution of the problem, follows directly from **29**, 3, since it was there pointed out that for a positive $\xi_1 \neq \xi$, necessarily $\xi_1^k \neq \xi^k$, i. e. $\neq a$.

If k is an even number, then $-\xi$ is also a solution of the problem. We shall not, however, take this into account in the following pages, but interpret the kth root of a positive number a as meaning *only the positive number* ξ, *completely and uniquely determined by* **30**[11]. — For $a = 0$, we may also put $\sqrt[k]{a} = 0$[12]

[10] g is the last of the numbers $0, 1, 2, \ldots, p$ whose kth power is $\leqq a$.

[11] In accordance with this we have, for instance, $\sqrt{x^2}$ *not* always $= x$, but always $= |x|$.

[12] For negative a's we will not define $\sqrt[k]{a}$ at all; we can, however, if k is odd, write $\sqrt[k]{a} = -\sqrt[k]{|a|}$.

We will not enter further into the rules for calculations with roots, but consider them as familiar to every one, and will only prove the following simple theorems:

29, 3 gives at once the

31. Theorem 1. *If* $a > 0$ *and* $a_1 > 0$, *then* $\sqrt[k]{a} \lesseqgtr \sqrt[k]{a_1}$, *according as* $a \lesseqgtr a_1$. — Further we have the

Theorem 2. *If* $a > 0$, *then* $\left(\sqrt[n]{a}\right)$ *is a monotone sequence; and we have, more precisely,*

$$a > \sqrt{a} > \sqrt[3]{a} > \cdots > 1, \quad \textit{if } a > 1,$$

but

$$a < \sqrt{a} < \sqrt[3]{a} < \cdots < 1, \quad \textit{if } a < 1.$$

(For $a = 1$, *the sequence is of course* $\equiv 1$.)

Proof. By **29**, 3, $a > 1$ involves $a^{n+1} > a^n > 1$, and therefore by the preceding theorem, taking $n(n+1)^{\text{th}}$ roots,

$$\sqrt[n]{a} > \sqrt[n+1]{a} > 1.$$

Since for $a < 1$ *all* the inequality signs are reversed, this proves the whole statement. — Hence finally we deduce the

Theorem 3. *If* $a > 0$, *then the numbers*

$$x_n = \sqrt[n]{a} - 1$$

form a null sequence (monotone by the preceding theorem).

Proof. For $a = 1$, the assertion is trivial, as then $x_n \equiv 0$. If $a > 1$, and therefore $\sqrt[n]{a} > 1$, i. e. $x_n = \sqrt[n]{a} - 1 > 0$, then we reason as follows: By the inequality of Bernoulli (v. **10**, 7), $\sqrt[n]{a} = 1 + x_n$ gives

$$a = (1 + x_n)^n > 1 + n x_n > n x_n.$$

Consequently $x_n = |x_n| < \frac{a}{n}$, therefore (x_n), by **26**, 1 or 2, is a null sequence.

If $0 < a < 1$, then $\frac{1}{a} > 1$, and so, by the result obtained,

$$\left(\sqrt[n]{\frac{1}{a}} - 1\right)$$

is a null sequence. If we multiply this term by term by the factors $\sqrt[n]{a}$, — which certainly form a bounded sequence, as $a \leqq \sqrt[n]{a} < 1$, — then it at once follows, by **26**, 2, that

$$\left(1 - \sqrt[n]{a}\right), \quad \text{and therefore also } (x_n),$$

is a null sequence, — q. e. d.

III. Powers with rational exponents.

We again regard as substantially known, in what manner one may pass from roots with integral exponents to powers with any rational exponent: *By* $a^{\frac{p}{q}}$, *with integral* $p \gtreqless 0$, $q > 0$, *we mean, for any positive a, the positive number uniquely defined by*

$$a^{\frac{p}{q}} = \left(\sqrt[q]{a}\right)^p.$$

If $p > 0$, then a may also be $= 0$; $a^{\frac{p}{q}}$ must then be taken to have the value 0.

With these definitions, the three fundamental rules **29**, 1, i. e. the formulae

$$a^r \cdot a^{r'} = a^{r+r'}; \quad a^r b^r = (a\,b)^r; \quad (a^r)^{r'} = a^{r r'}$$

remain unaltered, for any rational exponents, and therefore calculations with these powers are formally the same as when the exponents are integers.

These formulae contain, at the same time, all the rules for working with roots, since every root may now be written as a power with a rational exponent. — Of the less known results we may prove, as they are particularly important for the sequel, these theorems:

Theorem 1. *When* $a > 1$, — *then* $a^r > 1$, *if, and only if,* $r > 0$. **32.**
Similarly, when $a < 1$ *(but positive), then* a^r *is* < 1 *if, and only if,* $r > 0$.

Proof. By **31**, 2, a and $\sqrt[q]{a}$ are either *both* greater or *both* less than 1; by **29** the same is true of a and $\left(\sqrt[q]{a}\right)^p = a^r$ if and only if $p > 0$.

Theorem 1a. *If the rational number* $r > 0$, *and both bases are positive, then* $a^r \lesseqgtr a_1{}^r$, *according as* $a \lesseqgtr a_1$.

The proof is at once obtained from **31**, 1 and **29**, 3.

Theorem 2. *If* $a > 0$, *and the rational number* r *lies between the rational numbers* r' *and* r'', *then* a^r *also always lies between* $a^{r'}$ *and* $a^{r''}$[13], *and conversely,* — *whether* a *be* $<$, $=$ *or* > 1, *and* $r' <$, $=$ *or* $> r''$.

Proof. If, firstly, $a > 1$ and $r' < r''$, then

$$a^r = a^{r'} \cdot a^{r-r'} = \frac{a^{r''}}{a^{r''-r}}.$$

[13] The term "between" may be taken, as we please, either to *include* or *exclude* equality on both sides, — excepting when $a = 1$, and therefore all the powers a^r also $= 1$.

By Theorem **1**, this already proves the validity of our statement for this case, and in the other possible cases the proof is quite as easy. — From this proof we deduce, indeed, more precisely, the

Theorem 2a. *If $a > 1$, then to the larger (rational) exponent also corresponds the larger value of the power. If $a < 1$ (but positive) then the larger exponent gives the smaller power. — In particular: If the (positive) base $a \neq 1$, then different exponents give different powers.* — Hence we deduce, further,

Theorem 3. *If (r_n) is any (rational) null sequence, then the numbers*

$$x_n = a^{r_n} - 1, \qquad (a > 0)$$

also form a null sequence. If (r_n) is monotone, then so is (x_n).

Proof. By **31**, 3, $\left(\sqrt[n]{a} - 1\right)$ and $\left(\sqrt[n]{\frac{1}{a}} - 1\right)$ are null sequences. If therefore $\varepsilon > 0$ be given, we can so choose n_1 and n_2 that

$$\text{for } n > n_1, \quad \left|\sqrt[n]{a} - 1\right| < \varepsilon,$$

$$\text{and for } n > n_2, \quad \left|\sqrt[n]{\frac{1}{a}} - 1\right| < \varepsilon.$$

If m is an integer larger than both n_1 and n_2, then the numbers $\left(a^{\frac{1}{m}} - 1\right)$ and $\left(a^{-\frac{1}{m}} - 1\right)$ *both* lie between $-\varepsilon$ and $+\varepsilon$, i. e.

$$a^{\frac{1}{m}} \text{ and } a^{-\frac{1}{m}} \text{ lie between } 1 - \varepsilon \text{ and } 1 + \varepsilon.$$

By Theorem 2, a^r then lies between the same bounds, if r lies between $-\frac{1}{m}$ and $+\frac{1}{m}$. By hypothesis we can, however, so choose n_0, that for every $n > n_0$,

$$|r_n| < \frac{1}{m} \quad \text{or} \quad -\frac{1}{m} < r_n < +\frac{1}{m};$$

for $n > n_0$, a^{r_n} is therefore between $1 - \varepsilon$ and $1 + \varepsilon$. Hence, for these n's,

$$\left|a^{r_n} - 1\right| < \varepsilon,$$

proving that $(a^{r_n} - 1)$ is a null sequence. — That it is monotone, if (r_n) is, follows immediately from Theorem 2a.

These theorems form the basis for the definition of

IV. Powers with arbitrary real exponents.

For this we first state the

33. **Theorem.** *If $(x_n \mid y_n)$ is any nest of intervals (with rational end-points) and a is positive, then*

$$\textit{for } a \geqq 1, \quad \sigma = (a^{x_n} \mid a^{y_n})$$

$$\textit{and for } a \leqq 1, \quad \sigma = (a^{y_n} \mid a^{x_n})$$

is also a nest of intervals. And if $(x_n \mid y_n)$ *is rational valued and* $= r$, *then* $\sigma = a^r$.

Proof. That in either case the left endpoints form a monotone ascending sequence, the right endpoints a monotone descending sequence, follows at once from **32**, 2a. By the same theorem, $a^{x_n} \leqq a^{y_n}$ in the one case ($a \geqq 1$) and $a^{y_n} \leqq a^{x_n}$ in the other ($a \leqq 1$), for every n. Finally, that in both cases the lengths of the intervals form a null sequence, follows, with the aid of **26**, from

$$\left| a^{y_n} - a^{x_n} \right| = \left| a^{y_n - x_n} - 1 \right| \cdot a^{x_n};$$

for here the first factor, by **32**, 3, is a null sequence, because $(y_n - x_n)$ is by hypothesis a null sequence with rational terms; and the second factor is bounded, because for every n

$$0 < a^{x_n} \leqq a^{y_1}$$

in the one case ($a \geqq 1$),

$$\leqq a^{x_1}$$

in the other ($a \leqq 1$).

Now if $(x_n \mid y_n) = r$, then r lies between x_n and y_n, for every n, and so by **32**, 2, a^r lies between a^{x_n} and a^{y_n}, for every n; hence by § 5, Theorem 4, necessarily $\sigma = a^r$.

In consequence of this theorem, we may agree to the following

Definition [14]. *If* $a > 0$, *and* $\varrho = (x_n \mid y_n)$ *is an arbitrary real number, then:*

$$a^\varrho = \sigma, \text{ i. e. } \begin{cases} = (a^{x_n} \mid a^{y_n}) & \text{if } a \geqq 1 \\ = (a^{y_n} \mid a^{x_n}) & \text{if } a \leqq 1. \end{cases}$$

This definition can of course only be regarded as appropriate, if the concept of a *general power* thereby determined obeys substantially the same laws as the type of power so far considered, that with rational exponents. That this is so, in the fullest sense, is shewn by the following considerations.

1. For rational exponents, the new definition gives the same result as the old. **34.**

2. If $\varrho = \varrho'$, then [15] $a^\varrho = a^{\varrho'}$.

---

[14] This combination **33** of theorem and definition is, from the point of view of method, of exactly the same kind as those set forth in **14**—**19**: What is demonstrable in the case of rational exponents is raised, in the case of arbitrary exponents, to the rank of a definition, — whose appropriateness has then to be verified.

[15] This assertion, formally rather trivial in appearance, when put somewhat more explicitly, runs thus: If $(x_n \mid y_n) = \varrho$ and $(x_n' \mid y_n') = \varrho'$ are two nests of intervals, which may be regarded as equal in the sense of **14**, then so are those nests of intervals equal (again in the sense of **14**), which by Definition **33** give the powers a^ϱ and $a^{\varrho'}$.

3. For two arbitrary real numbers ϱ and ϱ', and positive a and b, the three fundamental rules

$$a^{\varrho} \cdot a^{\varrho'} = a^{\varrho+\varrho'}; \qquad (a^{\varrho} \cdot b^{\varrho}) = (a\,b)^{\varrho}; \qquad (a^{\varrho})^{\varrho'} = a^{\varrho\varrho'},$$

hold, so that with the general powers now introduced we may calculate formally in precisely the same way as with the special types hitherto used.

Into the extremely simple proofs of these facts we will, as emphasized on p. 49, not enter further[16]; we will also, so far as concerns the extension of theorems **32**, 1—3 to general powers, now immediately possible, content ourselves with the statement and a few indications of the proof. We have therefore the theorems, generalized from **32**, 1—3:

35. **Theorem 1.** *When $a > 1$, we have $a^{\varrho} > 1$ if, and only if, $\varrho > 0$. Similarly, when $a < 1$, (but positive), we have $a^{\varrho} < 1$ if, and only if, $\varrho > 0$.*

For by **32**, **1**, we have e. g. for $a > 1$, $a^{x_n} > 1$ if, and only if, $x_n > 0$.

Theorem 1a. *If the real number ϱ* is > 0, *and both bases are positive, then $a^{\varrho} \lesseqgtr a_1^{\varrho}$, according as $a \lesseqgtr a_1$.*

Proof by **32**, 1a and **15**.

Theorem 2. *If $a > 0$ and ϱ is between ϱ' and ϱ'', then a^{ϱ} is always between $a^{\varrho'}$ and $a^{\varrho''}$.* — The proof is precisely the same as **32**, 2. It yields, more exactly, the

Theorem 2a. *If $a > 1$, then to the larger exponent corresponds the larger value of the power; if $a < 1$ (but positive), then the larger exponent gives the smaller power. In particular: If $a \neq 1$, then different exponents give different powers.* — And from this theorem, exactly as in **32**, 3, follows the final

---

[16] As a model we may sketch the proof of the first of the three fundamental rules: If $\varrho = (x_n \,|\, y_n)$ and $\varrho' = (x_n' \,|\, y_n')$, then by **16**, $\varrho + \varrho' = (x_n + x_n' \,|\, y_n + y_n')$ and therefore — we assume $a \geqq 1$ —:

$$a^{\varrho} = \left(a^{x_n} \,\middle|\, a^{y_n}\right), \qquad a^{\varrho'} = \left(a^{x_n'} \,\middle|\, a^{y_n'}\right), \qquad a^{\varrho+\varrho'} = \left(a^{x+x_n'} \,\middle|\, a^{y_n+y_n'}\right).$$

Since all endpoints (as powers with rational exponents) are positive, we have, by **18**,

$$a^{\varrho} \cdot a^{\varrho'} = \left(a^{x_n} \cdot a^{x_n'} \,\middle|\, a^{y_n} \cdot a^{y_n'}\right).$$

Since, however, for rational exponents, the first of the three fundamental rules has already been seen to hold, this last nest of intervals is not only equal, in the sense of **14**, to that defining $a^{\varrho+\varrho'}$, but even coincides with it term by term.

Theorem 3. *If* (ϱ_n) *is any null sequence, then the numbers*

$$x_n = a^{\varrho_n} - 1 \qquad (a > 0)$$

form a null sequence. If (ϱ_n) *is monotone, then so is* (x_n).

As a special application, we may mention the

Theorem 4. *If* (x_n) *is a null sequence with all its terms positive, then for every positive* α,

$$x_n' = x_n^{\alpha},$$

is also the term of a null sequence. — Thus $\left(\frac{1}{n^\alpha}\right)$ for every $\alpha > 0$ is a null sequence.

Proof. If $\varepsilon > 0$ be given arbitrarily, $\varepsilon^{\frac{1}{\alpha}}$ is also a positive number. By hypothesis, we can choose n_0 so that, for every $n > n_0$ (cf. **10**, 4 and 12),

$$|x_n| = x_n < \varepsilon^{\frac{1}{\alpha}}.$$

For $n > n_0$, by **35**, 1a, we then also have, however,

$$x_n^{\alpha} = |x_n'| < \varepsilon$$

which at once proves the whole statement.

The above theorems comprise the main principles used in calculations with generalized powers.

V. Logarithms.

The foundation for the definition of logarithms lies in the

Theorem. *If* $a > 0$ *and* $b > 1$ *are two real, and in all further* **36.**
respects quite arbitrary numbers, then one and only one real number ξ *always exists, for which*

$$b^{\xi} = a.$$

Proof. That *at most one* such number can exist, already follows from **35**, 2a, because the base b with different exponents cannot give the same value a. That such a number *does* exist, we show constructively, by assigning a nest of intervals which determines it, — thus for instance by the method of decimal sections: Since $b > 1$, $(b^{-n}) = \left(\frac{1}{b^n}\right)$ is a null sequence, by **10**, 7, and there exists, consequently, since a and $\frac{1}{a}$ are positive, natural numbers p and q for which

$$b^{-p} < a \quad \text{and} \quad b^{-q} < \frac{1}{a} \quad \text{or} \quad b^q > a.$$

If, now, we consider the various integers between $-p$ and $+q$ in succession, as exponents of b, there must be one, and can be *only one* — call it g — for which

$$b^g \leqq a, \quad \text{but} \quad b^{g+1} > a.$$

The interval $J_0 = g \ldots (g+1)$ thereby determined we divide into 10 equal parts and obtain, just as on p. 51, a "digit" z_1, for which

$$b^{g+\frac{z_1}{10}} \leqq a, \quad \text{but} \quad b^{g+\frac{z_1+1}{10}} > a.$$

By repetition of the process of subdivision we find a perfectly definite nest of intervals

$$\xi = (x_n \mid y_n), \quad \text{with} \quad \begin{cases} x_n = g + \frac{z_1}{10} + \cdots + \frac{z_{n-1}}{10^{n-1}} + \frac{z_n}{10^n}, \\ y_n = g + \frac{z_1}{10} + \cdots + \frac{z_{n-1}}{10^{n-1}} + \frac{z_n+1}{10^n}, \end{cases}$$

for which

$$b^{x_n} \leqq a < b^{y_n}$$

for every n, — for which, therefore, in accordance with **33**,

$$b^{\xi} = a.$$

This theorem justifies us in the following

Definition. *If $a > 0$ and $b > 1$ are arbitrarily given, then the real number ξ, uniquely determined by*

$$b^{\xi} = a$$

is called the logarithm of a to the base b; and, symbolically,

$$\xi = \log_b a.$$

(g is also called the characteristic, and the set of the digits $z_1, z_2, z_3 \ldots$ the mantissa, of the logarithm.)

We speak of a system of logarithms, when the base b is assumed fixed once for all and the logarithms of all possible numbers are taken to this base b. The suffix b in $\log_b$ is then usually omitted as superfluous. Very soon a particular real number, usually denoted by e, appears quite naturally as the most convenient for all theoretical considerations; the system of logarithms built up on this base is usually called the system of *natural* logarithms. For practical purposes, however, the base 10 is, as we know, the most convenient; logarithms to this base are called *common* or *Briggs' logarithms*. These are the logarithms found in all the ordinary tables[17].

The rules for working with logarithms we assume, as we did with powers, to be already known, and content ourselves with a mere mention of the most important of them. If the base $b > 1$ is arbitrary,

[17] As a matter of course, a system of logarithms may also be built up on a positive base *less* than 1. This, however, is not usual. The first logarithms calculated by *Napier* in 1614 were, however, built up on a base $b < 1$, which presents some small advantages, particularly for logarithms of trigonometrical functions. Neither *Napier* nor *Briggs*, however, really used any base. The idea of logarithms as the inverse of powers only developed in the course of the 18th century.

but assumed fixed in what follows, and if $a, a', a'' \ldots$ denote any positive numbers, then

1. $\log(a' a'') = \log a' + \log a''$. **37.**
2. $\log 1 = 0$; $\log \frac{1}{a} = -\log a$; $\log b = 1$.
3. $\log a^\varrho = \varrho \log a$ (ϱ arbitrary, real).
4. $\log a \lesseqgtr \log a'$, according as $a \lesseqgtr a'$; in particular,
5. $\log a \gtreqless 0$, according as $a \gtreqless 1$.
6. If b and b_1 are two different bases (> 1), and ξ and ξ_1 the logarithms of the same number a to these two bases, i. e.

$$\xi = \log_b a, \qquad \xi_1 = \log_{b_1} a,$$

then

$$\xi = \xi_1 \cdot \log_b b_1, \text{ —}$$

as follows at once from $(a =) b^\xi = b_1{}^{\xi_1}$, by taking logarithms on both sides to the base b and taking account of **37**, 2 and 3.

7. $\left(\frac{1}{\log n}\right)$, $n = 2, 3, 4, \ldots$ is a null sequence. In fact $\frac{1}{\log n} < \varepsilon$, provided $\log n > \frac{1}{\varepsilon}$, that is, $n > b^{\frac{1}{\varepsilon}}$.

VI. Circular functions.

To introduce the so-called circular functions (the sine of a given angle[18], with the cosine, tangent, cotangent etc.) in an equally strict manner, i. e. avoiding on principle all reference to geometrical intuition as *element of proof* and founding solely on the concept of the real number, is at this stage not yet possible. This question will be resumed later (§ 24). In spite of this, we will unhesitatingly enlist them to enrich our applications and enliven our examples (but of course never to prove general propositions), in so far as their knowledge may be presupposed from elementary work.

Thus e. g. the following two simple facts can at once be ascertained: **37a.**

1. If $\alpha_1, \alpha_2, \ldots, \alpha_n, \ldots$ are any angles (that is to say, any numbers), then

$$(\sin \alpha_n) \quad \text{and} \quad (\cos \alpha_n)$$

are bounded sequences; and

[18] Angles will in general be measured in radians. If in a circle of radius unity we imagine the radius to turn from a definite initial position, then we measure the angle of turning by the length of the path which the extremity of the moving radius has traversed — taking it as *positive* when the sense of turning is *counterclockwise*, otherwise as *negative*. An angle is accordingly a pure number; a straight angle has the measure $+\pi$ or $-\pi$, a right angle the measure $+\frac{\pi}{2}$ or $-\frac{\pi}{2}$. To every definitely placed angle there belongs an infinite number of measures which, however, differ from one another only by integral multiples of 2π, i. e. by whole turns. The measure 1 belongs to the angle, the arc corresponding to which is equal to the radius, and which therefere in degrees is $57^\circ 17' 44''{\cdot}8$ nearly.

2. the sequences

$$\left(\frac{\sin \alpha_n}{n}\right) \quad \text{and} \quad \left(\frac{\cos \alpha_n}{n}\right)$$

are (by **26**) null sequences, for their terms are derived from those of the null sequence $\left(\frac{1}{n}\right)$ by multiplication by bounded factors.

VII. Special null sequences.

As a further application of the concepts now defined, we will examine a number of *special sequences*:

38. °1. *If* $|a| < 1$, *then besides* (a^n) *even* $(n\,a^n)$ *is a null sequence.*

Proof. Our reasoning is analogous to that of **10**, 7[19]: For $a = 0$, the assertion is trivial; for $0 < |a| < 1$, we may write, with $\varrho > 0$,

$$|a| = \frac{1}{1+\varrho}, \quad \text{and therefore} \quad |a^n| = \frac{1}{1+\binom{n}{1}\varrho + \cdots + \binom{n}{n}\varrho^n}.$$

Since here in the denominator each term of the sum is positive, we have for every $n > 1$,

$$|a^n| < \frac{1}{\binom{n}{2}\varrho^2}, \quad \text{therefore} \quad |n\,a^n| < \frac{1\cdot 2}{(n-1)\,\varrho^2}.$$

Thus we have

$$|n\,a^n| < \varepsilon, \quad \text{as soon as} \quad \frac{1\cdot 2}{(n-1)\,\varrho^2} < \varepsilon$$

i. e. for every

$$n > 1 + \frac{2}{\varepsilon\cdot\varrho^2}.$$

The result thus proved is very remarkable: it asserts, in fact, that for a large n the fraction $\frac{n}{(1+\varrho)^n}$ is very small, and its denominator therefore very much greater than its numerator. This denominator is however constant $(=1)$ for $\varrho = 0$, and when ϱ is very small (and positive), it only increases very slowly with n. Nevertheless, our result shows that provided only n be taken *sufficiently* large, the denominator is very much larger than the numerator[20]. The point n_0, from which $|n\,a^n| = \frac{n}{(1+\varrho)^n}$ lies below a given ε — we found $n_0 = 1 + \frac{2}{\varepsilon\cdot\varrho^2}$ — does indeed lie very far to the right, not only when ε, but also when $\varrho = \frac{1}{|a|} - 1$, is very small (i. e. $|a|$ very near to 1). Substantially this

---

[19] Except that a and ϱ need no longer be rational.

and only this is true: However $|a| < 1$ and $\varepsilon > 0$ may be given, we have always, from a readily assignable point onwards, $|n a^n| < \varepsilon$.

From this result many others may be deduced, e. g. the still more paradoxical fact:

°2. *If* $|a| < 1$ *and* α *real and arbitrary, then* $(n^\alpha a^n)$ *is also a null sequence.*

Proof. If $\alpha \leqq 0$, then this is evident from **10**, 7, because of **26**, 2; if $\alpha > 0$, write $|a|^{\frac{1}{\alpha}} = a_1$, so that by **35**, 1a, the positive number a_1 is also < 1. By the preceding result, $(n\, a_1{}^n)$ is a null sequence. By **35**, 4

$$[n\, a_1{}^n]^\alpha, \quad \text{i. e.} \quad n^\alpha |a|^n \quad \text{or} \quad |n^\alpha a^n|,$$

therefore, finally, (by **10**, 5), $n^\alpha a^n$ itself is also the term of a null sequence [21].

3. *If* $\sigma > 0$, then $\left(\frac{\log n}{n^\sigma}\right)$ *is a null sequence* [22], *to whatever base* $b > 1$ *the logarithms are taken.*

Proof. Since $b > 1$, $\sigma > 0$, we have (by **35**, 1a), $b^\sigma > 1$. Therefore $\left(\frac{n}{(b^\sigma)^n}\right)$ is a null sequence, by 1. Given $\varepsilon > 0$, we have consequently from a certain point onwards, — say for every $n > m$ —

$$\frac{n}{(b^\sigma)^n} < \varepsilon' = \frac{\varepsilon}{b^\sigma}.$$

But, in any case,

$$\frac{\log n}{n^\sigma} < \frac{g+1}{(b^g)^\sigma} = b^\sigma \cdot \frac{g+1}{(b^\sigma)^{g+1}},$$

if g denote the characteristic of $\log n$ (so that $g \leqq \log n < g + 1$). If, therefore, we take $n > b^m$, $\log n$, and *a fortiori* $g + 1$, is $> m$. Hence the last value above, with our choice of m, is

$$< b^\sigma \cdot \frac{\varepsilon}{b^\sigma} = \varepsilon, \quad \text{i. e.} \quad \frac{\log n}{n^\sigma} < \varepsilon \quad \text{for every} \quad n > n_0 = b^m.$$

---

[20] Writing as above $|a| = \frac{1}{1+\varrho}$, $|n a^n| = \frac{n}{(1+\varrho)^n}$, we may also say: $(1+\varrho)^n$ becomes — for a positive ϱ — *more pronouncedly large*, or, also *more pronouncedly infinite*, than n itself, — by which we again (cf. 7, 3) mean nothing more and nothing less than that our sequence is precisely a null sequence. — For future reference we remark here that the results proved in 1 and 2 are also valid for a complex a, provided only $|a| < 1$.

[21] With the same change of notation as above, we may say here: "$(1+\varrho)^n$ becomes *more pronouncedly infinite* than every (fixed) power however large of n itself".

[22] Or, in words, "$\log n$ becomes *less pronouncedly large* than every power, however small (but determinate and positive), of n itself".

4. *If α and β are arbitrary positive numbers, then*

$$\left(\frac{(\log n)^\alpha}{n^\beta}\right)$$

is a null sequence —, however large α and however small β may be[23].

Proof. By 3., $\left(\frac{\log n}{n^{\beta/\alpha}}\right)$ is a null sequence, because $\frac{\beta}{\alpha} > 0$; by **35**, 4, therefore, so is the given sequence.

5. $(x_n) \equiv (\sqrt[n]{n} - 1)$ *is a null sequence.* (This result is also very remarkable. For when n is large, we have a large number under the $\sqrt{}$; the exponent of the $\sqrt{}$ is, it is true, also large; but it is not at all evident *a priori* which of the two — radicand or exponent — will, so to speak, prove the stronger.)

Proof. For $n > 1$, we certainly have $\sqrt[n]{n} > 1$, therefore $x_n = \sqrt[n]{n} - 1$ certainly > 0. Hence in

$$n = (1 + x_n)^n = 1 + \binom{n}{1} x_n + \cdots + \binom{n}{n} x_n^n$$

all the terms of the sum are positive. Consequently we have, in particular,

$$n > \binom{n}{2} x_n^2 = \frac{n(n-1)}{1 \cdot 2} x_n^2$$

or[24]

$$x_n^2 < \frac{2}{n-1} \leqq \frac{2}{n - \frac{n}{2}} = \frac{4}{n}.$$

Hence

$$|x_n| < \frac{2}{n^{\frac{1}{2}}},$$

so that $(x_n) = \sqrt[n]{n} - 1$ is in fact by **26**, 1 and **35**, 4 a null sequence.

6. *If (x_n) is a null sequence whose terms are all > -1, then for every (fixed) integer k, the numbers*

$$x_n' = \sqrt[k]{1 + x_n} - 1$$

also form a null sequence[25].

---

[23] "Every power of $\log n$, however large, (but fixed) becomes less pronouncedly large than every power of n itself, however small (but fixed).

[24] The substitution, when $n > 1$, of the value $n - \frac{n}{2}$ for $(n-1)$ which it cannot exceed, is an artifice often useful in simplifying calculations.

[25] By the assumption that *all* x_n's > -1, we merely wish to ensure that the numbers x_n' are defined for *every* n. From a definite point onwards this is automatically the case, since (x_n) is assumed to be a null sequence and therefore from some point certainly $|x_n| < 1$, and hence $x_n > -1$.

Proof. From the formulae set forth on p. 22, Footnote 13, where we put $a = \sqrt[k]{1 + x_n}$ and $b = 1$, it follows that[26]

$$x_n' = \frac{x_n}{\left(\sqrt[k]{1+x_n}\right)^{k-1} + \left(\sqrt[k]{1+x_n}\right)^{k-2} + \ldots + 1};$$

therefore, since the terms in the denominator are all positive and the last is 1,

$$|x_n'| \leqq |x_n|;$$

whence, by **26**, the statement at once follows.

7. *If* (x_n) *is a null sequence of the same kind as in* 6., *then the numbers*

$$y_n = \log(1 + x_n)$$

also form a null sequence.

Proof. If $b > 1$ is the base to which the logarithms are taken, and $\varepsilon > 0$ is given, we write

$$b^\varepsilon - 1 = \varepsilon_1, \qquad 1 - b^{-\varepsilon} = \varepsilon_2$$

so that we have $\varepsilon_1 = b^\varepsilon \cdot \varepsilon_2 > \varepsilon_2 > 0$. We then choose n_0 so large, that for every $n > n_0$, $|x_n| < \varepsilon_2$. For those n's we have, *a fortiori*,

$$-\varepsilon_2 < x_n < \varepsilon_1, \text{ i.e. } b^{-\varepsilon} < 1 + x_n < b^{+\varepsilon};$$

therefore (by **35**, 2 or **37**, 4)

$$|y_n| = |\log(1 + x_n)| < \varepsilon;$$

with which the statement is proved.

8. *If* (x_n) *is again a null sequence of the same kind as in* 6., *then the numbers*

$$z_n = (1 + x_n)^\varrho - 1$$

also form a null sequence, if ϱ *denote any real number.*

Proof. By 7. and **26**, 3, the numbers

$$\varrho_n = \varrho \cdot \log(1 + x_n)$$

form a null sequence. By **35**, 3 and **37**, 3 the same is true of the numbers

$$b^{\varrho_n} - 1 = (1 + x_n)^\varrho - 1 = z_n, \qquad \text{q. e. d.}$$

---

[26] We assume $k \geqq 2$, since for $k = 1$ the assertion is trivial.

§ 8. Convergent sequences.

Definitions.

So far, when considering the behaviour of a given sequence, we have been chiefly concerned to discover whether it was a null sequence or not. By extending this point of view somewhat, in a manner which readily suggests itself, we reach the most important concept of all with which we shall have to deal, namely, that of the *convergence of a sequence.*

We have already (cf. **10**, 10) described the property which a sequence (x_n) may have, of being a null sequence, by saying that its members *become* small, become *arbitrarily small*, with increasing n. We may also say: Its terms, as n increases, approach the value 0, — without, in general, ever reaching it, it is true; but they approach *arbitrarily* near to this value in the sense that the values of its terms (that is to say, their *differences from* 0) sink below every number ε (> 0), however small. If we substitute for the value 0 in this conception any other real number ξ, we shall be concerned with a sequence (x_n) for which the differences of the various terms from the definite number ξ — that is to say, by **3**, II, 6, the values $|x_n - \xi|$, — sink, with increasing n, below every number $\varepsilon > 0$, however small.

We state the matter more precisely in the following:

39. °Definition. *If (x_n) is a given sequence, and if it is related to a definite number ξ in such a way that*

$$(x_n - \xi)$$

forms a **null sequence**[1], *then we say that the sequence (x_n)* **converges** *to ξ, or that it is* **convergent**. *The number ξ is called the limiting value or* **limit** *of this sequence; the sequence is also said to* **converge to ξ**, *and we say that its terms approach the (limiting) value ξ, tend to ξ, have the limit ξ. This fact is expressed by the symbols*

$$x_n \to \xi \quad \textit{or} \quad \lim x_n = \xi.$$

To make it plainer that the approach to ξ is effected by taking the index n larger and larger, we also frequently write[2]

$$x_n \to \xi \quad \textit{for} \quad n \to \infty \quad \textit{or} \quad \lim_{n \to \infty} x_n = \xi.$$

Including the definition of a null sequence in the new definition, we may also say:

$x_n \to \xi$ for $n \to \infty$ (or $\lim\limits_{n\to\infty} x_n = \xi$) if for every chosen $\varepsilon > 0$, we can always assign a number $n_0 = n_0(\varepsilon)$, so that for every $n > n_0$, we have

$$|x_n - \xi| < \varepsilon.$$

---

[1] Or $(\xi - x_n)$ or $|x_n - \xi|$; by **10**, 5 the result is exactly the same.

[2] Read: "x_n (tends) towards ξ for n tending to infinity" in the one case, and "Limit x_n for n tending to infinity equals ξ" in the other. In view of the definitions **40**, 2 and 3, it would be more correct to write here "$n \to +\infty$"; but for simplicity the $+$ sign is usually omitted.

Remarks and Examples.

1. Instead of saying "(x_n) is a null sequence", we may now, more shortly, write "$x_n \to 0$". Null sequences are convergent sequences with the special limiting value 0.

2. Substantially, all remarks made in **10** therefore hold here, since we are concerned only with a very obvious generalisation of the concept of a null sequence.

3. By **31**, 3 and **38**, 5, we have for $a > 0$

$$\sqrt[n]{a} \to 1 \quad \text{and} \quad \sqrt[n]{n} \to 1.$$

4. If $(x_n \mid y_n) = \sigma$, then $x_n \to \sigma$ and $y_n \to \sigma$. For both

$$|x_n - \sigma| \quad \text{and also} \quad |y_n - \sigma| \quad \text{are} \quad \leqq |y_n - x_n|,$$

so that both, by **26**, 1, form null sequences together with $(y_n - x_n)$.

5. For $x_n = 1 - \frac{(-1)^n}{n}$, that is, for the sequence $2, \frac{1}{2}, \frac{4}{3}, \frac{3}{4}, \frac{6}{5}, \frac{5}{6}, \ldots, x_n \to 1$, for $|x_n - 1| = \frac{1}{n}$ forms a null sequence.

6. In geometrical language, $x_n \to \xi$ means that all terms with sufficiently large indices lie in the neighbourhood of the fixed point ξ. Or more precisely (cf. **10**, 13), in every ε-neighbourhood of ξ, the whole of the terms, *with at most a finite number of exceptions*, are to be found [3]. — In applying the mode of representation of **7**, 6, we draw parallels to the axis of abscissae, through the two points $(0, \xi \pm \varepsilon)$ and may say: $x_n \to \xi$, if the whole graph of the sequence (x_n), with the exception of a finite initial portion, lies in every ε-strip (however narrow).

7. The lax mode of expression: "for $n = \infty$, $x_n = \xi$" instead of $x_n \to \xi$, should be most emphatically rejected. — For *an integer* $n = \infty$ *does not exist and* x_n *need never be* $= \xi$. We are concerned merely with a process of approximation, sufficiently clear from all that precedes, which there is no ground whatever for imagining *completed* in any form. (In older text books and writings we frequently find, however, the symbolical mode of writing: "$\lim\limits_{n=\infty} x_n = \xi$", to which, since it is after all meant only symbolically, no objection can be taken, — excepting that it is clumsy, and that writing "$n = \infty$" must necessarily create some confusion regarding the concept of the infinite in mathematics.

8. If $x_n \to \xi$, then the isolated terms of the sequence (x_n) are also called *approximations* to ξ, and the difference $\xi - x_n$ is called the *error* corresponding to the approximation x_n.

9. The name "*convergent*" appears to have been first used by *J. Gregory* (*Vera circuli et hyperbolae quadratura*, Padua 1667), and "*divergent*" (**40**) by *Bernoulli* (Letter to *Leibniz* of 7. 4. 1713). It was through the publications of *A. L. Cauchy* (see p. 72, footnote 18) that a limiting value came to be denoted generally by the prefixed symbol "lim". The arrow sign ($\to$), which is so particularly appropriate, came into common use after 1906, through the works of *G. H. Hardy*, who himself referred it back to *J. G. Leatham* (1905).

To the definition of convergence we at once append that of divergence:

°Definition 1. *Every sequence which is not convergent in the sense of* **39** *is called* ***divergent****.* **40.**

---

[3] Frequently this is expressed more briefly: In every ε-neighbourhood of ξ "almost all" terms of the sequence are situated. The expression "almost all" has, however, other meanings, e. g. in the Theory of Sets of Points.

With this definition, the sequences **6**, 2, 4, 7, 8, **11** are certainly divergent.

Among divergent sequences, one type is distinguished by its particularly simple and transparent behaviour, e. g. the sequences (n^2), (n), (a^n) for $a > 1$, $(\log n)$, and others. Their common property is evidently that the terms increase with increasing n beyond every bound, however high. For this reason, we may also say that they tend to $+\infty$, or that they (or their terms) become infinitely large. This we put more precisely in the following

Definition 2. *If the sequence (x_n) has the property that, given an arbitrary (large) positive number G, another number n_0 can always be assigned such that for* **every** *$n > n_0$*

$$x_n > G,$$

then[4] *we shall say that (x_n) diverges to $+\infty$, tends to $+\infty$, or is definitely divergent*[5] *with the limit $+\infty$; and we then write*

$$x_n \to +\infty \ (\text{for } n \to \infty) \quad \text{or} \quad \lim x_n = +\infty \quad \text{or} \quad \lim_{n\to\infty} x_n = +\infty.$$

We are merely interchanging right and left by defining further:

Definition 3. *If the sequence (x_n) has the property that, given an arbitrary negative number $-G$ (large in absolute value), another number n_0 can always be assigned such that for* **every** *$n > n_0$*

$$x_n < -G,$$

then we shall say that (x_n) diverges to $-\infty$, tends to $-\infty$ or is definitely divergent[5] *with the limit $-\infty$, and we write*

$$x_n \to -\infty \ (\text{for } n \to \infty) \quad \text{or} \quad \lim x_n = -\infty \quad \text{or} \quad \lim_{n\to\infty} x_n = -\infty.$$

Remarks and Examples.

1. The sequences (n), (n^2), (n^α) for $\alpha > 0$, $(\log n)$, $(\log n)^\alpha$ for $\alpha > 0$, tend to $+\infty$; those whose terms have these values with the negative sign tend to $-\infty$.

2. In general: If $x_n \to +\infty$, then $x_n' = -x_n \to -\infty$, and conversely. — It is therefore sufficient, substantially, to consider divergence to $+\infty$ in what follows.

3. In geometrical language, $x_n \to +\infty$ means, of course, that however a point G (very far to the right) may be chosen, all points x_n, except at most a finite number of them, remain beyond it on the right. — With the mode of

[4] Notice that here not merely the *absolute values* $|x_n|$, but the numbers x_n themselves, are required to be $> G$.

[5] It is sometimes even said, — with apparent distortion of facts, — that the sequence converges to $+\infty$. The reason for this is that the behaviour described in Definition 2 resembles in many respects that of convergence (**39**). We will not, however, subscribe to this mode of exp·ession, although a misunderstanding would never have to be feared. — Similarly for $-\infty$.

representation in **7**, 6, it means that: however far above the axis of abscissae we may have drawn the parallel to it, the whole graph of the sequence (x_n) — excepting a finite initial portion, lies still further above it.

4. The divergence to $\pm\infty$ need not be monotone; thus for instance the sequence $1, 2^1, 2, 2^2\ 3, 2^3, 4, 2^4, \ldots, k, 2^k, \ldots$ also diverges to $+\infty$.

5. The succession $1, -2, +3, -4, \ldots, (-1)^{n-1}n, \ldots$ does not diverge to $+\infty$ or to $-\infty$. — This leads us to the further

Definition 4. *A sequence (x_n), which either converges in the sense of definition* **39**, *or diverges definitely in the sense of the definitions* **40**, *2 and 3, will be said to behave definitely (for $n \to \infty$). All other sequences, which therefore neither converge, nor diverge definitely, will be called* ***indefinitely divergent*** *or, for short,* ***indefinite***[6].

Remarks and Examples.

1. The sequences $[(-1)^n]$, $[(-2)^n]$, (a^n) for $a \leqq -1$, and likewise the sequences $0, 1, 0, 2, 0, 3, 0, 4, \ldots$ and $0, -1, 0, -2, 0, -3, \ldots$, as also the sequences **6**, 4, 8 are obviously indefinitely divergent.

2. On the contrary, the sequence $(|a^n|)$ for arbitrary a, and, in spite of all irregularities in detail, the sequences $(3^n + (-2)^n)$, $(n + (-1)^n \log n)$, $(n^2 + (-1)^n n)$, show *definite* behaviour.

3. The geometrical interpretation of indefinite behaviour follows immediately from the fact that there is neither convergence (v. **39**, 6) nor definite divergence (v. **40**, 3, rem. 3).

4. Both from $x_n \to +\infty$ and from $x_n \to -\infty$ it follows, provided every term $\neq 0$[7], that $\frac{1}{x_n} \to 0$; for $|x_n| > G = \frac{1}{\varepsilon}$ evidently implies $\left|\frac{1}{x_n}\right| < \varepsilon$. — On the other hand, $x_n \to 0$ in no way involves definite behaviour of $\left(\frac{1}{x_n}\right)$.

Example: For $x_n = \frac{(-1)^n}{n}$, we have $x_n \to 0$, but $\left(\frac{1}{x_n}\right)$ indefinitely divergent. — We have however, as is easily proved, the

Theorem: *If (x_n) is a null sequence whose terms all have the same sign, then the sequence $\left(\frac{1}{x_n}\right)$ is definitely divergent; — and of course to $+\infty$ or $-\infty$, according as the x_n's are all positive or all negative.*

---

[6] We have therefore to consider three typical modes of behaviour of a sequence, namely: a) Convergence to a number ξ, in accordance with **39**; b) divergence to $\pm\infty$, in accordance with **40**, 2 and 3; c) neither of the two. — Since the behaviour b) shows some analogy with a) and some with c), modes of expressions in use for it vary. Usually, it is true, b) is reckoned as divergence (the mode of expression mentioned in the last footnote cannot be consistently maintained) but "limiting values" $+\infty$ and $-\infty$ are at the same time spoken of. — We therefore speak, in the cases a) and b), of a definite, in the case c) of an indefinite, behaviour; in case a), and only in this case, we speak of convergence, in the cases b) and c) of divergence. — Instead of "definitely and indefinitely divergent", the words "properly and improperly divergent" are also used Since, however, as remarked, definite divergence still shows many analogies to convergence and a limit is still spoken of in this case, it does not seem advisable to designate this case precisely as that of proper divergence.

[7] From some place onwards this is certainly the case.

To facilitate the understanding of certain cases which frequently occur, we finally introduce the following further mode of expression:

○ **Definition 5.** *If two sequences (x_n) and (y_n), not necessarily convergent, are so related to one another that the quotient*

$$\frac{x_n}{y_n}$$

tends, for $n \to +\infty$, to a definite finite limit different from zero[8], *then we shall say that the two sequences are* ***asymptotically proportional*** *and write briefly*

$$x_n \sim y_n.$$

If in particular this limit is 1, then we say that the two sequences are ***asymptotically equal*** *and write, more expressively*

$$x_n \cong y_n.$$

Thus for instance

$$\sqrt{n^2+1} \cong n, \quad \log(5\,n^9+23) \sim \log n, \quad \sqrt{n+1}-\sqrt{n} \sim \frac{1}{\sqrt{n}},$$

$$1+2+\cdots+n \sim n^2, \quad 1^2+2^2+\cdots+n^2 \cong \tfrac{1}{3}\,n^3.$$

These designations are due substantially to *P. du Bois-Reymond* (Annali di matematica pura ed appl. (2) IV, p. 338, 1870/71).

To these definitions we now attach a series of simple, but quite fundamental

Theorems on convergent sequences.

41. ○ Theorem 1. *A convergent sequence determines its limit quite uniquely*[9].

Proof. If $x_n \to \xi$, and simultaneously $x_n \to \xi'$, then $(x_n - \xi)$ and $(x_n - \xi')$ are null sequences. By **28**, 2,

$$((x_n - \xi) - (x_n - \xi')) = (\xi' - \xi)$$

is then also a null sequence, i. e. $\xi = \xi'$, q. e. d.[10]

---

[8] x_n and y_n must then necessarily be $\neq 0$ *from some place onwards.* This is not required for *every* n in the above definition.

[9] A convergent sequence therefore defines (determines, gives ...) its limit quite as uniquely as any nest of intervals or *Dedekind* section defines the number to which it corresponds. Thus from this point we may consider a real number as *given* if we know a *sequence converging to it.* And as formerly we said for brevity that a nest of intervals $(x_n \mid y_n)$ or a *Dedekind* section $(A \mid B)$ or a radix fraction *is* a real number, so we may now with equal right say that a sequence (x_n) converging to ξ *is* the real number ξ, or symbolically: $(x_n) = \xi$. For further details of this conception, which was used by *G. Cantor* to construct his theory of real numbers, see pp. 79 and 95.

[10] The last step in our reasoning, by which the reader may at first sight be taken aback, amounts simply to this: If with respect to a definite numerical value α we know that, for every $\varepsilon > 0$, we always have $|\alpha| < \varepsilon$, then we

° Theorem 2. *A convergent sequence* (x_n) *is invariably bounded. And if* $|x_n| \leqq K$, *then for the limit* ξ *we have*[11] $|\xi| \leqq K$.

Proof. If $x_n \to \xi$, then we can, given $\varepsilon > 0$, assign a number m, such that for every $n > m$

$$\xi - \varepsilon < x_n < \xi + \varepsilon.$$

If therefore K_1 is a number greater than the m values $|x_1|, |x_2|, \ldots, |x_m|$, and greater than $|\xi| + \varepsilon$, then obviously

$$|x_n| < K_1$$

for every n. Now let K be any bound of the numbers $|x_n|$. If we had $|\xi| > K$, then $|\xi| - K > 0$ and therefore, from some place onwards in the sequence,

$$|\xi| - |x_n| \leqq |x_n - \xi| < |\xi| - K$$

and therefore $|x_n| > K$, which is contrary to the meaning of K.

° Theorem 2a. $x_n \to \xi$ *implies* $|x_n| \to |\xi|$.

Proof. We have (v. 3, II, 4)

$$\big|\,|x_n| - |\xi|\,\big| \leqq |x_n - \xi|;$$

therefore $(|x_n| - |\xi|)$ is by 26, 2 a null sequence when $(x_n - \xi)$ is.

° Theorem 3. *If a convergent sequence* (x_n) *has all its terms different from zero, and if its limit* ξ *is also* $\neq 0$, *then the sequence* $\left(\frac{1}{x_n}\right)$ *is bounded; or in other words, a number* $\gamma > 0$ *exists, such that* $|x_n| \geqq \gamma > 0$ *for every* n; *the numbers* $|x_n|$ *possess a positive lower bound.*

Proof. By hypothesis, $\frac{1}{2}|\xi| = \varepsilon > 0$, and there exists an integer m, such that for every $n > m$, $|x_n - \xi| < \varepsilon$ and therefore $|x_n| > \frac{1}{2}|\xi|$[12]. If the smallest of the $(m + 1)$ positive numbers $|x_1|, |x_2|, \ldots, |x_m|$ and $\frac{1}{2}|\xi|$ be denoted by γ, then $\gamma > 0$, and for every n, $|x_n| \geqq \gamma$, $\left|\frac{1}{x_n}\right| \leqq K = \frac{1}{\gamma}$, q. e. d.

If, given a sequence (x_n) converging to ξ, we apply to the null sequence $(x_n - \xi)$ the theorems 27, 1 to 5, then we immediately obtain the theorems:

---

necessarily have $\alpha = 0$. For 0 is the only number whose absolute value is less than *every* positive ε. (In fact $|0| < \varepsilon$ is true for every $\varepsilon > 0$. But if $\alpha \neq 0$, so that $|\alpha| > 0$, then $|\alpha|$ is certainly *not* less than the positive number $\varepsilon = \frac{1}{2}|\alpha|$.) Similarly, if we know of a definite numerical value α that, for every $\varepsilon > 0$, we always have $\alpha \leqq K + \varepsilon$, then we must have further $\alpha \leqq K$. The method of reasoning involved here: "*If for every* $\varepsilon > 0$, *we always have* $|\alpha| < \varepsilon$, *then necessarily* $\alpha = 0$" is precisely the same as was constantly applied by the Greek mathematicians (cf. *Euclid*, Elements X) and later called the *method of exhaustion.*

[11] Here the sign of equality in "$|\xi| \leqq K$" must not be omitted, even when, for every n, $|x_n| < K$.

[12] For $n > m$, all the x_n's are therefore necessarily $\neq 0$.

°**Theorem 4.** *If* (x_n') *is a sub-sequence of* (x_n), *then*

$$x_n \to \xi \quad \textit{implies} \quad x_n' \to \xi .$$

°**Theorem 5.** *If the sequence* (x_n) *can be divided into two sub-sequences*[13] *of which each converges to* ξ, *then* (x_n) *itself converges to* ξ.

°**Theorem 6.** *If* (x_n') *is an arbitrary rearrangement of* x_n, *then*

$$x_n \to \xi \quad \textit{implies} \quad x_n' \to \xi .$$

°**Theorem 7.** *If* $x_n \to \xi$ *and* (x_n') *results from* (x_n) *by a finite number of alterations, then* $x_n' \to \xi$.

°**Theorem 8.** *If* $x_n' \to \xi$ *and* $x_n'' \to \xi$, *and if the sequence* (x_n) *is so related to the sequences* (x_n') *and* (x_n'') *that from some place onwards, (i. e. for every* $n \geqq m$, *say,)*

$$x_n' \leqq x_n \leqq x_n'',$$

then $x_n \to \xi$.

Calculations with convergent sequences are based on the following four theorems:

°**Theorem 9.** $x_n \to \xi$ *and* $y_n \to \eta$ *always implies* $(x_n + y_n) \to \xi + \eta$, *and the corresponding statement holds for term by term addition of any fixed number — say* p *— of convergent sequences.*

Proof. If $(x_n - \xi)$ and $(y_n - \eta)$ are null sequences, then so, by **28**, 1, is $((x_n + y_n) - (\xi + \eta))$. In the same way, **28**, 2 gives the

°**Theorem 9a.** $x_n \to \xi$ *and* $y_n \to \eta$, *always implies* $(x_n - y_n) \to \xi - \eta$.

°**Theorem 10.** $x_n \to \xi$ *and* $y_n \to \eta$, *always implies* $x_n y_n \to \xi \eta$, *and the corresponding statement holds for term by term multiplication of any fixed number — say* p *— of convergent sequences.*

In particular: $x_n \to \xi$ *implies* $c x_n \to c \xi$, *whatever number* p *denote.*

Proof. We have

$$x_n y_n - \xi \eta = (x_n - \xi) y_n + (y_n - \eta) \xi ;$$

and since here on the right hand side two null sequences are multiplied term by term by bounded factors and then added, the whole expression is itself the term of a null sequence, q. e. d.

°**Theorem 11.** $x_n \to \xi$ *and* $y_n \to \eta$ *always implies, if every* $x_n \neq 0$ *and also* $\xi \neq 0$,

$$\frac{y_n}{x_n} \to \frac{\eta}{\xi}.$$

Proof. We have

$$\frac{y_n}{x_n} - \frac{\eta}{\xi} = \frac{y_n \xi - x_n \eta}{x_n \cdot \xi} = \frac{(y_n - \eta)\xi - (x_n - \xi)\eta}{x_n \cdot \xi}.$$

---

[13] Or *three*, or any *definite* number.

Here the numerator, for the same reasons as above, represents a null sequence, and the factors $\frac{1}{\xi \cdot x_n}$ are, by theorem 3, bounded. Therefore the whole expression is again the term of a null sequence. — Only a particular case of this is the

°Theorem 11a[14]. *$x_n \to \xi$ always implies, if every x_n and also ξ are* $\neq 0$,

$$\frac{1}{x_n} \to \frac{1}{\xi}.$$

These fundamental theorems 8—11 lead, by repeated application, to the following more comprehensive

°**Theorem 12.** *Let $R = R(x^{(1)}, x^{(2)}, x^{(3)}, \ldots, x^{(p)})$ denote an expression built up, by a finite number of additions, subtractions, multiplications, and divisions, from the letters $x^{(1)}, x^{(2)}, \ldots, x^{(p)}$, and arbitrary numerical coefficients*[15]*; and let*

$$(x_n^{(1)}), \quad (x_n^{(2)}), \quad \ldots, \quad (x_n^{(p)})$$

be p given sequences, converging respectively to $\xi^{(1)}, \xi^{(2)}, \ldots, \xi^{(p)}$. Then the sequence of the numbers

$$R_n = R(x_n^{(1)}, x_n^{(2)}, \ldots, x_n^{(p)}) \to R(\xi^{(1)}, \xi^{(2)}, \ldots, \xi^{(p)})$$

provided neither in the evaluation of the terms R_n, nor in that of the number $R(\xi^{(1)}, \xi^{(2)}, \ldots, \xi^{(p)})$, division by 0 is anywhere required.

These theorems give us all that is required for the formal manipulation of convergent sequences: We give a few more

Examples.

1. $x_n \to \xi$ implies, if $a > 0$, *invariably*, **42.**

$$a^{x_n} \to a^{\xi}.$$

For

$$a^{x_n} - a^{\xi} = a^{\xi}\left(a^{x_n - \xi} - 1\right)$$

is a null sequence by **35**, 3.

2. $x_n \to \xi$ implies, if every x_n and also ξ are > 0, that

$$\log x_n \to \log \xi.$$

Proof. We have

$$\log x_n - \log \xi = \log \frac{x_n}{\xi} = \log\left(1 + \frac{x_n - \xi}{\xi}\right)$$

which by **38**, 7 is a null sequence, since $x_n > 0$ implies $\frac{x_n - \xi}{\xi} > -1$.

---

[14] In theorems 3, 11 and 11a, it is sufficient to postulate that the limit of the denominators is $\neq 0$, for then the denominators are, from some index m onwards, necessarily $\neq 0$, and only "a finite number of alterations" need be made, or the new sequence need only be considered for $n > m$, to ensure this being the case for all.

[15] More shortly: a rational function of the p variables $x^{(1)}, x^{(2)}, \ldots, x^{(p)}$ with arbitrary numerical coefficients.

3. Under the same hypotheses as in 2., we also have, for arbitrary real ϱ,

$$x_n^\varrho \to \xi^\varrho.$$

Proof. We have

$$x_n^\varrho - \xi^\varrho = \xi^\varrho\left(\frac{x_n^\varrho}{\xi^\varrho} - 1\right) = \xi^\varrho\left[\left(1 + \frac{x_n - \xi}{\xi}\right)^\varrho - 1\right],$$

which by 38, 8 is a null sequence[16], since $\frac{x_n - \xi}{\xi} > -1$ and tends to 0 as $n \to \infty$. (This is to a certain extent further completed by 35, 4.)

Cauchy's theorem of limits and its generalisations.

There is a group of theorems on limits[17] essentially more profound than the above, and of great significance for later work, which originated in their simplest form with *Cauchy*[18] and have in recent times been extended in different directions. We have first the simple

43. °Theorem 1. *If* $(x_0, x_1, \ldots)$ *is a null sequence, then the arithmetic means*

$$x_n' = \frac{x_0 + x_1 + \cdots + x_n}{n+1}, \qquad n = 0, 1, 2, \ldots,$$

also form a null sequence.

Proof. If ε is given > 0, then m can be so chosen, that for every $n > m$ we have $|x_n| < \frac{\varepsilon}{2}$. For these n's, we then have

$$|x_n'| \leqq \frac{|x_0 + x_1 + \cdots + x_m|}{n+1} + \frac{\varepsilon}{2}\,\frac{n-m}{n+1}.$$

Since the numerator of the first fraction on the right hand side now contains a fixed number, we can further determine n_0, so that for $n > n_0$ that fraction remains $< \frac{\varepsilon}{2}$. But then, for every $n > n_0$, we have $|x_n'| < \varepsilon$, — and our theorem is proved. — Somewhat more general, but nevertheless an immediate corollary of this, is the

°Theorem 2. *If* $x_n \to \xi$, *then so do the arithmetic means*

$$x_n' = \frac{x_0 + x_1 + \cdots + x_n}{n+1} \to \xi.$$

---

[16] Examples 1. to 3. mean — in the language of the theory of functions — that the function a^x is continuous at every point, the functions $\log x$ and x^ϱ at every positive point.

[17] The reader may defer the study of these theorems until, in the later chapters, they come into use.

[18] *Augustin Louis Cauchy*, born 1789 in Paris, died 1857 in Sceaux. In his work *Analyse algébrique*, Paris 1821 (German edition, Berlin 1885, Julius Springer) the foundations of higher analysis are for the first time developed with full rigour, and among them the theory of infinite series. In what follows we shall frequently have to refer to it; the above theorem 2 may be found on p. 59 of that treatise.

Proof. By theorem 1,

$$\left(\frac{(x_0-\xi)+(x_1-\xi)+\dots+(x_n-\xi)}{n+1}\right)=(x_n'-\xi)$$

is a null sequence when $(x_n-\xi)$ is, q. e d.

From this theorem, the corresponding one for geometric means now follows quite easily.

Theorem 3. *Let the sequence* $(y_1, y_2, \dots)\to\eta$, *and have all its members and its limit* η *positive. Then also the sequence of geometric means*

$$y_n'=\sqrt[n]{y_1\,y_2\dots y_n}\to\eta.$$

Proof. From $y_n\to\eta$, since all the numbers are positive, we deduce, by **42**, 2, that

$$x_n=\log y_n\to\xi=\log\eta.$$

By theorem 2, it follows that

$$x_n'=\frac{x_1+x_2+\dots+x_n}{n}=\log\sqrt[n]{y_1\,y_2\dots y_n}=\log y_n'\to\log\eta.$$

By **42**, 1, this at once proves the truth of our statement.

Examples.

1. $$\frac{1+\frac{1}{2}+\dots+\frac{1}{n}}{n}\to 0,\quad\text{because }\frac{1}{n}\to 0.$$

2. $$\sqrt[n]{n}=\sqrt[n]{1\cdot\frac{2}{1}\cdot\frac{3}{2}\cdots\frac{n}{n-1}}\to 1,\quad\text{because }\frac{n}{n-1}\to 1.$$

3. $$\frac{1+\sqrt{2}+\sqrt[3]{3}+\dots+\sqrt[n]{n}}{n}\to 1,\quad\text{because }\sqrt[n]{n}\to 1.$$

4. Because $\left(1+\frac{1}{n}\right)^n\to e$ (v. **46a** in the next §), we have by theorem **3**,

$$\sqrt[n]{\left(\frac{2}{1}\right)^1\cdot\left(\frac{3}{2}\right)^2\cdot\left(\frac{4}{3}\right)^3\cdots\left(\frac{n+1}{n}\right)^n}=\sqrt[n]{\frac{(n+1)^n}{n!}}=\frac{n+1}{\sqrt[n]{n!}}\quad\text{also}\to e$$

or, therefore,

$$\frac{1}{n}\sqrt[n]{n!}\to\frac{1}{e},$$

a relation which may also be noted in the form "$\sqrt[n]{n!}\cong\frac{n}{e}$".

Essentially more far-reaching, and yet as easily proved, is the following generalisation of *Cauchy*'s theorems 1 and 2, due to *O. Toeplitz*[19]:

○**Theorem 4.** *Let $(x_0, x_1, \ldots)$ be a null sequence and suppose the coefficients $a_{\mu\nu}$ of the system*

$$(A)\quad \begin{cases} a_{00} \\ a_{10} \quad a_{11} \\ a_{20} \quad a_{21} \quad a_{22} \\ \cdots\cdots\cdots\cdots \\ a_{n0} \quad a_{n1} \quad a_{n2} \quad \cdots \quad a_{nn} \\ \cdots\cdots\cdots\cdots\cdots\cdots \end{cases}$$

satisfy the two conditions:

(a) *Every column contains a null sequence, i. e. for fixed $p \geqq 0$*

$$a_{np} \to 0 \quad \textit{when} \quad n \to +\infty.$$

(b) *There exists a constant K, such that the sum of the absolute values of the terms in any one row, i. e., for every n, the sum*

$$|a_{n0}| + |a_{n1}| + \cdots + |a_{nn}| \textit{ remains } < K.$$

— *Then the sequence formed by the numbers*

$$x_n' = a_{n0} x_0 + a_{n1} x_1 + a_{n2} x_2 + \cdots + a_{nn} x_n$$

is also a null sequence.

Proof. If ε is given > 0, determine m so that for every $n > m$ $|x_n| < \frac{\varepsilon}{2K}$. Then for those n's,

$$|x_n'| < |a_{n0} x_0 + \cdots + a_{nm} x_m| + \frac{\varepsilon}{2}.$$

By the hypothesis (a), we may now (as m is *fixed*) choose $n_0 > m$, so that for every $n > n_0$, we have $|a_{n0} x_0 + \cdots + a_{nm} x_m| < \frac{\varepsilon}{2}$. Since for these n's $|x_n'|$ is then $< \varepsilon$, our theorem is proved.

In applications it is useful to have the following

○**Complement.** *If, for the coefficients $a_{\varkappa\lambda}$, are substituted other numbers $a'_{\varkappa\lambda} = a_{\varkappa\lambda} \cdot \alpha_{\varkappa\lambda}$, obtained from the numbers $a_{\varkappa\lambda}$ by multiplication*

[19] *Cauchy*'s Theorem 1 has been generalised in several ways, in particular by *J. L. W. V. Jensen* (Om en Sätning af Cauchy, Tidskrift for Mathematik, (5) Vol. 2, pp. 81—84. 1884) and *O. Stolz* (Über eine Verallgemeinerung eines Satzes von Cauchy, Mathemat. Annalen, Vol. 33, p. 237. 1889). The above formulation, due to *O. Toeplitz* (Über lineare Mittelbildungen, Prace matematycznofizyczne, Vol. 22, p. 113—119. 1911), is in a certain sense a final generalisation, for this reason that it shows (l. c.) the conditions, recognised in Theorem 5 as sufficient, to be also *necessary*, for $x_n \to \xi$ *to imply* $x_n' \to \xi$ *in all cases* (cf. **221**, and the work of *I. Schur*: Über lineare Transformationen in der Theorie der unendlichen Reihen, Jour. f.d. reine u. angew. Math., Vol. 151, pp. 79—111. 1920).

by factors $\alpha_{\varkappa\lambda}$ — *all in absolute value less than a fixed constant* α, — *then the numbers*

$$x_n'' = a_{n0}' x_0 + a_{n1}' x_1 + \cdots + a_{nn}' x_n$$

also form a null sequence.

Proof. The $a_{\varkappa\lambda}'$'s also satisfy the conditions (a) and (b) of theorem 4; for, if p is fixed, $a_{np}' \to 0$ by **26**, 1, and the sums

$$|a_{n0}'| + |a_{n1}'| + \cdots + |a_{nn}'| \quad \text{remain} \quad < K' = \alpha K.$$

From Theorem 4 we may now deduce the

°**Theorem 5.** *If* $x_n \to \xi$, *and the coefficients* $a_{\mu\nu}$ *satisfy, besides the conditions* (a) *and* (b) *of Theorem* 4, *the further condition*

$$\text{(c)} \qquad a_{n0} + a_{n1} + \cdots + a_{nn} = A_n \to 1,$$ [20]

then also the sequence formed by the numbers

$$x_n' = a_{n0} x_0 + a_{n1} x_1 + \cdots + a_{nn} x_n \to \xi.$$

Proof. We now have

$$x_n' = A_n \cdot \xi + a_{n0}(x_0 - \xi) + a_{n1}(x_1 - \xi) + \cdots + a_{nn}(x_n - \xi),$$

whence our statement at once follows, in consequence of condition (c), by theorem 4.

Before giving examples and applications of these important theorems, we may prove the following further generalisation, which points in a new direction.

°**Theorem 6.** *If the coefficients* $a_{\mu\nu}$ *of the system* (A) *satisfy, besides the conditions* (a), (b) *and* (c) *mentioned in Theorems* 4 *and* 5, *the further condition, that*

(d) *the numbers in each of the "diagonals" of* A *form a null sequence, i. e. for fixed* p, $a_{n\,n-p} \to 0$ *when* $n \to +\infty$,

then it follows from $x_n \to \xi$ *and* $y_n \to \eta$ *that the numbers*

$$z_n = a_{n0} x_0 y_n + a_{n1} x_1 y_{n-1} + \cdots + a_{nn} x_n y_0 \to \xi \cdot \eta.$$ [21]

Proof. Since

$$x_\nu y_{n-\nu} = (x_\nu - \xi) y_{n-\nu} + \xi \cdot y_{n-\nu},$$

we have

$$z_n = \sum_{\nu=0}^{n} a_{n\nu} y_{n-\nu} (x_\nu - \xi) + \xi \cdot \sum_{\nu=0}^{n} a_{n\nu} y_{n-\nu}.$$

---

[20] In the applications, we shall generally have $A_n = 1$.

[21] For positive $a_{\mu\nu}$, this theorem may be found in a paper by the author "Über Summen der Form $a_0 b_n + a_1 b_{n-1} + \cdots + a_n b_0$" (Rend. del circolo mat. di Palermo, Vol. 32, p. 95–110. 1911).

Here the first sum tends to zero, by Theorem 4 and its complement, for $(x_\nu - \xi)$ is a null sequence and the factors $y_{n-\nu}$ are bounded. And if the second sum be written in the form

$$\xi \cdot \sum_{\nu=0}^{n} a_{n\,n-\nu}\, y_\nu \equiv \xi \cdot \sum_{\nu=0}^{n} a'_{n\nu}\, y_\nu$$

we see, by theorem 5, that this, and thereby also z_n, tends $\to \xi\eta$; for the numbers $a'_{n\nu} = a_{n\,n-\nu}$ satisfy, in consequence of (d), precisely the condition (a) there stipulated.

44. Remarks, applications and examples.

1. Theorem 1 is a particular case of Theorem 4; we need only put, in the latter,

$$a_{n0} = a_{n1} = \ldots = a_{nn} = \frac{1}{n+1}, \qquad (n = 0, 1, 2, \ldots)$$

Theorem 2 is derived in the same way from Theorem 5. The conditions (a), (b), (c) are fulfilled.

2. If $\alpha_0, \alpha_1, \ldots$ are any *positive* numbers, for which the sums

$$\alpha_0 + \alpha_1 + \ldots + \alpha_n = \sigma_n \to +\infty,$$

it follows [22] from $x_n \to \xi$ that

$$x_n' = \frac{\alpha_0 x_0 + \alpha_1 x_1 + \ldots + \alpha_n x_n}{\alpha_0 + \alpha_1 + \ldots + \alpha_n} \text{ also } \to \xi.$$

In fact, we need only put, in theorem 5,

$$a_{n\nu} = \frac{\alpha_\nu}{\sigma_n} \qquad \begin{cases} n = 0, 1, 2, \ldots \\ \nu = 0, 1, \ldots, n \end{cases}$$

to see that the statement is correct. The conditions (a), (b), (c) are fulfilled. — For $\alpha_n \equiv 1$, we again obtain Theorem 2.

2a. The theorem of no. 2. remains true for $\xi = +\infty$ or $\xi = -\infty$. The same remark holds for the general theorem 5, provided all the $a_{\mu\nu}$'s are $\geqq 0$ there. For if $x_n \to +\infty$ and, as in the proof of Theorem 4, m be so chosen, given $G > 0$, that for every $n > m$ we have $x_n > G + 1$, then for those n's we have

$$x_n' > (G+1)(a_{n\,m+1} + \ldots + a_{nn}) - a_{n0}\,|x_0| - \ldots - a_{nm}\,|x_m|.$$

In consequence of the conditions (a) and (c) in Theorems 4 and 5, we may therefore so choose n_0 that for every $n > n_0$ we have $x_n' > G$. Hence $x_n' \to +\infty$.

°3. Instead of assuming the α_n's positive and $\sigma_n \to +\infty$, it suffices [by (b)] to require only that $|\alpha_0| + |\alpha_1| + \ldots + |\alpha_n| \to +\infty$, with the proviso, however, that a constant K exists, such that [23] for every n

$$|\alpha_0| + |\alpha_1| + \ldots + |\alpha_n| \leqq K \cdot |\alpha_0 + \alpha_1 + \ldots + \alpha_n|.$$

(For positive α_n, $K = 1$ gives all that is here required.)

---

[22] *O. Stolz*, loc. cit. — Of course it also suffices, that the α_n's be *from some point onwards* $\geqq 0$, provided only $\sigma_n \to +\infty$. The x_n''s must then be considered from that point onwards, after which σ_n is > 0.

[23] *Jensen*, loc. cit. — If α_m is the first of the α's to be $\neq 0$, then the x_n''s are defined only for $n \geqq m$.

4. If in 2. or 3. we put, for brevity, $\alpha_n x_n = y_n$, then we obtain:

$$\frac{y_0 + y_1 + \dots + y_n}{\alpha_0 + \alpha_1 + \dots + \alpha_n} \to \xi, \quad \text{provided} \quad \frac{y_n}{\alpha_n} \to \xi,$$

and provided the α_n's satisfy the conditions given in 2. or 3.

5. If we write further $y_0 + y_1 + \dots + y_n = Y_n$, and $\alpha_0 + \alpha_1 + \dots + \alpha_n = A_n$, then the last result takes the form:

$$\frac{Y_n}{A_n} \to \xi, \quad \text{provided} \quad \frac{Y_n - Y_{n-1}}{A_n - A_{n-1}} \to \xi,$$

and provided the numbers $\alpha_n = A_n - A_{n-1}$ $(n \geqq 1,\ \alpha_0 = A_0)$ satisfy the conditions given in 2. or 3.

6. Thus we have, for instance, by 5.:

$$\lim \frac{1 + 2 + \dots + n}{n^2} = \lim \frac{n}{n^2 - (n-1)^2} = \lim \frac{n}{2n - 1} = \frac{1}{2}.$$

Similarly we have

$$\lim \frac{1^2 + 2^2 + \dots + n^2}{n^3} = \lim \frac{n^2}{n^3 - (n-1)^3} = \frac{1}{3},$$

and generally

$$\lim \frac{1^p + 2^p + \dots + n^p}{n^{p+1}} = \lim \frac{n^p}{n^{p+1} - (n-1)^{p+1}}$$

$$= \lim \frac{n^p}{(p+1)\, n^p - \binom{p+1}{2} n^{p-1} + \dots} = \frac{1}{p+1},$$

if p denotes a positive integer.

7. Similarly we find, if we anticipate the proof in **46a** of the convergence of the sequence of numbers $\left(1 + \frac{1}{n}\right)^{n+1}$:

$$\frac{\log 1 + \log 2 + \dots + \log n}{n \log n} = \frac{\log n!}{\log n^n} \to 1, \quad \text{i. e.} \quad \log n! \cong \log n^n.$$

8. The numbers

$$a_{n\nu} = \frac{1}{2^n} \binom{n}{\nu} \qquad \begin{cases} n = 0, 1, 2, \dots \\ \nu = 0, 1, \dots, n \end{cases}$$

fulfil the conditions (a), (b) and (c) of the theorems 4 and 5; for if p be fixed, $a_{np} \to 0$, seeing that it is

$$= \frac{1}{2^n} \binom{n}{p}, \quad \text{and therefore} \quad < \frac{n^p}{2^n} \qquad (\text{v. } \mathbf{38}, 2)$$

while

$$|a_{n0}| + \dots + |a_{nn}| = a_{n0} + \dots + a_{nn} = 1,$$

for every n. Therefore $x_n \to \xi$ always implies

$$\frac{x_0 + \binom{n}{1} x_1 + \binom{n}{2} x_2 + \dots + \binom{n}{n} x_n}{2^n} \to \xi.$$

9. The same specialisations as were given in 1., 2., 3. and 8. for theorem 5 may of course also be applied to theorem 6. We merely mention the two following theorems:

(a) From $x_n \to \xi$ and $y_n \to \eta$ it always follows that

$$\frac{x_0 y_n + x_1 y_{n-1} + x_2 y_{n-2} + \cdots + x_n y_0}{n+1} \to \xi \eta.$$

(b) If (x_n) and (y_n) are two null sequences, the second of which fulfils the extra condition that for every n

$$|y_0| + |y_1| + \cdots + |y_n|$$

remains less than a fixed number K, then the numbers

$$z_n = x_0 y_n + x_1 y_{n-1} + \cdots + x_n y_0$$

form a null sequence. (For the proof we put $a_{n\nu} = y_{n-\nu}$ in theorem 4.)

10. The reader will have noticed that it is in no wise essential that the rows of the system (A) of theorem 4 should break off exactly at the nth term. On the contrary, these rows may contain any number of terms. Indeed, after we have mastered the first principles of the theory of infinite series, we shall see that these rows may contain even an infinity of terms $(a_{n0}, a_{n1}, \ldots, a_{n\nu}, \ldots)$, provided only the other conditions imposed on the system be fulfilled. The theorem hereby indicated will be formulated and proved in **221**.

§ 9. The two main criteria.

We are now sufficiently prepared to attack the actual problems of convergence. There are two main points of view from which we propose, in what follows, to examine the sequences which come before us. We have above all to consider the

Problem A. *Is a given sequence (x_n) convergent, or definitely or indefinitely divergent?* (Briefly: How does the sequence behave with respect to convergence?) — And if a sequence has proved to be convergent, so that the existence of a limiting value is ensured, we have further to consider the

Problem B. *To what limit ξ does the sequence (x_n), recognized to be convergent, tend?*

A few examples may make the significance of these problems clearer: If for instance we are given the sequences

$$\left(n(\sqrt[n]{2}-1)\right), \quad \left(n(\sqrt[n]{n}-1)\right), \quad \left(\left(1+\frac{1}{n}\right)^n\right), \quad \left(\left(1+\frac{1}{n^2}\right)^n\right),$$

$$\left(\frac{1+2^2+3^3+\cdots+n^n}{n^2}\right), \qquad \left(\frac{1+\frac{1}{2}+\cdots+\frac{1}{n}}{\log n}\right), \text{ etc.}$$

examination of their construction shows that there are always two (or more) forces which here, so to speak, oppose one another and thereby call forth the variation of the terms. One force tends to increase, the other to diminish them, and it is not clear at a glance which of the two will get the upper hand or in what degree this will happen. Every means which enables us to decide the question of convergence or divergence of a given sequence, we call a *criterion* of *convergence* or of *divergence*; these serve, therefore, to solve the problem A.

The problem B is in general much more difficult. In fact, we might almost say that it is insoluble, — or else is trivial. The latter, because a convergent sequence (x_n), by theorem **41**, 1, entirely determines its limit ξ, which may therefore be regarded as "given" by the sequence itself (cf. footnote to **41**, 1). On account, however, of the boundless complexity and multiplicity of form which sequences show, this conclusion does not seem very satisfactory. We shall wish, rather, not to consider the limit ξ as "known", until we have before us a Dedekind section, or still better a nest of intervals, for instance a radix fraction, in particular a decimal fraction. These latter especially are the methods of representing a real number with which we have always been most familiar. If we regard the problem in this light, we may call it *the question of numerical calculation of the limit*[1].

This question, one of great practical significance, is usually in theoretical considerations of very second-rate importance, for from a theoretical point of view, all modes of representation for a real number (nests, sections, sequences, ...) are precisely equivalent. If we observe further, that the representation of a real number by a sequence may be considered as the most general mode of representation, our problem B may be stated in the following form:

Problem B′. Two convergent sequences (x_n) and (x_n') are given, — how may we determine whether or not both define the same limit, or whether or not the two limits stand in a simple relation to one another?

A few examples will serve to illustrate the kind of question referred to:

1. Let **45.**

$$x_n = \left(1 + \frac{1}{n}\right)^n \text{ and } x_n' = \left(1 + \frac{\alpha}{n}\right)^n.$$

Both sequences are quite easily (v. **46**a and **111**) seen to be convergent. But it is not so apparent that if ξ denotes the limit of the first sequence, that of the second is $= \xi^\alpha$.

2. Given the sequence

$$\frac{1}{1}, \frac{3}{2}, \frac{7}{5}, \frac{17}{12}, \frac{41}{29}, \ldots$$

in which the numerator of each fraction is formed by adding twice the numerator of the last fraction preceding to the numerator of the last fraction but one (e. g. $41 = 2 \cdot 17 + 7$), and similarly for the denominators. — The question of

[1] Numerical calculation of a real number = representation of that number by a decimal fraction. For further details, see chapter VIII.

convergence again gives no trouble, nor does the numerical evaluation of the limit, — but how are we to recognise that this limit $=\sqrt{2}$?

3. Let $$x_n = \left(1 - \frac{1}{3} + \frac{1}{5} - \frac{1}{7} + \cdots + \frac{(-1)^{n-1}}{2n-1}\right) \qquad (n = 1, 2, \ldots$$

and let x_n' be the perimeter of the regular polygon with n sides inscribed in the circle of radius 1. Here also both sequences are easily seen to be convergent. If ξ and ξ' are their limits, — how does one see that here $\xi' = 8 \cdot \xi$?

These examples make it seem sufficiently probable, that Problem B or B′ is considerably harder to attack than Problem A. We therefore confine our attention in the first instance entirely to the latter, and to begin with make ourselves acquainted with two criteria, from which *all others* may be deduced.

First main criterion (for monotone sequences).

46. *A monotone bounded sequence is invariably convergent; a monotone sequence which is not bounded is always definitely divergent.* (Or, therefore: A monotone sequence always behaves definitely, and is then and only then convergent, when it is bounded, and then and only then divergent, when it is *not* bounded. In the latter case the divergence is towards $+\infty$ or $-\infty$ according as the monotone sequence is ascending or descending.)

Proof. a) Let the sequence (x_n) be monotone ascending and not bounded. Since it is then (because $x_n \geqq x_1$) certainly bounded on the left, it cannot be bounded on the right; given any arbitrary (large) positive number G, there is then always an index n_0, for which

$$x_{n_0} > G.$$

But then, since the sequence is monotone increasing, we have for every $n > n_0$, *a fortiori*, $x_n > G$, and so, by Definition **40**, 2, actually $x_n \to +\infty$. Interchanging right and left, we see in the same way that a monotone descending sequence which is not bounded must diverge to $-\infty$. Thus the second part of the proposition is also proved.

b) Now let (x_n) be a monotone ascending, but bounded sequence. There is then a number K, such that $|x_n| \leqq K$ for every n, so that

$$x_1 \leqq x_n \leqq K$$

for every n. The interval $J_1 = x_1 \ldots K$ therefore contains all the terms of (x_n); to this interval we apply the method of successive bisection: We denote the *right* or the *left* half of J_1 by J_2, according as the right half does or does not still contain points of (x_n). From J_2 we select one half by the *same* rule, and call this J_3; and so on. The intervals of the nest so constructed have the property[2], that *no* point

[2] The reader should illustrate the circumstances on the number-axis.

of the sequence lies to the *right* of them, but at least *one* lies *inside* each of them. Or in other words: the points of the sequence (while monotonely progressing towards the right) penetrate *into* each interval, but do not emerge from it again; in each of these intervals, therefore, *all* points from a certain index onwards come to lie. We may therefore, if we suppose the numbers $n_1, n_2, \ldots$ properly chosen, say that:

In J_k lie all x_n's with $n > n_k$, but to the right of J_k lie no more x_n's.

If ξ is now the number determined by the nest (J_n), it can at once be shewn that $x_n \to \xi$. For if ε is given > 0, choose the index p so that the length of J_p is less than ε. For $n > n_p$, all the x_n's lie, together with ξ, in J_p, so that for these n's we must have

$$|x_n - \xi| < \varepsilon.$$

$(x_n - \xi)$ is therefore a null sequence, and $x_n \to \xi$, q. e. d.

By a suitable interchange of right and left, we see that monotone descending bounded sequences must also be convergent. Thus every part of the theorem is proved.

Remarks and Examples.

1. We first draw attention again to the fact that (cf. **41**, 1) even when $|x_n| < K$, we *may* have for the limiting value ξ the equality $|\xi| = K$.

2. Let

$$x_n = \frac{1}{n+1} + \frac{1}{n+2} + \cdots + \frac{1}{2n} \qquad (n = 1, 2, \ldots).$$

As

$$x_{n+1} - x_n = \frac{1}{2n+1} + \frac{1}{2n+2} - \frac{1}{n+1} = \frac{1}{2n+1} - \frac{1}{2n+2} > 0,$$

the sequence is monotone increasing, and as $x_n \leqq n \cdot \frac{1}{n+1} < 1$, it is also bounded. *It is therefore convergent.* Of its limit ξ we know no more, so far, than that

$$x_n < \xi \leqq 1$$

for every n, which e. g. for $n = 3$ becomes $\frac{37}{60} < \xi \leqq 1$. Whether it has a rational value, or whether ξ bears a close relation to a number appearing in any other connection — in short: an answer to problem B — cannot here be perceived at once. Later on we shall see that ξ is equal to the natural logarithm of 2. I. e. the logarithm of 2 whose base is the number e introduced in **46a** below.

3. Let $x_n = \left(1 + \frac{1}{2} + \frac{1}{3} + \cdots + \frac{1}{n}\right)$, so that the sequence (x_n) is monotone increasing (cf. **6**, 12). Is it bounded or not? — If G is given arbitrarily > 0, chose $m > 2G$; then for $n > 2^m$

$$x_n > \left(1 + \frac{1}{2}\right) + \left(\frac{1}{3} + \frac{1}{4}\right) + \left(\frac{1}{5} + \cdots + \frac{1}{8}\right) + \cdots + \left(\frac{1}{2^{m-1}+1} + \cdots + \frac{1}{2^m}\right)$$

$$> \frac{1}{2} + 2 \cdot \frac{1}{4} + 4 \cdot \frac{1}{8} + 8 \cdot \frac{1}{16} + \cdots + 2^{m-1} \cdot \frac{1}{2^m} = \frac{m}{2} > G.$$

The sequence is therefore not bounded and consequently diverges $\to +\infty$.

4. If $\sigma = (x_n \mid y_n)$ is an arbitrary nest of intervals, the left and right end-points of the intervals respectively form two monotone, bounded and therefore convergent sequences. We then have

$$\lim x_n = \lim y_n = (x_n \mid y_n) = \sigma.$$

46a. As a particularly important example, we will consider the two sequences whose terms are

$$x_n = \left(1 + \frac{1}{n}\right)^n \quad \text{and} \quad y_n = \left(1 + \frac{1}{n}\right)^{n+1}. \quad (n = 1, 2, 3, \ldots)$$

We have no means of perceiving immediately (cf. the general remark on p. 78) how the sequences behave as n increases.

We proceed to show first that the second sequence is monotone descending, that is to say that for $n \geqq 2$

$$y_{n-1} > y_n \quad \text{or} \quad \left(1 + \frac{1}{n-1}\right)^n > \left(1 + \frac{1}{n}\right)^{n+1}.$$

This inequality is in fact equivalent[3] to

$$\left(\frac{1 + \frac{1}{n-1}}{1 + \frac{1}{n}}\right)^n > 1 + \frac{1}{n}$$

or to

$$\left(\frac{n^2}{n^2 - 1}\right)^n > 1 + \frac{1}{n}, \quad \text{i. e. to} \quad \left(1 + \frac{1}{n^2 - 1}\right)^n > 1 + \frac{1}{n}.$$

But the truth of *this* inequality is evident, since, by Bernoulli's inequality **10**, 7 we have, for $\alpha > -1$, $\alpha \neq 0$ and every $n > 1$,

$$(1 + \alpha)^n > 1 + n\alpha,$$

or in particular

$$\left(1 + \frac{1}{n^2 - 1}\right)^n > 1 + \frac{n}{n^2 - 1} > 1 + \frac{n}{n^2} = 1 + \frac{1}{n}.$$

As, moreover, $y_n > 1$ for every n, the sequence (y_n) is monotone descending and bounded, *and therefore convergent*. Its limit will often occur later on; it is, since *Euler*'s time, denoted by the special[4] letter *e*. As regards this number, we can only deduce for the present that

$$1 \leqq e < y_n$$

which for e. g. $n = 5$ becomes

$$1 \leqq e < \frac{6^6}{5^6} < 3.$$

---

[3] That is to say, each inequality follows from all the others.

[4] *Euler* uses this letter to designate the above limit in a letter to *Goldbach* (25. Nov. 1731) and in 1736 in his work: Mechanica sive motus scientia analytice exposita, II, p. 251.

The first of our two sequences, on the contrary, is *monotone ascending.* In fact, $x_{n-1} < x_n$ here means [5]

$$\left(1+\frac{1}{n-1}\right)^{n-1} < \left(1+\frac{1}{n}\right)^n$$

or

$$\left(1+\frac{1}{n-1}\right)^{-1} < \left(\frac{1+\frac{1}{n}}{1+\frac{1}{n-1}}\right)^n$$

i. e.

$$1-\frac{1}{n} < \left(\frac{n^2-1}{n^2}\right)^n = \left(1-\frac{1}{n^2}\right)^n.$$

But, again by **10**, 7, we have actually for every $n > 1$,

$$\left(1-\frac{1}{n^2}\right)^n > 1-\frac{n}{n^2} = 1-\frac{1}{n}.$$

The sequence (x_n) is therefore monotone increasing.

As, in any case,

$$\left(1+\frac{1}{n}\right)^n < \left(1+\frac{1}{n}\right)^{n+1}, \quad \text{i. e.} \quad x_n < y_n,$$

we have, for every n, $x_n < y_1$, i. e. (x_n) is also bounded and *hence convergent.* As, finally, the numbers

$$y_n - x_n = \left(1+\frac{1}{n}\right)^n \cdot \left(1+\frac{1}{n}-1\right) = \frac{1}{n}\cdot x_n$$

are all positive and (by **26**, 1) form a null sequence, we conclude at once that (x_n) has *the same* limit as (y_n). Thus

$$\lim x_n = \lim y_n = e.$$

And for this number e we have furthermore, as has appeared in the proof, in

$$\boldsymbol{e} = (\boldsymbol{x}_n \,|\, \boldsymbol{y}_n) = \left(\left(\mathbf{1}+\frac{\mathbf{1}}{\boldsymbol{n}}\right)^n \middle| \left(\mathbf{1}+\frac{\mathbf{1}}{\boldsymbol{n}}\right)^{n+1}\right)$$

a nest of intervals defining it. (It provides, for instance taking $n = 3$, the inequality $\frac{64}{27} < e < \frac{256}{81}$; we shall however become acquainted later on (§ 23) with other sequences converging to e, which are more convenient for numerical calculation.)

This is the number e that (cf. p. 58) forms the *base of the natural* logarithms. We shall accordingly agree to use the symbol log to mean this natural logarithm to the base e, unless the contrary is expressly stated.

The fruitfulness of the first main criterion is due above all to the fact that it allows us to deduce the convergence of a sequence of numbers from very few hypotheses, and these such as are usually very easy to verify — namely, from monotony and boundedness alone. On the other hand, however, it still relates only to a special, even though particularly frequent and important kind of sequence, and therefore

[5] Cf. footnote 3.

appears theoretically insufficient. We shall therefore ask for a criterion which enables us to decide *quite generally* as to the convergence or divergence of any sequence. This is accomplished by the following

47. °Second main criterion (1st form).

An arbitrary sequence (x_n) *is* **convergent** *if and only if, given* $\varepsilon > 0$, *a number* $n_0 = n_0(\varepsilon)$ *can always be assigned, such that for any two indices* n *and* n' *both greater than* n_0, *we have in every case*

$$|x_n - x_{n'}| < \varepsilon. \text{ —}$$

We first give a few

Explanations and Examples.

1. The remarks **10**, 1, 3, 4 and 9 are also substantially applicable here; and the reader is recommended to read them through once more in this connection.

2. The criterion states — to put it in intuitive language: all x_n's with very high indices must lie very close together.

3. Let $x_0 = 0$, $x_1 = 1$, and let every term after these be the arithmetic mean between the two terms which precede it, i. e. for $n \geqq 2$

$$x_n = \frac{x_{n-1} + x_{n-2}}{2}$$

so that $x_2 = \frac{1}{2}$, $x_3 = \frac{3}{4}$, $x_4 = \frac{5}{8}$, ... In this evidently *not* monotone sequence it is clear, on the one hand, that the differences between consecutive terms form a null sequence; for it may be verified quite easily by induction that[6]

$$x_{n+1} - x_n = \frac{(-1)^n}{2^n}$$

and so tends to 0. On the other hand, between these two consecutive numbers all the following ones lie. If therefore, after ε has been assigned > 0, we choose p so large that $\frac{1}{2^p} < \varepsilon$, we have

$$|x_n - x_n'| < \varepsilon$$

provided only n and n' are $> p$. By the 2nd main criterion the sequence (x_n) is *therefore convergent*. The limit ξ also happens to be easily obtainable. A little reflection in fact leads to the surmise that $\xi = \frac{2}{3}$. In point of fact, the formula

$$x_n - \frac{2}{3} = \frac{2}{3} \cdot \frac{(-1)^{n+1}}{2^n}$$

can immediately be proved by induction and shows that $x_n - \frac{2}{3}$ is actually a null sequence.

Before trying to fathom the meaning of the 2nd main criterion further, we proceed to give its

Proof. a) That the condition of the theorem — let us call it for brevity its ε-condition — is *necessary*, i. e. that it is always fulfilled

---

[6] This is true for $n = 0$ and 1. From $x_{k+2} - x_{k+1} = \frac{x_{k+1} - x_k}{2} + \frac{x_k - x_{k-1}}{2}$ it follows that if proved for every $n \leqq k$, it is true for $n = k + 1$.

if (x_n) is convergent, is seen thus: If $x_n \to \xi$, then $(x_n - \xi)$ is a null sequence; given $\varepsilon > 0$, we can so choose n_0 that for every $n > n_0$, $|x_n - \xi|$ is $< \frac{\varepsilon}{2}$. If besides n, we also have $n' > n_0$, then $|x_{n'} - \xi|$ is also $< \frac{\varepsilon}{2}$, and so

$$|x_n - x_{n'}| = |(x_n - \xi) - (x_{n'} - \xi)| \leqq |x_n - \xi| + |x_{n'} - \xi| < \frac{\varepsilon}{2} + \frac{\varepsilon}{2} = \varepsilon;$$

which proves this part of the theorem.

b) That the ε-condition is also sufficient is not so easy to see. We again prove it constructively, by deducing from the sequence (x_n) a nest of intervals (J_n) and then showing that the number determined thereby is the limit of the sequence. This is done as follows:

Any $\varepsilon > 0$ being chosen, $|x_n - x_{n'}|$ must always be $< \varepsilon$ provided only the indices n and n' *both* exceed some sufficiently large value. If we suppose the one fixed and denote it by p, then we may also say: Given any $\varepsilon > 0$, we can always assign an index p (actually, as far to the right as we please) so that for every $n > p$

$$|x_n - x_p| < \varepsilon.$$

If we choose successively $\varepsilon = \frac{1}{2}, \frac{1}{4}, \ldots, \frac{1}{2^k}, \ldots$ then we get:

1) There is an index p_1 such that

$$\text{for every } n > p_1, \quad \text{we have} \quad |x_n - x_{p_1}| < \frac{1}{2}.$$

2) There is an index p_2, which we may assume $> p_1$, such that

$$\text{for every } n > p_2, \quad \text{we have} \quad |x_n - x_{p_2}| < \frac{1}{2^2},$$

and so on. A kth step of this kind gives:

k) There is an index p_k, which we may assume $> p_{k-1}$, such that

$$\text{for every } n > p_k, \quad \text{we have} \quad |x_n - x_{p_k}| < \frac{1}{2^k}.$$

Accordingly we form the intervals J_k:

1. The interval $x_{p_1} - \frac{1}{2} \ldots x_{p_1} + \frac{1}{2}$ call J_1; it contains all the x_n's for $n > p_1$, in particular, therefore, the point x_{p_2}. It therefore contains in whole or part the interval $x_{p_2} - \frac{1}{4} \ldots x_{p_2} + \frac{1}{4}$, in which all x_n's with $n > p_2$ lie. As these points also lie in J_1, they lie in the *common part of the two intervals*. This common part we denote

2) by J_2 and may state: J_2 lies in J_1 and contains all points x_n with $n > p_2$. If in this result we replace p_1 and p_2 by p_{k-1} and p_k, and denote therefore

k) by J_k the portion of the interval $x_{p_k} - \frac{1}{2^k} \ldots x_{p_k} + \frac{1}{2^k}$ which lies in J_{k-1}, we may then state: J_k lies in J_{k-1} and contains all points x_n with $n > p_k$.

But (J_k) is then a nest of intervals; for each interval lies in the preceding and the length of J_k is $\leqq \frac{2}{2^k}$.

Now if ξ is the number thus determined, we assert, finally, that

$$x_n \to \xi.$$

In fact, if an arbitrary $\varepsilon > 0$ be now given, we choose an index r so large that $\frac{2}{2^r} < \varepsilon$. We then have

$$\text{for every } n > p_r, \quad |x_n - \xi| \text{ is } < \varepsilon,$$

since ξ, together with all x_n's for $n > p_r$, lies in J_r and the length of J_r is $< \varepsilon$. This proves all that was required[7].

Further examples and remarks.

48. 1 The sequence **45**, 3 can easily now be seen to be convergent. For we have here, if $n' > n$:

$$x_{n'} - x_n = \pm\left(\frac{1}{2n+1} - \frac{1}{2n+3} + \cdots + \frac{(-1)^{n'-n-1}}{2n'+1}\right).$$

If inside the bracket, we take the successive terms in pairs, we see (cf. later **81c**, 3) that the value of the bracket is positive, so that

$$|x_{n'} - x_n| = \frac{1}{2n+1} - \frac{1}{2n+3} + \cdots + \frac{(-1)^{n'-n-1}}{2n'+1}.$$

If we now let the first term stand by itself and take the following terms in pairs, we see further that

$$|x_{n'} - x_n| < \frac{1}{2n+1}.$$

Therefore $|x_{n'} - x_n|$ is $< \varepsilon$, provided n and n' are both $> \frac{1}{2\varepsilon}$. The sequence is therefore convergent.

2. If $x_n = \left(1 + \frac{1}{2} + \cdots + \frac{1}{n}\right)$, we have already seen in **46**, 3 that (x_n) is not convergent. With the aid of the 2nd main criterion, this is deducible from the fact that here the ε-condition is *not* satisfied for $\varepsilon < \frac{1}{2}$. For however n_0 may be chosen, we have for $n > n_0$ and $n' = 2n$ (also therefore $> n_0$)

$$x_{n'} - x_n = \frac{1}{n+1} + \frac{1}{n+2} + \cdots + \frac{1}{2n} > n\,\frac{1}{2n} = \frac{1}{2},$$

not therefore $< \varepsilon$. The sequence is therefore divergent, and in fact *definitely* divergent, since it is evidently monotone ascending.

3. The previous example shows at the same time that the contrary of the fulfilment of the ε-condition is the following (cf. also **10**, 12): Not for *every* choice of $\varepsilon > 0$ can n_0 be so assigned that the ε-condition is then fulfilled; there exists on the contrary (at least) one particular number $\varepsilon_0 > 0$ such that,

---

[7] We shall become acquainted with other proofs of this fundamental criterion. The proof given above leads immediately to the definition of the limit by the aid of a nest of intervals. — A critical account of earlier proofs of the criterion may be found in *A. Pringsheim* (Sitzungsber. d. Akad. München, Vol. 27, p. 303. 1897).

above every number n_0, *however large* (therefore infinitely often) **two** positive integers n and n' may be found for which

$$|x_{n'} - x_n| \geqq \varepsilon_0 > 0 .$$

4. The 2nd main criterion is now usually, after *P. du Bois Reymond* (Allgemeine Funktionentheorie, Tübingen 1882), called the *general principle of convergence.* In substance, it originated with *B. Bolzano* (1817, cf. *O. Stolz*, Mathem. Ann. Vol. 18, p. 259, 1881) but was first made a starting point, as an expressly formulated principle, by *A. L. Cauchy* (Analyse algébrique, p. 125).

Our main criterion may also be given somewhat different forms, which are sometimes more convenient in applications. We suppose the notation for the numbers n and n' so chosen that $n' > n$, and therefore we may write $n' = n + k$, where k is again a positive integer. We then formulate thus the

○ Second main criterion (Form 1a). 49.

The necessary and sufficient condition for the convergence of the sequence (x_n) *is that, given any* $\varepsilon > 0$, *a number* $n_0 = n_0(\varepsilon)$ *can always be assigned so that for every* $n > n_0$ *and every* $k \geqq 1$ *we always have*

$$|x_{n+k} - x_n| < \varepsilon .$$

From this statement of the criterion we can draw further conclusions. If we suppose *quite arbitrary natural* numbers $k_1, k_2, \ldots, k_n, \ldots$ chosen, then we must have, in view of the above, for every $n > n_0$

$$|x_{n+k_n} - x_n| < \varepsilon .$$

But this implies that the sequence of differences

$$d_n = (x_{n+k_n} - x_n)$$

forms a *null sequence.* — In order to make ourselves more readily understood, we will call the sequence (d_n) for short a *difference-sequence* of (x_n). In it, d_n is therefore the difference between x_n and *some definite later term.* Our criterion may then be formulated thus:

○ Second main criterion (2nd form). 50.

The sequence (x_n) *is convergent if and only if* ***every one*** *of its difference-sequences is a null sequence.*

Proof. The necessity of this condition we have just proved; we have still to show that it is *sufficient.* We accordingly *assume that* every difference-sequence tends to 0, and have to show that (x_n) converges. But if (x_n) were divergent, there would, by **48**, 3, exist a particular number ε_0 such that above every number n_0, however large, two numbers n and $n' = n + k$ would always lie, for which the difference

$$|x_{n+k} - x_n| \text{ was } \geqq \varepsilon_0 .$$

Since this must be the case infinitely often, there would — in contradiction to the hypothesis — exist difference-sequences [8] which did not tend to 0; (x_n) *must therefore converge*, q. e. d.

Remark. If (x_n) is convergent, and we choose a *particular* difference-sequence (d_n), we therefore certainly have $d_n \to 0$. But it should be expressly emphasized that from $d_n \to 0$ alone the convergence of (x_n) *need not* follow. On the contrary, for this, it is only sufficient that *every arbitrary* difference-sequence (not merely a particular one) should prove to be a null sequence.

If for instance the sequence $(1, 0, 1, 0, 1, \ldots)$ is considered, *every* difference-sequence for which all k_n's (from some point onwards) are *even* numbers is a null sequence. Nevertheless the sequence in question is not convergent. Similarly in the divergent sequence (x_n) with $x_n = 1 + \frac{1}{2} + \ldots + \frac{1}{n}$ *every* difference-sequence for which the indices k_n are *bounded* forms a null sequence.

Extending somewhat further the last obtained formulation of the criterion, we may finally formulate it thus:

51. ### °Second main criterion (3rd form).

If $\nu_1, \nu_2, \ldots, \nu_n, \ldots$ is *any* sequence of positive integers [9] which diverges to $+\infty$, and $k_1, k_2, \ldots, k_n, \ldots$ are *any* positive integers (without any restriction), and if we again call the sequence of differences

$$d_n = x_{\nu_n+k_n} - x_{\nu_n}$$

for short a *difference-sequence* of (x_n), then for the convergence of (x_n) it is again necessary and sufficient that (d_n) is *in every case* a null sequence.

Proof. That this condition is *sufficient* is obvious from the preceding form of the criterion, since (d_n) must, in the present case also, always be a null sequence when ν_n is chosen $= n$. And that it is *necessary* may at once be seen. For if ε is chosen > 0, there certainly exists, if (x_n) is convergent (v. Form 1a), a number m, such that for every $n > m$ and *every* $k \geqq 1$, we have

$$|x_{n+k} - x_n| < \varepsilon.$$

As ν_n diverges $\to +\infty$, there must be a number n_0 such that

$$\text{for} \quad n > n_0, \quad \text{we have always} \quad \nu_n > m.$$

But then, by the preceding, we have, for $n > n_0$, always

$$|x_{\nu_n+k_n} - x_{\nu_n}| = |d_n| < \varepsilon,$$

i. e. (d_n) is a null sequence, q. e. d.

---

[8] For if we denote by $n_1, n_2, n_3, \ldots$ the infinite number of values of n for which that inequality (each time with a suitable choice of k) is assumed to be possible, a difference-sequence would exist whose $n_1^{\text{th}}, n_2^{\text{th}}, n_3^{\text{th}}, \ldots$ terms were all in absolute value $\geqq \varepsilon_0 > 0$. This could not then be a null sequence.

[9] Equal or unequal, monotone or not monotone.

§ 10. Limiting points and upper and lower limits.

The concept of the convergence of a sequence of numbers as defined in the two preceding paragraphs admits of another, some· what more general mode of treatment, by which we shall at the same time become acquainted with some other concepts, of the utmost importance for all that comes after.

In **39**, 6, we have already illustrated the fact of a given sequence (x_n) being convergent by saying that every ε-neighbourhood (however small) of ξ must contain all the terms of the sequence — with the possible exception of a finite number at most. — There is therefore in every neighbourhood of ξ, however small, certainly *an infinite number* of terms of the sequence. For this reason, ξ may be called a *limiting point* or *point of accumulation* of the given sequence. Such points may, as we shall at once see, occur also in the case of divergent sequences, and we define therefore quite generally:

°**Definition.** *A number ξ shall be called a* ***limiting point**** *of a given sequence (x_n) if every neighbourhood of ξ, however small, contains an infinite number of the terms of the sequence;* or, therefore, if, for any chosen $\varepsilon > 0$, there is always an infinite number of indices n for which **52.**

$$|x_n - \xi| < \varepsilon.$$

Remarks and examples.

1. The distinction between this definition and the definition of limit given in **39** lies, as already indicated, in the fact that here $|x_n - \xi| < \varepsilon$ needs to be fulfilled not for *every* n after a certain point, but only for *any* infinite number of n's, and therefore in particular for at least *one* n beyond every n_0. On the other hand, in accordance with **39**, the limit ξ of a convergent sequence (x_n) is always a limiting point of the sequence. **53.**

2. The sequence **6**, 1 has the limiting point 0; **6**, 4, the limiting points 0 and 1. (Every number which occurs an infinite number of times in a sequence (x_n) is *ipso facto* a limiting point.) **6**, 2, 7 and 11 have no limiting point; **6**, 9 and 10 have the limiting point 1.

3. We now form an example of more than illustrative significance: If p is an integer $\geqq 2$, there is obviously only a finite number of positive fractions for which the sum of numerator and denominator $= p$, namely the fractions $\frac{p-1}{1}, \frac{p-2}{2}, \ldots, \frac{1}{p-1}$. Of these we suppose left out all those which are not in their lowest terms, and now consider in succession all the fractions thus formed for $p = 2, 3, 4, \ldots$. This gives the sequence, beginning with

$$\text{(a)} \qquad 1, 2, \frac{1}{2}, 3, \frac{1}{3}, 4, \frac{3}{2}, \frac{2}{3}, \frac{1}{4}, \ldots,$$

which contains *all positive rational numbers.* If after each of these numbers we insert the same number with sign changed and start with 0 as first term, we have in the sequence

---

* German: ***Häufungswert,*** *Häufungspunkt or Häufungsstelle.* (Tr.)

(b) $$0,\ 1,\ -1,\ 2,\ -2,\ \frac{1}{2},\ -\frac{1}{2},\ 3,\ -3,\ \frac{1}{3},\ -\frac{1}{3},\ 4,\ -4,\ \frac{3}{2},\ -\frac{3}{2},\ \frac{2}{3},\ -\frac{2}{3},\ \frac{1}{4},\ \ldots$$

thus formed obviously *all rational numbers* occurring, each exactly once.

For this remarkable sequence *every* real number is a limiting point; for *every* neighbourhood of *every* real number contains an infinity of rational numbers (cf. p. 12).

4. We shall frequently make use of the principle of arrangement in order applied in this example. We therefore formulate it somewhat more generally: Suppose that for every k of the series $k = 0, 1, 2, \ldots$ a sequence

$$x_0^{(k)},\ x_1^{(k)},\ x_2^{(k)},\ \ldots \qquad (k = 0, 1, 2, \ldots)$$

is given. *We can then, in many different ways, form a sequence* (x_n) *which contains every term of each of these sequences and contains it exactly once.*

The proof consists simply in assigning a sequence (x_n) which fulfils what is required. For this purpose we write the given sequences in rows one below the other:

$$\left\{\begin{array}{llllll} x_0^{(0)}, & x_1^{(0)}, & x_2^{(0)}, & \ldots, & x_n^{(0)}, & \ldots \\ x_0^{(1)}, & x_1^{(1)}, & x_2^{(1)}, & \ldots, & x_n^{(1)}, & \ldots \\ \cdots & \cdots & \cdots & \cdots & \cdots & \cdots \\ \cdots & \cdots & \cdots & \cdots & \cdots & \cdots \\ x_0^{(k)}, & x_1^{(k)}, & x_2^{(k)}, & \ldots, & x_n^{(k)}, & \ldots \\ \cdots & \cdots & \cdots & \cdots & \cdots & \cdots \end{array}\right.$$

The "diagonal" of this system which joins the element $x_0^{(p)}$ to the element $x_p^{(0)}$ then contains all elements $x_n^{(k)}$ for which $k + n = p$, and no others. They are $p + 1$ in number. These terms we write down in succession, taking $p = 0, 1, 2, \ldots$, and describe each of the diagonals say from bottom to top. Thus we obtain the sequence

$$x_0^{(0)},\ x_0^{(1)},\ x_1^{(0)},\ x_0^{(2)},\ x_1^{(1)},\ x_2^{(0)},\ x_0^{(3)},\ x_1^{(2)},\ \ldots,$$

which evidently fulfils the requirements. (*Arrangement by diagonals**).

Another arrangement frequently used is that "by squares". Here we first write the elements $x_0^{(p)}, x_1^{(p)}, \ldots, x_p^{(p)}$ of the pth row, then the elements standing vertically above $x_p^{(p)}$ in the above system: $x_p^{(p-1)}, \ldots, x_p^{(0)}$. These groups of $2p+1$ terms are then written down in succession for $p = 0, 1, 2, \ldots$, and this gives, beginning with

$$x_0^{(0)},\ x_0^{(1)},\ x_1^{(1)},\ x_1^{(0)},\ x_0^{(2)},\ x_1^{(2)},\ x_2^{(2)},\ x_2^{(1)},\ x_2^{(0)},\ x_0^{(3)},\ \ldots$$

the *arrangement by squares***.

If some or all of the rows in the above system consist of only a finite number of terms, or if the system consists of only a finite number of rows, then the arrangements described above undergo slight and immediately obvious modifications.

---

* German: *Anordnung nach Schräglinien.* (Tr.)
** German: *Anordnung nach Quadraten.* (Tr.)

5. An example similar to 3. is the following: For every $p \geqq 2$ there are exactly $p-1$ numbers of the form $\frac{1}{k}+\frac{1}{m}$ for which the sum of the positive integers k and m is equal to p. If we suppose these written down in succession, for $p = 2, 3, 4, \ldots$, we obtain the sequence

$$2,\ \frac{3}{2},\ \frac{3}{2},\ \frac{4}{3},\ 1,\ \frac{4}{3},\ \frac{5}{4},\ \frac{5}{6},\ \frac{5}{6},\ \ldots.$$

We find that this sequence has the limiting points $0,\ 1,\ \frac{1}{2},\ \frac{1}{3},\ \frac{1}{4},\ \ldots$ and no others.

6. As in the case of the limit of a convergent sequence, the limiting points of an arbitrary sequence may very well not belong to the sequence itself. Thus in 3. the irrational numbers, and in 5. the value 0, certainly do not belong to the sequence concerned. On the other hand, in both cases the value $\frac{1}{2}$, for instance, is both a limiting point and a term of the sequence.

We proceed to give a theorem which is fundamental for our purpose, due originally to *B. Bolzano*[10], though its significance was first fully recognised by *K. Weierstrass*[11].

°**Theorem.** *Every bounded sequence possesses at least one limiting point.* **54.**

Proof. We again determine the number in question by a suitable nest of intervals. By hypothesis there exists an interval J_0 which contains all the terms of the given sequence (x_n). To this interval we apply the method of successive bisection and designate as J_1 its *left* or *right* half *according as the left half contains an infinite number of the terms of the sequence or not.* By the *same* rule we designate a definite half of J_1 as J_2, and so on. Then the intervals of the nest (J_n) so formed all have the property that an infinite number of terms is contained *in* each, whilst to the left of their left endpoint there is always at most a finite number of points of the sequence. The point ξ thus defined is obviously a limiting point; for if $\varepsilon > 0$ is given arbitrarily, choose from the succession of intervals J_n one, say J_p, whose length is $< \varepsilon$. The terms of (x_n), in number infinite, which belong to the interval J_p then lie *ipso facto* in the ε-neighbourhood of ξ, — which proves all that we require.

The similarity of the definitions of *limiting point* and *limit* (or *limiting value*) in spite of the difference emphasized in **53**, 1 ("every limit is also a limiting point, but not conversely") naturally creates a certain relationship between them. This is elucidated by the following

---

[10] Rein analytischer Beweis des Lehrsatzes, daß zwischen je zwey Werthen, die ein entgegengesetztes Resultat gewähren, wenigstens eine reelle Wurzel der Gleichung liege, Prag 1817.

[11] In his lectures.

55. °Theorem. *Every limiting point ξ of a sequence (x_n) may be regarded as the limit of a suitable sub-sequence of (x_n).*

Proof. Since for every $\varepsilon > 0$, we have, for an infinite number of indices, $|x_n - \xi| < \varepsilon$, we have, in particular, for a suitable $n = k_1$, $|x_{k_1} - \xi| < 1$; for a suitable $n = k_2 > k_1$, we have similarly $|x_{k_2} - \xi| < \frac{1}{2}$, and in general, for a suitable $n = k_\nu > k_{\nu-1}$

$$|x_{k_\nu} - \xi| < \frac{1}{\nu} \qquad (\nu = 2, 3, \ldots).$$

For the subsequence $(x_n') \equiv (x_{k_n})$ thus picked out, we have $x_n' \to \xi$, as $(x_{k_n} - \xi)$, by **26**, 2, forms a null sequence.

The proof of the theorem of *Bolzano-Weierstrass* gives occasion for a further most important remark: The intervals J_n of the nest there constructed not only had the property that *within them* lay an *infinite* number of terms of the sequence (x_n), but as we noticed, they had the further property that to the left of the left endpoint of any definite one of the intervals there lay *always a finite number only* of the terms of the sequence. From this, however, it follows at once that no further limiting point can lie to the left of the limiting point ξ already determined. For if we choose any real number $\xi' < \xi$, we have $\varepsilon = \frac{1}{2}(\xi - \xi') < 0$; choosing an interval J_q of length $< \varepsilon$, we have the whole of the ε-neighbourhood of the point ξ' lying to the left of the left endpoint of J_q and therefore containing only a finite number of terms of the sequence. Therefore no point ξ' to the left of ξ can be a limiting point of the sequence (x_n), and we have the

56. Theorem. *Every bounded sequence has a well-defined least limiting point* (i. e. one farthest to the *left*).

If we interchange right and left in these considerations, we obtain[12] quite similarly the

57. Theorem. *Every bounded sequence has a well-defined greatest limiting point*[13] (i. e. one farthest to the *right*).

These two special limiting points we will designate by a special name.

58. Definition. *The least limiting point of a (bounded) sequence will be called* its* ***lower limit or limes inferior.*** *Denoting it by $\varkappa$, we write*

$$\varliminf_{n\to\infty} x_n = \varkappa \quad \text{or} \quad \liminf_{n\to\infty} x_n = \varkappa$$

---

[12] Or by reflection at the origin.

[13] These theorems are again obvious except in the case in which the sequence (x_n) has an *infinite number* of limiting points, like e. g. the sequence **53**, 5. For among a *finite* number of values there must always be both a greatest and a least.

* The German text has "*untere Häufungsgrenze, unterer Limes, Limes inferior*". (Tr.)

(possibly omitting the subscript $n \to \infty$). If μ is the greatest limiting point of the sequence, we write

$$\overline{\lim_{n\to\infty}}\, x_n = \mu \quad \textit{or} \quad \limsup_{n\to\infty} x_n = \mu$$

and call μ the* **upper limit** *or* **limes superior** *of the sequence (x_n). We have necessarily always $\varkappa \leqq \mu$.*

Since every ε-neighbourhood of the point $\varkappa$ contains an infinite number of terms of the sequence (x_n), and since on the other hand only a finite number of terms of the sequence can lie to the left of the left endpoint of any such neighbourhood, $\varkappa$ (or similarly μ) is also characterised by the following conditions:

Theorem. *The number $\varkappa$ (or μ) is the lower (or upper) limit of the sequence (x_n) if and only if, given an arbitrary $\varepsilon > 0$, we have still for an infinite number of n's,* **59.**

$$x_n < \varkappa + \varepsilon \quad (\textit{or} > \mu - \varepsilon),$$

but for at most a finite number[14] *of n's,*

$$x_n < \varkappa - \varepsilon \quad (\textit{or} > \mu + \varepsilon).$$

Before we give a few examples and explanations of this theorem, let us complete our definitions for the case of unbounded sequences.

Definitions. 1. *If a sequence is unbounded on the left, then we will say that $-\infty$ is a limiting point of the sequence; and if it is unbounded on the right, we will say that $+\infty$ is a limiting point of the sequence.* In these cases, however large we choose the number $G > 0$, the sequence has an infinity of terms[15] below $-G$ or above $+G$. **60.**

2. *If therefore the sequence (x_n) is unbounded on the left, then $-\infty$ is the least limiting point,* so that we have to write

$$\varkappa = \underline{\lim_{n\to+\infty}}\, x_n = -\infty.$$

Similarly we have to write

$$\mu = \overline{\lim_{n\to+\infty}}\, x_n = +\infty$$

if the sequence is unbounded on the right. In these cases, however large we choose the number $G > 0$, we have, for an infinity of indices, $x_n < -G$ or $x_n > +G$.

---

* The German text has "*obere Häufungsgrenze, oberen Limes, Limes superior*". (Tr.)

[14] Or: *There is an index n_0* from and after which we never have $x_n < \varkappa - \varepsilon$ $(> \mu + \varepsilon)$ but beyond *every* index n, there is always another n for which $x_n < \varkappa + \varepsilon$ $(> \mu - \varepsilon)$.

[15] Here therefore — and similarly in the following definitions — the portion of the straight line to the right of $+G$ plays the part of an ε-neighbourhood of $+\infty$, the portion to the left of $-G$ that of an ε-neighbourhood of $-\infty$.

3. If, finally, the sequence is *bounded on the left, but not on the right* and (besides $+\infty$) has *no other* limiting point, then $+\infty$ is not only its *greatest*, but at the same time its *least* limiting point, and *we shall therefore equate the lower limit also to* $+\infty$:

$$\varkappa = \underline{\lim}_{n\to+\infty} x_n = +\infty;$$

and correspondingly we shall have to *equate the upper limit to* $-\infty$,

$$\mu = \overline{\lim}_{n\to+\infty} x_n = -\infty$$

if the sequence is *bounded on the right, but not on the left*, and (besides $-\infty$) has *no other* limiting point. The former (latter) case occurs if and only if, given any $G > 0$, *the inequality*

$$x_n > G \quad (x_n < -G)$$

holds for an infinite number of n's, but the inequality

$$x_n < G \quad (x_n > -G)$$

for at most a finite number of n's, that is to say therefore when $x_n \to +\infty$ $(-\infty)$, Cf. **63**, Theorem 2.

Examples and explanations.

61. 1. In consequence of the preceding definitions, every sequence of numbers now of itself defines, absolutely uniquely, two determinate symbols $\varkappa$ and μ, (which may now, it is true, stand for $+\infty$ or $-\infty$, and which bear the relation $\varkappa \leqq \mu$ to one another[16]. And the following examples show that $\varkappa$ and μ may actually assume all finite or infinite values compatible with the inequality $\varkappa \leqq \mu$.

In fact, for the sequence $(x_1, x_2, x_3, \ldots)$	we have $\varkappa =$	$\mu =$
1. $(n) \equiv 1, 2, 3, 4, \ldots$	$+\infty$	$+\infty$
2. $(a + n^{(-1)^n}) \equiv a+1,\ a+2,\ a+\frac{1}{3},\ a+4, \ldots$	a	$+\infty$
3. $a, b, a, b, a, b, \ldots \quad (a < b)$	a	b
4. $\left(a + \frac{(-1)^n}{n}\right) \equiv a-1,\ a+\frac{1}{2},\ a-\frac{1}{3},\ a+\frac{1}{4}, \ldots$	a	a
5. $((-1)^n \cdot n) \equiv -1, +2, -3, +4, \ldots$	$-\infty$	$+\infty$
6. $(a - n^{(-1)^n}) \equiv a-1,\ a-2,\ a-\frac{1}{3},\ a-4, \ldots$	$-\infty$	a
7. $(-n) \equiv -1, -2, -3, \ldots$	$-\infty$	$-\infty$

2. The reader should note particularly that it is not contradictory to theorem **59** that an *infinite* number of terms of the sequence should lie to the left of $\varkappa$ or to the right of μ. Thus for instance we have, for the sequence $\left((-1)^n \frac{n+1}{n}\right)$, i. e. for the sequence $-2, +\frac{3}{2}, -\frac{4}{3}, +\frac{5}{4}, -\frac{6}{5}, \ldots$ evidently

[16] We say of every real number that it is $< +\infty$ and $> -\infty$, and for this reason we occasionally designate it expressly as "finite".

$\varkappa = -1$, $\mu = +1$, and both to the left of $\varkappa$ and to the right of μ lies an infinite number of terms of the sequence (and *between* $\varkappa$ and μ lies *no* term of the sequence!). It is therefore not at all necessary that there should be only a finite number of terms of the sequence outside the interval $\varkappa \ldots \mu$. Theorem **59** only asserts in fact that at most a finite number of terms of the sequence can lie to the left of $\varkappa - \varepsilon$ or to the right of $\mu + \varepsilon$.

3. "A finite number of alterations" has no effect on the limiting points of a sequence — none, in particular, on its upper and lower limits. These therefore represent an *ultimate* property of the sequence.

4. Since a sequence (x_n) determines both the numbers $\varkappa$ and μ with complete uniqueness, and since their value, in connection with our definition, was also enclosed by a well defined nest of intervals, we have herein a new legitimate means of defining (determining, giving) real numbers: *a real number shall henceforth also be regarded as "given", if it is the upper or lower limit of a given sequence.* This means of determining real numbers is evidently still more general than the one mentioned in **41**, 1 since now the sequence utilised need not even be convergent, or be subject to any restriction whatever[17].

As may be seen, in the light of **55**, we have also the following

Theorem. *The upper limit* μ *of the sequence* (x_n), $\mu = \overline{\lim}\, x_n$, *is* **62.**
also, in the case $\mu \neq \pm\infty$, *characterised by the two following conditions:*

a) *the limit* ξ' *of every convergent sub-sequence* (x_n') *of* (x_n) *is invariably* $\leqq \mu$*; but there exists*

b) *at least one such sub-sequence, whose limit is equal to* μ*; — and correspondingly for the lower limit.*

A concept related to that of the upper and lower limits, though one which must be sharply distinguished from it, is the concept of *upper and lower bounds* of a sequence (x_n), which is derived from the following consideration: If no term of the sequence lies to the right of $\mu = \overline{\lim}\, x_n$, so that *for every* n, $x_n \leqq \mu$, then μ is a bound above (**8**, 4) of the sequence, — but one which cannot be replaced by any smaller one; μ is therefore in this case the *least bound above.* But such a least bound also exists if there is a term of the sequence $> \mu$. For if for instance x_p is $> \mu$, then by **59** there is certainly only a finite number of terms in the sequence which are $\geqq x_p$, and among these there is necessarily (**8**, 5) a largest one, say x_q. We then have, *for every* n, $x_n \leqq x_q$, i. e. x_q is a bound above of the sequence, — but again one, which cannot be replaced by any smaller one. *Every sequence bounded on the right therefore possesses a definite least bound above.* Since, in the same way, every sequence bounded

---

[17] Whereas therefore a nest of intervals (with rational endpoints) was at first to count as the only means of defining a real number, we have now deduced quite a series of other means which we now admit as equally legitimate: Radix fractions, Dedekind sections, nests of intervals with arbitrary real endpoints, convergent sequences, upper and lower limits of a sequence. In all these cases, however, we saw how at once to assign a nest of intervals (with rational endpoints) which encloses the given number.

on the left must have a definite greatest bound below, we are justified in the following

Definition. *We define as* ***the upper bound*** * *of a sequence bounded on the right the least of its bounds above (invariably determinate by our preliminary remarks), and similarly as* ***the lower bound*** * *of a sequence bounded on the left the greatest of its bounds below. A sequence unbounded on the right is said to possess the upper bound* $+\infty$, *one unbounded on the left, to possess the lower bound* $-\infty$.

The concepts of upper and lower limits are due to *A. L. Cauchy* (Analyse algébrique, p. 132. Paris 1821) but were first made generally known by *P. du Bois-Reymond* (Allgemeine Funktionentheorie, Tübingen 1882). Both nomenclature and notation have remained variable up to the present day. The particularly convenient notation $\overline{\lim}$ and $\underline{\lim}$ used in the text was introduced by *A. Pringsheim* (Sitzungsber. d. Akad. zu München, vol. 28, p. 62. 1898), to whom the designations of *upper and lower limits* are also due **.

It should be expressly pointed out again that the upper (and similarly the lower) bound is not necessarily determined by the tail-end of the sequence. Thus the upper bound of the sequence $\left(\frac{1}{n}\right)$ is 1, and is obviously altered if the first term of the sequence is altered.

The previous investigations of this paragraph were carried out quite independently of the considerations on convergence of §§ 8 and 9, and give us, for this very reason, a new means of attacking the problem of convergence A of § 9. It may be shewn that the knowledge of the lower and upper limits $\varkappa$ and μ of a sequence — the knowledge, therefore, of two numbers whose *existence* is *a priori* ensured — entirely suffices to decide whether or how the sequence converges or diverges. We have in fact the theorems

63. **Theorem 1.** *The sequence* (x_n) *is convergent if and only if its lower and upper limits* $\varkappa$ *and* μ *are equal and finite. If* λ is the common value (different, therefore, from $+\infty$ or $-\infty$) of $\varkappa$ and μ, then $x_n \to \lambda$.

Proof. a) Let $\varkappa = \mu$ and their common value $= \lambda$. Then, by **59**, given ε, there is at most a finite number of n's for which

$$x_n \leqq \kappa - \varepsilon = \lambda - \varepsilon,$$

---

* German: *Obere, untere Grenze* (frontier). The word "frontier" is not usual in English writings, though sometimes found in French. The distinction between *any* bounds and the *narrowest* bounds is emphasized chiefly by the article *the* in the latter case; *the* upper bound and *the* lower bound always denoting the latter. For fear of ambiguity, however, the word "bound" in the general sense is avoided as much as possible in English text-books. (Tr.)

** We have omitted reference here to the untranslated term "Häufungsgrenze" of the German text: "Die im Texte benutzte ausführlichere Bezeichnung *Häufungsgrenze* soll nur den Unterschied zu der soeben definierten unteren und oberen *Grenze* stärker betonen". (Tr.)

and similarly at most a finite number of n's for which

$$x_n \geqq \mu + \varepsilon = \lambda + \varepsilon.$$

For *every* $n \geqq$ some n_0, we therefore have

$$\lambda - \varepsilon < x_n < \lambda + \varepsilon, \quad \text{or} \quad |x_n - \lambda| < \varepsilon,$$

i. e. the sequence is convergent and λ is its limit.

b) If, conversely, $\lim x_n = \lambda$, then, given $\varepsilon > 0$, we have, for every $n > n_0(\varepsilon)$, $\lambda - \varepsilon < x_n < \lambda + \varepsilon$. Therefore the inequality

$$x_n < \lambda + \varepsilon \quad (> \lambda - \varepsilon)$$

is satisfied for an infinite number of n's, but the inequality

$$x_n < \lambda - \varepsilon \quad (> \lambda + \varepsilon)$$

for at most a finite number of n's. The former inequalities (with $<$) imply $\varkappa = \lambda$, the latter $\mu = \lambda$. This proves all that we required.

Theorem 2. *The sequence* (x_n) *is definitely divergent if, and only if, its upper and lower limits are equal, but have the common value*[18] $+\infty$ *or* $-\infty$. In the former case it diverges to $+\infty$, in the latter to $-\infty$.

Proof. a) If $\varkappa = \mu = +\infty$ (or $-\infty$), then this signifies, by **60**, 2 and 3, that, given $G > 0$, we have from and after a certain n_0

$$x_n > +G \quad (< -G);$$

we therefore then have $\lim x_n = +\infty$ $(-\infty)$.

b) If, conversely, $\lim x_n = +\infty$, then, given $G > 0$, we have for every n after a certain n_0, $x_n > +G$; therefore

the inequality $x_n < +G$ is satisfied for at most a finite number of n's, whereas

the inequality $x_n > +G$ is satisfied for an infinite number of n's. But this implies, by **60**, that $\varkappa = +\infty$ and *ipso facto* also $\mu = +\infty$. Therefore $\varkappa = \mu = +\infty$. And in precisely the same way we show that if $\lim x_n = -\infty$, then $\varkappa = \mu = -\infty$.

From these two theorems we at once deduce further:

Theorem 3. *The sequence* (x_n) *is indefinitely divergent if and only if its upper and lower limits are distinct.*

The content of these three theorems provides us with the following

Third main criterion *for the convergence or divergence of a sequence:* **64.**

The sequence (x_n) *behaves definitely or indefinitely, according as its upper and lower limits are equal or distinct. In the case of definite behaviour, it is convergent or divergent, according as the common value of the upper and lower limits is finite or infinite.*

---

[18] In occasionally speaking of the symbols $+\infty$ and $-\infty$ (which are certainly not numbers) as "values", we make use of a mere verbal licence, to which no importance should be attached.

The following table gives a summary of possibilities as regards the convergence or divergence of a sequence and of the designations used in this connection.

$\varkappa = \mu$, both $= \lambda \neq \pm\infty$	$\varkappa = \mu = +\infty$ or $-\infty$	$\varkappa < \mu$
convergent (with limit λ) $\lim\limits_{(n \to +\infty)} x_n = \lambda$ $x_n \to \lambda$ (for $n \to +\infty$)	divergent (or possibly: convergent) towards (or: with limit) $+\infty$ or $-\infty$; in both cases: definitely divergent. $\lim x_n = +\infty$ or $-\infty$ $x_n \to +\infty$ or $-\infty$	indefinitely divergent
convergent	divergent	
definite behaviour		indefinite behaviour

§ 11. Infinite series, infinite products, and infinite continued fractions.

A numerical sequence can be specified in the most diverse ways; this is sufficiently evident from the examples which have been given. In these, however, for the most part, the nth term x_n was for convenience given by an explicit formula, enabling us to calculate it at once. This is by no means the rule, however, in the applications of sequences in all parts of mathematics. On the contrary, the sequences to be examined generally present themselves indirectly. Besides several less important kinds, three types especially come into consideration; of these we will now give a brief discussion.

66. I. **Infinite series.** These are sequences given in the following way. A sequence is at first assigned in any manner (usually by direct indication of its terms), but without being intended itself to form the object of discussion. From it a new sequence is to be deduced, whose terms we now denote by s_n, writing

$$s_0 = a_0; \quad s_1 = a_0 + a_1; \quad s_2 = a_0 + a_1 + a_2;$$

and generally

$$s_n = a_0 + a_1 + a_2 + \ldots + a_n \qquad (n = 0, 1, 2, \ldots).$$

It is the sequence (s_n) of *these numbers* which then forms the object of investigation. For this sequence (s_n) we use the symbolical expression

67. a) $$a_0 + a_1 + a_2 + \ldots + a_n + \ldots$$

or more shortly

b) $$a_0 + a_1 + a_2 + \ldots$$

or still more shortly and more expressively:

$$\sum_{n=0}^{\infty} a_n$$

and this new symbol we call an ***infinite series***; the numbers s_n are called the *partial sums* or *sections* * of the series. — We may therefore state the

° **Definition.** *An infinite series is a symbol of the form* **68.**

$$\sum_{n=0}^{\infty} a_n \quad \text{or} \quad a_0 + a_1 + a_2 + \ldots$$

or

$$a_0 + a_1 + a_2 + \ldots + a_n + \ldots$$

by which is meant the sequence (s_n) of the partial sums

$$s_n = a_0 + a_1 + \ldots + a_n \qquad (n = 0, 1, 2, \ldots).$$

Remarks and Examples.

1. The symbols

$$a_0 + \sum_{n=1}^{\infty} a_n; \quad a_0 + a_1 + \sum_{n=2}^{\infty} a_n; \quad a_0 + a_1 + \ldots + a_m + \sum_{n=m+1}^{\infty} a_n$$

shall be entirely equivalent to $\sum_{n=0}^{\infty} a_n$. The index n is called the *index of summation.* Of course any other letter may take its place

$$\sum_{\nu=0}^{\infty} a_\nu; \quad a_0 + a_1 + a_2 + \sum_{\varrho=3}^{\infty} a_\varrho; \quad \text{etc.}$$

The numbers a_n are the *terms* of the series. They need not be indexed from 0 onwards. Thus the symbol

$$\sum_{\lambda=1}^{\infty} a_\lambda \quad \text{denotes the sequence} \quad (a_1, a_1 + a_2, a_1 + a_2 + a_3, \ldots)$$

and more generally,

$$\sum_{k=p}^{\infty} a_k$$

denotes the sequence of numbers $s_p, s_{p+1}, s_{p+2}, \ldots$ given by

$$s_n = a_p + a_{p+1} + \ldots + a_n \quad \text{for} \quad n = p, p+1, \ldots$$

Here p may be any integer $\gtreqless 0$. Finally we also write quite shortly

$$\sum a_\nu$$

when there is no ambiguity as to the values which the index of summation has to assume, — or when this is a matter of indifference.

2. For $n = 0, 1, 2, \ldots$ let a_n be

a) $= \dfrac{1}{2^n}$; b) $= \dfrac{1}{(n+1)(n+2)}$; c) $= 1$; d) $= n$;

e) $= \dfrac{(-1)^n}{n+1}$; f) $= (-1)^n$; g) $= (-1)^n (2n+1)$;

h) $= \dfrac{1}{(\alpha+n)(\alpha+n+1)}$ $\qquad \alpha =$ a real number $\neq 0, -1, -2, \ldots$

---

* German: *Teilsummen* oder *Abschnitte.*

We are then concerned with the infinite series

$$\text{a) } \sum_{n=0}^{\infty} \frac{1}{2^n} \equiv 1 + \frac{1}{2} + \frac{1}{4} + \frac{1}{8} + \cdots;$$

$$\text{b) } \sum_{n=0}^{\infty} \frac{1}{(n+1)(n+2)} \equiv \frac{1}{1\cdot 2} + \frac{1}{2\cdot 3} + \frac{1}{3\cdot 4} + \cdots;$$

$$\text{c) } 1 + 1 + 1 + \cdots; \qquad \text{d) } 0 + 1 + 2 + 3 + \cdots;$$

$$\text{e) } \sum_{\nu=0}^{\infty} \frac{(-1)^\nu}{\nu+1} \equiv 1 - \frac{1}{2} + \frac{1}{3} - \frac{1}{4} + - \cdots;$$

$$\text{f) } \sum_{\lambda=0}^{\infty} (-1)^\lambda \equiv 1 - 1 + 1 - 1 + - \cdots; \qquad \text{g) } 1 - 3 + 5 - 7 + 9 - + \cdots;$$

$$\text{h) } \sum_{k=0}^{\infty} \frac{1}{(\alpha+k)(\alpha+k+1)} \equiv \frac{1}{\alpha(\alpha+1)} + \frac{1}{(\alpha+1)(\alpha+2)} + \frac{1}{(\alpha+2)(\alpha+3)} + \cdots.$$

And we have in these simply a new — and as will be seen, very con venient — symbol for the sequences $(s_0, s_1, s_2 \ldots)$ for which s_n is

$$\text{a) } = 1 + \frac{1}{2} + \frac{1}{4} + \cdots + \frac{1}{2^n} = 2 - \frac{1}{2^n};$$

$$\text{b) } = \frac{1}{1\cdot 2} + \frac{1}{2\cdot 3} + \frac{1}{3\cdot 4} + \cdots + \frac{1}{(n+1)(n+2)}$$

$$= \left(1 - \frac{1}{2}\right) + \left(\frac{1}{2} - \frac{1}{3}\right) + \cdots + \left(\frac{1}{n+1} - \frac{1}{n+2}\right) = 1 - \frac{1}{n+2};$$

$$\text{c) } = n + 1; \qquad \text{d) } = \frac{n(n+1)}{2};$$

$$\text{e) } = 1 - \frac{1}{2} + \frac{1}{3} - + \cdots + \frac{(-1)^n}{n+1} \quad \text{(cf. } \mathbf{45}, 3 \text{ and } \mathbf{48}, 1);$$

$$\text{f) } = \tfrac{1}{2}[1 - (-1)^{n+1}] \text{ (see footnote 19)};$$

$$\text{g) } = (-1)^n (n+1);$$

$$\text{h) } = \frac{1}{\alpha(\alpha+1)} + \frac{1}{(\alpha+1)(\alpha+2)} + \cdots + \frac{1}{(\alpha+n)(\alpha+n+1)}$$

$$= \left(\frac{1}{\alpha} - \frac{1}{\alpha+1}\right) + \left(\frac{1}{\alpha+1} - \frac{1}{\alpha+2}\right) + \cdots + \left(\frac{1}{\alpha+n} - \frac{1}{\alpha+n+1}\right)$$

$$= \frac{1}{\alpha} - \frac{1}{\alpha+n+1}.$$

3. We emphasise above all that the new symbols have *no significance in themselves*. Addition, it is true, is a well-defined operation, always possible, with regard to two or any particular number of values, in one and only one way. The partial sums s_n therefore, however the terms a_n may be given, have under all circumstances definite values. But the symbol $\sum_{n=0}^{\infty} a_n$ has *in itself* no meaning whatever, — not even in a case as transparent, seemingly, as 2 a; for the addition of an *infinite number* of terms is something quite undefined, something perfectly meaningless. It must be considered substantially as a convention that we are to take the new symbol to mean the sequence of its partial sums.

[19] Equal to 1 or 0, according as n is even or odd.

4. The reader should take particular care to distinguish a series from a sequence [20]: A *series* is a new symbol for a *sequence* deducible by a definite rule from it.

5. The symbol with the sign of summation "Σ" can of course only be used when the terms of the series are formed by an explicitly assigned law, or when a particular notation is available for them. If for instance the numbers

$$\frac{1}{2}, \frac{1}{3}, \frac{1}{5}, \frac{1}{7}, \frac{1}{11}, \frac{1}{13}, \frac{1}{17}, \ldots$$

or the numbers

$$\frac{1}{3}, \frac{1}{7}, \frac{1}{8}, \frac{1}{15}, \frac{1}{24}, \frac{1}{26}, \frac{1}{31}, \ldots$$

are to be the terms of a series, we shall have to use the explicit symbols

$$\frac{1}{2}+\frac{1}{3}+\frac{1}{5}+\frac{1}{7}+\frac{1}{11}+\frac{1}{13}+\cdots$$

and

$$\frac{1}{3}+\frac{1}{7}+\frac{1}{8}+\frac{1}{15}+\frac{1}{24}+\frac{1}{26}+\frac{1}{31}+\cdots$$

and write down as many terms as necessary, till we may assume that the reader has recognised the law of formation. For the first of these two series, this may be expected after the term $\frac{1}{13}$: the terms are the reciprocals of the successive prime numbers. In the second example it will not be known even after the term $\frac{1}{31}$ how to proceed: the denominators of the terms are meant to be the integers of the form

$$p^q - 1 \qquad (p, q = 2, 3, 4, \ldots)$$

in order of magnitude.

We now adopt the further convention that all expressions used to describe the behaviour, in respect of convergence, of a sequence are to be carried over from the sequence (s_n) to the infinite series Σa_n itself. Thereby we obtain in particular the following

69. Definition. *An infinite series Σa_n is said to be convergent, definitely divergent or indefinitely divergent, according as the sequence of its partial sums shows the behaviour indicated by those names. If, in the case of convergence, $s_n \to s$, then we say that s is the* **value** *or the* **sum** *of the convergent infinite series and we write for brevity*

$$\sum_{\nu=0}^{\infty} a_\nu = s,$$

so that $\sum\limits_{\nu=0}^{\infty} a_\nu$ denotes not only the sequence (s_n) of the partial sums, as laid down in the preceding definition, but also the limit $\lim s_n$, *when this exists* [21]. *In the case of definite divergence of (s_n), we also say that the series is definitely divergent and that it diverges to $+\infty$ or $-\infty$ according as $s_n \to +\infty$ or $\to -\infty$. If finally, in the case of indefinite divergence of (s_n), $\varkappa$ and μ are the lower and upper limits of the sequence, then we also say that the series is indefinitely divergent and oscillates between the (lower and upper) limits $\varkappa$ and μ.*

---

[20] The additional epithet of "infinite" may be omitted when obvious.

[21] Exactly as we may now, in accordance with the footnote 9 to **41**, 1, write $(s_n) = s$.

Remarks and examples.

1. It is at once obvious that the series **68**, 2a, b and h converge and have for sums $+2$, 1 and $\frac{1}{\alpha}$ respectively; 2c and d are definitely divergent towards $+\infty$; 2e is convergent and has for sum the number s defined by the nest[22] $(s_{2k-1} \mid s_{2k})$; 2f, finally, oscillates between 0 and 1, and 2g between $-\infty$ and $+\infty$.

2. As regards the term *sum* the reader must be expressly cautioned about a possible misunderstanding: The number s ***is not*** a sum in any sense previously in use, but only *the limit of an infinite sequence of sums*; the equation

$$\sum_{n=0}^{\infty} a_n = s \quad \text{or} \quad a_0 + a_1 + \cdots + a_n + \cdots = s$$

is therefore neither more nor less than another way of writing

$$\lim s_n = s \quad \text{or} \quad s_n \to s.$$

It would therefore seem more appropriate to speak not of the *sum* but of the *limit* or *value* of the series. However the term "sum" has remained in use from the time when infinite series first appeared in mathematical science and when no one had a clear notion of the underlying limiting processes or, generally, of the "infinite" at all.

3. The number s *is* therefore no sum, but is only so named, for the sake of brevity. In particular, calculations involving series will in no wise obey all the rules for calculating with sums. Thus for instance in an (actual) sum we may introduce or omit brackets in any manner, so that for instance,

$$1-1+1-1=(1-1)+(1-1)=1-(1-1)-1=0.$$

But on the contrary

$$\sum_{n=0}^{\infty}(-1)^n \equiv 1-1+1-1+-\cdots$$

is *not* the same thing as

$$(1-1)+(1-1)+(1-1)+\cdots \equiv 0+0+0+\cdots$$

or as

$$1-(1-1)-(1-1)-(1-1)-\cdots \equiv 1-0-0-0-\cdots.$$

Nevertheless, calculations involving series will have many analogies with those involving (actual) sums. The existence of such an analogy has, however, *in every particular case to be first established.*

4. It is also, perhaps, not superfluous to remark that it is really quite paradoxical that an infinite series, say $\sum_{n=0}^{\infty} \frac{1}{2^n}$, should possess anything at all

---

[22] In fact $s_{2k-1} = \left(1-\frac{1}{2}\right)+\left(\frac{1}{3}-\frac{1}{4}\right)+\cdots+\left(\frac{1}{2k-1}-\frac{1}{2k}\right)=\frac{1}{1\cdot 2}+\frac{1}{3\cdot 4}+\cdots+\frac{1}{(2k-1)\,2k}$, so that $s_1 < s_3 < s_5 < \cdots$; similarly from $s_{2k} = 1-\left(\frac{1}{2}-\frac{1}{3}\right)-\cdots-\left(\frac{1}{2k}-\frac{1}{2k+1}\right)$ we deduce that $s_0 > s_2 > s_4 > \cdots$. Finally $s_{2k}-s_{2k-1}=+\frac{1}{2k+1}$, i. e. positive and tending to 0. By **46**, 4 and **41**, 5, we have $s_n \to (s_{2k-1} \mid s_{2k})$. Cf. **81c**, 3 and **82**, 5 where these considerations are generalised.

capable of being called its sum. Let us interpret it in fourth-form fashion by shillings and pence: I give some one first 1 s., then $^1/_2$ s., then $^1/_4$ s., then $^1/_8$ s., and so on. If now I never come to an end with these gifts, the question arises, whether the fortune of the recipient must thereby necessarily increase beyond all bounds, or not At first one has the feeling that the former must occur; for if I continue constantly adding something, the sum must — it seems — ultimately exceed every value. In the case under consideration this is not so, since for every n

$$s_n = 1 + \frac{1}{2} + \frac{1}{4} + \ldots + \frac{1}{2^n} = 2 - \frac{1}{2^n} \text{ remains } < 2.$$

The total gift therefore never reaches even the amount of 2 s. And if we now, in spite of this, say that $\sum \frac{1}{2^n}$ *is equal* to 2, then we are really only using an abbreviated expression for the fact that the sequence of partial sums tends to the *limit* 2. — Cf. the well-known paradox of Achilles and the tortoise (Zenon's paradox).

5. In the case of definite divergence we can also, in an extended sense, speak of a sum of the series, which then has the "value" $+\infty$ or $-\infty$. Thus for instance the series

$$\sum_{n=1}^{\infty} \frac{1}{n} \equiv 1 + \frac{1}{2} + \frac{1}{3} + \frac{1}{4} + \cdots$$

is definitely divergent, and has the "sum" $+\infty$, because by **46**, 3 its partial[23] sums $\to +\infty$. We write for short

$$\sum_{n=1}^{\infty} \frac{1}{n} = +\infty$$

which is only another mode of writing for

$$\lim \left(1 + \frac{1}{2} + \cdots + \frac{1}{n}\right) = +\infty.$$

6. In the case of an indefinitely divergent series however, the word "sum" loses all significance. If in this case $\underline{\lim}\, s_n = \varkappa$ and $\overline{\lim}\, s_n = \mu\,(> \varkappa)$, then we said, in the above, that the series *oscillates between* $\varkappa$ *and* μ. But it must be carefully noted (cf. **61**, 2), that this refers only to a description of the *ultimate* behaviour of the series. In fact the partial sums s_n *need not* lie *between* $\varkappa$ and μ. Thus, for instance, if $a_0 = 2$, and for $n > 0$,

$$a_n = (-1)^n \left[\frac{n+1}{n} + \frac{n+2}{n+1}\right]$$

we can at once verify that

$$s_n = a_0 + a_1 + \cdots + a_n = (-1)^n \frac{n+2}{n+1} \qquad (n = 0, 1, 2, \ldots)$$

and therefore $\underline{\lim}\, s_n = -1$, $\overline{\lim}\, s_n = +1$. But *all* the terms of the sequence (s_n)

---

[23] If therefore the payments discussed in 4. have the values 1 s., $^1/_2$ s., $^1/_3$ s., $^1/_4$ s., ... the fortune of the recipient now does increase beyond all bounds. It is not at first at all obvious to what it is due that in the case 4, the sum does not exceed a modest amount, whereas in the present case it exceeds every bound. The divergence of this series was discovered by *John Bernoulli* and published by *James Bernoulli* in 1689; but seems to have been already known to *Leibniz* in 1673.

lie *outside* the interval $-1 \ldots +1$, alternately on the left and on the right, so that an infinite number of terms of the sequence lies on both sides of the interval.

7. As we emphasized above that a series Σa_n represents *merely* the sequence (s_n) of its partial sums, — and therefore is merely another mode of symbolising a *sequence*, so we may easily convince ourselves that conversely every sequence $(x_0, x_1, \ldots)$ may be written as a series. We need only write

$$a_0 = x_0, \quad a_1 = x_1 - x_0, \quad a_2 = x_2 - x_1, \ldots, \quad a_n = x_n - x_{n-1}, \ldots \quad (n \geqq 1).$$

For then the series $\sum_{n=0}^{\infty} a_n \equiv x_0 + \sum_{k=1}^{\infty} (x_k - x_{k-1})$ has for partial sums $s_0 = x_0$, $s_1 = x_0 + (x_1 - x_0) = x_1$ and generally for $n \geqq 1$

$$s_n = x_0 + (x_1 - x_0) + (x_2 - x_1) + \ldots + (x_{n-1} - x_{n-2}) + (x_n - x_{n-1}) = x_n,$$

so that the above written series does actually stand for the sequence (x_n). The new symbol of the infinite series is therefore neither more special nor more general than that of the infinite sequence. Its significance resides principally in the fact that the emphasis is on the *difference* $a_n = s_n - s_{n-1}$ of each term of the sequence (s_n) from the preceding, rather than on these terms themselves.

The convention laid down in **68**, 1, by which for instance

$$\sum_{n=0}^{\infty} a_n \text{ and } a_0 + a_1 + \ldots + a_m + \sum_{n=m+1}^{\infty} a_n$$

are to mean the same thing, now becomes the theorem (cf. **70** and **82**, 4) that the two series involved converge and diverge together, and that, when convergent, the two expressions have the same value.

8. With regard to the *History of Infinite Series*, an excellent account is given in a little book by *R. Reiff* (Tübingen 1889). Here it may suffice to mention the following facts: The first example of an infinite series is usually ascribed to *Archimedes* (Opera, ed. J. L. Heiberg, Vol. 2, pp. 310 seqq., Leipzig 1913). He, however, merely shows that $1 + \frac{1}{4} + \ldots + \frac{1}{4^n}$ remains less than $\frac{4}{3}$, whatever value n may have, and that the difference between the two values is $\frac{1}{3} \cdot \frac{1}{4^n}$, and consequently less than a given positive number, provided n be taken sufficiently large. He therefore proves — in our phraseology — that the series $\sum_{n=0}^{\infty} \frac{1}{4^n}$ is convergent, and shows that its sum equals $\frac{4}{3}$. A more general use of infinite series does not, however, begin till the second half of the 17th Century, when *N. Mercator* and *W. Brouncker*, in 1668, while engaged on the quadrature of the hyperbola, discovered the logarithmic series **120**, and when *I. Newton*, in 1669, in his work *De analysi per aequationes numero terminorum infinitas* placed their use on a firmer basis. In the 18th Century the consideration of principles was, it is true, entirely neglected, but the practice of series, on the other hand, was developed, above all by *Euler*, in a magnificent manner. In the 19th Century, finally, the theory was established by *A. L. Cauchy* (Analyse algébrique, Paris 1821) in an irreproachable manner, except for the want of clearness which then still attached to the concept of number as such. (For further historical remarks, see Introduction to § 59.)

II. Infinite products.

Here we are concerned with products of the form

$$u_1 \cdot u_2 \cdot u_3 \ldots u_n \ldots \quad \text{or} \quad \prod_{n=1}^{\infty} u_n;$$

they must be taken, in a precisely similar manner to the infinite series just considered, simply as a new symbolic form for the well-defined sequence of the *partial products*

$$p_1 = u_1; \quad p_2 = u_1 \cdot u_2; \quad \ldots; \quad p_n = u_1 \cdot u_2 \ldots u_n; \quad \ldots.$$

However we shall later, with reference to the exceptional part played by the number 0 in multiplication, have to make a few special conventions in this connection.

1. If for instance we have, for every $n \geqq 1$, $u_n = \frac{(n+1)^2}{n(n+2)}$, then the infinite product

$$\prod_{n=1}^{\infty} \frac{(n+1)^2}{n(n+2)} \quad \text{or} \quad \frac{2^2}{1\cdot 3} \cdot \frac{3^2}{2\cdot 4} \cdot \frac{4^2}{3\cdot 5} \cdot \frac{5^2}{4\cdot 6} \cdots \frac{(n+1)^2}{n(n+2)} \cdots$$

represents the sequence of numbers

$$p_1 = \frac{4}{3}; \quad p_2 = \frac{2\cdot 3}{4}; \quad p_3 = \frac{2\cdot 4}{5}; \quad \ldots; \quad p_n = \frac{2(n+1)}{n+2}; \quad \ldots.$$

2. The additions and remarks just made in I retain *mutatis mutandis* their significance here. All further details will be considered later (Chapter VII).

III. Infinite continued fractions. Here the sequence (x_n) under examination is formed by means of *two* other sequences $(a_1, a_2 \ldots)$ and $(b_0, b_1, \ldots)$, by writing:

$$x_0 = b_0, \quad x_1 = b_0 + \frac{a_1}{b_1}, \quad x_2 = b_0 + \cfrac{a_1}{b_1 + \cfrac{a_2}{b_2}}, \quad x_3 = b_0 + \cfrac{a_1}{b_1 + \cfrac{a_2}{b_2 + \cfrac{a_3}{b_3}}}$$

and so on, x_n, in the general case, being deduced from x_{n-1} by substituting for the last denominator b_{n-1} of x_{n-1} the value $b_{n-1} + \frac{a_n}{b_n}$, and proceeding thus *ad infinitum*. For the "infinite continued fraction" so formed the notation

$$b_0 + \frac{a_1|}{|b_1} + \frac{a_2|}{|b_2} + \cdots + \frac{a_n|}{|b_n} + \cdots$$

is fairly usual. The most natural notation for it would be

$$b_0 + \mathop{K}_{n=1}^{\infty} \frac{a_n}{b_n}.$$

Here also a few special conventions have to be made, to take the fact into account that in division the number 0 again plays an exceptional part. The subject of continued fractions we shall not, however, enter into in this treatise [24].

Of the three modes of assigning a sequence discussed above, that by infinite series is by far the most important for all applications in higher mathematics. We shall therefore have to deal mainly with these. — Since series merely represent sequences, the introductory developments of § 9 provide us with the points of view from which a given series will have to be investigated: Together with the *problem A* which concerns the convergence or divergence of a given series, we have again the harder *problem B*, which relates to the sum of a series already seen to be convergent. And for exactly the same

[24] A complete account of their theory and applications is given by *O. Perron*, Die Lehre von den Kettenbrüchen, 2nd Edition, Leipzig 1929.

reasons as we there explained, the second problem will generally present itself in the form: *A series Σa_n is known to be convergent; does its sum coincide with that of any other series or with the limit of any other sequence, or does it stand in any assignable relation to such another sum or limit?*[25]

Since the problem A is the easier and since — in contradistinction to problem B — it admits of a methodical solution, we will proceed in the first place to give our attention to this in detail.

Exercises on Chapter II[26].

9. Prove Theorems **15** to **19** of Chapter I by the method indicated in the footnote to **14**.

10. Prove in all details that the ordered arrangement, defined by **14** and **15**, of the system of all nests of intervals, obeys each of the theorems of order **1**. (For this cf. **14**, 4 and **15**, 2.)

11. Carry out the details of the proof required on p. 32; i. e. prove that the four modes of combining nests of intervals, defined by **16** to **19**, obey all the fundamental laws **2**.

12. For fixed ϱ, with $\varrho < 1$,

$$x_n = (n+1)^{\varrho} - n^{\varrho} \to 0.$$

13. For arbitrary positive α and β,

$$\frac{(\log \log n)^{\alpha}}{(\log n)^{\beta}} \to 0.$$

14. Which of the two numbers $\left(\frac{e}{2}\right)^{\sqrt{3}}$ and $\left(\sqrt{2}\right)^{\frac{\pi}{2}}$ is the larger?

---

[25] Thus e. g. the series $1 + 1 + \frac{1}{2!} + \frac{1}{3!} + \cdots + \frac{1}{n!} + \cdots$ will easily be shown to converge. How do we see that its sum coincides with the number e given by the sequence $\left(1 + \frac{1}{n}\right)^n$? Similarly we may very soon convince ourselves of the convergence of the two series

$$1 + \frac{1}{4} + \frac{1}{9} + \cdots + \frac{1}{n^2} + \cdots \quad \text{and} \quad 1 - \frac{1}{3} + \frac{1}{5} - \frac{1}{7} + - \cdots.$$

But how do we discover that if s and s' are their sums, $s = \frac{8}{3} s'^2$ and $4s' = \pi$ (i. e. equal to the limit in a third limiting process, which occurs in relation to the circle; cf. pp. 200 and 214)?

[26] In several of the following exercises, a few of the simplest results with regard to logarithms, and the numbers e and π, are assumed known, although they are only deduced later on in the text.

15. Prove the following limiting relations:

a) $$\left[\frac{1}{n^2}+\frac{2}{n^2}+\cdots+\frac{n}{n^2}\right]\to\frac{1}{2}.$$

b) $$\left[\log\left(1+\frac{1}{n^2}\right)+\log\left(1+\frac{2}{n^2}\right)+\cdots+\log\left(1+\frac{n}{n^2}\right)\right]\to\frac{1}{2};$$

c) $$\left[\frac{1}{\sqrt{n^2+1}}+\frac{1}{\sqrt{n^2+2}}+\cdots+\frac{1}{\sqrt{n^2+n}}\right]\to 1;$$

d) $$\left[\frac{n}{n^2+1^2}+\frac{n}{n^2+2^2}+\cdots+\frac{n}{n^2+n^2}\right]\to\frac{\pi}{4};$$

e) $$\left[\left(\frac{n}{n}\right)^n+\left(\frac{n-1}{n}\right)^n+\cdots+\left(\frac{1}{n}\right)^n\right]\to\frac{e}{e-1};$$

f) $$\frac{1}{n}\sqrt[n]{(n+1)(n+2)\ldots(n+n)}\to\frac{4}{e}.$$

Note that in examples a) to d) a *term by term* passage to the limit gives a wrong result, whereas in e) it gives a correct result.

16. Let a be >0, $x_1>0$ and the sequence $(x_1, x_2, \ldots)$ defined by the convention that for $n\geqq 2$

a) $$x_n=\sqrt{a+x_{n-1}},$$

b) $$x_n=\frac{a}{1+x_{n-1}}.$$

Shew that in case a) the sequence tends monotonely to the positive root of $x^2-x-a=0$; that in case b) it tends to that of $x^2+x-a=0$, but with x_n lying alternately to the left and to the right of the limit.

17 Investigate the convergence or divergence of the following sequences:

a) x_0, x_1 arbitrary; for every $n\geqq 2$, $x_n=\frac{1}{2}(x_{n-1}+x_{n-2})$;

b) $x_0, x_1, \ldots, x_{p-1}$ arbitrary; for every $n\geqq p$

$$x_n=a_1x_{n-1}+a_2x_{n-2}+\cdots+a_px_{n-p}$$

$$\left(a_1, a_2, \ldots, a_p \text{ given constants, e. g. all equal to } \frac{1}{p}\right);$$

c) x_0, x_1 positive; for every $n\geqq 2$, $x_n=\sqrt{x_{n-1}\cdot x_{n-2}}$;

d) x_0, x_1 arbitrary; for every $n\geqq 2$, $x_n=\dfrac{2x_{n-1}x_{n-2}}{x_{n-1}+x_{n-2}}$.

18 If in Ex. 17, c we put, in particular, $x_0=1$, $x_1=2$, then the limit of the sequence is $=\sqrt[3]{4}$.

19. Let $a_1, a_2, \ldots, a_p$ be arbitrary given positive quantities and let us write, for $n=1, 2, \ldots$

$$\frac{a_1^n+a_2^n+\cdots+a_p^n}{p}=s_n \quad\text{and}\quad \sqrt[n]{s_n}=x_n.$$

Show that x_n always *increases monotonely* and if one, say a_1, of the given numbers is greater than all the others, then $x_n \to a_1$ as limit.

(Hint: First show that

$$s_1 \leqq \frac{s_2}{s_1} \leqq \frac{s_3}{s_2} \leqq \cdots).$$

20. Somewhat similarly to last Ex., write

$$\frac{\sqrt[n]{a_1} + \sqrt[n]{a_2} + \cdots + \sqrt[n]{a_p}}{p} = s_n' \quad \text{and} \quad (s_n')^n = x_n'$$

and show that x_n' *decreases monotonely* and $\to \sqrt[p]{a_1 a_2 \ldots a_p}$.

21. Divide the interval $a \ldots b$ $(0 < a < b)$ into n equal parts; let $x_0 = a$, $x_1, x_2, \ldots, x_n = b$ denote the points of division. Show that the geometric mean

$$\sqrt[n+1]{x_0 \cdot x_1 \cdot x_2 \ldots x_n} \to \frac{1}{e}\left(\frac{b^b}{a^a}\right)^{\frac{1}{b-a}} = \exp\left(\frac{1}{b-a}\int_a^b \log x\, dx\right)$$

and the harmonic mean $\dfrac{n+1}{\dfrac{1}{x_0} + \dfrac{1}{x_1} + \cdots + \dfrac{1}{x_n}} \to \dfrac{b-a}{\log b - \log a}$.

22. Show that in the case of the general sequence of Ex. 5

$$\frac{x_n}{\alpha^n} \to \frac{x_1 - \beta x_0}{(\alpha - \beta)}.$$

23. Set $x > 0$ and let the sequence (x_n) be defined by

$$x_1 = x, \quad x_2 = x^{x_1}, \quad x_3 = x^{x_2}, \quad \ldots, \quad x_n = x^{x_{n-1}}, \quad \ldots$$

For what values of x is the sequence convergent? (Answer: If and only if

$$\left(\frac{1}{e}\right)^e \leqq x \leqq e^{\frac{1}{e}}.)$$

24. Let $\underline{\lim}\, x_n = \varkappa$, $\overline{\lim}\, x_n = \mu$, $\underline{\lim}\, x_n' = \varkappa'$, $\overline{\lim}\, x_n' = \mu'$. What may be said of the position of the limits for the sequences

$$(-x_n), \quad \left(\frac{1}{x_n}\right), \quad (x_n + x_n'), \quad (x_n - x_n'), \quad (x_n \cdot x_n'), \quad \left(\frac{x_n}{x_n'}\right)?$$

Discuss all possible cases.

25. Let (α_n) be bounded and (with the possible exception of a few initial terms) let us put

$$\log\left(1 + \frac{\alpha_n}{n}\right) = \frac{\beta_n}{n}.$$

Then (α_n) and (β_n) have the same upper and lower limits. The same holds if we put

$$\log\left(1 + \frac{1}{n} + \frac{\alpha_n}{n \log n}\right) = \frac{1}{n} + \frac{\beta_n}{n \log n}.$$

26. Does Theorem **43**, 3 still hold if $\eta = 0$ or $= +\infty$?

27. If the sequences (x_n) and (y_n) given in **43**, 2 and 3 are *monotone*, then so are the sequences (x_n') and (y_n') mentioned there.

28. If the sequence $\left(\frac{a_n}{b_n}\right)$ is monotone and $b_n > 0$, then the sequence having n^{th} term

$$\frac{a_1 + a_2 + \cdots + a_n}{b_1 + b_2 + \cdots + b_n}$$

is also monotone.

29. We have

$$\lim \frac{a_n}{b_n} = \lim \frac{a_n - a_{n+1}}{b_n - b_{n+1}},$$

provided the limit *on the right* exists and (a_n) and (b_n) are null sequences, with (b_n) monotone.

30. For positive, monotone c_n's,

$$\frac{x_0 + x_1 + \cdots + x_n}{n+1} \to \xi$$

implies

$$\frac{c_0 x_0 + c_1 x_1 + \cdots + c_n x_n}{c_0 + c_1 + \cdots + c_n} \to \xi$$

provided $\left(\frac{n\, c_n}{C_n}\right)$ is bounded and $C_n \to +\infty$. (Here $C_n = c_0 + c_1 + \cdots + c_n$.)

31. If $b_n > 0$, and $b_0 + b_1 + \cdots + b_n = B_n \to +\infty$, and $x_n \to +\infty$, then

$$\frac{B_n}{b_n}(x_{n+1} - x_n) \to \xi$$

implies

$$\left[x_{n+1} - \frac{b_0 x_0 + b_1 x_1 + \cdots + b_n x_n}{b_0 + b_1 + \cdots + b_n}\right] \to \xi.$$

32. For every sequence (x_n), we invariably have

$$\underline{\lim}\, x_n \leq \underline{\lim} \frac{x_0 + x_1 + \cdots + x_n}{n+1} \leq \overline{\lim} \frac{x_0 + x_1 + \cdots + x_n}{n+1} \leq \overline{\lim}\, x_n.$$

(Cf. Theorem **161**.)

33. Show that if the coefficients $a_{\lambda\mu}$ of the Theorem of *Toeplitz* **43**, 5 are *positive*, then for *every* sequence (x_n) the relation

$$\underline{\lim}\, x_n \leq \underline{\lim}\, x_n' \leq \overline{\lim}\, x_n' \leq \overline{\lim}\, x_n$$

holds, where $x_n' = a_{n0} x_0 + a_{n1} x_1 + \cdots + a_{nn} x_n$.

Part II.

Foundations of the theory of infinite series.

Chapter III.

Series of positive terms.

§ 12. The first principal criterion and the two comparison tests.

In this chapter we shall be concerned exclusively with series, all of whose terms are positive or at least non-negative numbers. If Σa_n is such a series, which we shall designate for brevity as a *series of positive terms*, then, since $a_n \geqq 0$, we have

$$s_n = s_{n-1} + a_n \geqq s_{n-1},$$

so that the sequence (s_n) of partial sums is a monotone increasing sequence. Its behaviour is therefore particularly simple, since it is then determined by the first main criterion **46**. This at once provides the following simple and fundamental

70. First principal criterion. *A series with positive terms either converges or else diverges to* $+\infty$. *And it is convergent if, and only if, its partial sums are bounded* [1].

Before indicating the first applications of this fundamental theorem, we may facilitate its use by the following additional propositions:

Theorem 1. *If p is any positive integer, then the two series*

$$\sum_{n=0}^{\infty} a_n \quad \textit{and} \quad \sum_{n=p}^{\infty} a_n$$

converge and diverge together [2], *and when both series converge*,

$$\sum_{n=0}^{\infty} a_n = a_0 + a_1 + \ldots + a_{p-1} + \sum_{n=p}^{\infty} a_n.$$

---

[1] Only boundedness *on the right* (boundedness above) comes into question, since an increasing sequence is invariably bounded on the *left*.

[2] More shortly: We "may" omit an arbitrary initial portion. — For this reason, it is often unnecessary to indicate the limits of summation (between which the index n is made to vary).

Proof. If s_n $(n = 0, 1, \ldots)$ are the partial sums of the first series, and s_n' $(n = p, p+1, \ldots)$ those of the second, then, for $n \geqq p$,

$$s_n = a_0 + a_1 + \ldots + a_{p-1} + s_n',$$

whence, for $n \to \infty$, both statements follow, — even without requiring the terms a_n to be non-negative.

Theorem 2. *If Σc_n is a convergent series with positive terms, then so is $\Sigma \gamma_n c_n$, if the factors γ_n are any positive, but bounded, numbers* [3].

Proof. If the partial sums of Σc_n remain constantly $< K$ and the factors $\gamma_n < \gamma$, then the partial sums of $\Sigma \gamma_n c_n$ obviously remain always $< \gamma K$, which, by the fundamental criterion, proves the theorem.

Theorem 3. *If Σd_n is a divergent series with positive terms, then so is $\Sigma \delta_n d_n$, if the factors δ_n are any numbers with a positive lower bound δ.*

Proof. If $G > 0$ be arbitrarily chosen, then by hypothesis the partial sums of Σd_n, from a suitable index onwards, are all $> G : \delta$. From the *same* index onwards, the partial sums of $\Sigma \delta_n d_n$ are then $> G$. Thus $\Sigma \delta_n d_n$ is divergent.

Both theorems are substantially contained in the following

Theorem 4. *If the factors α_n satisfy the inequalities*

$$0 < \alpha' \leqq \alpha_n \leqq \alpha'',$$

then the two series with positive terms Σa_n and $\Sigma \alpha_n a_n$ converge and diverge together. Or otherwise expressed: Two series with positive terms Σa_n and $\Sigma a_n'$ converge and diverge together if two positive numbers α' and α'' can be assigned for which, constantly, (or at least from some n onwards) [4]

$$\alpha' \leqq \frac{a_n'}{a_n} \leqq \alpha''$$

in particular therefore if $a_n' \sim a_n$ or, a fortiori, if $a_n' \cong a_n$ (v. **40**, 5).

Examples and Remarks. **71.**

1. If K is a bound above for the partial sums of the series Σa_n with positive terms, then the sum s of this series is $\leqq K$ (v. **46**, 1).

2. ***The geometric series.*** Given $a > 0$, and the so-called *geometric series*

$$\sum_{n=0}^{\infty} a^n \equiv 1 + a + a^2 + \cdots + a^n + \cdots,$$

we have, if $a \geqq 1$, then $s_n > n$ and so (s_n) is certainly not bounded; the series

---

[3] We shall in future usually denote by c_n the terms of a series *assumed convergent*, and by d_n those of a series assumed divergent.

[4] Since, in this formulation of the hypotheses, division by a_n occurs, the assumption is of course implied that $a_n > 0$ and never $= 0$. — Corresponding restrictions should be observed in the more frequent cases in the sequel.

is therefore in that case divergent. But if $a < 1$, then

$$s_n = 1 + a + a^2 + \cdots + a^n = \frac{1 - a^{n+1}}{1 - a}, \text{ (cf. p. 22, footnote 13)}$$

and therefore we have, for every n,

$$s_n < \frac{1}{1-a},$$

so that the series is then convergent. Since further

$$\left| s_n - \frac{1}{1-a} \right| = \frac{1}{1-a} \cdot a^{n+1}$$

forms a null sequence, by **10**, 7 and **26**, 1, we at the same time obtain — this is rarely the case — a simple expression for the *sum* of the series:

$$\sum_{n=0}^{\infty} a^n = \frac{1}{1-a}.$$

3. The series $\sum_{n=1}^{\infty} \frac{1}{n(n+1)} \equiv \frac{1}{1 \cdot 2} + \frac{1}{2 \cdot 3} + \frac{1}{3 \cdot 4} + \cdots$ has the partial sums

$$s_n = \left(1 - \frac{1}{2}\right) + \left(\frac{1}{2} - \frac{1}{3}\right) + \cdots + \left(\frac{1}{n} - \frac{1}{n+1}\right) = 1 - \frac{1}{n+1}.$$

These are constantly < 1, the series is therefore convergent. As it happens, we can see at once that $s_n \to 1$, so that $s = 1$.

4. ***Harmonic series.*** $\sum_{n=1}^{\infty} \frac{1}{n} \equiv 1 + \frac{1}{2} + \cdots + \frac{1}{n} + \cdots$ is *divergent*, for, as we saw in **46**, 3, its partial sums

$$s_n = 1 + \frac{1}{2} + \cdots + \frac{1}{n}$$

diverge[5] to $+\infty$. But the series

$$\sum_{n=1}^{\infty} \frac{1}{n^2} = 1 + \frac{1}{4} + \frac{1}{9} + \frac{1}{16} + \cdots$$

is *convergent*. For its nth partial sum is

$$s_n = 1 + \frac{1}{2 \cdot 2} + \frac{1}{3 \cdot 3} + \cdots + \frac{1}{n \cdot n} < 1 + \frac{1}{1 \cdot 2} + \frac{1}{2 \cdot 3} + \cdots + \frac{1}{(n-1)n}$$

hence

$$= 1 + \left(1 - \frac{1}{2}\right) + \left(\frac{1}{2} - \frac{1}{3}\right) + \cdots + \left(\frac{1}{n-1} - \frac{1}{n}\right) = 2 - \frac{1}{n},$$

and therefore s_n is constantly < 2, so that the given series is convergent. — The sum s is not so readily obtainable in this case; we have however at any rate $s < 2$, indeed certainly $s < \frac{7}{4}$. We shall find later (see **136**, **156**, **189** and **210**) that $s = \frac{\pi^2}{6}$. — A series of the form $\sum \frac{1}{n^\alpha}$ is called an *harmonic series*.

5. The series $\sum_{n=0}^{\infty} \frac{1}{n!} \equiv 1 + 1 + \frac{1}{2!} + \frac{1}{3!} + \ldots$ has the partial sums $s_0 = 1$, $s_1 = 2$, and for $n \geqq 2$,

[5] Cf. footnote 23, p. 103.

$$s_n = 2 + \frac{1}{2} + \frac{1}{2\cdot 3} + \dots + \frac{1}{2\cdot 3 \dots n}.$$

Replacing each factor in the denominators by the least, namely 2, we deduce that

$$s_n \leqq 2 + \frac{1}{2} + \frac{1}{2\cdot 2} + \dots + \frac{1}{2\cdot 2 \dots 2}$$
$$= 2 + \frac{1}{2} + \frac{1}{2^2} + \dots + \frac{1}{2^{n-1}} = 3 - \frac{1}{2^{n-1}} < 3.$$

The series is therefore convergent, with sum $\leqq 3$. We shall see later that this sum coincides with the limit e of the numbers $\left(1 + \frac{1}{n}\right)^n$.

6. As we remarked above that every series with positive terms represents a monotone increasing sequence, so we see, conversely, that *every* monotone increasing sequence $(x_0, x_1, \dots)$ may be expressed as a series with positive terms, provided x_0 is positive. We need only write

$$a_0 = x_0, \quad a_1 = x_1 - x_0, \dots, \quad a_n = x_n - x_{n-1}, \dots;$$

for, actually,

$$s_n = x_0 + (x_1 - x_0) + \dots + (x_n - x_{n-1}) = x_n$$

and all the a_n's are $\geqq 0$.

From our fundamental theorem we shall in due course deduce criteria which are more special, but are also easier to manipulate. This we shall be enabled to do chiefly by the instrumentality of the two following "comparison tests" *:

Comparison test of the 1st kind. **72.**

Let Σc_n and Σd_n be two series with positive terms, already known to be the first convergent, the second divergent. If the terms of a given series Σa_n, also with positive terms, satisfy, for every $n >$ a certain m,

a) *the condition*

$$a_n \leqq c_n,$$

then the series Σa_n is also convergent. — If, however, for every $n >$ a certain m,

b) *we have constantly*

$$a_n \geqq d_n,$$

then the series Σa_n must also diverge [6].

Proof. By **70**, 1, it suffices to establish the convergence or divergence of $\sum_{n=m+1}^{\infty} a_n$. In case a) the convergence of this series results at once, by **70**, 2, from that of $\sum_{n=m+1}^{\infty} c_n$, because by hypothesis we may,

* German: Vergleichskriterien. (Tr.)

[6] *Gauss* used this criterion in 1812 (v. Werke III, p. 140). It was not, however, formulated explicitly, nor was the following test of the 2nd kind, before *Cauchy*, Analyse algébrique (Paris 1821).

for every $n > m$, write $a_n = \gamma_n c_n$, with $\gamma_n \leqq 1$. In case b) the divergence results similarly[7] from that of $\sum_{n=m+1}^{\infty} d_n$, because here we may write $a_n = \delta_n d_n$, with $\delta_n \geqq 1$.

73. *Comparison test of the 2nd kind.*

Let Σc_n and Σd_n again denote respectively a convergent and a divergent series of positive terms. If the terms of a given series Σa_n of positive terms satisfy, for every $n \geqq$ a certain m,

a) *the conditions*

$$\frac{a_{n+1}}{a_n} \leqq \frac{c_{n+1}}{c_n},$$

then the series Σa_n is also convergent. If, however, for every $n \geqq$ a certain m, we have

b) *constantly*

$$\frac{a_{n+1}}{a_n} \geqq \frac{d_{n+1}}{d_n},$$

then Σa_n must also diverge.

Proof. In case a), we have for every $n \geqq m$

$$\frac{a_{n+1}}{c_{n+1}} \leqq \frac{a_n}{c_n}.$$

The sequence of the ratio $\gamma_n = \frac{a_n}{c_n}$ is, from a certain point onwards, monotone descending, and consequently, since all its terms are positive, it is necessarily *bounded*. Theorem **70**, 2 now establishes the convergence. In case b) we have, analogously, $\frac{a_{n+1}}{d_{n+1}} \geqq \frac{a_n}{d_n}$, so that the ratios $\delta_n = \frac{a_n}{d_n}$ *increase* monotonely from a point onwards. But as they are constantly positive, they then have a positive lower bound. Theorem **70**, 3 now proves the divergence.

These comparison tests or criteria can of course only be useful to us if we are already acquainted with a large number of convergent and divergent series with positive terms. We shall therefore have to lay in as large a stock as possible, so to speak, of series whose convergence or divergence is known. For this purpose the following examples may form a nucleus:

---

[7] Or else — almost more concisely —: In case a) every bound above of the partial sums of Σc_n is also one for the partial sums of Σa_n; and in case b), the partial sums of Σa_n must ultimately exceed *every* bound, since those of Σd_n do so.

Examples.

1. $\sum_{n=1}^{\infty}\frac{1}{n}$ was seen to be divergent, $\sum_{n=1}^{\infty}\frac{1}{n^2}$ convergent. By the first comparison **74.**
test, the so-called *harmonic series*

$$\sum_{n=1}^{\infty}\frac{1}{n^\alpha}$$

is therefore certainly divergent for $\alpha \leqq 1$, convergent for $\alpha \geqq 2$. It is, however, only known in the case α = even integer how its sum may be related to numbers occurring in other connections; for instance we shall see later on that for $\alpha = 4$ the sum is $\frac{\pi^4}{90}$.

2. By the preceding, the convergence or divergence of $\sum\frac{1}{n^\alpha}$ only remains questionable in case $1 < \alpha < 2$. We may prove as follows that the series *converges* for every $\alpha > 1$: To obtain a bound above for any partial sum s_n of the series, choose k so large that $2^k > n$. Then

$$s_n \leqq s_{2^k-1} = 1 + \left(\frac{1}{2^\alpha}+\frac{1}{3^\alpha}\right) + \left(\frac{1}{4^\alpha}+\frac{1}{5^\alpha}+\frac{1}{6^\alpha}+\frac{1}{7^\alpha}\right) + \dots + \left(\frac{1}{(2^{k-1})^\alpha}+\dots+\frac{1}{(2^k-1)^\alpha}\right).$$

Here we group in one parenthesis those terms whose indices run from a power of 2 (inclusive) to the next power of 2 (exclusive). Replace, in each pair of parentheses, every separate term by the first; this involves an increase of value and we have therefore

$$s_n \leqq 1 + \frac{2}{2^\alpha} + \frac{4}{4^\alpha} + \dots + \frac{2^{k-1}}{(2^{k-1})^\alpha}.$$

If we now write for brevity $\frac{1}{2^{\alpha-1}} = \vartheta$, — a positive number certainly < 1, since $\alpha > 1$, — then we have

$$s_n \leqq 1 + \vartheta + \vartheta^2 + \dots + \vartheta^{k-1} = \frac{1-\vartheta^k}{1-\vartheta} < \frac{1}{1-\vartheta};$$

and since this holds for every n, the partial sums of our series are bounded, and the series itself is convergent, q. e. d. (Cf. **77.**)

All harmonic series $\sum\frac{1}{n^\alpha}$ *for* $\alpha \leqq 1$ *are divergent, and for* $\alpha > 1$, *convergent.* In these, with the geometric series, we have already quite a useful stock of comparison series.

3. Series of the type

$$\sum_{n=1}^{\infty}\frac{1}{(a\,n+b)^\alpha},$$

where a and b are given positive numbers, also diverge for $\alpha \leqq 1$, converge for $\alpha > 1$. For since

$$\frac{n^\alpha}{(a\,n+b)^\alpha} = \left(\frac{1}{a+\frac{b}{n}}\right)^\alpha \to \frac{1}{a^\alpha}, \quad \text{we have} \quad \frac{1}{(a\,n+b)^\alpha} \sim \frac{1}{n^\alpha};$$

and **70**, 4 proves the truth of our statement.

Accordingly the series

$$1+\frac{1}{3^\alpha}+\frac{1}{5^\alpha}+\cdots \equiv \sum_{n=0}^{\infty}\frac{1}{(2n+1)^\alpha},$$

in particular, are convergent for $\alpha > 1$, divergent for $\alpha \leqq 1$.

4. If $\sum\limits_{n=0}^{\infty} c_n$ is a convergent series with positive terms, and we deduce from it a new series $\Sigma c_n'$ by omitting any (possibly an infinite number) of its terms, or by inserting in any way terms with the value 0, thus "diluting" the series, then the resulting "sub-series" $\Sigma c_n'$ is also convergent. For every number which is a bound above for the partial sums of Σc_n is then also a bound above for those of the new series.

In accordance with this, the series $\Sigma \frac{1}{p^\alpha}$, where p runs through all prime integral values, i. e. the series

$$\frac{1}{2^\alpha}+\frac{1}{3^\alpha}+\frac{1}{5^\alpha}+\frac{1}{7^\alpha}+\frac{1}{11^\alpha}+\cdots$$

is certainly convergent for $\alpha > 1$. (On the other hand, of course, we cannot conclude without further examination that it diverges for $\alpha \leqq 1$!)

5. Since Σa^n is already recognised as convergent for $0 \leqq a < 1$, we infer in particular the convergence of

$$\sum_{n=1}^{\infty}\frac{1}{10^n}=\frac{1}{10}+\frac{1}{10^2}+\cdots+\frac{1}{10^n}+\cdots,$$

If $z_1, z_2, \ldots, z_n, \ldots$ denote any "digits", i. e. if each of them be one of the numbers $0, 1, 2, \ldots, 9$, and if z_0 is any integer $\gtreqless 0$, then, by **70**, 2, the series

$$\sum_{n=0}^{\infty}\frac{z_n}{10^n}$$

is also convergent. — Thus we see that an infinite decimal fraction may also be regarded as an infinite series. In this sense we may say that *every* infinite decimal fraction is convergent and therefore represents a definite real number. — In this form of series we also have, according to our customary order of ideas, an immediate conception of the value of its sum.

§ 13. The root test and the ratio test.

We prepare the way for a more systematic use of these two comparison tests, by the two following theorems. If we take as comparison series, to begin with, the geometric series Σa^n, with $0 < a < 1$, then we immediately obtain the

75. Theorem 1. *If, given a series Σa_n of positive terms, we have, from some place onwards in the series, $a_n \leqq a^n$ with $0 < a < 1$, i. e.*

$$\sqrt[n]{a_n} \leqq a < 1,$$

then the series is convergent. *If however, from some place onwards,*

$$\sqrt[n]{a_n} \geqq 1,$$

then the series is divergent. (Cauchy's root test[8].)

Supplementary note. For divergence it clearly suffices that $\sqrt[n]{a_n} \geqq 1$ should be known to hold for *infinitely many* distinct values of n. For we then also have, for those values of n, $a_n \geqq 1$; and a particular partial sum s_m will consequently exceed a given (positive integral) number G, if m is chosen so large that the inequality $a_n \geqq 1$ occurs at least G times while $0 \leqq n \leqq m$. The sequence (s_n) is therefore certainly not bounded.

The second comparison test gives immediately:

Theorem 2. *If, from some place onwards in the series,* $a_n > 0$, *and*

$$\frac{a_{n+1}}{a_n} \leqq \alpha < 1,$$

then the series Σa_n *is* convergent. *If however, from some place onwards,*

$$\frac{a_{n+1}}{a_n} \geqq 1,$$

then the series Σa_n *is* divergent. (Cauchy's ratio test[9].)

Remarks and Examples. 76.

1. In both these theorems, it is *essential* for convergence that $\sqrt[n]{a_n}$ and $\frac{a_{n+1}}{a_n}$ respectively should be ultimately less than a fixed proper fraction α. It does not at all suffice for convergence that we should have

$$\sqrt[n]{a_n} < 1, \quad \text{or} \quad \frac{a_{n+1}}{a_n} < 1$$

for every n. An example presents itself at once in the harmonic series $\sum \frac{1}{n}$, for which we certainly *always* have

$$\sqrt[n]{\frac{1}{n}} < 1 \quad \text{and also} \quad \frac{1}{n+1} : \frac{1}{n} = 1 - \frac{1}{n+1} < 1,$$

though the series diverges. It is quite essential that the root and ratio should *not approach arbitrarily near to* 1.

2. If one of the sequences $\left(\sqrt[n]{a_n}\right)$ or $\left(\frac{a_{n+1}}{a_n}\right)$ is *convergent*, say with limit α, then theorems 1 and 2 show that the series $\sum a_n$ is convergent if $\alpha < 1$,

[8] Analyse algébrique, p. 132 seqq.

[9] Analyse algébrique, p. 134 seqq.

divergent if $\alpha > 1$. For suppose $\sqrt[n]{a_n} \to \alpha < 1$, for instance; then $\varepsilon = \frac{1-\alpha}{2} > 0$ and m may be determined so that, for every $n > m$, we have

$$\sqrt[n]{a_n} < \alpha + \varepsilon = \frac{1+\alpha}{2} = a.$$

And since this value a is < 1, theorem 1 proves the convergence. If on the contrary $\alpha > 1$, then $\varepsilon' = \frac{\alpha - 1}{2} > 0$, and m' may be so determined that, for every $n > m'$, we have

$$\sqrt[n]{a_n} > \alpha - \varepsilon' = \frac{1+\alpha}{2} = a.$$

And since this value a is > 1, theorem 1 proves the divergence. — The proof in the case of the ratio is quite analogous.

If $\alpha = 1$, these two theorems prove nothing.

3. The reasoning just applied in 2. is obviously also legitimate when $\overline{\lim} \sqrt[n]{a_n}$ or $\overline{\lim} \frac{a_{n+1}}{a_n}$ is < 1, in the one case, and $\underline{\lim} \sqrt[n]{a_n}$ or $\underline{\lim} \frac{a_{n+1}}{a_n}$ is > 1, in the other. If one of these upper or lower limits is $= 1$, or the upper limit > 1, the lower < 1, then we can infer *almost nothing* as to the convergence or divergence of Σa_n. The supplementary note to **75**, 1, however, shows that, in the root test, it is sufficient for divergence[10] that $\overline{\lim} \sqrt[n]{a_n} > 1$.

4. The remarks just made in 2. and 3. are so obvious that, in similar cases in future, we shall not specially mention them.

5. The root and ratio tests are by far the most important tests used in practice. For most of the series which occur in applications, the question of convergence or divergence can be solved by their means. We append a few examples, in which x, for the present, represents a positive number.

a) $\Sigma n^\alpha x^n$ (α arbitrary).

Here we have

$$\frac{a_{n+1}}{a_n} = \left(\frac{n+1}{n}\right)^\alpha \cdot x \to x,$$

as $\frac{n+1}{n} = 1 + \frac{1}{n} \to 1$ and is permanently positive (v. **38**, 8). The series is therefore — and this without reference to the value of α — convergent if $x < 1$, divergent if $x > 1$. For $x = 1$ our two tests are inconclusive; however we then get the harmonic series, with which we are already acquainted.

b) $$\sum_{n=0}^{\infty} \binom{n+p}{n} x^n \equiv \sum_{n=0}^{\infty} (-1)^n \binom{-p-1}{n} x^n \qquad (p \text{ an integer } \geqq 1).$$

Here we have

$$\frac{a_{n+1}}{a_n} = \frac{(n+p+1)(n+p)\ldots(n+2)\cdot p!}{(n+p)(n-1+p)\ldots(n+1)\cdot p!} \cdot x = \frac{n+p+1}{n+1} x \to x.$$

---

[10] Thereby the criterion obtains a *disjunctive* form. Σa_n is convergent or divergent according as $\overline{\lim} \sqrt[n]{a_n}$ is < 1 or > 1. (Further details in §§ 36 and 42.)

Hence this series too is convergent for $x < 1$, divergent for $x > 1$, whatever be the value of p. For $x = 1$ and $p \geqq 0$ it obviously diverges, since then $\frac{a_{n+1}}{a_n} \geqq 1$ for every n. In the case of convergence we shall later on find for its sum the value $\left(\frac{1}{1-x}\right)^{p+1}$.

c) $\sum_{n=0}^{\infty} \frac{x^n}{n!} \equiv 1 + x + \frac{x^2}{2!} + \cdots + \frac{x^n}{n!} + \cdots.$

Here we have for *every* $x > 0$

$$\frac{a_{n+1}}{a_n} = \frac{x}{n+1} \to 0 \ (< 1);$$

the series is therefore convergent for every $x \geqq 0$. For the sum we shall later on find the value e^x.

d) $\sum \frac{x^n}{n^n}$ is convergent for $x \geqq 0$, as $\sqrt[n]{\frac{x^n}{n^n}} = \frac{x}{n} \to 0$.

e) $\sum \frac{1}{(\log n)^n}$ is convergent[11], as again $\sqrt[n]{a_n} \to 0$.

f) $\sum \frac{1}{1+n^2}$ convergent, because $a_n < \frac{1}{n^2}$;

$\sum \frac{n!}{n^n}$ convergent, because $a_n = \frac{1 \cdot 2 \ldots n}{n \cdot n \ldots n} \leqq \frac{2}{n^2}$ for every $n \geqq 2$;

$\sum \frac{1}{\sqrt{n(n+1)}}$ divergent, because $a_n > \frac{1}{n+1}$;

$\sum \frac{1}{\sqrt{n(1+n^2)}}$ convergent, because $a_n < \frac{1}{n^{\frac{3}{2}}}$.

g) $\sum \frac{1}{(\log n)^p}$ (p fixed > 0), is divergent, since by 38, 4 from some n onwards $(\log n)^p < n$.

h) $\sum \frac{1}{(\log n)^{\log n}}$ is convergent, as we may at once recognize by writing the generic term in the form

$$\frac{1}{n^{\log \log n}}.$$

---

[11] In this series, summation may only begin with $n = 2$, since $\log 1 = 0$. Such and similar obvious restrictions we shall in future not always expressly mention; it suffices, for the question of convergence or divergence, that the indicated terms of the series, *from some place onwards*, have determinate values. — In all that follows, as already agreed on p. 83, the sign "log" will always stand for the *natural* logarithm, i. e. that to the base e (**46a**).

On the other hand

$$\sum \frac{1}{(\log n)^{\log \log n}} \equiv \sum e^{-(\log \log n)^2}$$

is divergent, because by **38**, 4 and Ex. 13, $(\log \log n)^2 < \log n$ from some n onwards, so that the generic term of the series is $> \frac{1}{n}$.

§ 14. Series of positive, monotone decreasing terms.

Before passing from these quite elementary considerations, we will mention a particularly simple class of series of positive terms, namely those series whose terms a_n, at least from some place onwards, form a monotone sequence. To this class belong nearly all the series given as examples above and also the majority of those which occur in applications. For such series we have the following:

77. **Cauchy's theorem of convergence**[12]. *If* $\sum_{n=1}^{\infty} a_n$ *is a series whose terms form a positive monotone decreasing sequence* (a_n), *then it converges and diverges with*

$$\sum_{k=0}^{\infty} 2^k a_{2^k} \equiv a_1 + 2a_2 + 4a_4 + 8a_8 + \cdots$$

Preliminary remark. What is particularly remarkable in this theorem is that it shows that a small proportion of all the terms of the series suffices to determine the convergence or divergence of the whole series. For this reason it is also called the *condensation theorem.*

It shows that the harmonic series $\sum \frac{1}{n}$, for instance, is certainly divergent, for it converges and diverges with the series

$$\sum \frac{2^k}{2^k} \equiv 1 + 1 + 1 + \cdots$$

which is unmistakably divergent. And speaking generally, the series $\frac{1}{n^\alpha}$ is inferred to converge and diverge with the series

$$\sum \frac{2^k}{(2^k)^\alpha} \equiv \sum \left(\frac{1}{2^{\alpha-1}}\right)^k;$$

but this is a geometric series and therefore converges or diverges according as $\alpha > 1$ or $\alpha \leqq 1$.

These examples also show us that the convergence or divergence of $\Sigma 2^k a_{2^k}$ is often more easily ascertained than that of the series Σa_n itself; it is just in this that the value of the theorem lies.

Proof. We denote the partial sums of the given series by s_n, those of the new series by t_k. Then we have (cf. **74**, 2)

[12] Analyse algébrique, p. 135.

a) for $n < 2^k$

$$s_n \leqq a_1 + (a_2 + a_3) + \cdots + (a_{2^k} + \cdots + a_{2^{k+1}-1})$$
$$\leqq a_1 + 2a_2 + 4a_4 + \cdots + 2^k a_{2^k} = t_k,$$

i. e.

$$s_n \leqq t_k;$$

b) for $n > 2^k$

$$s_n \geqq a_1 + a_2 + (a_3 + a_4) + \cdots + (a_{2^{k-1}+1} + \cdots + a_{2^k})$$
$$\geqq \tfrac{1}{2}a_1 + a_2 + 2a_4 + \cdots + 2^{k-1} a_{2^k} = \tfrac{1}{2} t_k,$$

i. e.

$$2s_n \geqq t_k.$$

Inequality a) shows that the sequence (s_n) is bounded if the sequence (t_k) is bounded; inequality b), conversely, that if (s_n) is bounded, so is (t_k). The two sequences are therefore either both bounded or both unbounded, and therefore the two series under consideration either *both* converge or *both* diverge, q. e. d.

Before given further examples illustrating this theorem, we may extend it somewhat[13]; for it is immediately evident that the number 2 plays no essential part in the theorem. In fact we have, more generally, the

Theorem. *If Σa_n is again a series whose terms form a positive monotone decreasing sequence (a_n), and if $(g_0, g_1, \ldots)$ is any monotone increasing sequence of integers, then the two series* **78.**

$$\sum_{n=0}^{\infty} a_n \quad \text{and} \quad \sum_{k=0}^{\infty} (g_{k+1} - g_k)\, a_{g_k}$$

are either both convergent or both divergent, provided g_k, for every $k > 0$, fulfils the conditions

$$g_k > g_{k-1} \geqq 0 \quad \text{and} \quad g_{k+1} - g_k \leqq M \cdot (g_k - g_{k-1})$$

in the second of which M stands for a positive constant[14].

Proof. Exactly as before we have

a) for $n < g_k$, — denoting by A the sum of the terms possibly preceding a_{g_0} (or otherwise 0), —

$$s_n \leqq s_{g_k} \leqq A + (a_{g_0} + \cdots + a_{g_1-1}) + \cdots + (a_{g_k} + \cdots + a_{g_{k+1}-1})$$
$$\leqq A + (g_1 - g_0)\, a_{g_0} + \cdots + (g_{k+1} - g_k)\, a_{g_k},$$

i. e.

$$s_n \leqq A + t_k,$$

[13] *Schlömilch, O.:* Zeitschr. f. Math. u. Phys., Vol. 18, p. 425. 1873.

[14] The second condition signifies that the gaps in the sequence (g_k), relatively to the sequence of *all* positive integers, must not increase at too great a rate.

b) for $n > g_k$

$$s_n \geqq s_{g_k} > (a_{g_0+1} + \cdots + a_{g_1}) + \cdots + (a_{g_{k-1}+1} + \cdots + a_{g_k})$$
$$\geqq (g_1 - g_0)\, a_{g_1} + \cdots + (g_k - g_{k-1})\, a_{g_k},$$
$$M s_n \geqq (g_2 - g_1)\, a_{g_1} + \cdots + (g_{k+1} - g_k)\, a_{g_k}$$

i. e.

$$M s_n \geqq t_k - t_0.$$

And from the two inequalities the statements in question follow in the same way as before.

79. Remarks.

1. It suffices of course that the conditions in either theorem be fulfilled from and after a definite place in the series. Therefore we may, in the extended theorem, suppose, as a particular case,

$$g_k = 3^k, \ = 4^k, \ldots, \qquad \text{or} \qquad = [g^k]$$

where g is any real number >1 and $[g^k]$ the largest integer not greater than g^k. We also satisfy the requirements of this theorem by taking

$$g_k = k^2, \ = k^3, \ = k^4, \ldots .$$

With $g_k = k^2$ we obtain, for instance, the theorem that the series $\sum_{n=0}^{\infty} a_n$, — if (a_n) is a positive monotone decreasing sequence, — converges and diverges with

$$\sum_{k=0}^{\infty} (2k+1)\, a_{k^2} \equiv a_0 + 3a_1 + 5a_4 + 7a_9 + \cdots$$

We may also replace this last series, according to **70**, 4, by the series $\sum k\, a_{k^2} \equiv a_1 + 2a_4 + 3a_9 + \cdots$.

2. $\sum_{n=2}^{\infty} \frac{1}{n \log n}$ is *divergent*, — although its terms are materially less than those of the harmonic series; for according to our theorem, this series converges and diverges with

$$\sum_{k=1}^{\infty} \frac{2^k}{2^k \cdot \log(2^k)} \equiv \sum_{k=1}^{\infty} \frac{1}{(\log 2)\cdot k}$$

and is therefore, by **70**, 2, like the harmonic series, divergent. The divergence of this series and of those considered in the next examples was first discovered by *N. H. Abel*[15] (v. Œuvres II, p. 200).

3. $\sum_{n=3}^{\infty} \frac{1}{n \log n \cdot \log\log n}$ is also still divergent, although its terms are again considerably less than those of the *Abel's* series just considered. For by Cauchy's theorem it converges and diverges with

$$\sum_{k=2}^{\infty} \frac{2^k}{2^k \cdot \log 2^k \cdot \log(\log 2^k)} \equiv \sum_{k=2}^{\infty} \frac{1}{k \log 2 \cdot \log(k \log 2)};$$

---

[15] *Niels Henrik Abel,* born Aug. 5th, 1802, at Findoe near Stavanger (Norway), died April 6th, 1829, at the Froland ironworks, near Arendal.

and this, since $\log 2 < 1$, has larger terms than *Abel's* series $\sum \frac{1}{k \log k}$ discussed above, and must therefore diverge.

4. Thus we may continue as long as we please. To abbreviate, let us denote by $\log_r x$ the r^{ple} repeated or *iterated* logarithm of a positive number x, so that

$$\log_0 x = x, \quad \log_1 x = \log x, \quad \log_2 x = \log(\log x), \ldots$$
$$\log_r x = \log(\log_{r-1} x).$$

We may also take $\log_{-1} x$ to denote the value e^x.

These iterated logarithms only have a meaning if x is sufficiently large; thus $\log x$ only for $x > 0$, $\log_2 x$ only for $x > 1$, $\log_3 x$ only for $x > e$, and so on; and we shall only place them in the denominators of the terms of our series if they are positive, i. e. $\log x$ only for $x > 1$, $\log_2 x$ only for $x > e$, $\log_3 x$ only for $x > e^e$, and so on. If therefore we wish to consider the series

$$\sum_n \frac{1}{n \log n \cdot \log_2 n \ldots \log_p n} \qquad (p \text{ integer} \geqq 1),$$

then the summation must only begin with a suitably large index, — whose exact value, however, (by **70**, 1), does not matter. Since the logarithms increase monotonely with n, and the terms therefore decrease monotonely, the series, by *Cauchy's* theorem, converges and diverges with

$$\sum_k \frac{1}{\log 2^k \cdot \log_2 2^k \ldots \log_p 2^k}$$

and this, since $2 < e$, must certainly diverge, if

$$\sum_k \frac{1}{k \log k \ldots \log_{p-1} k}$$

diverges. Since the divergence of the latter series was proved for $p = 1$ (and $p = 2$), it follows by Mathematical Induction (**2**, V) that it diverges for *every* $p \geqq 1$.

5. The series above considered, however, become convergent if we raise the last factor in the denominator to a power > 1. That $\sum \frac{1}{n^\alpha}$ converges for $\alpha > 1$, we already know. If we assume proved for a particular (integer) $p \geqq 1$, that the series[16]

$$(*) \qquad \sum_k \frac{1}{k \cdot \log k \ldots \log_{p-2} k \cdot (\log_{p-1} k)^\alpha} \qquad (\alpha > 1)$$

is convergent, it follows just as before that the series

$$\sum_n \frac{1}{n \cdot \log n \ldots \log_{p-1} n \cdot (\log_p n)^\alpha} \qquad (\alpha > 1)$$

is also convergent. For this, by the extended *Cauchy's* theorem **78**, converges and diverges with the series — we choose $g_k = 3^k$ —

$$\sum_k \frac{3^{k+1} - 3^k}{3^k \log 3^k \ldots (\log_p 3^k)^\alpha}.$$

---

[16] For $p = 1$, this reduces to the series $\sum \frac{1}{k^\alpha}$.

As $3 > e$, this series has its terms *less* than those of the series (*) (assumed convergent), if the terms of the latter are multiplied by 2 (which by **70**, 2 leaves the convergence undisturbed).

The series brought forward in the two last examples will later on render us most valuable services as comparison series.

We will prove one more remarkable theorem on series of positive *monotone decreasing* terms, although it anticipates to a certain extent the general considerations on convergence of the following chapter (v. **82**, Theorem 1).

80. Theorem. *If the series Σa_n of positive monotone decreasing terms is to converge, then we must have not only $a_n \to 0$, but*[17]

$$n a_n \to 0.$$

Proof. By hypothesis, the sequence of partial sums $a_0 + a_1 + \cdots + a_n = s_n$ is convergent. Having chosen $\varepsilon > 0$, we can therefore so choose m that for every $\nu > m$ and every $\lambda \geqq 1$ we have

$$|s_{\nu+\lambda} - s_\nu| < \frac{\varepsilon}{2},$$

i. e.

$$a_{\nu+1} + a_{\nu+2} + \cdots + a_{\nu+\lambda} < \frac{\varepsilon}{2}.$$

If we now choose $n > 2m$, then, taking $\nu = [\frac{1}{2} n]$, the largest integer not greater than $\frac{1}{2} n$, we have $\nu \geqq m$ and therefore

$$a_{\nu+1} + a_{\nu+2} + \cdots + a_n < \frac{\varepsilon}{2};$$

a fortiori, therefore,

$$(n - \nu) a_n < \frac{\varepsilon}{2}$$

and

$$\frac{n}{2} \cdot a_n < \frac{\varepsilon}{2}, \quad \text{i. e.} \quad n a_n < \varepsilon.$$

Therefore $n a_n \to 0$, q. e. d.

Remark. We must expressly emphasize the fact that the condition $n a_n \to 0$ is only a *necessary*, not a *sufficient* one for the convergence of our present type of series, i. e. if $n a_n$ *does not* tend to 0, then the series in question is certainly divergent[18], while $n a_n \to 0$ does not necessarily imply anything as to the possible convergence of the series. In point of fact, the *Abel's* series $\sum \frac{1}{n \log n}$ diverges, although it has monotone decreasing terms and

$$n a_n = \frac{1}{\log n} \to 0.$$

---

[17] *Olivier, L.*: Journ. f. d. reine u. angew. Math., Vol. 2, p. 34. 1827.

[18] Accordingly, the harmonic series $\sum \frac{1}{n}$, for instance, must diverge because it has monotone decreasing terms, but $n \cdot \frac{1}{n}$ does *not* tend to 0.

Exercises on Chapter III.

34. Investigate the behaviour (convergence or divergence) of a series Σa_n, for which a_n, from some index onwards, has the following values:

$$\frac{1}{n^{1+\frac{1}{n}}},\quad \frac{n^a}{n!},\quad \frac{n+\sqrt{n}}{n^2-n},\quad \frac{(n!)^2}{(2n)!},\quad \frac{2^n\cdot n!}{n^n},\quad \frac{3^n\cdot n!}{n^n},\quad \binom{a+n}{n},$$

$$\frac{1}{1+a^n},\quad \left(\sqrt[n]{a}-1\right),\quad \left(\sqrt[n]{a}-1-\frac{1}{n}\right),\quad \left(\sqrt{n+1}-\sqrt{n}\right),\quad \left(\sqrt[3]{n+1}-\sqrt[3]{n}\right),$$

$$\frac{\sqrt{n+1}-\sqrt{n}}{n},\quad \frac{1}{(\log\log n)^{\log n}},\quad \frac{1}{(\log_3 n)^{\log n}},\quad a^{\sqrt{n}},\quad a^{\log n},\quad a^{\log\log n},$$

$$\left(1-\frac{\log n}{n}\right)^n,\quad \left(\sqrt[n]{n}-1\right)^a,\quad \left(\sqrt[n]{n}-1\right)^n.$$

35. If Σd_n diverges, so also does $\Sigma \frac{d_n}{1+d_n}$. What is the behaviour of $\Sigma \frac{d_n}{1+n\,d_n}$ and $\Sigma \frac{d_n}{1+n^2 d_n}$? $(d_n > 0)$.

36. Under the same assumption that Σd_n diverges and $d_n > 0$, what is the behaviour of the series $\Sigma \frac{d_n}{1+d_n^2}$?

37. Suppose $p_n \to +\infty$. What is the behaviour of the series

$$\Sigma \frac{1}{p_n^{\,n}},\qquad \Sigma \frac{1}{p_n^{\,\log n}},\qquad \Sigma \frac{1}{p_n^{\,\log\log n}}?$$

38. Suppose $p_n \to +\infty$, but with

$$0 < \underline{\overline{\lim}}\,(p_{n+1}-p_n) < +\infty.$$

What must be the upper and lower limits of the sequence (ϱ_n) so that

$$\Sigma \frac{1}{p_n^{\,\varrho_n}}$$

converge or so that it diverge?

39. For every $n > 1$,

$$\frac{n}{2} < 1+\frac{1}{2}+\frac{1}{3}+\frac{1}{4}+\cdots+\frac{1}{2^n-1} < n.$$

40. The sequence of numbers

$$x_n = \left[1+\frac{1}{2}+\cdots+\frac{1}{n}-\log n\right]$$

is monotone descending.

41. If Σa_n has positive terms and is convergent, then $\Sigma \sqrt{a_n\, a_{n+1}}$ is also convergent. Show by an example that the converse of this theorem is not true in general, and prove that it does nevertheless hold when (a_n) is monotone.

42. If Σa_n converges, and $a_n \geqq 0$, then $\Sigma \frac{\sqrt{a_n}}{n}$ also converges, and also indeed the series $\Sigma \frac{\sqrt{a_n}}{(\sqrt{n})^{1+\delta}}$, for every $\delta > 0$.

43. Every positive real number x_1 is, in one and only one way, expressible in the form

$$x_1 = a_1 + \frac{a_2}{2!} + \frac{a_3}{3!} + \frac{a_4}{4!} + \cdots$$

where a_n is a non-negative integer with $a_n \leqq n-1$ for $n > 1$, subject to the condition of not being $= n-1$ for *every* n after a definite n_0. If x_1 is rational, and only then, the series terminates.

44. If $0 < x \leqq 1$, then there is one and only one sequence of positive integers (k_ν), with

$$1 < k_1 \leqq k_2 \leqq k_3 \leqq \cdots,$$

for which

$$x = \frac{1}{k_1} + \frac{1}{k_1 k_2} + \cdots + \frac{1}{k_1 k_2 \ldots k_n} + \cdots.$$

x is rational if, and only if, the k_ν's are all equal after some index ν_0.

Chapter IV.

Series of arbitrary terms.

§ 15. The second principal criterion and the algebra of convergent series.

An infinite series $\sum_{n=0}^{\infty} a_n$, — whose terms are now no longer assumed subjected to any restriction, but may be arbitrary real numbers, — was, we agreed, to be considered as essentially a new symbol for the sequence (s_n) of its partial sums

$$s_n = a_0 + a_1 + \cdots + a_n \qquad (n = 0, 1, 2, \ldots)$$

and we proposed to transfer immediately to the series itself the designations introduced to characterise the convergence or divergence of (s_n). The case of convergence again occupies our main attention. The second main criterion (**47**—**51**), expressing the necessary and sufficient condition for convergence, at once provides the following

81. °**Fundamental theorem (First form).** *The necessary and sufficient condition for the convergence of the series $\sum a_n$ is that, having chosen any $\varepsilon > 0$, we can assign a number $n_0 = n_0(\varepsilon)$ such that for every $n > n_0$ and every $k \geqq 1$, we have*

$$|s_{n+k} - s_n| < \varepsilon,$$

that is to say, in the present case, that

$$|a_{n+1} + a_{n+2} + \cdots + a_{n+k}| < \varepsilon.$$

Starting with the second form of the main criterion, we also obtain for the present fundamental theorem the following

°Second form. *The series $\sum a_n$ converges if, and only if, given a perfectly arbitrary sequence (k_n) of positive integers, — the sequence of numbers* **81a.**

$$T_n = (a_{n+1} + a_{n+2} + \cdots + a_{n+k_n})$$

invariably proves to be a null sequence[1]. *And as before we can extend this somewhat to the*

°Third form. *The series $\sum a_n$ converges if, and only if, given two perfectly arbitrary sequences (ν_n) and (k_n) of positive integers, of which the first, at least, tends to $+\infty$, — the sequence of numbers* **81b.**

$$T_n = (a_{\nu_n+1} + a_{\nu_n+2} + \cdots + a_{\nu_n+k_n})$$

invariably proves to be a null sequence.

Remarks.

1. A series represents essentially a new symbolic expression for sequences of numbers, and in particular, as we remarked, not only every series represents a sequence, but every sequence is also expressible as a series; all remarks and examples given on p. 84 have their parallels here.

2. The contents of the fundamental theorem may be formulated as follows: Given $\varepsilon > 0$, *every portion* of the series, *however long*, provided only its initial index be sufficiently large, must have a sum whose absolute value is $< \varepsilon$. Or: Given $\varepsilon > 0$, we must be able to assign an index m so that for $n > m$ the addition, to s_n, of an arbitrary number of terms immediately consecutive to a_n can only alter this partial sum by less than ε.

3. Our present theorems and remarks of course also hold for series of positive terms. This the reader should verify in each separate case.

A finite part of the series, such as

$$a_{\nu+1} + a_{\nu+2} + \cdots + a_{\nu+\lambda},$$

we may for brevity call a ***portion*** of the series, denoting it by T_ν if it begins immediately after the νth term. When required, we may further explicitly indicate the number of terms in the portion by denoting this by $T_{\nu,\lambda}$. If we are considering an arbitrary sequence of such portions *whose initial index* $\to +\infty$, we shall refer to it for short as a ***"sequence of portions"*** of the given series. The second and third form of the fundamental theorem may then also be expressed thus:

°4th form. *The series $\sum a_n$ converges if, and only if, every "sequence of portions" of the series is a null sequence.* **81c.**

[1] It is substantially in this form that *N. H. Abel* establishes the criterion in his fundamental memoir on the *Binomial series* (Journ f. die reine u. angew. Math., Vol. 1, p. 311. 1826).

Remarks and examples.

1. Σa_n is thus divergent if, and only if at least *one* sequence of portions can be assigned which is *not* a null sequence. For the harmonic series $\sum \frac{1}{n}$, for instance, we have

$$T_n = T_{n,n} = \frac{1}{n+1} + \frac{1}{n+2} + \cdots + \frac{1}{2n} > n \cdot \frac{1}{2n} = \frac{1}{2}.$$

The sequence (T_n) is therefore certainly not a null sequence, and therefore $\sum \frac{1}{n}$ is divergent.

2. For $\sum \frac{1}{n^2}$ we have

$$\begin{aligned} T_\nu = T_{\nu,\lambda} &= \frac{1}{(\nu+1)^2} + \cdots + \frac{1}{(\nu+\lambda)^2} \\ &< \frac{1}{\nu(\nu+1)} + \frac{1}{(\nu+1)(\nu+2)} + \cdots + \frac{1}{(\nu+\lambda-1)(\nu+\lambda)} \\ &= \left(\frac{1}{\nu} - \frac{1}{\nu+1}\right) + \left(\frac{1}{\nu+1} - \frac{1}{\nu+2}\right) + \cdots + \left(\frac{1}{\nu+\lambda-1} - \frac{1}{\nu+\lambda}\right) = \frac{1}{\nu} - \frac{1}{\nu+\lambda}, \end{aligned}$$

therefore $T_\nu < \frac{1}{\nu}$, so that $T_\nu \to 0$, when $\nu \to +\infty$. The series therefore converges.

3. For the sequence

$$\sum_{n=1}^{\infty} \frac{(-1)^{n-1}}{n} \equiv 1 - \frac{1}{2} + \frac{1}{3} - \frac{1}{4} + \frac{1}{5} - + \cdots$$

we have

$$T_n = T_{n,k} = (-1)^n \left[\frac{1}{n+1} - \frac{1}{n+2} + \frac{1}{n+3} - + \cdots + \frac{(-1)^{k-1}}{n+k}\right].$$

Whether k is even or odd, the expression in brackets is certainly positive and $< \frac{1}{n+1}$. For if we take together, in pairs, each positive term and the following negative term, the sum of the two is in each case positive. If *k is even* all terms are exhausted in this manner, if *k is uneven* a positive term remains, so that in either case the complete expression is seen to be positive. If, on the other hand, we write it in the form

$$\frac{1}{n+1} - \left(\frac{1}{n+2} - \frac{1}{n+3}\right) - \left(\frac{1}{n+4} - \frac{1}{n+5}\right) - \cdots$$

all the terms are now exhausted when k is *odd* and a negative term remains over if k is even, so that in both cases only subtractions from $\frac{1}{n+1}$ occur, and thus the expression is $< \frac{1}{n+1}$. As we now have

$$|T_n| = |T_{n,k}| < \frac{1}{n+1}$$

this involves

$$T_n \to 0,$$

and our series converges. — We shall see later (cf. **120**) that its sum coincides with the limit of the sequence **46**, 2 and has the value log 2.

To these four separate forms of the second fundamental criterion we may at once attach the following simple but important considerations: Since in the second form, by putting $k_n = 1$, we obtain $a_{n+1} \to 0$, we have also (by **27**, 4), $a_n \to 0$, i. e. we have the

° Theorem 1. *In a convergent series, the terms a_n necessarily form a null sequence: $a_n \to 0$.* **82.**

That this condition is not sufficient for convergence, we know already, from the example of the harmonic series.

If, on the other hand, we already know that $\sum a_n$ converges, then so does the series $a_{n+1} + a_{n+2} + a_{n+3} + \cdots \equiv \sum_{\nu=n+1}^{\infty} a_\nu$, whose sum is usually, as the so called ***remainder*** of the series $\sum a_n$, denoted by r_n (so that $s_n + r_n = s =$ the sum of the complete series). Now we may, in the inequality

$$|a_{n+1} + a_{n+2} + \cdots + a_{n+k}| < \varepsilon,$$

valid for $n > n_0$ and every $k \geqq 1$, allow k to increase beyond all bounds and so obtain, for every $n > n_0$, $r_n \leqq \varepsilon$. Thus we have the

°Theorem 2. *The remainders $r_n = \sum_{\nu=n+1}^{\infty} a_\nu$ of a convergent series $\sum_{n=0}^{\infty} a_n = s$, — i. e. the numbers*

$$(r_{-1} = s), r_0, r_1, r_2, \ldots, r_n, \ldots, -$$

always form a null sequence.

In **80**, we saw further that if the terms of a convergent series $\sum a_n$ (of positive terms) are *monotone* decreasing, then, over and above the theorem just proved, the condition $n a_n \to 0$ must hold. That this need no longer be the case in series of arbitrary terms is already shewn by the series given in **81 c**, 3. We can, however, show that we must have

$$\frac{a_1 + 2a_2 + \cdots + n a_n}{n} \to 0$$

i. e. that the terms of the sequence $(n a_n)$ are small *on the average.* In fact we have [2] the more general

°Theorem 3. *If $\sum_{n=0}^{\infty} a_n$ is a convergent series of arbitrary terms and if $(p_0, p_1, \ldots)$ denotes an arbitrary monotone increasing sequence of positive numbers tending to $+\infty$, then the ratio*

$$\frac{p_0 a_0 + p_1 a_1 + \ldots + p_n a_n}{p_n} \to 0.$$

[2] *L. Kronecker,* Comptes rendus de l'Ac. de Paris, Vol. 103, p. 980. 1886. — Moreover, this condition is not only necessary, but also, in a quite determinate sense, sufficient, for the convergence of the series Σa_n (cf. Ex. 58 a).

Proof. By **44**, 2, $s_n \to s$ implies

$$\frac{p_1 s_0 + (p_2 - p_1) s_1 + \cdots + (p_n - p_{n-1}) s_{n-1}}{p_n} \to s.$$

Since $\frac{p_0 s_0}{p_n} \to 0$ and $s_n \to s$, we must therefore have

$$s_n - \frac{(p_1 - p_0) s_0 + (p_2 - p_1) s_1 + \cdots + (p_n - p_{n-1}) s_{n-1}}{p_n} \to 0.$$

But this is precisely the relation we had to prove, as may be seen at once by reducing to the common denominator p_n and grouping in succession the terms which contain $p_0, p_1, \ldots, p_n$ respectively[3].

As regards any condition for convergence whatsoever, we have to repeat expressly that the stipulations made therein always concern — or only need concern — those terms of the series which follow on some determinate one, whose index may moreover be replaced by any larger index. In deciding whether a series is or is not convergent, *the beginning of the series,* — as it is usually put for brevity, — *does not come into account.* This we express more exactly in the following

°**Theorem 4.** *If we deduce, from a given series $\sum_{n=0}^{\infty} a_n$, a new series $\sum_{n=0}^{\infty} a_n'$ by omitting a finite number of terms, prefixing a finite number of terms, or altering a finite number of terms (or doing all three things at once) and now designating afresh the terms of the series so produced by $a_0', a_1', \ldots,$[4] then either both series converge or both diverge.*

Proof. The hypotheses imply that a definite integer $q \gtreqless 0$ exists such that from some place onwards, say for every $n > m$, we have

$$a_n' = a_{n+q}.$$

Every portion of the one series is therefore also a portion of the other, provided only its initial index be $> m + |q|$. The fundamental theorem **81a** immediately proves the correctness of our statement.

---

[3] Instead of the positive p_n we may (cf. **44**, 3 and 5) take *any* sequence (p_n), for which, on the one hand, $|p_n| \to +\infty$ and, on the other, a constant K is assignable for which

$$|p_0| + |p_1 - p_0| + \cdots + |p_n - p_{n-1}| < K |p_n|$$

for every n.

[4] I. e. in short: "... by making a finite number of alterations (**27**, 4) in the sequence (a_n) of the *terms* of the series ..."

Remark.

It should be expressly noted that for series of arbitrary terms, comparison tests of every kind become entirely powerless. In particular, of two series Σa_n and $\Sigma a_n'$ whose terms are asymptotically equal $(a_n \cong a_n')$, the one may quite well converge and the other diverge. Take for instance $a_n = \frac{(-1)^n}{n}$ and $a_n' = a_n + \frac{1}{n \log n}$.

Finally we prove the following criterion of convergence, which appears almost unique in consequence of its particularly elementary character, and relates to the so-called ***alternating series***, i. e. to series whose terms have alternately positive and negative signs:

Theorem 5. [*Leibniz's rule* [5].] *An alternating series, for which the absolute values of the terms form a monotone null sequence, is invariably convergent.*

The proof proceeds on quite similar lines to that of **81 c**, 3. For if $\sum a_n$ is the given alternating series, then a_n has either the sign $(-1)^n$, for every n, or the sign $(-1)^{n+1}$, for every n. If we write, therefore, $|a_n| = \alpha_n$, we have

$$T_n = T_{n,k} = \pm [\alpha_{n+1} - \alpha_{n+2} + \alpha_{n+3} - + \cdots + (-1)^{k-1} \alpha_{n+k}].$$

As the α's are monotone decreasing, we may convince ourselves precisely as in the example referred to, that the value of the square bracket is always positive, but less than its first term α_{n+1}. Thus

$$|T_n| = |T_{n,k}| < \alpha_{n+1},$$

which, since α_n forms a null sequence by hypothesis, involves $T_n \to 0$ and therefore convergence of $\sum a_n$, by **81 c**.

The algebra of convergent series.

Already in **69**, 2, 3, it has been emphasized that the term "sum", to designate the limit of the sequence of partial sums of a series, is misleading in so far as it arouses a belief that an infinite series may be operated on by the same rules as an (actual) sum of a definite number of terms, e. g. of the form $(a + b + c + d)$, say. This is not the case, however, and the presumption is therefore fundamentally erroneous, although *some* of the rules in question do actually remain valid for infinite series. The principal laws in the algebra of (actual) sums are (according to **2**, I and III) the associative, distributive and commutative laws. The following theorems are intended to show how far these laws remain true for infinite series.

---

[5] Letters to *J. Hermann* of 26. VI. 1705 and to *John Bernoulli* of 10. I. 1714.

83. ° Theorem 1. *The associative law holds for convergent infinite series unrestrictedly in the following sense only:*

$$a_0 + a_1 + a_2 + \cdots = s$$

implies

$$(a_0 + a_1 + \cdots + a_{\nu_1}) + (a_{\nu_1+1} + a_{\nu_1+2} + \cdots + a_{\nu_2}) + \cdots = s,$$

if $\nu_1, \nu_2, \ldots$ *denote any increasing sequence of different integers and the sum of the terms enclosed in each bracket is considered as one term of a new series*

$$A_0 + A_1 + \cdots + A_k + \cdots$$

where, therefore, for $k = 0, 1, 2, \ldots,$

$$A_k = a_{\nu_k+1} + a_{\nu_k+2} + \cdots + a_{\nu_{k+1}}$$

$(\nu_0 = -1)$. The converse is however not always true.

Proof. The succession of partial sums S_k of ΣA_k is obviously the sub-sequence $s_{\nu_1}, s_{\nu_2}, \ldots, s_{\nu_k}, \ldots$ of the sequence of partial sums s_n of Σa_n. By **41**, 4, S_n therefore tends to the same limit as s_n.

Remarks and examples.

1. The convergence of $\sum_{n=1}^{\infty} \frac{(-1)^{n-1}}{n} \equiv 1 - \frac{1}{2} + \frac{1}{3} - \frac{1}{4} \cdots$ therefore implies that of

$$\sum_{k=1}^{\infty}\left(\frac{1}{2k-1} - \frac{1}{2k}\right) \equiv \sum_{k=1}^{\infty} \frac{1}{(2k-1)\cdot 2k} \equiv \frac{1}{1\cdot 2} + \frac{1}{3\cdot 4} + \frac{1}{5\cdot 6} + \cdots$$

and also, similarly, of

$$1 - \left(\frac{1}{2} - \frac{1}{3}\right) - \left(\frac{1}{4} - \frac{1}{5}\right) - \cdots \equiv 1 - \frac{1}{2\cdot 3} - \frac{1}{4\cdot 5} - \frac{1}{6\cdot 7} - \cdots,$$

and all three series have the same sum. If we denote this by s, the second series shows that in any case, $s > \frac{1}{1\cdot 2} + \frac{1}{3\cdot 4} = \frac{7}{12}$, and the third, that $s < 1 - \frac{1}{2\cdot 3} = \frac{10}{12}$. Thus

$$\frac{7}{12} < s < \frac{10}{12}.$$

2. That we may *introduce* brackets, but may not without consideration *omit* brackets occurring in a series, the following simple example shows: The series $0 + 0 + 0 + \ldots$ is certainly convergent and has the sum 0. If we substitute everywhere $(1 - 1)$ for 0, we obtain the correct equality

$$(1 - 1) + (1 - 1) + \ldots \equiv \Sigma(1 - 1) = 0.$$

But by omitting the brackets we obtain the *divergent* series

$$1 - 1 + 1 - 1 + - \ldots,$$

which therefore may not be put "$= 0$". For we should then by again grouping the terms, though in a slightly different way, obtain

$$1 - (1 - 1) - (1 - 1) - \ldots \equiv 1 - 0 - 0 - 0 \ldots,$$

which again converges and has the sum 1. We should therefore finally deduce that $0 = 1$!![6].

We proceed at once to complete Theorem 1 by the following

° Theorem 2. *If the terms of a convergent infinite series* $\sum_{k=0}^{\infty} A_k$ *are themselves actual sums (say, as above,* $A_k = a_{\nu_k+1} + \ldots + a_{\nu_{k+1}}$; $k = 0, 1, \ldots$; $\nu_0 = -1$), *then we "may" omit the brackets enclosing these if, and only if, the new series* $\sum_{n=0}^{\infty} a_n$ *thus obtained also converges.*

In fact in that case, by the preceding theorem, $\Sigma a_n = \Sigma A_k$, while in the case of divergence of Σa_n, this equality would become meaningless.

A usually sufficient indication as to *whether* the new series converges is provided by the following

° Supplementary theorem. *The new series* Σa_n *deduced from* ΣA_k *in accordance with the preceding theorem is certainly convergent if the quantities*

$$A_k' = |a_{\nu_k+1}| + |a_{\nu_k+2}| + \ldots + |a_{\nu_{k+1}}|$$

form a null sequence[7].

Proof. If ε be given > 0, choose m_1 so large that, for every $k > m_1$, we have

$$|S_{k-1} - s| < \frac{\varepsilon}{2}$$

and choose m_2 so large that, for every $k > m_2$, we have $A_k' < \frac{\varepsilon}{2}$. If m is larger than both these numbers m_1 and m_2, then we have, for every $n > \nu_m$,

$$|s_n - s| < \varepsilon.$$

---

[6] In former times — before the strict foundation of the algebra of infinite series (v. Introduction) — mathematicians found themselves fairly at a loss when confronted with paradoxes such as this. And even though the better mathematicians instinctively avoided arguments such as the above, the lesser brains had all the more opportunity of indulging in the boldest speculations. — Thus e. g. *Guido Grandi* (according to *R. Reiff*, v. 69, 8) believed that in the above erroneous train of argument which turns 0 into 1, he had obtained a mathematical proof of the possibility of the creation of the world from nothing!

[7] As $A_k \to 0$, this is of itself the case if the terms which constitute A_k have one and the same sign — in particular, therefore, if by omission of the brackets we obtain a series of positive terms. Furthermore, this is always the case if the terms a_n form a null sequence and if the number $\nu_{k+1} - \nu_k$ of terms grouped together in A_k forms a bounded sequence for $k = 0, 1, 2, \ldots$ (An example is afforded by the series $\Sigma (a_n + b_n)$ in the next Theorem 3.)

For to each such n corresponds a perfectly definite number k, for which

$$\nu_k < n \leqq \nu_{k+1}$$

and this number k must be $\geqq m$. In that case, however,

$$s_n = S_{k-1} + a_{\nu_k+1} + \cdots + a_n, \qquad |s_n - S_{k-1}| \leqq A_k' < \frac{\varepsilon}{2}.$$

And since

$$s_n - s = (s_n - S_{k-1}) + (S_{k-1} - s)$$

we then have, effectually,

$$|s_n - s| < \varepsilon, \quad \text{i. e.} \quad \sum_{n=0}^{\infty} a_n = s, \qquad \text{q. e. d.}$$

Example.

$$\sum_{k=1}^{\infty} A_k \equiv \left(1 + \frac{1}{3} - \frac{1}{2}\right) + \left(\frac{1}{5} + \frac{1}{7} - \frac{1}{4}\right) + \cdots + \left(\frac{1}{4k-3} + \frac{1}{4k-1} - \frac{1}{2k}\right) + \cdots$$

is convergent; for A_k is positive, and, for every $k > 1$, is

$$< \frac{2}{4k-4} - \frac{1}{2k} = \frac{1}{2(k-1) \cdot k} \leqq \frac{1}{k^2}.$$

Since similarly, for every $k > 1$,

$$A_k' < \frac{2}{4k-4} + \frac{1}{2k} < \frac{1}{k-1},$$

(A_k') is a null sequence. Therefore the series

$$1 + \frac{1}{3} - \frac{1}{2} + \frac{1}{5} + \frac{1}{7} - \frac{1}{4} + + - \cdots$$

is also convergent. — Its sum — call it S — is certainly $> A_1 + A_2 > \frac{11}{12}$, as the series in its first form had only positive terms.

○ **Theorem 3.** *Convergent series may be added term by term. More precisely,*

$$\sum_{n=0}^{\infty} a_n = s \quad \text{and} \quad \sum_{n=0}^{\infty} b_n = t$$

imply both

$$\sum_{n=0}^{\infty} (a_n + b_n) = s + t$$

and also — without brackets! —

$$a_0 + b_0 + a_1 + b_1 + a_2 + \cdots = s + t.$$

Proof. If s_n and t_n are the partial sums of the first two series, then $(s_n + t_n)$ are those of the third. By **41**, 9, it therefore follows at once that $(s_n + t_n) \to s + t$. That the brackets may be omitted, in the series thereby proved convergent, follows from the supplementary theorem of Theorem 2, since $(|a_n|)$ and $(|b_n|)$ and therefore also $(|a_n| + |b_n|)$ are null sequences.

°**Theorem 4.** *Convergent series may in the same sense be subtracted term by term.* The proof is identical.

°**Theorem 5.** *Convergent series may be multiplied by a constant, that is to say, from $\sum a_n = s$ it follows, if c is an arbitrary number, that*

$$\sum (c\, a_n) = c\, s\,.$$

Proof. The partial sums of the new series are $c\, s_n$, if those of the old are s_n. Theorem **41**, 10 at once proves the statement. — This theorem, to some extent, provides the extension to infinite series of the *distributive law.*

Remarks and Examples. **84.**

1. These simple theorems are all the more important, as they not only allow us to deduce the *convergence* of the new series from the convergence of known series, but also set up a relation between its sum and that of the known series. They form therefore the foundation for actual calculation in terms of infinite series.

2. The series $\sum_{n=1}^{\infty} \frac{(-1)^{n-1}}{n}$ was convergent. Let s denote its sum. By theorem 1, the series

$$\sum_{k=1}^{\infty} \left(\frac{1}{2k-1} - \frac{1}{2k}\right) \quad \text{and} \quad \sum_{k=1}^{\infty} \left(\frac{1}{4k-3} - \frac{1}{4k-2} + \frac{1}{4k-1} - \frac{1}{4k}\right)$$

are then also convergent with the sum s. Multiply the first by $\frac{1}{2}$, in accordance with Theorem 5, — this giving

$$\sum_{k=1}^{\infty} \left(\frac{1}{4k-2} - \frac{1}{4k}\right) = \frac{s}{2}$$

— and add this term by term to the second; we obtain

$$\sum_{k=1}^{\infty} \left(\frac{1}{4k-3} + \frac{1}{4k-1} - \frac{1}{2k}\right) = \frac{3}{2}\, s\,,$$

or more precisely: we obtain the convergence of the series on the left hand side *and* the value of its sum, — the latter expressed in terms of the sum of the series from which we started. The *convergence* was also proved directly in connection with theorem 2; the present considerations have led however appreciably further, since they afford a definite statement as to the sum of the series.

Before we examine the validity of the commutative and distributive laws and investigate, in relation to the latter, the possibility of forming the product of two series, we still require an important preliminary.

§ 16. Absolute convergence. Derangement of series.

The series $1 - \frac{1}{2} + \frac{1}{3} - \frac{1}{4} + \cdots$ proved (**81c**, 3) to be *convergent.* But if we replace each term by its absolute value, the series becomes the *divergent* harmonic series $1 + \frac{1}{2} + \frac{1}{3} + \cdots$. In all that follows, it will usually make a very material difference whether a convergent series Σa_n remains convergent or becomes divergent, when all its terms are replaced by their absolute values. Here we have, to begin with, the

85. °**Theorem.** *A series Σa_n is certainly convergent if the series (of positive terms) $\Sigma |a_n|$ converges*[8]. *And if $\Sigma a_n = s$, $\Sigma |a_n| = S$ then $|s| \leqq S$.*

Proof. Since

$$|a_{n+1} + a_{n+2} + \cdots + a_{n+k}| \leqq |a_{n+1}| + \cdots + |a_{n+k}|$$

the left hand side is here certainly $< \varepsilon$ if the right hand side is, whence by the fundamental theorem **81** our first statement at once follows. Since further

$$|s_n| \leqq |a_0| + |a_1| + \cdots + |a_n| < S,$$

we have also, by **41**, 2, $|s| \leqq S$.

By this theorem, all convergent series are divided into two classes and Σa_n belongs to the one or the other according as $\Sigma |a_n|$ is or is not also convergent. We define

86. ° Definition. *If a convergent series Σa_n is such that $\Sigma |a_n|$ also converges, then the first series will be called absolutely convergent, and otherwise non-absolutely convergent*[9].

Examples.

The series

$$\sum_{n=1}^{\infty} \frac{(-1)^n}{n^2}; \qquad \sum_{n=1}^{\infty} \frac{(-1)^{\frac{1}{2}n(n-1)}}{n^{\alpha}}, \quad (\alpha > 1); \qquad \sum_{n=0}^{\infty} a^n, \quad 0 > a > -1;$$

$$\sum_{n=0}^{\infty} \frac{x^n}{n^n}, \quad x < 0; \qquad \sum_{n=1}^{\infty} \frac{x^n}{n!}, \quad x < 0;$$

are absolutely convergent. — Every convergent series of positive terms is of course absolutely convergent.

The very great significance of the concept of absolute convergence will first appear in this: the convergence of absolutely convergent series is much more easy to recognise than that of non-absolutely convergent series, — usually, in fact, by comparison with series of positive terms,

---

[8] *Cauchy,* Analyse algébrique, p. 142. (The proof is inadequate.) — On the other hand, the example just given showed that the convergence of Σa_n need not involve that of $\Sigma |a_n|$.

[9] A series is thus "non-absolutely convergent" if it converges, but not absolutely. The designation "non-absolutely convergent" applies therefore to *convergent* series only.

so that the simple and far-reaching theorems of the preceding chapter become available for the purpose. But this significance will immediately become further visible in that we may operate on absolutely convergent series, on the whole, precisely as we operate on (actual) sums of a definite number of terms, whereas in the case of non-absolutely convergent series this is in general no longer the case. — The following theorems will show this in detail.

°**Theorem 1.** *If Σc_n is a convergent series of positive terms and if the terms of a given series Σa_n, for every $n > m$, satisfy the condition* **87.**

$$|a_n| \leqq c_n \quad \text{or the condition} \quad \left|\frac{a_{n+1}}{a_n}\right| \leqq \frac{c_{n+1}}{c_n},$$

then Σa_n is (absolutely) convergent.[10]

Proof. By the 1st and 2nd comparison tests, **72** and **73**, respectively, $\Sigma |a_n|$ is in either case convergent[11], and so therefore, by **85**, is Σa_n.

In consequence of this simple theorem the complete store of convergence tests relating to series of positive terms becomes available for series of arbitrary terms. We infer at once from it the following

°**Theorem 2.** *If Σa_n is an absolutely convergent series and if the factors α_n form a bounded sequence, then the series*

$$\Sigma \alpha_n a_n$$

is also (absolutely) convergent.

Proof. Since $(|\alpha_n|)$ is a bounded sequence simultaneously with (α_n), it follows from **70**, 2 that $\Sigma |\alpha_n| \cdot |a_n| \equiv \Sigma |\alpha_n a_n|$ is convergent simultaneously with $\Sigma |a_n|$.

Examples.

1. If Σc_n is any convergent series of positive terms and if the α_n 's are *bounded*, then $\Sigma \alpha_n c_n$ is also convergent, for then Σc_n is also absolutely convergent. We may thus, for instance, instead of joining the terms $c_0, c_1, c_2, \ldots$ with the invariable sign $+$, replace this by quite arbitrary $+$ and $-$ signs, — in every case we get a convergent series; for the factors ± 1 certainly form a bounded sequence. Thus for instance the series

$$\Sigma (-1)^n c_n, \qquad \Sigma (-1)^{[\sqrt{n}]} c_n, \qquad \Sigma (-1)^{[\log n]} c_n, \ldots$$

are all convergent, where $[z]$, as usual, stands for the largest integer not greater than z.

---

[10] In the second condition, it is tacitly assumed that, for every $n > m$, $a_n \neq 0$ and $c_n \neq 0$.

[11] The corresponding criteria of divergence,

$$\text{“}|a_n| \geqq d_n\text{”}, \quad \text{and} \quad \text{“}\left|\frac{a_{n+1}}{a_n}\right| \geqq \left|\frac{d_{n+1}}{d_n}\right|\text{”}$$

are of course abolished, since the divergence of $\Sigma |a_n|$, not necessarily of Σa_n, is all that follows. Cf. Footnote 8.

2. If Σa_n is absolutely convergent, then the series obtained from it by an arbitrary alteration in the signs of its terms, is invariably an absolutely convergent series.

We shall now — returning thereby to the questions put aside at the end of last section (§ 15), — show that for absolutely convergent series the fundamental laws of the algebra of (actual) sums are in all essentials maintained, but that for non-absolutely convergent series this is no longer the case.

Thus the commutative law "$a + b = b + a$" does not in general hold for infinite series. The meaning of this statement is as follows: If $(\nu_0, \nu_1, \nu_2, \ldots)$ is any rearrangement (**27**, 3) of the sequence (0, 1, 2, ...) then the series

$$\sum_{n=0}^{\infty} a_n' \equiv \sum_{n=0}^{\infty} a_{\nu_n} \qquad \text{(i. e. with } a_n' = a_{\nu_n} \text{ for } n = 0, 1, 2, \ldots)$$

will be said, for brevity, to result from the given series $\sum_{n=0}^{\infty} a_n$ *by rearrangement* or ***derangement*** of the latter. The value of (actual) sums of a definite number of terms remains unaltered, however the terms may be rearranged (permuted). *For infinite series this is no longer the case*[12]. This is shown already by the two series considered as examples in **81c**, 3 and **83** Theorems 1 and 2, namely

$$1 - \tfrac{1}{2} + \tfrac{1}{3} - \tfrac{1}{4} + - \cdots \quad \text{and} \quad 1 + \tfrac{1}{3} - \tfrac{1}{2} + \tfrac{1}{5} + \tfrac{1}{7} - \tfrac{1}{4} + + - \cdots$$

which are evidently rearrangements of one another, but have *different* sums. The sum of the first was in fact $s < \frac{10}{12}$, while that of the second was $s' > \frac{11}{12}$; and indeed the considerations of **84**, 2 showed more precisely that $s' = \frac{3}{2} s$.

This circumstance of course enforces the greatest care in working with infinite series, since we must — to put it shortly — take account of the order of the terms[13]. It is therefore all the more valuable to know in which cases we may not need to be so careful, and for this we have the

88. °**Theorem 1.** *For absolutely convergent series, the commutative law holds unrestrictedly*[14].

Proof. Let Σa_n be any *absolutely convergent* series (i. e. $\Sigma |a_n|$ is convergent as well), and let $\Sigma a_n' \equiv \Sigma a_{\nu_n}$ be a derangement of Σa_n.

---

[12] This was first remarked by *Cauchy* (Résumés analytiques, Turin 1833).

[13] As Σa_n merely represents the sequence (s_n), and a rearrangement of Σa_n produces a series $\Sigma a_n'$ with *entirely different* partial sums s_n', — these not merely forming a rearrangement of (s_n), but representing *entirely different* numbers!! — it seems *a priori* most improbable that such a derangement will be without effect on the behaviour of the series.

[14] *Lejeune-Dirichlet, G.*: Abh. Akad. Berlin 1837, p. 48 (Werke I, p. 319). Here we also find the example given in the text, of the alteration in the sum of the series by derangement.

Then every bound for the partial sums of $\Sigma\,|\,a_n\,|$ is clearly also a bound for the partial sums of $\Sigma\,|\,a_n'\,|$. So $\Sigma\, a_n'$ is absolutely convergent with $\Sigma\, a_n$. Let s_n denote the partial sums of $\Sigma\, a_n$, and s_n' those of $\Sigma\, a_n'$. Then if ε is arbitrarily given > 0, we may first choose m, in accordance with **81**, so large, that for every $k \geqq 1$

$$|\,a_{m+1}\,| + |\,a_{m+2}\,| + \ldots + |\,a_{m+k}\,| < \varepsilon$$

and now choose n_0 so large that the numbers $\nu_0, \nu_1, \nu_2, \ldots, \nu_{n_0}$ comprise [15] at least all the numbers $0, 1, 2, \ldots, m$. Then the terms $a_0, a_1, a_2, \ldots, a_m$ evidently cancel in the difference $s_n' - s_n$, for every $n > n_0$, and only terms of index $> m$ remain, — that is, only (a finite number of the) terms $a_{m+1}, a_{m+2}, \ldots$ Since, however, the sum of the absolute values of any number of these terms is always $< \varepsilon$, we have, for every $n > n_0$,

$$|\,s_n' - s_n\,| < \varepsilon,$$

and therefore $(s_n' - s_n)$ is a null sequence. But this implies that $s_n' = s_n + (s_n' - s_n)$ has the same limit as s_n, i. e. $\Sigma\, a_n'$ is convergent and has the same sum as $\Sigma\, a_n$, q. e. d.

This property of absolutely convergent series is so essential that it deserves a special designation:

° Definition. *A convergent infinite series which obeys the commutation law without any restriction, — i. e. remains convergent, with unaltered sum, under every rearrangement, — shall be called* ***unconditionally convergent.*** *A convergent series, on the other hand, whose behaviour as to convergence can be altered by rearrangement, for which therefore the order of the terms must be taken into account, shall be called* ***conditionally convergent.*** **89.**

The theorem proved just above can now be expressed as follows: "*Every absolutely convergent series is unconditionally convergent.*" —

The *converse* of this theorem also holds, namely

° Theorem 2. *Every non-absolutely convergent series is only conditionally convergent* [16]. In other words, the validity of the equality

$$\sum_{n=0}^{\infty} a_n = s$$

in the case of a non-absolutely convergent series $\Sigma\, a_n$ depends *essentially* on the order of the terms of the series on the left, and may therefore, by a suitable rearrangement, be disturbed.

---

[15] That such a number n_0 exists follows from the very definition of derangement.

[16] Cf. Fundamental theorem of § 44.

Proof. It obviously suffices to prove that, by a suitable rearrangement, we can deduce from $\Sigma\, a_n$ a divergent series $\Sigma\, a_n'$. This we may do as follows: The terms of the series $\Sigma\, a_n$ which are $\geqq 0$, we denote, in the order in which they occur in $\Sigma\, a_n$, by $p_1, p_2, p_3, \ldots$; those which are < 0 we denote similarly by $-q_1, -q_2, -q_3, \ldots$ Then $\Sigma\, p_n$ and $\Sigma\, q_n$ are series of positive terms. Of these, one at least must diverge. For if both were convergent, with sums P and Q say, then we should obviously have, for each n,

$$|a_0| + |a_1| + \ldots + |a_n| \leqq P + Q,$$

hence $\Sigma\, a_n$ would, by **70**, be absolutely convergent, in contradiction with our assumption [17]. If for instance $\Sigma\, p_n$ diverges, then we consider a series of the form

$$p_1 + p_2 + \ldots + p_{m_1} - q_1 + p_{m_1+1} + p_{m_1+2} + \ldots + p_{m_2} - q_2 + p_{m_2+1} + \ldots,$$

in which, therefore, we have alternately a group of positive terms followed by a single negative term. This series is clearly a rearrangement of the given series $\Sigma\, a_n$ and will, as such, be denoted by $\Sigma\, a_n'$. Now since the series $\Sigma\, p_n$ was assumed to diverge, and its partial sums are therefore unbounded, we can, in the above, first choose m_1 so large that $p_1 + p_2 + \ldots + p_{m_1} > 1 + q_1$, then $m_2 > m_1$ so large that

$$p_1 + p_2 + \ldots + p_{m_1} + \ldots + p_{m_2} > 2 + q_1 + q_2$$

and, generally, $m_\nu > m_{\nu-1}$ so large that

$$p_1 + p_2 + \ldots + p_{m_\nu} > \nu + q_1 + q_2 + \ldots + q_\nu$$

$(\nu = 3, 4, \ldots)$. But $\Sigma\, a_n'$ is then clearly divergent; for each of those partial sums of this series whose last term is a negative term $-q_\nu$ of $\Sigma\, a_n$, is by the above $> \nu$ $(\nu = 1, 2, \ldots)$. And since ν may stand for every positive integer, the partial sums of $\Sigma\, a_n'$ are certainly not bounded, and $\Sigma\, a_n'$ itself is divergent, q. e. d. [18].

If $\Sigma\, q_n$ is divergent, we need only interchange $\Sigma\, p_n$ and $\Sigma\, q_n$ suitably in the above to reach the same conclusion.

---

[17] It is not difficult to see that actually *both* the series $\Sigma\, p_n$ and $\Sigma\, q_n$ must diverge (cf. § 44); but this is for the moment superfluous.

[18] $\Sigma\, a_n'$ clearly diverges to $+\infty$.

Example.

$\sum_{n=1}^{\infty} \frac{(-1)^n}{n} \equiv -1 + \frac{1}{2} - \frac{1}{3} + \frac{1}{4} - \frac{1}{5} + \ldots$ was seen to be non-absolutely convergent. Since (cf. **46**, 3), for $\lambda = 1, 2, \ldots,$

$$\frac{1}{2} + \frac{1}{4} + \frac{1}{6} + \ldots + \frac{1}{2^\lambda} > \frac{\lambda}{4}$$

we have, for $\nu = 1, 2, \ldots,$

$$\frac{1}{2} + \frac{1}{4} + \frac{1}{6} + \ldots + \frac{1}{2^{8\nu}} > 2\nu.$$

If therefore we apply to the series $\sum \frac{(-1)^n}{n}$ the procedure described above, we need only put $m_\nu = 2^{8\nu}$, to deduce from it by rearrangement the *divergent* series

$$\frac{1}{2} + \frac{1}{4} + \frac{1}{6} + \ldots + \frac{1}{2^8} - 1 + \frac{1}{2^8 + 2} + \ldots + \frac{1}{2^{16}} - \frac{1}{3} + \ldots$$

For the partial sums of this series terminating with the νth negative term is greater than 2ν minus ν proper fractions, — i. e. certainly $> \nu$.

Theorem **88**, 1 on the derangement of absolutely convergent series may still be considerably extended. For the purpose, we first prove the following simple

° Theorem 3. *If $\sum a_n$ is absolutely convergent, then every "sub-series" $\sum a_{\lambda_n}$ — for which the indices λ_n denote, therefore, any monotone increasing sequence of different positive integers, — is again convergent and in fact again absolutely convergent.*

Proof. By **74**, 4, $\sum |a_{\lambda_n}|$ converges with $\sum (a_n)$. By **85**, the statement at once follows.

We may now extend the rearrangement theorem **88**, 1 in the following manner. We begin by picking out a first sub-series $\sum a_{\lambda_n}$ of the given absolutely convergent series $\sum a_n$, and arranging this first sub-series in any order, denote it by

$$a_0^{(0)} + a_1^{(0)} + \ldots + a_n^{(0)} + \ldots;$$

let $z^{(0)}$ be the sum of this series, certainly existing, by the preceding theorem, and independent of the chosen arrangement by **88**, 1 [19]. We may also allow this and the following sub-series to consist of only a finite number of terms, — i. e. not to be an *infinite* series at all. From the remaining

[19] The letter z is intended as a reference to the *rows* of the following doubly infinite array.

terms — as far as is possible — we again pick out a (finite or infinite) sub-series, and denote it, arranged in any order, by

$$a_0^{(1)} + a_1^{(1)} + a_2^{(1)} + \ldots + a_n^{(1)} + \ldots,$$

its sum by $z^{(1)}$; from the remaining terms we again pick out a sub-series, and so on. In this manner, we obtain, in general, an *infinite* series of finite or (absolutely) convergent infinite series:

$$\begin{aligned} a_0^{(0)} + a_1^{(0)} + a_2^{(0)} + \ldots + a_n^{(0)} + \ldots &= z^{(0)} \\ a_0^{(1)} + a_1^{(1)} + a_2^{(1)} + \ldots + a_n^{(1)} + \ldots &= z^{(1)} \\ a_0^{(2)} + a_1^{(2)} + a_2^{(2)} + \ldots + a_n^{(2)} + \ldots &= z^{(2)} \\ \cdots\cdots\cdots\cdots\cdots\cdots\cdots \end{aligned}$$

If the process was such as to give each non-zero term [20] of the series Σa_n a place in one (and only one) of these sub-series, then the series

$$z^{(0)} + z^{(1)} + z^{(2)} + \ldots,$$

or, that is to say, the series

$$(\Sigma a_n^{(0)}) + (\Sigma a_n^{(1)}) + (\Sigma a_n^{(2)}) + \ldots$$

may in a further extended sense be called a *rearrangement* of the given series [21]. For this again we have, corresponding to theorem **88**, 1:

° Theorem 4. *An absolutely convergent series "may" also in the extended sense be rearranged.* More precisely: *The series*

$$z^{(0)} + z^{(1)} + z^{(2)} + \ldots$$

is again (absolutely) convergent, and its sum is equal to that of Σa_n.

Proof. If $\varepsilon > 0$ be given, first determine m so that, for every $k \geqq 1$, the remainder $|a_{m+1}| + |a_{m+2}| + \ldots < \varepsilon$, and then choose n_0 so that in the first $n_0 + 1$ sub-series $\Sigma a_n^{(\nu)}$, $\nu = 0, 1, \ldots, n_0$, the terms $a_0, a_1, a_2, \ldots, a_m$ of the given series certainly appear. If $n > n_0$ and $> m$, then the series

$$(z^{(0)} + z^{(1)} + \ldots + z^{(n)}) - s_n$$

---

[20] The introduction or omission of zero terms in Σa_n or in the partial sums is obviously without influence on the present considerations.

[21] Put into the first sub-series, besides a_0 and a_1, all those terms a_n, for instance, in any order, whose indices n are divisible by 2; into the next all those of the remaining terms whose indices are divisible by 3; into the next again all remaining terms whose indices are divisible by 5; and so on, using the prime numbers 7, 11, 13 . . . as divisors.

contains only terms $\pm a_n$ whose indices are $> m$. Hence, by the choice of m, the absolute value of this difference is $< \varepsilon$, and tends therefore, with increasing n, to zero, so that

$$\lim_{n\to\infty} (z^{(0)} + z^{(1)} + \ldots + z^{(n)}) = \lim_{n\to\infty} s_n = s = \Sigma a_n.$$

Moreover, the convergence of $\Sigma z^{(k)}$ which is thus established is also absolute, since for each n we have obviously

$$|z^{(0)}| + |z^{(1)}| + \ldots + |z^{(n)}| \leqq S = \Sigma |a_\nu|.$$

The converse of this theorem is, of course, even less valid than that of theorem **83, 1**, without further consideration. Given, for $k = 0$, 1, 2, ..., the convergent series

$$z^{(k)} = \sum_{n=0}^{\infty} a_n^{(k)},$$

if the aggregate of terms $a_n^{(k)}$ be arranged in any way as a sequence (cf. **53, 4**), then Σa_n need not at all converge, — even should $\sum_{k=0}^{\infty} z^{(k)}$ be convergent. To show this is possible we have only to take, for *each* of the series $z^{(k)}$, the series $1 - 1 + 0 + 0 + 0 + \ldots$. And even *if* Σa_n converges, the sum need not be equal to that of $\Sigma z^{(k)}$.

A general discussion of the question under what circumstances this converse of our theorem does hold, belongs to the theory of double series. However, we may even here prove the following case, which is a particularly important one for applications:

° **Main rearrangement theorem** [22]. *We suppose given an infinite number of convergent series* **90.**

$$\text{(A)}\quad \begin{cases} z^{(0)} = a_0^{(0)} + a_1^{(0)} + \ldots + a_n^{(0)} + \ldots \\ z^{(1)} = a_0^{(1)} + a_1^{(1)} + \ldots + a_n^{(1)} + \ldots \\ \cdot\quad\cdot\quad\cdot\quad\cdot\quad\cdot\quad\cdot\quad\cdot\quad\cdot\quad\cdot\quad\cdot\quad\cdot\quad\cdot\quad\cdot\quad\cdot \\ z^{(k)} = a_0^{(k)} + a_1^{(k)} + \ldots + a_n^{(k)} + \ldots \\ \cdot\quad\cdot\quad\cdot\quad\cdot\quad\cdot\quad\cdot\quad\cdot\quad\cdot\quad\cdot\quad\cdot\quad\cdot\quad\cdot\quad\cdot\quad\cdot \end{cases}$$

and assume that these series are not only absolutely convergent, but satisfy the stricter condition that, if we write

$$\sum_{n=0}^{\infty} |a_n^{(k)}| = \zeta^{(k)} \qquad (k = 0, 1, 2, \ldots, \textit{fixed}),$$

the series

$$\sum_{k=0}^{\infty} \zeta^{(k)} = \sigma$$

[22] Also called *Cauchy's Double Series Theorem.*

is convergent. Then the terms standing vertically one below the other also form (absolutely) convergent series; and if we write[23]

$$\sum_{k=0}^{\infty} a_n^{(k)} = s^{(n)} \qquad (n = 0, 1, 2, \ldots, \textit{fixed}),$$

then $\Sigma s^{(n)}$ is again absolutely convergent and we have

$$\sum_{n=0}^{\infty} s^{(n)} = \sum_{k=0}^{\infty} z^{(k)};$$

in other words, the two series formed by the sums of the rows and by the sums of the columns, respectively, are both absolutely convergent and have the same sum.

The proof is extremely simple: Suppose all the terms in (A) arranged anyhow (in accordance with **53**, 4) in a simple sequence, and denoted, as terms of this sequence, by $a_0, a_1, a_2, \ldots$. Then Σa_n is absolutely convergent. For every partial sum of $\Sigma |a_n|$, for instance

$$|a_0| + |a_1| + \ldots + |a_m|,$$

must still be $\leqq \sigma$, since by choosing k so large that the terms $a_0, a_1, a_2, \ldots, a_m$ all occur in the k first rows of (A), we certainly have

$$|a_0| + |a_1| + \ldots + |a_m| \leqq \zeta^{(0)} + \zeta^{(1)} + \ldots + \zeta^{(k)},$$

i. e. $\leqq \sigma$. A different arrangement of the terms $a_n^{(k)}$ in (A) as a simple sequence $a_0', a_1', a_2', \ldots$ would produce a series $\Sigma a_n'$ which would be a mere rearrangement of Σa_n, and therefore again absolutely convergent, with the same sum. Let this invariable sum be denoted by s.

Now both $\Sigma z^{(k)}$ and also $\Sigma s^{(n)}$ are rearrangements of $\Sigma a_n = s$, in the extended sense of theorem 4, just proved. Therefore these two series are both absolutely convergent and have the same sum s, q. e. d.

This rearrangement theorem may be expressed in somewhat more general form as follows:

° **Supplementary theorem.** *If M is a countable set of numbers and there exists a constant K such that the sum of the absolute values of any finite number of the elements of M remains invariably $< K$,*

---

[23] Here the letter s is intended as a reference to the *columns* of (A).

then we can assert the absolute convergence — with the invariable sum s — of every series ΣA_k *whose terms* A_k *represent sums of a finite or infinite number of elements of M (provided each element of M occurs in one and only one of the terms* A_k*). And this remains true if we allow a repetition of the elements of M, provided each element occurs exactly the same number of times in all the* A_k*'s taken together, as in M itself*[24].

Examples of these important theorems will occur at several crucial points in what follows. Here we may give one or two obvious applications:

1. Let $\Sigma a_n = s$ be an absolutely convergent series and put

$$\frac{a_0 + 2a_1 + 4a_2 + \cdots + 2^n a_n}{2^{n+1}} = a_n' \qquad (n = 0, 1, 2, \ldots)$$

Then we also have $\Sigma a_n' = s$. The proof results immediately, by the previous rearrangement theorem, from the consideration of the array

$$\begin{cases} a_0 = \frac{a_0}{2} + \frac{a_0}{4} + \frac{a_0}{8} + \frac{a_0}{16} + \cdots \\ a_1 = 0 + 2\frac{a_1}{4} + 2\frac{a_1}{8} + 2\frac{a_1}{16} + \cdots \\ a_2 = 0 + 0 + 4\frac{a_2}{8} + 4\frac{a_2}{16} + \cdots \\ \cdots\cdots\cdots\cdots\cdots \end{cases}$$

2. Similarly, from $\frac{1}{p(p+1)} + \frac{1}{(p+1)(p+2)} + \cdots = \frac{1}{p}$ (v. **68**, 2h), and the array

$$\begin{cases} a_0 = \frac{a_0}{1\cdot 2} + \frac{a_0}{2\cdot 3} + \frac{a_0}{3\cdot 4} + \cdots \\ a_1 = 0 + 2\frac{a_1}{2\cdot 3} + 2\frac{a_1}{3\cdot 4} + \cdots \\ a_2 = 0 + 0 + 3\frac{a_2}{3\cdot 4} + \cdots \\ \cdots\cdots\cdots\cdots\cdots \end{cases}$$

we deduce the equality, valid for any absolutely convergent series Σa_n:

$$\sum_{n=0}^{\infty} a_n = \frac{a_0}{1\cdot 2} + \frac{a_0 + 2a_1}{2\cdot 3} + \frac{a_0 + 2a_1 + 3a_2}{3\cdot 4} + \cdots$$

3. The preceding rearrangement theorem evidently holds whenever every $a_n^{(k)}$ is $\geqq 0$ and at least one of the two series $\Sigma z^{(k)}$ and $\Sigma s^{(n)}$ converges; it holds further whenever it is possible to construct a second array (A′) similar to (A), whose terms are positive and $\geqq$ the absolute values of the corresponding terms in (A), and such that, in (A′), either the sums of the rows or the sums of the columns form convergent series.

---

[24] An infinite number of repetitions of a term different from zero is excluded from the outset, since otherwise the constant K of the theorem would certainly not exist. And the number 0 can produce no disturbance.

§ 17. Multiplication of infinite series.

We finally enquire to what extent the distributive law "$a(b+c) = ab + ac$" holds for infinite series. That a convergent infinite series Σa_n may be multiplied term by term by a constant, we have already seen in **83**, 5. In the simplest form

$$c(\Sigma a_n) = \Sigma(c\,a_n)$$

the distributive law is therefore valid for all convergent series. In the case of actual sums, it at once follows further, from the distributive law, that $(a+b)(c+d) = ac + ad + bc + bd$, and more generally, that $(a_0 + a_1 + \cdots + a_l)(b_0 + b_1 + \cdots + b_m) = a_0 b_0 + a_0 b_1 + \cdots + a_l b_m$, or in short, that

$$\left(\sum_{\lambda=0}^{l} a_\lambda\right) \cdot \left(\sum_{\mu=0}^{m} b_\mu\right) = \sum_{\substack{\lambda=0,\ldots,l \\ \mu=0,\ldots,m}} a_\lambda b_\mu$$

where the notation on the right is intended to convey that the indices λ and μ assume, *independently of one another*, all the integral values from 0 to l and 0 to m respectively, and that all $(l+1)(m+1)$ such products $a_\lambda b_\mu$ are to be added, in any order we please.

Does this result continue to hold for infinite series? If $\Sigma a_n = s$ and $\Sigma b_n = t$ are two given convergent infinite series of sum s and t, is it possible to multiply out in the product

$$s \cdot t = \left(\sum_{\lambda=0}^{\infty} a_\lambda\right) \cdot \left(\sum_{\mu=0}^{\infty} b_\mu\right)$$

in any similar way, and in what sense is this possible? More precisely: Let the products

$$a_\lambda b_\mu \qquad \begin{pmatrix} \lambda = 0, 1, 2, \ldots \\ \mu = 0, 1, 2, \ldots \end{pmatrix}$$

be denoted, in any order we choose [25], by $p_0, p_1, p_2, \ldots$; is the series Σp_n convergent, and if convergent, does it have the sum $s \cdot t$? — Here again absolutely convergent series behave like actual sums. In fact we have the

91. ° Theorem [26]. *If the series $\Sigma a_n = s$ and $\Sigma b_n = t$ are absolutely convergent, then the series Σp_n also converges absolutely and has the sum $s \cdot t$.*

---

[25] We suppose, for this, that the products $a_\lambda b_\mu$ are written down exactly in the same way as $a_n^{(k)}$ or $a_\lambda^{(\mu)}$ for **53**, 4 and **90**, to form a doubly infinite array (A). We can then suppose in particular the arrangement by *diagonals* or the arrangement by *squares* carried out for these products.

[26] *Cauchy*: Analyse algébrique, p. 147.

Proof. 1. Let n be a definite integer > 0 and let m be the largest of the indices λ and μ of the products $a_\lambda b_\mu$ which have been denoted by $p_0, p_1, \ldots, p_n$. Evidently

$$|p_0| + |p_1| + \cdots + |p_n| \leqq \left(\sum_{\lambda=0}^{m} |a_\lambda|\right) \cdot \left(\sum_{\mu=0}^{m} |b_\mu|\right),$$

i. e. $< \sigma \cdot \tau$, if σ and τ denote the sums of the series $\Sigma |a_\lambda|$ and $\Sigma |b_\mu|$. The partial sums of $\Sigma |p_n|$ are therefore bounded and Σp_n is absolutely convergent.

2. The absolute convergence of Σp_n having been proved, we need only determine its sum — call it S — for a special arrangement of the products $a_\lambda b_\mu$, for instance the arrangement "by squares". For this we have, however, obviously,

$$a_0 b_0 = p_0, \qquad (a_0 + a_1)(b_0 + b_1) = p_0 + p_1 + p_2 + p_3$$

and in general

$$(a_0 + \cdots + a_n)(b_0 + \cdots + b_n) = p_0 + \cdots + p_{(n+1)^2-1},$$

an equality which, by **41**, 10 and 4, becomes, when $n \to \infty$,

$$s \cdot t = S$$

which was the relation to be proved.

Remarks and Examples.

1. As remarked, for the validity of the relation $\Sigma p_n = s \cdot t$ under the hypotheses made, it is perfectly indifferent in what manner the products $a_\lambda b_\mu$ are enumerated, that is to say arranged in order as a simple sequence (p_n). The arrangement by diagonals is particularly important in applications, and leads, if the products in each diagonal are grouped together (**83**, 1), to the following relation:

$$\sum_{n=0}^{\infty} a_n \cdot \sum_{n=0}^{\infty} b_n = a_0 b_0 + (a_0 b_1 + a_1 b_0) + (a_0 b_2 + a_1 b_1 + a_2 b_0) + \cdots$$

$$\equiv \sum_{n=0}^{\infty} c_n$$

writing for brevity $a_0 b_n + a_1 b_{n-1} + a_2 b_{n-2} + \cdots + a_n b_0 = c_n$. The validity of this relation is therefore secured when both series on the left converge absolutely.

We are also led to this form or arrangement of the "product series", sometimes called *Cauchy's product* of the two given series[27], by the consideration of products of *rational integral functions* and those of *power series*, which latter will be discussed in the following chapter: If in fact, we form the product of two rational integral functions (polynomials)

$$a_0 + a_1 x + a_2 x^2 + \cdots + a_l x^l \quad \text{and} \quad b_0 + b_1 x + b_2 x^2 + \cdots + b_m x^m$$

[27] *Cauchy* loc. cit. examines the product series in this special form only.

and arrange the result again in order of increasing powers of x, then the first terms are

$$a_0 b_0 + (a_0 b_1 + a_1 b_0) x + (a_0 b_2 + a_1 b_1 + a_2 b_0) x^2 + \cdots,$$

so that we have the numbers $c_0, c_1, c_2, \cdots$, above introduced, appearing as coefficients. It is precisely due to this connection that *Cauchy's product* of two series occurs particularly often.

2. Since Σx^n is convergent for $|x| < 1$, we have for such an x

$$\left(\frac{1}{1-x}\right)^2 = \sum_{n=0}^{\infty} x^n \cdot \sum_{n=0}^{\infty} x^n = \sum_{n=0}^{\infty} (n+1) x^n.$$

3. The series $\sum \frac{x^n}{n!}$, cf. **76**, 5c and **85**, is absolutely convergent for every real number x. If therefore x_1 and x_2 are any two real numbers, we may form the product of

$$\sum a_n \equiv \sum \frac{x_1^n}{n!} \quad \text{and} \quad \sum b_n \equiv \sum \frac{x_2^n}{n!}$$

according to *Cauchy*'s rule. We get

$$c_n = \sum_{\nu=0}^{n} a_\nu b_{n-\nu} = \sum_{\nu=0}^{n} \frac{x_1^\nu x_2^{n-\nu}}{\nu!\,(n-\nu)!} = \frac{1}{n!} \sum_{\nu=0}^{n} \frac{n!}{\nu!\,(n-\nu)!} x_1^\nu x_2^{n-\nu} = \frac{(x_1+x_2)^n}{n!}.$$

Therefore we have — for arbitrary x_1 and x_2 — putting $x_1 + x_2 = x_3$:

$$\sum_{n=0}^{\infty} \frac{x_1^n}{n!} \cdot \sum_{n=0}^{\infty} \frac{x_2^n}{n!} = \sum_{n=0}^{\infty} \frac{x_3^n}{n!}.$$

By our theorem, we have now established that the distributive law may at any rate be extended without change to infinite series, — and this, moreover, with an arbitrary arrangement of the products $a_\lambda b_\mu$ —, if both the two given series are absolutely convergent. It is conceivable that this restricting assumption is unnecessarily strict. On the other hand, the following example, given already by *Cauchy*[28] for the purpose, shows that some restriction is necessary, or the theorem no longer holds: Let

$$a_0 = b_0 = 0 \quad \text{and} \quad a_n = b_n = \frac{(-1)^{n-1}}{\sqrt{n}} \quad \text{for } n \geqq 1,$$

so that Σa_n and Σb_n are convergent in accordance with Leibnitz's rule **82**, 5. Then $c_0 = c_1 = 0$, and for $n \geqq 2$,

$$c_n = (-1)^n \left[\frac{1}{\sqrt{1}\,\sqrt{n-1}} + \frac{1}{\sqrt{2}\,\sqrt{n-2}} + \cdots + \frac{1}{\sqrt{n-1}\,\sqrt{1}}\right].$$

Replacing each root in the denominators by the largest, $\sqrt{n-1}$, it follows that, for $n \geqq 2$,

$$|c_n| \geqq \frac{n-1}{\sqrt{n-1} \cdot \sqrt{n-1}} = 1$$

and therefore the product series $\Sigma c_n = \Sigma (a_0 b_n + a_1 b_{n-1} + \cdots + a_n b_0)$

[28] Analyse algébrique, p. 149.

is certainly divergent in accordance with **82**, 1. This is therefore *a fortiori* the case when we omit the brackets.

Nevertheless, the question remains open, whether we may not be able, under less stringent conditions than that of absolute convergence of both the series Σa_n and Σb_n, to prove the convergence of the product series Σp_n — at least for some special arrangement of the terms $a_\lambda b_\mu$, for instance as in the series Σc_n above. To this question we shall return in § 45.

Exercises on Chapter IV.

45. Examine the convergence or divergence of the series $\Sigma(-1)^n a_n$, for which a_n, from some n onwards, has one of the following values:

$$\frac{1}{a+n},\quad \frac{1}{an+b},\quad \frac{1}{\sqrt{n}},\quad \frac{1}{\log n},\quad \frac{1}{\log\log n},\quad \frac{1}{\sqrt[n]{n}},\quad \frac{1}{\sqrt{n}}+\frac{(-1)^n}{n}.$$

46. What alterations have to be made in the answers to Ex. 34, when the behaviour of $\Sigma(-1)^n a_n$ is required?

47. Let

$$\varepsilon_n = \begin{cases} +1 \text{ for } 2^{2k} \leqq n < 2^{2k+1}, \\ -1 \text{ for } 2^{2k+1} \leqq n < 2^{2k+2}. \end{cases} \qquad (k=0, 1, 2, \ldots)$$

Then the series

$$\sum_{k=2}^{\infty} \frac{\varepsilon_n}{n \log n}$$

converges. What is the behaviour of $\sum \frac{\varepsilon_n}{n}$?

48. $\sum_{n=1}^{\infty} (-1)^{n-1} \cdot \frac{2n+1}{n(n+1)}$ is convergent and has the sum **1**.

49. Let the partial sums of the series $1-\frac{1}{2}+\frac{1}{3}-\frac{1}{4}+-\cdots$ be denoted by s_n, and its sum by s, and put $\frac{1}{n+1}+\frac{1}{n+2}+\cdots+\frac{1}{2n}=x_n$. Show that, for every n,

$$x_n = s_{2n}$$

so that $\lim x_n = \sum_{n=1}^{\infty} \frac{(-1)^{n-1}}{n} = s\,(=\log 2)$.

50. Let $s\,(=\log 2)$ denote as above the sum of the series $1-\frac{1}{2}+\frac{1}{3}-+\cdots$ Prove the following relations:

a) $\frac{1}{3}-\frac{1}{5}-\frac{1}{7}+\frac{1}{5}-\frac{1}{9}-\frac{1}{11}+\frac{1}{7}-\frac{1}{13}-\frac{1}{15}+--\cdots=\frac{1}{3}-\frac{1}{2}\log 2;$

b) $1-\frac{1}{2}-\frac{1}{4}+\frac{1}{3}-\frac{1}{6}-\frac{1}{8}+\frac{1}{5}-\frac{1}{10}-\frac{1}{12}+--\cdots=\frac{1}{2}\log 2;$

c) $1-\frac{1}{2}-\frac{1}{4}+\frac{1}{5}+\frac{1}{7}-\frac{1}{8}-\frac{1}{10}++--\cdots=\frac{2}{3}\log 2;$

d) $1+\frac{1}{3}+\frac{1}{5}-\frac{1}{2}-\frac{1}{4}+++--\cdots=\frac{1}{2}\log 6;$

e) $1+\frac{1}{3}+\frac{1}{5}-\frac{1}{2}-\frac{1}{4}-\frac{1}{6}+++---\cdots=\log 2.$

51. With reference to the last two questions, show generally that the series remains convergent when we alternately write throughout p positive terms and q negative ones, and that the sum is then $=\log 2+\frac{1}{2}\log\frac{p}{q}$.

52. The harmonic series $1+\frac{1}{2}+\frac{1}{3}+\frac{1}{4}+\cdots$ remains divergent, when the signs are so changed that we have throughout alternately p positive terms and q negative ones, with $p \neq q$. If $p=q$ the resulting series is convergent.

53. Consider the rearrangements of the series $\sum\limits_{n=1}^{\infty}\frac{(-1)^{n-1}}{\sqrt{n}}$ exactly corresponding to those of the series $\sum\frac{(-1)^{n-1}}{n}$ in Ex. 50 and 51. When is the resulting series convergent and when is it not? When is the sum expressible in terms of the sum of the given series?

54. Consider, with the series $\sum\frac{1}{\sqrt{n}}$, the same alterations in signs as in Ex. 52, for the series $\sum\frac{1}{n}$. When is a convergent series obtained?

55. For which values of α do the following two series converge:

$$1-\frac{1}{2^\alpha}+\frac{1}{3}-\frac{1}{4^\alpha}+\frac{1}{5}-\frac{1}{6^\alpha}+-\cdots,$$

$$1+\frac{1}{3^\alpha}-\frac{1}{2^\alpha}+\frac{1}{5^\alpha}+\frac{1}{7^\alpha}-\frac{1}{4^\alpha}++-\cdots?$$

56. The sum of the series $1-\frac{1}{2^\alpha}+\frac{1}{3^\alpha}-\frac{1}{4^\alpha}+-\cdots$ lies between $\frac{1}{2}$ and 1, for every $\alpha>0$.

57. Given

$$\sum_{n=1}^{\infty}\frac{1}{n^2}=s\left(=\frac{\pi^2}{6}\right)$$

show that

$$1+\frac{1}{3^2}+\frac{1}{5^2}+\frac{1}{7^2}+\cdots=\frac{3}{4}s,$$

$$1+\frac{1}{5^2}+\frac{1}{7^2}+\frac{1}{11^2}+\frac{1}{13^2}+\cdots=\frac{2}{3}s,$$

$$1-\frac{1}{2^2}-\frac{1}{4^2}+\frac{1}{5^2}+\frac{1}{7^2}-\frac{1}{8^2}-\frac{1}{10^2}++--\cdots=\frac{4}{9}s.$$

(With the latter equality cf. Ex. 50c.)

58. In every (conditionally) convergent series the terms can be grouped together in such a manner that the new series converges absolutely.

58a. The following complement to *Kronecker's* theorem **82**, 3 holds good: If a series Σa_n is so constituted that for *every* positive monotone sequence (p_n) tending to $+\infty$, the quotients

$$\frac{p_0 a_0+p_1 a_1+\cdots+p_n a_n}{p_n}$$

tend to 0, *then* Σa_n *is convergent.* — In this sense, therefore, *Kronecker's* condition is *necessary and sufficient* for the convergence.

59. If from a given series Σa_n, with the partial sums s_n, we deduce, by association of terms, a new series ΣA_k with the partial sums S_k, then the inequalities

$$\underline{\lim}\, s_n \leqq \underline{\lim}\, S_k \leqq \overline{\lim}\, S_k \leqq \overline{\lim}\, s_n$$

invariably hold good, whether Σa_n converges or not.

60. If Σa_n, with the partial sums s_n, diverges indefinitely, and s' is a value of accumulation (**52**) of the sequence (s_n), then we can always deduce from Σa_n, by association of terms, a series ΣA_k converging to s' as sum.

61. If Σa_n, with the partial sums s_n, diverges indefinitely, and $a_n \to 0$, then *every* point of the stretch between the upper and lower limits of s_n is a point of accumulation of this sequence.

62. If *every* sub-series of Σa_n converges, then the series itself is *absolutely* convergent.

63. *Cauchy's* product of the two definitely divergent series

$$1 - \frac{3}{2} - \left(\frac{3}{2}\right)^2 - \left(\frac{3}{2}\right)^3 - \cdots$$

and

$$1 + \left(2 + \frac{1}{2^2}\right) + \frac{3}{2}\left(2^2 + \frac{1}{2^3}\right) + \left(\frac{3}{2}\right)^2\left(2^3 + \frac{1}{2^4}\right) + \cdots$$

is

$$1 + \frac{3}{4} + \left(\frac{3}{4}\right)^2 + \left(\frac{3}{4}\right)^3 + \cdots;$$

that of the two series $3 + \sum_{n=1}^{\infty} 3^n$ and $-2 + \sum_{n=1}^{\infty} 2^n$ is $-6 + 0 + 0 + 0 + \cdots$. In both cases it is *absolutely convergent.* How can this paradox be explained?

Chapter V.

Power series.

§ 18. The radius of convergence.

The terms of the series which we have examined so far were, for the most part, determinate numbers. In such cases the series may be more particularly characterised as having constant terms. This however was not everywhere the case. In the geometric series Σa^n, for instance, the terms only become determinate when the value of a is assigned. Our investigation of the behaviour of this series did not, consequently, terminate with a mere statement of convergence or divergence, — the result was: Σa^n converges *if* $|a| < 1$, but diverges *if* $|a| \geqq 1$. The solution of the question of convergence or divergence thus depends, as do the terms of the series themselves, on the value of a quantity left undetermined — a *variable.* Series which have their terms, — and accordingly their convergence or divergence, — depending on a variable quantity (such a quantity will usually be denoted by x

and we shall speak of *series of variable terms*[1]) will be investigated later in more detail. For the moment we propose only to consider series of the above type whose generic term, instead of being a number a_n, has the form

$$a_n x^n,$$

i. e. we shall consider series of the form [2]

$$a_0 + a_1 x + a_2 x^2 + \cdots + a_n x^n + \cdots \equiv \sum_{n=0}^{\infty} a_n x^n.$$

Such series are called ***power series*** (in x), and the numbers a_n are their *coefficients*. For such power series, we are thus not concerned simply with the alternatives "convergent" or "divergent", but with the more precise question: For what values of x is the series convergent, and for what values divergent?

92. Simple examples have already come before us:

1. The geometric series Σx^n is convergent for $|x| < 1$, divergent for $|x| \geqq 1$. For $|x| < 1$, indeed, we have absolute convergence.

2. $\sum \frac{x^n}{n!}$ is (absolutely) convergent for every real x; likewise the series

$$\sum_{k=0}^{\infty} (-1)^k \frac{x^{2k}}{(2k)!} \quad \text{and} \quad \sum_{k=0}^{\infty} (-1)^k \frac{x^{2k+1}}{(2k+1)!}$$

3. $\sum \frac{x^n}{n}$, because $\left|\frac{x^n}{n}\right| \leqq |x|^n$, is absolutely convergent for $|x| < 1$. For $|x| > 1$, the series is divergent, because in that case (by **38**, 1 and **40**), $\left|\frac{x^n}{n}\right| \to +\infty$. For $x = 1$ it reduces to the *divergent* harmonic series, and for $x = -1$, to a series *convergent* by **82**, Theorem 5.

4. $\sum_{n=1}^{\infty} \frac{x^n}{n^2 \cdot 2^n}$ is (absolutely) convergent for $|x| \leqq 2$, but divergent for $|x| > 2$.

5. $\sum_{n=1}^{\infty} n^n x^n$ is convergent for $x = 0$; but for *every* value of $x \neq 0$ it is divergent, for if $x \neq 0$, $|nx| \to +\infty$ and *a fortiori* $|n^n x^n| \to +\infty$, so that (by **82**, Theorem 1) there can be no question of the series converging.

For $x = 0$, obviously *every* power series $\Sigma a_n x^n$ is convergent, whatever be the values of the coefficients a_n. The general case is evidently that in which the power series converges for some values of x, and diverges for others, while, in special instances, the two extreme cases may occur, in which the series converges for *every* x (Example 2), or for *none* $\neq 0$ (Example 5).

---

[1] The harmonic series $\sum \frac{1}{n^x}$ is also of this type: it converges for $x > 1$, diverges for $x \leqq 1$.

[2] We here write, for convenience, $x^0 = 1$, even when $x = 0$.

In the first of these special cases we say that the power series *is everywhere convergent*, in the second — leaving out of account the self-evident point of convergence $x = 0$ — we say that it is *nowhere convergent*. In general, the totality of points x for which the given series $\Sigma a_n x^n$ converges is called its *region of convergence*.

In 2. this consists therefore of the whole axis of x, in 5. of the single point 0; in the other examples, it consists of a stretch bisected at the origin, — sometimes with, sometimes without one or both of its endpoints.

In this we may see already the behaviour of the series in the most general case, for we have the

°**Fundamental theorem.** *If $\Sigma a_n x^n$ is any power series which* **93.** *does not merely converge everywhere or nowhere, then a definite positive number r exists such that $\Sigma a_n x^n$ converges for every $|x| < r$ (indeed absolutely), but diverges for every $|x| > r$. The number r is called the* ***radius of convergence****, or for short the* ***radius****, and the stretch $-r \ldots +r$ the* ***interval of convergence****, of the given power series*[3]. — Fig. 2 schematizes the typical situation established by this theorem.

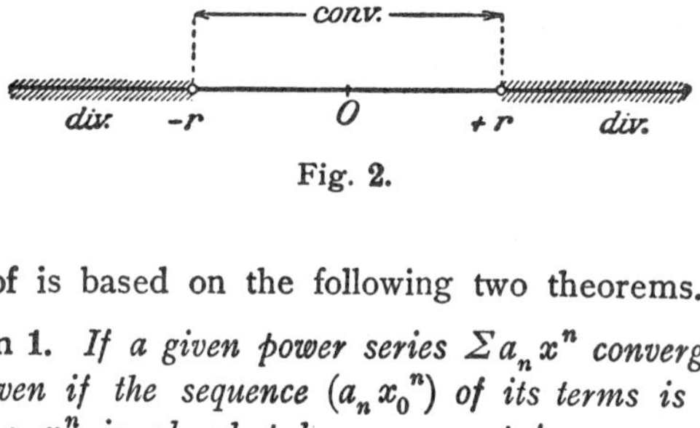

Fig. 2.

The proof is based on the following two theorems.

°**Theorem 1.** *If a given power series $\Sigma a_n x^n$ converges for $x = x_0$ ($x_0 \neq 0$), or even if the sequence $(a_n x_0^n)$ of its terms is only bounded there, then $\Sigma a_n x^n$ is* a b s o l u t e l y *convergent for every $x = x_1$ nearer to the origin than x_0, i. e. with $|x_1| < |x_0|$.*

Proof. If $|a_n x_0^n| < K$, say, then

$$|a_n x_1^n| = |a_n x_0^n| \cdot \left|\frac{x_1}{x_0}\right|^n \leqq K \vartheta^n,$$

where $\vartheta =$ the proper fraction $\frac{x_1}{x_0}$. By **87**, 1 the result stated follows immediately.

°**Theorem 2.** *If the given power series $\Sigma a_n x^n$ diverges for $x = x_0$ then it diverges a fortiori for every $x = x_1$ further from the origin than x_0, i. e. with $|x_1| > |x_0|$.*

---

[3] In the two extreme cases we may also say that the radius of convergence of the series is $r = 0$ or $r = +\infty$, respectively.

Proof. If the series were convergent for x_1, then by theorem 1 it would have to converge for the point x_0, nearer 0 than x_1, — which contradicts the hypothesis.

Proof of the fundamental theorem. By hypothesis, there exists at least one point of divergence, and one point of convergence $\neq 0$. We can therefore choose a *positive* number x_0 nearer 0 than the point of convergence and a *positive* number y_0 further from 0 than the point of divergence. By theorems 1 and 2, the series $\sum a_n x^n$ is convergent for $x = x_0$, divergent for $x = y_0$, and therefore we certainly have $x_0 < y_0$. To the interval $J_0 = x_0 \dots y_0$, we apply the method of successive bisection: we denote by J_1 the *left* or the *right* half of J_1 according as $\sum a_n x^n$ *diverges* or *converges* at the middle point of J_0. By the *same* rule, we designate a particular half of J_1 by J_2, and so on. The intervals of this nest (J_n) all have the property that $\sum a_n x^n$ *converges* at their *left* end point (say x_n) but diverges at their right end point (say y_n). The number r (necessarily positive), which this nest determines, is the number required for the theorem.

In fact, if $x = x'$ is any real number for which $|x'| < r$ (equality excluded), then we have $|x'| < x_k$, for a sufficiently large k, i. e. such that the length of J_k is less than $r - |x'|$. By theorem 1, x' is a point of convergence at the same time as x_k is; and indeed at x' we have absolute convergence. If, on the contrary, x'' is a number for which $|x''| > r$, then $|x''| > y_m$, provided m is large enough for the length of J_m to be less than $|x''| - r$. By theorem 2, x'' is then a point of divergence at the same time as y_m is. This proves all that was desired.

This proof, which appeals to the mind by its extreme simplicity, is yet not entirely satisfying, in that it merely establishes the *existence* of the radius of convergence without supplying any information as to its magnitude. We will therefore prove the fundamental theorem by an alternative method, this time obtaining the magnitude of the radius itself. For this purpose, we proceed — quite independently of our previous theorem, — to prove the more precise

94. °**Theorem**[4]: *If the power series $\sum a_n x^n$ is given and μ denotes the upper limit of the (positive) sequence of numbers*

$$|a_1|, \quad \sqrt{|a_2|}, \quad \sqrt[3]{|a_3|}, \dots, \sqrt[n]{|a_n|}, \dots,$$

i. e.

$$\mu = \overline{\lim} \sqrt[n]{|a_n|},$$

[4] *Cauchy*: Analyse algébrique p. 151. — This beautiful theorem remained for the time entirely unnoticed, till *J. Hadamard* (J. de math. pures et appl., (4) Vol. 8, p. 107. 1892) rediscovered it and made use of it in important applications.

then

a) *if* $\mu = 0$, *the power series is everywhere convergent;*
b) *if* $\mu = +\infty$, *the power series is nowhere convergent;*
c) *if* $0 < \mu < +\infty$, *the power series*

$$\textit{converges absolutely for every } |x| < \frac{1}{\mu},$$
$$\textit{but diverges for every } |x| > \frac{1}{\mu}.$$

Thus — with the suitable interpretation,

$$r = \frac{1}{\mu} = \frac{1}{\overline{\lim}\sqrt[n]{|a_n|}}$$

is the radius of convergence of the given power series[5].

Proof. If in case a) x_0 is an arbitrary real number $\neq 0$, then $\frac{1}{2|x_0|} > 0$ and therefore by **59**,

$$\sqrt[n]{|a_n|} < \frac{1}{2|x_0|} \quad \text{or} \quad |a_n x_0^n| < \frac{1}{2^n}$$

for every $n > m$. By **87**, 1, this shows that $\Sigma a_n x_0^n$ converges absolutely, — which proves a).

If conversely $\Sigma a_n x^n$ converges for $x = x_1 \neq 0$, then the sequence $(a_n x_1^n)$ and, *a fortiori*, the sequence $\left(\sqrt[n]{|a_n x_1^n|}\right)$, are bounded. If $\sqrt[n]{|a_n x_1^n|} < K_1$, say, for every n, then $\sqrt[n]{|a_n|} < \frac{K_1}{|x_1|} = K$, for every n, i. e. $\left(\sqrt[n]{|a_n|}\right)$ is a bounded sequence. In case b), in which the sequence is assumed unbounded above, the series therefore cannot converge for any $x \neq 0$.

Finally, in case c), if x' is any number for which $|x'| < \frac{1}{\mu}$, then choose a positive ϱ for which $|x'| < \varrho < \frac{1}{\mu}$, and so $\frac{1}{\varrho} > \mu$. By the definition of μ, we must have, for every $n >$ some n_0,

$$\sqrt[n]{|a_n|} < \frac{1}{\varrho} \quad \text{and consequently} \quad \sqrt[n]{|a_n x'^n|} < \frac{x'}{\varrho} < 1.$$

By **75**, 1, $\Sigma a_n x'^n$ is therefore (absolutely) convergent.

---

[5]) For convenience of exposition, we here exceptionally write $\frac{1}{0} = +\infty$, $\frac{1}{+\infty} = 0$. — Furthermore it should be noticed that $\frac{1}{\overline{\lim}\sqrt[n]{|a_n|}}$ is not for instance the same as $\overline{\lim}\frac{1}{\sqrt[n]{|a_n|}}$, — as the student should verify by means of obvious examples. (Cf. Ex. 24.)

On the other hand, if $|x''| > \frac{1}{\mu}$, so that $\left|\frac{1}{x''}\right| < \mu$, then we must have, for an infinite number of n's (again and again; v. **59**)

$$\sqrt[n]{|a_n|} > \left|\frac{1}{x''}\right| \quad \text{or} \quad |a_n x''^n| > 1.$$

By **82**, Theorem **1**, therefore the series certainly cannot converge[6].

Thus the theorem is proved in all its parts.

Remarks and Examples.

1. Since the three parts a), b), c) of the preceding theorem are mutually exclusive, it follows that the conditions are not merely sufficient, but also necessary for the corresponding behaviour of $\sum a_n x^n$.

2. In particular, we have $\sqrt[n]{|a_n|} \to 0$ for any power series everywhere convergent. For by the remark above, $\mu = 0$, and since we are concerned with a sequence of positive numbers, these certainly have their *lower* limit $\varkappa \geqq \mu$. Since on the other hand $\varkappa$ must be $\leqq \mu$, we must have $\varkappa = \mu = 0$. By **63** the sequence $\left(\sqrt[n]{|a_n|}\right)$ is therefore convergent with limit **0**

Thus for instance

$$\sqrt[n]{\frac{1}{n!}} \to 0, \quad \text{or} \quad \sqrt[n]{n!} \to \infty,$$

because $\sum \frac{x^n}{n!}$ converges everywhere. (Cf. **43**, Example 4.)

3. Theorems **93** and **94** give us no information as to the behaviour of the series for $x = +r$ and for $x = -r$; this differs from case to case: $\sum x^n$, $\sum \frac{x^n}{n}$, $\sum \frac{x^n}{n^2}$ all have the radius 1. The first converges neither at **1** nor at -1, the second only at one of the two, the third at both.

4 Further examples of power series will occur continually in the course of the next paragraphs, so that we need not indicate any particular examples here.

We saw that the convergence of a power series in the interior of the interval of convergence is, indeed, absolute convergence. We proceed to show further that the convergence is so pronounced as to be undisturbed by the introduction of decidedly large factors. We have in fact the

95. °Theorem. *If* $\sum_{n=0}^{\infty} a_n x^n$ *has the radius of convergence* r, *then the power series* $\sum_{n=0}^{\infty} n\, a_n x^{n-1}$, *or what is the same thing,* $\sum_{n=0}^{\infty} (n+1)\, a_{n+1} x^n$, *has precisely the same radius.*

---

[6] Case c) may be dealt with somewhat more concisely: If

$$\overline{\lim} \sqrt[n]{|a_n|} = \mu, \text{ then } \overline{\lim} \sqrt[n]{|a_n x^n|} = \overline{\lim} \sqrt[n]{|a_n|} \cdot |x| = \mu \cdot |x|$$

(for what reason?). By **76**, 3 the series is therefore absolutely convergent for $\mu \cdot |x| < 1$, and certainly divergent for $\mu \cdot |x| > 1$, q. e. d.

Proof. This theorem may be immediately inferred from Theorem **94**. For if we write $n a_n = a_n'$, then

$$\sqrt[n]{|a_n'|} = \sqrt[n]{a_n} \cdot \sqrt[n]{n}.$$

Since (by **38**, 5), $\sqrt[n]{n} \to 1$, it follows at once from Theorem **62** that the sequences $\left(\sqrt[n]{|a_n'|}\right)$ and $\left(\sqrt[n]{|a_n|}\right)$ have the *same* upper limits. For if we pick out the same sub-sequences from both, as corresponding terms only differ by the factor $\sqrt[n]{n}$, which $\to +1$, these sub-sequences either both diverge or both converge to the *same* limit[7].

Examples.

1. By repeated application of the theorem, we deduce that the series $\Sigma n a_n x^{n-1}$, $\Sigma n(n-1) a_n x^{n-2}, \ldots, \Sigma n(n-1) \cdots (n-k+1) a_n x^{n-k}$ or, what is exactly the same thing, the series

$$\Sigma (n+1) a_{n+1} x^n, \quad \Sigma (n+1)(n+2) a_{n+2} x^n, \ldots,$$
$$\sum (n+1)(n+2) \cdots (n+k) a_{n+k} x^n \equiv k! \sum \binom{n+k}{k} a_{n+k} x^n$$

all have the same radius as $\Sigma a_n x^n$, whatever positive integer be chosen for k.

2. The same of course is true of the series

$$\sum \frac{a_n}{n+1} x^{n+1}, \quad \sum \frac{a_n}{(n+1)(n+2)} x^{n+2}, \ldots, \quad \sum \frac{a_n}{(n+1)(n+2)\cdots(n+k)} x^{n+k}.$$

Thus far we have only considered power series of the form $\Sigma a_n x^n$. These considerations are scarcely altered, if we take the more general type

$$\sum_{n=0}^{\infty} a_n (x - x_0)^n.$$

Putting $x - x_0 = x'$, we see that these series converge absolutely for

$$|x'| = |x - x_0| < r,$$

but diverge for $|x - x_0| > r$, if r again denotes the number determined by Theorem **94**. The region of convergence of this series — except in the extreme cases, in which it converges only for $x = x_0$, or for every x, — is therefore a stretch bisected by the point x_0, sometimes with, sometimes without one or both of its end-points. Except for this displacement of the interval of convergence, all our considerations remain valid. The point x_0 will for brevity be called the *centre of the series*. If $x_0 = 0$, we have the previous form of the series again.

---

[7] Alternative proof. By **76**, 5a or **91**, 2, the series $\Sigma n \vartheta^{n-1}$ is convergent for every $|\vartheta| < 1$. If $|x_0| < r$, and ϱ is so chosen that $|x_0| < \varrho < r$, then $\Sigma a_n \varrho^n$ converges, $(a_n \varrho^n)$ is therefore bounded, say $|a_n \varrho^n| < K$. We infer that $|n a_n x_0^{n-1}| < \frac{K}{\varrho} \cdot n \left|\frac{x_0}{\varrho}\right|^{n-1}$, which, since $\left|\frac{x_0}{\varrho}\right| < 1$, proves the convergence.

In the interval of convergence, the power series $\Sigma a_n (x - x_0)^n$ has a definite sum s, for each x, and usually of course a different sum for a different x. In order to express this dependence on x, we write

$$\sum_{n=0}^{\infty} a_n (x - x_0)^n = s(x)$$

and say that the power series defines, in its interval of convergence, a *function* of x.

The foundations of the theory of real functions, that is to say the foundations of the differential and integral calculus, we assume, as remarked in the Introduction, to be already known to the reader in all that is essential. It is only to avoid any possible uncertainty as to the extent of the facts required from these domains, that we shall rapidly indicate, in the following section, all the definitions and theorems which we shall need, without going into more exact elucidations or proofs.

§ 19. Functions of a real variable.

Definition 1 (Function). If to each value x of an interval of the x-axis, by any prescribed rule, a definite value y is made to correspond, then we say that y is a ***function of x*** defined in that interval and write, for short,

$$y = f(x),$$

where "f" symbolises the prescribed rule in virtue of which each x has corresponding to it the relevant value of y.

The interval, which may be closed or open on one or both sides, bounded or unbounded, is called the *interval of definition* of $f(x)$.

Definition 2 (Boundedness). If there exists a constant K_1 such that for every x of the interval of definition we have

$$f(x) \geqq K_1,$$

then the function $f(x)$ is said to be *bounded on the left* (or *below*) in the interval, and K_1 is *a bound below* (or *left hand bound*) of $f(x)$. If there exists a constant K_2 such that for every x of the interval of definition $f(x) \leqq K_2$, then $f(x)$ is said to be *bounded on the right* (or *above*) and K_2 is *a bound above* (or *right hand bound*) of $f(x)$. A function bounded on both sides is said simply to be *bounded*. There then exists a constant K such that for every x of the interval of definition, we have

$$|f(x)| \leqq K.$$

Definition 3 (Upper and lower bound, oscillation). There is always a *least* one among all the bounds above of a bounded function, and always a *greatest* among all its bounds below[8]. The former we call *the upper bound*, the latter *the lower bound*, and their difference *the oscillation* of the function $f(x)$ in its interval of definition. Corresponding designations are defined for a sub-interval $a' \ldots b'$ of the interval of definition.

Definition 4 (Limit of a function). If ξ is a point of the interval of definition of a function $f(x)$, or one of the endpoints of that interval, then the notation

$$\lim_{x \to \xi} f(x) = c$$

or

$$f(x) \to c \quad \text{for} \quad x \to \xi$$

means that

a) for *every* sequence of numbers x_n of the interval of definition which converges to ξ, *but with all its terms different from* ξ, the sequence of the corresponding values

$$y_n = f(x_n) \qquad (n = 1, 2, 3, \ldots)$$

of the function converges to c; or

b) an arbitrary positive number ε being chosen, another positive number $\delta = \delta(\varepsilon)$ can always be assigned, such that for all values of x in the interval of definition with

$$|x - \xi| < \delta \quad \text{but} \quad x \neq \xi,$$

we have[9]

$$|f(x) - c| < \varepsilon.$$

The two forms of definition a) and b) mean precisely the same thing.

Definition 5 (Right hand and left hand limits). If, in the case of definition 4, it is stipulated besides that all points x_n or x taken into account lie to the *right* of ξ (which must not of course be the right hand endpoint of the interval of definition of $f(x)$), then we speak of a *right hand limit* (or *limit on the right*) and write

$$\lim_{x \to \xi + 0} f(x) = c;$$

similarly we write

$$\lim_{x \to \xi - 0} f(x) = c,$$

and speak of a *left hand limit* (or *limit on the left*), if ξ is not the left hand endpoint of the interval of definition of $f(x)$, and if points x_n or x to the *left* of ξ are alone taken into account.

---

[8] Cf. 8, 2, and also **62**.

[9] The older notation $\lim_{x = \xi} f(x)$ for $\lim_{x \to \xi} f(x)$ should be *absolutely discarded* since the whole point is that x is to remain $\neq \xi$.

Definition 5a (Further types of limits). Besides the three types of limit already defined, the following may also occur[10]:

$$\left.\begin{array}{r}\lim f(x) = \\ \text{or} \qquad \\ f(x) \to \end{array}\right\} \quad c, \quad +\infty, \quad -\infty$$

with one of the five supplementary indications ("*motions of x*")

for $\quad x \to \xi, \quad \to \xi + 0, \quad \to \xi - 0, \quad \to +\infty, \quad \to -\infty.$

With reference to 2 and 3 there will be no difficulty in formulating precisely the definitions — in the form a) or b) — which correspond to the definitions just discussed.

Since, as remarked, we assume these matters to be familiar to the reader, in all essentials, we suppress all elucidations of detail and examples, and only emphasize that the value c to which a function tends, for instance for $x \to \xi$, need bear no relation whatever to the value of the function *at* ξ. Only for this we will give an *example*: let $f(x)$ be defined for *every* x by putting $f(x) = 0$ if x is an irrational number, but $f(x) = \frac{1}{q}$ if x is a rational number which in its lowest terms is of the form $\frac{p}{q}$ $(q > 0)$. Thus e. g. $f\left(\frac{3}{4}\right) = \frac{1}{4}$, $f(0) = f\left(\frac{0}{1}\right) = 1$, $f(\sqrt{2}) = 0$, etc.

Here we have for every ξ

$$\lim_{x \to \xi} f(x) = 0 .$$

For if ε is an arbitrary positive number and m is so large that $\frac{1}{m} < \varepsilon$, then there are not more than a finite number of rational points whose (least positive) denominator is $\leqq m$. These we imagine marked in the interval $\xi - 1 \ldots \xi + 1$. As there are only a finite number of them, we can find one nearest of all to ξ; (if ξ itself is one of these points we of course should *not* take it into account here). Let δ denote its (positive) distance from ξ. Then *every* x, for which

$$0 < |x - \xi| < \delta,$$

is either irrational, or a rational number whose least positive denominator q is $> m$. In the one case, $f(x) = 0$; in the other, $= \frac{1}{q} < \frac{1}{m} < \varepsilon$. Therefore we have, for *every* x in $0 < |x - \xi| < \delta$,

$$|f(x) - 0| < \varepsilon$$

i. e., as asserted,

$$\lim_{x \to \xi} f(x) = 0 .$$

If therefore ξ is in particular a rational number, then this *limit* differs decidedly from the *value* $f(\xi)$ itself.

Calculations with limits are rendered possible by the following theorem:

---

[10] In the first of these three cases we say that $f(x)$ *tends* or *converges* to c; in the second and third cases: $f(x)$ *tends* or *diverges* (definitely) to $+\infty$ or $-\infty$; and in all three, we speak of a *definite behaviour* or also of a *limit in the wider sense*. If $f(x)$ shows none of these three modes of behaviour, then we say that $f(x)$ *diverges indefinitely* for the motion of x under consideration.

Theorem 1. If $f_1(x), f_2(x), \dots f_p(x)$ are given functions (p some determinate positive integer), each of which, for one and the same motion of x of the types mentioned in Definition 5a, tends to a *finite* limit, say $f_1(x) \to c_1, \dots, f_p(x) \to c_p$, then

a) the function
$$f(x) = [f_1(x) + f_2(x) + \cdots + f_p(x)] \to c_1 + c_2 + \cdots + c_p;$$
b) the function
$$f(x) = [f_1(x) \cdot f_2(x) \cdots f_p(x)] \to c_1 \cdot c_2 \cdots c_p;$$

c) in particular, therefore, the function $a f_1(x) \to a c_1$, (a = arbitrary real number) and the function $f_1(x) - f_2(x) \to c_1 - c_2$;

d) the function $\frac{1}{f_1(x)} \to \frac{1}{c_1}$, provided $c_1 \neq 0$.

Theorem 2. If $\lim\limits_{x \to \xi} f(x) = c \ (\neq \pm\infty)$, then $f(x)$ is *bounded* in a neighbourhood of ξ, i. e. two positive numbers δ and K exist such that

$$|f(x)| < K, \quad \text{when } |x - \xi| < \delta,$$

and corresponding statements hold in the case of a (finite) $\lim f(x)$ for $x \to \xi + 0$, $\xi - 0$, $+\infty$, $-\infty$.

Definition 6 (Continuity at a point). If ξ is a point of the interval of definition of $f(x)$, then $f(x)$ is said to be *continuous* at ξ if

$$\lim_{x \to \xi} f(x)$$

exists and coincides with the value $f(\xi)$ of the function at ξ:

$$\lim_{x \to \xi} f(x) = f(\xi).$$

If we include the definition of lim in this new definition, we may also state:

Definition 6a. $f(x)$ is said to be *continuous* at a point ξ, if for *every* sequence of x_n's of the interval of definition, which tends to ξ, the corresponding values of the function

$$y_n = f(x_n) \to f(\xi).$$

Definition 6b. $f(x)$ is said to be continuous at ξ, if, having chosen an arbitrary $\varepsilon > 0$, we can always assign $\delta = \delta(\varepsilon) > 0$, such that for every x of the interval of definition with

$$|x - \xi| < \delta \quad \text{we have} \quad |f(x) - f(\xi)| < \varepsilon$$

Definition 7 (Right hand and left hand continuity). $f(x)$ is said to be *continuous on the right* (*right-handedly*) or *on the left* (*left-handedly*) if $\lim f(x)$ exists at least for $x \to \xi + 0$ or $x \to \xi - 0$ respectively, and coincides with $f(\xi)$.

Corresponding to Theorem 1 we have here the

Theorem 3. If $f_1(x), f_2(x), \dots, f_p(x)$ are given functions (p a particular positive integer), all continuous at ξ, then the functions

a) $f_1(x) + f_2(x) + \dots + f_p(x)$,

b) $f_1(x) \cdot f_2(x) \dots f_p(x)$,

c) $a f_1(x)$ (a = an arbitrary real number), $f_1(x) - f_2(x)$, and

d) if $f_1(\xi) \neq 0$, also $\dfrac{1}{f_1(x)}$

are all continuous at ξ. Corresponding statements hold, when only right hand or only left hand continuity is assumed.

By repeated application of this theorem to the function $f(x) \equiv x$, certainly continuous everywhere (since for $x \to \xi$ we have precisely $x \to \xi$), we at once deduce:

All *rational* functions are continuous everywhere, with the exception of (at most a finite number of) points where the denominator $= 0$. In particular: Rational integral functions are continuous everywhere.

Similarly, the limiting relations **42**, 1—3, showed that: a^x, $(a > 0)$ is continuous for every real x; $\log x$ is continuous for every $x > 0$; x^α (α = arbitrary real number) is continuous for every $x > 0$.

Definition 8 (Continuity in an interval). If a function is continuous at every individual point of an interval J, then we say that it is *continuous in this interval.* Continuity at an endpoint of the interval is here taken to be continuity "inwards", i. e. right handed continuity at the left hand endpoint, and left handed continuity at the right hand endpoint. These endpoints of J may or may not, according to the circumstances, be reckoned as in the interval. Functions which are continuous in a *closed* interval give rise to a series of important theorems, of which we may mention the following:

Theorem 4. If $f(x)$ is continuous in the closed interval $a \leqq x \leqq b$ and if $f(a) > 0$, but $f(b) < 0$, then there exists, between a and b, at least one point ξ for which $f(\xi) = 0$.

Theorem 4a. If $f(x)$ is continuous in the closed interval $a \leqq x \leqq b$ and η is any real number between $f(a)$ and $f(b)$, then there exists, between a and b, at least one point ξ for which $f(\xi) = \eta$. Or: The equation $f(x) = \eta$ has at least one solution in that interval.

Theorem 5. If $f(x)$ is continuous in the closed interval $a \leqq x \leqq b$, then, having chosen any $\varepsilon > 0$, we can always assign *some* number $\delta > 0$ so that, if x' and x'' are any two points of the interval in question whose distance $|x'' - x'|$ is $< \delta$, the difference of the corresponding values of the function, $|f(x'') - f(x')|$, is $< \varepsilon$. (The property, established by this theorem, of a function continuous in a closed interval is called *uniform continuity* of the function in the interval.)

Definition 9 (Monotony). A function defined in the interval $a \dots b$ is said to be *monotone increasing* or *decreasing* in the interval, if for every pair of points x_1 and x_2 of that interval, with $x_1 < x_2$, we invariably have $f(x_1) \leqq f(x_2)$ in the one case, or invariably $f(x_1) \geqq f(x_2)$, in the other. We also speak of *strictly* increasing and *strictly* decreasing functions, when the equality signs, in the inequalities between the values of the function just written down, are excluded.

Theorem 6. The point ξ, certainly existing under the hypotheses of Theorems 4 and 4a, is necessarily unique of its kind if the function $f(x)$ under consideration is *strictly monotone* in the interval $a \ldots b$. Thus in that case, to each η between $f(a)$ and $f(b)$ corresponds one and only one ξ for which $f(\xi) = \eta$. We say in this case: *The inverse function of $y = f(x)$ is everywhere existent and one-valued (or $y = f(x)$ is reversible) in the interval.*

Definition 10 (Differentiability). A function $f(x)$ defined at a point ξ and in a certain neighbourhood of ξ is said to be *differentiable at ξ* if the limit

$$\lim_{x \to \xi} \frac{f(x) - f(\xi)}{x - \xi}$$

exists. Its value is called the (unique derivative or) *differential coefficient* of $f(x)$ at ξ and is denoted by $f'(\xi)$. If the limit in question only exists on the left or on the right (that is, only for $x \to \xi + 0$ or $x \to \xi - 0$ respectively), then we speak of *right hand* or *left hand differentiability, differential coefficient,* etc.

If a function is differentiable at each individual point of an interval J, then we say for brevity that the function is *differentiable in this interval.*

The rules for differentiation of a sum or product of a particular (fixed) number of functions, of a difference or quotient of two functions, of functions of a function, as also the rules for differentiation of the elementary functions and of their combinations, we regard as known to the reader.

All means necessary to their construction have been developed in the above, if we anticipate a knowledge of the limit defined in **112** and there determined in a perfectly direct manner. If, for instance, it is inquired whether a^x ($a > 0$ and $\neq 1$) is differentiable, and, if so, what is its differential coefficient, at the point ξ, then, following Defs. 10 and 4, we have to choose a null sequence (x_n) with terms all $\neq 0$ and to examine the sequence of numbers

$$X_n = \frac{a^{\xi + x_n} - a^{\xi}}{x_n} = a^{\xi} \cdot \frac{a^{x_n} - 1}{x_n}.$$

If we write y_n for the numerator in the last fraction, then by **35**, 3 we know that (y_n) is also a null sequence, and indeed one for which none of the terms is *equal* to 0. X_n may then be written in the form

$$X_n = a^{\xi} \cdot \frac{y_n \cdot \log a}{\log(1 + y_n)}.$$

But since, as remarked, y_n is a null sequence, we have by **112**

$$\frac{\log(1 + y_n)}{y_n} \to 1.$$

Since the same then holds for the reciprocal values, by **41**, 11a, we deduce $X_n \to a^{\xi} \cdot \log a$. The function a^x is thus differentiable for every x and has the differential coefficient $a^x \cdot \log a$.

In the same way, as regards differentiability and differential coefficient of $\log x$ for $\xi > 0$, we deduce, by consideration of

$$X_n = \frac{\log(\xi + x_n) - \log \xi}{x_n} = \frac{\log\left(1 + \frac{x_n}{\xi}\right)}{x_n} = \frac{1}{\xi}\cdot\log\left(1 + \frac{x_n}{\xi}\right)^{\frac{\xi}{x_n}}$$

that the differential coefficient exists here and $= \frac{1}{\xi}$.

Of the properties of differentiable functions we shall for the present require scarcely more than is contained in the following simple theorems:

Theorem 7. If a function $f(x)$ is differentiable in an interval J and its differential coefficient is there constantly equal to 0, then $f(x)$ is constant in J, that is to say is $\equiv f(x_0)$, where x_0 is any point of J.

If two functions $f_1(x)$ and $f_2(x)$ are differentiable in J and their differential coefficients constantly coincide there, then the difference of the two functions is constant in J; therefore we have

$$f_2(x) = f_1(x) + c = f_1(x) + [f_2(x_0) - f_1(x_0)]$$

where x_0 is any point of J.

Theorem 8. (*First mean value theorem of the differential calculus.*) If $f(x)$ is continuous in the closed interval $a \leqq x \leqq b$ and differentiable in at least the open interval $a < x < b$, then there is, in the latter interval, at least one point ξ for which

$$\frac{f(b) - f(a)}{b - a} = f'(\xi).$$

(In words: The finite difference quotient relative to the endpoints of the intervals is equal to the differential coefficient at a suitable interior point.)

Theorem 9. If $f(x)$ is differentiable at ξ and $f'(\xi)$ is > 0 (< 0) then $f(x)$ "increases" ("decreases") at ξ, i. e. the difference

$$f(x) - f(\xi) \text{ has } \begin{Bmatrix} \text{the same} \\ \text{(the opposite)} \end{Bmatrix} \text{ sign as (to) } (x - \xi),$$

provided $|x - \xi|$ be less than a suitable number δ.

Theorem 10. If $f(x)$ is differentiable at an interior point ξ of its interval of definition, then unless $f'(\xi) = 0$ the functional value $f(\xi)$ cannot be $\geqq$ every other functional value $f(x)$ in a neighbourhood of ξ of the form $|x - \xi| < \delta$, i. e. ξ cannot be a (relative) *maximum* point. Similarly the condition $f'(\xi) = 0$ is necessary for ξ to be a (relative) *minimum* point, i. e. such that $f(\xi)$ is not greater than any other functional value $f(x)$, as long as x remains in a suitable neighbourhood of ξ.

Definition 11 (**Differential coefficients of higher orders.** If $f(x)$ is differentiable in J, then (in accordance with Def. 1) $f'(x)$ is again a function defined in J*. If this function is again differentiable in J,

---

* and called the *derived function* of $f(x)$.

then its differential coefficient is called the second differential coefficient of $f(x)$ and is denoted by $f''(x)$. Correspondingly, we obtain the third and, generally, the kth differential coefficient of $f(x)$, which is denoted by $f^{(k)}(x)$. For the existence of the kth differential coefficient at ξ it is thus (v. Def. 10) necessary that the $(k-1)^{\text{th}}$ differential coefficient should exist both at ξ *and at all points of a certain neighbourhood of* ξ. — The lth differential coefficient of $f^{(k)}(x)$ is $f^{(k+l)}(x)$, $k \geqq 0$, $l \geqq 0$. (As 0th differential coefficient of $f(x)$ we then take the function itself.)

Of the integral calculus we shall, in the sequel, require only the simplest concepts and theorems, except in the two paragraphs on Fourier series, where rather deeper material has to be brought in.

Definition 12 (Indefinite integral). If a function $f(x)$ is given in an interval $a \ldots b$ and if a differentiable function $F(x)$ can be found such that, for all points of the interval in question, $F'(x) = f(x)$, then we say that $F(x)$ *is an indefinite integral of* $f(x)$ in that interval. (Besides $F(x)$, the functions $F(x) + c$ are then also indefinite integrals of $f(x)$, if c denotes any real number. Besides these, however, there are no others). We write

$$F(x) = \int f(x)\,dx.$$

In the simplest cases, indefinite integrals are obtained by inverting the elementary formulae of the differential calculus. E. g. from $(\sin \alpha x)' = \alpha \cos \alpha x$ it follows that $\int \cos \alpha x\,dx = \frac{\sin \alpha x}{\alpha}$, and so on. These elementary rules we assume known. Special integrals of this kind, excepting the very simplest, are little used in the sequel; we mention

$$\int \frac{dx}{1+x^3} = \frac{1}{3}\log(1+x) - \frac{1}{6}\log(1-x+x^2) + \frac{1}{\sqrt{3}}\tan^{-1}\frac{2x-1}{\sqrt{3}},$$

$$\int \frac{dx}{1+x^4} = \frac{\sqrt{2}}{8}\log\frac{x^2+x\sqrt{2}+1}{x^2-x\sqrt{2}+1} + \frac{\sqrt{2}}{4}[\tan^{-1}(x\sqrt{2}-1) + \tan^{-1}(x\sqrt{2}+1)],$$

$$\int \left[\cot x - \frac{1}{x}\right] dx = \log\frac{\sin x}{x}.$$

Though in indefinite integrals, we find no more than a new *mode of writing* for formulae of the differential calculus, the *definite integral* introduces an essentially new concept.

Definition 13 (Definite integral). A function defined in a closed interval $a \ldots b$ and there *bounded* is said to be *integrable over this interval* if it fulfils the following condition:

Divide the interval $a \ldots b$ in any manner into n equal or unequal parts ($n \geqq 1$, a positive integer), and denote by $x_1, x_2, \ldots, x_{n-1}$ the points of division between $a = x_0$ and $b = x_n$. Next in each of these n parts (in which *both* endpoints may be reckoned) choose any point, and denote the chosen points in corresponding order by $\xi_1, \xi_2, \ldots, \xi_n$. Then form the sum[11]

$$S_n = \sum_{\nu=1}^{n} (x_\nu - x_{\nu-1}) f(\xi_\nu).$$

Let such sums S_n be evaluated for each $n = 1, 2, 3, \ldots$ independently (that is to say, at each stage x_ν and ξ_ν may be chosen afresh). But, at the same time, l_n, the length of *the longest* of the n parts into which the interval is divided when forming S_n, shall *tend to* 0[12].

If the sequence of numbers $S_1, S_2, \ldots$, in whatever way they may have been formed, invariably proves to be convergent and always gives the same[13] *limit S, then $f(x)$ will be called integrable in Riemann's sense and the limit S will be called the definite integral of $f(x)$ over $a \ldots b$, and written*

$$\int_a^b f(x)\,dx.$$

x is called the variable of integration and may of course be replaced by any other letter. — Instead of $f(\xi_\nu)$ we may also take, to form S_n, *the lower bound* α_ν or *the upper bound* β_ν of all the functional values[14] in the interval $x_{\nu-1} \leqq x \leqq x_\nu$.

Theorem 11 (*Riemann's test of integrability*). The necessary and sufficient condition for a function $f(x)$, defined in the closed interval $a \ldots b$ and there bounded, to be integrable over $a \ldots b$, is as follows: Given $\varepsilon > 0$, a choice of n and of the points $x_1, x_2, \ldots, x_{n-1}$ must be possible, for which

$$\sum_{\nu=1}^{n} i_\nu \sigma_\nu < \varepsilon$$

if $i_\nu = |x_\nu - x_{\nu-1}|$ is the length of the νth part of $a \ldots b$ and σ_ν the oscillation of $f(x)$ in this sub-interval.

This criterion may also be expressed as follows, assuming the notation chosen so that $a < b$: After choosing ε, we must be able to assign two "step-functions" (functions constant in stretches) such that in $a \leqq x \leqq b$ we have always

$$g(x) \leqq f(x) \leqq G(x)$$

---

[11] If $f(x) > 0$, $a > b$, and we consider a plane portion S bounded on the one side by the axis of abscissae, on the other by the verticals through a and b and by the curve $y = f(x)$, then S_n is an approximate value of the area of S. This however only provides a satisfactory representation if $y = f(x)$ is a curve in the intuitive sense.

[12] We may then also say that the subdivisions, with increasing n, become indefinitely closer.

[13] It is easily shewn that if the sequence (S_n) is *invariably* convergent it also *ipso facto* always gives *the same* limit.

[14] In these cases S_n gives the area of a ("step-") polygon inscribed or circumscribed to the plane portion S.

as well as [15] $$\int_a^b (G(x) - g(x))\,dx < \varepsilon.$$

It suffices in fact to put, in $x_{\nu-1} \leqq x \leqq x_\nu$,

$$g(x) = \alpha_\nu, \quad G(x) = \beta_\nu, \quad \nu = 1, 2, \ldots, n,$$

together with $$g(b) = \alpha_n, \quad G(b) = \beta_n.$$

From this criterion, the following particular theorems are deduced:

Theorem 12. Every function *monotone* in $a \leqq x \leqq b$, and also every function *continuous* in $a \leqq x \leqq b$, is integrable over $a \ldots b$.

Theorem 13. The function $f(x)$ is integrable over $a \ldots b$, if, in $a \ldots b$, it is bounded and has only *a finite number* of discontinuities.

Riemann's test of integrability may also be given the following form:

Theorem 14. The function $f(x)$ is integrable over $a \ldots b$ if, and only if, it is bounded there and if, two arbitrary positive numbers δ and ε being assigned, the subdivision of $a \ldots b$ into n sub-intervals described in theorem 11 can be so carried out that the sub-intervals i_ν in which the oscillation of $f(x)$ exceeds δ add up to a total length less than ε.

Theorem 15. The function $f(x)$ is certainly *not* integrable over $a \ldots b$ if it is discontinuous at *every* point of that interval.

Theorem 16. If $f(x)$ is integrable over $a \ldots b$, then $f(x)$ is also integrable over every sub-interval $a' \ldots b'$ of $a \ldots b$.

Theorem 17. If the function $f(x)$ is integrable over $a \ldots b$, then every other function $f_1(x)$ is integrable over $a \ldots b$, and has the same integral, which results from $f(x)$ by an arbitrary change in a *finite number* of its values.

Theorem 18. If $f(x)$ and $f_1(x)$ are two functions integrable over $a \ldots b$, then they have the same integral provided that they coincide at least at all points of a set everywhere dense in $a \ldots b$ (e. g. all rational points).

For calculations with integrals we have the following simple theorems, where $f(x)$ denotes a function integrable over the interval $a \ldots b$.

Theorem 19. We have $\int_b^a f(x)\,dx = -\int_a^b f(x)\,dx$ and if a_1, a_2, a_3 are three arbitrary points of the interval $a \ldots b$,

$$\int_{a_1}^{a_2} f(x)\,dx + \int_{a_2}^{a_3} f(x)\,dx + \int_{a_3}^{a_1} f(x)\,dx = 0.$$

Theorem 20. If $f(x)$ and $g(x)$ are two functions integrable over $a \ldots b$, $(a < b)$, and if in $a \ldots b$ we have constantly $f(x) \leqq g(x)$, then we also have

$$\int_a^b f(x)\,dx \leqq \int_a^b g(x)\,dx.$$

---

[15] It is immediately obvious from the first form of the criterion that a step-function such as $G(x) - g(x)$ is integrable.

Theorem 20a. $|f(x)|$ is integrable with $f(x)$ and we have, if $a < b$,

$$\left|\int_a^b f(x)\,dx\right| \leqq \int_a^b |f(x)|\,dx.$$

Theorem 21. (***First mean value theorem of the integral calculus.***) We have

$$\int_a^b f(x)\,dx = \mu \cdot (b-a)$$

if μ is a suitable number between the lower bound α and the upper bound β of $f(x)$ in $a \ldots b$ ($\alpha \leqq \mu \leqq \beta$). In particular we have

$$\left|\int_a^b f(x)\,dx\right| \leqq K \cdot (b-a)$$

if K denotes a bound above of $|f(x)|$ in $a \ldots b$.

Theorem 22. If the functions $f_1(x), f_2(x), \ldots, f_p(x)$ are all integrable over $a \ldots b$ ($p =$ *fixed* positive integer), then so are their sum and their product and for the integral of the sum we have the formula

$$\int_a^b (f_1(x) + \ldots + f_p(x))\,dx = \int_a^b f_1(x)\,dx + \ldots + \int_a^b f_p(x)\,dx;$$

i. e. the sum of a *fixed* number of functions may be integrated *term* by *term*.

Theorem 22a. If $f(x)$ is integrable over $a \ldots b$ and if the lower bound of $|f(x)|$ in $a \ldots b$ is > 0, then $\frac{1}{f(x)}$ is also integrable over $a \ldots b$.

Theorem 23. If $f(x)$ is integrable over $a \ldots b$, then the function

$$F(x) = \int_a^x f(t)\,dt$$

is *continuous* in the interval $a \ldots b$ and is also *differentiable* at every point of the interval, where $f(x)$ itself is continuous. If x_0 is such a point, then $F'(x_0) = f(x_0)$ there.

Theorem 24 (***Fundamental theorem of the differential and integral calculus***). If $f(x)$ is integrable over $a \ldots b$, and if $f(x)$ has an indefinite integral $F(x)$ in that interval, then

$$\int_a^b f(x)\,dx = F(b) - F(a).$$

Theorem 25 (***Change of the variable of integration***). If $f(x)$ is integrable over $a \ldots b$ and $x = \varphi(t)$ is a function differentiable in $\alpha \ldots \beta$, with $\varphi(\alpha) = a$ and $\varphi(\beta) = b$, if further, when t varies from α to β, $\varphi(t)$ varies *monotonely* (in the stricter sense) from a to b, and if $\varphi'(t)$, the differential coefficient of $\varphi(t)$, is integrable [16] over $\alpha \ldots \beta$, then

$$\int_a^b f(x)\,dx = \int_\alpha^\beta f(\varphi(t)) \cdot \varphi'(t)\,dt.$$

---

[16] The derivative of a differentiable function need not be integrable. Examples of this fact are, however, not very easily constructed (cf. e. g. *H. Lebesgue*, Leçons sur l'intégration, 2nd Edition, Paris 1928, pp. 93—94).

Theorem 26 (*Integration by parts*). If $f(x)$ is integrable over $a \dots b$ and $F(x)$ is the indefinite integral of $f(x)$, if further $g(x)$ is a function, differentiable in $a \dots b$, whose differential coefficient is integrable over $a \dots b$, then [17]

$$\int_a^b f(x)\, g(x)\, d\,x = [F(x) \cdot g(x)]_a^b - \int_a^b F(x) \cdot g'(x)\, d\,x.$$

The following penetrates considerably further than all the above simple theorems:

Theorem 27 (*Second mean value theorem of the integral calculus*). If $f(x)$ and $\varphi(x)$ are integrable over $a \dots b$ and $\varphi(x)$ is monotone in that interval, then a number ξ, with $a \leqq \xi \leqq b$, can be so chosen that

$$\int_a^b \varphi(x) \cdot f(x)\, d\,x = \varphi(a) \int_a^\xi f(x)\, d\,x + \varphi(b) \int_\xi^b f(x)\, d\,x.$$

Here $\varphi(a)$ may also be replaced by the limit, certainly existing under the hypotheses, $\varphi_a = \lim\limits_{x \to a+0} \varphi(x)$, and similarly $\varphi(b)$ by $\varphi_b = \lim\limits_{x \to b+0} \varphi(x)$; but in this case a different value may have to be chosen for ξ.

We mention only the following of the applications of the concept of integral above considered:

Theorem 28 (*Area*). If $f(x)$ is integrable over $a \dots b$, $(a < b)$ and, let us suppose, always positive in the interval [18], then the portion of plane surface bounded by the axis of abscissae, the ordinates through a and b, and the curve $y = f(x)$ — or more precisely, the set of points (x, y) for which $a \leqq x \leqq b$, and at the same time, for each such x, $0 \leqq y \leqq f(x)$, — has a *measurable area* and its measure is $\int_a^b f(x)\, d\,x$.

Theorem 29 (*Length*). If $x = \varphi(t)$ and $y = \psi(t)$ are two functions differentiable in $\alpha \leqq t \leqq \beta$, and if $\varphi'(t)$ and $\psi'(t)$ themselves are continuous in $\alpha \dots \beta$, then the path traced out by the point $x = \varphi(t), y = \psi(t)$ in the plane of a rectangular coordinate system $O\,x$, $O\,y$, when t describes the interval from α to β, has a *measurable length* and this is given by the integral

$$\int_\alpha^\beta \sqrt{\varphi'(t)^2 + \psi'(t)^2}\, d\,t.$$

Finally we may say a few words on the subject of so-called *improper integrals*.

---

[17] Here $[h(x)]_a^b$ denotes the difference $h(b) - h(a)$.

[18] — which may always be arranged by the addition of a suitable constant.

Definition 14. If $f(t)$ is defined for $t \geqq a$ and is integrable over $a \leqq t \leqq x$, for every x, so that the function

$$F(x) = \int_a^x f(t)\,dt$$

is also defined for every $x \geqq a$, then, if $\lim_{x \to +\infty} F(x)$ exists and $= c$, we say that the improper integral

$$\int_a^{+\infty} f(t)\,dt$$

converges and has the value c.

Theorem 30. If $f(t)$ is constantly $\geqq 0$ or constantly $\leqq 0$ for every $t \geqq a$, then $\int_a^{\infty} f(t)\,dt$ converges if and only if the function $F(x)$ of Def. 14 is *bounded* for $x > a$. If $f(t)$ is capable of both signs for $t \geqq a$, then the same integral converges if, and only if, given an arbitrary $\varepsilon > 0$, $x_0 > a$ can be so determined that

$$\left| \int_{x'}^{x''} f(t)\,dt \right| < \varepsilon$$

for every x' and x'' both $> x_0$.

And quite analogously:

Definition 15. If $f(x)$ is defined, but not bounded, in the interval $a < t \leqq b$, *open on the left*, and is integrable, for every x of $a < x < b$, over the interval $x \leqq t \leqq b$, so that the function

$$F(x) = \int_x^b f(t)\,dt$$

is defined for each of these x's, then, if $\lim_{x \to a+0} F(x)$ exists and $= c$, we say that the *improper integral* (improper *at* a)

$$\int_a^b f(t)\,dt$$

is convergent and has the value c.

Exactly analogous conventions are made for an interval *open on the right.* The case of an interval *open on both sides* is reduced to the two preceding cases by dividing it at an interior point into two half-open intervals, and then taking theorem 19 as a definition.

Theorem 31. If in the case of Def. 15, we further have $f(t) \geqq 0$ everywhere or $\leqq 0$ everywhere, then the improper integral in question exists if, and only if, $F(x)$ remains *bounded* in $a < x \leqq b$. If $f(t)$ assumes both signs, then the integral exists if, and only if, given $\varepsilon > 0$, we can choose $\delta > 0$ so that

$$\left| \int_{x'}^{x''} f(t)\,dt \right| < \varepsilon$$

for every x' and x'' *both* between a (excl.) and $a + \delta$.

§ 20. Principal properties of functions represented by power series.

We interrupted our discussion of power series at the observation, terminating § 18, that the sum of a power series, in the interior of its interval of convergence, defines a *function*, which we will now denote by $f(x)$:

$$f(x) = \sum_{n=0}^{\infty} a_n (x - x_0)^n, \qquad |x - x_0| < r.$$

We resume it at that point, and agree in this connection, unless special remark to the contrary is made, to leave the interval of convergence *open at both ends*, even should the power series converge at one or both of the endpoints.

Now if, as is the case here, an infinite series defines a function in a certain interval, then the most important problem is, in general, to deduce from the series the principal properties of the function represented by it — interpreting these for instance in the sense of the summary of the preceding section.

In the case of power series, this presents no great difficulty. We shall see, on the whole, that a function represented by a power series possesses all the properties which we may consider particularly important and that the algebra of power series assumes a peculiarly simple form. For this reason, power series play a prominent part, and it is precisely on this account that their discussion belongs to the elements of the theory of infinite series.

In these investigations, we may, without thereby restricting the scope of the results, assume $x_0 = 0$, i. e. assume the series to be of the simplified form $\Sigma a_n x^n$. Its radius of convergence is of course assumed positive (> 0), but may be $+\infty$, i. e. the series may be everywhere convergent. — We then have, first, the

°**Theorem.** *The function $f(x)$ defined, in its interval of convergence, by the power series $\sum_{n=0}^{\infty} a_n (x - x_0)^n$, is* c o n t i n u o u s *at $x = x_0$; that is to say, we have* **96.**

$$\lim_{x \to x_0} f(x) = \lim_{x \to x_0} \sum_{n=0}^{\infty} a_n (x - x_0)^n = a_0 = f(x_0).$$

Proof. If $0 < \varrho < r$, then by **83**, 5,

$$\sum_{n=1}^{\infty} |a_n| \varrho^{n-1} \quad \text{converges with} \quad \sum_{n=0}^{\infty} |a_n| \varrho^n.$$

If we write $K (> 0)$ for the sum of the former, then we have, for every $|x - x_0| \leqq \varrho$,

$$|f(x) - a_0| = \left|(x - x_0) \cdot \sum_{n=1}^{\infty} a_n (x - x_0)^{n-1}\right| \leqq |x - x_0| K.$$

If therefore $\varepsilon > 0$ is arbitrarily given and if $\delta > 0$ is less than both ϱ and $\frac{\varepsilon}{K}$, then we have, for every $|x - x_0| < \delta$,

$$|f(x) - a_0| < \varepsilon;$$

which by § 19, Def. 6b, proves all that was required.

From this theorem, we immediately deduce the extremely far-reaching and very frequently applied:

97. °**Identity Theorem for power series.** *If the two power series*

$$\sum_{n=0}^{\infty} a_n x^n \quad \text{and} \quad \sum_{n=0}^{\infty} b_n x^n$$

have the same sum in an interval $|x| < \varrho$ *in which both of them converge*[19], *then the two series are entirely identical, that is to say, for every* $n = 0, 1, 2, \ldots$, *we then have*

$$a_n = b_n.$$

Proof. From

(a) $$a_0 + a_1 x + a_2 x^2 + \cdots = b_0 + b_1 x + b_2 x^2 + \cdots$$

it follows, by the preceding theorem, letting $x \to 0$ on both sides of the equation, that

$$a_0 = b_0.$$

Leaving out these terms and dividing by x, we infer that for $0 < |x| < \varrho$

(b) $$a_1 + a_2 x + a_3 x^2 + \cdots = b_1 + b_2 x + b_3 x^2 + \cdots,$$

an equation from which we deduce, in exactly the same way[20], that

$$a_1 = b_1$$

and

$$a_2 + a_3 x + \cdots = b_2 + b_3 x + \cdots.$$

Proceeding in this manner, we infer successively (more precisely: by complete induction) that for *every* n the statement is fulfilled.

Examples and illustrations.

1. This identity theorem will often appear both in the theory and in the applications. We may also interpret it thus: if a function can be represented by a power series in the neighbourhood of the origin, then this is only possible *in one way*. In this form, the theorem may also be called the *theorem of uniqueness*. It of course holds, in the corresponding statement, for the general power series $\Sigma a_n (x - x_0)^n$.

2. Since the assertion in the theorem culminates in the fact that the corresponding coefficients on both sides of the equation (a) are equal, we may also speak, when applying the theorem, of the *method of equating coefficients*.

---

[19] Or even for every $x = x_\nu$ of a *null sequence* (x_ν) whose terms are all $\neq 0$. — In the proof we have then to carry out the limiting processes in accordance with § 19, Def. 4a.

[20] For $x = 0$, equation (b) is not in the first instance secured, since it was established by means of division by x. But for the limiting process $x \to 0$ this is quite immaterial (cf. § 19, Def. 4).

3. A simple example of this form of application is the following: We certainly have, for every x,

$$(1+x)^k (1+x)^k = (1+x)^{2k}$$

or

$$\sum_{\nu=0}^{k} \binom{k}{\nu} x^\nu \cdot \sum_{\nu=0}^{k} \binom{k}{\nu} x^\nu = \sum_{\lambda=0}^{2k} \binom{2k}{\lambda} x^\lambda .$$

If we multiply out on the left, by **91**, Rem. 1, and equate the coefficients on both sides, then we obtain, for instance, by equating the coefficients of x^k:

$$\binom{k}{0}\binom{k}{k} + \binom{k}{1}\binom{k}{k-1} + \cdots + \binom{k}{k}\binom{k}{0} \equiv \binom{k}{0}^2 + \binom{k}{1}^2 + \cdots + \binom{k}{k}^2 = \binom{2k}{k},$$

a relation between the binomial coefficients which would not have been so easy to prove by other methods.

4. If $f(x)$ is defined for $|x| < r$ and we have, for all such x's,

$$f(-x) = f(x),$$

then $f(x)$ is called an *even function*. If it is representable by a power series, then we at once obtain by equating coefficients,

$$a_1 = a_3 = a_5 = \cdots = a_{2k+1} = 0$$

so that in the power series of $f(x)$, only *even* powers of x can have coefficients different from 0.

5. If on the other hand, $f(-x) = -f(x)$, then the function is said to be *odd*. Its expansion in power series can then only contain *odd* powers of x. In particular, $f(0) = 0$.

We now proceed one step further and prove a number of theorems which must be regarded as in every respect the most important in the theory of infinite series:

°**Theorem 1.** *If* **98.**

$$\sum_{n=0}^{\infty} a_n (x - x_0)^n$$

is a power series with (positive) radius r, then the function $f(x)$ thereby represented, for $|x - x_0| < r$, may also be expanded in a power series with any other point x_1 of the interval of convergence as centre; we have, in fact,

$$f(x) = \sum_{k=0}^{\infty} b_k (x - x_1)^k$$

where

$$b_k = \sum_{n=0}^{\infty} \binom{n+k}{k} a_{n+k} (x_1 - x_0)^n,$$

and the radius r of this new series is at least equal to the positive number $r - |x_1 - x_0|$.

Proof. If x_1 lies in the interval of convergence of the series, so that $|x_1 - x_0| < r$, then

$$f(x) = \sum_{n=0}^{\infty} a_n [(x_1 - x_0) + (x - x_1)]^n$$

i. e.

(a) $$f(x) = \sum_{n=0}^{\infty} a_n \left[(x_1 - x_0)^n + \binom{n}{1} (x_1 - x_0)^{n-1} (x - x_1) + \cdots \right.$$
$$\left. \cdots + \binom{n}{n} (x - x_1)^n \right];$$

and all that we have to show is that we may here group together all terms with the same power of $(x - x_1)$, i. e. that the main rearrangement theorem **90** may be applied. If, however, to test its validity, we replace, in the latter series, every term by its absolute value, then we obtain the series

$$\sum_{n=0}^{\infty} |a_n| [|x_1 - x_0| + |x - x_1|]^n;$$

and this is certainly still convergent, if

$$|x_1 - x_0| + |x - x_1| < r, \quad \text{or} \quad |x - x_1| < r - |x_1 - x_0|.$$

If therefore x is nearer to x_1 than either of the endpoints of the original interval of convergence, then the projected rearrangement is allowed, and we obtain for $f(x)$, as asserted, a representation of the form

$$f(x) = \sum_{k=0}^{\infty} b_k (x - x_1)^k \qquad (|x - x_1| < r - |x_1 - x_0|).$$

If we proceed in detail to group the terms containing $(x - x_1)^k$ together, by writing the terms of the series (a) in successive rows one below the other, then the kth column gives

$$b_k = \binom{k}{k} a_k + \binom{k+1}{k} a_{k+1} (x_1 - x_0) + \ldots = \sum_{n=0}^{\infty} \binom{n+k}{k} a_{n+k} (x_1 - x_0)^n,$$

which completes the required proof[21].

From this theorem we deduce the most diverse consequences. First we have the

°**Theorem 2.** *A function represented by a power series*

$$f(x) = \sum_{n=0}^{\infty} a_n (x - x_0)^n$$

is continuous at every point x_1 interior to the interval of convergence.

Proof. By the preceding theorem, we may write, for a certain neighbourhood of x_1,

$$f(x) = \sum_{n=0}^{\infty} a_n (x - x_0)^n = \sum_{n=0}^{\infty} b_n (x - x_1)^n$$

with

$$b_0 = \sum_{n=0}^{\infty} a_n (x_1 - x_0)^n = f(x_1).$$

For $x \to x_1$, the second of the representations of $f(x)$, by **96**, at once gives the required relation (v. § 19, Def. 6):

$$\lim_{x \to x_1} f(x) = f(x_1).$$

°**Theorem 3.** *A function represented by a power series*

$$f(x) = \sum_{n=0}^{\infty} a_n (x - x_0)^n$$

is differentiable at every interior point x_1 of the interval of convergence

[21] We thus have, quite incidentally, a fresh proof of the convergence, already established in **95**, of the different series obtained for the coefficients b_k.

(*v.* § 19, *Def.* 10) *and its differential coefficient at that point,* $f'(x_1)$, *may be obtained by means of term-by-term differentiation, i. e. we have*

$$f'(x_1) = \sum_{n=1}^{\infty} n\, a_n (x_1 - x_0)^{n-1} = \sum_{n=0}^{\infty} (n+1)\, a_{n+1} (x_1 - x_0)^n.$$

Proof. Since $f(x) = \sum_{n=0}^{\infty} b_n (x - x_1)^n$, we have for every x sufficiently near x_1:

$$\frac{f(x) - f(x_1)}{x - x_1} = b_1 + b_2 (x - x_1) + \cdots,$$

whence for $x \to x_1$, by **96**, taking into account the meaning of b_1, we at once deduce the required result: $f'(x_1) = b_1 = \sum n\, a_n (x_1 - x_0)^{n-1}$.

Theorem 4. *A function represented by a power series,*

$$f(x) = \sum_{n=0}^{\infty} a_n (x - x_0)^n,$$

has, at every interior point x_1 *of its interval of convergence, differential coefficients of every order and we have*

$$f^{(k)}(x_1) = k!\, b_k = \sum_{n=0}^{\infty} (n+1)(n+2)\ldots(n+k)\, a_{n+k} (x_1 - x_0)^n.$$

Proof. For every x of the interval of convergence we have, as we have just shown,

$$f'(x) = \sum_{n=0}^{\infty} (n+1)\, a_{n+1} (x - x_0)^n.$$

$f'(x)$ is thus again a function represented by a power series, — and in fact by one which, in accordance with **95**, has the same interval of convergence as the original series. Hence the same result may be again applied to $f'(x)$, giving

$$f''(x) = \sum_{n=1}^{\infty} n(n+1)\, a_{n+1} (x - x_0)^{n-1} = \sum_{n=0}^{\infty} (n+1)(n+2)\, a_{n+2} (x - x_0)^n.$$

By a repetition of this simple process, we obtain for every k,

$$f^{(k)}(x) = \sum_{n=0}^{\infty} (n+1)(n+2)\ldots(n+k)(a_{n+k} (x - x_0)^n, \text{ —}$$

valid for every x of the original interval of convergence. Putting in particular $x = x_1$, we therefore at once deduce the required statement.

If we substitute, for the coefficients b_k in the expansion of theorem 1, the values $\frac{1}{k!} f^{(k)}(x_1)$ now obtained, then we finally infer from all the above the so-called

°**Taylor series**[22]. *If for* $|x - x_0| < r$, *we have* **99.**

$$f(x) = \sum_{n=0}^{\infty} a_n (x - x_0)^n,$$

and if x_1 *is an interior point of the interval of convergence, then we*

---

[22] *Brook Taylor*: Methodus incrementorum directa et inversa, London 1715. — Cf. *A. Pringsheim*, Geschichte des Taylorschen Lehrsatzes, Bibl. math. (3) Vol. 1, p. 433. 1900.

have, for every x for which[23] $|x - x_1| < r_1 = r - |x_1 - x_0|$,

$$f(x) = f(x_1) + \frac{f'(x_1)}{1}(x - x_1) + \frac{f''(x_1)}{2!}(x - x_1)^2 + \cdots$$
$$\cdots + \frac{f^{(k)}(x_1)}{k!}(x - x_1)^k + \cdots.$$

With Theorem 3 for the differentiation of our series, we couple the corresponding theorem for integration. Since a function represented by a power series is continuous in the interior of its interval of convergence, it is also, by § 19, Theorem 12, integrable over every interval contained, together with its endpoints, in the interior of this interval of convergence. For this we have the

°**Theorem 5.** *The integral of the (continuous) function $f(x)$ represented by $\sum_{n=0}^{\infty} a_n (x - x_0)^n$ in the interval of convergence, may be obtained by means of term by term integration, with the formula*

$$\int_{x_1}^{x_2} f(t)\,dt = \sum_{n=0}^{\infty} \frac{a_n}{n+1}\left[(x_2 - x_0)^{n+1} - (x_1 - x_0)^{n+1}\right],$$

provided x_1 and x_2 are both interior to the interval of convergence.

Proof. By **95**, 2, the power series

$$F(x) = \sum_{n=0}^{\infty} \frac{a_n}{n+1}(x - x_0)^{n+1}$$

has the same interval of convergence as the given series

$$f(x) = \sum_{n=0}^{\infty} a_n (x - x_0)^n.$$

By **98**, 3, the first series is an indefinite integral of the second. Hence by § 19, Theorem 24, the statement follows at once.

These theorems on power series we may complete in a special direction by the following important addition: Theorem 2 on the continuity of the function represented by a power series was, as we may again expressly observe, only valid for the *open* interval of convergence. Thus, for instance, in the case of the geometric series $\sum x^n$,

---

[23] The number $r_1 = r - |x_1 - x_0|$ of the text need not be the exact radius of convergence of the new series. On the contrary, the latter may prove considerably larger. Thus for $f(x) = \sum x^n = \frac{1}{1-x}$ and $x_1 = -\frac{1}{2}$ we obtain, by an easy calculation,

$$f(x) = \sum_{k=0}^{\infty} \left(\frac{2}{3}\right)^{k+1} \cdot \left(x + \frac{1}{2}\right)^k$$

and the radius of this series is not $= r - |x_1 - x_0| = \frac{1}{2}$, but is $= \frac{3}{2}$.

of sum $\frac{1}{1-x}$, we can deduce from our considerations neither its continuity at the point $x=-1$, nor its discontinuity at $x=+1$, by immediate inspection of the series. Even if the power series converged at one of the endpoints of the intervals $\left(\text{as here } \sum \frac{x^n}{n} \text{ for } x=-1\right)$, we should not be able to conclude this fact directly. That however, in this last particular case, the presumption is, at least to some extent, justified, we learn from the following:

Abel's limit theorem[24]. *Let the power series* $f(x)=\sum_{n=0}^{\infty} a_n x^n$ **100.**
have radius of convergence r *and still converge for* $x=+r$.

Then
$$\lim_{x\to r-0} f(x) \quad \textit{exists and} \quad = \sum_{n=0}^{\infty} a_n r^n.$$

Or in other words: If $\sum_{n=0}^{\infty} a_n x^n$ *still converges for* $x=+r$, *then the function* $f(x)$ *defined by the series in* $-r<x\leqq +r$, *is also continuous on the left at the endpoint* $x=+r$.

Proof. There is no restriction[25] in assuming $r=+1$. For if $\sum a_n x^n$ has radius r, then the series $\sum a_n' x^n$, in which $a_n'=a_n r^n$, obviously has radius 1; and the latter series is convergent at $+1$ or -1, if, and only if, the former was at $+r$ or $-r$ respectively.

We therefore in future assume $r=+1$. Our hypothesis is, therefore, that $f(x)=\sum a_n x^n$ has radius 1 and that $\sum a_n = s$ converges; and our statement is that

$$\lim_{x\to 1-0} f(x) = s, \quad \text{i. e.} \quad = \sum_{n=0}^{\infty} a_n.$$

Now by **91** (v. also later, **102**), we have for $|x|<1$,

$$\frac{1}{1-x}\sum_{n=0}^{\infty} a_n x^n = \sum_{n=0}^{\infty} x^n \sum_{n=0}^{\infty} a_n x^n = \sum_{n=0}^{\infty} s_n x^n,$$

if by s_n we denote the partial sum of $\sum a_n$. Consequently $f(x) = (1-x)\sum s_n x^n$ and since $1=(1-x)\sum x^n$, we therefore deduce, for $|x|<1$,

$$\text{(a)} \qquad s - f(x) = (1-x)\sum_{n=0}^{\infty}(s-s_n)\,x^n \equiv (1-x)\sum_{n=0}^{\infty} r_n x^n.$$

---

[24] Journal f. d. reine u. angew. Math. Vol. 1, p. 311. 1826. cf. **233** and § 62. — The theorem had already been stated and used by *Gauss* (Disquis. generales circa seriem ..., 1812; Werke III, p. 143) and in fact precisely in the form proved further on, that $r_n \to 0$ involves $(1-x)\sum r_n x^n \to 0$ if $x\to 1$ from the left (v. eq. (a)). The proof given by *Gauss* loc. cit. is however *incorrect*, as he interchanged the two limiting processes which come under consideration for this theorem, without at all testing whether he was justified in so doing.

[25] This remark holds in general for all discussions of (not everywhere convergent) power series of positive radius r.

Here we have written $s - s_n = r_n$, the "remainder" of the series; these remainders, by **82**, Theorem 2, form a null sequence.

If now $\varepsilon > 0$ is arbitrarily given, then we first choose m so large that, for every $n > m$, we have $|r_n| < \frac{\varepsilon}{2}$. We then have, for $0 \leqq x < 1$,

$$|s - f(x)| \leqq |(1 - x) \sum_{n=0}^{m} r_n x^n| + \frac{\varepsilon}{2}(1 - x) \cdot \sum_{n=m+1}^{\infty} x^n,$$

hence, if p denotes a positive number greater than $|r_0| + |r_1| + \ldots + |r_m|$, this is

$$\leqq p \cdot (1 - x) + \frac{\varepsilon}{2}(1 - x) \cdot \frac{x^{m+1}}{1-x}.$$

If we now write $\delta =$ the smaller of the two numbers 1 and $\frac{\varepsilon}{2p}$, then we have, for $1 - \delta < x < 1$,

$$|s - f(x)| < \frac{\varepsilon}{2} + \frac{\varepsilon}{2} = \varepsilon,$$

which, by § 19, Def. 5, proves the required statement "$f(x) \to s$ for $x \to 1 - 0$".

We have of course, quite similarly, Abel's limit theorem for the *left* endpoint of the interval of convergence:

If $\sum_{n=0}^{\infty} a_n x^n$ *still converge for* $x = -r$, *then*

$$\lim_{x \to -r+0} f(x) \quad \textit{exists and} \quad = \sum_{n=0}^{\infty} (-1)^n a_n r^n.$$

The continuity theorem **98**, 2 and Abel's theorem **100** together assert that

101. $$\lim_{x \to \xi} (\Sigma a_n x^n) = \Sigma a_n \xi^n$$

if the series on the right converges and x *tends to* ξ *from the side on which lies the origin.*

If the series $\Sigma a_n \xi^n$ diverges, we cannot assert anything, without further assumptions, as to the behaviour of $\Sigma a_n x^n$ when $x \to \xi$. We have however in this connection the following somewhat more definite:

Theorem *If* Σa_n *is a divergent series of positive terms, and* $\Sigma a_n x^n$ *has radius* 1, *then*

$$f(x) = \sum_{n=0}^{\infty} a_n x^n \to +\infty$$

when x *tends towards* $+1$ *from the origin.*

Proof. A divergent series of positive terms can only diverge to $+\infty$. If therefore $G > 0$ is arbitrarily given, we can choose m so large that $a_0 + a_1 + \ldots + a_m > G + 1$, and then by § 19, Theorem 3, choose $\delta < 1$ so small that for every $1 > x > 1 - \delta$, we continue to have

$$a_0 + a_1 x + \ldots + a_m x^m > G.$$

But then we have, *a fortiori*,

$$f(x) = \sum_{n=0}^{\infty} a_n x^n > G,$$

which is all that required proof.

Remarks and examples for the theorems of the present paragraph will be given in detail in the next chapter.

§ 21. The algebra of power series.

Before we make use of the far-reaching theorems of the preceding section (§ 20), which lead to the very centre of the wide field of application of the theory of infinite series, we will enter into a few questions whose solution should facilitate our *operations on power series*.

That power series, as long as they converge, may be added and subtracted term by term already follows from **83**, 3 and 4. That we may immediately multiply out term by term, in the product of two power series, provided we remain *in the interior* of the intervals of convergence, follows at once from **91**, since power series always converge absolutely *in the interior* of their intervals of convergence. We therefore have, with

$$\Sigma a_n x^n \pm \Sigma b_n x^n = \Sigma (a_n \pm b_n) x^n$$

also $$\sum_{n=0}^{\infty} a_n x^n \cdot \sum_{n=0}^{\infty} b_n x^n = \sum_{n=0}^{\infty} (a_0 b_n + a_1 b_{n-1} + \dots + a_n b_0) x^n,$$

provided x is interior to the intervals of convergence of *both* series [26].

The formulae **91**, Rem. 2 and 3 were themselves a first application of this theorem. If the second series is, in particular, the geometric series, then we find

$$\sum_{n=0}^{\infty} a_n x^n \cdot \sum_{n=0}^{\infty} x^n = \sum_{n=0}^{\infty} s_n x^n,$$

i. e. $$\frac{1}{1-x} \sum_{n=0}^{\infty} a_n x^n = \sum_{n=0}^{\infty} s_n x^n$$

or $$\sum_{n=0}^{\infty} a_n x^n = (1-x) \sum_{n=0}^{\infty} s_n x^n,$$ **102.**

where $s_n = a_0 + a_1 + \dots + a_n$, and $|x| < 1$ and also less than the radius of $\Sigma a_n x^n$.

We infer in as simple a manner that every series may be multiplied — and in fact, arbitrarily often — by itself. Thus

$$\left(\sum_{n=0}^{\infty} a_n x^n\right)^2 = \sum_{n=0}^{\infty} (a_0 a_n + a_1 a_{n-1} + \dots + a_n a_0) x^n ;$$

and generally, for every positive integral exponent k,

$$\left(\sum_{n=0}^{\infty} a_n x^n\right)^k = \sum_{n=0}^{\infty} a_n^{(k)} x^n$$ **103.**

[26] Here we see the particular importance of *Cauchy*'s product (v. **91**, 1).

where the coefficients $a_n^{(k)}$ are constructed from the coefficients a_n in a perfectly determinate manner — even though not an extremely obvious one[27] for larger k's. And these series are all absolutely convergent, so long as $\Sigma a_n x^n$ itself is.

This result makes it seem probable that we "may" also divide by power series, — that for instance we may also write

$$\frac{1}{a_0 + a_1 x + a_2 x^2 + \cdots} = c_0 + c_1 x + c_2 x^2 + \cdots$$

and that the coefficients c_n may again be constructed in a perfectly determinate manner from the coefficients a_n. For we may first, writing $-\frac{a_n}{a_0} = a_n'$, for $n = 1, 2, 3, \ldots$, replace the left hand ratio by

$$\frac{1}{a_0} \cdot \frac{1}{1 - (a_1' x + a_2' x^2 + \cdots)}$$

and then by

$$\frac{1}{a_0}[1 + (a_1' x + a_2' x^2 + \cdots) + (a_1' x + \cdots)^2 + (a_1' x + \cdots)^3 + \cdots]$$

which must actually result in a power series of the form $\Sigma c_n x^n$, if the powers are expanded by **103** and like powers of x then grouped together.

Our justification for writing the above may at once be tested from a somewhat more general point of view:

We suppose given a power series $\Sigma a_n x^n$ (in the above, the series $\sum\limits_{n=0}^{\infty} a_n' x^n$), whose sum we denote by $f(x)$ or more shortly by y. We further suppose given a power series in y, for instance $g(y) = \Sigma b_n y^n$ (in the above, the geometric series Σy^n) and in this we substitute for y the former power series:

$$b_0 + b_1 (a_0 + a_1 x + \cdots) + b_2 (a_0 + a_1 x + \cdots)^2 + \cdots.$$

Under what conditions do we, by expanding all the powers, in accordance with **103**, and grouping like powers of x together, obtain a new power series $c_0 + c_1 x + c_2 x^2 + \cdots$ which converges and has for sum the value of the function of a function $g(f(x))$? We assert the

104. °**Theorem.** *This certainly holds for every x for which $\sum\limits_{n=0}^{\infty} |a_n x^n|$ converges and has a sum less than the radius of $\Sigma b_n y^n$.*

---

[27] Recurrence formulae for the evoluation of $a_n^{(k)}$ are to be found in *J. W. L. Glaisher*, Note on Sylvester's paper: Development of an idea of Eisenstein (Quarterly Journal, Vol. 14, p. 79—84. 1875), where further references to the bibliography may also be obtained. See also *B. Hansted*, Tidskrift for Mathematik, (4) Vol. 5, pp. 12—16, 1881.

If *both* series converge *everywhere*, then the theorem holds without restriction for every x.

2. If $a_0 = 0$ and both series have a positive radius, then the theorem certainly holds for every "sufficiently" small x, that is to say, there is then certainly a positive number ϱ, such that the theorem holds for every $|x| < \varrho$. For if $y = a_1 x + a_2 x^2 + \cdots$, then $\eta = |a_1| \cdot |x| + |a_2| \cdot |x^2| + \cdots$; and since for $x \to 0$, we now also, by **96**, have $\eta \to 0$, η is certainly less than the radius of $\Sigma b_k y^k$ for all x whose absolute value is less than a suitable number ϱ.

3. In the series $\sum \frac{y^n}{n!}$, we "may" for instance substitute $y = \Sigma x^n$ for $|x| < 1$, or $y = \sum \frac{x^n}{n!}$ for *every* n, and then rearrange in powers of x.

4. To write, as we did above:

$$\frac{1}{a_0 + a_1 x + a_2 x^2 + \cdots} = c_0 + c_1 x + c_2 x^2 + \cdots$$

is, we now see, certainly allowed if $a_0 \neq 0$ and further x is in absolute value so small that

$$\eta = \left|\frac{a_1}{a_0} x\right| + \left|\frac{a_2}{a_0} x^2\right| + \cdots < 1,$$

which by Rem. 2 is certainly the case for every $|x| < \varrho$ with a suitable choice of ϱ. We may therefore say: *We "may" divide by a power series of positive radius if its constant term is* $\neq 0$ *and provided we restrict ourselves to sufficiently small*[29] *values of* x.

To determine the coefficients c_n by the general method used to prove their existence, would, — even for the first few indices, — be an extremely laborious process. But once we have established the *possibility* of the expansion — which is at the same time necessarily unique by **97**, — we may determine the c_n's more rapidly by remarking that

$$\Sigma a_n x^n \cdot \Sigma c_n x^n \equiv 1,$$

so that we have successively

$$\begin{aligned} &a_0 c_0 = 1 \\ &a_0 c_1 + a_1 c_0 = 0 \\ &a_0 c_2 + a_1 c_1 + a_2 c_0 = 0 \\ &a_0 c_3 + a_1 c_2 + a_2 c_1 + a_3 c_0 = 0 \\ &\cdots\cdots\cdots\cdots\cdots \end{aligned}$$

From these relations, since $a_0 \neq 0$, the successive coefficients $c_0, c_1, c_2, \ldots$ may be uniquely determined, the simplest method being with the aid of determinants, by *Cramer*'s Rule, which immediately yields a closed expression[30] for c_n in terms of $a_0, a_1, \ldots, a_n$.

5. As a particularly important example for many subsequent investigations we may set the following question[31]:

Expand $\quad \dfrac{1}{1 + \frac{x}{2!} + \frac{x^2}{3!} + \cdots} \quad$ *or* $\quad \dfrac{x}{\left(1 + x + \frac{x^2}{2!} + \frac{x^3}{3!} + \cdots\right) - 1}$

[29] *How* small x has to be, is usually immaterial. But what is essential, is that some positive radius ϱ exists, such that the relation holds for every $|x| < \varrho$. — The determination of the *precise* region of validity requires deeper methods of function theory.

[30] Explicit formulae for the coefficients of the expansion, in the case of the quotient of two power series, may be found e. g. in *J. Hagen*, On division of series, Americ. Journ. of Math., Vol. 5, p. 236, 1883.

[31] *Euler*: Institutiones calc. diff., Vol. 2, § 122. 1755.

Proof. We have obviously here a case of the main rearrang theorem **90**, and we have only to verify that the hypotheses c theorem are fulfilled. If we first write

$$y^k = (a_0 + a_1 x + \cdots)^k = a_0^{(k)} + a_1^{(k)} x + a_2^{(k)} x^2 + \cdots,$$

forming the powers by **103**, and also suppose this notation[28] ad for $k = 0$ and $k = 1$, then we have, in

$$(\mathrm{A}) \quad \begin{cases} b_0 = b_0 (a_0^{(0)} + a_1^{(0)} x + \cdots + a_0^{(0)} x^n + \cdots) \\ b_1 y = b_1 (a_0^{(1)} + a_1^{(1)} x + \cdots + a_n^{(1)} x^n + \cdots) \\ \cdot \quad \cdot \quad \cdot \quad \cdot \quad \cdot \quad \cdot \quad \cdot \quad \cdot \quad \cdot \quad \cdot \\ b_k y^k = b_k (a_0^{(k)} + a_1^{(k)} x + \cdots + a_n^{(k)} x^n + \cdots) \\ \cdot \quad \cdot \quad \cdot \quad \cdot \quad \cdot \quad \cdot \quad \cdot \quad \cdot \quad \cdot \quad \cdot \end{cases}$$

the series $z^{(k)}$ occurring in the theorem **90**. If we now take, inst of $y = \Sigma a_n x^n$, the series $\eta = \Sigma | a_n x^n |$, and, writing $| x | = \xi$, fo quite similarly,

$$(\mathrm{A}') \quad \begin{cases} | b_0 | = | b_0 | (\alpha_0^{(0)} + \alpha_1^{(0)} \xi + \cdots + \alpha_n^{(0)} \xi^n + \cdots) \\ | b_1 | \eta = | b_1 | (\alpha_0^{(1)} + \alpha_1^{(1)} \xi + \cdots + \alpha_n^{(1)} \xi^n + \cdots) \\ \cdot \quad \cdot \quad \cdot \quad \cdot \quad \cdot \quad \cdot \quad \cdot \quad \cdot \quad \cdot \quad \cdot \\ | b_k | \eta^k = | b_k | (\alpha_0^{(k)} + \alpha_1^{(k)} \xi + \cdots + \alpha_n^{(k)} \xi^n + \cdots) \\ \cdot \quad \cdot \quad \cdot \quad \cdot \quad \cdot \quad \cdot \quad \cdot \quad \cdot \quad \cdot \quad \cdot \end{cases}$$

then all the numbers in this array (A′) are $\geqq 0$ and since furthermo $\Sigma | b_k | \eta^k$ was assumed to converge, the main rearrangement theore is applicable to (A′). But obviously every number of the array A in absolute value $\leqq$ the corresponding number in (A′); hence ou theorem is *a fortiori* applicable to (A) (cf. **90**, Rem. 3). In particula therefore, the coefficients standing vertically one below the other i (A) always form (absolutely) convergent series

$$\sum_{k=0}^{\infty} b_k a_n^{(k)} = c_n \qquad \text{(for every definite } n = 0, 1, 2, \ldots$$

and the power series formed with these numbers as coefficients, i. e

$$\sum_{n=0}^{\infty} c_n x^n$$

is again, for the considered values of x, (absolutely) convergent and has the same sum as $\Sigma b_n y^n$. We therefore have, as asserted,

$$g(f(x)) = \sum_{n=0}^{\infty} c_n x^n$$

with the indicated meaning of c_n.

Remarks and Examples.

1. If the "outer" series $g(y) = \Sigma b_k x^k$ converges *everywhere*, then our theorem evidently holds for every x for which $\Sigma a_n x^n$ converges absolutely.

---

[28] We have therefore to write $a_0^{(0)} = 1$, $a_1^{(0)} = a_2^{(0)} = \cdots = 0$, and $a_n^{(1)} = a_n$, the latter for $n = 0, 1, 2, \ldots$.

in powers of x. Here the determination of the new coefficients becomes peculiarly elegant if we denote them, not by c_n, but by $\frac{c_n}{n!}$, or as we shall do, for historic reasons, by $\frac{B_n}{n!}$. Then the above equation is

$$\left(1+\frac{x}{2!}+\frac{x^2}{3!}+\cdots\right)\left(B_0+\frac{B_1}{1!}x+\frac{B_2}{2!}x^2+\cdots\right)\equiv 1$$

and the equations for determining B_n are, in succession,

$$B_0=1\,,\quad \frac{1}{2!}\frac{B_0}{0!}+\frac{1}{1!}\cdot\frac{B_1}{1!}=0\,,$$

and, in general, for $n=2, 3, \ldots$,

$$\frac{1}{n!}\cdot\frac{B_0}{0!}+\frac{1}{(n-1)!}\cdot\frac{B_1}{1!}+\frac{1}{(n-2)!}\cdot\frac{B_2}{2!}+\cdots+\frac{1}{1!}\cdot\frac{B_{n-1}}{(n-1)!}=0\,.$$

If we multiply by $n!$, we may write this more concisely:

$$\binom{n}{0}B_0+\binom{n}{1}B_1+\binom{n}{2}B_2+\cdots+\binom{n}{n-1}B_{n-1}=0\,.$$

Now if we here had B^ν in place of B_ν, for each ν, then we could write instead

$$(B+1)^n-B^n=0\,; \qquad \textbf{106.}$$

and the recurring formula under consideration also may be borne in mind under this convenient form, as a *symbolic equation*, i. e. one which is not intended to be interpreted literally, but only becomes valid with a particular convention, — here the convention that after expanding the nth power of the binomial $(B+1)$, we replace each B^ν by B_ν. Our formula now yields, for $n=2, 3, 4, 5, \ldots$ successively, the equations

$$\begin{aligned}&2B_1+1=0\,,\\&3B_2+3B_1+1=0\,,\\&4B_3+6B_2+4B_1+1=0\,,\\&5B_4+10B_3+10B_2+5B_1+1=0\,,\\&\ldots\ldots\ldots\ldots\ldots\ldots\end{aligned}$$

from which we deduce

$$B_1=-\frac{1}{2}\,,\quad B_2=\frac{1}{6}\,,\quad B_3=0\,,\quad B_4=-\frac{1}{30}$$

and then

$$B_5=B_7=B_9=B_{11}=B_{13}=B_{15}=0\,,$$

and

$$B_6=\frac{1}{42}\,,\quad B_8=-\frac{1}{30}\,,\quad B_{10}=\frac{5}{66}\,,\quad B_{12}=-\frac{691}{2730}\,,\quad B_{14}=\frac{7}{6}\,.$$

These are called ***Bernoulli's numbers*** and will be mentioned repeatedly later on (§ 24, 4; § 32, 4; § 55, IV; § 64). For the moment, we are able to infer only that the numbers B_n are definite *rational* numbers. They do not, however, conform to any apparent or superficial law, and have formed the subject of many elaborate discussions[32].

---

[32] Bernoulli's numbers are frequently indexed somewhat differently, B_0, B_1, B_3, B_5, B_7, ... being omitted and $(-1)^{k-1}B_k$ written instead of B_{2k}, for $k=1, 2, \ldots$ A table of the numbers B_2, B_4, ..., to B_{124} may be found in *J. C. Adams*, Journ. f. d. reine u. angew. Math., Vol. 85, 1878. We may mention in passing that B_{120} has for numerator a number with 113 digits, and for denominator the number 2 358 255 930; while B_{122} has the denominator 6 and,

Finally we will prove one more general theorem on power series: Given the power series $y = \sum_{n=0}^{\infty} a_n (x - x_0)^n$, convergent for $|x - x_0| < r$, we have, for every x in the neighbourhood of x_0, a determinate corresponding value of y, in particular for $x = x_0$ the value $y = a_0$, which we will accordingly denote by y_0. Then we have

$$y - y_0 = a_1 (x - x_0) + a_2 (x - x_0)^2 + \dots .$$

Because of the continuity of the function, to *every* x near x_0 also corresponds a value of y near y_0. We would now enquire *whether or how far every value of y near y_0 is obtained and whether it is obtained once only.* If the latter was the case, not merely y would be determined by x, but conversely x would be determined by y, and therefore x would be a function of y. The given function $y = f(x)$ would, as we say for brevity, be ***reversible*** in the neighbourhood of x_0 (cf. § 19, Theorem 6). The question of reversibility is dealt with by:

107. ° **Reversion theorem for power series.** *Given the expansion*

$$y - y_0 = a_1 (x - x_0) + a_2 (x - x_0)^2 + \dots,$$

convergent for $|x - x_0| < r$, the function $y = f(x)$ thereby determined is reversible in the neighbourhood of x_0, under the sole hypothesis that $a_1 \neq 0$; i. e. there then exists one and only one function $x = \varphi(y)$ which is expressible by a power series, convergent in a certain neighbourhood of y_0, of the form

$$x - x_0 = b_1 (y - y_0) + b_2 (y - y_0)^2 + \dots$$

and for which, in that neighbourhood, we have (in the sense of **104**)

$$f(\varphi(y)) \equiv y.$$

Moreover $b_1 = 1 : a_1$.

Proof. As we have already done more than once, we assume in the proof that x_0 and y_0 are $= 0$, — which implies no restriction [33]. But we will then further assume that $a_1 = 1$, so that the expansion

(a) $$y = x + a_2 x^2 + a_3 x^3 + \dots$$

is the one to be reversed. That too implies no restriction, for since $a_1 \neq 0$, by hypothesis, we can write $a_1 x + a_2 x^2 + \dots$ in the form

$$(a_1 x) + \frac{a_2}{a_1^2}(a_1 x)^2 + \frac{a_3}{a_1^3}(a_1 x)^3 + \dots .$$

---

in the numerator, a number with 107 digits. The numbers $B_2, B_4, \dots$, to B_{62} had previously been calculated by *Ohm*, ibid., Vol. 20, p. 111, 1840. — The numbers B_p first occur in *James Bernoulli*, Ars conjectandi, 1713, p. 96. — A comprehensive account is given by *L. Saalschütz*, "Vorlesungen über die *Bernoulli*schen Zahlen", Berlin (J. Springer) 1893, and by *N. E. Nörlund*, "Vorlesungen über Differenzenrechnung", Berlin (J. Springer) 1924. New investigations, which chiefly concern the arithmetical part of the theory, are given by *G. Frobenius*, Sitzgsber. d. Berl. Ak., 1910, p. 809—847.

[33] Or: we write for brevity $x - x_0 = x'$ and $y - y_0 = y'$ and then, for simplicity's sake, omit the accents.

If we write for brevity $a_1 x = x'$ and, for $n \geqq 2$,

$$\frac{a_n}{a_1^n} = a_n',$$

and subsequently, for simplicity's sake, omit the accents, then we obtain precisely the above form of expansion. It suffices therefore to consider this. But we can then show that a power series, convergent in a certain interval, of the form

$$\text{(b)} \qquad x = y + b_2 y^2 + b_3 y^3 + \cdots$$

exists which represents the inverse function of the former, so that

$$\text{(c)} \quad (y + b_2 y^2 + \cdots) + a_2 (y + b_2 y^2 + \cdots)^2 + a_3 (y + b_2 y^2 + \cdots)^3 + \cdots$$

is identically $= y$, if this series is arranged in powers of y, in accordance with **104**, — i. e. all the coefficients must be $= 0$ except that of y^1, which is $= 1$.

Since we have written, for brevity, x instead of $a_1 x$, we see that the series on the right hand side of (b) has still to be divided by a_1 to represent the inverse of the series $a_1 x + a_2 x^2 + \cdots$, where a_1 has no specialised value. In this general case we shall therefore have $b_1 = \frac{1}{a_1}$ as coefficient of y^1.

If we assume, provisionally, that the statement (b) is correct, then the coefficients b_ν are *quite uniquely* determined by the condition that the coefficients of $y^2, y^3, \ldots$ in (c) after the rearrangement, have all to be $= 0$. In fact, this stipulation gives the equations

$$\text{(d)} \qquad \begin{cases} b_2 + a_2 = 0 \\ b_3 + 2 b_2 a_2 + a_3 = 0 \\ b_4 + (b_2^2 + 2 b_3) a_2 + 3 b_2 a_3 + a_4 = 0 \\ \ldots\ldots\ldots\ldots\ldots\ldots\ldots, \end{cases}$$

from which, as is immediately evident, the coefficients b_ν may be determined in succession, without any ambiguity. Thus we obtain, the values

$$\text{(e)} \qquad \begin{cases} b_2 = -a_2 \\ b_3 = -2 b_2 a_2 - a_3 = 2 a_2^2 - a_3 \\ b_4 = -(b_2^2 + 2 b_3) a_2 - 3 b_2 a_3 - a_4 \\ b_5 = \cdots \\ \ldots\ldots\ldots\ldots\ldots\ldots\ldots \end{cases}$$

but the calculation soon becomes too complicated to convey any clear idea of the whole. Nevertheless, the equations we have written down show that if there *exists at all* an inverse function of $y = f(x)$, capable of expansion in form of a power series, then there exists *only one.*

Now the calculation just indicated shows that whatever may have been the original given series (a), we can *invariably* obtain perfectly determinate values b_ν, so that we can *invariably* construct a power series $y + b_2 y^2 + \cdots$ which *at least formally* satisfies the conditions of the problem, the series (c) becoming identically $= y$. It only remains to be seen whether the power series has a *positive* radius of convergence. If that can be proved, then the reversion is completely carried out.

The required verification may, as *Cauchy* first showed, actually be attained, in the general case, as follows: Choose any positive numbers α_n for which we have

$$|a_\nu| \leqq \alpha_\nu$$

and $\Sigma \alpha_\nu x^\nu$ has a positive radius of convergence. Proceeding in the above manner, for the series:

$$y = x - \alpha_2 x^2 - \alpha_3 x^3 + \cdots$$

whose inverse is, then, say,

$$x = y + \beta_2 y^2 + \beta_3 y^3 + \cdots$$

we obtain, for the coefficients β_ν, the equations

$$\begin{aligned} &\beta_2 = \alpha_2 \\ &\beta_3 = 2\beta_2 \alpha_2 + \alpha_3 \\ &\beta_4 = (\beta_2{}^2 + 2\beta_3)\alpha_2 + 3\beta_2\alpha_3 + \alpha_4 \\ &\ldots\ldots\ldots\ldots\ldots\ldots, \end{aligned}$$

in which all the terms are now positive. Thus for every ν,

$$\beta_\nu \geqq |b_\nu|.$$

If, therefore, it is possible so to choose the α_ν that the series $\Sigma \beta_\nu y^\nu$ has a positive radius of convergence, it would follow that $\Sigma b_\nu y^\nu$ also had a positive radius and our proof would be complete.

We choose the α_ν's as follows: There is certainly a positive number ϱ, for which the original series $x + a_2 x^2 + \cdots$ converges absolutely. A positive number K must, however, then exist (by **82**, Theorem **1** and **10**, 11) such that we have, for every $\nu = 2, 3, \ldots$,

$$|a_\nu| \varrho^\nu \leqq K \quad \text{or} \quad |a_\nu| \leqq \frac{K}{\varrho^\nu}.$$

We then choose, for $\nu = 2, 3, \ldots$,

$$\alpha_n = \frac{K}{\varrho^\nu},$$

so that we are concerned with reversing the series, convergent for $|x| < \varrho$,

$$y = x - K \cdot \frac{x^2}{\varrho^2}\left(1 + \frac{x}{\varrho} + \frac{x^2}{\varrho^2} + \cdots\right) = x - \frac{K \cdot x^2}{\varrho(\varrho - x)}.$$

But this function is immediately reversible. For we may at once see by differentiation — we are dealing, in fact, with a simple hyperbola, of which the student should draw a graph for himself —, that in

$$-\infty < x < x_1 = \varrho\left(1 - \sqrt{\frac{K}{K+\varrho}}\right),$$

the function increases monotonely (in the stricter sense) from $-\infty$ to the value

$$y_1 = 2K + \varrho - 2\sqrt{K(K+\varrho)}$$

and therefore possesses, for $y < y_1$, a uniquely determined inverse whose values are $< x_1$. For this, since

$$y = x - \frac{K \cdot x^2}{\varrho(\varrho - x)} \quad \text{or} \quad (K+\varrho)x^2 - \varrho(\varrho + y)x + \varrho^2 y = 0,$$

we have, *uniquely*,

$$x = \frac{\varrho}{2(K+\varrho)}\left[\varrho + y - \sqrt{y^2 - 2(2K+\varrho)y + \varrho^2}\right].$$

Further

$$y^2 - 2(2K+\varrho)y + \varrho^2 \equiv (y - y_1)(y - y_2),$$

if we write for brevity, with the above defined value of y_1,

$$y_2 = 2K + \varrho + 2\sqrt{K(K+\varrho)}, \text{ —}$$

and *both* y_1 and y_2 are > 0, since the second is and the two have product $= \varrho^2$. But

$$x = \frac{\varrho^2}{2(K+\varrho)}\left[1 + \frac{y}{\varrho} - \left(1 - \frac{y}{y_1}\right)^{\frac{1}{2}} \cdot \left(1 - \frac{y}{y_2}\right)^{\frac{1}{2}}\right].$$

In the following chapter we shall see that, for $|z| < 1$, the power $(1-z)^{\frac{1}{2}}$ can actually be expanded in a power series — beginning with $1 - \frac{z}{2} + \cdots$. Assuming this result, it follows immediately that x also may be expanded in a power series, convergent at least for $|y| < y_1$:

$$\begin{aligned} x &= \frac{\varrho^2}{2(K+\varrho)}\left[1 + \frac{y}{\varrho} - \left(1 - \frac{y}{2y_1} + \cdots\right)\left(1 - \frac{y}{2y_2} + \cdots\right)\right] \\ &= y + \beta_2 y^2 + \cdots. \end{aligned}$$

By our first remarks the proof is hereby entirely completed.

The actual construction of the series

$$y + b_2 y^2 + \cdots$$

from the series

$$x + a_2 x^2 + \cdots$$

here also involves in general considerable difficulties and necessitates the use of special artifices in each particular case[34]. Examples of this will occur in §§ 26, 27.

We only note further, a fact which will be of use later on, — that if (b) is the inverse of (a), then the inverse of the series

(a′) $$y = x - a_2 x^2 + a_3 x^3 - + - \cdots$$

where the signs are alternated, is obtained from (b) by similarly alternating the signs, i. e.

(b′) $$x = y - b_2 y^2 + b_3 y^3 - + - \cdots$$

---

[34] The general values of the coefficients of expansion b_n are worked out as far as b_{13} by *C. E. van Orstrand*, Reversion of power series, Philos. Magazine (6), Vol. 19, p. 366, 1910.

This is at once evident, if we first actually expand the powers of

$$(y + b_2 y^2 + \dots)$$

in (c), obtaining, say,

(c)'
$$(y + b_2 y^2 + \dots) + a_2 (y^2 + b_3^{(2)} y^3 + \dots) + a_3 (y^3 + b_4^{(3)} y^4 + \dots) + \dots .$$

Under the new assumption, the same process, since the product of two series with alternating coefficients is again a series with alternating coefficients, gives

(c')
$$(y - b_2 y^2 + \dots) - a_2 (y^2 - b_3^{(2)} y^3 + \dots) + a_3 (y^3 - b_4^{(3)} y^4 + \dots) - \dots .$$

And from this we immediately infer that on equating to zero the coefficients of $y^2, y^3, \dots$, we must obtain the identical equations (d), thus deducing for b_ν precisely the same values as before.

The exact analogue holds good when the two power series contain, from the first, only *odd* powers of x. Thus, if the inverse series of

$$y = x + a_3 x^3 + a_5 x^5 + \dots$$

is
$$x = y + b_3 y^3 + b_5 y^5 + \dots,$$

then the inverse series of

$$y = x - a_3 x^3 + a_5 x^5 - + \dots$$

is necessarily
$$x = y - b_3 y^3 + b_5 y^5 - + \dots .$$

Exercises on Chapter V.

64. Determine the radius of convergence of the power series $\Sigma a_n x^n$, when a_n has, from some point onwards, the values given in Ex. 34 or 45.

65. Determine the radii of the power series

$$\Sigma \vartheta^{n^2} \cdot x^n. \quad 0 < \vartheta < 1; \qquad \Sigma \left(\frac{1 \cdot 2 \dots n}{3 \cdot 5 \dots (2n+1)}\right)^2 x^n;$$

$$\Sigma \frac{n!}{n^n} x^n; \quad \Sigma \frac{n!}{a^{n^2}} x^n, \quad a > 1; \quad \Sigma \frac{(n!)^3}{(3n)!} x^n.$$

66. Denoting by $\varkappa$ and μ the lower and upper limits of $\left|\frac{a_n}{a_{n+1}}\right|$, the radius r of the power series $\Sigma a_n x^n$ *invariably* satisfies the relation $\varkappa \leqq r \leqq \mu$. In particular: If $\lim \left|\frac{a_n}{a_{n+1}}\right|$ exists, it has for value the radius of $\Sigma a_n x^n$.

67. $\Sigma a_n x^n$ has radius r, $\Sigma a_n' x^n$ radius r'. What may be said of the radius of the power series

$$\Sigma (a_n \pm a_n') x^n, \qquad \Sigma a_n a_n' x^n, \qquad \Sigma \frac{a_n}{a_n'} x^n ?$$

67 a. What is the radius of $\Sigma a_n x^n$ if $0 < \overline{\lim} |a_n| < +\infty$?

68. The power series $\sum_{n=2}^{\infty} \frac{\varepsilon_n}{n \log n} x^n$, where ε_n has the same value as in Ex. 47, converges at both ends of the interval of convergence, but in either case only conditionally.

69. Prove, with reference to 97, example 3, that

$$\sum_{\nu=0}^{n} \binom{n}{\nu}^2 = (-1)^n \sum_{\nu=0}^{2n} (-1)^\nu \binom{2n}{\nu}^2 = \binom{2n}{n}.$$

70. As a complement to *Abel's* theorem **100**, it may be shewn that in every case in which $\Sigma a_n x^n$ has a radius $r \geqq 1$, we have

$$\underline{\lim}\, s_n \leqq \overline{\lim_{x \to 1-0}} \left(\sum_{n=0}^{\infty} a_n x^n \right) \leqq \overline{\lim}\, s_n$$

$(s_n = a_0 + a_1 + \ldots + a_n)$.

71. The converse of *Abel's* theorem **100**, not in general true, holds, however, if the coefficients a_n are $\geqq 0$; if therefore, in that case,

$$\lim_{x \to r-0} \Sigma a_n x^n$$

exists, then $\Sigma a_n r^n$ converges and its sum is equal to that limit.

72. Let $\sum_{n=1}^{\infty} a_n x^n = f(x)$ and $\sum_{n=1}^{\infty} b_n x^n = g(x)$,

both series converging for $|x| < \varrho$. We then have (for what values of x?)

$$\sum_{n=1}^{\infty} b_n f(x^n) = \sum_{n=1}^{\infty} a_n g(x^n).$$

(By specialising the coefficients many interesting identities may be obtained. Write e. g. $b_n \equiv 1, (-1)^{n-1}, \frac{1}{n}$, etc.)

73. What are the first terms of the series, obtained by division, for

$$\frac{1}{1 - \frac{x^2}{2!} + \frac{x^4}{4!} - + \ldots}, \qquad \frac{1}{1 + \frac{x}{2} + \frac{x^2}{3} + \ldots}?$$

(Further exercises on power series will be found in the following Chapter.)

Chapter VI.

The expansions of the so-called elementary functions.

The theorems of the two preceding sections (§§ 20, 21) afford us the means of mastering completely a large number of series. We proceed to explain this in the most important cases.

A certain — not very large — number of power series, or functions represented thereby, have a considerable bearing on the whole of Analysis and are therefore frequently referred to as the elementary functions. These will occupy us first of all.

§ 22. The rational functions.

From the geometric series

$$1 + x + x^2 + \ldots = \sum_{n=0}^{\infty} x^n = \frac{1}{1-x}, \qquad |x| < 1,$$

which forms the groundwork for many of the following special investigations, we deduce, by repeated differentiation, in accordance with **98**, 4:

$$\sum_{n=0}^{\infty} (n+1) x^n = \frac{1}{(1-x)^2}, \quad \sum_{n=0}^{\infty} \binom{n+2}{2} x^n = \frac{1}{(1-x)^3}, \ldots$$

and generally, for any positive p:

$$\sum_{n=0}^{\infty} \binom{n+p}{n} x^n = \frac{1}{(1-x)^{p+1}}, \qquad |x| < 1. \quad \textbf{108.}$$

If we multiply this equation once more, in accordance with **91**, by $\sum x^n = \frac{1}{1-x}$, we obtain, by **91** and **108**:

$$\sum_{n=0}^{\infty}\left[\binom{p}{p}+\binom{p+1}{p}+\cdots+\binom{p+n}{p}\right]x^n = \sum_{n=0}^{\infty}\binom{n+p+1}{p+1}x^n.$$

By comparing coefficients (in accordance with **97**), we deduce from this that

$$\binom{p}{p}+\binom{p+1}{p}+\cdots+\binom{p+n}{p}=\binom{n+p+1}{p+1},$$

which may of course be proved quite easily directly (by induction). If we do this, we may also deduce the equality **108** by repeated multiplication of $\sum x^n = \frac{1}{1-x}$ with itself, by **103**.

Since we have

$$\binom{n+p}{p}=\binom{n+p}{n}=(-1)^n\binom{-p-1}{n}$$

we obtain from **108**, if we there write $-x$ for x and $-k$ for $p+1$, the formula

109. $$(1+x)^k = \sum_{n=0}^{\infty}\binom{k}{n}x^n,$$

valid for $|x|<1$ and negative integral k. This formula is evidently an extension of the binomial theorem (**29**, 4) to *negative* integral exponents; for this theorem may for positive integral k (or for $k=0$), also be written in the form **109**, as the terms of the series for $n>k$ are in that case all $=0$.

Formulae such as those we have just deduced have — as we may observe immediately, and once for all — a two-fold meaning; if we read them from left to right, they give the expansion or representation of a function by a power series; if we read them from right to left, they give us a closed expression for the sum of an infinite series. According to circumstances, the one interpretation or the other may occupy the foremost place in our attention.

By means of these simple formulae we may often succeed in expanding, in a power series, an arbitrary given rational function

$$f(x)=\frac{a_0+a_1x+\cdots+a_mx^m}{b_0+b_1x+\cdots+b_kx^k},$$

namely whenever $f(x)$ may be split up into partial fractions, i. e. expressed as a sum of fractions of the form

$$\frac{A}{(x-a)^p}.$$

Every separate fraction of this kind, and therefore the given function also, can be expanded in a power series by **108**. And in fact this expansion can be carried out for the neighbourhood of every point x_0 distinct from a. We only have to write

$$\left(\frac{1}{x-a}\right)^p=\frac{1}{(x_0-a)^p}\cdot\left(\frac{1}{1-\left(\frac{x-x_0}{a-x_0}\right)}\right)^p$$

and then expand the last fraction by **108**. By this means we see, at the same time, that the expansion will converge for $|x - x_0| < |a - x_0|$ and only for these values of x.

This method, however, only assumes fundamental importance when we come to use complex numbers.

Examples. **110.**

1. $\sum_{n=0}^{\infty} \frac{1}{2^n} = 2$
2. $\sum_{n=0}^{\infty} \frac{n+1}{2^n} = 4$
3. $\sum_{n=0}^{\infty} \frac{(n+1)(n+2)}{2^{n+1}} = 8$
4. $\sum_{n=0}^{\infty} \binom{n+p}{p} \cdot \left(\frac{2}{3}\right)^n = 3^{p+1}$.

§ 23. The exponential function.

1. Besides the geometric series, the so-called *exponential series*

$$\sum_{n=0}^{\infty} \frac{x^n}{n!} \equiv 1 + x + \frac{x^2}{2!} + \frac{x^3}{3!} + \cdots + \frac{x^n}{n!} + \cdots$$

plays a specially fundamental part in the sequel. We proceed now to examine in more detail the function which it represents. This so-called ***exponential function*** we denote provisionally by $E(x)$. As the series converges *everywhere* by **92**, 2, $E(x)$ is certainly, by **98**, defined, continuous and differentiable any number of times, for every x. For its derived function, we at once find

$$E'(x) = E(x),$$

so that for all derived functions of higher order we must also have

$$E^{(\nu)}(x) = E(x).$$

We shall attempt to deduce all further properties *from the series itself*. We have already shown in **91**, 3 that if x_1 and x_2 are any two real numbers, we have in all cases

$$\text{(a)} \qquad \boldsymbol{E(x_1 + x_2) = E(x_1) \cdot E(x_2)}.$$

This fundamental formula is referred to briefly as the *addition theorem for the exponential function*[1]. It gives further

$$E(x_1 + x_2 + x_3) = E(x_1 + x_2) \cdot E(x_3) = E(x_1) \cdot E(x_2) \cdot E(x_3)$$

and by repetition of this process, we find that for any number of real numbers $x_1, x_2, \ldots, x_k$,

$$\text{(b)} \qquad E(x_1 + x_2 + \cdots + x_k) = E(x_1) \cdot E(x_2) \ldots E(x_k).$$

---

[1] Alternative proof. The *Taylor*'s series **99** for $E(x)$ is

$$E(x) = E(x_1) + \frac{E'(x_1)}{1!}(x - x_1) + \cdots,$$

valid for *all* values of x and x_1. If we observe that $E^{(\nu)}(x_1) = E(x_1)$, then it at once follows, replacing x by $x_1 + x_2$, that

$$E(x_1 + x_2) = E(x_1) \cdot \left[1 + \frac{x_2}{1!} + \frac{x_2^{\,2}}{2!} + \cdots\right] = E(x_1) \cdot E(x_2),$$

q. e. d.

If we here write $x_\nu = 1$ for each ν, we deduce in particular that

$$E(k) = [E(1)]^k$$

holds for every positive integer k. Since $E(0) = 1$, it also holds for $k = 0$. If we now write, in (b), $x_\nu = \frac{m}{k}$ for each ν, denoting by m a second integer $\geqq 0$, then it follows that

$$E\left(k \cdot \frac{m}{k}\right) = \left[E\left(\frac{m}{k}\right)\right]^k$$

or, — since $E(m) = [E(1)]^m$, — that

$$E\left(\frac{m}{k}\right) = [E(1)]^{\frac{m}{k}}.$$

If we write for brevity $E(1) = E$, we have thus shewn that the equation

(c) $$E(x) = E^x$$

holds for *every rational* $x \geqq 0$.

If ξ is any *positive irrational* number, then we can in any number of ways form a sequence (x_n), of positive rational terms, converging to ξ. For each n, we have, by the above,

$$E(x_n) = E^{x_n}.$$

When $n \to +\infty$, the left hand side, by **98**, 2, tends to $E(\xi)$, and the right hand side, by **42**, 1, to E^ξ, so that we obtain

$$E(\xi) = E^\xi.$$

Thus equation (c) is proved for *every real* $x \geqq 0$.

But, finally, (a) gives

$$E(-x) \cdot E(x) = E(x - x) = E(0) = 1,$$

whence we first conclude that $E(x) = 0$ cannot hold [2] for any real x and that for $x \geqq 0$

$$E(-x) = \frac{1}{E(x)} = \frac{1}{E^x} = E^{-x}.$$

But this implies that equation (c) is also valid for every negative real x.

We have thus proved that the equation holds for *every real* x; and at the same time the function $E(x)$ has justified its designation of exponential function; $E(x)$ is the xth power of a fixed base, namely of

$$E = E(1) = 1 + \frac{1}{1!} + \frac{1}{2!} + \frac{1}{3!} + \cdots + \frac{1}{n!} + \cdots$$

---

[2] This may of course, for $x \geqq 0$, be deduced immediately from the series, by inspection, since this is a series of positive terms whose term of rank 0 is $= 1$.

2. It will next be required to obtain some further information about this base. We shall show that it is identical with the number e already met in **46 a**, so that[3]

$$\lim\left(1+\frac{1}{n}\right)^n=\sum_{\nu=0}^{\infty}\frac{1}{\nu!}.$$

The proof may be made somewhat more comprehensive, by at once establishing the following theorem, and thus completing the investigation of **46, a**:

○ Theorem. *For every real x,* **111.**

$\lim\limits_{n\to\infty}\left(1+\frac{x}{n}\right)^n$ *exists and is equal to the sum*[4] *of the series* $\sum\limits_{\nu=0}^{\infty}\frac{x^\nu}{\nu!}$.

Proof. We write for brevity

$$\left(1+\frac{x}{n}\right)^n=x_n \quad\text{and}\quad \sum_{\nu=0}^{\infty}\frac{x^\nu}{\nu!}=s(x)=s.$$

It then suffices to prove that $(s-x_n)\to 0$. Now if, — given, first, a definite value for x, — ε is chosen >0, we can assume p so large that the remainder

$$\frac{|x|^{p+1}}{(p+1)!}+\frac{|x|^{p+2}}{(p+2)!}+\cdots<\frac{\varepsilon}{2}.$$

Further, for $n>2$,

$$x_n=1+\binom{n}{1}\frac{x}{n}+\cdots+\binom{n}{k}\frac{x^k}{n^k}+\cdots+\binom{n}{n}\frac{x^n}{n^n}$$
$$=1+x+\frac{1}{2!}\left(1-\frac{1}{n}\right)x^2+\frac{1}{3!}\left(1-\frac{1}{n}\right)\left(1-\frac{2}{n}\right)x^3+\cdots$$
$$\cdots+\frac{1}{k!}\left[\left(1-\frac{1}{n}\right)\left(1-\frac{2}{n}\right)\cdots\left(1-\frac{k-1}{n}\right)\right]x^k+\cdots,$$

a series which terminates of itself at the nth term. The term in x^k, $k=0,1,\ldots,$ evidently has a coefficient $\geqq 0$, but not greater than the coefficient $1/k!$ of the corresponding term of the exponential series. The same is also true, therefore, of the difference of the former and the latter term. Accordingly we have, for $n>p$ — from the manner in which p was chosen[5] —

$$|s-x_n|\leqq\frac{1}{2!}\left[1-\left(1-\frac{1}{n}\right)\right]\cdot|x|^2+\cdots$$
$$\cdots+\frac{1}{p!}\left[1-\left(1-\frac{1}{n}\right)\cdots\left(1-\frac{p-1}{n}\right)\right]\cdot|x|^p+\frac{\varepsilon}{2}.$$

Every individual term of the $(p-1)$ first terms on the right hand side

---

[3] We have here, therefore, a significant example of problem B. Cf. introduction to § 9.

[4] First proved — if not in an entirely irreproachable manner — by *Euler*, Introductio in analysin infinitorum, Lausanne 1748, p. 86. — The exponential series and its sum e^x were already known to *Newton* (1669) and *Leibniz* (1676).

[5] We assume $p>2$ from the first.

is now obviously the nth term of a null sequence[6]; hence their sum — for p is a fixed number — also tends to 0, and we may choose $n_0 > p$ so large that this sum remains $< \frac{\varepsilon}{2}$ for every $n > n_0$. But we then have, for every $n > n_0$,

$$|s - x_n| < \varepsilon,$$

which proves our statement[7]. — For $x = 1$, we deduce in particular

$$E = \sum_{\nu=0}^{\infty} \frac{1}{\nu!} = \lim_{\nu \to \infty} \left(1 + \frac{1}{n}\right)^n = e;$$

and more generally, for *every real* x,

$$E(x) = \sum_{\nu=0}^{\infty} \frac{x^\nu}{\nu!} = e^x.$$

The new representation thus obtained for the number e, by the exponential series, is a very much more convenient one for the further discussion of this number. In the first place, we can, by this means, easily obtain a good approximation to e. For, since all the terms of the series are positive, we evidently have, for every n,

$$s_n < e < s_n + \frac{1}{(n+1)!} + \frac{1}{(n+1)!\,(n+1)} + \frac{1}{(n+1)!\,(n+1)^2} + \cdots$$

or

$$s_n < e < s_n + \frac{1}{(n+1)!} \cdot \frac{n+1}{n}$$

i. e.

(a) $$s_n < e < s_n + \frac{1}{n!\,n},$$

---

[6] We have $\left(1 - \frac{1}{n}\right) \to 1$, $\left(1 - \frac{2}{n}\right) \to 1$, $\ldots$, $\left(1 - \frac{p-1}{n}\right) \to 1$, and so their product (by **41**, 10), also $\to 1$, or $\left[1 - \left(1 - \frac{1}{n}\right)\ldots\left(1 - \frac{p-1}{n}\right)\right] \to 0$; so, as x and p are fixed numbers, the product of this last expression by $\frac{1}{p!}|x|^p$ also $\to 0$; and similarly for the other terms. — We can also infer the result directly from **41**, 12.

[7] The artifice here adopted is not one imagined *ad hoc*, but one which is frequently used: The terms of a sequence are represented as a sum $x_n = x_0^{(n)} + x_1^{(n)} + \cdots + x_{k_n}^{(n)}$, where the terms summed not only depend individually on n, but also increase *in number* with n: $k_n \to \infty$. If we know how each individual term behaves for $n \to \infty$, as for instance, that $x_\nu^{(n)}$ for fixed ν tends to ξ_ν, then we may often attain our end by separating out a fixed number of terms, say $x_0^{(n)} + x_1^{(n)} + \cdots + x_p^{(n)}$ with fixed p; this tends, when $n \to \infty$, to $\xi_0 + \xi_1 + \cdots + \xi_p$, by **41**, 9. The remaining terms, $x_{p+1}^{(n)} + \cdots + x_{k_n}^{(n)}$ we then endeavour to estimate in the bulk directly, by finding bounds above and below for them, which often presents no difficulties, provided p was suitably chosen.

where s_n denotes a partial sum of the new series for e. If we calculate these simple values e. g. for $n = 9$ (v. p. 251) then we find

$$2{\cdot}718\,281 < e < 2{\cdot}718\,282,$$

which already gives us a good idea of the value[8] of the number e. — From the formula (a) we may, however, draw further important inferences. A number is not completely before us unless it is rational and is written in the form $\frac{p}{q}$. *Is e perhaps a rational number?* The inequalities (a) show quite easily that this is unfortunately *not* the case. For if we had $e = \frac{p}{q}$, then for $n = q$, formula (a) would give:

$$s_q < \frac{p}{q} < s_q + \frac{1}{q!\,q}$$

where $s_q = 2 + \frac{1}{2!} + \cdots + \frac{1}{q!}$. If we multiply this inequality by $q!$, then $q!\,s_q$ is an integer, which we will denote for the moment by g, and it follows that

$$g < p\cdot(q-1)! < g + \frac{1}{q} \leqq g + 1.$$

But this is impossible; for between the two *consecutive* integers g and $g+1$ there cannot be another integer $p\cdot(q-1)!$ distinct from either: *e is an irrational number.*

3. The above investigations give us all the information, with regard to the limit of $\left(1+\frac{x}{n}\right)^n$, which we, in the first instance, require; the two problems A and B (§ 9) are both satisfactorily solved. In spite of this, we propose, in view of the fundamental importance of these matters, to determine the same limit again and in a different way, — entirely independent of the preceding.

We use only the fact, previously established, that $\left(1+\frac{1}{n}\right)^n \to e$. This we will first extend by showing that

$$\left(1+\frac{1}{y_n}\right)^{y_n} \to e$$

also, when (y_n) *is any sequence of positive numbers tending to* $+\infty$. When $y_n =$ a positive integer, for every n, this is an immediate consequence of the previous result[9].

---

[8] The number e has been calculated to 346 places of decimals by *J. M. Boormann* (Math. magazine, Vol. 1, No. 12, p. 204, 1884).

[9] For if ε is given > 0, and n_0 is determined, by **46a**, so that $\left|\left(1+\frac{1}{n}\right)^n - e\right|$ remains $< \varepsilon$ for every $n > n_0$, then we shall also have $\left|\left(1+\frac{1}{y_n}\right)^{y_n} - e\right| < \varepsilon$ for every $n > n_1$, provided n_1 is so chosen that for every $n > n_1$ we have $y_n > n_0$.

If the numbers y_n are not integers, there will still be for each n one (and only one) integer k_n such that

$$k_n \leqq y_n < k_n + 1,$$

and the sequence of these integers k_n must evidently also tend to $+\infty$. Now, however, if $k_n \geqq 1$,

$$\left(1+\frac{1}{k_n+1}\right)^{k_n} < \left(1+\frac{1}{y_n}\right)^{y_n} < \left(1+\frac{1}{k_n}\right)^{k_n+1}.$$

And since the numbers k_n are integers, the sequence

$$\left(1+\frac{1}{k_n}\right)^{k_n+1} = \left(1+\frac{1}{k_n}\right)^{k_n} \cdot \left(1+\frac{1}{k_n}\right)$$

and the sequence

$$\left(1+\frac{1}{k_n+1}\right)^{k_n} = \left(1+\frac{1}{k_n+1}\right)^{k_n+1} \cdot \frac{1}{1+\frac{1}{k_n+1}}$$

both tend to e, by our first remark. Hence, by **41**, 8, we also have

$$\left(1+\frac{1}{y_n}\right)^{y_n} \to e.$$

We may next show that when $y_n' \to -\infty$, we also have

$$\left(1+\frac{1}{y_n'}\right)^{y_n'} \to e,$$

or, otherwise, that when $y_n \to +\infty$, we have

$$\left(1-\frac{1}{y_n}\right)^{-y_n} \to e.$$

All the numbers y_n' must, however, be assumed < -1, i. e. $y_n > 1$, so that the base of the power does not reduce to 0 or a negative value; this can always be brought about by "a finite number of alterations". Since

$$\left(1-\frac{1}{y_n}\right)^{-y_n} = \left(\frac{y_n}{y_n-1}\right)^{y_n} = \left(1+\frac{1}{y_n-1}\right)^{y_n-1} \cdot \left(1+\frac{1}{y_n-1}\right),$$

and since, with y_n, $y_n - 1$ also $\to +\infty$, the statement to be proved is an immediate consequence of the preceding one.

Writing $\frac{1}{y_n} = z_n$, we may couple the two results thus:

$$(1+z_n)^{\frac{1}{z_n}} \to e$$

provided (z_n) is any null sequence with only positive or only negative terms, — the terms in the latter case being all > -1. From this we finally obtain the theorem, including all the above results:

12. **Theorem:** *If (x_n) is an arbitrary null sequence whose terms are different from 0 and > -1 from the first* [10], *then* [11]

(a) $$\lim_{n\to\infty} (1+x_n)^{\frac{1}{x_n}} = e.$$

---

[10] The latter may always be effected by "a finite number of alterations" (cf. **38**, 6).

[11] *Cauchy*: Résumé des leçons sur le calcul infinit., Paris 1823, p. 81.

Proof. Since all the x_n's $\neq 0$, the sequence (x_n) may be divided into two sub-sequences, one with only positive and one with only negative terms. Since, for both sub-sequences, the limit in question, as we have proved, exists [12] and $= e$, it follows by **41**, 5 that the given sequence also converges, with limit e.

By **42**, 2, the result thus obtained may also be expressed in the form

(b) $$\frac{\log_e (1 + x_n)}{x_n} \to 1,$$

which will frequently be used.

By § 19, Def. 4, the result also signifies that, *invariably*:

$$\lim_{x \to 0} (1 + x)^{\frac{1}{x}} = e.$$

From these results, it again follows, — quite independently, as we announced, of our investigations of 1. and 2. —, that

$$\left(1 + \frac{x}{n}\right)^n \to e^x;$$

for $\left(\frac{x}{n}\right)$ is certainly a null sequence[13], so that we have, by the preceding theorem,

$$\left(1 + \frac{x}{n}\right)^{\frac{n}{x}} \to e \quad \text{and therefore} \quad \left(1 + \frac{x}{n}\right)^n \to e^x, \text{ —}$$

which was what we required[14].

4. If $a > 0$, and x is an arbitrary real number, then, denoting by log the *natural* logarithm (v. p. 211),

$$a^x = e^{x \log a} = 1 + \frac{\log a}{1!} x + \frac{(\log a)^2}{2!} x^2 + \frac{(\log a)^3}{3!} x^3 + \ldots$$

is an expansion in power series of an arbitrary power. We deduce the limiting relation [15]

$$\frac{a^x - 1}{x} \to \log a \quad \text{for} \quad x \to 0, \qquad a > 0. \qquad \textbf{113.}$$

---

[12] If one of the two sub-series breaks off after a finite number of terms, then we can, by a finite number of alterations, leave it out of account.

[13] We consider this null sequence for $n > |x|$ only, so that we may always have $\frac{x}{n} > -1$.

[14] Combining this with the result deduced in 2., that the above limit has the same value as the sum of the exponential series, we have a second proof of the fact that the sum of the exponential series is $= e^x$.

[15] Direct proof: If the x_n's form a null sequence, then by **35**, **3**, so do the numbers $y_n = a^{x_n} - 1$; and consequently, by **112** (b),

$$\frac{a^{x_n} - 1}{x_n} = \frac{y_n \cdot \log a}{\log (1 + y_n)} \to \frac{\log a}{1} = \log a.$$

This formula provides us with a first means of calculating logarithms, which is already to a certain extent practicable. For it gives, e. g. (cf. § 9, p. 78)

$$\log a = \lim_{n\to\infty} n(\sqrt[n]{a} - 1)$$
$$= \lim_{k\to\infty} 2^k(\sqrt[2^k]{a} - 1).$$

As roots whose exponent is a power of 2 can be calculated directly by repeated taking of square roots, we have in this a means (though still a primitive one) for the evaluation of logarithms.

5. We have already noted that e^x is everywhere continuous and differentiable up to any order, — with $e^x = (e^x)' = (e^x)'' = \cdots$. It also shares with the general power a^x, of base $a > 1$, the property of being *everywhere positive* and *monotone increasing* with x.

More noteworthy than these are the properties expressed by a series of simple inequalities, of which we shall make use repeatedly in the sequel, and which are mostly obtained by comparison of the exponential with the geometric series. The proofs we will leave to the reader.

114. α) For every[16] x, $\quad e^x > 1 + x$,

β) for $x < 1$, $\quad e^x < \dfrac{1}{1-x}$,

γ) for $x > -1$, $\quad \dfrac{x}{1+x} < 1 - e^{-x} < x$,

δ) for $x < +1$, $\quad x < e^x - 1 < \dfrac{x}{1-x}$,

ε) for $x > -1$, $\quad 1 + x > e^{\frac{x}{1+x}}$,

ζ) for $x > 0$, $\quad e^x > \dfrac{x^p}{p!}$, $\quad (p = 0, 1, 2, \ldots)$,

η) for $x > 0$ and $y > 0$, $\quad e^x > \left(1 + \dfrac{x}{y}\right)^y > e^{\frac{xy}{x+y}}$,

ϑ) for every $x \neq 0$, $\quad |e^x - 1| < e^{|x|} - 1 < |x|\, e^{|x|}$.

§ 24. The trigonometrical functions.

We are now in a position to introduce the circular functions rigorously, i. e. employing purely arithmetical methods. For this purpose, we consider the series, everywhere convergent by **92**, 2:

$$C(x) = 1 - \frac{x^2}{2!} + \frac{x^4}{4!} - + \cdots + (-1)^k \frac{x^{2k}}{(2k)!} + \cdots$$

---

[16] Only for $x = 0$ do these and the following inequalities reduce to equalities. — The reader should illustrate the meaning of the inequalities on the relative curves.

and $$S(x) = x - \frac{x^3}{3!} + \frac{x^5}{5!} - + \cdots + (-1)^k \frac{x^{2k+1}}{(2k+1)!} + \cdots.$$
Each of these series represents a function everywhere continuous and differentiable any number of times in succession. The properties of these functions will be established, taking as starting point their expansions in series form, and it will be seen finally that they coincide with the functions $\cos x$ and $\sin x$ with which we are familiar from elementary studies.

1. We first find, by **98**, 3, that their derived functions have the following values:
$$C' = -S, \quad C'' = -C, \quad C''' = S, \quad C'''' = C;$$
$$S' = C, \quad S'' = -S, \quad S''' = -C, \quad S'''' = S;$$
— relations valid for every x (which symbol is for brevity omitted). Since, here, the 4th derived functions are seen to coincide with the original functions, the same series of values repeats itself, in the same order, from that point onwards in the succession of differentiations. Further, we see at once that $C(x)$ is an even, and $S(x)$ an odd, function:
$$C(-x) = C(x), \qquad S(-x) = -S(x).$$
These functions also, like the exponential function, satisfy simple ad dition theorems, by means of which they can then be further examined. They are most easily obtained by *Taylor*'s expansion (cf. p. 191, footnote 1). This gives, for any two values x_1 and x_2, — since the two series converge everywhere (absolutely), —
$$C(x_1 + x_2) = C(x_1) + \frac{C'(x_1)}{1!} x_2 + \frac{C''(x_1)}{2!} x_2^2 + \cdots,$$
and as this series converges absolutely, we may, by **89**, 4, rearrange it in any order we please, in particular we may group together all those terms for which the derived functions which they contain have the same value. This gives
$$C(x_1 + x_2) = C(x_1)\left[1 - \frac{x_2^2}{2!} + \frac{x_2^4}{4!} - + \cdots\right]$$
$$- S(x_1)\left[x_2 - \frac{x_2^3}{3!} + \frac{x_2^5}{5!} - + \cdots\right]$$
or

(a) $$C(x_1 + x_2) = C(x_1)\,C(x_2) - S(x_1)\,S(x_2);$$
and we find [17] quite similarly

(b) $$S(x_1 + x_2) = S(x_1)\,C(x_2) + C(x_1)\,S(x_2).$$

---

[17] Second proof. By multiplying out and rearranging in series form, we obtain from
$$C(x_1)\,C(x_2) - S(x_1)\,S(x_2)$$
the series $C(x_1 + x_2)$, — as in **91**, 3 for the exponential series.

Third proof. The derived function of $f(x) =$
$$[C(x_1 + x) - C(x_1)\,C(x) + S(x_1)\,S(x)]^2 + [S(x_1 + x) - S(x_1)\,C(x) - C(x_1)\,S(x)]^2$$
is, as may at once be seen, $\equiv 0$. Consequently (by § 19, theorem 7), $f(x) \equiv f(0) = 0$. Hence each of the square brackets must be separately $\equiv 0$, which at once gives both the addition theorems.

From these theorems, — whose form coincides with that of the addition theorems, with which we are already acquainted from an elementary standpoint, for the functions cos and sin, — it easily follows that our functions C and S also satisfy all the other so-called purely goniometrical formulae. We note, in particular:

From (a), writing $x_2 = -x_1$, we deduce that, for every x,

(c) $$C^2(x) + S^2(x) = 1;$$

from (a) and (b), replacing both x_1 and x_2 by x:

(d) $$\begin{aligned} C(2x) &= C^2(x) - S^2(x) \\ S(2x) &= 2\,C(x)\,S(x). \end{aligned}$$

2. It is a little more troublesome to infer the properties of periodicity directly from the series. This may be done as follows: We have

$$C(0) = 1 > 0.$$

On the other hand, $C(2) < 0$; for

$$C(2) = 1 - \frac{2^2}{2!} + \frac{2^4}{4!} - \left(\frac{2^6}{6!} - \frac{2^8}{8!}\right) - \left(\frac{2^{10}}{10!} - \frac{2^{12}}{12!}\right) - \cdots$$

where the expressions in brackets are all positive, — since for $n \geqq 2$,

$$\frac{2^n}{n!} - \frac{2^{n+2}}{(n+2)!} > 0 \quad —,$$

and therefore $C(2) < 1 - \frac{4}{2} + \frac{16}{24} = -\frac{1}{3}$, i. e. certainly negative. By § 19, Theorem 4, the function $C(x)$ therefore vanishes at least *once* between 0 and 2. Since further, as may be again easily verified,

$$S(x) = x\left(1 - \frac{x^2}{2\cdot 3}\right) + \frac{x^5}{5!}\left(1 - \frac{x^2}{6\cdot 7}\right) + \cdots$$

is positive for all values of x between 0 and 2, and therefore $C'(x) = -S(x)$ constantly negative there, — it follows that $C(x)$ is (strictly) monotone decreasing in this interval and can only vanish at *one single* point ξ in that interval. *The least positive zero of* $C(x)$, i. e. ξ, is accordingly a well-defined real number. We shall immediately see that it is equal to a quarter of the perimeter of a circle of radius 1 and we accordingly at once denote it[18] by $\frac{\pi}{2}$:

$$\xi = \frac{\pi}{2}, \qquad C\left(\frac{\pi}{2}\right) = 0.$$

From (c), it then follows that $S^2\left(\frac{\pi}{2}\right) = 1$, i. e. since $S(x)$ was seen to be positive between 0 and 2, that

$$S\left(\frac{\pi}{2}\right) = 1.$$

[18] The situation is thus that π is to stand for the moment as a mere abbreviation for 2ξ; only subsequently shall we show that this number π has the familiar meaning for the circle.

The formulae (d) show further that

$$C(\pi) = -1, \qquad S(\pi) = 0,$$

and by a second application, that

$$C(2\pi) = 1, \qquad S(2\pi) = 0.$$

It then finally follows from the addition theorems that, for every x,

$$\text{(e)} \quad \begin{cases} C\left(x + \frac{\pi}{2}\right) = -S(x), & S\left(x + \frac{\pi}{2}\right) = C(x), \\ C(x + \pi) = -C(x), & S(x + \pi) = -S(x), \\ C(\pi - x) = -C(x), & S(\pi - x) = S(x), \\ \boldsymbol{C(x + 2\pi) = C(x)}, & \boldsymbol{S(x + 2\pi) = S(x)}. \end{cases}$$

Our two functions thus possess[19] the period 2π.

3. It therefore only remains to show that the number π, introduced by us in a purely arithmetical way, has the familiar geometrical significance for the circle. Thereby we shall have also established the complete identity of our functions $C(x)$ and $S(x)$ with the functions $\cos x$ and $\sin x$ respectively.

Let a point P (fig. 3) of the plane of a rectangular coordinate system OXY, be assumed to move in such a manner that, at the time t, its two coordinates are given by

$$x = C(t) \quad \text{and} \quad y = S(t);$$

then its distance $|OP| = \sqrt{x^2 + y^2}$ from the origin of coordinates is constantly $= 1$, by (c). The point P therefore moves along the perimeter of a circle of radius 1 and centre O. If, in particular, t increases from 0 to 2π, then the point P starts from the point A of the positive x-axis and describes the perimeter of the circle exactly once, in the mathematically positive (i. e. anticlockwise) sense. In fact, as t increases from 0 to π, $x = C(t)$ decreases, as is now evident, from $+1$ to -1, monotonely, and the abscissa of P thus assumes each of the values between $+1$ and -1, exactly once. At the same time, $S(t)$ remains constantly positive; this therefore implies that P describes the upper half of the circle from A to B steadily, and passes through

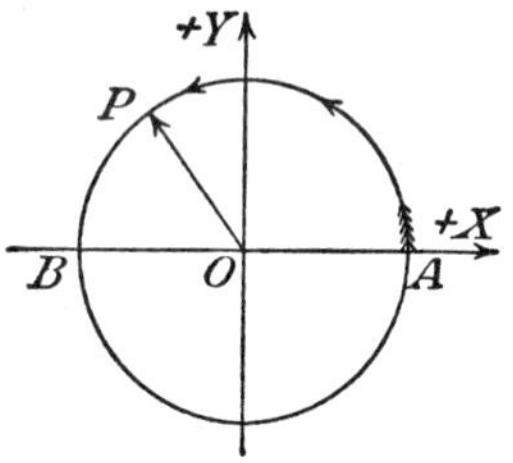

Fig. 3.

[19] 2π is also a so-called primitive period of our functions, i. e. a period, no (proper) fraction of which is itself a period. For the formulae (e) show that $\frac{2\pi}{2} = \pi$ is certainly not a period. And a fraction $\frac{2\pi}{m}$, with $m > 2$, cannot be a period, as then e. g. $S\left(\frac{2\pi}{m}\right) = S(0) = 0$, which is impossible since $S(x)$ was seen to be positive between 0 and 2 and in fact, as $S(\pi - x) = S(x)$, is positive between 0 and π. Similarly for $C(x)$, $\frac{2\pi}{m}$ $(m > 1)$ cannot be a period.

each of its points exactly once. The formulae (e) then show further that when t increases from π to 2π, the lower semi-circle is described in exactly the same way from B to A. These considerations provide us first with the

Theorem. *If x and y are any two real numbers for which $x^2+y^2=1$, then there exists one and only one number t between 0 (incl.) and 2π (excl.), for which, simultaneously,*

$$C(t)=x \quad \text{and} \quad S(t)=y.$$

If we next require the length of the path described by P when t has increased from 0 to a value t_0, the formula of § 19, Theorem 29 gives at once, for this, the value

$$\int_0^{t_0} \sqrt{C'^2+S'^2}\,dt=\int_0^{t_0} dt=t_0.$$

In particular, the complete perimeter of the circle is

$$=\int_0^{2\pi} \sqrt{C'^2+S'^2}\,dt=\int_0^{2\pi} dt=2\pi.$$

The connection which we had in view between our original considerations and the geometry of the circle, is thus completely established: $C(t)$, as abscissa of the point P for which the arc $\widehat{AP}=t$, coincides with the cosine of that arc, or of the corresponding angle at the centre, and $S(t)$, as ordinate of P, coincides with the sine of that angle. From now on we may therefore write $\cos t$ for $C(t)$ and $\sin t$ for $S(t)$. — Our mode of treatment differs from the elementary one chiefly in that the latter introduces the two functions from geometrical considerations, making use naively, as we might say, of measurements of length, angle, arc and area, and from this the expansion of the functions in power series is only reached as the ultimate result. We, on the contrary, started from these series, examined the functions defined by them, and finally established — using a concept of length elucidated by the integral calculus — the familiar interpretation in terms of the circle.

4. The functions $\cot x$ and $\tan x$ are defined as usual by the ratios

$$\cot x=\frac{\cos x}{\sin x}, \qquad \tan x=\frac{\sin x}{\cos x};$$

as functions, they therefore represent nothing essentially new.

The expansions in power series for these functions are however not so simple. A few of the coefficients of the expansions could of course easily be obtained by the process of division described in **105**, 4. But this gives us no insight into any relationships. We proceed as follows: In **105**, 5, we became acquainted with the expansion[20]

[20] The expression on the left hand side is defined in a neighbourhood of 0 *exclusive* of this point; the right hand side is also defined in such a neighbourhood, but *inclusive* of 0, *and moreover is continuous for* $x=0$. In such case we usually make no special mention of the fact that we define the left hand side for $x=0$ by the value of the right hand side at the point.

$$\frac{x}{e^x - 1} = \sum_{\nu=0}^{\infty} \frac{B_\nu x^\nu}{\nu!} \equiv 1 - \frac{x}{2} + \frac{B_2 x^2}{2!} + \frac{B_3 x^3}{3!} + \cdots$$

where the *Bernoulli*'s numbers B_ν are, it is true, not explicitly known, but still are easily obtainable by the very lucid recurrence formula **106**. These numbers we may, and accordingly will, in future, regard as entirely known [21]. We have therefore, — for every "sufficiently" small x (cf. **105**, 2, 4) —

$$\frac{x}{e^x - 1} + \frac{x}{2} = 1 + \frac{B_2 x^2}{2!} + \cdots.$$

The function on the left hand side is however equal to

$$= \frac{x}{2}\left(\frac{2}{e^x - 1} + 1\right) = \frac{x}{2}\frac{e^x + 1}{e^x - 1} = \frac{x}{2}\frac{e^{\frac{x}{2}} + e^{-\frac{x}{2}}}{e^{\frac{x}{2}} - e^{-\frac{x}{2}}}$$

and from this we see that it is an *even* function. *Bernoulli's* numbers B_3, B_5, B_7 are therefore, by **97**, 4, all $= 0$, as already seen in **106**, and we have, using the exponential series for $e^{\frac{x}{2}}$ and writing for brevity $\frac{x}{2} = z$:

$$z \cdot \frac{1 + \frac{z^2}{2!} + \frac{z^4}{4!} + \cdots}{z + \frac{z^3}{3!} + \frac{z^5}{5!} + \cdots} = 1 + \frac{B_2}{2!}(2z)^2 + \frac{B_4}{4!}(2z)^4 + \cdots$$

If on the left hand side, we had the signs $+$ and $-$ occurring alternately, both in the numerator and denominator, we should have precisely the function $z \cot z$. Dividing out on the left hand side by the factor z, so that only even powers of z occur, may we then deduce straight away that the relation

$$z \cot z = \frac{1 - \frac{z^2}{2!} + \frac{z^4}{4!} - + \cdots}{1 - \frac{z^2}{3!} + \frac{z^4}{5!} - + \cdots} = 1 - \frac{B_2}{2!}(2z)^2 + \frac{B_4}{4!}(2z)^4 - + \cdots$$

obtained from our equality by alternating the signs throughout, is also valid? Clearly we may. For if, to take the general case, we have for every sufficiently small z:

$$\frac{1 + a_2 z^2 + a_4 z^4 + \cdots}{1 + b_2 z^2 + b_4 z^4 + \cdots} = 1 + c_2 z^2 + c_4 z^4 + \cdots,$$

the same relation holds good when the $+$ signs throughout are replaced by alternate $+$ and $-$ signs. In *either* case, in fact, the coefficients $c_{2\nu}$ are obtained, according to **105**, 4, from the equations:

$$c_2 + b_2 = a_2; \quad c_4 + c_2 b_2 + b_4 = a_4; \quad c_6 + c_4 b_2 + c_2 b_4 + b_6 = a_6; \quad \ldots;$$
$$c_{2\nu} + c_{2\nu-2} b_2 + \cdots + c_2 b_{2\nu-2} + b_{2\nu} = a_{2\nu}; \quad \ldots$$

---

[21] As appears from the definition, they are certainly all rational.

We therefore, as presumed, — now writing x for z, — have the formula[22]:

115. $$x\cot x = 1 - \frac{2^2 B_2}{2!}x^2 + \frac{2^4 B_4}{4!}x^4 - + \cdots + (-1)^k \frac{2^{2k} B_{2k}}{(2k)!}x^{2k} + \cdots$$
$$= 1 - \frac{1}{3}x^2 - \frac{1}{45}x^4 - \frac{2}{945}x^6 - \frac{1}{4725}x^8 - \cdots$$

The expansion for $\tan x$ is now most simply obtained by means of the addition theorem

$$2\cot 2x = \frac{\cos^2 x - \sin^2 x}{\cos x \cdot \sin x} = \cot x - \tan x,$$

from which we deduce

$$\tan x = \cot x - 2\cot 2x$$

and therefore[23]

116. (a) $$\tan x = \sum_{k=1}^{\infty} (-1)^{k-1} \frac{2^{2k}(2^{2k}-1) B_{2k}}{(2k)!} x^{2k-1}$$
$$= x + \frac{1}{3}x^3 + \frac{2}{15}x^5 + \frac{17}{315}x^7 + \cdots.$$

From the two expansions, with the help of the formula

$$\cot x + \tan\frac{x}{2} = \frac{1}{\sin x}$$

we obtain further

(b) $$\frac{x}{\sin x} = \sum_{k=0}^{\infty} (-1)^{k-1} \frac{(2^{2k}-2) B_{2k}}{(2k)!} x^{2k}$$
$$= 1 + \frac{1}{6}x^2 + \frac{7}{360}x^4 + \frac{31}{15120}x^6 + \cdots$$ [23]

(An expansion for $1/\cos x$ will be found on p. 239.) — These expansions, at the present point, are still unsatisfactory, as their interval of validity cannot be assigned; we only know *that* the series have a positive radius of convergence, not, however, what its value is.

5. From another quite different starting point, *Euler* obtained an interesting expansion for the cotangent which we proceed to deduce, especially as it is of great importance for many problems in series[24]. At the same time, it will give us the radius of convergence of the series **115** and **116** (v. **241**).

---

[22] This and the following expansions are almost all due to *Euler* and are found in the 9th and 10th chapters of his *Introductio in analysin infinitorum*, Lausanne 1748.

[23] We shall afterwards see that B_{2k} has the sign $(-1)^{k-1}$ (v. **136**), so that the expansion of $x\cot x$, after the initial term 1, has only *negative* coefficients, those of $\tan x$ and $\frac{x}{\sin x}$ only *positive* coefficients.

[24] The following considerable simplification of *Euler*'s method for obtaining the expansion is due to *Schröter* (Ableitung der Partialbruch- und Produkt-entwicklungen für die trigonometrischen Funktionen. Zeitschrift für Math. u. Phys., Vol. 13, p. 254. 1868).

We have, as was just shewn,

$$\cot y = \frac{1}{2}\left(\cot\frac{y}{2} - \tan\frac{y}{2}\right)$$

or

$$(*) \qquad \cot\pi x = \frac{1}{2}\left\{\cot\frac{\pi x}{2} + \cot\frac{\pi(x\pm 1)}{2}\right\},$$

a formula in which we may, on the right, take either of the signs $\pm$. Let x be *an arbitrary real number distinct from* $0, \pm 1, \pm 2, \ldots,$ whose value will remain fixed in what follows. Then

$$\pi x\cot\pi x = \frac{\pi x}{2}\left\{\cot\frac{\pi x}{2} + \cot\frac{\pi(x+1)}{2}\right\}$$

and applying the formula (*) once more to both functions on the right hand side, taking for the first the $+$ and for the second, the $-$ sign, we obtain

$$\pi x\cot\pi x = \frac{\pi x}{4}\left\{\cot\frac{\pi x}{4} + \left[\cot\frac{\pi(x+1)}{4} + \cot\frac{\pi(x-1)}{4}\right] + \cot\frac{\pi(x+2)}{4}\right\}.$$

A third similar step gives, for $\pi x \cot \pi x$, the value

$$\frac{\pi x}{8}\left\{\cot\frac{\pi x}{8}\begin{matrix} +\cot\frac{\pi(x+1)}{8} + \cot\frac{\pi(x+2)}{8} + \cot\frac{\pi(x+3)}{8} \\ \bullet \\ +\cot\frac{\pi(x-1)}{8} + \cot\frac{\pi(x-2)}{8} + \cot\frac{\pi(x-3)}{8}\end{matrix} + \cot\frac{\pi(x+4)}{8}\right\},$$

since here each pair of terms which occupy symmetrical positions relatively to the centre (●) of the aggregate in the curly brackets give, except for a factor $\frac{1}{2}$, a term of the preceding aggregate, in accordance with the formula (*). If we proceed thus through n stages, we obtain for $n > 1$

$$(\dagger) \quad \pi x\cot\pi x = \frac{\pi x}{2^n}\left\{\cot\frac{\pi x}{2^n} + \sum_{\nu=1}^{2^{n-1}-1}\left[\cot\frac{\pi(x+\nu)}{2^n} + \cot\frac{\pi(x-\nu)}{2^n}\right] - \tan\frac{\pi x}{2^n}\right\}$$

Now by **115**,

$$\lim_{z\to 0} z\cot z = 1$$

and hence for each $\alpha \neq 0$

$$\lim_{n\to\infty}\frac{1}{2^n}\cot\frac{\alpha}{2^n} = \frac{1}{\alpha};$$

if in the above expression we let $n \to \infty$ and, *at first tentatively*, carry out the limiting process for each term separately, we obtain the expansion

$$\pi x\cot\pi x = 1 + x\sum_{\nu=1}^{\infty}\left(\frac{1}{x+\nu} + \frac{1}{x-\nu}\right) - 0 = 1 + 2x^2\sum_{\nu=1}^{\infty}\frac{1}{x^2-\nu^2}.$$

We proceed to show that this in general *faulty mode of passage to the limit* has, however, led in this case to a right result.

We first note that the series converges absolutely for every $x \neq \pm 1, \pm 2, \ldots,$ by **70**, 4, since the absolute values of its terms are asymptotically equal to those of the series $\sum \frac{1}{\nu^2}$. Now choose an arbitrary integer $k > 6\,|x|$, to be kept provisionally fixed. If n is so large that the number $2^{n-1} - 1$, which we will denote for short by m, is $> k$, we then split up the expression (†) for $\pi x \cot x$, as follows[25]:

$$\pi x \cot \pi x = \frac{\pi x}{2^n}\left\{\cot \frac{\pi x}{2^n} - \tan \frac{\pi x}{2^n} + \sum_{\nu=1}^{k}[\cdots]\right\} + \frac{\pi x}{2^n}\left\{\sum_{\nu=k+1}^{m}[\cdots]\right\}.$$

(In the square brackets we have of course to insert the same expression as occurs in (†).) The two parts of this expression we denote by A_n and B_n. Since A_n consists of a finite number of terms, the passage to the limit term by term is certainly allowed there, by **41**, 9, and we have

$$\lim_{n\to\infty} A_n = 1 + 2x^2 \sum_{\nu=1}^{k} \frac{1}{x^2 - \nu^2}.$$

Also B_n is precisely $\pi x \cot \pi x - A_n$, hence $\lim B_n$ certainly exists. Let r_k denote its value, depending as it does upon the chosen value k; thus

$$\lim_{n\to\infty} B_n = r_k = \pi x \cot \pi x - \left[1 + 2x^2 \sum_{\nu=1}^{k} \frac{1}{x^2 - \nu^2}\right].$$

Bounds above for the numbers B_n, for their limit r_k and so finally for the difference on the right hand side, may now quite easily be estimated:

We have

$$\cot(a+b) + \cot(a-b) = \frac{-2\cot a}{\dfrac{\sin^2 b}{\sin^2 a} - 1}$$

and hence

$$\cot \frac{\pi(x+\nu)}{2^n} + \cot \frac{\pi(x-\nu)}{2^n} = \frac{-2\cot\alpha}{\dfrac{\sin^2\beta}{\sin^2\alpha} - 1},$$

writing for the moment $\frac{\pi x}{2^n} = \alpha$ and $\frac{\pi\nu}{2^n} = \beta$, for short.

As $2^n > k > 6\,|x|$, we certainly have $|\alpha| = \left|\frac{\pi x}{2^n}\right| < 1$ and so[26]

$$|\sin\alpha| = \left|\alpha - \frac{\alpha^3}{3!} + \cdots\right| \leq |\alpha|\left(1 + \frac{1}{3!} + \frac{1}{5!} + \cdots\right) < 2\,|\alpha|.$$

Since, further, $0 < \beta < \frac{\pi}{2} < 2$, we have[27]

$$\sin\beta = \beta\left(1 - \frac{\beta^2}{2\cdot 3}\right) + \frac{\beta^5}{5!}\left(1 - \frac{\beta^2}{6\cdot 7}\right) + \cdots > \frac{\beta}{3}.$$

---

[25] Cf. Footnote 7, p. 194.

[26] For the sake of later applications we make these estimates in the above rough form.

[27] Cf. p. 200.

Hence
$$\left|\frac{\sin\beta}{\sin\alpha}\right| > \frac{\beta}{6\,|\alpha|} = \frac{\nu}{6\,|x|} > 1,$$
the latter, because $\nu > k > 6\,|x|$. It therefore follows that (for $\nu > k$)
$$\left|\cot\frac{\pi(x+\nu)}{2^n} + \cot\frac{\pi(x-\nu)}{2^n}\right| \leqq \frac{2\left|\cot\frac{\pi x}{2^n}\right|}{\frac{\nu^2}{36\,x^2}-1}$$
and hence
$$|B_n| \leqq \left|\frac{\pi x}{2^n}\cot\frac{\pi x}{2^n}\right| \cdot \sum_{\nu=k+1}^{m}\frac{72\,x^2}{\nu^2-36\,x^2}.$$
The factor outside the sign of summation is — quite roughly estimated — certainly < 3; for $\left|\frac{\pi x}{2^n}\right|$ was < 1, and for $|z| < 1$ we have [26]
$$|z\cot z| = \left|\frac{1-\frac{z^2}{2!}+\frac{z^4}{4!}-+\cdots}{1-\frac{z^2}{3!}+\frac{z^4}{5!}-+\cdots}\right| < \frac{1+\frac{1}{2!}+\frac{1}{4!}+\cdots}{1-\frac{1}{3!}-\frac{1}{5!}-\cdots} < 3.$$
Accordingly,
$$|B_n| < 216\,x^2 \cdot \sum_{\nu=k+1}^{m}\frac{1}{\nu^2-36\,x^2} < 216\,x^2 \cdot \sum_{\nu=k+1}^{\infty}\frac{1}{\nu^2-36\,x^2}.$$
But this is a number quite independent of n, so that we may also write
$$|\lim B_n| = |r_k| \leqq 216\,x^2 \cdot \sum_{\nu=k+1}^{\infty}\frac{1}{\nu^2-36\,x^2}.$$
But the bound above which we have thus obtained for r_k is equal to the remainder, after the kth term, of a convergent series [28]. Hence $r_k \to 0$ as $k \to +\infty$. If we refer back to the meaning of r_k, we see that this implies
$$\lim_{k\to\infty}\left\{\pi x\cot\pi x - \left[1 + 2\,x^2\sum_{n=1}^{k}\frac{1}{x^2-\nu^2}\right]\right\} = 0,$$
or, as asserted,
$$\boldsymbol{\pi x\cot\pi x = 1 + 2\,x^2\sum_{\nu=1}^{\infty}\frac{1}{x^2-\nu^2},}$$
— a formula which is thus proved valid for every $x \neq 0, \pm 1, \pm 2, \ldots$. **117.**

6. We shall in the next chapter but one make important applications (p. 236 seqq.) of this most remarkable *expansion in partial fractions*, as it is called, *of the function* cot. We can of course easily deduce many further such expressions from it; we make note of the following:

[28] The convergence is obtained just as simply as, previously, that of the series $\Sigma\frac{1}{x^2-\nu^2}$.

The formula

$$\pi \cot \frac{\pi x}{2} - 2\pi \cot \pi x = \pi \tan \frac{\pi x}{2}$$

first of all gives[29]

$$\pi \tan \frac{\pi x}{2} = \sum_{\nu=0}^{\infty} \frac{4x}{(2\nu+1)^2 - x^2}, \qquad x \neq \pm 1, \pm 3, \pm 5, \ldots$$

$$= 2 \sum_{\nu=0}^{\infty} \left(\frac{1}{(2\nu+1) - x} - \frac{1}{(2\nu+1)+x} \right).$$

The formula

$$\cot z + \tan \frac{z}{2} = \frac{1}{\sin z}$$

then gives further, for $x \neq 0, \pm 1, \pm 2, \ldots$

$$\frac{\pi}{\sin \pi x} = \frac{1}{x} + \frac{2x}{1^2 - x^2} - \frac{2x}{2^2 - x^2} + - \cdots$$

$$= \frac{1}{x} + \left(\frac{1}{1-x} - \frac{1}{1+x} \right) - \left(\frac{1}{2-x} - \frac{1}{2+x} \right) + - \cdots.$$

Finally if we here replace x by $\frac{1}{2} - x$, we deduce

$$\frac{\pi}{\cos \pi x} = \frac{2}{1-2x} + \left(\frac{2}{1+2x} - \frac{2}{3-2x} \right) - \left(\frac{2}{3+2x} - \frac{2}{5-2x} \right) + - \cdots.$$

By **83**, 2, Supplementary theorem, the brackets may here be omitted. But if we then take the terms together again in pairs, starting from the beginning, we obtain, — provided $x \neq \pm \frac{1}{2}, \pm \frac{3}{2}, \pm \frac{5}{2}, \ldots,$

118. $$\frac{\pi}{\cos \pi x} = \frac{1}{(\frac{1}{2})^2 - x^2} - \frac{3}{(\frac{3}{2})^2 - x^2} + \frac{5}{(\frac{5}{2})^2 - x^2} - + \cdots$$

With these *expansions in partial fractions* for the functions cot, tan, $\frac{1}{\sin}$ and $\frac{1}{\cos}$, we will terminate our discussion of the trigonometrical functions.

§ 25. The binomial series.

We have already, in § 22, seen that the binomial theorem for positive integral exponents, if written in the form

$$(1+x)^k = \sum_{n=0}^{\infty} \binom{k}{n} x^n,$$

remains unaltered in the case[30] of a negative integral k. But we have then to stipulate $|x| < 1$. We will now show that with this restriction

[29] The formula first follows only for $x \neq 0, \pm 1, \pm 2, \ldots$ but can then be verified without any difficulty for $x = 0, \pm 2, \pm 4, \ldots$. (The series has the sum 0, as is most easily seen from the second expression, for an even integral x.)

[30] In the former case the series is infinite only in form, in the latter it is actually so.

the theorem holds[31] even for any real exponent α, i. e.

$$(1+x)^\alpha = \sum_{n=0}^{\infty} \binom{\alpha}{n} x^n \qquad \begin{cases} |x| < 1 \\ \alpha \text{ any real number.} \end{cases}$$ **119.**

As in the preceding cases, we will start from the series and shew that it represents the function in question.

The convergence of the series for $|x| < 1$ may be at once established; for the absolute value of the ratio of the $(n+1)^{\text{th}}$ to the nth term is

$$= \left| \frac{\alpha - n}{n-1} x \right| \quad \text{and therefore} \quad \to |x|,$$

which by **76, 2** proves that the exact radius of convergence of the binomial series is 1. It is not quite so easy to see that its sum is equal to the — of course positive — value of $(1+x)^\alpha$. If we denote provisionally by $f_\alpha(x)$ the function represented by the series for $|x| < 1$, the proof may be carried out as follows.

Since $\sum \binom{\alpha}{n} x^n$ converges absolutely for $|x| < 1$, whatever may be the value of α, it follows, by **91**, Rem. 1, that for any α and β, and every $|x| < 1$, we have

$$\sum_{n=0}^{\infty} \binom{\alpha}{n} x^n \cdot \sum_{n=0}^{\infty} \binom{\beta}{n} x^n = \sum_{n=0}^{\infty} \left[\binom{\alpha}{0}\binom{\beta}{n} + \binom{\alpha}{1}\binom{\beta}{n-1} + \cdots + \binom{\alpha}{n}\binom{\beta}{0} \right] x^n.$$

Now
$$\binom{\alpha}{0}\binom{\beta}{n} + \binom{\alpha}{1}\binom{\beta}{n-1} + \cdots + \binom{\alpha}{n}\binom{\beta}{0} = \binom{\alpha+\beta}{n}$$

as may quite easily be verified e. g. by induction[32]. Hence — for

---

[31] The symbol $\binom{\alpha}{n}$ is defined for an arbitrary real α and integral $n \geqq 0$ by the two conventions

$$\binom{\alpha}{0} = 1, \qquad \binom{\alpha}{n} = \frac{\alpha(\alpha-1)\dots(\alpha-n+1)}{1 \cdot 2 \cdot \;\dots\; n} \qquad \text{for } n \geqq 1$$

and for every real α and every $n \geqq 1$, it satisfies the relation, which may at once be verified by calculation:

$$\binom{\alpha-1}{n-1} + \binom{\alpha-1}{n} = \binom{\alpha}{n}.$$

[32] For this, give the statement, by multiplying by $n!$, the form

$$\binom{n}{0} \cdot \overline{\beta(\beta-1)\dots(\beta-n+1)} + \cdots$$
$$+ \binom{n}{k} \overline{\alpha(\alpha-1)\dots(\alpha-k+1)} \cdot \overline{\beta(\beta-1)\dots(\beta-n+k+1)} + \cdots$$
$$+ \binom{n}{n} \overline{\alpha(\alpha-1)\dots(\alpha-n+1)} = (\alpha+\beta)(\alpha+\beta-1)\dots(\alpha+\beta-n+1).$$

Then multiply each of the $(n+1)$ terms on the left hand side first by the corresponding term of

$$\alpha,\ (\alpha-1),\ \dots,\ (\alpha-k),\ \dots,\ (\alpha-n),$$

then by the corresponding term of

$$(\beta-n),\ (\beta-n+1),\ \dots,\ (\beta-n+k),\ \dots,\ \beta$$

and add, so that in all we multiply by $(\alpha+\beta-n)$; grouping together the similar terms on the left, we obtain precisely the asserted equality, where n is replaced by $n+1$. — The above formula is usually called the *addition theorem for the binomial coefficients.*

fixed $|x| < 1$, — we have, for any α and β,

$$f_\alpha \cdot f_\beta = f_{\alpha+\beta}.$$

By precisely the same method as we used to deduce from the addition theorem of the exponential function, — $E(x_1)\cdot E(x_2) = E(x_1 + x_2)$, — that for every real x we had $(E(x) = (E(1))^x$, — so we could here conclude that for every α,

$$f_\alpha = (f_1)^\alpha,$$

if we knew here also that f_α was for every real α (with fixed x) a continuous function of α. As $f_1 = 1 + x$, the equality

$$f_\alpha = (1 + x)^\alpha$$

would then be established generally for the stated values of x.

The proof of the continuity results quite simply from the main rearrangement theorem **90**: If we write the series for f_α in the more explicit form

$$\text{(a)} \qquad f_\alpha = 1 + \alpha x + \left(\frac{\alpha^2}{2} - \frac{\alpha}{2}\right) x^2 + \left(\frac{\alpha^3}{6} - \frac{\alpha^2}{2} + \frac{\alpha}{3}\right) x^3 + \cdots$$

and then replace each term by its absolute value, we obtain the series

$$\sum_{n=0}^{\infty} \frac{|\alpha|(|\alpha|+1)\cdots(|\alpha|+n-1)}{1\cdot\ 2\cdot\ \ \cdots\ \ n} |x|^n = \sum_{n=0}^{\infty} \binom{|\alpha|+n-1}{n} |x|^n,$$

also convergent for $|x| < 1$ by the ratio test. We may accordingly rearrange the above series (a) in powers of α, obtaining

$$\text{(b)} \quad f_\alpha = 1 + \left(x - \frac{x^2}{2} + \frac{x^3}{3} - + \cdots + \frac{(-1)^{n-1}}{n} x^n + \cdots\right)\alpha + \cdots,$$

i. e. certainly a *power series in* α. Since this — still for fixed x in $|x| < 1$ — converges, by the manner in which it was obtained, for *every* α, we have an everywhere convergent power series in α, hence certainly a *continuous* function of α.

This completes the proof [33] of the validity of the expansion **119** and at the same time fills the gap left in the proof of the reversion theorem § 21.

---

[33] An alternative proof, perhaps still easier than the above, but using the differential calculus, is as follows: From $f_\alpha(x) = \sum_{n=0}^{\infty} \binom{\alpha}{n} x^n$ it follows that

$$f_\alpha'(x) = \sum_{n=1}^{\infty} n \binom{\alpha}{n} x^{n-1} = \sum_{n=0}^{\infty} (n+1)\binom{\alpha}{n+1} x^n.$$

Since however $(n+1)\binom{\alpha}{n+1} = \alpha\binom{\alpha-1}{n}$, it follows further that

$$f_\alpha'(x) = \alpha \cdot f_{\alpha-1}(x).$$

But

$$(1+x) f_{\alpha-1}(x) = (1+x)\cdot \sum_{n=0}^{\infty} \binom{\alpha-1}{n} x^n$$

$$= 1 + \sum_{n=1}^{\infty} \left[\binom{\alpha-1}{n} + \binom{\alpha-1}{n-1}\right] x^n = \sum_{n=0}^{\infty} \binom{\alpha}{n} x^n;$$

The binomial series provides, like the exponential series, an expansion of the general power a^ϱ: Choose a (positive) number c for which, on the one hand, c^ϱ may be regarded as known, and on the other, $0 < \frac{a}{c} < 2$. Then we may write $\frac{a}{c} = 1 + x$ with $|x| < 1$ and so obtain, as the required expansion,

$$a^\varrho = c^\varrho \cdot (1+x)^\varrho = c^\varrho \left[1 + \binom{\varrho}{1} x + \binom{\varrho}{2} x^2 + \dots\right].$$

Thus e. g.

$$\sqrt{2} = 2^{\frac{1}{2}} = \left(\frac{49}{25}\right)^{\frac{1}{2}} \cdot \left(\frac{50}{49}\right)^{\frac{1}{2}} = \frac{7}{5} \cdot \left(\frac{49}{50}\right)^{-\frac{1}{2}} = \frac{7}{5} \cdot \left(1 - \frac{1}{50}\right)^{-\frac{1}{2}}$$

$$= \frac{7}{5}\left[1 - \binom{-\frac{1}{2}}{1}\frac{1}{50} + \binom{-\frac{1}{2}}{2}\frac{1}{50^2} - \binom{-\frac{1}{2}}{3}\frac{1}{50^3} + - \dots\right]$$

is a convenient expansion of $\sqrt{2}$.

The discovery of the binomial series by *Newton*[34] forms one of the landmarks in the development of mathematical science. Later *Abel*[35] made this series the subject of researches which represent a perhaps equally important landmark in the development of the theory of series (cf. below **170**, 1 and **247**).

§ 26. The logarithmic series.

As already observed on pp. 58 and 83, in theoretical investigations it is convenient to employ exclusively the so-called natural logarithms, that is to say, those with the base e. *In the sequel,* $\log x$ *shall therefore always stand for* $\log_e x$ $(x > 0)$.

If $y = \log x$, then $x = e^y$ or

$$x - 1 = y + \frac{y^2}{2!} + \frac{y^3}{3!} + \dots.$$

By the theorem for the reversion of power series (**107**), $y = \log x$ is there-

---

thus, for every $|x| < 1$, we have the equation

$$(1+x) f_\alpha'(x) - \alpha f_\alpha(x) = 0.$$

Since $(1+x)^\alpha > 0$, this shows that the quotient

$$\frac{f_\alpha(x)}{(1+x)^\alpha}$$

has everywhere the differential coefficient 0, i. e. is identically equal to one and the same constant. For $x = 0$ the value is at once calculated and $= +1$; thus the assertion $f_\alpha(x) = (1+x)^\alpha$ is proved afresh.

[34] Letter to *Oldenburg*, 13 June 1676. — *Newton* at that time possessed no proof of the formula; the first proof was found in 1774 by *Euler*.

[35] J. f. d. reine u. angew. Math., Vol. 1, p. 311, 1826.

fore expansible in powers of $(x - 1)$ for all values of x sufficiently near to $+1$, or $y = \log(1 + x)$ in powers of x, for every sufficiently small $|x|$:

$$y = \log(1 + x) = x + b_2 x^2 + b_3 x^3 + \dots .$$

The coefficients b_n may actually be evaluated by the process indicated, provided the working is skilfully set out [36]. But it is advisable to seek more convenient methods: For this purpose, the developments of the preceding section suffice. For $|x| < 1$ and arbitrary α, the function $f_\alpha = f_\alpha(x)$ there examined is

$$= (1 + x)^\alpha = e^{\alpha \log(1+x)}.$$

Using, for the left hand side, the expression (b) of the former paragraph and for the right hand side, the exponential series, we obtain the two power series everywhere convergent:

$$1 + \left[x - \frac{x^2}{2} + \frac{x^3}{3} - + \dots + \frac{(-1)^{n-1}}{n} x^n + \dots\right] \alpha + \dots$$
$$= 1 + [\log(1 + x)]\, \alpha + \dots .$$

By the identity theorem for power series **97**, the coefficients of corresponding powers of α must here coincide. Thus, in particular [37], — and for every $|x| < 1$ —

120. (a) $$\log(1 + x) = x - \frac{x^2}{2} + \frac{x^3}{3} - + \dots + \frac{(-1)^{n-1}}{n} x^n + \dots .$$

Thus we have obtained the desired expansion, which, we also see *a posteriori*, cannot hold for $|x| > 1$. If we replace in this *logarithmic series*, as it is called, x by $-x$ and change the signs on both sides of the equality, we obtain, — equally for every $|x| < 1$, —

(b) $$\log \frac{1}{1 - x} = x + \frac{x^2}{2} + \frac{x^3}{3} + \dots + \frac{x^n}{n} + \dots .$$

By addition we deduce, — again for every $|x| < 1$, —

(c) $$\log \frac{1 + x}{1 - x} = 2\left[x + \frac{x^3}{3} + \frac{x^5}{5} + \dots + \frac{x^{2k+1}}{2k+1} + \dots\right].$$

There are of course various other ways of obtaining these expansions; but they either do not follow so immediately from the definition of the log as inverse function of the exponential function, or make more extensive use of the differential and integral calculus [38].

---

[36] *Herm. Schmidt*, Jahresber. d. Deutsch. Math. Ver., Vol. 48, p. 56. 1938.

[37] Cf. the historical remarks in **69**, 8.

[38] We may indicate the following two ways:

1. We know from the reversion theorem that we may write

$$\log(1 + x) = x + b_2 x^2 + b_3 x^3 + \dots ;$$

it follows from *Taylor*'s series **99** that

$$b_k = \frac{1}{k!}\left(\frac{d^k \log(1 + x)}{d x^k}\right)_{x=0} = \frac{(-1)^{k-1}}{k}.$$

Our mode of obtaining the logarithmic series — also the two modes mentioned in the footnote — do not enable us to determine whether the representation remains valid for $x = +1$ or $x = -1$. Since however **120a** reduces, for $x = +1$, to the convergent series (v. **81c**, 3)

$$1 - \frac{1}{2} + \frac{1}{3} - \frac{1}{4} + - \cdots,$$

the value of this series, by *Abel*'s theorem of limits, is

$$= \lim_{x \to 1-0} \log(1+x) = \log 2.$$

Our representation (a) there remains valid for $x = +1$; but for $x = -1$ it certainly no longer holds, as the series is then divergent.

§ 27. The cyclometrical functions.

Since the trigonometrical functions sin and tan are expansible in power series in which the first power of the variable has the coefficient 1, different from 0, this is also true of their inverses, the so-called cyclometrical functions $\sin^{-1}$ and $\tan^{-1}$. We have therefore to write, for every sufficiently small $|x|$,

$$y = \sin^{-1} x = x + b_3 x^3 + b_5 x^5 + \cdots$$
$$y = \tan^{-1} x = x + b_3' x^3 + b_5' x^5 + \cdots$$

where we have left out the even powers at once, since our functions are odd. Here too it would be tedious to seek to evaluate the coefficients b and b' by the general process of **107**. We again choose more convenient methods: The series for $\tan^{-1} x$ is the inverse of

(a) $$x = \tan y = \frac{\sin y}{\cos y} = \frac{y - \frac{y^3}{3!} + \frac{y^5}{5!} - + \cdots}{1 - \frac{y^2}{2!} + \frac{y^4}{4!} - + \cdots},$$

or of the series obtained by **105**, 4 after carrying out the process of division in the last quotient. If here all the signs, in numerator and denominator, were $+$, then we should be concerned with reversing the function

$$x = \frac{e^y - e^{-y}}{e^y + e^{-y}} = \frac{e^{2y} - 1}{e^{2y} + 1}.$$

---

2. $\frac{d \log(1+x)}{dx} = \frac{1}{1+x} = 1 - x + x^2 - + \cdots + (-1)^k x^k + \cdots \equiv \sum_{n=0}^{\infty} (-1)^k x^k.$
Integrating, it follows at once, by **99**, theorem 5, — since $\log 1 = 0$, — that

$$\log(1+x) = \sum_{k=0}^{\infty} \frac{(-1)^k}{k+1} x^{k+1} = \sum_{k=1}^{\infty} \frac{(-1)^{k-1}}{k} x^k.$$

The method in the text is so far simpler that it proceeds entirely without the use of the differential and integral calculus.

But the inverse of *this* function is, as we immediately find,

$$\frac{1}{2}\log\frac{1+x}{1-x} = x + \frac{x^3}{3} + \frac{x^5}{5} + \cdots.$$

By the general remark at the end of § 21, the reverse series of the series for $x = \tan y$ actually before us is obtained from the series last written down by alternating the signs [39] again, i. e.

121. $$\tan^{-1} x = x - \frac{x^3}{3} + \frac{x^5}{5} - + \cdots.$$

If therefore this power series, which obviously has the radius of convergence 1, is substituted for y in the quotient on the right of (a), and this is then rearranged, as is certainly allowed, — we obtain the terminating power series x. Hence its sum for an arbitrary given $|x| < 1$ is a solution of $\tan y = x$, and is precisely the so-called principal value of the function $\tan^{-1} x$ hereby defined, i. e. the value which is $= 0$ for $x = 0$ and then varies continuously with x. Hence for $-1 < x < +1$, it satisfies the condition

$$-\frac{\pi}{4} < y < +\frac{\pi}{4}$$

and is defined, in the interior of this interval, without any ambiguity.

For $|x| > 1$ the expansion obtained is certainly no longer valid; but *Abel*'s theorem of limits shews that it does still hold for $x = \pm 1$. For the series remains convergent at both endpoints of the interval of convergence and $\tan^{-1} x$ is continuous at both these points. We have therefore in particular the series, peculiarly remarkable for clearness and simplicity:

122. $$\frac{\pi}{4} = 1 - \frac{1}{3} + \frac{1}{5} - \frac{1}{7} + - \cdots$$

giving at the same time a first means of determining π of some practical value. This beautiful equation is usually named after *Leibniz* [40]; it may be said to reduce the treatment of the number π to pure arithmetic. It is as if, by this expansion, the veil which hung over that strange number had been drawn aside.

---

[39] A different method is the following: We have

$$\frac{d\tan^{-1} x}{dx} = \frac{1}{\frac{d\tan y}{dy}} = \frac{1}{1+\tan^2 y} = \frac{1}{1+x^2} = 1 - x^2 + x^4 - + \cdots,$$

the latter for $|x| < 1$. As $\tan^{-1} 0 = 0$, it follows by **99**, theorem 5, that for $|x| < 1$,

$$\tan^{-1} x = x - \frac{x^3}{3} + \frac{x^5}{5} - + \cdots.$$

A method corresponding to that given first in the preceding footnote is somewhat more troublesome here, as the differential coefficients of higher order of $\tan^{-1} x$ — even at the single point 0 — are not easy to find directly. — The expansion of $\tan^{-1} x$ was found in 1671 by *J. Gregory*, but did not become known till 1712.

[40] He probably discovered it in 1673 from geometrical considerations and without reference to the inverse tan-series.

For the deduction of a series for $\sin^{-1} x$, the method which we have just used for $\tan^{-1} x$ is not available. The process indicated in the last footnote, however, provides the desired series: We have for $|x| < 1$

$$\frac{d\sin^{-1} x}{dx} = \frac{1}{\left(\frac{d\sin y}{dy}\right)} = \frac{1}{\cos y} = \frac{1}{\sqrt{1-x^2}} = (1-x^2)^{-\frac{1}{2}},$$

the positive sign being given to the radical since the derived function of $\sin^{-1} x$ is constantly positive in the interval $-1 \ldots +1$. From

$$(\sin^{-1} x)' = 1 - \binom{-\frac{1}{2}}{1} x^2 + \binom{-\frac{1}{2}}{2} x^4 - + \cdots$$

it at once follows, however, by **99**, theorem 5, as $\sin^{-1} 0 = 0$, that for $|x| < 1$

$$\sin^{-1} x = x + \frac{1}{2}\frac{x^3}{3} + \frac{1\cdot 3}{2\cdot 4}\cdot\frac{x^5}{5} + \frac{1\cdot 3\cdot 5}{2\cdot 4\cdot 6}\cdot\frac{x^7}{7} + \cdots. \qquad \textbf{123.}$$

This power series also has radius 1, and on quite similar grounds to the above we conclude that for $|x| < 1$ its sum is the principal value of $\sin^{-1} x$, i. e. that uniquely determined solution y of the equation $\sin y = x$ which lies between $-\frac{\pi}{2}$ and $+\frac{\pi}{2}$.

For $x = \pm 1$, the equality is not yet secured. By *Abel's* theorem of limits it will hold there if, and only if, the series converges there. As we have a mere change of sign in passing from $+x$ to $-x$, this only needs testing for the point $+1$. There we have a series of positive terms and it suffices to show that its partial sums are bounded. Now for $0 < x < 1$, if we denote by $s_n(x)$ the partial sums of **123**,

$$s_n(x) < \sin^{-1} x < \sin^{-1} 1 = \frac{\pi}{2}.$$

And as this holds (with fixed n) for every positive $x < 1$, we also have

$$\lim_{x\to 1} s_n(x) = s_n(1) \leqq \frac{\pi}{2};$$

and as this holds for every n, we have proved what we required. Thus

$$\frac{\pi}{2} = 1 + \frac{1}{2}\cdot\frac{1}{3} + \frac{1\cdot 3}{2\cdot 4}\cdot\frac{1}{5} + \frac{1\cdot 3\cdot 5}{2\cdot 4\cdot 6}\cdot\frac{1}{7} + \cdots. \qquad \textbf{124.}$$

§§ 22 to 27 have thus put us in possession of all the power series which are most important for applications.

Exercises on Chapter VI.

74. Show that the expansions in power series of the following functions have the form indicated in each case:

a) $e^x \sin x = \sum_{n=0}^{\infty} \frac{s_n}{n!} x^n$ with $s_n = \sqrt{2^n}\cdot\sin n\frac{\pi}{4}$, i. e. $s_{4k} = 0$, $s_{4k+1} = (-1)^k 2^{2k}$, $s_{4k+2} = (-1)^k 2^{2k+1}$, $s_{4k+3} = (-1)^k 2^{2k+1}$;

b) $\frac{1}{2}(\tan^{-1} x)^2 = \sum_{k=1}^{\infty} (-1)^{k-1} b_k\cdot\frac{x^{2k}}{2k}$ with $b_k = 1 + \frac{1}{3} + \frac{1}{5} + \cdots + \frac{1}{2k-1}$;

c) $\frac{1}{4}\tan^{-1}x\cdot\log\frac{1+x}{1-x}=\sum_{k=0}^{\infty}c_k\cdot\frac{x^{4k+2}}{4k+2}$ with $c_k=1-\frac{1}{3}+\frac{1}{5}+\cdots+\frac{1}{4k+1}$;

d) $\frac{1}{2}\tan^{-1}x\cdot\log(1+x^2)=\sum_{k=1}^{\infty}(-1)^{k-1}h_{2k}\cdot\frac{x^{2k+1}}{2k+1}$,

with $h_n=1+\frac{1}{2}+\cdots+\frac{1}{n}$;

e) $\frac{1}{2}\left[\log\frac{1}{1-x}\right]^2=\sum_{n=2}^{\infty}\frac{h_{n-1}}{n}x^n$, with the same meaning of h_n as in d).

75. Show that the expansions in power series of the following functions begin with the terms indicated:

a) $\dfrac{x}{\log\frac{1}{1-x}}=1-\frac{x}{2}-\frac{x^2}{12}-\frac{x^3}{24}-\cdots$;

b) $(1-x)\,e^{x+\frac{x^2}{2}+\cdots+\frac{x^m}{m}}=1-\frac{x^{m+1}}{m+1}+\cdots,\quad(m\geqq 1)$;

c) $\tan(\sin x)-\sin(\tan x)=\frac{1}{30}x^7+\frac{29}{756}x^9+\cdots$;

d) $\frac{1}{e}(1+x)^{\frac{1}{x}}=1-\frac{x}{2}+\frac{11}{24}x^2-\frac{7}{16}x^3+\frac{2447}{5760}x^4-\frac{959}{2304}x^5+\cdots$;

e) $\dfrac{x^2}{x-\log(1+x)}=2+\frac{4}{3}x-\frac{1}{9}x^2+\frac{8}{135}x^3+\cdots$.

76. Deduce, with reference to **105**, 5, **115** and **116**, the expansions in power series of the following functions

a) $\log\cos x$;

b) $\log\frac{\sin x}{x}$;

c) $\log\frac{\tan x}{x}$;

d) $\frac{x}{\sin x}$;

e) $\frac{x^2}{1-\cos x}$;

f) $\frac{1}{\cos x}$;

g) $\log\dfrac{x}{2\sin\frac{1}{2}x}$;

h) $\frac{1}{e^x+1}$;

i) $\frac{e^x}{e^x+1}$;

k) $\frac{1}{\cos x-\sin x}$.

77. Show that, for $a\neq 0,-2,-4,\ldots$

$$\frac{1}{\sqrt{1-x}}\cdot\left[\frac{1}{a}+\frac{1}{2}\frac{x}{a+2}+\frac{1\cdot 3}{2\cdot 4}\cdot\frac{x^2}{a+4}+\cdots\right]=$$
$$=\frac{1}{a}\left[1+\frac{a+1}{a+2}x+\frac{(a+1)(a+3)}{(a+2)(a+4)}x^2+\cdots\right].$$

78. We have $\left(\frac{2n+1}{2n-1}\right)^n\to e$. Is the sequence monotone? Increasing or decreasing? What, in this respect, is the behaviour of the sequences

$$\left(1+\frac{1}{n}\right)^{n+\alpha},\quad 0<\alpha<1?$$

79. From $x_n\to\xi$ it invariably follows that

$$\left(1+\frac{x_n}{n}\right)^n\to e^{\xi}$$

and also, if x_n and ξ are positive, that

$$n\left(\sqrt[n]{x_n}-1\right)\to\log\xi.$$

80. If (x_n) is an arbitrary real sequence, for which $\frac{x_n^2}{n}\to 0$, and we write $\left(1-\frac{x_n}{n}\right)^n=y_n$, then, in every case,

$$y_n \cong e^{-x_n}.$$

81. Prove the inequalities of **114**.

82. Express the sums of the following series by closed expressions in terms of the elementary functions:

a) $\frac{1}{2}+\frac{x}{5}+\frac{x^2}{8}+\frac{x^3}{11}+\cdots.$

(Hint: If $f(x)$ be the required function, then obviously

$$(x^2\cdot f(x^3))'=\frac{x}{1-x^3}$$

whence $f(x)$ may be determined. Similarly in the following examples.)

b) $\frac{x^3}{1\cdot3}-\frac{x^5}{3\cdot5}+\frac{x^7}{5\cdot7}-\frac{x^9}{7\cdot9}+-\cdots;$

c) $\frac{1}{1\cdot2\cdot3}+\frac{x}{2\cdot3\cdot4}+\frac{x^2}{3\cdot4\cdot5}+\cdots;$

d) $x+\frac{x^3}{3}-\frac{x^5}{5}-\frac{x^7}{7}++--\cdots.$

83. Obtain the sums of the following series as particular values of elementary functions:

a) $\frac{1}{2!}+\frac{2}{3!}+\frac{3}{4!}+\frac{4}{5!}+\cdots=1;$

b) $\frac{1}{2}+\frac{1}{2\cdot4}+\frac{1\cdot3}{2\cdot4\cdot6}+\frac{1\cdot3\cdot5}{2\cdot4\cdot6\cdot8}+\cdots=1;$

c) $\frac{1}{2}+\frac{1\cdot3}{2\cdot4\cdot6}+\frac{1\cdot3\cdot5\cdot7}{2\cdot4\cdot6\cdot8\cdot10}+\frac{1\cdot3\cdot5\cdot7\cdot9\cdot11}{2\cdot4\cdot6\cdot8\cdot10\cdot12\cdot14}+\cdots=\frac{1}{2}\sqrt{2};$

d) $\frac{1}{2}-\frac{1\cdot3}{2\cdot4\cdot6}+\frac{1\cdot3\cdot5\cdot7}{2\cdot4\cdot6\cdot8\cdot10}-\frac{1\cdot3\cdot5\cdot7\cdot9\cdot11}{2\cdot4\cdot6\cdot8\cdot10\cdot12\cdot14}+-\cdots=\sqrt{\frac{\sqrt{2}-1}{\sqrt{2}}}.$

84. Deduce from the expansion in partial fractions **117** seq. the following expressions for π:

$$\pi=\alpha\cdot\tan\frac{\pi}{\alpha}\cdot\left[1-\frac{1}{\alpha-1}+\frac{1}{\alpha+1}-\frac{1}{2\alpha-1}+\frac{1}{2\alpha+1}-+\cdots\right],$$

$$\pi=\alpha\cdot\sin\frac{\pi}{\alpha}\cdot\left[1+\frac{1}{\alpha-1}-\frac{1}{\alpha+1}-\frac{1}{2\alpha-1}+\frac{1}{2\alpha+1}+--++\cdots\right],$$

where $\alpha\neq 0, \pm1, \pm\frac{1}{2}, \pm\frac{1}{3},\ldots$ Substitute in particular $\alpha=3, 4, 6$.

Chapter VII.

Infinite products.

§ 28. Products with positive terms.

An infinite product

$$u_1 \cdot u_2 \cdot u_3 \cdot \ldots \cdot u_n \cdot \ldots$$

is, by § 11, II, to be taken merely as representing a new symbol for the sequence of the partial products

$$u_1 \cdot u_2 \cdot \ldots \cdot u_n.$$

Accordingly such an infinite product should be called convergent, with value U,

$$\prod_{n=1}^{\infty} u_n = U,$$

if the sequence of the partial products tends to the number U as limit. But this is particularly inconvenient, owing to the fact that then every product would have to be called convergent for which a single factor was $= 0$. For if u_m were 0, then the sequence of partial products also would tend to $U = 0$, since its terms would all be equal to 0 for $n \geqq m$. Similarly *every* product would be convergent — again with the value 0 — for which from some m onwards

$$|u_n| \leqq \vartheta < 1.$$

In order to exclude these trivial cases, we do *not* describe the behaviour of an infinite product by that of the sequence of its partial products, but adopt the following more suitable definition, which takes into account the peculiar part played by the number 0 in multiplication:

125. ∘ **Definition.** *The infinite product*

$$\prod_{n=1}^{\infty} u_n \equiv u_1 \cdot u_2 \cdot u_3 \cdot \ldots$$

will be called **convergent** *(in the stricter sense) if from some point onwards — say for every $n > m$ — no factor vanishes, and if the partial products, beginning immediately beyond this point*

$$p_n = u_{m+1} \cdot u_{m+2} \cdot \ldots \cdot u_n, \qquad (n > m)$$

tend, as n increases, to a limit, finite and different from 0. *If this be $= U_m$, then the number*

$$U = u_1 \cdot u_2 \cdot \ldots \cdot u_m \cdot U_m,$$

obviously independent of m, is regarded as the **value** *of the product*[1].

[1] Infinite products are first found in *F. Vieta* (Opera, Leyden 1646, p. 400) who gives the product

$$\frac{2}{\pi} = \sqrt{\frac{1}{2}} \cdot \sqrt{\frac{1}{2} + \frac{1}{2}\sqrt{\frac{1}{2}}} \cdot \sqrt{\frac{1}{2} + \frac{1}{2}\sqrt{\frac{1}{2} + \frac{1}{2}\sqrt{\frac{1}{2}}}},$$

We then have first, as for finite products, the

○ **Theorem 1.** *A convergent infinite product has the value* 0 *if, and only if, one of its factors is* $=0$.

As further $p_{n-1} \to U_m$ with $p_n \to U_m$, and as U_m is $\neq 0$, we have (by **41**, 11)

$$u_n = \frac{p_n}{p_{n-1}} \to 1$$

and we have the

○ **Theorem 2.** *The sequence of the factors in a convergent infinite product always tends* $\to 1$.

On this account, it will be more convenient to denote the factors by $u_n = 1 + a_n$, so that the products considered have the form

$$\prod_{n=1}^{\infty} (1 + a_n).$$

For these, the condition $a_n \to 0$ *is then a necessary condition for convergence.* The numbers a_n — as the most essential parts of the factors — will be called the ***terms*** of the product. If they are all $\geqq 0$, then as in the case of infinite series, we speak of *products with positive terms.* We will first concern ourselves with these.

The question of convergence is entirely answered here by the

Theorem 3. *A product* $\Pi(1 + a_n)$ *with positive terms* a_n *is convergent if, and only if, the series* $\sum a_n$ *converges.*

Proof. The partial products $p_n = (1 + a_1) \ldots (1 + a_n)$, since $a_n \geqq 0$, increase monotonely; hence the First main criterion (**46**) is available and we only have to show that the partial products p_n are bounded if, and only if, the partial *sums* $s_n = a_1 + a_2 + \cdots + a_n$ are bounded. Now by **114** α, $1 + a_\nu \leqq e^{a_\nu}$ and so for each n

$$p_n \leqq e^{s_n};$$

on the other hand

$$p_n = (1 + a_1) \ldots (1 + a_n) = 1 + a_1 + a_2 + \cdots + a_n + a_1 a_2 + \cdots > s_n,$$

the latter because in the product, after expansion, we have, besides the terms of s_n, many others, but all non-negative ones, occurring. Thus for each n

$$s_n < p_n.$$

(cf. Ex. 89) and in *J. Wallis* (Opera I, Oxford 1695, p. 468) who in 1656 gives the product

$$\frac{\pi}{2} = \frac{2}{1} \cdot \frac{2}{3} \cdot \frac{4}{3} \cdot \frac{4}{5} \cdot \frac{6}{5} \cdot \frac{6}{7} \cdots$$

But infinite products first secured a footing in mathematics through *Euler*, who established a number of important expansions in infinite product form. The first criteria of convergence are due to *Cauchy*.

The former inequality shows that p_n remains bounded when s_n does, the latter, conversely, that s_n remains bounded when p_n does, — which proves the statement.[2]

Examples.

1. As we are already acquainted with a number of examples of convergent series Σa_n with positive terms, we may obtain, by theorem 3, as many examples of convergent products $\Pi(1+a_n)$. We may mention:

$\Pi\left(1+\frac{1}{n^\alpha}\right)$ is convergent for $\alpha > 1$, divergent for $\alpha \leqq 1$. — The latter is more easily recognised here than in the corresponding series[3], for

$$\left(1+\frac{1}{1}\right)\left(1+\frac{1}{2}\right)\cdots\left(1+\frac{1}{n}\right)=\frac{2}{1}\cdot\frac{3}{2}\cdot\frac{4}{3}\cdots\frac{n+1}{n}=n+1\to+\infty.$$

2. $\Pi(1+x^n)$ is convergent for $0 \leqq x < 1$; similarly $\Pi(1+x^{2^n})$.

3. $$\prod_{n=2}^{\infty}\left(1-\frac{2}{n(n+1)}\right)\equiv\prod_{n=2}^{\infty}\frac{(n-1)(n+2)}{n(n+1)}=\frac{1}{3}.$$

With theorem 3 we may at once couple the following very similar

Theorem 4. *If, for every n, $a_n \geqq 0$, then the product $\Pi(1-a_n)$ also is convergent if, and only if, $\sum a_n$ converges.*

Proof. If a_n does not tend to 0, both the series and the product certainly diverge. But if $a_n \to 0$, then from some point onwards, say for every $n > m$, we have $a_n < \frac{1}{2}$, or $1 - a_n > \frac{1}{2}$. We consider the series and product from this point onwards only.

Now if the product converges, then the monotone decreasing sequence of its partial products $p_n = (1-a_{m+1})\cdots(1-a_n)$ tends to a *positive* (>0) number U_m, and, for every $n > m$,

$$(1-a_{m+1})\cdots(1-a_n) \geqq U_m > 0.$$

Since, for $0 < a_\nu < 1$, we always have

$$(1+a_\nu) \leqq \frac{1}{1-a_\nu}$$

(as is at once seen by multiplying up), we certainly have

$$(1+a_{m+1})(1+a_{m+2})\cdots(1+a_n) \leqq \frac{1}{U_m}.$$

---

[2] In the first part of the proof of this elementary theorem, we use the transcendental exponential function. We can avoid this as follows: If $\sum a_n = s$ converges, choose m so that for every $n > m$

$$a_{m+1}+a_{m+2}+\cdots+a_n < \frac{1}{2}.$$

As, obviously, for these n's, we now have

$$(1+a_{m+1})\cdots(1+a_n) \leqq 1 + (a_{m+1}+\cdots+a_n) + (a_{m+1}+\cdots+a_n)^2 + \cdots + (a_{m+1}+\cdots+a_n)^n < 2,$$

we certainly have, for all n's,

$$p_n < 2\cdot(1+a_1)\cdots(1+a_m) = K,$$

hence (p_n) is bounded.

[3] In this we have therefore, on account of theorem 3, a new proof of the divergence of $\sum\frac{1}{n}$.

Accordingly the convergence of the product $\Pi(1+a_n)$, and hence of the series Σa_n, results from that of $\Pi(1-a_n)$. — If, conversely, Σa_n converges, then so does $\Sigma 2a_n$, and consequently by Theorem 3 the product $\Pi(1+2a_n)$ also does. Hence, with a suitable choice of K, the products $(1+2a_{m+1})\ldots(1+2a_n)$ remain $< K$. If we now use the fact that, for $0 \leqq a_\nu \leqq \frac{1}{2}$,

$$1-a_\nu \geqq \frac{1}{1+2a_\nu}$$

— as may again be seen by multiplying up — we infer

$$(1-a_{m+1})\ldots(1-a_n) > \frac{1}{K} > 0;$$

and the partial products on the left hand side, as they form a monotone decreasing sequence, therefore tend to a *positive* limit: i. e. the product $\Pi(1-a_n)$ is convergent.

Remarks and Examples.

126.

1. $\prod_{n=2}^{\infty}\left(1-\frac{1}{n^\alpha}\right)$ is convergent for $\alpha > 1$, divergent for $\alpha \leqq 1$.

2. If $a_n < 1$ and if Σa_n diverges, then $\Pi(1-a_n)$ is not convergent, with our definition. As however the partial products p_n decrease monotonely and remain > 0, they have a limit, but one which is necessarily $= 0$. We say that the product *diverges* to 0. The exceptional part played by the number 0 thus involves us in some slight incongruity of expression. A product is called *divergent* whose partial products form a decidedly convergent sequence, namely a null sequence, (p_n). The addition "in the stricter sense" to the word "convergent" in Def. **125** is intended to serve as a reminder of this fact.

3. That e. g. $\prod_{n=2}^{\infty}\left(1-\frac{1}{n}\right)$ diverges to 0 is again very easily seen from

$$p_n = \left(1-\frac{1}{2}\right)\left(1-\frac{1}{3}\right)\ldots\left(1-\frac{1}{n}\right) = \frac{1}{2}\cdot\frac{2}{3}\cdot\frac{3}{4}\ldots\frac{n-1}{n} = \frac{1}{n} \to 0.$$

§ 29. Products with arbitrary terms.[4] Absolute convergence.

If the terms a_n of a product have arbitrary signs, then the following theorem — corresponding to the second principal criterion **81** for series — holds:

∘ **Theorem 5.** *The infinite product* $\Pi(1+a_n)$ *converges if, and*

[4] A full and systematic account of the theory of convergence of infinite products may be found in *A. Pringsheim:* Über die Konvergenz unendlicher Produkte, Math. Annalen, Vol. 33, p. 119—154, 1889.

only if, given $\varepsilon > 0$, *we can determine*[5] n_0 *so that for every* $n > n_0$ *and every* $k \geqq 1$,

$$[(1+a_{n+1})(1+a_{n+2})\cdots(1+a_{n+k})-1] < \varepsilon.$$

Proof. a) If the product converges, then from some point onwards, say for every $n > m$, we have $a_n \neq -1$, and the partial products

$$p_n = (1+a_{m+1})\cdots(1+a_n), \qquad (n > m)$$

tend to a limit $\neq 0$. Hence there exists (v. **41**, 3) a positive number β such that, for every $n > m$, $|p_n| \geqq \beta > 0$. By the second principal criterion **49** we may now, given $\varepsilon > 0$, determine n_0 so that for every $n > n_0$ and every $k \geqq 1$,

$$|p_{n+k} - p_n| < \varepsilon \cdot \beta.$$

But then, for the same n and k,

$$\left|\frac{p_{n+k}}{p_n} - 1\right| = |(1+a_{n+1})(1+a_{n+2})\cdots(1+a_{n+k}) - 1| < \varepsilon,$$

which is precisely what we asserted.

b) Conversely, if the ε-condition of the theorem is fulfilled, first choose $\varepsilon = \frac{1}{2}$, and determine m so that, for every $n > m$,

$$|(1+a_{m+1})\cdots(1+a_n) - 1| = |p_n - 1| < \tfrac{1}{2}.$$

For these n's we then have

$$\tfrac{1}{2} < |p_n| < \tfrac{3}{2},$$

showing that, for every $n > m$, we must have $1 + a_n \neq 0$; and further, that *if* p_n tends to a limit at all, this certainly cannot be 0. But we may now, given $\varepsilon > 0$, choose the number n_0 so that for every $n > n_0$ and every $k \geqq 1$,

$$\left|\frac{p_{n+k}}{p_n} - 1\right| < \frac{\varepsilon}{2}$$

or

$$|p_{n+k} - p_n| < |p_n| \cdot \frac{\varepsilon}{2} < \varepsilon.$$

And this shows that p_n really *has* a unique limit. Thus the convergence of the product is established.

As in the case of infinite series, so similarly in that of infinite products, those are the most easily dealt with which converge "absolutely". By this we do not mean products Πu_n for which $\Pi |u_n|$ also converges, — such a definition would be valueless, since then *every* convergent product would also be absolutely convergent, — but we define, on the contrary, as follows:

127. ∘ Definition. *The product* $\Pi(1+a_n)$ *is said to be absolutely convergent if the product* $\Pi(1+|a_n|)$ *converges.*

[5] Or — v. **81**, 2nd form — if *invariably*

$$[(1+a_{n+1})(1+a_{n+2})\cdots(1+a_{n+k_n})] \to 1;$$

or — v. **81**, 3rd form — if *invariably*

$$[(1+a_{\nu_n+1})\cdots(1+a_{\nu_n+k_n})] \to 1.$$

This definition only gains significance through the theorem:

∘**Theorem 6.** *The convergence of* $\Pi(1+|a_n|)$ *involves that of* $\Pi(1+a_n)$.

Proof. We have invariably

$$|(1+a_{n+1})(1+a_{n+2})\dots(1+a_{n+k})-1|$$
$$\leqq(1+|a_{n+1}|)(1+|a_{n+2}|)\dots(1+|a_{n+k}|)-1,$$

as is at once verified by multiplying out. If therefore the necessary and sufficient condition for the convergence of Theorem 5 is satisfied by $\Pi(1+|a_n|)$, it is *ipso facto* satisfied by $\Pi(1+a_n)$, q. e. d.

In consequence of Theorem 3, we may therefore at once state

∘**Theorem 7.** *A product* $\Pi(1+a_n)$ *is absolutely convergent if, and only if,* $\sum a_n$ *converges absolutely.*

As we have an already sufficiently developed theory for the determination of the absolute convergence of a series, Theorem 7 solves the problem of convergence in a satisfactory manner for absolutely convergent products. In all other cases, the following theorem reduces the problem of convergence of products completely to the corresponding one for series:

Theorem 8. *The product* $\Pi(1+a_n)$ *converges if, and only if the series*

$$\sum_{n=m+1}^{\infty}\log(1+a_n)$$

commencing with a suitable index[6], *converges. And the convergence of the product is absolute if, and only if, that of the series is so. Furthermore, if L is the sum of the series, then*

$$\prod_{n=1}^{\infty}(1+a_n)=(1+a_1)\dots(1+a_m)\cdot e^L.$$

Proof. a) If $\Pi(1+a_n)$ converges, then $a_n\to 0$ and hence from some point onwards, say for every $n>m$, we have $|a_n|<1$. Since, further, the partial products

$$p_n=(1+a_{m+1})\dots(1+a_n),\qquad (n>m),$$

tend to a limit $U_m\neq 0$ (hence positive), we have by (**42**, 2),

$$\log p_n\to\log U_m.$$

But $\log p_n$ is the partial sum, ending with the nth term, of the series in question. This, therefore, converges to the sum $L=\log U_m$. As $U_m=e^L$, we thus have

$$\Pi(1+a_n)=(1+a_1)\dots(1+a_m)\cdot e^L.$$

b) If, conversely, the series is known to converge, and to have the sum L, then we have precisely $\log p_n\to L$, and consequently (by **42**, 1)

$$p_n=e^{\log p_n}\to e^L.$$

This completes the proof of the first part of the theorem, since $e^L\neq 0$.

---

[6] It suffices to choose m so that for every $n>m$ we have $|a_n|<1$.

To deduce, finally, that the series and product are, in every possible case, either both or neither absolutely convergent, we use with theorem 7 and **70**, 4, the fact that (**112**, b), when $a_n \to 0$

$$\left|\frac{\log(1+a_n)}{a_n}\right| \to 1.$$

(Here any terms a_n which $= 0$ may be simply omitted from consideration.)

Although we have thus completely reduced the problem of the convergence of infinite products to that of infinite series, yet the result cannot entirely satisfy us, because of the difficulties usually involved in the practical determination of the convergence of a series of the form $\sum \log(1+a_n)$. The want here felt may, at least partially, be supplied by the following

°**Theorem 9.** *The series (starting with a suitable initial index) $\Sigma \log(1+a_n)$ and with it the product $\Pi(1+a_n)$, is certainly convergent, if Σa_n converges and if Σa_n^2 is absolutely convergent*[7].

Proof. We choose m so that for every $n > m$, we have $|a_n| < \frac{1}{2}$, and consider $\Pi(1+a_n)$ and $\Sigma \log(1+a_n)$, starting with the $(m+1)^{\text{th}}$ terms. If we write

$$\log(1+a_n) = a_n + \vartheta_n \cdot a_n^2 \quad \text{or} \quad \frac{\log(1+a_n) - a_n}{a_n} = \vartheta_n,$$

then the numbers ϑ_n so determined certainly form a bounded sequence, for[8], as $a_n \to 0$, $\vartheta_n \to -\frac{1}{2}$. If therefore Σa_n and $\Sigma |a_n|^2$ are convergent, $\Sigma \log(1+a_n)$, and hence also $\Pi(1+a_n)$, is convergent.

This simple theorem leads easily to the following further theorem:

°**Theorem 10.** *If Σa_n^2 is absolutely convergent, and $|a_n|$ is < 1 for every $n > m$, then the partial products*

$$p_n = \prod_{\nu=m+1}^{n} (1+a_\nu) \quad \textit{and the partial sums} \quad s_n = \sum_{\nu=m+1}^{n} a_\nu, \qquad (n > m),$$

are so related that

$$p_n \sim e^{s_n}$$

i. e. the ratio of the two sides of this relation tends to a definite limit, finite and $\neq 0$, — whether or no Σa_n converges.

---

[7] Σa_n^2, if convergent at all, is certainly absolutely convergent. We adopt the above wording so that the theorem may remain true for complex a_n's, for which a_n^2 is not necessarily > 0 (cf. § 57).

[8] For $0 < |x| < 1$ we have in fact

$$\log(1+x) = x + x^2\left[-\frac{1}{2} + \frac{x}{3} - \frac{x^2}{4} + - \cdots\right]$$

or

$$\frac{\log(1+x) - x}{x^2} = -\frac{1}{2} + \frac{x}{3} - + \cdots$$

And those terms which are possibly $= 0$ may be again simply neglected, as they have no influence on the question under consideration.

Proof. If we adopt the notation of the *preceding* proof, then, as $\log(1+a_n) = a_n + \vartheta_n \cdot a_n^2$, we have for every $n > m$

$$(1+a_{m+1})\cdots(1+a_n) = \prod_{\nu=m+1}^{n} e^{a_\nu + \vartheta_\nu a_\nu^2} = e^{\Sigma a_\nu} \cdot e^{\Sigma \vartheta_\nu a_\nu^2},$$

if the sums in the last two exponents are taken also from $\nu = m+1$ to $\nu = n$.

And as $\Sigma \vartheta_n a_n^2$, the ϑ_n's being bounded, converges absolutely when Σa_n^2 does so, we can, from the above equation, at once infer the result stated. — This theorem also provides the following, often useful

Supplementary theorem. *If Σa_n^2 converges absolutely, then Σa_n and $\Pi(1+a_n)$ converge and diverge together.*

Remarks and examples. **128.**

1. The conditions of Theorem 9 are only *sufficient*; the product $\Pi(1+a_n)$ may converge, without Σa_n converging. But in that case, by Theorem 10, $\Sigma |a_n|^2$ must also diverge.

2. If we apply theorem 10 to the (divergent) product $\prod_{n=1}^{\infty}\left(1+\frac{1}{n}\right)$, then it follows that

$$e^{h_n} \sim n$$

if h_n denotes the nth partial sum of the harmonic series $h_n = 1 + \frac{1}{2} + \cdots + \frac{1}{n}$. Accordingly the limits $\lim \frac{e^{h_n}}{n} = c$ and $\lim[h_n - \log n] = \log c = C$ exist, the latter because $c \neq 0$, hence > 0. The number C defined by the second limit is called *Euler*'s or *Mascheroni*'s constant. Its numerical value is $C = 0{\cdot}5772156649\ldots$ (cf. Ex. 86a, **176**, 1 and § 64, B, 4). The latter result gives us further valuable information as to the degree of divergence of the harmonic series, as it gives

$$h_n \cong \log n.$$

Further the estimates of bounds above made for the proof of Theorem 3 show, even more precisely, if we there put $a_\nu = \frac{1}{\nu}$, that

$$e^{h_n} > e^{h_{n-1}} > n \quad \text{or} \quad h_n > h_{n-1} > \log n$$

so that *Euler*'s constant cannot be negative.

3. $\prod_{n=1}^{\infty}\left(1+\frac{(-1)^{n-1}}{n}\right)$ is convergent. Its value may, as it happens, be found at once by forming the partial products, and is $= 1$.

4. $\prod_{n=1}^{\infty}\left(1+\frac{x}{n}\right)$ diverges for $x \neq 0$. However, theorem 10 shows that

$\prod_{\nu=1}^{\infty}\left(1+\frac{x}{\nu}\right) \sim e^{x\left(1+\frac{1}{2}+\cdots+\frac{1}{n}\right)}$, or — what is the same thing by 2, — $\sim n^x$, i. e. (v. **40**, def. 5) the ratio

$$n^{-x} \cdot \prod_{\nu=1}^{n}\left(1+\frac{x}{\nu}\right) = \frac{(x+1)(x+2)\ldots(x+n)}{n!\, n^x}$$

has, for every (fixed) x, when $n \to \infty$, a *determinate* (finite) limit which is also *different from* 0 if x is taken $\neq -1, -2, \ldots$ (cf. below, **219**, 4).

5. $\prod_{n=1}^{\infty}\left(1-\frac{x^2}{n^2}\right)$ is absolutely convergent for every x.

6. $\prod_{n=2}^{\infty}\left(1-\frac{1}{n^2}\right)=\frac{1}{2}$.

§ 30. Connection between series and products. Conditional and unconditional convergence.

We have more than once observed that an infinite series Σa_n is merely another symbol for the sequence (s_n) of its partial sums. Apart from the fact that we have to take into account the exceptional part played by the value 0 in multiplication, the corresponding remark holds good for infinite products. It follows that, with this reservation, every series may be written as a product and every product as a series. As regards detail, this has to be done as follows:

129. 1. If $\prod_{n=1}^{\infty}(1+a_n)$ is given, then this product, if we write

$$\prod_{\nu=1}^{n}(1+a_\nu)=p_n,$$

represents essentially the sequence (p_n). This sequence, on the other hand, is represented by the series

$$p_1+(p_2-p_1)+(p_3-p_2)+\cdots \equiv p_1+\sum_{n=2}^{\infty}(1+a_1)\ldots(1+a_{n-1})\cdot a_n,$$

This and the given product have the same meaning — if the product converges in accordance with our definition. But the series *may* also have a meaning without this being the case for the product (e. g. if the factor $(1+a_5)$ is $=0$ and all other factors are $=2$).

2. If conversely the series $\sum_{n=1}^{\infty} a_n$ is given, then it represents the sequence for which $s_n=\sum_{\nu=1}^{n} a_\nu$. This is also what is meant by the product

$$s_1\cdot\frac{s_2}{s_1}\cdot\frac{s_3}{s_2}\cdots \equiv s_1\cdot\prod_{n=2}^{\infty}\frac{s_n}{s_{n-1}}\equiv a_1\cdot\prod_{n=2}^{\infty}\left(1+\frac{a_n}{a_1+a_2+\cdots+a_{n-1}}\right),$$

— provided it has a meaning at all. And for this obviously all that we require is that *each* $s_n \neq 0$. In general the convergence of the product implies the convergence of the series, and conversely. In the case, however, of $s_n \to 0$, although we call the series *convergent* with sum 0, we say that the product *diverges* to 0.

Thus e. g. the symbols $\sum_{n=1}^{\infty}\frac{1}{2^n}$ and $\frac{1}{2}\cdot\prod_{n=2}^{\infty}\left(1+\frac{1}{2^n-2}\right)$,

or $\sum_{n=1}^{\infty}\frac{1}{n(n+1)}$ and $\frac{1}{2}\cdot\prod_{n=2}^{\infty}\left(1+\frac{1}{n^2-1}\right)$,

have precisely the same meaning.

It is, however, only in rare cases that a passage such as this from the one symbol to the other will be advantageous for actual investigations. The connection between series and products which is theoretically conclusive was, moreover, established by Theorem 8 alone, — or by Theorem 7, if we are concerned with the mere question of absolute convergence. In order to show the bearing of these theorems on general questions, we may prove — as analogue of Theorem **88**, 1, and **89**, 2, — the following:

° **Theorem 11.** *An infinite product $\Pi(1+a_n)$ is unconditionally convergent — i. e. remains convergent, with value unaltered, however its factors be rearranged* (v. **27**, 3) — *if, and only if, it converges absolutely*[9]. **130**

Proof. We suppose given a convergent infinite product $\Pi(1+a_n)$. The terms a_n, certainly finite in number, for which $|a_n| \geqq \frac{1}{2}$, we replace by 0. In so doing, we only make a "finite number of alterations" and we ensure $|a_n| < \frac{1}{2}$ for every n. The number m in the proof of theorem 8 may then be taken $= 0$. We first prove the theorem for the altered product.

Now, with the present values of a_n,

$$\Pi(1+a_n) \qquad \text{and} \qquad \Sigma \log(1+a_n)$$

are convergent together, and their values U and L stand in the relation $U = e^L$ to one another. It follows that a rearrangement of the factors of the product leaves this convergent, with the same value U, if and only if the corresponding rearrangement of the terms of the series also leaves this convergent, with the same sum. But this, for a series, is the case if, and only if, it converges *absolutely*. By theorem 8 the same therefore holds for the product.

Now if, before the rearrangement, we have made a finite number of alterations, and then after the rearrangement make them again in the opposite sense, this can have no influence on the present question. The theorem is therefore true for *all* products.

Additional remark. Using the theorem of *Riemann* proved later (**187**) we can of course say, more precisely: If the product is not absolutely convergent and has no factor $= 0$, then we can by *suitable* rearrangement of its factors, always arrange that the sequence of its partial products has *prescribed lower and upper limits* $\varkappa$ and μ, provided they have the same sign as the value of the given product[10]. Here $\varkappa$ and μ may also be 0 or $\pm\infty$.

---

[9] *Dini, U.*: Sui prodotti infiniti, Annali di Matem., (2) Vol. 2, pp. 28—38. 1868.

[10] For a convergent infinite product has certainly only a finite number of negative factors; and their number is not altered by the rearrangement.

Exercises on Chapter VII.

85. Prove that the following products converge and have the values indicated:

a) $\prod_{n=2}^{\infty} \frac{n^3-1}{n^3+1} = \frac{2}{3}$; b) $\prod_{n=0}^{\infty} \left(1 + \left(\frac{1}{2}\right)^{2^n}\right) = 2$;

c) $\prod_{n=2}^{\infty} \left(1 + \frac{2n+1}{(n^2-1)(n+1)^2}\right) = \frac{4}{3}$.

85a. By **128**, 2 the sequences

$$x_n = 1 + \tfrac{1}{2} + \ldots + \frac{1}{n-1} - \log n \quad \text{and} \quad y_n = 1 + \tfrac{1}{2} + \ldots + \frac{1}{n} - \log n$$

have *positive* terms for $n > 1$. Show that $(x_n \mid y_n)$ is a nest of intervals. The value so defined is *Euler's constant.*

86. Determine the behaviour of the following products:

a) $\prod_{n=2}^{\infty} \left(1 + \frac{(-1)^n}{\sqrt{n}}\right)$; b) $\prod_{n=2}^{\infty} \left(1 + \frac{(-1)^n}{\log n}\right)$;

c) $\left(1 - \frac{1}{\sqrt{3}}\right)\left(1 + \frac{1}{\sqrt{2}}\right)\left(1 - \frac{1}{\sqrt{5}}\right)\left(1 - \frac{1}{\sqrt{7}}\right)\left(1 + \frac{1}{\sqrt{4}}\right)\left(1 - \frac{1}{\sqrt{9}}\right)\cdots$;

d) $\left(1 + \frac{1}{\alpha-1}\right)\left(1 - \frac{1}{2\alpha-1}\right)\left(1 + \frac{1}{3\alpha-1}\right)\left(1 - \frac{1}{4\alpha-1}\right)\cdots$,

for $\alpha \neq 1, \frac{1}{2}, \frac{1}{3}, \ldots$

87. Show that $\prod \cos x_n$ converges if $\sum |x_n|^2$ converges.

88. The product in Ex. 86 d has, for positive integral values of α, the value $\sqrt[\alpha]{2}$.

[Hint: The partial product with last factor $\left(1 - \frac{1}{2k\alpha - 1}\right)$ is $= \prod_{\nu=k+1}^{2k} \left(1 - \frac{1}{\nu\alpha}\right)^{-1}$.]

89. Prove, with reference to Ex. 87, that $\cos\frac{\pi}{4} \cdot \cos\frac{\pi}{8} \cdot \cos\frac{\pi}{16} \cdots = \frac{2}{\pi}$.

(We recognise *Vieta*'s product mentioned in footnote 1, p. 218.)

90. Show, more generally, that for every x

$$\cos\frac{x}{2} \cdot \cos\frac{x}{4} \cdot \cos\frac{x}{8} \cdot \cos\frac{x}{16} \cdots = \frac{\sin x}{x}.$$

$$\cosh\frac{x}{2} \cdot \cosh\frac{x}{4} \cdot \cosh\frac{x}{8} \cdot \cosh\frac{x}{16} \cdots \frac{\sinh x}{x},$$

in which latter formula $\cosh x = \frac{e^x + e^{-x}}{2}$, $\sinh x = \frac{e^x - e^{-x}}{2}$ denote the hyperbolic cosine and sine of x.

91. With the help of Ex. 90, show that the number defined by the nest of intervals in Ex. 8c is $= \frac{\sin\vartheta}{\vartheta} y_1$, where ϑ is defined as the acute angle for which $\cos\vartheta = \frac{x_1}{y_1}$. Similarly the number defined by Ex. 8d is $= \frac{\sinh\vartheta}{\vartheta} x_1$, if ϑ is defined by $\cosh\vartheta = \frac{y_1}{x_1}$.

92. In a similar way, show that the numbers defined in Ex. 8e and 8f have the values:

e) $\frac{\sinh 2\vartheta}{2\vartheta} x_1$ with $\cosh^2\vartheta = \frac{y_1}{x_1}$,

f) $\frac{\sin 2\vartheta}{2\vartheta} y_1$ with $\cos^2\vartheta = \frac{x_1}{y_1}$.

93. We have

$$1 - \frac{x}{a_1} + \frac{x(x-a_1)}{a_1 \cdot a_2} - + \cdots + (-1)^n \frac{x(x-a_1)\ldots(x-a_{n-1})}{a_1 a_2 \ldots a_n}$$

$$= \left(1 - \frac{x}{a_1}\right)\left(1 - \frac{x}{a_2}\right)\cdots\left(1 - \frac{x}{a_n}\right).$$

What can you deduce for the series and product of which we here have the initial portions?

94. With the help of theorem 10 of § 29, show that

$$\frac{1 \cdot 3 \cdot 5 \ldots (2n-1)}{2 \cdot 4 \cdot 6 \ldots 2n} \sim \frac{1}{\sqrt{n}}.$$

95. Similarly, show that, for $0 \leqq x < y$,

$$\frac{x(x+1)(x+2)\ldots(x+n)}{y(y+1)(y+2)\ldots(y+n)} \to 0.$$

96. Similarly, show that if a and b are positive, and A_n and G_n are respectively the arithmetic and geometric means, of the n quantities

$$a,\quad a+b,\quad a+2b,\quad \ldots,\quad a+(n-1)b,$$

$(n = 2, 3, 4, \ldots)$ then

$$\frac{A_n}{G_n} \to \frac{e}{2}.$$

97. What can be deduced, from the convergence of $\Pi(1+a_n)$ and $\Pi(1+b_n)$, as to that of

$$\Pi(1+a_n)(1+b_n) \quad \text{and} \quad \Pi\frac{1+a_n}{1+b_n}?$$

(Cf. **83**, 3 and 4.)

98. Given (u_n) monotone decreasing and $\to 1$, is

$$u_1 \cdot \frac{1}{u_2} \cdot u_3 \cdot \frac{1}{u_4} \cdot u_5 \cdots$$

always convergent? (Cf. **82**, theorem 5.)

99. To complete § 29, theorem 9, prove that $\Pi(1+a_n)$ certainly converges if the two series

$$\Sigma(a_n - \tfrac{1}{2}a_n^2) \quad \text{and} \quad \Sigma|a_n|^3$$

converge. — How may this be generalized? — On the other hand, show, by the example of the product

$$\left(1-\frac{1}{2^\alpha}\right)\left(1+\frac{1}{2^\alpha}+\frac{1}{2^{2\alpha}}\right)\left(1-\frac{1}{3^\alpha}\right)\left(1+\frac{1}{3^\alpha}+\frac{1}{3^{2\alpha}}\right)\left(1-\frac{1}{4^\alpha}\right)\cdots,$$

where we assume $\frac{1}{3} < \alpha \leqq \frac{1}{2}$, that $\Pi(1+a_n)$ may converge even when Σa_n and Σa_n^2 *both* diverge.

Chapter VIII.

Closed and numerical expressions for the sums of series.

§ 31. Statement of the problem.

In Chapters III and IV, we were concerned mainly with our problem A, the question of the convergence of series, and it was not till the last few chapters that we considered also the sum of the series. This latter point of view we shall now place in the foreground. It is necessary, however, in order to supplement our developments of pp. 78–79 and 105, that we should make it quite clear once more what is the significance of the questions which arise in this connection. If, for instance, we have proved the relation **122**:

$$\frac{\pi}{4} = 1 - \frac{1}{3} + \frac{1}{5} - \frac{1}{7} + - \cdots,$$

we may interpret it in two ways. On the one hand, the equation indicates that the sum of the series on the right has the value $\frac{\pi}{4}$, one quarter of the value of a number[1] which we meet with in many other connections and to which approximations are well-known. In this sense, it may be claimed that we have specified the sum of the series written down above. But such a statement can only hold in a very relative sense; for it is not possible to give a complete specification of the number π, otherwise than by a nest of intervals or some equivalent symbol, and such a symbol is precisely furnished by the *series*; i. e. the expression on the *right*, in the above equation. We are therefore equally justified in claiming the exact opposite, namely that the equation provides an (extremely simple) *expression for the number* π in series form, — that is to say, by means of a convergent sequence of numbers, — which happens indeed in our case to have a peculiarly straightforward and convenient form and may also (**69**, 1) be immediately expressed as a nest of intervals[2].

The circumstances are entirely altered when we come to the equation (cf. **68**, 2b):

$$\frac{1}{1\cdot 2} + \frac{1}{2\cdot 3} + \frac{1}{3\cdot 4} + \cdots = \mathbf{1}.$$

---

[1] In former times, when these matters were all interpreted rather geometrically, $\frac{\pi}{4}$ was always thought of as the ratio of the area of a circle to that of the circumscribed square.

[2] Namely:

$$\frac{\pi}{4} = (s_{2k} \,|\, s_{2k+1}),$$

where

$$s_n = 1 - \frac{1}{3} + \frac{1}{5} - + \cdots + \frac{(-1)^{n-1}}{2n-1}, \qquad (n = 2, 3, \ldots)$$

Here we are perfectly satisfied with the statement that the sum of the series is $=1$, precisely because the number 1 (and similarly every rational number) can be fully and literally assigned. In such cases, we have a perfect right to assert that we have a closed expression for the sum of the series. But in all other cases, where the sum of the series is not a rational number, or at any rate not known to be one[3], we cannot strictly speak of evaluating the sum of the series by means of a closed expression. On the contrary, the series ought then to be regarded as a (more or less imperfect) means of *representing* or *approximating to* its sum. By proceeding to express these *approximations* (usually in the form of decimal fractions) and estimating the errors involved, we form what is called a *numerical evaluation* of the sum.

Lastly, as above in the case of the series for $\frac{\pi}{4}$, we may have ascertained merely that the given series has for sum a number related in some simple (or at any rate specifiable) manner to a number which we meet with in other connections; as e. g. it follows from **122** and **124** that

$$1+\frac{1}{2}\cdot\frac{1}{3}+\frac{1\cdot 3}{2\cdot 4}\cdot\frac{1}{5}+\cdots=2\left[1-\frac{1}{3}+\frac{1}{5}-+\cdots\right].$$

In that case, we should still welcome the information so obtained, since it establishes a connection between results where formerly we saw none. It is usual, in such cases, still to say — though in an extended sense — that we have evaluated the sum by means of a closed expression; in fact, the number concerned is then regarded as "known" through those other connections, and we simply express the sum of the series "by means of a closed expression" involving this number. Here the student must, however, guard against self-delusion. If it has been ascertained, for instance (v. p. 211) that the sum of the series

$$1+\frac{1}{2}\cdot\frac{1}{50}+\frac{1\cdot 3}{2\cdot 4}\cdot\frac{1}{50^2}+\frac{1\cdot 3\cdot 5}{2\cdot 4\cdot 6}\cdot\frac{1}{50^3}+\cdots$$

has the value $\frac{5}{7}\sqrt{2}$, it is still only in a very relative sense "determined in the form of a closed expression". The number $\sqrt{2}$ is not *per se* any better known than the sum of any *arbitrary* convergent series. It is only because $\sqrt{2}$ occurs in so many hundreds of other connections and has, for practical purposes, been so often evaluated numerically, that we are in the habit of considering its value as almost as perfectly "known" as any literally specified rational number. If

[3] For instance, if we have determined the sum of a series to be equal to Euler's constant, we do not know to this day whether we are confronted with a rational number or not.

instead of the above series, we consider, for instance, the following binomial series:

$$\frac{5}{2}\left[1+\frac{1}{5}\cdot\frac{24}{1000}-\frac{4}{5\cdot 10}\cdot\frac{24^2}{1000^2}+\frac{4\cdot 9}{5\cdot 10\cdot 15}\cdot\frac{24^3}{1000^3}-+\cdots\right]$$

and its sum has been ascertained to be equal to $\sqrt[5]{100}$, we shall be less inclined to regard the sum as fully determined thereby; on the contrary, we shall prefer to accept the series as a most useful means of evaluating $\sqrt[5]{100}$ to a degree of approximation not so easily attainable by other means. In other words, — with the exception of those few cases in which the sum of a series can be specified as a *definite rational number*, — when we consider equalities of the form "$s=\Sigma a_n$", the emphasis will be laid sometimes on the right hand side and sometimes on the left, according to the circumstances of the case. If s may be considered as known through other connections, we shall still (though in an extended sense) say that the sum of the series has been evaluated *in the form of a closed expression.* If this is not the case, we shall say that the series is a means of evaluating the number s (of which it provides the definition). (Obviously both points of view may be taken with regard to the same equality.) In the former of the two cases, we shall, so to speak, have achieved our object, since the problem B (v. p. 105) also is then solved to our satisfaction. In the latter case, however, a new task now begins, that of actually expressing the approximations, provided by the series itself, to its sum, in a convenient and simple form (e. g. in decimal fraction form, as the most desirable for our purposes), and of estimating the errors involved in these approximations.

§ 32. Evaluation of the sum of a series by means of a closed expression.

1. Direct evaluation. It is obvious that we may without difficulty *construct* series with any *assigned* sum. If s be the assigned sum, construct, by any one of the many processes at our disposal, a sequence (s_n) converging to s, and consider the series

$$s_0+(s_1-s_0)+(s_2-s_1)+\cdots+(s_n-s_{n-1})+\cdots.$$

Since its nth partial sum is precisely $=s_n$, this series is convergent and has the sum s. This simple procedure affords an inexhaustible means of *constructing* series capable of summation in the form of a closed expression; e. g. we need only assume one of the numerous null sequences (x_n) known to us, and write $s_n=s-x_n$, $n=0, 1, 2, \ldots$.

Examples of series of sum 1.

1. $(x_n)\equiv\left(\frac{1}{n+1}\right)$ gives $\frac{1}{1\cdot 2}+\frac{1}{2\cdot 3}+\frac{1}{3\cdot 4}+\cdots=1$

2. $(x_n) \equiv \left(\frac{(-1)^n}{n+1}\right)$ „ $\frac{3}{1\cdot 2} - \frac{5}{2\cdot 3} + \frac{7}{3\cdot 4} - \frac{9}{4\cdot 5} + - \cdots = 1$

3. $(x_n) \equiv \left(\frac{1}{n+1}\right)^2$ „ $\sum_{n=1}^{\infty} \frac{2n+1}{n^2(n+1)^2} = 1$

4. $(x_n) \equiv \left(\frac{1}{n+1}\right)^3$ „ $\sum_{n=1}^{\infty} \frac{3n^2+3n+1}{n^3(n+1)^3} = 1$

5. $(x_n) \equiv \left(\frac{1}{2^n}\right)$ „ $\sum_{n=1}^{\infty} \frac{1}{2^n} = 1$

6. $(x_n) \equiv \left(\frac{1}{\sqrt{(n+1)}}\right)$ „ $\sum_{n=1}^{\infty} \frac{1}{\sqrt{n(n+1)}\,(\sqrt{n}+\sqrt{n+1})} = 1.$

7. If we multiply the terms of one of these series by s, we obtain a convergent series of sum s.

It is not superfluous to be able to construct such examples, as we shall see that the power to provide series with known sum is an advantage in the discussion of further series.

The converse of the principle just treated is expressed by the

° **Theorem.** *Given a series* $\sum_{n=0}^{\infty} a_n$, *whose terms* a_n *are expressible in the form* $a_n = x_n - x_{n+1}$, *where* x_n *is the term of a convergent sequence of known limit* ξ, *the sum of the series can be specified, for we have* **131.**

$$\sum_{n=0}^{\infty} a_n = x_0 - \xi.$$

Proof. We may write

$$s_n = (x_0 - x_1) + (x_1 - x_2) + \cdots + (x_n - x_{n+1}) = x_0 - x_{n+1}.$$

Since $x_n \to \xi$, the statement follows.

Examples. **132.**

1. If α be any real number $\neq 0, -1, -2, \ldots$, then (v. **68**, 2h):

$$\sum_{n=0}^{\infty} \frac{1}{(\alpha+n)(\alpha+n+1)} = \frac{1}{\alpha}, \quad \text{as here} \quad a_n = \left[\frac{1}{\alpha+n} - \frac{1}{\alpha+n+1}\right].$$

2. Similarly

$$\sum_{n=0}^{\infty} \frac{1}{(\alpha+n)(\alpha+n+1)(\alpha+n+2)} = \frac{1}{2\alpha(\alpha+1)}$$

as here

$$a_n = \frac{1}{2}\left[\frac{1}{(\alpha+n)(\alpha+n+1)} - \frac{1}{(\alpha+n+1)(\alpha+n+2)}\right].$$

3. Generally, if p denotes any positive integer,

$$\sum_{n=0}^{\infty} \frac{1}{(\alpha+n)(\alpha+n+1)\ldots(\alpha+n+p)} = \frac{1}{p}\,\frac{1}{\alpha(\alpha+1)\ldots(\alpha+p-1)}.$$

4. Putting $\alpha = \frac{1}{3}$, we thus obtain, for instance, from 2.:

$$\frac{1}{1\cdot 4\cdot 7} + \frac{1}{4\cdot 7\cdot 10} + \frac{1}{7\cdot 10\cdot 13} + \cdots = \frac{1}{24}.$$

5. Putting $\alpha = 1$ in 3. we obtain

$$\frac{1}{1\cdot 2\ldots(p+1)} + \frac{1}{2\cdot 3\ldots(p+2)} + \cdots = \frac{1}{p\cdot p!}$$

or

$$\sum_{n=0}^{\infty} \frac{1}{\binom{p+n+1}{p+1}} = \frac{p+1}{p}.$$

The following is a somewhat more general theorem.

133. ○ **Theorem.** *If the term a_n of a given series Σa_n is expressible in the form $x_n - x_{n+q}$, where x_n is the term of a convergent sequence of known limit ξ, and q denotes a fixed integer > 0, then*

$$\sum_{n=0}^{\infty} a_n = x_0 + x_1 + \cdots + x_{q-1} - q\,\xi.$$

Proof. We have, for $n > q$,

$$s_n = (x_0 - x_q) + (x_1 - x_{q+1}) + \cdots + (x_{q-1} - x_{2q-1}) + (x_q - x_{2q}) + \cdots + (x_n - x_{n+q})$$
$$= (x_0 + x_1 + \cdots + x_{q-1}) - (x_{n+1} + x_{n+2} + \cdots + x_{n+q}).$$

Since $x_\nu \to \xi$ (by **41**, 9), the statement at once follows.

Examples.

1. $$\sum_{n=0}^{\infty} \frac{1}{(\alpha+n)(\alpha+n+q)} = \frac{1}{q}\left(\frac{1}{\alpha} + \frac{1}{\alpha+1} + \cdots + \frac{1}{\alpha+q-1}\right),$$

since here we have

$$a_n = \frac{1}{q}\left(\frac{1}{\alpha+n} - \frac{1}{\alpha+n+q}\right).$$

In particular, writing $\alpha = \frac{1}{2}$,

$$\sum_{n=0}^{\infty} \frac{1}{(2n+1)(2n+2q+1)} = \frac{1}{2q}\left(1 + \frac{1}{3} + \cdots + \frac{1}{2q-1}\right).$$

2. For $\alpha = 1$ and $q = 2$ we have accordingly:

$$\frac{1}{1\cdot 3} + \frac{1}{2\cdot 4} + \frac{1}{3\cdot 5} + \cdots = \frac{3}{4};$$

and for $\alpha = \frac{1}{2}$, $q = 3$:

$$\frac{1}{1\cdot 7} + \frac{1}{3\cdot 9} + \frac{1}{5\cdot 11} + \cdots = \frac{23}{90}.$$

3. Somewhat more generally, if k, as well as q, denotes a fixed integer > 0:

$$\sum_{n=0}^{\infty} \frac{1}{(\alpha+n)(\alpha+n+q)\ldots(\alpha+n+kq)} =$$
$$= \frac{1}{kq}\sum_{\nu=0}^{q-1} \frac{1}{(\alpha+\nu)(\alpha+\nu+q)\ldots(\alpha+\nu+\overline{k-1}\,q)}.$$

4. Thus for $\alpha = \frac{1}{2}$, $q = 2$, $k = 2$ we find

$$\frac{1}{1\cdot 5\cdot 9} + \frac{1}{3\cdot 7\cdot 11} + \frac{1}{5\cdot 9\cdot 13} + \cdots = \frac{13}{420}.$$

The artifices here employed may be extended to obtain, finally, the following considerably further reaching

134. ○ **Theorem.** *If the terms of a series Σa_n are expressible, for every n, in the form*

$$a_n = c_1 x_{n+1} + c_2 x_{n+2} + \cdots + c_k x_{n+k} \quad (k \text{ constant}, \geqq 2)$$

where (x_n) denotes a convergent sequence of known limit ξ, and the coefficients c_λ satisfy the condition

$$c_1 + c_2 + \cdots + c_k = 0,$$

then Σa_n *is convergent and has for sum:*

$$\sum_{n=0}^{\infty} a_n = c_1 x_1 + (c_1 + c_2) x_2 + \cdots + (c_1 + c_2 + \cdots + c_{k-1}) x_{k-1}$$
$$+ (c_2 + 2 c_3 + \cdots + \overline{k-1}\, c_k) \xi .$$

The proof is at once obtained by writing the expressions for $a_1, a_2, \ldots, a_m$, one below the other so that terms involving x_ν occupy the same vertical. Carrying out the addition in columns, — which of course is allowed even without reference to the main rearrangement theorem — we find, for $m > k$, taking into account the condition fulfilled by the coefficients c_λ,

$$\sum_{n=0}^{m} a_n = \sum_{\lambda=1}^{k-1} (c_1 + c_2 + \cdots + c_\lambda) x_\lambda + \sum_{\lambda=1}^{k-1} (c_{\lambda+1} + \cdots + c_k) x_{m+\lambda+1},$$

which is again the sum of a *finite* number of terms. Letting $m \to \infty$, we at once obtain the required relation.

Examples.

1. Putting $x_n = \frac{n^2}{n^2+1}$, $k = 2$, $c_1 = -1$, $c_2 = +1$, we obtain

$$\frac{3}{2\cdot 5} + \frac{5}{5\cdot 10} + \frac{7}{10\cdot 17} + \cdots + \frac{2n+1}{(n^2+1)(\overline{n+1}^2+1)} + \cdots = \frac{1}{2}.$$

2. $$\sum_{n=0}^{\infty} \frac{1}{(3n+1)(3n+10)} = \frac{1}{27} \sum_{n=0}^{\infty} \left(\frac{1}{-\frac{2}{3}+n+1} - \frac{1}{-\frac{2}{3}+n+4} \right) = \frac{13}{84}.$$

These examples may of course easily be multiplied to any extent desired.

2. **Application to the elementary functions.** The above few theorems have, speaking generally, made us familiar with all types of series which may, without requiring any more refined artifices, be summed in the form of a closed expression.

By far the most frequent series, in all applications, are those obtained by substituting particular values for x in series expansions of elementary functions and in series derived from these by every species of transformation or combination, or other known processes of deduction. Examples, obtained in this manner, of summation by closed expressions are innumerable. We must content ourselves with referring the reader to the particularly ample selection of examples at the end of this chapter, in the working out of which the student will rapidly become familiar with all the main artifices used in this connection. The developments in this and the following section will afford further guidance in this part of the subject. Let us merely observe quite generally, for the moment, that it is often possible to deal with a given series by splitting it up into two or more parts, each of which again represents a convergent series; or else by adding to or subtracting from Σa_n, term by term, a second series of known sum. In particular, if a_n is a rational function of n its *expansion in partial fractions* will frequently be a considerable help.

3. Application of *Abel*'s theorem of limits. A further means of evaluating the sum of a series, — one of great theoretical importance, differing from that just indicated in the principle it involves, though in most cases intimately connected with it in virtue of **101**, — consists in applying *Abel*'s theorem of limits. Given a convergent series Σa_n, the power series $f(x) = \Sigma a_n x^n$ converges at least for $-1 < x \leqq +1$, and hence, by **101**,

$$\Sigma a_n = \lim_{x \to 1-0} f(x).$$

If we suppose that the function $f(x)$ which the power series represents is so far known, that the latter limit can be evaluated, the summation of the series is achieved. The developments of Chapter VI offer a wide basis for this mode of procedure, and in fact *Abel*'s theorem has already been used there more than once in the sense now explained.

We shall give here only a few relatively obvious examples, with a reference to the exercises at the end of this chapter.

135. Examples. We are already acquainted with the series:

$$1.\quad \sum_{x=0}^{\infty} \frac{(-1)^n}{n+1} = \lim_{x\to 1-0} \sum_{n=0}^{\infty} (-1)^n \frac{x^{n+1}}{n+1} = \lim_{x\to 1-0} \log(1+x) = \log 2,$$

$$2.\quad \sum_{n=0}^{\infty} \frac{(-1)^n}{2n+1} = \lim_{x\to 1-0} \sum_{n=0}^{\infty} (-1)^n \frac{x^{2n+1}}{2n+1} = \lim_{x\to 1-0} \tan^{-1} x = \frac{\pi}{4}.$$

We have the further example

$$3.\quad \sum_{n=0}^{\infty} \frac{(-1)^n}{3n+1} = \lim_{x\to 1-0} \left(x - \frac{x^4}{4} + \frac{x^7}{7} - + \cdots\right).$$

The series inside the bracket has for derived series

$$1 - x^3 + x^6 - + \cdots = \frac{1}{1+x^3}$$

and therefore represents the function (v. § 19, Def. 12)

$$\int_0^x \frac{dx}{1+x^3} = \frac{1}{6}\log\frac{(x+1)^2}{x^2-x+1} + \frac{1}{\sqrt{3}}\tan^{-1}\frac{2x-1}{\sqrt{3}} + \frac{\pi}{6\sqrt{3}}.$$

Accordingly, the sum of the given series is $= \frac{1}{3}\log 2 + \frac{\pi}{3\sqrt{3}}$.

4. Similarly we find (v. § 19, Def. 12)

$$\sum_{n=0}^{\infty} \frac{(-1)^n}{4n+1} \equiv 1 - \frac{1}{5} + \frac{1}{9} - \frac{1}{13} + - \cdots = \frac{1}{8}\sqrt{2}\left[\pi + \log(3+2\sqrt{2})\right].$$

For further series constructed on the same lines, the formulae of course become more and more complicated.

4. Application of the main rearrangement theorem. Equally great theoretical and practical significance attaches, in our present problem, to the application of the main rearrangement theorem. This application we proceed at once to illustrate by one of the most important cases; additional examples will again be furnished by the exercises.

In **115** and **117**, we obtained two entirely distinct expansions of the function $x \cot x$, both valid at least for every sufficiently small $|x|$.

If, in the first of these, we replace x by πx, we obtain, certainly for every sufficiently small $|x|$,

$$1 + \sum_{n=1}^{\infty} (-1)^n \frac{2^{2n} B_{2n}}{(2n)!} (\pi x)^{2n} = 1 - \sum_{k=1}^{\infty} \frac{2x^2}{k^2 - x^2}.$$

Each term of the series on the right may obviously be expanded in powers of x:

$$-\frac{2x^2}{k^2 - x^2} = -\sum_{n=1}^{\infty} 2\left(\frac{x^2}{k^2}\right)^n. \quad (k = 1, 2, \ldots \textit{ fixed})$$

These are the series $z^{(k)}$ of the main rearrangement theorem; since the series $\zeta^{(k)}$ of that theorem in our case only differ in sign from the series $z^{(k)}$ themselves, the conditions of that theorem are all fulfilled, and we may sum *in columns*. The coefficient of x^{2p} on the right then becomes

$$= -2\sum_{k=1}^{\infty} \frac{1}{k^{2p}} \qquad (p \textit{ fixed})$$

and since, by **97**, it has to coincide with that on the left, we obtain the important result (once more denoting the index of summation by n)

$$\sum_{n=1}^{\infty} \frac{1}{n^{2p}} = (-1)^{p-1} \frac{B_{2p} (2\pi)^{2p}}{2(2p)!}. \qquad (p \textit{ fixed}) \quad \textbf{136.}$$

This gives us the sum of the series

$$1 + \frac{1}{2^{2p}} + \frac{1}{3^{2p}} + \cdots + \frac{1}{n^{2p}} + \cdots \qquad (p \textit{ fixed})$$

in the form of a closed expression, since the number π and the (rational) *Bernoulli*'s numbers may be regarded as known[4].

In particular,

$$\sum_{n=1}^{\infty} \frac{1}{n^2} = \frac{\pi^2}{6}, \qquad \sum_{n=1}^{\infty} \frac{1}{n^4} = \frac{\pi^4}{90}, \qquad \sum_{n=1}^{\infty} \frac{1}{n^6} = \frac{\pi^6}{945}.$$

---

[4] Quite incidentally, formula **136** shows that *Bernoulli*'s numbers B_{2n} are of alternating signs and that $(-1)^{n-1} B_{2n}$ is positive; further, that they increase with extreme rapidity as n increases; for since the value of $\sum_{k=1}^{\infty} \frac{1}{k^{2n}}$ lies between 1 and 2, whatever be the value of n, we necessarily have

$$2\frac{2(2n)!}{(2\pi)^{2n}} > (-1)^{n-1} B_{2n} > \frac{2(2n)!}{(2\pi)^{2n}},$$

whence it follows that $\left|\frac{B_{2n+2}}{B_{2n}}\right| \to +\infty$. Finally, as the above transformation holds for $|x| < 1$, it also follows that the series **115** converges absolutely at least for $|x| < \pi$. But for $|x| > \pi$ it certainly cannot converge absolutely, for then $\cot x$ would be continuous for $x = \pi$, by **98**, 2, which we know is not the case; thus the series **115** has exactly the radius π. It follows from this that **116a** has the radius $\frac{\pi}{2}$, **116b** the radius π.

It is not superfluous to try to realise all that was needed to obtain even the first of these elegant formulae[5]. This will be seen to involve much of our investigations up to this point.

The above provides us with the sum of every harmonic series with an *even integral* exponent; we know nothing yet of the sum of a harmonic series with *odd* exponent (>1); that is to say, we have not succeeded as yet in finding any obvious relations that might result in connecting such a sum $\left(\text{e. g. } \Sigma \frac{1}{k^3}\right)$ with any numbers occurring elsewhere. (There is of course no obstacle to our evaluating the sum of any harmonic series numerically, to any degree of approximation[6]; v. § 35). On the other hand, our results readily yield the following further formulae: We have

$$\sum_{n=1}^{\infty} \frac{1}{n^{2p}} = \sum_{\nu=1}^{\infty} \frac{1}{(2\nu-1)^{2p}} + \sum_{\nu=1}^{\infty} \frac{1}{(2\nu)^{2p}}.$$

The latter series is precisely the same thing as $\frac{1}{2^{2p}} \sum_{n=1}^{\infty} \frac{1}{n^{2p}}$. Subtracting this from both sides, we obtain

$$\sum_{n=1}^{\infty} \frac{1}{(2n-1)^{2p}} = \left(1 - \frac{1}{2^{2p}}\right) \sum_{n=1}^{\infty} \frac{1}{n^{2p}}$$

or

137. $$1 + \frac{1}{3^{2p}} + \frac{1}{5^{2p}} + \cdots = (-1)^{p-1} \frac{2^{2p}-1}{2\,(2p)!} B_{2p} \cdot \pi^{2p}.$$

For $p = 1, 2, 3, \ldots$, the sums are in particular

$$\frac{\pi^2}{8}, \quad \frac{\pi^4}{96}, \quad \frac{\pi^6}{960}, \quad \ldots$$

If we again subtract the same series $\frac{1}{2^{2p}} \sum_{n=1}^{\infty} \frac{1}{n^{2p}}$, we obtain

$$\sum_{n=1}^{\infty} \frac{(-1)^{n-1}}{n^{2p}} = \left(1 - \frac{2}{2^{2p}}\right) \sum_{n=1}^{\infty} \frac{1}{n^{2p}}$$

or

138. $$1 - \frac{1}{2^{2p}} + \frac{1}{3^{2p}} - \frac{1}{4^{2p}} + - \cdots = (-1)^{p-1} \frac{2^{2p-1}-1}{(2p)!} B_{2p} \cdot \pi^{2p}.$$

---

[5] *James* and *John Bernoulli* did their utmost to sum the series

$$1 + \frac{1}{4} + \frac{1}{9} + \frac{1}{16} + \cdots.$$

The former of the two did not live to see the solution of the problem, which was found by *Euler* in 1736. *John Bernoulli*, to whom it became known soon after, wrote in this connection (Werke, Vol. 4, p. 22): Atque ita satisfactum est ardenti desiderio Fratris mei, qui agnoscens summae huius pervestigationem *difficiliorem quam quis putaverit*, ingenue fassus est omnem suam industriam fuisse elusam ... Utinam Frater superstes esset! A second proof, of a quite different kind, will be found in **156**, a third in **189**, and a fourth in **210**.

[6] *T. J. Stieltjes* (Tables des valeurs des sommes $S_k = \sum_{n=1}^{\infty} \frac{1}{n^k}$, Acta mathematica, Vol. 10, p. 299, 1887) evaluated the sums of these series, up to the exponent 70, to 32 places of decimals.

In particular, for $p = 1, 2, 3, \ldots$ the sums are

$$\frac{1}{12}\pi^2, \quad \frac{7}{720}\pi^4, \quad \frac{31}{30240}\pi^6, \ldots.$$

Here again, however, we know nothing of the corresponding series with odd exponents. — The last two results might of course also have been obtained by starting with the expansions in partial fractions of the functions tan or $\frac{1}{\sin}$, and reasoning as above for that of the function cot. We may deduce further results by treating the expansion in partial fractions, given in **118**, of the function $\frac{1}{\cos}$, i. e.

$$\frac{\pi}{4\cos\frac{\pi x}{2}} = \frac{1}{1^2 - x^2} - \frac{3}{3^2 - x^2} + \frac{5}{5^2 - x^2} - + \cdots \equiv \sum_{\nu=0}^{\infty} \frac{(-1)^\nu (2\nu+1)}{(2\nu+1)^2 - x^2}.$$

The νth term is here expressible by the power series

$$(-1)^\nu \sum_{k=0}^{\infty} \frac{x^{2k}}{(2\nu+1)^{2k+1}};$$

after rearranging, the coefficient of x^{2p} thus becomes:

$$\sum_{\nu=0}^{\infty} \frac{(-1)^\nu}{(2\nu+1)^{2p+1}} \equiv 1 - \frac{1}{3^{2p+1}} + \frac{1}{5^{2p+1}} - + \cdots.$$

Let us denote these sums provisionally by σ_{2p+1}; then

$$\frac{\pi}{4\cos\frac{\pi x}{2}} = \sigma_1 + \sigma_3 x^2 + \sigma_5 x^4 + \cdots$$

or

$$\frac{1}{\cos z} = \frac{4}{\pi}\left[\sigma_1 + \sigma_3\left(\frac{2z}{\pi}\right)^2 + \sigma_5\left(\frac{2z}{\pi}\right)^4 + \cdots\right].$$

On the other hand, this power series may be obtained by direct division and its coefficients — just like *Bernoulli*'s numbers in **105**, 5 — by simple recurring formulae. We usually write

$$\frac{1}{\cos z} = \sum_{n=0}^{\infty} (-1)^n \frac{E_{2n}}{(2n)!} z^{2n},$$

so that

$$\left(1 - \frac{x^2}{2!} + \frac{x^4}{4!} - + \cdots\right)\left(E_0 - \frac{E_2}{2!}x^2 + \frac{E_4}{4!}x^4 - + \cdots\right) \equiv 1.$$

This gives $E_0 = 1$, and, for every $n \geqq 1$, recurring formulae[7] which may be written as follows (after multiplication by $(2n)!$):

$$E_{2n} + \binom{2n}{2} E_{2n-2} + \binom{2n}{4} E_{2n-4} + \cdots + E_0 = 0, \qquad \textbf{139.}$$

[7] The numbers determined by these formulae (which are moreover *rational integral numbers*) are usually referred to as ***Euler's numbers***. The numbers E_ν up to $\nu = 30$ have been calculated by *W. Scherk*, Mathem. Abh., Berlin 1825.

or in the shorter symbolical form (cf. **106**):

$$(E+1)^k+(E-1)^k=0,$$

now holding for every $k\geqq 1$.

We deduce without difficulty:

$$E_1=E_3=E_5=\cdots=0$$

and

$$E_0=1,\quad E_2=-1,\quad E_4=5,\quad E_6=-61,\quad E_8=1385,\ldots.$$

In terms of these numbers, which we are perfectly justified in considering as known, we have, finally,

$$(-1)^p\frac{E_{2p}}{(2p)!}=\frac{4}{\pi}\sigma_{2p+1}\cdot\frac{2^{2p}}{\pi^{2p}},$$

i. e.

140. $$1-\frac{1}{3^{2p+1}}+\frac{1}{5^{2p+1}}-+\cdots\equiv(-1)^p\frac{E_{2p}}{2^{2p+2}(2p)!}\pi^{2p+1}.$$

In particular, for $p=0,1,2,3,\ldots$, this gives the values

$$\frac{1}{4}\pi,\quad\frac{1}{32}\pi^3,\quad\frac{5}{1536}\pi^5,\quad\frac{61}{2^{12}\cdot 45}\pi^7,\ldots$$

for the sums of the corresponding series.

§ 33. Transformation of series.

In the preceding section (§ 32), we became acquainted with the most important types of series which can be summed by means of a *closed expression* — either in the stricter or in the wider sense of the term. In the evaluations last made, which are really of a profound nature, the main rearrangement theorem played an essential part; indeed, in virtue of this theorem, the original series was changed, so to speak, into a completely different series which then yielded further information. We were therefore principally concerned with a special *transformation of series*[8]. Such transformations are frequently of the greatest use, and indeed even more so in the numerical calculations which form the subject of the following two sections, than in the determination of closed expressions for the sums of series. To these transformations we will now turn our attention, and we start at once with a more general conception of the transformation deduced from the main rearrangement theorem and repeatedly applied to advantage already in the preceding section.

[8] Such transformations were first indicated by *J. Stirling* (Methodus differentialis, London 1730); they are based, in his case, on similar lines to the above, excepting that he fails to verify the fulfilment of the conditions under which the processes are valid.

Given a convergent series $\sum_{k=0}^{\infty} z^{(k)}$, let each of its terms be expressed, *in any manner*, (e. g. by § 32, p. 232) as the sum of an infinite series:

$$\text{(A)} \qquad \begin{cases} z^{(0)} = a_0^{(0)} + a_1^{(0)} + a_2^{(0)} + \cdots + a_n^{(0)} + \cdots \\ z^{(1)} = a_0^{(1)} + a_1^{(1)} + a_2^{(1)} + \cdots + a_n^{(1)} + \cdots \\ \cdot \quad \cdot \quad \cdot \quad \cdot \quad \cdot \quad \cdot \quad \cdot \quad \cdot \quad \cdot \quad \cdot \\ z^{(k)} = a_0^{(k)} + a_1^{(k)} + a_2^{(k)} + \cdots + a_n^{(k)} + \cdots \\ \cdot \quad \cdot \quad \cdot \quad \cdot \quad \cdot \quad \cdot \quad \cdot \quad \cdot \quad \cdot \quad \cdot \end{cases}$$

We shall assume further that the vertical columns in this array themselves constitute convergent series, and denote their sums by $s^{(0)}, s^{(1)}, \ldots, s^{(n)}, \ldots$. *Under what conditions may the series* $\sum_{n=0}^{\infty} s^{(n)}$ *formed by these numbers be expected to converge, with*

$$\sum_{k=0}^{\infty} z^{(k)} = \sum_{n=0}^{\infty} s^{(n)}?$$

If this equality is justified, we have certainly effected a transformation of the given series. The *main rearrangement theorem* immediately gives the

°**Theorem.** *If the horizontal rows of the array* (A) *all constitute* **141.** *absolutely convergent series and — denoting by* $\zeta^{(k)}$ *the sum,* $\sum_{n=0}^{\infty} |a_n^{(k)}|$, *of the absolute values of the terms in one row —, if the series* $\sum \zeta^{(k)}$ *is convergent, the series* $\sum s^{(n)}$ *also converges and* $= \sum z^{(k)}$.

It is this theorem that we have applied in the preceding paragraph. The question arises whether its requirements are not unnecessarily stringent, whether the transformation is not allowed under very much wider conditions.

A. In this direction, an extremely far-reaching theorem was proved by *A. Markoff*[9]. He assumes first only that the series constituted by the vertical columns of the array (A) converge, as well as the original series and the series constituted by the horizontal rows of the array. The numbers $s^{(n)}$ have thus determinate values. Since $\sum_{k=0}^{\infty} z^{(k)}$ and $\sum_{k=0}^{\infty} a_0^{(k)}$ converge, so does $\sum_{k=0}^{\infty} (z^{(k)} - a_0^{(k)})$; and also, similarly, for any fixed m, the series

$$\sum_{k=0}^{\infty} \left(z^{(k)} - a_0^{(k)} - a_1^{(k)} - \cdots - a_{m-1}^{(k)}\right) \qquad (m \text{ fixed}).$$

[9] Mémoire sur la transformation de séries (Mém. de l'Acad. Imp. de St. Pétersburg, (7) Vol. 37. 1891). Cf. a note by the author, "Einige Bemerkungen zur *Kummer*schen und *Markoff*schen Reihentransformation", Sitzungsberichte der Berl. Math. Ges., Vol. 19. pp. 4—17, 1919.

The terms of this series are, however, precisely the remainders, each with the initial[10] index m, of the series constituted by the individual rows of the array. If, for brevity, we denote these remainders by $r_m^{(k)}$, so that

$$r_m^{(k)} = \sum_{n=m}^{\infty} a_n^{(k)} \qquad (k \text{ and } m \text{ fixed}),$$

the series

$$\sum_{k=0}^{\infty} r_m^{(k)} = R_m \qquad (m \text{ fixed})$$

is convergent. The further assumption is then made that

$$R_m \to 0 \quad \text{when} \quad m \to \infty.$$

It may be shown that under these hypotheses $\Sigma s^{(n)}$ converges and $= \Sigma z^{(k)}$. The theorem obtained will thus be as follows:

142. ° ***Markoff's transformation of series.*** *Let a convergent series $\sum_{k=0}^{\infty} z^{(k)}$ be given with each of its terms itself expressed as a convergent series:*

$$\text{(A)} \qquad z^{(k)} = a_0^{(k)} + a_1^{(k)} + \ldots + a_n^{(k)} + \ldots \quad (k = 0, 1, 2, \ldots).$$

Let the individual columns $\sum_{k=0}^{\infty} a_n^{(k)}$ of the array (A) *so formed represent convergent series with sum $s^{(n)}$, $n = 0, 1, 2, \ldots$, so that the remainders*

$$r_m^{(k)} = \sum_{n=m}^{\infty} a_n^{(k)} \qquad (m \geqq 0)$$

of the series in the horizontal rows also constitute a convergent series

$$\sum_{k=0}^{\infty} r_m^{(k)} = R_m \qquad (m \text{ fixed}).$$

In order that the sums by vertical columns should form a convergent series $\Sigma s^{(n)}$, it is necessary and sufficient that $\lim R_m = R$ should exist; and in order that the relation

$$\sum_{n=0}^{\infty} s^{(n)} = \sum_{k=0}^{\infty} z^{(k)}$$

should hold as well, it is necessary and sufficient that this limit R should be 0.

The proof is almost trivial, for we have

$$\text{(a)} \qquad s^{(0)} + s^{(1)} + \ldots + s^{(n)} = R_0 - R_{n+1},$$

whence the first statement is immediate. Since it follows that

$$\sum_{n=0}^{\infty} s^{(n)} = R_0 - R,$$

and since R_0 is simply $\sum_{k=0}^{\infty} r_0^{(k)} = \sum_{k=0}^{\infty} z^{(k)}$, the second statement now follows also.

[10] Here we of course take $m = 0$ to give the *whole* series, i. e. $z^{(k)}$ itself.

B. The superiority of *Markoff*'s transformation over Theorem **141** consists, of course, in the absence of any mention of absolute convergence, only convergence pure and simple being required throughout. Its applications are numerous and fruitful: those bearing on numerical evaluations will be considered in §35, and we shall only indicate in this place one of the prettiest of its applications, which consists in obtaining a transformation given by *Euler* [11] — of course, in his case, without any considerations of convergence.

It is advantageous here to use the notation of the calculus of finite differences, and this we will accordingly first elucidate in brief. Given any sequence $(x_0, x_1, x_2, \ldots)$, the numbers

$$x_0 - x_1, \quad x_1 - x_2, \quad \ldots, \quad x_k - x_{k+1}, \quad \ldots$$

are called the *first differences* of (x_n) and are denoted by

$$\Delta x_0, \quad \Delta x_1, \quad \ldots, \quad \Delta x_k, \quad \ldots$$

The differences of the first order of (Δx_n), i. e. the numbers $\Delta x_k - \Delta x_{k+1}$, $k = 0, 1, 2, \ldots$, are called the *second differences* of (x_n), denoted by

$$\Delta^2 x_0, \quad \Delta^2 x_1, \ldots, \Delta^2 x_k, \ldots$$

In general, we write for $n \geqq 1$

$$\Delta^{n+1} x_k = \Delta^n x_k - \Delta^n x_{k+1} \qquad (k = 0, 1, 2, \ldots)$$

and this formula may also be taken to comprise the case $n = 0$ if we interpret $\Delta^0 x_k$ as being the number x_k itself. It is convenient to imagine the numbers x_k and $\Delta^n x_k$ arranged in rows so as to form the following triangular array, in which each difference occupies the place in its own row immediately below the space, in the row above, between the two terms whose difference it is:

$$(\Delta) \qquad \begin{array}{ccccccccccc}
x_0, & & x_1, & & x_2, & & x_3, & & x_4, & \ldots\ldots \\
& \Delta x_0, & & \Delta x_1, & & \Delta x_2, & & \Delta x_3, & \ldots\ldots & \\
& & \Delta^2 x_0, & & \Delta^2 x_1, & & \Delta^2 x_2, & \ldots\ldots & & \\
& & & \Delta^3 x_0, & & \Delta^3 x_1, & \ldots\ldots & & & \\
& & & & \Delta^4 x_0, & \ldots\ldots & & & & \\
& & & & & \ldots\ldots & & & &
\end{array}$$

The difference $\Delta^n x_k$ may be expressed in terms of the given numbers x_k directly. In fact

$$\begin{aligned}\Delta^2 x_k = \Delta x_k - \Delta x_{k+1} &= (x_k - x_{k+1}) - (x_{k+1} - x_{k+2}) \\ &= x_k - 2 x_{k+1} + x_{k+2}\end{aligned}$$

and similarly

$$\Delta^3 x_k = x_k - 3 x_{k+1} + 3 x_{k+2} - x_{k+3} ;$$

---

[11] Institutiones calculi differentialis, 1755, p. 281.

143. the formula

$$\Delta^n x_k = x_k - \binom{n}{1} x_{k+1} + \binom{n}{2} x_{k+2} - + \dots + (-1)^n \binom{n}{n} x_{k+n}$$

for fixed k, is thus established in the cases $n = 1, 2, 3$. By induction, its validity for every n follows. For, supposing **143** proved for a particular positive integer n, we have for $n + 1$:

$$\begin{aligned} \Delta^{n+1} x_k &= \Delta^n x_k - \Delta^n x_{k+1} \\ &= x_k - \binom{n}{1} x_{k+1} + \binom{n}{2} x_{k+2} - + \dots + (-1)^n \binom{n}{n} x_{k+n} \\ &\quad - \binom{n}{0} x_{k+1} + \binom{n}{1} x_{k+2} - + \dots + (-1)^n \binom{n}{n-1} x_{k+n} \\ &\quad + (-1)^{n+1} \binom{n}{n} x_{k+n+1}, \end{aligned}$$

whence by addition, since $\binom{n}{\nu} + \binom{n}{\nu-1} = \binom{n+1}{\nu}$, we have the formula **143** for $n + 1$ instead of n. This proves all that is required.

Making use of the above simple facts and notation, we may now state the following theorem:

144. ° ***Euler's transformation of series.*** *Given an arbitrary convergent series* [12]

$$\sum_{k=0}^{\infty} (-1)^k a_k \equiv a_0 - a_1 + a_2 - + \dots,$$

we invariably have:

$$\sum_{k=0}^{\infty} (-1)^k a_k = \sum_{n=0}^{\infty} \frac{\Delta^n a_0}{2^{n+1}},$$

i. e. *the series on the right also converges and has the same sum as the given series* [13].

---

[12] The series need not be an alternating series, i. e. the numbers a_n need not all be positive. There are however small, though by no means essential, advantages in *writing* the series in alternating form as above, when effecting the transformation.

[13] This general transformation is due to *Euler* (Inst. calc. diff., pp. 281 seq., 1755). The particular transformation given below in example 2 is to be found already in a letter to *Leibniz* dated 2. 8. 1704, from *J. Bernoulli*, who attributed the discovery to *N. Fatzius*. (Cf. also *J. Hermann*, letter to Leibniz of 21. 1. 1705.) An early investigation of a more searching kind, using remainder terms, was undertaken by *J. V. Poncelet*, Journ. f. d. reine u. angew. Math., Vol. 13, pp. 1 seq., 1835. The proof that the transformation is always valid, provided only the series $\Sigma (-1)^k a_k$ is assumed also convergent, was first given by *L. D. Ames* (Annals of Math., (2) Vol. 3, p. 185. 1901). Cf. also *E. Jacobsthal* (Mathem. Zeitschr., Vol. 6, p. 100. 1920) and the note bearing on that by the author (ibid. p. 118).

Proof. In the array (A) of p. 241, we substitute for $a_n^{(k)}$:

(b) $$a_n^{(k)} = (-1)^k \left[\frac{1}{2^n} \Delta^n a_k - \frac{1}{2^{n+1}} \Delta^{n+1} a_k\right].$$

By **131**, if we now sum for every n, keeping k fixed (i. e. form the sum of the kth horizontal row), we obtain

$$z^{(k)} = \sum_{n=0}^{\infty} a_n^{(k)} = (-1)^k \frac{1}{2^0} \Delta^0 a_k = (-1)^k a_k \qquad (k \text{ fixed}).$$

For

$$\lim_{n\to\infty} \frac{\Delta^n a_k}{2^n} \lim_{n\to\infty} \frac{\binom{n}{0} a_k - \binom{n}{1} a_{k+1} + - \ldots + (-1)^n \binom{n}{n} a_{n+k}}{2^n}$$

is equal to zero by **44**, 8, because a_k, $-a_{k+1}$, a_{k+2}, ... certainly form a null sequence. Accordingly (b) gives an expression for the individual terms of the given series $\Sigma(-1)^k a_k$ in infinite series. Forming the sum of the nth column, we obtain the series

$$\sum_{k=0}^{\infty} (-1)^k \left[\frac{1}{2^n} \Delta^n a_k - \frac{1}{2^{n+1}} \Delta^{n+1} a_k\right] \qquad (n \text{ fixed});$$

the generic term of this series, as $\Delta^{n+1} a_k = \Delta^n a_k - \Delta^n a_{k+1}$, can be written in the form

$$\frac{(-1)^k}{2^{n+1}} [\Delta^n a_k + \Delta^n a_{k+1}] = \frac{1}{2^{n+1}} [(-1)^k \Delta^n a_k - (-1)^{k+1} \Delta^n a_{k+1}],$$

so that the series under consideration may again be summed directly, by **131**. We obtain

$$\sum_{k=0}^{\infty} a_n^{(k)} = \frac{1}{2^{n+1}} [\Delta^n a_0 - \lim_{k\to\infty} (-1)^k \Delta^n a_k] \qquad (n \text{ fixed}).$$

Since, however, the numbers a_k form a null sequence, so do the first differences and the nth differences generally, for any fixed n. The vertical columns are thus seen to constitute convergent series of sums

$$s^{(n)} = \frac{\Delta^n a_0}{2^{n+1}}.$$

The validity of *Euler*'s transformation will accordingly be established when we have shown that $R_m \to 0$. Now the horizontal remainders are seen to have the values

$$r_m^{(k)} = (-1)^k \frac{\Delta^m a_k}{2^m},$$

— following precisely the same line of argument as was used above for the entire horizontal rows. Thus

$$R_m = \frac{1}{2^m} \sum_{k=0}^{\infty} (-1)^k \Delta^m a_k \qquad \text{(fixed } m\text{)}.$$

If we write for brevity

$$(-1)^k (a_k - a_{k+1} + a_{k+2} - + \cdots) = r_k,$$

this series for R_m may be thought of as obtained by term-by-term addition from the $(m+1)$ series:

$$r_0, \quad \binom{m}{1} r_1, \quad \binom{m}{2} r_2, \quad \ldots, \quad \binom{m}{m} r_m.$$

Hence

$$R_m = \frac{r_0 + \binom{m}{1} r_1 + \binom{m}{2} r_2 + \cdots + \binom{m}{m} r_m}{2^m};$$

therefore, as r_m is the term of a null sequence, so is R_m, by **44**, 8. This proves the validity of *Euler*'s transformation with full generality.

Examples.

1. Take

$$s = 1 - \frac{1}{2} + \frac{1}{3} - \frac{1}{4} + - \cdots.$$

The triangular array (Δ) takes the form

$$\begin{array}{ccccccccc}
1, & & \frac{1}{2}, & & \frac{1}{3}, & & \frac{1}{4}, & & \frac{1}{5}, \ \ldots \\
& \frac{1}{1\cdot 2}, & & \frac{1}{2\cdot 3}, & & \frac{1}{3\cdot 4}, & & \frac{1}{4\cdot 5}, \ \ldots & \\
& & \frac{1\cdot 2}{1\cdot 2\cdot 3}, & & \frac{1\cdot 2}{2\cdot 3\cdot 4}, & & \frac{1\cdot 2}{3\cdot 4\cdot 5}, \ \ldots & & \\
& & & \frac{1\cdot 2\cdot 3}{1\cdot 2\cdot 3\cdot 4}, & & \frac{1\cdot 2\cdot 3}{2\cdot 3\cdot 4\cdot 5}, \ \ldots & & & \\
& & & & \cdots\cdots\cdots & & & &
\end{array}$$

The general expression of the nth difference is found to be

$$\Delta^n a_k = \frac{n!}{(k+1)(k+2)\ldots(k+n+1)},$$

so that in particular

$$\Delta^n a_0 = \frac{1}{n+1}.$$

This is easily verified by induction. Accordingly we have

$$s = \log 2 = 1 - \frac{1}{2} + \frac{1}{3} - \frac{1}{4} + - \cdots = \frac{1}{1\cdot 2^1} + \frac{1}{2\cdot 2^2} + \frac{1}{3\cdot 2^3} + \frac{1}{4\cdot 2^4} + \cdots.$$

The significance of this transformation e. g. for purposes of numerical calculation (§ 34) is at once apparent.

2. With equal facility, we may deduce

$$\frac{\pi}{4} = 1 - \frac{1}{3} + \frac{1}{5} - \frac{1}{7} + - \cdots = \frac{1}{2}\left[1 + \frac{1}{3} + \frac{1\cdot 2}{3\cdot 5} + \frac{1\cdot 2\cdot 3}{3\cdot 5\cdot 7} + \cdots\right].$$

In what cases this transformation is particularly advantageous for purposes of numerical calculation will be seen in the following section.

C. °*Kummer's* transformation of series. Another very obvious transformation consists simply in subtracting from a given series one whose sum is capable of representation by means of a known closed expression and which at the same time has terms as similar in construction as possible to those of the given series. By this means, subtracting for instance from $s = \Sigma \frac{1}{n^2}$ the known series (v. **68**, 2b)

$$1 = \sum_{n=1}^{\infty} \frac{1}{n(n+1)},$$

we deduce the transformation

$$s = \sum_{n=1}^{\infty} \frac{1}{n^2} = 1 + \sum_{n=1}^{\infty} \frac{1}{n^2(n+1)}.$$

The advantage of this transformation for numerical purposes is at once clear.

Simple and obvious as this transformation is, it yet forms what is really the kernel of *Kummer's transformation of series*[14]; the only difference being that a particular emphasis is now laid on a suitable choice of the series to be subtracted. This choice is regulated as follows: Let $\Sigma a_n = s$ be the given series (of course, by hypothesis, convergent). Let $\Sigma c_n = C$ be a convergent series of known sum C. *Let us suppose that the terms of the two series are asymptotically proportional,* say

$$\lim_{n \to \infty} \frac{a_n}{c_n} = \gamma \neq 0.$$

In that case

$$s = \sum_{n=0}^{\infty} a_n = \gamma C + \sum_{n=0}^{\infty} \left(1 - \gamma \frac{c_n}{a_n}\right) a_n, \qquad \textbf{145.}$$

and the new series occuring on the right may be regarded as a transformation of the given series. The advantage of this transformation lies mainly in the fact that the new series has terms *less* in absolute value than those of the given series, as in fact $\left(1 - \gamma \frac{c_n}{a_n}\right) \to 0$. Consequently its field of application belongs for the most part to the domain of numerical calculations and examples illustrating it will be found in the following paragraph.

§ 34. Numerical evaluations.

1. General considerations. As repeatedly explained already, it is only on very rare occasions that a closed expression, properly so-called, exists for the sum of a series. In the general case, the real

---

[14] *Kummer, E. E.*: Journ. f. d. reine u. angew. Math., Vol. 16, p. 206. 1837. Cf. also *Leclert* and *Catalan*, Mémoires couronnés et de savants étrangers de l'Ac. Belgique, Vol. 33, 1865—67, and the note by the author mentioned in footnote 9.

number to which a given convergent series, or the sequence of numbers for which it stands, converges, is, so to speak, first defined (given, determined, ...) by the series itself, in the only sense in which a number can be given, according to the discussion of Chapters I and II[15]. In this sense, we may boldly affirm that the convergent series *is* the number to which its partial sums converge. But for most practical purposes we gain very little by this assertion. In practice, we usually require to know something more precise about the *magnitude* of the number and to compare different numbers among themselves, etc. For this purpose, we require to be able to reduce all numbers, defined by *any kind* of limiting process, *to one and the same typical form.* The form of a decimal fraction is that most familiar to us to-day, and the expression, in this form, of numbers represented by series accordingly interests us first and foremost[16]. The student should, however, get it quite clear in his own mind that by obtaining such an expression we have merely, at bottom, substituted for the definition of a number by a given limiting process, a representation by means of another limiting process. The advantages of the latter, namely of the decimal form, are mainly that numbers so represented are easily compared with one another and that the error involved in terminating an infinite decimal at any given place is easily evaluated. Opposed to this there are, however, considerable disadvantages: the complete obscurity of the mode of succession of the digits in by far the greater number of cases and the consequent labour involved in their successive evaluation.

These advantages and disadvantages may be conveniently illustrated by the two following examples:

$$\left(\frac{\pi}{4}=\right)1-\frac{1}{3}+\frac{1}{5}-\frac{1}{7}+-\cdots=0{\cdot}785\,398\ldots$$

$$(\log 2=)\,1-\frac{1}{2}+\frac{1}{3}-\frac{1}{4}+-\cdots=0{\cdot}693\,147\ldots$$

By the *series*, distinct laws of formation are given; but they afford us no means of recognizing which of the two numbers is the larger of the two, for instance, or what is its excess over the smaller number. The *decimal fractions*, on the other hand, exhibit no such laws, but give us a direct sense of the relative and absolute magnitudes of both numbers.

---

15 Indeed an infinite series — our previous considerations give ample confirmation of the fact — is one of the most useful modes of so defining a number, one of the most significant both for theoretical and practical purposes.

16 And only in special cases the expression in ordinary fractional form. The reason is always that of convenience of comparison; which, of $\frac{11}{17}$ or $\frac{25}{39}$, is the larger, we cannot say at once, whereas the answer to the same question for 0·647 and 0·641 requires no calculation whatever.

We shall therefore henceforth reserve the term ***numerical*** *evaluation for the expression of a number in decimal form.*

As no infinite decimal fraction can be specified *in toto,* it will be necessary to break it off after a definite number of digits. We have still a few words to say as to the significance of this process of *breaking off* decimal fractions. If it be desired, for instance, to indicate the number e by a two-digit decimal fraction, we may with equal justification write 2·71 and 2·72, — the former, because the two first decimals are actually 7 and 1, — the latter, because it appears to involve a lesser error. We shall therefore make the following convention: when the n specified digits after the decimal point are the *actual* first n digits of the complete infinite decimal which expresses a given number, we shall insert a few dots after the nth digit, writing for instance $e = 2{\cdot}71\ldots$; when, however, the number is indicated by the nearest possible decimal fraction of n digits, we insert no dots after the nth digit, but write[17] e. g. $e \approx 2{\cdot}72$, in the latter case the nth digit written down is thus the nth digit of the actual infinite fraction raised or not by unity according as the succeeding part of the infinite fraction represents more or less than one half of a unit in the nth decimal place.

In point of fact, either specification has the effect of assigning an *interval of length* $1/10^n$ containing the required number. In the one case, the left hand endpoint is indicated, in the other, the centre of the interval. The margin, for the actual value, is the same in both cases. On the other hand, the error attaching to the indicated value, relatively to the true value of the number considered, is in the former case only known to be $\geqq 0$ and $\leqq 1/10^n$, in the latter to have modulus $\leqq \frac{1}{2}/10^n$. We may therefore describe the first indication as theoretically the clearer, and the second as practically the more useful. The *difficulty of actual determination* of the digits is also in all essential particulars the same in both cases. For in either, it may become necessary, when a specially unfavourable case is considered, to diminish the error of calculation to very appreciably less than $1/10^n$ before the nth digit can be properly determined. If we are, for instance, concerned with a number $\alpha = 5{\cdot}279\,999\,993\,26\ldots$, — to determine whether $\alpha = 5{\cdot}27\ldots$ or $5{\cdot}28\ldots$ (retaining two decimals), we have to diminish the error to less than a unit in the 8th decimal place. On the other hand, if we are concerned with a number $\beta = 2{\cdot}385\,000\,002\,6\ldots$, the choice between $\beta \approx 2{\cdot}38$ and $2{\cdot}39$ would be influenced by an uncertainty of one unit in the 8th decimal place[18].

---

[17] In $e = 2{\cdot}71\ldots$, the sign of equality may be justified as representing a limiting relation.

[18] The probability of such cases occurring is of course extremely small. By mentioning them, we have merely wished to draw attention to the significance of these facts. In Ex. 131, however, a particularly crude case is indicated.

2. Evaluation of errors and remainders. When given a convergent series $\Sigma a_n = s$, we shall of course assume that the individual terms of the series are "known", i. e. that their expressions in decimal form can easily be obtained to any number of digits. By addition, every partial sum s_n may accordingly also be evaluated. The question[19] remains: what is the magnitude of the error attaching to a given s_n? Here the word *error* designates the (positive or negative) number which has to be *added* to s_n to obtain the required value s. Since this error is $s - s_n$, i. e. is equal to the remainder of the series, starting immediately *after* the nth term, we will denote it by r_n, and the process of determining this error will also be designated by the term *evaluation of remainders.*

In practical problems, evaluations of remainders almost invariably reduce to one of the two following types:

A. *Remainders of absolutely convergent series.* If $s = \Sigma a_n$ converges absolutely, determine a series $\Sigma a_n'$ of positive terms, capable of summation in a convenient closed expression, and with terms *not less than* the absolute values of the corresponding terms of the given series (though also exceeding these by *as little as possible*). Obviously

$$|r_n| \leqq |a_{n+1}| + |a_{n+2}| + \cdots \leqq a'_{n+1} + a'_{n+2} + \cdots = r_n'$$

and the number r_n', which is assumed known, thus provides a means of estimating the magnitude of the remainder r_n, i. e. $|r_n| \leqq r_n'$, and this all the more closely the less a_n' exceeds $|a_n|$.

A particularly frequent case is that in which, for some fixed m, and every $k \geqq 1$:

$$|a_{m+k}| \leqq |a_m| \cdot \alpha^k \qquad \text{with } 0 < \alpha < 1;$$

in that case, of course,

$$|r_m| \leqq |a_m| \frac{\alpha}{1-\alpha},$$

and in particular, if $0 < \alpha \leqq \frac{1}{2}$:

$$|r_m| \leqq |a_m|.$$

The absolute value of the remainder is in this case *not greater than* that of the *term last calculated*[20].

B. *Remainders of alternating series.* Given a series of the form $s = \Sigma(-1)^n a_n$, and supposing that the (positive) numbers a_n form a *monotone* (decreasing) null sequence, we have (cf. **82**, Theorem 5):

$$0 < (-1)^{n+1} r_n = (a_{n+1} - a_{n+2}) + (a_{n+3} - a_{n+4}) + \cdots$$
$$= a_{n+1} - (a_{n+2} - a_{n+3}) - \cdots < a_{n+1}.$$

---

[19] Or in more practical form: Up to what order of decimal does s_n coincide with the required value s?

[20] In forming these estimates, it should be noticed that they give no indications as to the sign of the remainder r_n, only as to its *absolute value.*

Hence we may assert that *the error r_n has the same sign as the first neglected term, but has a smaller absolute value.*

When neither of these two modes of procedure is applicable, the evaluation of remainders is usually more troublesome, and it becomes necessary to adopt special artifices in each particular case. We shall, then, designate the series considered as *rapidly or slowly convergent,* according as r_n does or does not fall within the desired limit of error for moderate values [21] of n.

A few further fundamental remarks may be elucidated by the

3. Evaluation of the number e. We found

$$e = 1 + \frac{1}{1!} + \frac{1}{2!} + \frac{1}{3!} + \cdots + \frac{1}{n!} + \cdots.$$

Already, on p. 194, we have mentioned that the (*positive*) remainder r_n was less than the nth part of the term immediately before, so that

$$s_n < e < s_n + \frac{1}{n!\,n}.$$

In effecting the numerical calculations, we have now to take into account the following fact: When we express the individual *terms* of the series in decimal form, we have even at that point to break off the decimals at some particular digit, and we therefore incur a certain error. Unless n remains comparatively small, these errors may accumulate to such an extent that the whole calculation is in danger of becoming illusory. The mode of procedure is then as follows: Supposing that we are retaining 9 digits, we write [22]

$$\begin{aligned}
a_0 + a_1 + a_2 = \tfrac{5}{2} &= 2{\cdot}500\,000\,000 \\
a_3 &= 0{\cdot}166\,666\,667^- \\
a_4 &= 0{\cdot}.41\,666\,667^- \\
a_5 &= 0{\cdot}..8\,333\,333^+ \\
a_6 &= 0{\cdot}..1\,388\,889^- \\
a_7 &= 0{\cdot}...\ 198\,413^- \\
a_8 &= 0{\cdot}....24\,802^- \\
a_9 &= 0{\cdot}.....2\,756^- \\
a_{10} &= 0{\cdot}......\ 276^- \\
a_{11} &= 0{\cdot}.......25^+ \\
a_{12} &= 0{\cdot}........2^+ \\
[r_{12} &< 0{\cdot}........0^+]
\end{aligned}$$

Here the small $+$ and $-$ signs are intended to indicate whether the error in the term in question is positive or negative. In either case it is in absolute value less than one half of a unit in the last decimal place. By addition, we obtain the number

$$2{\cdot}718\,281\,830\,.$$

---

[21] A more precise definition of rapid convergence will be given in § 37.

[22] a_n is deduced from a_{n-1} by simple division by n.

But s_{12} itself may *possibly* (namely if all positive errors are nearly 0 and all negative ones nearly $\frac{1}{2}$ of a unit in the last decimal place) fall short of the number required by as much as $\frac{7}{2}$ of a unit in the last decimal place; or it may, on the other hand, be as much as $\frac{3}{2}$ of a unit in excess, since there are 7 negative and 3 positive errors. Taking also into account the remainder, we can only deduce with certainty, since $s_n < e = s_n + r_n$, that

$$2{\cdot}718281826 < e < 2{\cdot}718281832\,.$$

Our calculation thus secures only the first seven *true decimals*, while the *approximate value* [23] is obtained with eight digits: $e \approx 2{\cdot}71828183$.

In practice it will generally suffice to proceed a few decimal places further (2 or 3 at most) with the evaluation of the terms than it is desired to proceed for the sum. The number n of terms taken into account will be chosen so large that the remainder r_n contributes at most *one* unit in the last decimal place considered. The error in the individual terms will then, in general, have no appreciable effect. But to obtain perfect security for the resulting digits, it is necessary to proceed as described above. For we may retain a large number of digits beyond the desired number in calculating the individual terms, — yet as an error attaches to each of the decimals broken off and these errors accumulate, they may, in particularly unfavourable cases (cf. the example on p. 249), influence some of the much earlier digits.

4. Evaluation of the number π. The chief means placed at our disposal, up to the present, for the evaluation of the number π, are the series expansions of the functions $\tan^{-1}$ and $\sin^{-1}$; of these, the former has the preference, owing to its simple mode of formation. From this series, we deduced the expansion

$$\frac{1}{4}\pi = 1 - \frac{1}{3} + \frac{1}{5} - + \cdots,$$

which for numerical purposes is practically valueless. In fact, by p. 250, we can say no more on inspection about the remainder r_n in this expansion, than that it has the sign $(-1)^{n+1}$ and is in absolute value $< \frac{1}{2n+3}$. In order to secure 6 decimals, we should therefore be obliged to take $n > 10^6$, but an evaluation of a million terms is, for practical purposes, quite impossible. The rapidity of the convergence may be increased very materially by *Euler*'s transformation **144**, 2. In the next paragraph, we shall discuss the utility of such transformations for purposes of numerical calculation. Our present object is to deduce more convenient series expressions for π directly from the $\tan^{-1}$ series itself.

The series expansion for $\tan^{-1}\frac{1}{\sqrt{3}} = \frac{\pi}{6}$ is already of appreciable use: this gives

$$\frac{\pi}{6} = \frac{1}{\sqrt{3}}\left[1 - \frac{1}{3\cdot 3} + \frac{1}{5\cdot 3^2} - \frac{1}{7\cdot 3^3} + - \cdots\right].$$

---

[23] Cf. p. 249.

The following mode of procedure, however, provides considerably more convenient series[24].

The number

$$\alpha = \tan^{-1}\frac{1}{5} = \frac{1}{5} - \frac{1}{3\cdot 5^3} + \frac{1}{5\cdot 5^5} - \frac{1}{7\cdot 5^7} + - \cdots$$

is easily calculated from the series itself (see below). For this value of α, $\tan\alpha = \frac{1}{5}$, and so

$$\tan 2\alpha = \frac{2\tan\alpha}{1-\tan^2\alpha} = \frac{5}{12}$$

and

$$\tan 4\alpha = \frac{120}{119}.$$

Consequently 4α exceeds $\frac{\pi}{4}$ by only a small amount. Writing

$$4\alpha - \frac{\pi}{4} = \beta,$$

we have

$$\tan\beta = \frac{\tan 4\alpha - \tan\frac{1}{4}\pi}{1+\tan 4\alpha\tan\frac{1}{4}\pi} = \frac{1}{239}.$$

Hence β can very easily be evaluated from the series

$$\beta = \tan^{-1}\frac{1}{239} = \frac{1}{239} - \frac{1}{3}\,\frac{1}{239^3} + - \cdots.$$

The two numbers α and β give us

$$\pi = 4(4\alpha - \beta)$$

$$= 16\cdot\left[\frac{1}{5} - \frac{1}{3\cdot 5^3} + \frac{1}{5\cdot 5^5} - + \cdots\right] - 4\left[\frac{1}{239} - \frac{1}{3\cdot 239^3} + - \cdots\right]. \qquad \textbf{146.}$$

If it be desired to obtain the *first seven true decimals of* π, we may endeavour to attain this end by taking, say, 9 decimals for each of the terms and for the remainder[25] — a scanty enough margin, for the errors incurred on the numbers α and β have ultimately to be multiplied by 16 and 4 respectively. Denoting the first series by $a_1 - a_3 + a_5 - + \cdots$, the second by $a_1' - a_3' + a_5' - + \cdots$, and the corresponding partial sums by s_ν and s_ν', the calculation proceeds as follows:

$a_1 = 0{\cdot}200\,000\,000$	$a_3 = 0{\cdot}002\,666\,667^-$
$a_5 = 0{\cdot}000\,064\,000$	$a_7 = 0{\cdot}000\,001\,829^-$
$a_9 = 0{\cdot}\ldots\ldots.57^-$	$a_{11} = 0{\cdot}\ldots\ldots..2^-$
$a_1 + a_5 + a_9 = 0{\cdot}200\,064\,057^-$	$a_3 + a_7 + a_{11} = 0{\cdot}002\,668\,498^{---}$

Hence, as the errors change signs in a subtraction,

$$s_{11} = 0{\cdot}197\,395\,559^{+++-}$$

and

$$0 < r_{11} < 10^{-10}$$

Accordingly

$$3{\cdot}158\,328\,936 < 16\,\alpha < 3{\cdot}158\,328\,970,$$

---

[24] *J. Machin* (in *W. Jones*: Synopsis, London 1706).

[25] The result alone can show whether this suffices. In fact we do not know *a priori* whether we are not in the presence of one of the particularly unfavourable cases described on p. 249.

for after multiplying by 16 we have to subtract $\frac{16}{2} = 8$ units of the 9th decimal place, or add $\frac{48}{2} = 24$ of these units, to obtain bounds on either side for $16\, s_{11}$. Since

$$0 < 16\, r_{11} < 2 \cdot 10^{-9},$$

we have finally to add 2 units to the bound above, to obtain the corresponding bounds of $16\,\alpha$. Further

$$\begin{aligned} a_1' &= 0{\cdot}004\,184\,100+ \\ a_3' &= 0{\cdot}\ldots\ldots 024+ \\ \hline a_1' - a_3' &= 0{\cdot}004\,184\,076\pm \end{aligned} \qquad 0 < r_3' < 10^{-12},$$

hence

$$-0{\cdot}016\,736\,307 < -4\beta < -0{\cdot}016\,736\,302.$$

Combining the two results, we get

$$3{\cdot}141\,592\,629 < \pi < 3{\cdot}141\,592\,668.$$

This brief calculation thus really gives us the seven first true decimals of π:

$$\pi = 3{\cdot}141\,5926\ldots$$

(The same procedure would only have secured six decimals for the approximate value; cf. calculation of e, where circumstances, in this respect, were the exact reverse.)

The series here utilized for the calculation of π are among the most convenient; by their means, a very much greater number of decimals may also be secured[26] with relatively small trouble, and we are therefore fully justified in regarding π henceforth as one of the "known" numbers.

147. **5. Calculation of logarithms.** The starting point for the calculation of logarithms resides in the series

$$\log \frac{1+x}{1-x} = 2\left[x + \frac{x^3}{3} + \frac{x^5}{5} + \cdots\right]. \qquad (|x| < 1).$$

This series converges with considerable rapidity for $x = \frac{1}{3}$, and at once gives

$$\log 2 = 2\left[\frac{1}{3} + \frac{1}{3\cdot 3^3} + \frac{1}{5\cdot 3^5} + \cdots\right].$$

Denoting by $a_0, a_1, \ldots,$ the terms of the series inside the square bracket, we have

$$a_n = \frac{1}{(2n+1)\cdot 3^{2n+1}}$$

and

$$0 < r_n < \frac{1}{(2n+3)\cdot 3^{2n+3}}\left[1 + \frac{1}{9} + \frac{1}{9^2} + \cdots\right],$$

or

$$0 < r_n < \frac{1}{(2n+1)\cdot 3^{2n+1}} \cdot \frac{1}{3^2} \cdot \frac{9}{8} = \frac{a_n}{8}.$$

---

[26] The number π has been evaluated to 810 places of decimals (Mathematical Gazette, Feb. 1948, p. 37).

Our calculations then proceed as follows, if we again take 9 decimals for each of the terms a_n:

$$\begin{aligned}
a_0 &= 0{\cdot}333\,333\,333+\\
a_1 &= 0{\cdot}012\,345\,679+\\
a_2 &= 0{\cdot}000\,823\,045+\\
a_3 &= 0{\cdot}\ldots 065\,321+\\
a_4 &= 0{\cdot}\ldots 005\,645+\\
a_5 &= 0{\cdot}\ldots\ldots 513+\\
a_6 &= 0{\cdot}\ldots\ldots 048+\\
a_7 &= 0{\cdot}\ldots\ldots 005-\\
[r_7 &< 0{\cdot}\ldots\ldots 001]\\
\hline
&\;\;0{\cdot}346\,573\,589
\end{aligned}$$

Whence it follows, taking into account the remainder and the small + and − signs:

$$\log 2 = 0{\cdot}693\,1471\ldots \quad \text{or} \quad \log 2 \approx 0{\cdot}693\,1472$$

with seven decimals secured[27].

Once log 2 is evaluated, the calculation of the logarithms of all other numbers involves very little further trouble. In fact, our series gives, for $x = \frac{1}{2p+1}$,

$$\log(p+1) = \log p + 2\left[\frac{1}{2p+1} + \frac{1}{3(2p+1)^3} + \frac{1}{5(2p+1)^5} + \cdots\right]; \quad \mathbf{148.}$$

therefore if $\log p$ is known $(p = 2, 3, \ldots)$, we obtain the value of $\log(p+1)$, by the above formula. Moreover, since $\frac{1}{2p+1} = \frac{1}{5}, \frac{1}{7}, \ldots$, the expression involves a series converging *very rapidly*. In fact (cf. above, case $p = 1$)

$$0 < r_n < \frac{1}{(2n+3)\cdot(2p+1)^{2n+3}} \cdot \frac{1}{1 - \frac{1}{(2p+1)^2}} < \frac{a_n}{4p(p+1)},$$

so that the remainder is already very small for quite moderate values of n. The rapidity of convergence of course increases when p is given somewhat larger values, i. e. as soon as the first few logarithms have been successfully determined. It is useful to observe that by **37, 1**, only logarithms of prime numbers 2, 3, 5, 7, 11, 13, ... need be evaluated; those of all other numbers follow by mere combination.

Now supposing that we have effected the calculations for the logarithms of the first four prime numbers, 2, 3, 5, 7, the labour involved in calculating the logarithms of further primes is small. Thus, for instance, taking $p = 10$, we have

$$\log 11 = \log 2 + \log 5 + 2\left[\frac{1}{21} + \frac{1}{3\cdot 21^3} + \frac{1}{5\cdot 21^5} + \cdots\right]$$

with

$$0 < r_n < \frac{a_n}{11\cdot 40}.$$

---

[27] The series $1 - \frac{1}{2} + \frac{1}{3} - \frac{1}{4} + \cdots$ for log 2 is of course inappropriate for the evaluation of this number; even its *Euler*'s transformation effected in **144, 1** is less convenient than the series utilized above.

Thus already for $n = 3$,

$$0 < r_n < \frac{1}{7\cdot 21^7\cdot 11\cdot 40} < \frac{1}{20^8\cdot 2\cdot 11\cdot 7} < \frac{1}{10^9\cdot 2^9\cdot 7} < \frac{1}{10^{12}}$$

ensuring a degree of approximation sufficient even for the most refined scientific needs.

It would accordingly appear desirable to possess somewhat more convenient methods of calculation for $\log 2$, $\log 3$, $\log 5$, and also, at any rate, $\log 7$. Diverse artifices may be applied for the purpose, all of which consist mainly in finding rational numbers $\frac{k}{m}$, as near as possible to **1**, whose numerators and denominators are products of powers of these first four primes. If q of these primes have been utilized, q fractions will be needed to deduce the logarithms of those q primes from those of the fractions. For actually effecting these calculations, it is convenient to follow the method indicated by *Adams*[28]: Evaluate the logarithms of $\frac{10}{9}, \frac{25}{24}, \frac{81}{80}$ by means, not of the series **120**, c just employed, but of the original series **120**, a and b, which here give

$$\log\frac{10}{9} = -\log\left(1-\frac{1}{10}\right) = \frac{1}{10}+\frac{1}{2\cdot 10^2}+\frac{1}{3\cdot 10^3}+\cdots$$

$$\log\frac{25}{24} = -\log\left(1-\frac{4}{100}\right) = \frac{4}{100}+\frac{1}{2}\cdot\frac{16}{100^2}+\frac{1}{3}\cdot\frac{64}{100^3}+\cdots$$

$$\log\frac{81}{80} = \log\left(1+\frac{1}{80}\right) = \frac{1}{8\cdot 10}-\frac{1}{2}\cdot\frac{1}{64\cdot 10^2}+\frac{1}{3\cdot 510\cdot 10^3}-+\cdots.$$

Owing to the occurrence, in the denominator, of powers of 10, the calculation here becomes extremely simple. With the aid of these logarithms, we then obtain, as may be verified immediately:

$$\log 2 = 7\log\frac{10}{9} - 2\log\frac{25}{24} + 3\log\frac{81}{80}$$

$$\log 3 = 11\log\frac{10}{9} - 3\log\frac{25}{24} + 5\log\frac{81}{80}$$

$$\log 5 = 16\log\frac{10}{9} - 4\log\frac{25}{24} + 7\log\frac{81}{80}$$

If we proceed further to evaluate, as we may with equal facility,[29]

$$\log\frac{126}{125} = \log\left(1+\frac{8}{1000}\right) = \frac{8}{10^3}-\frac{1}{2}\cdot\frac{8^2}{10^6}+\frac{1}{3}\cdot\frac{8^3}{10^9}-+\cdots,$$

we also obtain

$$\log 7 = 19\log\frac{10}{9} - 4\log\frac{25}{24} + 8\log\frac{81}{80} + \log\frac{126}{125}.$$

---

[28] Proc. of the Royal Society, Vol. 27, p. 88, 1878.

[29] The facility with which this calculation is effected may be seen by the following, which in 5 simple lines provides $\log\frac{126}{125}$ with 10 decimals secured:

$$\left.\begin{array}{l} +0{\cdot}008\,000\,000\,000 \\ -0{\cdot}\ldots 032\,000\,000 \\ +0{\cdot}\ldots\ldots 170\,667^- \\ -0{\cdot}\ldots\ldots 001\,024 \\ +0{\cdot}\ldots\ldots\ldots 007^- \end{array}\right\};\quad \log\frac{126}{125} = 0{\cdot}007\,968\,1696\ldots$$

We have thus, for the actual calculation of natural logarithms, a method which is convenient and easily applicable in practice. Into further details of the computation of logarithmic tables we cannot enter in this place.

Having obtained $\log 2$ and $\log 5$, we have also the value of $\log 10$; and hence, in

$$M = \frac{1}{\log 10} = 0{\cdot}434\,294\,48190\ldots,$$

the "modulus" of ***Briggs'*** system of logarithms to the base 10, or factor by which the natural logarithm of a number must be multiplied to give the *Briggian* logarithm[30].

6. Calculation of roots. Once logarithms have been mastered no great practical importance attaches to the problem of obtaining simple methods of calculation for the roots of natural numbers. We shall therefore be quite brief in the following explanations. The rapidity of convergence of the binomial series

$$(1+x)^{\alpha} = \sum_{n=0}^{\infty} \binom{\alpha}{n} x^n$$

increases as $|x|$ diminishes. Now the calculation of a power $\sqrt[p]{q} = q^{\frac{1}{p}}$ can always be reduced to that of a power of the form $(1+x)^{\frac{1}{p}}$, with some small value of $|x|$.

A few examples may serve to illustrate the above. On p. 211, we gave **149.**
for $\sqrt{2}$ the series expansion of $\frac{7}{5}\left(1-\frac{1}{50}\right)^{-\frac{1}{2}}$:

$$\sqrt{2} = \frac{7}{5}\left[1 + \frac{1}{2}\cdot\frac{1}{50} + \frac{1\cdot 3}{2\cdot 4}\cdot\frac{1}{50^2} + \frac{1\cdot 3\cdot 5}{2\cdot 4\cdot 6}\cdot\frac{1}{50^3} + \cdots\right].$$

Since $(-1)^n\binom{-\frac{1}{2}}{n}$ is constantly positive and forms a monotone decreasing sequence, the remainder r_n may be estimated by means of the inequality

$$0 < r_n < a_n\cdot\left(\frac{1}{50} + \frac{1}{50^2} + \cdots\right) = \frac{a_n}{49},$$

showing that, even for small values of n, a considerable degree of approximation is attained[31]. The method is even more effective if we write

$$\sqrt{2} = \frac{99}{70}\left(1 + \frac{1}{9800}\right)^{-\frac{1}{2}}, \quad \text{or} \quad \sqrt{2} = \frac{141}{100}\left(1 - \frac{119}{20\,000}\right)^{-\frac{1}{2}},$$

---

[30] We may remark in passing that we have certainly found ample justification, by this time, for what seemed at first the rather arbitrary designation of the logarithms with the remarkable base e as the "natural" logarithms.

[31] How simply the calculation proceeds is shewn by the following details:

$$\left.\begin{aligned} a_0 + a_1 &= 1{\cdot}010\ldots\ldots\ldots 0\\ a_2 &= 0{\cdot}\ldots 15\ldots\ldots\ldots 0\\ a_3 &= 0{\cdot}\ldots\ldots 25\ldots\ldots 0\\ a_4 &= 0{\cdot}\ldots\ldots\ldots 4375.0\\ a_5 &= 0{\cdot}\ldots\ldots\ldots\ldots 7875 \end{aligned}\right\} \quad \begin{gathered}\text{hence — indeed without any error! —}\\ s_5 = 1{\cdot}010\,1525445375\\ 0 < r_5 < 17\cdot 10^{-12}\\ \sqrt{2} = 1{\cdot}4142135623\ldots,\end{gathered}$$

by which the first 10 decimals are thus already secured.

or other similar expressions, obtained by taking any rough approximation a to $\sqrt{2}$ $\left(\frac{99}{70}\right.$ in the first case, 1·41 in the second$\left.\right)$, and putting

$$\sqrt{2} = a\sqrt{\frac{2}{a^2}}.$$

Since a^2 is chosen to be very near 2, the quantity under the $\sqrt{\ }$ is of the form $1+x$, with small $|x|$. — Similarly, if we are already aware that $\sqrt{3} = 1{\cdot}732\ldots$, we have only to write

$$\sqrt{3} = 1{\cdot}732\sqrt{\frac{3}{(1{\cdot}732)^2}} = 1{\cdot}732\left[1 - \frac{176}{3000000}\right]^{-\frac{1}{2}}$$

to obtain, with the greatest ease, an expansion of $\sqrt{3}$ to 50 or more places of decimals.

We may, without further explanation, indicate the examples:

$$\sqrt{11} = \frac{10}{3}\left(1 - \frac{1}{100}\right)^{\frac{1}{2}}, \qquad \sqrt{13} = \frac{18}{5}\left(1 - \frac{1}{325}\right)^{-\frac{1}{2}}$$

$$\sqrt[3]{2} = \frac{5}{4}\left(1 + \frac{3}{125}\right)^{\frac{1}{3}}, \qquad \sqrt[3]{3} = \frac{10}{7}\left(1 + \frac{29}{1000}\right)^{\frac{1}{3}}.$$

150. **7. Calculation of trigonometrical functions.** The series expansions of $\sin x$ and $\cos x$ converge with even greater rapidity than the exponential series, since only the even or only the odd powers occur in them, and these have, moreover, alternating signs. Accordingly, no special artifices are required; for angles of no excessive magnitude, the series furnish all that can possibly be desired.

To determine, for instance, $\sin 1^0$, we have first to express 1^0 in circular measure. We have $1^0 = \frac{\pi}{180} = 0{\cdot}017453292\ldots$, i. e. certainly $< \frac{1}{50}$. Denoting this quantity by α,

$$\sin 1^0 = \alpha - \frac{\alpha^3}{3!} + \frac{\alpha^5}{5!} - + \cdots \equiv a_0 - a_1 + a_2 - + \cdots,$$

and the error r_n may at once be estimated (p. 250, B) by

$$0 < (-1)^{n+1} r_n < \frac{\alpha^{2n+3}}{(2n+3)!},$$

which last expression is already less than $\frac{1}{4}\cdot 10^{-15}$ for $n = 2$.

Circumstances are similar in the case of $\cos 1^0$; this quantity may also, however, since $\sin^2 1^0 < \frac{1}{2500}$, be obtained easily from the relation

$$\cos 1^0 = (1 - \sin^2 1^0)^{\frac{1}{2}}$$

by means of the binomial series: — $\tan x$ and $\cot x$ are then obtained by division, or from their expansions **116** and **115**, whose convergence is still quite sufficiently rapid when $|x|$ is small.

These latter series also lead to useful expansions for the *logarithms* of $\sin x$ and $\cos x$, — which for practical purposes are of greater importance than the values of $\sin x$ and $\cos x$ themselves. We have[32] (cf. § 19, Def. 12)

$$\log \sin x = \log x + \log \frac{\sin x}{x} = \log x + \int_0^x \left[\cot x - \frac{1}{x}\right] dx$$

$$= \log x + \sum_{k=1}^{\infty} (-1)^k \frac{2^{2k} \cdot B_{2k}}{2k \cdot (2k)!} x^{2k} \qquad \mathbf{151.}$$

and similarly from **116**

$$-\log \cos x = \int_0^x \tan x \, dx = \sum_{k=1}^{\infty} (-1)^{k-1} \frac{2^{2k}(2^{2k}-1) B_{2k}}{2k \cdot (2k)!} x^{2k}; \qquad \mathbf{152.}$$

$\log \tan x$ and $\log \cot x$ may be obtained from these by simple addition. As regards the convergence of these series, we can only state in the first instance that they certainly do converge *for all sufficiently small values of* $|x|$. However, the remarks of p. 237, footnote 4, show further that the series in **151** has the radius π, that in **152** the radius $\frac{\pi}{2}$.

Further details in the computation of trigonometrical tables will not be entered into here, as they do not concern the theory of infinite series.

8. More accurate evaluation of remainders. In the cases previously considered, the sum of a given series was invariably deduced by evaluating suitable partial sums and estimating the error involved in the corresponding remainder. It is obvious that this method is impracticable unless the convergence of the series is relatively rapid. If it be desired to evaluate, with some degree of approximation, for instance

$$s = \sum_{n=1}^{\infty} \frac{1}{n^2},$$

this direct method is pretty hopeless[33]. Even if we are very cautious in the margin we allow, we can only deduce, as an upper estimate of the remainder

$$r_n = \frac{1}{(n+1)^2} + \frac{1}{(n+2)^2} + \cdots,$$

[32] The function in the square bracket has to be understood to stand for the series **115** after division by x and subtraction of the foremost term $\frac{1}{x}$. The function is therefore defined and continuous also for $x = 0$.

[33] As we happen to know that the sum is $\frac{\pi^2}{6}$, its evaluation indirectly by means of the value of π is of course quite simple. But for the moment we are assuming that we know as little about this sum as e. g. about the sum of $\sum \frac{1}{n^3}$.

the inequality

$$r_n < \frac{1}{n(n+1)} + \frac{1}{(n+1)(n+2)} + \cdots = \frac{1}{n};$$

according to this, it would become necessary to calculate a million terms, in order to secure 6 places of decimals. This of course is out of the question.

This state of things may frequently be improved to some extent, if it is possible to supplement the upper estimate of the remainder r_n by a lower estimate, i. e. to deduce an inequality for r_n of opposite sense to the above in our case. In our example, the same principle as that already used gives

$$r_n > \frac{1}{(n+1)(n+2)} + \frac{1}{(n+2)(n+3)} + \cdots = \frac{1}{n+1};$$

we are thus able to assert that our sum s satisfies the conditions

$$1 + \frac{1}{2^2} + \cdots + \frac{1}{n^2} + \frac{1}{n+1} < s < 1 + \frac{1}{2^2} + \cdots + \frac{1}{n^2} + \frac{1}{n},$$

for every n. To secure 6 decimals, we may accordingly need only 1000 terms. This is still too large a number for practical purposes. But in special examples this method of *upper and lower estimates of the remainder* (cf. Ex. 131) may lead to a satisfactory result.

These cases are, however, so rare, that they do not come into account for practical purposes. Greater importance attaches to methods for transformation of slowly convergent into rapidly convergent series, because they admit of a far wider range of applications. To these methods we proceed to give our attention.

§ 35. Applications of the transformation of series to numerical evaluations.

In cases of slow convergence, one naturally attempts to change the given series into one with a more rapid convergence, by means of some suitable modification. We proceed to examine in this light the transformations discussed in § 33, so as to see how far they will be of use to us here.

A. *Kummer*'s transformation. For this transformation it is immediately obvious whether and to what extent an increase in the rapidity of the convergence can be obtained by it. In fact, using the notation of **145**, we have

$$\sum_{n=0}^{\infty} a_n = \gamma C + \sum_{n=0}^{\infty} \left(1 - \gamma \frac{c_n}{a_n}\right) a_n;$$

as $\left(1 - \gamma \frac{c_n}{a_n}\right) \to 0$, the terms of the new series (from some index onwards) are less than those of the given series. The method will accordingly be all the more effective the smaller the factors $\left(1-\gamma\frac{c_n}{a_n}\right)$ are, from the first; or in other words, the nearer the terms of Σc_n are to those of Σa_n.

Examples. **153.**

1. We found on p. 247 that $\sum\frac{1}{n^2}=1+\sum\frac{1}{n^2(n+1)}$. The terms of the new series are asymptotically equal to those of the series

$$\sum_{n=1}^{\infty}\frac{1}{n(n+1)(n+2)}=\frac{1}{2}\sum_{n=1}^{\infty}\left(\frac{1}{n(n+1)}-\frac{1}{(n+1)(n+2)}\right)=\frac{1}{4};$$

thus here $C=\frac{1}{4}$ and $\gamma=1$, and so

$$\sum_{n=1}^{\infty}\frac{1}{n^2}=1+\frac{1}{4}+\sum_{n=1}^{\infty}\frac{1}{n^2(n+1)(n+2)}.$$

Proceeding in this manner, we obtain, at the pth stage:

$$\sum_{n=1}^{\infty}\frac{1}{n^2}=1+\frac{1}{2^2}+\frac{1}{3^2}+\cdots+\frac{1}{p^2}+p!\sum_{n=1}^{\infty}\frac{1}{n^2(n+1)(n+2)\ldots(n+p)},$$

The latter series, even for moderate values of p, shows an appreciably rapid convergence.

2. Consider the somewhat more general series

$$\sum_{n=0}^{\infty}a_n\equiv\sum_{n=0}^{\infty}\frac{1}{(n+\alpha)^2(n+\alpha+1)^2\ldots(n+\alpha+p-1)^2},\quad\begin{cases}\alpha \text{ arbitrary} \neq 0,-1,\ldots\\ p,\text{ integer}\geqq 1\end{cases}$$

Here we take

$$c_n=(n+\gamma)\,a_n-(n+1+\gamma)\,a_{n+1},\qquad n=0,1,2,\ldots,$$

and we try to determine γ (independent of n) so that c_n is as near a_n as possible[34]. Here we have $C=\gamma\cdot a_0$ and a simple calculation gives $\gamma=\frac{1}{2p-1}$. Hence we obtain

$$\sum_{n=0}^{\infty}a_n=\frac{\gamma a_0}{2p-1}+\sum_{n=0}^{\infty}\left(1-\frac{(n+\gamma)\,a_n-(n+1+\gamma)\,a_{n+1}}{(2p-1)\,a_n}\right)a_n.$$

The expression in the large bracket is

$$1-\frac{(n+\gamma)(n+\alpha+p)^2-(n+1+\gamma)(n+\alpha)^2}{(2p-1)(n+\alpha+p)^2}.$$

Since, by simplification, the terms in n^3 and n^2 must disappear of themselves, this gives

$$\frac{(2\alpha+3p-2-2\gamma)\,pn+(2p-1-\gamma)(\alpha+p)^2+\alpha^2(1+\gamma)}{(2p-1)(n+\alpha+p)^2}.$$

If we now choose γ so that the terms in n also disappear, i. e. take $\gamma=\alpha+\frac{3}{2}p-1$: then the expression in the large bracket above now becomes

$$\frac{p^3}{2(2p-1)}\cdot\frac{1}{(n+\alpha+p)^2},$$

[34] The choice of a number c_n of the form x_n-x_{n+1} will, by **131**, always prove most convenient, as in that case C at any rate may be specified at once and the choice still be so arranged that the c_n's are near to the a_n's.

and accordingly

$$\sum_{n=0}^{\infty} \frac{1}{(n+\alpha)^2 \ldots (n+\alpha+p-1)^2}$$

$$= \frac{\left(\alpha+\frac{3}{2}p-1\right)\cdot\frac{1}{2p-1}}{\alpha^2(\alpha+1)^2\ldots(\alpha+p-1)^2} + \frac{p^3}{2(2p-1)} \sum_{n=0}^{\infty} \frac{1}{(n+\alpha)^2\ldots(n+\alpha+p)^2}.$$

The transformation thus has the effect of introducing an additional quadratic factor in the denominator. — Particular cases:

a) $\alpha = 1$.

$$\sum_{n=0}^{\infty} \frac{1}{(n+1)^2(n+2)^2\ldots(n+p)^2}$$

$$= \frac{\frac{3}{2}\cdot p\cdot\frac{1}{2p-1}}{1^2\cdot 2^2\ldots p^2} + \frac{p^3}{2(2p-1)} \sum_{n=0}^{\infty} \frac{1}{(n+1)^2\ldots(n+p)^2(n+p+1)^2}$$

Write for brevity

$$\sum_{k=1}^{\infty} \frac{1}{k^2(k+1)^2\ldots(k+p-1)^2} \equiv \sum_{n=0}^{\infty} \frac{1}{(n+1)^2\ldots(n+p)^2} = S_p;$$

the result then takes the form

$$S_p = \frac{3p}{2(2p-1)\cdot 1^2\cdot 2^2\ldots p^2} + \frac{p^3}{2(2p-1)}\cdot S_{p+1}.$$

This formula enables us easily to obtain very rapidly convergent series for $s = S_1 = \sum \frac{1}{n^2}$.

b) Similarly, for $\alpha = \frac{1}{2}$:

$$\sum_{n=0}^{\infty} \frac{1}{(2n+1)^2(2n+3)^2\ldots(2n+2p-1)^2}$$

$$= \frac{3p-1}{2(2p-1)}\cdot\frac{1}{1^2\cdot 3^2\ldots(2p-1)^2} + \frac{2p^3}{2p-1} \sum_{n=0}^{\infty} \frac{1}{(2n+1)^2\ldots(2n+2p+1)^2}.$$

This formula similarly leads to rapidly convergent series for $\sum \frac{1}{(2n+1)^2}$.

For further examples, see Exercises 127 seq.

B. *Euler*'s transformation.

Euler's transformation **144** need not by any means involve an increase in the rapidity of convergence[35] of the series to which it is applied.

154. For instance the transformation of $\sum_{n=0}^{\infty}\left(\frac{1}{2}\right)^n$ gives the series $\frac{1}{2}\sum_{n=0}^{\infty}\left(\frac{3}{4}\right)^n$, which evidently has a less rapid convergence. But even

[35] The explicit definition of what we mean by more or less rapid convergence will be given in § 37: $\Sigma a_n'$ is said to be *more or less rapidly convergent* than Σa_n, according as

$$\left|\frac{r_n'}{r_n}\right| = \left|\frac{a_{n+1}'+a_{n+2}'+\cdots}{a_{n+1}+a_{n+2}+\cdots}\right| \to 0 \quad \text{or} \quad \to +\infty.$$

in the case of alternating series, the effect need not be an increased rapidity of convergence; indeed the following three examples show that all conceivable cases may actually occur here:

1. $\sum_{n=0}^{\infty}(-1)^n \frac{1}{2^n}$ gives a *more* rapidly convergent series, $\frac{1}{2}\sum_{n=0}^{\infty}\frac{1}{4^n}$.

2. $\sum_{n=0}^{\infty}(-1)^n \frac{1}{3^n}$ " " series with the *same* rapidity of convergence, $\frac{1}{2}\sum_{n=0}^{\infty}\frac{1}{3^n}$.

3. $\sum_{n=0}^{\infty}(-1)^n \frac{1}{4^n}$ " " *less* rapidly convergent series, $\frac{1}{2}\sum_{n=0}^{\infty}\left(\frac{3}{8}\right)^n$.

We shall now show, however, that such an increase in the rapidity of the convergence does result, in the case of those *alternating series* $\Sigma(-1)^n a_n$, $a_n > 0$, whose terms, though not showing rapidity of convergence, still tend to zero in a particular *regular manner,* which we proceed to describe. These are the only types of alternating series of any practical importance.

The hypothesis required will be that not only the numbers a_n form a monotone decreasing sequence, i. e. have positive *first differences* Δa_n, but that the same is true of all differences of every order. A (positive) sequence $a_0, a_1, a_2, \ldots$ is said to have *p-fold monotony*[36] if its first, second, ..., p^{th} differences are all positive, and it is said to be *fully monotone* if *all* the differences $\Delta^k a_n$, $(k, n = 0, 1, 2, \ldots)$ are positive. With these designations, the theorem referred to is:

Theorem 1. *If* $\sum_{n=0}^{\infty}(-1)^n a_n$ *is an alternating series for which* **155.**
the (positive) numbers $a_0, a_1, \ldots$ *form a fully monotone null sequence, while, from the first,* $\frac{a_{n+1}}{a_n} \geqq a > \frac{1}{2}$ *(for every* n*)*[37], *then the transformed series* $\Sigma \frac{1}{2^{n+1}} \Delta^n a_0$ *converges more rapidly than the given series.*

The proof is very simple. As $\frac{a_{n+1}}{a_n} \geqq a$, we have $a_n \geqq a_0 \cdot a^n$. Further, for the remainder r_n of the given series we have

$$(-1)^{n+1} r_n = a_{n+1} - a_{n+2} + - \ldots = \Delta a_{n+1} + \Delta a_{n+3} + \Delta a_{n+5} + \ldots,$$

hence, since (Δa_ν) is itself a monotone null sequence,

$$|r_n| \geqq \frac{1}{2}(\Delta a_{n+1} + \Delta a_{n+2} + \Delta a_{n+3} + \ldots) = \frac{1}{2} a_{n+1} \geqq \frac{1}{2} a_0 \cdot a^{n+1}.$$

---

[36] Cf. Memoir of *E. Jacobsthal* referred to in **144.**

[37] This assumption is the precise formulation of the expression used above, that the given series should not converge particularly rapidly. The series will in fact, as the example shows more distinctly, converge less rapidly than $\Sigma\left(\frac{1}{2}\right)^n$. Cf. further the work by *F. V. Poncelet* quoted in footnote 13.

On the other hand, as $\Delta^n a_0 - \Delta^{n+1} a_0 = \Delta^n a_1 \geqq 0$, the numerators of the transformed series also form a monotone null sequence, and in particular are all $\leqq a_0$. Consequently the remainders r_n' of the transformed series, — which, moreover, is a series of positive terms, — satisfy

$$r_n' = \frac{\Delta^{n+1} a_0}{2^{n+2}} + \cdots \leqq \frac{a_0}{2^{n+2}}\left(1 + \frac{1}{2} + \frac{1}{4} + \cdots\right) = \frac{a_0}{2^{n+1}}.$$

Consequently, we have

$$\left|\frac{r_n'}{r_n}\right| \leqq \frac{1}{a}\left(\frac{1}{2a}\right)^n,$$

which proves our statement completely. Further, we see that the larger a is, the greater will be the increase in the rapidity of convergence, i. e. the more rapidly will $\frac{r_n'}{r_n} \to 0$. In particular, we may transform into series which converge with practically the same rapidity as $\sum\left(\frac{1}{2}\right)^n$, all alternating series for which the ratio of two consecutive terms tends to 1 in absolute value; such series have usually a slow convergence.

Examples. The two most striking examples of *Euler*'s transformation, that of $\sum \frac{(-1)^n}{n+1}$ and $\sum \frac{(-1)^n}{2n+1}$, were anticipated in **144**. For further applications it is essential to know which null sequences are *fully monotone*. We may prove, in this connection, by repeated application of the first mean value theorem of the differential calculus (§ 19, Theorem 8), the following theorem:

Theorem 2. *A (positive) sequence $a_0, a_1, \ldots$ is fully monotone decreasing if a function $f(x)$ exists, defined for $x \geqq 0$, and possessing differential coefficients of all orders for $x > 0$, for which $f(n) = a_n$ while the kth derived function has the constant sign $(-1)^k$, $(k = 0, 1, 2, \ldots)$.*

Accordingly the numbers

$$a^n, \ (0 < a < 1); \quad \frac{1}{(n+p)^\alpha}, \ (p > 0, \alpha > 0); \quad \frac{1}{\log(n+p)}, \ p > 1; \ \ldots$$

for instance, form fully monotone decreasing sequences; and from these many further sequences of this kind may be deduced, by means of the

Theorem 3. *If the numbers $a_0, a_1, \ldots$ and $b_0, b_1, \ldots$ constitute fully monotone decreasing sequences, the same is true of the products $a_0 b_0, a_1 b_1, a_2 b_2, \ldots$*

Proof. The following formula holds, and is easily verified by induction relatively to the index k:

$$\Delta^k a_n b_n = \sum_{\nu=0}^{k} \binom{k}{\nu} \Delta^{k-\nu} a_{n+\nu} \cdot \Delta^\nu b_n.$$

It shows that, as required, all the differences of $(a_n b_n)$ are positive, if those of (a_n) and (b_n) are so.

The following may be sketched as a particular numerical example:

The series

$$\sum_{n=0}^{\infty} (-1)^n a_n \equiv \frac{1}{\log 10} - \frac{1}{\log 11} + \frac{1}{\log 12} - + \cdots$$

has extraordinarily slow convergence; in fact, it converges with practically as small a rapidity as *Abel*'s series $\Sigma\, 1/n\,(\log n)^2$. Yet by means of *Euler*'s transformation, its sum may be calculated with relative ease. If we use only the first seven terms $\left(\text{to } \frac{1}{\log 16} \text{ inclusive}\right)$, we can deduce the first seven terms of the transformed series. If we use logarithms to seven places of decimals, we find, with 6 decimals secured, the value 0·221840... for the sum of the series[38].

C. *Markoff*'s transformation.

As the choice of the array (A), p. 241, from which *Markoff*'s transformation was deduced, is largely arbitrary, it is not surprising that we should be unable to formulate general theorems as to the effect of the transformation on the rapidity of the convergence. We shall therefore have to be content with laying down somewhat wider directing lines for its effective use, and with illustrating this by a few examples:

Denoting as before by $\Sigma z^{(k)}$ the given series (assumed convergent), we choose the terms of the 0th column in our array (A) to be as near as possible to those of the given series, and at the same time to possess a sum $s^{(0)}$ which we can indicate by a convenient closed expression; this is analogous to the condition of *Kummer*'s transformation. The series $\Sigma\left(z^{(k)} - a_0^{(k)}\right)$ now certainly converges more rapidly than $\Sigma z^{(k)}$; proceed with this new series in the same way, for the choice of the next column in our array, and so on. The effect of the transformation will be similar to that of an *indefinitely repeated Kummer*'s transformation, — the possibility of which was already indicated in the examples **153**, 2a (cf. Ex. 130).

As an example, we may take the series $\sum\limits_{k=1}^{\infty} \frac{1}{k^2}$, which is practically useless **156.** for the direct evaluation of its sum $\frac{\pi^2}{6}$. Here we think of the 0th row and column as consisting entirely of noughts, which we do not write down. The choice of the series $\sum \frac{1}{k(k+1)}$ for the *first column*, which was already used on p. 247, then appears obvious enough. This gives

$$\Sigma\left(z^{(k)} - a_1^{(k)}\right) \equiv \Sigma \frac{1}{k^2(k+1)}.$$

As second column, we shall then, as in **153**, 1, choose the series

$$\Sigma \frac{1}{k(k+1)(k+2)},$$

and so on. The kth row of the array thus takes the form

$$\frac{1}{k^2} = \frac{0!}{k(k+1)} + \frac{1!}{k(k+1)(k+2)} + \frac{2!}{k(k+1)(k+2)(k+3)} + \dots \quad (k \text{ fixed}).$$

The further calculations are, however, simplified by breaking off this series at the $(k-1)^{\text{th}}$ term and adding as k^{th} term the missing remainder r_k, after

[38] This example is taken from the work of *A. A. Markoff*: "Differenzenrechnung", Leipzig, p. 184. 1896.

which the series is regarded as consisting entirely of noughts. The kth row now has the form:

$$\frac{1}{k^2}=\frac{0!}{k(k+1)}+\frac{1!}{k(k+1)(k+2)}+\cdots+\frac{(k-2)!}{k(k+1)\dots(2k-1)}+r_k.$$

Subtracting the terms of the right hand side from the left *in succession*, we easily find

$$r_k=\frac{(k-1)!}{k^2(k+1)\dots(2k-1)}.$$

In our case, the process of splitting up the series $\sum\frac{1}{k^2}$ into an array of the form (A) of p. 241 thus gives:

$$\begin{array}{llll}
1 = 1 & & & \\
\frac{1}{2^2}=\frac{0!}{2\cdot 3} & +\frac{1!}{2^2\cdot 3} & & \\
\frac{1}{3^2}=\frac{0!}{3\cdot 4} & +\frac{1!}{3\cdot 4\cdot 5} & +\frac{2!}{3^2\cdot 4\cdot 5} & \\
\dots\dots & & & \\
\frac{1}{k^2}=\frac{0!}{k(k+1)} & +\frac{1!}{k(k+1)(k+2)}+\cdots & +\frac{(k-2)!}{k(k+1)\dots(2k-1)} & +\frac{(k-1)!}{k^2(k+1)\dots(2k-1)} \\
\dots\dots & & &
\end{array}$$

Since all the terms of this array are $\geqq 0$, the main rearrangement theorem **90** itself shows that we may sum in columns and must obtain $\frac{\pi^2}{6}$ as ultimate result. Now in the nth column we have the series

$$r_n+(n-1)!\left[\frac{1}{(n+1)\dots(2n+1)}+\frac{1}{(n+2)\dots(2n+2)}+\cdots\right], \quad (n \text{ fixed}).$$

By **132,** 3 for $\alpha=n+1$ and $p=n$, the series in the square brackets has the sum

$$\frac{1}{n(n+1)\dots 2n}.$$

Hence the nth column has for sum

$$s^{(n)}=(n-1)!\left[\frac{1}{n^2(n+1)\dots(2n-1)}+\frac{1}{n(n+1)\dots 2n}\right]$$
$$=3\,\frac{(n-1)!}{n(n+1)\dots(2n)}=3\,\frac{(n-1)!^2}{(2n!)}.$$

Therefore we have

$$\sum_{k=1}^{\infty}\frac{1}{k^2}=3\cdot\sum_{n=1}^{\infty}\frac{(n-1)!^2}{(2n)!}.$$

This formula is significant not only for *numerical* purposes, in view of the appreciable increase in the rapidity of the convergence, but almost more so because it provides a new means of obtaining the *closed* expression for the sum of the series $\sum\frac{1}{k^2}$, which we only succeeded in determining indirectly by using the expansion in partial fractions *as well as* the series expansion of the function cot. In fact we can easily establish directly (cf. Ex. 123), that **123** implies the expansion, for $|x|\leqq 1$:

$$(\sin^{-1}x)^2=\frac{1}{2}\sum_{n=1}^{\infty}\frac{(n-1)!^2}{(2n)!}(2x)^{2n}.$$

Putting $x=\frac{1}{2}$, we at once deduce[39]

$$\sum_{k=1}^{\infty}\frac{1}{k^2}=3\sum_{n=1}^{\infty}\frac{(n-1)^2}{(2n)!}=3\cdot 2\cdot\left(\frac{\pi}{6}\right)^2=\frac{\pi^2}{6}.$$

A further application of fundamental importance of *Markoff*'s transformation we have already come across (v. **144**) in *Euler*'s transformation, which was indeed deduced from *Markoff*'s.

For further applications of *Markoff*'s transformation we must refer to the accounts of *Markoff* himself (v. p. 265, footnote 38) and of *E. Fabry* (Théorie des séries à termes constants, Paris 1910). Their success depends for the most part on special artifices, but they are sometimes surprisingly effective. Numerous examples will be found, completely worked out, in the writings referred to.

Exercises on Chapter VIII.

I. Direct formation of the sequence of partial sums.

100. a) $\frac{x}{1+x}+\frac{2x^2}{1+x^2}+\frac{4x^4}{1+x^4}+\frac{8x^8}{1+x^8}+\cdots=\frac{x}{1-x}$ for $|x|<1$.

b) $\frac{x}{1-x^2}+\frac{x^2}{1-x^4}+\frac{x^4}{1-x^8}+\frac{x^8}{1-x^{16}}+\cdots=\begin{cases}\frac{x}{1-x} & \text{for } |x|<1,\\ -\frac{1}{x-1} & \text{for } |x|>1.\end{cases}$

101. $\sum_{n=1}^{\infty}\frac{a_n}{(1+a_1)(1+a_2)\ldots(1+a_n)}$ is, for a_n positive, *invariably* convergent. When does the series still continue to converge for arbitrary a_n, and what is its sum?

102. a) $\sum_{n=1}^{\infty}\tan^{-1}\frac{2}{n^2}=\frac{3\pi}{4}$;

$\left(\text{hint: } \tan^{-1}\frac{1}{n-1}-\tan^{-1}\frac{1}{n+1}=\tan^{-1}\frac{2}{n^2}\right).$

b) $\sum_{n=1}^{\infty}\tan^{-1}\frac{1}{n^2+n+1}=\frac{\pi}{4}.$

103. a) $\sum_{n=1}^{\infty}\frac{n}{(x+1)(2x+1)\ldots(nx+1)}=\frac{1}{x}$, if $x\neq 0,\ -1,\ -\frac{1}{2},\ -\frac{1}{3},\ldots$

b) $\frac{1}{y}+\frac{x}{y(y+1)}+\frac{x(x+1)}{y(y+1)(y+2)}+\cdots=\frac{1}{y-x}$, if $y>x>0$.

c) $1+\frac{a}{b}+\frac{a(a+1)}{b(b+1)}+\frac{a(a+1)(a+2)}{b(b+1)(b+2)}+\cdots=\frac{b-1}{b-a-1}$, if $b>a+1>1$.

104. $\frac{1}{k_1}+\frac{k_1-b}{k_1}\cdot\frac{1}{k_2}+\frac{(k_1-b)(k_2-b)}{k_1\cdot k_2}\cdot\frac{1}{k_3}+\cdots=\frac{1}{b}$, if $b\neq 0$, every $k_\nu>0$ and $\sum\frac{1}{k_\nu}$ is divergent.

---

[39] Cf. a note by *I. Schur* and the author: "Über die Herleitung der Gleichung $\sum\frac{1}{n^2}=\frac{\pi^2}{6}$", Archiv der Mathematik und Physik, Ser. 3, Vol. 27, p. 174. 1918.

105. a) $\sum\limits_{n=0}^{\infty} \frac{1}{2^n} \tan \frac{\pi}{2^{n+2}} = \frac{4}{\pi}$;

b) $\sum\limits_{n=1}^{\infty} \frac{1}{2^n} \tan \frac{x}{2^n} = \frac{1}{x} - \cot x$;

(hint: $\cot y - \tan y = 2 \cot 2y$).

106. In $\sum\limits_{n=0}^{\infty} \frac{g(n)}{(\alpha + p_1 + n)(\alpha + p_2 + n) \ldots (\alpha + p_k + n)}$, let $p_1, p_2, \ldots, p_k$ denote fixed given natural numbers, all different, and $\alpha \neq 0, -1, -2, \ldots$ any real number, while $g(x)$ denotes an integral rational function (polynomial) of degree $\leqq k - 2$. We assume the expansion in partial fractions:

$$\frac{g(x - \alpha)}{(x + p_1) \ldots (x + p_k)} = \frac{c_1}{x + p_1} + \cdots + \frac{c_k}{x + p_k}.$$

The given series then has the sum

$$-\sum_{\nu=1}^{k} c_\nu \left[\frac{1}{\alpha} + \frac{1}{\alpha + 1} + \cdots + \frac{1}{\alpha + p_\nu - 1}\right].$$

107. a) $\frac{1}{1 \cdot 2 \cdot 6 \cdot 7} - \frac{1}{3 \cdot 4 \cdot 8 \cdot 9} + \frac{1}{5 \cdot 6 \cdot 10 \cdot 11} - + \cdots = \frac{1}{60}\left(\pi - \frac{149}{60}\right)$;

b) $\frac{1}{1 \cdot 2 \cdot 4 \cdot 5} - \frac{1}{3 \cdot 4 \cdot 6 \cdot 7} + \frac{1}{5 \cdot 6 \cdot 8 \cdot 9} - + \cdots = \frac{5}{36} - \frac{1}{6} \log 2$;

c) $\frac{1}{1 \cdot 2 \cdot 4 \cdot 5} + \frac{1}{3 \cdot 4 \cdot 6 \cdot 7} + \frac{1}{5 \cdot 6 \cdot 8 \cdot 9} + \cdots = \frac{1}{36}$;

d) $\frac{1^3}{1^4 + 4} - \frac{3^3}{3^4 + 4} + \frac{5^3}{5^4 + 4} - + \cdots = 0$;

e) $\frac{1}{1 \cdot (1^4 + 4)} - \frac{1}{3 \cdot (3^4 + 4)} + \frac{1}{5 \cdot (5^4 + 4)} - + \cdots = \frac{\pi}{16}$;

f) $\frac{1}{1 \cdot (4 \cdot 1^4 + 1)} - \frac{1}{2 \cdot (4 \cdot 2^4 + 1)} + \frac{1}{3 \cdot (4 \cdot 3^4 + 1)} - + \cdots = \log 2 - \frac{1}{2}$;

g) $\frac{1}{1 \cdot 2 \cdot 3 \cdot 4} + \frac{1}{5 \cdot 6 \cdot 7 \cdot 8} + \cdots = \frac{1}{4} \log 2 - \frac{\pi}{24}$;

h) $\frac{1}{1 \cdot 2 \cdot 3} + \frac{1}{4 \cdot 5 \cdot 6} + \frac{1}{7 \cdot 8 \cdot 9} + \cdots = \frac{\pi}{12}\sqrt{3} - \frac{1}{4} \log 3$.

II. Determination of closed expressions by means of the expansions of elementary functions.

108. a) $1 - \frac{1}{5 \cdot 3^2} - \frac{1}{7 \cdot 3^3} + \frac{1}{11 \cdot 3^5} + \frac{1}{13 \cdot 3^6} - - + + \cdots = \log \sqrt{7}$;

b) $\frac{1}{2} \cdot \frac{1}{2} + \frac{1 \cdot 3}{2 \cdot 4} \cdot \frac{1}{4} + \frac{1 \cdot 3 \cdot 5}{2 \cdot 4 \cdot 6} \cdot \frac{1}{6} + \cdots = \log 2$;

c) $1 + \frac{1}{3} - \frac{1}{5} - \frac{1}{7} + \frac{1}{9} + \frac{1}{11} - - + + \cdots = \frac{\pi}{4}\sqrt{2}$;

d) $\frac{1}{x} + \frac{1}{x - y} + \frac{1}{x + y} + \frac{1}{x - 2y} + \frac{1}{x + 2y} + \cdots = \frac{\pi}{y} \cot \frac{\pi x}{y}$

gives for $y = 7$ and $x = 1, 2, 3$:

$$1 + \frac{1}{2} - \frac{1}{3} + \frac{1}{4} - \frac{1}{5} - \frac{1}{6} + \overset{7}{\smile}\!\!\!0 + + - + - - + \overset{14}{\smile}\!\!\!0 \cdots = \frac{\pi}{\sqrt{7}}$$

109. a) $\frac{1}{1\cdot2\cdot3}+\frac{1}{3\cdot4\cdot5}+\frac{1}{5\cdot6\cdot7}+\cdots=\log 2-\frac{1}{2}$;

b) $\frac{1}{1\cdot2\cdot3}-\frac{1}{3\cdot4\cdot5}+\frac{1}{5\cdot6\cdot7}-+\cdots=\frac{1}{2}(1-\log 2)$;

c) $\frac{1}{2\cdot3\cdot4}-\frac{1}{4\cdot5\cdot6}+\frac{1}{6\cdot7\cdot8}-+\cdots=\frac{1}{4}(\pi-3)$;

d) $\sum_{n=1}^{\infty}\frac{1}{4n^2-1}=\frac{1}{2}, \qquad \sum_{n=1}^{\infty}\frac{1}{(4n^2-1)^2}=\frac{\pi^2-8}{16},$

$$\sum_{n=1}^{\infty}\frac{1}{(4n^2-1)^3}=\frac{32-3\pi^2}{64};$$

e) $1-\frac{1}{5}+\frac{1}{7}-\frac{1}{11}+\frac{1}{13}-\frac{1}{17}+-\cdots=\frac{\pi}{2\sqrt{3}}$.

110. a) $\frac{1\cdot2}{1\cdot3}+\frac{1\cdot2\cdot3}{1\cdot3\cdot5}+\frac{1\cdot2\cdot3\cdot4}{1\cdot3\cdot5\cdot7}+\cdots=\frac{\pi}{2}$;

b) $\frac{1}{2\cdot3}+\frac{1\cdot2}{3\cdot4\cdot5}+\frac{1\cdot2\cdot3}{4\cdot5\cdot6\cdot7}+\cdots=\frac{2\pi}{3\sqrt{3}}-1$;

c) $\frac{1}{2\cdot3\cdot4}+\frac{1\cdot2}{3\cdot4\cdot5\cdot6}+\frac{1\cdot2\cdot3}{4\cdot5\cdot6\cdot7\cdot8}+\cdots=\frac{\pi^2}{18}-\frac{1}{2}$;

d) $\sum_{n=1}^{\infty}\frac{1}{n(4n^2-1)}=2\log 2-1$;

e) $\sum_{n=1}^{\infty}\frac{1}{n(4n^2-1)^2}=\frac{3}{2}-2\log 2$;

f) $\sum_{n=1}^{\infty}\frac{1}{2^n\cdot n^2}=\frac{\pi^2}{12}-\frac{1}{2}(\log 2)^2$.

111. If we write $\sum_{n=0}^{\infty}\left(\frac{n!}{(p+n)!}\right)^2=T_p$, then

$$T_2=\frac{\pi^2}{3}-3, \qquad T_3=\frac{\pi^2}{4}-\frac{39}{16}, \qquad T_4=\frac{5}{54}\pi^2-\frac{197}{216}.$$

112. If we write $\sum_{n=1}^{\infty}\frac{n^p}{n!}=g_p\cdot e$, $(p=1, 2, \ldots)$, then the numbers g_p are integers obtainable by the symbolical formula $g^{p+1}=(1+g)^p$. We have $g_1=1$ $g_2=2$, $g_3=5, \ldots$.

113.
$$\frac{1}{x}-\frac{1}{x+y}+\frac{1}{x+2y}-\frac{1}{x+3y}+-\cdots$$

may be summed in the form of a closed expression by means of elementary functions when x/y is a rational number. Special cases are:

$$1-\frac{1}{4}+\frac{1}{7}-\frac{1}{10}+-\cdots=\frac{1}{3}\left(\frac{\pi}{\sqrt{3}}+\log 2\right),$$

$$\frac{1}{2}-\frac{1}{5}+\frac{1}{8}-\frac{1}{11}+-\cdots=\frac{1}{3}\left(\frac{\pi}{\sqrt{3}}-\log 2\right),$$

$$1-\frac{1}{5}+\frac{1}{9}-\frac{1}{13}+-\cdots=\frac{1}{4\sqrt{2}}\left(\pi+2\log(\sqrt{2}+1)\right).$$

114. Writing $\sum_{n=1}^{\infty} \frac{x^n}{1-x^n} = L(x)$, $(|x|<1)$, we have, if (x_n) denotes *Fibonacci*'s sequence **6**, 7,

$$\sum_{k=1}^{\infty} \frac{1}{x_{2k}} \equiv 1 + \frac{1}{3} + \frac{1}{8} + \frac{1}{21} + \cdots = \sqrt{5}\left[L\left(\frac{3-\sqrt{5}}{2}\right) - L\left(\frac{7-3\sqrt{5}}{2}\right)\right].$$

And if we write $\sum_{k=1}^{\infty} \frac{1}{x_{2k-1}^2} = s$, and $\sum_{k-1}^{\infty} \frac{(-1)^{k-1}.k}{x_{2k}} = s'$, we have $\frac{s}{s'} = \sqrt{5}$.

III. Exercises on *Euler*'s transformation.

115. We have (for what values of x?)

a) $$\sum_{n=0}^{\infty} \frac{(-1)^n}{x+n} = \sum_{k=0}^{\infty} \frac{1}{2^{k+1}} \cdot \frac{k!}{x(x+1)\ldots(x+k)};$$

b) $$\sum_{n=0}^{\infty} \frac{(-1)^n}{\alpha+n} x^n = \frac{1}{\alpha(1+x)}\left[1 + \frac{1}{\alpha+1}\left(\frac{x}{1+x}\right) + \frac{1\cdot 2}{(\alpha+1)(\alpha+2)}\left(\frac{x}{1+x}\right)^2 + \cdots\right];$$

c) $$\sum_{n=0}^{\infty} \frac{(-1)^n}{(n+1)(n+2)\ldots(n+p+1)} = \frac{1}{p!}\sum_{k=0}^{\infty} \frac{1}{2^{k+1}(p+k+1)}.$$

116. If we put

$$e^{-x}\cdot \sum_{n=0}^{\infty} a_n \frac{x^n}{n!} = \sum_{n=0}^{\infty} (-1)^n b_n \frac{x^n}{n!},$$

we have $b_n = \Delta^n a_0$. In particular, therefore,

a) $$e^{-x}\left[1 + x + \frac{\alpha+2}{\alpha+1}\cdot\frac{x^2}{2!} + \frac{(\alpha+2)(\alpha+4)}{(\alpha+1)(\alpha+2)}\frac{x^3}{3!} + \cdots\right]$$
$$= 1 + \frac{1}{\alpha+1}\,\frac{x^2}{2^1\cdot 1!} + \frac{1}{(\alpha+1)(\alpha+3)}\,\frac{x^4}{2^2\cdot 2!}\cdots$$

b) $$e^x\cdot \sum_{n=1}^{\infty} \frac{(-1)^{n-1}}{n\cdot n!} x^n = \sum_{n=1}^{\infty} h_n \frac{x^n}{n!}, \quad \left(h_n = 1 + \frac{1}{2} + \cdots + \frac{1}{n}\right).$$

117. Quite special cases are:

a) $$\binom{n}{1} - \frac{1}{2}\binom{n}{2} + \frac{1}{3}\binom{n}{3} - + \cdots + \frac{(-1)^{n-1}}{n}\binom{n}{n} = h_n = 1 + \frac{1}{2} + \cdots + \frac{1}{n};$$

b) $$1 - \frac{1}{3}\binom{n}{1} + \frac{1}{5}\binom{n}{2} - + \cdots + (-1)^n \frac{1}{2n+1}\binom{n}{n} = \frac{2\cdot 4\ldots\ (2n)}{3\cdot 5\ldots(2n+1)}.$$

118. If $\Delta^n a_0 = b_n$, then $\Delta^n b_0 = a_n$. What accordingly are the inverse equations to those of the preceding exercise?

119. If (a_n) be a null sequence with $(p+1)$-fold decreasing monotony $(p \geqq 1)$, the sum s of the series $\sum_{n=0}^{\infty} (-1)^n a_n$ satisfies the inequalities

$$\frac{a_0}{2} + \frac{\Delta a_0}{2^2} + \cdots + \frac{\Delta^{p-1} a_0}{2^p} < s < \frac{a_0}{2} + \frac{\Delta a_0}{2^2} + \cdots + \frac{\Delta^{p-1} a_0}{2^p} + \frac{\Delta^p a_0}{2^p}.$$

Use this to prove the equality

$$\lim_{x \to 1-0} \left[\frac{1}{2} - \frac{x}{1+x} + \frac{x^2}{1+x^2} - + \cdots\right] = \frac{1}{4}.$$

120. If s_k and S_n denote the partial sums of both the series in **144**, we have

$$S_n = \frac{\binom{n+1}{1} s_0 + \binom{n+1}{2} s_1 + \cdots + \binom{n+1}{n+1} s_n}{2^{n+1}}.$$

Use this relation to prove the validity of *Euler*'s transformation.

121. The following relations hold, if the summation on either side is taken to start with the index 0 and the difference-symbols Δ operate on the coefficients on the left, a_k, a_{2k}, a_{2k+1} respectively:

a) $\sum (-1)^k a_k x^k = (1-y) \sum \Delta^n a_0 \cdot y^n$ with $(1+x)(1-y) = 1$;

b) $\sum (-1)^k a_{2k} x^{2k} = (1-y^2) \sum \Delta^n a_0 \cdot y^{2n}$ with $(1+x^2)(1-y^2) = 1$;

c) $\sum (-1)^k a_{2k+1} x^{2k+1} = \sqrt{1-y^2} \cdot \sum \Delta^n a_0 \cdot y^{2n+1}$ with $(1+x^2)(1-y^2) = 1$.

122. Thus e. g.

$$\tan^{-1} x = \frac{x}{1+x^2}\left[1 + \frac{2}{3}\,\frac{x^2}{1+x^2} + \frac{2\cdot 4}{3\cdot 5}\left(\frac{x^2}{1+x^2}\right)^2 + \cdots\right].$$

Putting $x = \frac{1}{2}, \frac{1}{3}, \frac{1}{7}, \frac{2}{11}, \frac{3}{79}, \ldots$, this provides peculiarly convenient series for π, as for instance

$$\frac{\pi}{4} = \tan^{-1}\frac{1}{2} + \tan^{-1}\frac{1}{3} = \frac{4}{10}\left[1 + \frac{2}{3}\left(\frac{2}{10}\right) + \cdots\right] + \frac{3}{10}\left[1 + \frac{2}{3}\left(\frac{1}{10}\right) + \cdots\right],$$

$$\frac{\pi}{4} = 2\tan^{-1}\frac{1}{3} + \tan^{-1}\frac{1}{7} = 5\tan^{-1}\frac{1}{7} + 2\tan^{-1}\frac{3}{79}, \text{ and others.}$$

123. The preceding series for $\tan^{-1} x$ may also be put in the form

$$\frac{\sin^{-1} y}{\sqrt{1-y^2}} = y + \frac{2}{3} y^3 + \frac{2\cdot 4}{3\cdot 5} y^5 + \cdots.$$

Hence deduce the expansion

$$2(\sin^{-1} y)^2 = \sum_{n=1}^{\infty} \frac{(n-1)!^2}{(2n)!}(2y)^{2n}.$$

IV. Other transformations of series.

124. Writing $\sum\limits_{n=2}^{\infty} \frac{1}{n^p} = S_p$, we have

a) $S_2 + S_3 + S_4 + \cdots = 1$; b) $S_2 + S_4 + S_6 + \cdots = \frac{3}{4}$;

c) $S_3 + S_5 + S_7 + \cdots = \frac{1}{4}$; d) $S_2 + \frac{1}{2} S_4 + \frac{1}{3} S_6 + \cdots = \log 2$;

e) $S_2 - S_3 + S_4 - + \cdots = \frac{1}{2}$; f) $S_2 - \frac{1}{2} S_4 + \frac{1}{3} S_6 - + \cdots = \log \frac{e^\pi - e^{-\pi}}{4\pi}$;

g) $\frac{1}{2} S_2 + \frac{1}{3} S_3 + \frac{1}{4} S_4 + \cdots = 1 - C$;

h) $\frac{1}{2} S_2 - \frac{1}{3} S_3 + - \cdots = \log 2 + C - 1$,

where C denotes *Euler*'s constant, defined in **128**, 2 and Ex. 85 a.

125. With the same meaning for S_p as in the preceding exercise, writing

$$\left(\frac{1}{k+1}-\frac{1}{2k}\right)=b_k \quad \text{and} \quad \lim \frac{n!\,e^n}{n^{n+\frac{1}{2}}}=\lambda,$$

we have

$$b_2 S_2 + b_3 S_3 + \cdots = 1 - \log \lambda.$$

(The *existence* of the limit λ results from the convergence of the series. We have $\lambda=\sqrt{2\pi}$.)

126.

a) $$\sum_{n=0}^{\infty} \frac{1}{x(x+1)\ldots(x+n)} = e\cdot\left[\frac{1}{x}-\frac{1}{1!}\frac{1}{x+1}+\frac{1}{2!}\frac{1}{x+2}-+\cdots\right];$$

b) $$\sum_{n=0}^{\infty} \frac{1}{n!}\frac{a^n}{1+x^2 a^{2n}} = e^{a} - x^2 e^{a^3} + x^4 e^{a^5} -+\cdots.$$

127. a) $$\frac{1}{\alpha}+\frac{1}{2}\frac{1!}{\alpha(\alpha+1)}+\frac{1}{3}\frac{2!}{\alpha(\alpha+1)(\alpha+2)}+\cdots=\frac{1}{\alpha^2}+\frac{1}{(\alpha+1)^2}+\cdots;$$

b) $$\sum_{n=0}^{\infty}\frac{1}{(n+1)^3(n+2)^3}=10-\pi^2.$$

128. With reference to § 35 A, establish the relation between

$$\sum_{n=1}^{\infty}\frac{1}{(n+\alpha)^3(n+\alpha+1)^3\ldots(n+\alpha+p-1)^3} \quad \text{and} \quad \sum_{n=1}^{\infty}\frac{1}{(n+\alpha-1)^3\ldots(n+\alpha+p)^3}$$

and, by giving special values to α and p, prove the following transformations:

a) $$\sum_{n=2}^{\infty}\frac{1}{n^3}=\frac{9}{8}+\frac{25}{2^2\cdot 3^4}-\frac{4}{3}\sum_{n=1}^{\infty}\frac{1}{(n+1)^3(n+2)^3(n+3)^3}$$

$$=\frac{9}{8}+\frac{133}{2^6\cdot 3^3}+\frac{3\cdot 4^4}{5}\sum_{n=1}^{\infty}\frac{1}{n^3(n+1)^3(n+2)^3(n+3)^3(n+4)^3};$$

b) $$\sum_{n=1}^{\infty}\frac{1}{n^3(n+1)^3}=\frac{83}{630}-\frac{2^3\cdot 3^4}{35}\sum_{n=1}^{\infty}\frac{1}{n^3(n+1)^3(n+2)^3(n+3)^3}.$$

Evaluate the sum of the first series to 6 places of decimals.

129. Prove similarly the transformations:

a) $$\sum_{n=1}^{\infty}\frac{1}{n^4}=\frac{7}{6}-\sum_{n=1}^{\infty}\frac{7n(n+1)+2}{6n^5(n+1)^5};$$

b) $$\sum_{n=1}^{\infty}\frac{1}{n^5}=\frac{7}{6}-\sum_{n=1}^{\infty}\frac{28n^2(n+\frac{3}{2})+24n+5}{12n^6(n+1)^6};$$

c) $$\sum_{n=1}^{\infty}\frac{(-1)^{n-1}}{n^2(n+1)^2\ldots(n+p-1)^2}$$

$$=\frac{5p+2}{4(p+1)}\frac{1}{(p!)^2}-\frac{p(p+1)^3}{4}\sum_{n=1}^{\infty}\frac{(-1)^{n-1}}{n^2(n+1)^2\ldots(n+p+1)^2}$$

Evaluate the sums of the series a) and b) to 6 places of decimals.

130. a) Denoting by T_p the sum of the series c) in the preceding exercise, we obtain relations between T_1 and T_{2k+1}, T_2 and T_{2k+2}. What are these relations? Is the process $k \to \infty$ allowed in them? What is the transformation thus obtained? Is it possible to deduce it directly as a *Markoff* transformation?

b) We have

$$\log 2 = \sum_{n=1}^{\infty} \frac{(-1)^{n-1}}{n} = \frac{1}{2} + \frac{1}{2} \sum_{n=1}^{\infty} \frac{(-1)^{n-1}}{n(n+1)} = \frac{3}{4} - \frac{1}{4} T_2.$$

Give the form now taken by the transformations of the series for $\log 2$ which were indicated in a).

c) Carry out the same process with the series **122** for $\frac{\pi}{4}$.

131. The sum of the series $\sum \frac{1}{n \log n \log_2 n (\log_3 n)^2}$, where n starts from the first integer satisfying $\log_3 n > 1$, evaluated to 8 decimal places, is exactly $\approx 1{\cdot}00000000$. — How may we determine whether the actual decimal expansion begins with $0 \cdot \ldots$ or with $1 \cdot \ldots$? — The solution of this problem requires a knowledge of the numerical value of $e''' = e^{(e^e)}$ to one decimal place at least: this is $e''' = 3814279{\cdot}1 \ldots$ It suffices, however, to know that $e''' - [e'''] = 0{\cdot}1 \ldots$. (Cf. remarks on p. 249.)

132. Arrange in order of magnitude all natural numbers of the form p^q. (p, q positive integers $\geqq 2$) and denote the nth of the numbers so arranged by p_n, so that

$$(p_1, p_2, \ldots) = (4, 8, 9, 16, 25, 27, 32, \ldots).$$

We then have

$$\sum_{n=1}^{\infty} \frac{1}{p_n - 1} = 1.$$

(Cf. **68**, 5.)

Part III.

Development of the theory.

Chapter IX.

Series of positive terms.

§ 36. Detailed study of the two comparison tests.

In the preceding chapters we contented ourselves with setting forth the fundamental facts of the theory of infinite series. Henceforth we shall aim somewhat further, and endeavour to penetrate deeper into the theory and proceed to give more extensive applications. For this purpose we first resume the considerations stated from a quite elementary standpoint in Chapters III and IV. We begin by examining in greater detail the two comparison tests of the first and second kinds (**72** and **73**), which were deduced immediately from the first main criterion (**70**), for the convergence or divergence of series of positive terms. These, and all related criteria, will in the sequel be expressed more concisely by using the notation Σc_n and Σd_n to denote any series of positive terms known *a priori* to be convergent and divergent respectively, whereas Σa_n shall denote a series — *also, in the present chapter, of positive terms only* — whose convergence or divergence is being examined. The criterion **72** can then be written in the simple form

157. (I) $$a_n \leqq c_n \quad : \quad \mathfrak{C}, \qquad a_n \geqq d_n \quad : \quad \mathfrak{D}.$$

This indicates that, *if* the terms of the series under consideration satisfy the first inequality *from and after a certain n*, then the series will *converge*; *if*, on the other hand, they satisfy the second inequality, *from and after a certain n*, then it must *diverge.*

The criterion **73** becomes in the same abbreviated notation

158. (II) $$\frac{a_{n+1}}{a_n} \leqq \frac{c_{n+1}}{c_n} \quad : \quad \mathfrak{C}, \qquad \frac{a_{n+1}}{a_n} \geqq \frac{d_{n+1}}{d_n} \quad : \quad \mathfrak{D}.$$

Before proceeding we may make a few remarks in this connexion. But let us first insist once more on one point: Neither these nor any of the analogous criteria to be established below will *necessarily solve* the question of convergence or divergence of any particular given series. They represent *sufficient* conditions only and may therefore very well fail in special cases. Their success will depend on the choice of the comparison series Σc_n and Σd_n (see below). The following pages will accordingly be devoted to establishing tests, as numerous and as efficacious as possible, so as to increase the probability of actually solving the problem in given special cases.

Remarks on the first comparison test (**157**). **159.**

1. Since for every positive number g the series $\Sigma g c_n$ and $\Sigma g d_n$ necessarily converge and diverge respectively with Σc_n and Σd_n, the first of our criteria may also be expressed in the form:

$$\frac{a_n}{c_n} \leqq g \, (< +\infty) \quad : \quad \mathfrak{C}, \qquad \frac{a_n}{d_n} \geqq g \, (> 0) \quad : \quad \mathfrak{D}$$

or, even more forcibly, in the form

$$\overline{\lim} \frac{a_n}{c_n} < +\infty \quad : \quad \mathfrak{C}, \qquad \underline{\lim} \frac{a_n}{d_n} > 0 \quad : \quad \mathfrak{D}.$$

2. Accordingly we must ***always*** have:

$$\overline{\lim} \frac{d_n}{c_n} = +\infty, \qquad \underline{\lim} \frac{c_n}{d_n} = 0$$

or, otherwise expressed:

$\overline{\lim} \frac{a_n}{c_n} = +\infty$ is a *necessary* condition for the divergence of Σa_n,

$\underline{\lim} \frac{a_n}{d_n} = 0$ " " *necessary* " " " convergence " Σa_n.

3. Here, as in all that follows, it is *not* necessary that actual unique limits should exist. This may be inferred, to take the question quite generally, from the fact that the convergence or divergence of a series of positive terms remains unaltered when the series is subjected to an arbitrary rearrangement (v. **88**). The latter can in every case be so chosen that the above limits do not exist. For instance Σc_n can be taken to be $1 + \frac{1}{2} + \frac{1}{4} + \frac{1}{8} + \cdots$, and Σa_n to be the series

$$\tfrac{1}{2} + 1 + \tfrac{1}{8} + \tfrac{1}{4} + \tfrac{1}{32} + \tfrac{1}{16} + \cdots,$$

obtained from the former by interchanging the terms in each successive pair; the ratio $\frac{a_n}{c_n}$ certainly tends to no unique limit; in fact, it has distinct upper and lower limits 2 and $\frac{1}{2}$. Similarly, let Σd_n be chosen to be the series $1 + \frac{1}{2} + \frac{1}{3} + \frac{1}{4} + \cdots$, and let Σa_n be the series

$$1 + \tfrac{1}{3} + \tfrac{1}{2} + \tfrac{1}{5} + \tfrac{1}{7} + \tfrac{1}{4} + \tfrac{1}{9} + \tfrac{1}{11} + \tfrac{1}{6} + \cdots,$$

deduced from the former by rearrangement, (in this series every *two* odd denominators are followed by *one* even one). Here $\frac{a_n}{d_n}$ has the two distinct upper and lower limits $\frac{3}{2}$ and $\frac{3}{4}$. In a similar manner we may convince ourselves by examples in the other cases that an actual unique limit need not exist. *If*, however, such an unique limit does exist, it necessarily satisfies the conditions indicated for $\overline{\lim}$ and $\underline{\lim}$, since it is then equal to both.

4. In particular: *No* condition of the form $\frac{a_n}{d_n} \to 0$ is necessary for the convergence of Σa_n — unless *all* the terms of the divergent series Σd_n remain greater than a fixed positive δ. For, even if we only have $\underline{\lim}\, d_n = 0$, by choosing

$$k_1 < k_2 < \cdots < k_\nu < \cdots,$$

so that

$$d_{k_\nu} < \frac{1}{2^\nu}$$

and writing $a_{k_\nu} = d_{k_\nu}$, $a_n = 0$ or $=$ the corresponding term c_n of any convergent series c_n for every other n, we evidently obtain a convergent series a_n, but it is equally evident that $\frac{a_n}{d_n}$ does *not* $\to 0$.

160. Remarks on the second comparison test (**158**).

1. The validity of the comparison test II may now be established more concisely as follows:

In the case marked ($\mathfrak{C}$), we have, from and after a definite n, $\frac{a_n}{c_n} \geqq \frac{a_{n+1}}{c_{n+1}}$, i. e. $\left(\frac{a_n}{c_n}\right)$ is a monotone descending sequence, whose limit γ is defined and $\geqq 0$. In particular $\overline{\lim}\,\frac{a_n}{c_n} = \gamma < +\infty$, and, by **159**, 1, Σa_n is convergent. In the case marked ($\mathfrak{D}$), $\left(\frac{a_n}{d_n}\right)$ is monotone ascending from and after a particular n, and accordingly also tends to a definite limit > 0, or to $+\infty$. In either case the condition $\underline{\lim}\,\frac{a_n}{d_n} > 0$ of **159**, 1 is fulfilled and this shows that Σa_n is divergent.

2. The comparison test II thus appears as an almost immediate *corollary* to the comparison test I. If the convergence or divergence of a series Σa_n can be inferred by comparison with a (definitely chosen) series Σc_n or Σd_n in accordance with **158**, then this may also be inferred by means of **157** (or **159**, 1), but not conversely, i. e. if I is decisive, II need not be so.

Examples of this have already occurred in the pairs of series of **159**, 3. For the first pair we have $\overline{\lim}\,\frac{a_n}{c_n} = 2$, while $\frac{a_{n+1}}{a_n}$ alternately $= 2$ and $= \frac{1}{8}$, i. e. it is sometimes greater, sometimes less than the corresponding ratio $\frac{c_{n+1}}{c_n}$, since this constantly $= \frac{1}{2}$. The second pair of series represents an equally simple case.

3. This relation between the two types of comparison tests becomes particularly interesting when we come to deal with the two tests to which we were led in § 13 as immediate applications of the first and second comparison tests. These were the root and ratio tests, inferred from I and II by the use of the geometric series as comparison series, and they may be stated thus: —

$$\sqrt[n]{a_n}\begin{cases}\leqq \vartheta < 1 & : \quad \mathfrak{C} \\ \geqq 1 & : \quad \mathfrak{D}\end{cases} \qquad \Bigg| \qquad \frac{a_{n+1}}{a_n}\begin{cases}\leqq \vartheta < 1 & : \quad \mathfrak{C} \\ \geqq 1 & : \quad \mathfrak{D}.\end{cases}$$

Our remark 2. shows that the ratio test may very well fail when the root test applies (the series Σa_n given there are obvious examples of this). On the other hand our remark 1. shows that the root test must *necessarily* work, if the ratio test does so. This relation between the two comparison tests is expressed in more significant form by the following theorem, which may be regarded as an extension of **43**, 3.

Theorem. *If $x_1, x_2, \ldots$ are arbitrary positive terms, we always have*[1] **161.**

$$\underline{\lim}\, \frac{x_{n+1}}{x_n} \leqq \underline{\lim}\, \sqrt[n]{x_n} \leqq \overline{\lim}\, \sqrt[n]{x_n} \leqq \overline{\lim}\, \frac{x_{n+1}}{x_n}.$$

Proof. The inner inequality is obvious[2], and the two outer inequalities are so closely similar that we may be content with proving one of them. Let us choose the right hand inequality and put

$$\overline{\lim}\, \sqrt[n]{x_n} = \mu, \qquad \overline{\lim}\, \frac{x_{n+1}}{x_n} = \mu',$$

so that the statement reduces to "$\mu \leqq \mu'$". Now if $\mu' = +\infty$, there is nothing to prove. But, if $\mu' < +\infty$, we may, given $\varepsilon > 0$, assign an integer p, such that, for every $\nu \geqq p$, we always have

$$\frac{x_{\nu+1}}{x_\nu} < \mu' + \frac{\varepsilon}{2}.$$

---

[1] This theorem is of the same character as **43**, 3. In fact, writing $y_1, y_2, y_3, \ldots$ for the ratios $\frac{x_1}{1}, \frac{x_2}{x_1}, \frac{x_3}{x_2}, \ldots$, we are concerned with a comparison of the upper and lower limits of y_n and of $y_n' = \sqrt[n]{y_1 y_2 \ldots y_n}$.

[2] For this reason, it is usual to write more shortly:

$$\underline{\lim}\, \frac{x_{n+1}}{x_n} \leqq \overline{\underline{\lim}}\, \sqrt[n]{x_n} \leqq \overline{\lim}\, \frac{x_{n+1}}{x_n},$$

implying that in the centre, either $\overline{\lim}$ or $\underline{\lim}$ may be considered indifferently. Such an abbreviated notation will frequently be used by us in the sequel.

This inequality may be supposed written down for every $\nu = p$, $p+1, \ldots, n-1$, and we then multiply all these inequalities together, deducing, for $n > p$,

$$x_n < x_p \cdot \left(\mu' + \frac{\varepsilon}{2}\right)^{n-p}.$$

Let us, for brevity, denote the *constant* number $x_p \cdot \left(\mu' + \frac{\varepsilon}{2}\right)^{-p}$ by A; then, for $n > p$, we always have

$$\sqrt[n]{x_n} < \sqrt[n]{A} \cdot \left(\mu' + \frac{\varepsilon}{2}\right).$$

But $\sqrt[n]{A} \to 1$, and hence $\left(\mu' + \frac{\varepsilon}{2}\right)\sqrt[n]{A} \to \mu' + \frac{\varepsilon}{2}$. We can therefore so choose $n_0 > p$ that, for every $n > n_0$, we have $\left(\mu' + \frac{\varepsilon}{2}\right)\sqrt[n]{A} < \mu' + \varepsilon$. We then have *a fortiori*, for every $n > n_0$,

$$\sqrt[n]{x_n} < \mu' + \varepsilon$$

and hence also $\mu \leqq \mu' + \varepsilon$, or, as asserted, since ε is arbitrary, $\mu \leqq \mu'$. (Cf. p. 68, footnote 10.) Moreover we can show by simple examples that the sign of equality need not hold in any of the three inequalities of **161**, which is now completely established.

4. The preceding theorem shows in particular that, if $\lim \frac{x_{n+1}}{x_n}$ exists, $\lim \sqrt[n]{x_n}$ must also exist and have the same value. Hence in particular: If the ratio test works in the form given in **76**, 2, then so will the root test, necessarily, (but not the converse!). To sum up: — The ratio test is *theoretically less powerful* than the root test. (Nevertheless it may frequently be preferred, as being easier of application.)

5. In this place we have also to refer to the remarks **75**, 1 and **76**, 3.

§ 37. The logarithmic scales.

We have already observed that such criteria as those just discussed only provide *sufficient* conditions and may accordingly fail in particular cases. Their efficiency will depend on the nature of the chosen comparison series Σc_n and Σd_n; in general terms we may say that a $\mathfrak{C}$-test will present a better prospect of success the *greater* the magnitude of the c_n's, a $\mathfrak{D}$-test, on the contrary, the *smaller* the magnitude of the d_n's. In order to express these circumstances more precisely, we proceed first to define the concept of the rapidity of convergence: A convergent series will be said to converge more or less rapidly according as its partial sums approach more or less rapidly to the sum of the series; and a divergent series will be said to diverge more or less rapidly in proportion to the rapidity with which its partial sums increase. More precisely:

Definition 1. *Given two convergent series* $\Sigma c_n = s$ *and* $\Sigma c_n' = s'$ *of positive terms, whose partial sums are denoted by* s_n *and* s_n', *the corresponding remainders by* $s - s_n = r_n$, $s' - s_n' = r_n'$, *we say that the second converges* ***more or less rapidly*** *(or* ***better or less well****) than the first, according as* **162.**

$$\lim \frac{r_n'}{r_n} = 0 \qquad \textit{or} \qquad \lim \frac{r_n'}{r_n} = +\infty.$$

If the limit of this ratio exists and has a finite positive value, or if it be known merely that its lower limit > 0 and its upper limit $< +\infty$, then the convergence of the two series will be said to be *of the same kind.* In any other case a comparison of the rapidity of convergence of the two series is impracticable[3].

Definition 2. *If* Σd_n *and* $\Sigma d_n'$ *are two divergent series of positive terms, whose partial sums are denoted by* s_n *and* s_n' *respectively, the second is said to* ***diverge more or less rapidly*** *(or* ***more or less markedly****) than the first according as*

$$\lim \frac{s_n'}{s_n} = +\infty \qquad \textit{or} \qquad \lim \frac{s_n'}{s_n} = 0.$$

If the upper and lower limits of this ratio are finite and positive, then the divergence of both series will be said to be *of the same kind.* In any other case we shall not compare the two series in respect of rapidity of divergence[4].

The two following theorems show that the rapidity of the convergence or divergence of two series may frequently be recognised from the terms themselves (without reference to partial sums or remainders):

Theorem 1. *If* $\frac{c_n'}{c_n} \to 0\,(+\infty)$, *then* $\Sigma c_n'$ *converges more (less) rapidly than* Σc_n.

---

[3] In the case $\underline{\lim} \frac{r_n'}{r_n} = 0\ (> 0)$ and $\overline{\lim} \frac{r_n'}{r_n} < +\infty\ (= +\infty)$, we might also speak of the series $\Sigma c_n'$ as "no less" ("no more") rapidly convergent than the series Σc_n; this however presents no particular advantages. In the case of the lower limit being 0 and the upper limit $+\infty$, the rapidity of the convergence of the two series is totally incommensurable. A similar remark holds for divergence. (The student should illustrate by examples the fact that all the cases mentioned can really occur.) — These definitions may be directly transferred to the case of series of arbitrary terms, replacing r_n and r_n' by their *absolute values.*

[4] The properties referred to in these definitions are obviously *transitive,* i. e. if a first given series converges more rapidly than a second, and this again more rapidly than a third, the first series will also converge more rapidly than the third.

Proof. In the first case, given ε, we choose n_0 so that for every $n > n_0$ we have $c_n' < \varepsilon c_n$. We then also have

$$\frac{r_n'}{r_n} = \frac{c_{n+1}' + c_{n+2}' + \cdots}{c_{n+1} + c_{n+2} + \cdots} < \varepsilon \frac{c_{n+1} + \cdots}{c_{n+1} + \cdots} = \varepsilon .$$

Consequently this ratio tends to 0. The second case reduces to the first by interchanging the two series (cf. the theorem of **40**, 4, Rem. 4). This proves all that was required.

Theorem 2. *If* $\frac{d_n'}{d_n} \to 0 (+\infty)$, *then* $\Sigma d_n'$ *diverges less (more) rapidly than* Σd_n.

Proof. By **44**, 4 it follows immediately from $\frac{d_n'}{d_n} \to 0$ that

$$\frac{d_1' + d_2' + \cdots + d_n'}{d_1 + d_2 + \cdots + d_n} = \frac{s_n'}{s_n} \to 0 .$$

This proves the statement.

163. Simple examples. 1. The series

$$\Sigma \frac{1}{n \log^2 n}, \quad \Sigma \frac{1}{n^2}, \quad \Sigma \frac{1}{n^3}, \quad \Sigma \frac{1}{2^n}, \quad \Sigma \frac{1}{3^n}, \quad \Sigma \frac{1}{n!}, \quad \Sigma \frac{1}{n^n},$$

are such that each converges more rapidly than the preceding. In fact we have e. g. for $n > 3$:

$$\frac{1}{n!} \div \frac{1}{3^n} = \frac{3^n}{n!} = \left(\frac{3 \cdot 3 \cdot 3}{1 \cdot 2 \cdot 3}\right) \frac{\cdot 3 \ldots 3}{\cdot 4 \ldots n} < \frac{9}{2} \cdot \left(\frac{3}{4}\right)^{n-3},$$

which tends to 0. Similarly $\frac{\log^2 n}{n} \to 0$ (by **38**,4); the other cases are even simpler.

2. The series

$$\Sigma 2^n, \quad \Sigma n, \quad \Sigma 1, \quad \Sigma \frac{1}{n}, \quad \Sigma \frac{1}{n \log n}, \quad \Sigma \frac{1}{n \log n \log_2 n}, \ldots$$

are such that each diverges less rapidly than the preceding.

Besides the above simple examples, the most important cases of series with rapidity of convergence forming a graduated scale are afforded by the series which we came across in § **14**. As we saw in that paragraph, the series

$$\Sigma \frac{1}{n^\alpha}, \quad \Sigma \frac{1}{n (\log n)^\alpha}, \quad \Sigma \frac{1}{n \log n (\log_2 n)^\alpha}, \quad \ldots, \quad \Sigma \frac{1}{n \log n \ldots \log_{p-1} n \cdot (\log_p n)^\alpha}$$

converge for $\alpha > 1$ and diverge for $\alpha \leqq 1$. Our theorems 1 and 2 now show more precisely that when p is fixed each of these series will converge or diverge less and less rapidly *as the exponent* α *approaches unity* (remaining > 1 in the first case and $\leqq 1$ in the second). Similarly each of these series will converge or diverge less and less rapidly, *as p increases,* whatever positive[5] value may be given to the exponent α (>1 in the first case, $\leqq 1$ in the second).

The second alone of these statements perhaps requires some justification. Divide the generic term of the $(p+1)^{\text{th}}$ series with the exponent α' by the corresponding term of the pth series taken with the exponent α. We obtain

$$\frac{(\log_p n)^{\alpha}}{\log_p n \cdot (\log_{p+1} n)^{\alpha'}}.$$

In the case of divergent series, α and α' are positive and $\leqq 1$; the ratio therefore tends to 0, q. e. d. In the case of convergence, i. e. α and α' both >1, the ratio tends to $+\infty$; in fact, — by reasoning analogous to that of **38**, 4, — we have the auxiliary theorem that the numbers

$$\frac{(\log_{p+1} n)^{\alpha'}}{(\log_p n)^{\beta}} = \frac{\{\log(\log_p n)\}^{\alpha'}}{(\log_p n)^{\beta}}$$

form a null sequence, $\beta = \alpha - 1$ denoting any positive exponent and p any positive integer. This proves all that was required.

The gradation in the rapidity of the convergence and divergence of these series enables us to deduce complete scales of convergence and divergence tests by introducing these series as comparison series in the tests I and II (p. 274). We first immediately obtain the following form of the criteria:

$$\text{(I)} \qquad \left.\begin{matrix} a_n \leqq \\ a_n \geqq \end{matrix}\right\} \frac{1}{n \log n \ldots \log_{p-1} n \, (\log_p n)^{\alpha}} \quad \text{with} \quad \left\{\begin{matrix} \alpha > 1 & : & \mathfrak{C} \\ \alpha \leqq 1 & : & \mathfrak{D} \end{matrix}\right.$$

164.

$$\text{(II)} \qquad \left.\begin{matrix} \dfrac{a_{n+1}}{a_n} \leqq \\ \geqq \end{matrix}\right\} \frac{n}{n+1} \cdot \frac{\log n}{\log(n+1)} \cdots \frac{\log_{p-1} n}{\log_{p-1}(n+1)} \cdot \left(\frac{\log_p n}{\log_p(n+1)}\right)^{\alpha}$$

$$\text{with} \quad \left\{\begin{matrix} \alpha > 1 & : & \mathfrak{C} \\ \alpha \leqq 1 & : & \mathfrak{D}. \end{matrix}\right.$$

These criteria will be referred to briefly as *the logarithmic tests of the first and second kinds* — also in the case $p = 0$. Their efficiency may be increased by the choice of p, and, for fixed p, by the choice of α, in accordance with our previous remarks[6].

---

[5] For $\alpha = -\beta < 0$, each series of course diverges more rapidly than the preceding one with the exponent replaced by 1; thus e. g. $\sum \frac{(\log n)^{\beta}}{n}$, with $\beta > 0$, diverges more rapidly than $\sum \frac{1}{n}$.

[6] The convergence and divergence of series of the above type was known to *N. H. Abel* in 1827, but was not published by him (Œuvres II, p. 200). *A. de Morgan* (The differential and integral calculus, London 1842) was the first

For practical purposes it is advantageous to give other forms of these criteria. Such transformations are given below with a few remarks appended, but without completely carrying out the necessary calculations.

165. Transformation of the logarithmic tests of the 1st kind.

1. When a and b are positive, the two inequalities $a \leqq b$ and $\log a \leqq \log b$ are equivalent; after a slight alteration the inequalities **164**, I accordingly become:

$$\text{(I}'\text{)} \qquad \frac{\log a_n + \log n + \log_2 n + \cdots + \log_p n}{\log_p n} \begin{cases} \leqq -\beta < 0 & : \quad \mathfrak{C} \\ \geqq 0 & : \quad \mathfrak{D}. \end{cases}$$

2. Denoting for a moment by A_n the expression on the left of (I′), we have, in

$$\text{(I}''\text{)} \qquad \overline{\lim}\, A_n < 0 \quad : \quad \mathfrak{C}, \qquad \underline{\lim}\, A_n > 0 \quad : \quad \mathfrak{D},$$

a test of practically the same effect. The parts relating to convergence are indeed completely equivalent in (I′) and (I″); that relating to divergence is not quite so powerful in (I″) as in (I′), since it is required in (I″) that A_n should remain, from some value of n onwards, not merely $= 0$ but greater than a fixed positive number [7].

3. If we use the somewhat more explicit notation $A_n = A_n^{(p)}$, and consider both $A_n^{(p)}$ and $A_n^{(p+1)}$, we obviously have

$$A_n^{(p+1)} = 1 + \frac{\log_p n}{\log_{p+1} n} \cdot A_n^{(p)}.$$

And since, by **38**, 4, $\dfrac{\log_p n}{\log_{p+1} n} = \dfrac{\log_p n}{\log\,(\log_p n)}$ tends as n increases to $+\infty$, this simple transformation leads to the following result: If for a particular p one of the limits of $A_n = A_n^{(p)}$ is different from zero, it is necessarily ∞ for the following p, in fact $+\infty$ or $-\infty$ according as the preceding p was positive or negative. More precisely, if we denote by μ_p and $\varkappa_p$ the upper and the lower limits of $A_n = A_n^{(p)}$, for every p, then if we have, for any particular p,

$$\varkappa_p \leqq \mu_p < 0, \quad \text{we have} \quad \varkappa_{p+1} = \mu_{p+1} = -\infty,$$

and if

$$\mu_p \geqq \varkappa_p > 0, \quad \text{we have} \quad \mu_{p+1} = \varkappa_{p+1} = +\infty.$$

If, however,

$$\varkappa_p < 0, \quad \mu_p > 0, \quad \text{we have} \quad \varkappa_{p+1} = -\infty, \ \mu_{p+1} = +\infty.$$

The scales of reference (I) thus lead to the solution of the question of convergence or divergence if, and only if, for a particular p, the values $\varkappa_p$ and μ_p have the *same* sign. If the sign is negative, the series converges; if positive, it

---

to use these series for the construction of criteria. Essentially, these criteria are consequences of **164**, I and II; numerous transformations of them were subsequently published as special criteria, e. g. by *J. Bertrand* (J. de math. pures et appl., (1) Vol. 7, p. 35. 1842), *O. Bonnet* (ibid., (1) Vol. 8, p. 78. 1843), *U. Dini* (Giornale di matematiche, Vol. 6, p. 166. 1868).

[7] It would clearly, however, be wrong to write the last $\mathfrak{D}$-test in the form $\underline{\lim}\, A_n \geqq 0$, since the lower limit may very well be 0 without a single term being positive.

diverges; if the two numbers have opposite signs for some value of p, then for *all* higher p's we have

$$\underline{\lim}\, A_n^{(p)} = -\infty, \qquad \overline{\lim}\, A_n^{(p)} = +\infty$$

and the scale therefore is not decisive. Similarly it fails when both numbers are zero for *every* p.

Transformation of the logarithmic tests of the 2nd kind. 166.

1. The following Lemmas are easily proved:

Lemma 1. *For every integer $p \geqq 0$, for every real α and every sufficiently large n, an equality of the form*

$$\left(\frac{\log_p(n-1)}{\log_p n}\right)^\alpha = 1 - \frac{\alpha}{n \log n \ldots \log_p n} - \frac{\vartheta_n}{n^2}$$

holds, where (ϑ_n) is bounded[8]. The index n is here assumed to start with a value from and after which all the denominators are defined and positive.

We immediately infer that, for every integer $p \geqq 0$, for every real α and every sufficiently large n,

$$\frac{n-1}{n} \cdot \frac{\log(n-1)}{\log n} \cdots \frac{\log_{p-1}(n-1)}{\log_{p-1} n} \cdot \left(\frac{\log_p(n-1)}{\log_p n}\right)^\alpha$$

$$= 1 - \frac{1}{n} - \frac{1}{n \log n} - \cdots - \frac{1}{n \log n \ldots \log_{p-1} n} - \frac{\alpha}{n \log n \ldots \log_p n} - \frac{\eta_n}{n^2}$$

where (η_n) is again certainly bounded[9].

Lemma 2. *Let Σa_n and $\Sigma a_n'$ be two series of positive terms; if the series whose n^{th} term is* 167.

$$\gamma_n = \left(\frac{a_{n+1}}{a_n} \cdot \frac{a_n'}{a_{n+1}'} - 1\right)$$

is absolutely convergent, the two given series are either both convergent or both divergent.

In fact, we have $\gamma_\nu > -1$ for every ν; taking, then, any positive integer m, writing down the relations

$$\frac{a_{\nu+1}}{a_\nu} \cdot \frac{a_\nu'}{a_{\nu+1}'} = 1 + \gamma_\nu$$

for $\nu = m, m+1, \ldots, n-1$, and multiplying them together, we at once deduce that the ratio a_n'/a_n for $n > m$ lies between two fixed positive numbers. —

---

[8] An equality of the above form of course holds *under any circumstances*. In fact we can consider the numbers ϑ_n as defined precisely by the equation:

$$\vartheta_n = n^2 \left[1 - \frac{\alpha}{n \log n \ldots \log_p n} - \left(\frac{\log_p(n-1)}{\log_p n}\right)^\alpha\right].$$

The emphasis lies on the statement that (ϑ_n) is *bounded*. — The proof is obtained inductively, with the help of the two remarks that if (ϑ_n') and (ϑ_n'') are defined, for every sufficiently large n, by

$$(1 - x_n)^\alpha = 1 - \alpha x_n - \vartheta_n' x_n^2 \quad \text{and} \quad \log\left(1 - \frac{1}{n y_n}\right) = -\frac{1}{n y_n} - \frac{\vartheta_n''}{n^2},$$

they are necessarily bounded, provided (x_n) is a null sequence and the numbers y_n are in absolute value $\geqq 1$, say.

[9] The interpretation in the case $p = 0$ is immediately obvious.

The conditions of the Lemma are fulfilled, in particular, when the ratios $\frac{a_{n+1}}{a_n}$ and $\frac{a'_{n+1}}{a'_n}$ lie between fixed positive bounds and the series $\sum\left|\frac{a_{n+1}}{a_n}-\frac{a'_{n+1}}{a'_n}\right|$ converges.

2. In accordance with the above we may express the logarithmic test of the second kind e. g. in the following form[10]:

168. $$\frac{a_{n+1}}{a_n}\begin{matrix}\leqq\\ \geqq\end{matrix}\Big\}\, 1-\frac{1}{n}-\frac{1}{n\log n}-\cdots-\frac{1}{n\log n\ldots\log_{p-1}n}-\frac{\alpha'}{n\log n\ldots\log_p n}$$

$$\text{with}\ \begin{cases}\alpha'>1 & : \quad \mathfrak{C}\\ \alpha'\leqq 1 & : \quad \mathfrak{D},\end{cases}$$

or, after a simple transformation,

169. $$\left[\frac{a_{n+1}}{a_n}-1+\frac{1}{n}+\cdots+\frac{1}{n\log n\ldots\log_p n}\right]\cdot n\log n\ldots\log_p n$$

$$\begin{cases}\leqq-\beta<0 & : \quad \mathfrak{C}\\ \geqq 0 & : \quad \mathfrak{D},\end{cases}$$

or, finally, denoting the expression on the left hand side for brevity by B_n, and slightly restricting the scope of the $\mathfrak{D}$-test (cf. **165**, 2),

$$\overline{\lim}\, B_n<0 \quad : \quad \mathfrak{C}, \qquad \underline{\lim}\, B_n>0 \quad : \quad \mathfrak{D}.$$

Remarks analogous to those of **165**, 2 hold here.

3. The developments of **165**, 3 also remain valid, with quite unessential alterations. For, if we use the more explicit notation $B_n=B_n^{(p)}$, we have obviously

$$B_n^{(p+1)}=1+B_n^{(p)}\cdot\log_{p+1}n.$$

And, as $\log_{p+1}n\to+\infty$, we may reason with this relation in precisely the same manner as with its analogue in **165**, 3. It is unnecessary to develop this in detail.

4. Still more generally, we may at once prove that a series of the form

$$\sum\frac{1}{e^{(\alpha-1)n}\cdot n^{\alpha_0}(\log n)^{\alpha_1}(\log_2 n)^{\alpha_2}\ldots(\log_q n)^{\alpha_q}}$$

converges *if, and only if,* the first of the exponents α, α_0, α_1, ..., α_q which differs from 1 is >1. The values of the subsequent exponents have no further influence. — When the comparison series is put into this form, *Raabe*'s test (§ 38) and *Cauchy*'s ratio test appear naturally as the 0th and the (—1)th terms of the logarithmic scale.

§ 38. Special comparison tests of the second kind.

The logarithmic tests deduced in the preceding article are undoubtedly of greater theoretical than practical interest. They afford indeed a more profound insight into the systematic theory of the convergence of series of positive terms, but are of little use in actually testing the convergence of such series as occur in applications of the

---

10 Here we make the nth term of the investigated series Σa_n correspond *to the* $(n-1)^{\text{th}}$ *term* of the comparison series, which, by **82**, theorem 4, is allowable.

theory. (For this reason we have only sketched the considerations relating to them.) For practical purposes the first two or three terms, at most, of the logarithmic scales may be turned to account; from these we proceed to deduce by specialization a number of simpler tests, which were discovered at various times, rather by chance, and each proved in its own way, but which may now be arranged in closer connexion with one another.

For $p = 0$ the logarithmic scale provides a criterion already established by *J. L. Raabe*[11]. We deduce it from **169**, first in the form

$$\left[\frac{a_{n+1}}{a_n} - 1 + \frac{1}{n}\right] n \begin{cases} \leqq -\beta < 0 & : \quad \mathfrak{C} \\ \geqq 0 & : \quad \mathfrak{D}, \end{cases}$$

or, as we may now write more advantageously,

$$\left[\frac{a_{n+1}}{a_n} - 1\right] n \begin{cases} \leqq -\alpha < -1 & : \quad \mathfrak{C} \\ \geqq -1 & : \quad \mathfrak{D}. \end{cases} \qquad \textbf{170.}$$

The very elementary nature and great practical utility of this criterion makes it worth while to give a direct *proof* of its validity: the $\mathfrak{C}$-condition means that, for every sufficiently large n,

$$\frac{a_{n+1}}{a_n} \leqq 1 - \frac{\alpha}{n} \qquad \text{or} \qquad n\, a_{n+1} \leqq (n-1)\, a_n - \beta\, a_n$$

where $\beta = \alpha - 1 > 0$. Hence

$$(n-1)\, a_n - n\, a_{n+1} \geqq \beta\, a_n > 0$$

and therefore $n\, a_{n+1}$ is the term of a monotone descending sequence, for a sufficiently large n. Since it is constantly positive, it tends to a limit $\gamma \geqq 0$. The series Σc_n with $c_n = (n-1)\, a_n - n\, a_{n+1}$ therefore converges, by **131**. Since $a_n \leqq \frac{1}{\beta} c_n$, the convergence of Σa_n immediately follows. — Similarly, if the $\mathfrak{D}$-condition is fulfilled, we have

$$\frac{a_{n+1}}{a_n} \geqq 1 - \frac{1}{n} \qquad \text{or} \qquad (n-1)\, a_n - n\, a_{n+1} \leqq 0.$$

Accordingly $n\, a_{n+1}$ is the term of a monotone increasing sequence and therefore remains greater than a fixed positive number γ. As $a_{n+1} > \frac{\gamma}{n}$, $\gamma > 0$, the divergence follows immediately.

If the expression on the left in **170** tends, when $n \to +\infty$, to a limit l, it follows from the reasoning already repeatedly applied (v. **76**, 2) that $l < -1$ involves the convergence of Σa_n, and $l > -1$, its divergence, while $l = -1$ leads to no immediate conclusion.

---

[11] Zeitschr. f. Phys. u. Math. von Baumgarten u. Ettinghausen, Vol. 10, p. 63, 1832. Cf. *Duhamel, J. M. C.*: Journ. de math. pures et appl., (1) Vol. 4, p. 214, 1839.

Examples.

1. In § 25 we examined the binomial series and were unable to decide there whether the series converged or not at the endpoints of the interval of convergence, that is, whether for given real α's the series

$$\sum_{n=0}^{\infty} \binom{\alpha}{n} \quad \text{and} \quad \sum_{n=0}^{\infty} (-1)^n \binom{\alpha}{n}$$

were, or were not, convergent. We are now able to decide this question.

For the *second* series we have

$$\frac{a_{n+1}}{a_n} = -\frac{\alpha - n}{n+1} = \frac{(n+1)-(\alpha+1)}{n+1}.$$

Since this ratio is positive from a certain stage on, it follows that the terms then maintain one sign; this we may assume to be the sign +, since changing the signs of all the terms does not, of course, affect the argument. Further, according to this,

$$\left(\frac{a_{n+1}}{a_n} - 1\right) n = -(\alpha+1)\cdot\frac{n}{n+1} \to -(\alpha+1),$$

from which we at once deduce, by *Raabe*'s test, that the second of our series converges for $\alpha > 0$, and diverges for $\alpha < 0$. For $\alpha = 0$, the series reduces to its initial term 1.

For the *first* series we have

$$\frac{a_{n+1}}{a_n} = -1 + \frac{\alpha+1}{n+1}$$

and, since this value becomes negative from some stage on, the terms of the series have an alternating sign from that stage on. If now we suppose $\alpha+1 \leqq 0$, we therefore have

$$\left|\frac{a_{n+1}}{a_n}\right| \geqq 1$$

whence we infer that ultimately the terms a_n are non-decreasing. The series must therefore diverge. If however we suppose $\alpha + 1 > 0$, we have ultimately, say for every $n \geqq m$,

(a) $$\left|\frac{a_{n+1}}{a_n}\right| = 1 - \frac{\alpha+1}{n+1} < 1,$$

and the terms ultimately decrease in absolute value. By *Leibniz*'s criterion for series with alternately positive and negative terms, our series must therefore converge, provided we can show that $\binom{\alpha}{n} \to 0$. If we write down the relations (a) for $m, m+1, \ldots, n-1$ and multiply them all together, we deduce for every $n > m$

$$|a_n| = |a_m| \cdot \prod_{\nu=m+1}^{n} \left(1 - \frac{\alpha+1}{\nu}\right).$$

Since, however, the product $\prod \left(1 - \frac{\alpha+1}{\nu}\right)$, by **126**, 2, 3, diverges to 0, a_n must also $\to 0$, and therefore $\sum \binom{\alpha}{n}$ must converge. Summing up, we therefore have the following results relating to the binomial series:

The series $\sum_{n=0}^{\infty} \binom{\alpha}{n} x^n$ *converges if, and only if, either* $|x| < 1$, *or* $x = -1$

and $\alpha > 0$[12], *or* $x = +1$ *and* $\alpha > -1$. *The sum of the series is then by Abel's theorem of limits always* $(1 + x)^\alpha$. *If* α *is an integer and is non-negative, then the series is finite and hence converges for each* x. *In all other cases the series is divergent.* (An appreciable addition to this theorem is provided by **247**.)

2. The following criterion does not differ essentially from that of *Raabe*; it is due to *O. Schlömilch*:

$$n \log \frac{a_{n+1}}{a_n} \begin{cases} \leqq -\alpha < -1 & : \quad \mathfrak{C} \\ \geqq -1 & : \quad \mathfrak{D}. \end{cases}$$

In fact, in the case ($\mathfrak{D}$), we have, by **114**, $\frac{a_{n+1}}{a_n} \geqq e^{-\frac{1}{n}} > 1 - \frac{1}{n}$,

from which the divergence follows by *Raabe*'s test. In case ($\mathfrak{C}$) we have, ultimately,

$$\frac{a_{n+1}}{a_n} \leqq e^{-\frac{\alpha}{n}} \leqq 1 - \frac{\alpha'}{n},$$

if $\alpha > \alpha' > 1$. By **170**, this involves convergence.

If, in the logarithmic scale, we choose $p = 1$, we obtain a criterion of the second kind which, omitting the limiting case $\alpha = 1$, we may write

$$\frac{a_{n+1}}{a_n} = 1 - \frac{1}{n} - \frac{\alpha_n}{n \log n} \quad \text{with} \quad \begin{cases} \alpha_n \geqq \alpha > 1 & : \quad \mathfrak{C} \\ \alpha_n \leqq \alpha < 1 & : \quad \mathfrak{D}. \end{cases}$$ **171.**

A *direct proof* of the validity of this criterion can be given as follows: As in the proof of *Raabe*'s test we first put the criterion in the following form:

$$[-1 + (n-1)\log n]\, a_n - [n \log n]\, a_{n+1} \begin{cases} \geqq \ \beta\, a_n \ \text{with} \ \beta > 0 & : \ \mathfrak{C} \\ \leqq -\beta' a_n \ \text{with} \ \beta' > 0 & : \ \mathfrak{D}. \end{cases}$$

If now the $\mathfrak{C}$*-condition is fulfilled,* since, as we may immediately verify by **114**, α,

$$(n-1)\log(n-1) > -1 + (n-1)\log n,$$

we have *a fortiori*

$$(n-1)\log(n-1)\cdot a_n - n \log n \cdot a_{n+1} \geqq \beta\, a_n.$$

Accordingly $n \log n \cdot a_{n+1}$ is the term of a monotone descending sequence and accordingly tends to a limit $\gamma \geqq 0$. By **131**, the series whose n^{th} term is

$$c_n = (n-1)\log(n-1)\cdot a_n - n \log n \cdot a_{n+1}$$

must converge. As $a_n \leqq \frac{1}{\beta} c_n$, the same is true of Σa_n.

If, on the other hand, the $\mathfrak{D}$*-condition is fulfilled,* we have

$$(n-1)\log(n-1)\cdot a_n - n \log n \cdot a_{n+1}$$
$$\leqq \left[-\beta' + 1 - (n-1)\log\left(1 + \frac{1}{n-1}\right)\right] a_n.$$

For $n \to +\infty$, however, the expression in square brackets $\to -\beta'$

[12] For $\alpha = 0$, see above

(by **112**, b), and is therefore negative for every sufficiently large n Hence for those n's the expression $n \log n \cdot a_{n+1}$ increases monotonely and consequently remains greater than a certain positive number γ As $a_{n+1} \geqq \frac{\gamma}{n \log n}$, $\gamma > 0$, it follows that Σa_n must diverge.

Here again we may observe, as repeatedly in previous instances, that, if α_n tends to a limit l, then $l > 1$ involves convergence, and $l < 1$ involves divergence, while, from $l = 1$, nothing can be directly inferred.

Even this, the first properly *logarithmic* criterion of the scale, will rarely be actually applied in practice. In fact, the series which are amenable to this test, and not already to a simpler one (*Raabe*'s test, or the ratio test), occur exceedingly seldom; and as their convergence is no more rapid than that of $\sum \frac{1}{n (\log n)^{\alpha}}$, $(\alpha > 1)$, these series are useless for numerical calculation.

It enables us, however, to deduce easily one or two other criteria. We will above all mention

172. Gauss's Test[13]: *If the ratio* $\frac{a_{n+1}}{a_n}$ *can be expressed in the form*

$$\frac{a_{n+1}}{a_n} = 1 - \frac{\alpha}{n} - \frac{\vartheta_n}{n^{\lambda}}$$

where $\lambda > 1$, *and* (ϑ_n) *is* bounded[14], *then* Σa_n *converges when* $\alpha > 1$ *and diverges when* $\alpha \leqq 1$.

The proof is immediate: when $\alpha \gtrless 1$, *Raabe*'s test itself proves the validity of the assertion. For $\alpha = 1$, we write

$$\frac{a_{n+1}}{a_n} = 1 - \frac{1}{n} - \frac{1}{n \log n}\left(\frac{\vartheta_n \cdot \log n}{n^{\lambda-1}}\right);$$

and as now the factor in brackets tends to zero since $(\lambda - 1) > 0$, the series certainly diverges, by **171**.

Gauss expressed this criterion in somewhat more special form as follows: "*If the ratio* $\frac{a_{n+1}}{a_n}$ *can be expressed in the form*

$$\frac{a_{n+1}}{a_n} = \frac{n^k + b_1 n^{k-1} + \cdots + b_k}{n^k + b_1' n^{k-1} + \cdots + b_k'} \qquad (k \text{ an integer } \geqq 1)$$

then Σa_n *will converge when* $b_1 - b_1' < -1$ *and diverge when* $b_1 - b_1' \geqq -1$." — The proof is obvious from the preceding.

[13] Werke, Vol. 3, p. 140. — This criterion was established by *Gauss* in 1812.

[14] Cf. footnote 8, p. 283.

Examples.

1. *Gauss* established this test in order to determine the convergence of the so-called ***hypergeometric series***

$$1+\frac{\alpha\cdot\beta}{1\cdot\gamma}x+\frac{\alpha(\alpha+1)}{1\cdot 2}\cdot\frac{\beta(\beta+1)}{\gamma(\gamma+1)}x^2+\frac{\alpha(\alpha+1)(\alpha+2)}{1\cdot 2\cdot 3}\cdot\frac{\beta(\beta+1)(\beta+2)}{\gamma(\gamma+1)(\gamma+2)}x^3+\cdots$$

$$\equiv\sum_{n=0}^{\infty}\frac{\alpha(\alpha+1)\ldots(\alpha+n-1)}{1\cdot 2\ldots n}\cdot\frac{\beta(\beta+1)\ldots(\beta+n-1)}{\gamma(\gamma+1)\ldots(\gamma+n-1)}x^n$$

where α, β, γ are any real numbers[15] different from $0, -1, -2, \ldots$. Here

$$\frac{a_{n+1}}{a_n}=\frac{(\alpha+n)(\beta+n)}{(1+n)(\gamma+n)}x$$

which shows in the first instance that the series converges (absolutely) for $|x|<1$, and diverges for $|x|>1$. Accordingly it only remains to examine the values $x=1$ and $x=-1$. This is analogous to the case of the binomial series, to which, of course, the present one reduces when we choose $\beta=\gamma\ (=1)$ and replace α and x by $-\alpha$ and $-x$.

For $x=1$, we have

$$\frac{a_{n+1}}{a_n}=\frac{n^2+(\alpha+\beta)n+\alpha\beta}{n^2+(\gamma+1)n+\gamma}.$$

This shows that for every sufficiently large n, the terms of the series have one and the same sign, which may be assumed positive. *Gauss*'s test now shows that the series converges for $\alpha+\beta-\gamma-1<-1$, i. e. for $\alpha+\beta<\gamma$, but diverges for $\alpha+\beta\geqq\gamma$.

For $x=-1$, the series has, from some stage on, alternately positive and negative terms, since $\frac{a_{n+1}}{a_n}\to -1$, i. e. is ultimately negative. The relation[16]

$$\frac{a_{n+1}}{a_n}=-\frac{n^2+(\alpha+\beta)n+\alpha\beta}{n^2+(\gamma+1)n+\gamma}=-\left[1+\frac{\alpha+\beta-\gamma-1}{n}+\frac{\vartheta_n}{n^2}\right],$$

with word for word the same reasoning as was employed in **170**, 1 for the binomial series, now shows that the hypergeometric series will

diverge when $\alpha+\beta-\gamma>1$
converge when $\alpha+\beta-\gamma<1$.

We have only to verify further that it also diverges when $\alpha+\beta-\gamma=1$, as this does not follow from precisely the same reasoning as before. If for every $n>p>1$ we have

$$\frac{a_{n+1}}{a_n}=-\left(1+\frac{\vartheta_n}{n^2}\right)\quad\text{with}\quad|\vartheta_n|\leqq\vartheta\quad\text{for every } n,$$

then, assuming p chosen so large that $p^2>\vartheta$,

$$|a_n|>|a_p|\left(1-\frac{\vartheta}{p^2}\right)\left(1-\frac{\vartheta}{(p+1)^2}\right)\cdots\left(1-\frac{\vartheta}{(n-1)^2}\right).$$

Since on the right hand side we have the product of the first $(n-p)$ factors of a *convergent* infinite product of positive factors, it follows that $|a_n|$, for *all these* values of n, remains greater than a certain positive number. The series can therefore only diverge.

---

[15] For these values, the series would terminate or become meaningless. For $n=0$, the general term of the series should be equated to 1.

[16] As before, (ϑ_n) denotes a bounded sequence of numbers.

2. *Raabe*'s $\mathfrak{C}$-test fails if the numbers α_n in the expression

$$\frac{a_{n+1}}{a_n} = 1 - \frac{\alpha_n}{n}$$

though constantly > 1, have the value 1 for lower limit. In that case, writing $\alpha_n = 1 + \beta_n$, the condition

$$\overline{\lim}\, n\, \beta_n = +\infty$$

is a *necessary* condition for the convergence of Σa_n. In fact, if $n\,\beta_n$ were bounded, we should have

$$\frac{a_{n+1}}{a_n} = 1 - \frac{1}{n} - \frac{\vartheta_n}{n^2}$$

and Σa_n would be divergent by *Gauss*'s test[17].

§ 39. Theorems of *Abel, Dini* and *Pringsheim* and their application to a fresh deduction of the logarithmic scale of comparison tests.

Our previous manner of deducing the logarithmic tests invests these, the most general criteria yet obtained, with something of a fortuitous character. In fact everything turned on the use, as comparison series, of *Abel*'s series, which were obtained themselves only as chance applications of *Cauchy*'s condensation test. This character of fortuitousness disappears to some extent if we approach the subject from a different direction, involving a greater degree of inevitableness. Our starting point for this is the following

173. Theorem of *Abel* and *Dini*[18]: *If* $\sum_{n=1}^{\infty} d_n$ *is an arbitrary divergent series of positive terms, and* $D_n = d_1 + d_2 + \cdots + d_n$ *denotes its partial sums, the series*

$$\sum_{n=1}^{\infty} a_n \equiv \sum_{n=1}^{\infty} \frac{d_n}{D_n^{\alpha}} \begin{cases} \textit{converges when } \alpha > 1 \\ \textit{diverges when } \alpha \leqq 1. \end{cases}$$

Proof. In the case $\alpha = 1$,

$$\frac{d_{n+1}}{D_{n+1}} + \cdots + \frac{d_{n+k}}{D_{n+k}} \geqq \frac{d_{n+1} + \cdots + d_{n+k}}{D_{n+k}} = 1 - \frac{D_n}{D_{n+k}}.$$

As $D_\nu \to +\infty$ by hypothesis, we can therefore choose $k = k_n$, for each n, so that

$$\frac{D_n}{D_{n+k_n}} < \frac{1}{2}, \quad \text{i. e.} \quad a_{n+1} + a_{n+2} + \cdots + a_{n+k_n} > \frac{1}{2};$$

17 *Cahen, E.*: Nouv. Annales de Math., (3) Vol. 5, p. 535.

18 *N. H. Abel* (J. f. d. reine u. angew. Math., Vol. 3, p. 81. 1828) only proved the divergence of $\sum \frac{d_n}{D_{n-1}}$; *U. Dini* (Sulle serie a termini positivi, Annali Univ. Toscana Vol. 9. 1867) established the theorem in the above complete form. It was not till 1881 that writings of *Abel* were discovered (Œuvres II, p. 197) which also contain the part relative to convergence of the theorem given above.

by **81**, 2, the series Σa_n must accordingly diverge when $\alpha = 1$, and *a fortiori* when $\alpha \leqq 1$.

The proof of its convergence in the case $\alpha > 1$ is slightly more troublesome. We may at the same time prove the following extension, due to *Pringsheim*[19].

Theorem of ***Pringsheim***: *The series* **174.**

$$\sum_{n=2}^{\infty} \frac{d_n}{D_n \cdot D_{n-1}^{\varrho}} \equiv \sum_{n=2}^{\infty} \frac{D_n - D_{n-1}}{D_n \cdot D_{n-1}^{\varrho}},$$

where d_n and D_n have the same meaning as before, converges for every $\varrho > 0$.

Proof. Choose a natural number p such that $\frac{1}{p} < \varrho$. It then suffices to prove the convergence of the above series when the exponent ϱ is replaced by $\tau = \frac{1}{p}$. Since, further, the series

$$\sum_{n=2}^{\infty} \left(\frac{1}{D_{n-1}^{\tau}} - \frac{1}{D_n^{\tau}} \right)$$

converges, by **131**, since $D_{n-1} \leqq D_n \to +\infty$, and since its terms are all positive, it would also suffice to establish the inequality

$$\frac{D_n - D_{n-1}}{D_n \cdot D_{n-1}^{\tau}} \leqq \frac{1}{\tau} \left(\frac{1}{D_{n-1}^{\tau}} - \frac{1}{D_n^{\tau}} \right) \quad \text{or} \quad 1 - \frac{D_{n-1}}{D_n} \leqq \frac{1}{\tau} \left(1 - \frac{D_{n-1}^{\tau}}{D_n^{\tau}} \right),$$

that is to say, to prove that

$$(1 - x^p) \leqq p(1 - x) \qquad \left[x = \left(\frac{D_{n-1}}{D_n} \right)^{\frac{1}{p}} \right]$$

for every x such that $0 < x \leqq 1$. But this is obvious at once, from

$$(1 - x^p) = (1 - x)(1 + x + \cdots + x^{p-1}).$$

Therefore the theorem is established.

Additions and Examples. **175.**

1. In the theorem of *Abel-Dini*, we may of course replace the quantities D_n by any other quantities D_n' asymptotically equal to them, or for which the ratio $\frac{D_n'}{D_n}$ lies between two fixed positive numbers, for every n (at least from some stage on). By **70**, 4 the convergence or divergence of the series Σa_n cannot be affected by this change.

2. By the theorem of *Abel-Dini*,

$$\Sigma d_n' \equiv \Sigma \frac{d_n}{D_n}$$

diverges with Σd_n. We may enquire what is the relation as to magnitude between the partial sums of the two series. Here we have the following elegant

[19] Math. Annalen, Vol. 35, p. 329. 1890.

Theorem[20]. *If* $\frac{d_n}{D_n} \to 0$, *we have*[21]

$$\frac{d_1}{D_1} + \frac{d_2}{D_2} + \cdots + \frac{d_n}{D_n} \cong \log D_n.$$

The new partial sums thus increase essentially like the l o g a r i t h m s *of the old ones.*

Proof. If $x_n = \frac{d_n}{D_n} \to 0$, we have, by **112**, b,

$$\frac{x_n}{\log \frac{1}{1-x_n}} = \frac{\frac{d_n}{D_n}}{\log \frac{D_n}{D_{n-1}}} \to 1.$$

The undefined number D_0 we here assume $= 1$, also replacing the above ratio by 1 for all indices n for which $x_n = 0$. By the theorem of limits **44**, 4, since $\log D_n \to +\infty$, we then have

$$\frac{\frac{d_1}{D_1} + \frac{d_2}{D_2} + \cdots + \frac{d_n}{D_n}}{\log D_1 + \log \frac{D_2}{D_1} + \cdots + \log \frac{D_n}{D_{n-1}}} = \frac{1}{\log D_n}\left[\frac{d_1}{D_1} + \frac{d_2}{D_2} + \cdots + \frac{d_n}{D_n}\right] \to 1.$$

This proves the theorem.

Further, it is at once clear that in the statement of this theorem, the numbers D_n may *on both sides* be replaced by others D_n' asymptotically equal to them.

3. These remarks now enable us to elaborate in the simplest manner the considerations indicated at the beginning of this section:

a) The series $\sum\limits_{n=1}^{\infty} d_n$, with $d_n = 1$, i. e. $D_n = n$, must be considered as the simplest of all divergent series, for the natural numbers $D_n = n$ form the prototype of divergence to $+\infty$. The theorem of *Abel-Dini* then shows at once that the harmonic series

$$\sum_{n=1}^{\infty} \frac{1}{n^\alpha} \begin{cases} \text{converges for } \alpha > 1 \\ \text{diverges for } \alpha \leqq 1, \end{cases}$$

and the theorem in 2. shows further that in the latter case we have for $\alpha = 1$,

$$1 + \frac{1}{2} + \frac{1}{3} + \cdots + \frac{1}{n} \cong \log n$$

(cf. **128**, 2).

b) Now choosing for Σd_n, in the theorem of *Abel-Dini*, the series $\sum \frac{1}{n}$ newly recognised to be divergent by a), and replacing, as we may by 1. and 2., D_n by $D_n' = \log n$, we conclude that

$$\sum_{n=2}^{\infty} \frac{1}{n(\log n)^\alpha} \begin{cases} \text{converges when } \alpha > 1 \\ \text{diverges when } \alpha \leqq 1. \end{cases}$$

The theorem in 2. shows further that

$$\frac{1}{2\log 2} + \frac{1}{3\log 3} + \cdots + \frac{1}{n \log n} \cong \log\log n = \log_2 n.$$

---

[20] v. *Cesàro, E.*: Nouv. Annales de Math., (3) Vol. 9, p. 353. 1890.

[21] This condition is certainly satisfied if the numbers d_n remain bounded, — hence in all the series which will occur in the sequel.

c) By repetition of this extremely simple method of inference, we obtain afresh, and quite independently of our previous results:

Starting from a suitably large index $(e_p + 1)$, *the series*[22]

$$\sum \frac{1}{n \log n \ldots \log_{p-1} n \,(\log_p n)^{\alpha}} \begin{cases} \text{converges when } \alpha > 1, \\ \text{diverges when } \alpha \leqq 1, \end{cases}$$

whatever value is given to the positive integer p. *The partial sums of the series for* $\alpha = 1$ *satisfy the asymptotic relation*

$$\sum_{\nu = e_p + 1}^{\infty} \frac{1}{\nu \log \nu \ldots \log_{p-1} \nu \cdot \log_p \nu} \cong \log_{p+1} n.$$

4. A theorem analogous to **173**, but starting from a convergent series, is the following:

Theorem of ***Dini***[23]. *If* $\sum c_n$ *is a convergent series of positive terms, and* $r_{n-1} = c_n + c_{n+1} + \cdots$ *denotes its remainder after the* $(n-1)^{\text{th}}$ *term, then*

$$\sum \frac{c_n}{r_{n-1}^{\alpha}} \equiv \sum \frac{c_n}{(c_n + c_{n+1} + \cdots)^{\alpha}} \begin{cases} \text{converges when } \alpha < 1, \\ \text{diverges when } \alpha \geqq 1. \end{cases}$$

Proof. The divergent case is again quite easily dealt with, since, for $\alpha = 1$,

$$\frac{c_n}{r_{n-1}} + \cdots + \frac{c_{n+k}}{r_{n+k-1}} \geqq \frac{c_n + \cdots + c_{n+k}}{r_{n-1}} = 1 - \frac{r_{n+k}}{r_{n-1}};$$

and for every (fixed) n, this value may be made $> \frac{1}{2}$ by a suitable choice of k, as $r_\lambda \to 0$. By **81**, 2 the series must therefore then diverge. For $\alpha > 1$ this will *a fortiori* also be the case, since r_n is < 1 for every sufficiently large n.

If, however, $\alpha < 1$, we may choose a positive integer p so that $\alpha < 1 - \frac{1}{p}$, and it now suffices — again because $r_n < 1$ for $n > n_1$ — to establish the convergence of the series

$$\sum \frac{c_n}{(r_{n-1})^{1-\tau}} \equiv \sum \frac{r_{n-1} - r_n}{r_{n-1}} \cdot r_{n-1}^{\tau}$$

where $\tau = \frac{1}{p}$.

Now r_n tends monotonely to 0 and consequently $\sum \left(r_{n-1}^{\tau} - r_n^{\tau}\right)$ is certainly convergent with positive terms. It therefore suffices to show that

$$\frac{r_{n-1} - r_n}{r_{n-1}} \cdot r_{n-1}^{\tau} \leqq \frac{1}{\tau} \left(r_{n-1}^{\tau} - r_n^{\tau}\right)$$

that is to say

$$(1 - y^p) \leqq p\,(1 - y) \qquad \left[y = \left(\frac{r_n}{r_{n-1}}\right)^{\frac{1}{p}}\right].$$

But the latter relation is evident, since $0 < y \leqq 1$.

---

[22] If we write $e = e'$, $e^{e'} = e''$, ..., $e^{e^{(\nu)}} = e^{(\nu+1)}$, ... and denote by $[e^{(\nu)}] = e_\nu$ the largest integer contained in $(\leqq)\, e^{(\nu)}$, we may say that the factors in the denominators of the terms of our series are all > 1, if n be taken to start from the value $(e_p + 1)$.

[23] v. footnote 18, p. 290.

§ 40. Series of monotonely diminishing positive terms.

Our previous investigations concerned for the most part series of quite arbitrary positive terms. The comparison series used for the construction of our criteria, however, were almost always of a much simpler nature; in particular, their terms decreased monotonely. It is clear that for such series simpler laws altogether will become valid and perhaps also simpler tests of convergence may be constructed.

We have already shown in **80** that if in a convergent series Σc_n the terms diminish monotonely to zero, we have necessarily $n c_n \to 0$, a fact which need not occur in the case of other convergent series (even with positive terms only). Again, *Cauchy*'s condensation test **77** belongs to the series we are considering.

We propose to institute one or two further investigations of this kind and, in the first instance, to deduce for such series a few very simple and at the same time very far-reaching criteria. Their convergence, as we shall see, is often very much more easily determined than that of more general types of series.

176. 1. **The integral test**[24]. *Let* $\sum_{n=1}^{\infty} a_n$ *be a given series of monotonely diminishing terms. If there exist a function* $f(x)$, *positive and monotone decreasing for* $x \geqq 1$, *for which*

$$f(n) = a_n \qquad \textit{for every } n,$$

then Σa_n *converges if, and only if, the numbers*

$$J_n = \int_1^n f(t)\,dt$$

are bounded[25].

Proof. Since, for $(k-1) \leqq t \leqq k$, we have $f(t) \geqq a_k$, and for $k \leqq t \leqq k+1$, $f(t) \leqq a_k$, (k an integer $\geqq 2$), it follows (by § 19, Theorem 20), that

$$\int_k^{k+1} f(t)\,dt \leqq a_k \leqq \int_{k-1}^{k} f(t)\,dt \qquad (k = 2, 3, \ldots).$$

Assuming these inequalities written down for $k = 2, 3, \ldots, n$ and added, we obtain

$$\int_2^{n+1} f(t)\,dt \leqq a_2 + a_3 + \cdots + a_n = s_n - a_1 < \int_1^n f(t)\,dt.$$

[24] *Cauchy*: Exercices mathém., Vol. 2, p. 221. Paris 1827.

[25] By **70**, 4 it is of course sufficient that $f(n)$ should be asymptotically proportional to the terms a_n, or that $f(n) = \alpha_n a_n$ with a *positive* lower limit for α_n. — Instead of requiring that J_n should remain bounded, we can of course also require that $\int_1^{\infty} f(t)\,dt$ should converge. The two conditions (by § 19, Def. 14) are exactly equivalent.

From the right hand inequality it follows, as the integrals J_n are bounded, that so are the partial sums of the series; from the left hand inequality the converse is inferred. This, by **70**, proves all that was required.

Supplement. *The differences* $(s_n - J_n)$ *at the same time form a monotone decreasing sequence with limit between* 0 *and* a_1. — In fact, we have

$$(s_n - J_n) - (s_{n+1} - J_{n+1}) = \int_n^{n+1} f(t)\,dt - a_{n+1} \geqq 0;$$

whence the statement follows, since $a_1 \geqq s_n - J_n \geqq a_1 - J_2 \geqq 0$. — The limit in question is therefore certainly positive, if $f(t)$ is *strictly* monotone decreasing.

Examples and Illustrations.

1. This test not only enables us to determine the convergence of numerous series, but is also frequently a means of conveniently estimating the rapidity of their convergence or divergence. Thus e. g. we can see at once that for $\alpha > 1$ the series

$$\sum_{n=1}^{\infty} \frac{1}{n^\alpha}, \quad \text{since } J_n = \int_1^n \frac{dt}{t^\alpha} = \frac{1}{\alpha-1}\left(1 - \frac{1}{n^{\alpha-1}}\right) < \frac{1}{\alpha-1},$$

must converge, whereas

$$\sum_{n=1}^{\infty} \frac{1}{n}, \quad \text{where } J_n = \int_1^n \frac{dt}{t} = \log n \to +\infty,$$

must diverge. But we learn further that, for $\alpha > 1$,

$$\int_{n+1}^{n+k+1} \frac{dt}{t^\alpha} < \sum_{\nu=n+1}^{n+k} \frac{1}{\nu^\alpha} < \int_n^{n+k} \frac{dt}{t^\alpha},$$

and therefore

$$\frac{1}{\alpha-1} \cdot \frac{1}{(n+1)^{\alpha-1}} < r_n < \frac{1}{\alpha-1} \cdot \frac{1}{n^{\alpha-1}}.$$

For $\alpha = 2$, this evaluation was already established on p. 260. In the same way the supplement to **176** gives a fresh proof of the fact that the difference

$$\left[1 + \frac{1}{2} + \cdots + \frac{1}{n} - \log n\right]$$

is the term of a monotone descending sequence tending to a positive limit between 0 and 1. This was Euler's constant mentioned in **128**, 2.

Similarly, the supplement also shows that when $0 < \alpha < 1$, the difference

$$1 + \frac{1}{2^\alpha} + \cdots + \frac{1}{n^\alpha} - \int_1^n \frac{dt}{t^\alpha}$$

is the term of a monotone descending sequence with a positive limit less than 1. Therefore, in particular (cf. **44**, 6), for $0 < \alpha < 1$:

$$1 + \frac{1}{2^\alpha} + \frac{1}{3^\alpha} + \cdots + \frac{1}{n^\alpha} \cong \frac{n^{1-\alpha}}{1-\alpha},$$

and it is easily seen that this relation holds equally when $\alpha \leqq 0$.

2. More generally, from

$$\int \frac{dt}{t\log t \ldots \log_{p-1} t \cdot (\log_p t)^\alpha} = \begin{cases} -\dfrac{1}{\alpha-1}\dfrac{1}{(\log_p t)^{\alpha-1}}, & \text{if } \alpha \neq 1, \\ \log_{p+1} t, & \text{if } \alpha = 1, \end{cases}$$

we can immediately deduce, by the same method, the known conditions of convergence and divergence of *Abel's* series. We have now three totally distinct methods of obtaining these. The supplement to **176** again affords us good evaluations of the remainders in the case of convergence, and of the partial sums in the case of divergence.

3. If $f(x)$ be *positive* for every sufficiently large x, and possesses, for those x's, a differential coefficient equal to a monotone decreasing (also positive) function with the limit 0 at infinity, the ratio $f'(x)/f(x)$ is also monotone decreasing. Since

$$\int \frac{f'(t)}{f(t)}\,dt = \log f(t),$$

it follows that the integrals

$$\int^x f'(t)\,dt \quad \text{and} \quad \int^x \frac{f'(t)}{f(t)}\,dt$$

are either both bounded or both unbounded. Hence we conclude that the series

$$\sum f'(n) \quad \text{and} \quad \sum \frac{f'(n)}{f(n)}$$

will either both converge or both diverge. In the case of divergence, when necessarily $f(n) \to +\infty$, we have

$$\sum \frac{f'(n)}{[f(n)]^\alpha} \qquad \text{convergent when } \alpha > 1.$$

In fact, here

$$\int \frac{f'(t)}{[f(t)]^\alpha} = -\frac{1}{\alpha-1}\frac{1}{[f(t)]^{\alpha-1}}$$

whence the validity of the statement can be directly inferred. — These theorems are closely connected with the theorem of *Abel-Dini*.

2. A test of practically the same scope, and independent of the integral calculus in its wording, is

***Ermakoff's test*[26].**

177\. *If $f(x)$ is related to a given series $\sum a_n$ of positive, monotonely diminishing terms, in the manner described in the integral test, and also satisfies the conditions there laid down, then*

$$\sum a_n \equiv \sum f(n) \begin{Bmatrix} \textit{converges} \\ \textit{diverges} \end{Bmatrix} \quad \textit{if} \quad \frac{e^x f(e^x)}{f(x)} \begin{Bmatrix} \leqq \vartheta < 1 \\ \geqq 1 \end{Bmatrix}$$

for every sufficiently large x.

Proof. If we suppose the first of these inequalities satisfied for $x \geqq x_0$, we have for these x's

$$\int_{e^{x_0}}^{e^x} f(t)\,dt = \int_{x_0}^{x} e^t f(e^t)\,dt \leqq \vartheta \int_{x_0}^{x} f(t)\,dt.$$

[26] Bulletin des sciences mathém., (1) Vol. 2, p. 250. 1871.

Consequently

$$(1-\vartheta)\int_{e^{x_0}}^{e^x} f(t)\,dt \leqq \vartheta\Big[\int_{x_0}^{x} f(t)\,dt - \int_{e^{x_0}}^{e^x} f(t)\,dt\Big]$$

$$\leqq \vartheta\Big[\int_{x_0}^{e^{x_0}} f(t)\,dt - \int_{x}^{e^x} f(t)\,dt\Big]$$

$$\leqq \vartheta\int_{x_0}^{e^{x_0}} f(t)\,dt.$$

Thus the integral on the left, and hence also $\int_{x_0}^{x} f(t)\,dt$, is, for every $x > x_0$, less than a certain fixed number. The series Σa_n must therefore converge, by the integral test.

If, on the other hand, we assume the second inequality satisfied for $x > x_1$, we have, for these x's,

$$\int_{e^{x_1}}^{e^x} f(t)\,dt = \int_{x_1}^{x} e^t f(e^t)\,dt \geqq \int_{x_1}^{x} f(t)\,dt.$$

A comparison of the first and third integrals shows further that

$$\int_{x}^{e^x} f(t)\,dt \geqq \int_{x_1}^{e^{x_1}} f(t)\,dt.$$

On the right hand side of this inequality, we have a fixed quantity $\gamma > 0$, and the inequality expresses the fact that for every $n\,(> x_1)$ we can assign k_n so that (with the same meaning for J_n as in **176**)

$$J_{n+k_n} - J_n = \int_{n}^{n+k_n} f(t)\,dt \geqq \gamma > 0.$$

By **46** and **50**, the numbers J_n cannot be bounded and Σa_n therefore cannot converge[27].

Remarks.

1. *Ermakoff*'s test bears a certain resemblance to *Cauchy*'s condensation test. It contains, in particular, like the latter, the complete logarithmic comparison scale, to which we have thus a fourth mode of approach. In fact, the behaviour of the series

$$\Sigma \frac{1}{n \log n \ldots \log_{p-1} n\,(\log_p n)^{\alpha}}$$

is determined by that of the ratic

$$\frac{\log_{p-1} x \cdot (\log_p x)^{\alpha}}{(\log_{p-1} x)^{\alpha}}.$$

[27] It is not difficult to carry out the proof without introducing integrals, but it makes it rather more clumsy.

As this ratio tends to zero, when $\alpha > 1$, but $\to +\infty$, when $\alpha \leqq 1$, *Ermakoff*'s test therefore provides the known conditions for convergence and divergence of these series, as asserted[28].

2. We may of course make use of other functions instead of e^x. If $\varphi(x)$ is any monotone increasing positive function, everywhere differentiable, for which $\varphi(x) > x$ always, the series Σa_n will converge or diverge according as we have

$$\frac{\varphi'(x) f(\varphi(x))}{f(x)} \begin{cases} \leqq \vartheta < 1 \\ \geqq 1 \end{cases}$$

for all sufficiently large x's.

With *Ermakoff*'s test and *Cauchy*'s integral test, we have command over the most important tests for our present series.

§ 41. General remarks on the theory of the convergence and divergence of series of positive terms.

Practically the whole of the 19th century was required to establish the convergence tests set forth in the preceding sections and to elucidate their meaning. It was not till the end of that century, and in particular by *Pringsheim*'s investigations, that the fundamental questions were brought to a satisfactory conclusion. By these researches, which covered an extremely extensive field, a series of questions were also solved, which were only timidly approached before his time, although now they appear to us so simple and transparent that it seems almost inconceivable that they should have ever presented any difficulty[29], still more so, that they should have been answered in a completely erroneous manner. How great a distance had to be traversed before this point could be reached is clear if we reflect that *Euler* never troubled himself at all about questions of convergence; when a series occurred, he would attribute to it, without any hesitation, the value of the expression which gave rise to the series[30]. *Lagrange* in 1770[31] was still of the opinion that a series represents a definite value, provided only that its terms decrease to 0[32]. To refute the latter

---

[28] This also holds for $p = 0$, if we interpret $\log_{-1} x$ to mean e^x.

[29] As a curiosity, we may mention that, as late as 1885 and 1889, several memoirs were published with the object of demonstrating the existence of convergent series Σc_n for which $\frac{c_{n+1}}{c_n}$ did not tend to a limit! (Cf. **159**, 3.)

[30] Thus in all seriousness he deduced from $\frac{1}{1-x} = 1 + x + x^2 + \cdots$, that

$$\frac{1}{2} = 1 - 1 + 1 - 1 + - \cdots$$

and

$$\frac{1}{3} = 1 - 2 + 2^2 - 2^3 + - \cdots.$$

Cf. the first few paragraphs of § 59.

[31] V. *Œuvres*, Vol. 3, p. 61.

[32] In this, however, some traces of a sense for convergence may be seen.

assumption expressly by referring to the fact (at that time already well known) of the divergence of $\sum \frac{1}{n}$, appears to us at present superfluous, and many other presumptions and attempts at proof current in previous times are in the same case. Their interest is therefore for the most part historical. A few of the questions raised, however, whether answered in the affirmative or negative, remain of sufficient interest for us to give a rapid account of them. A considerable proportion of these are indeed of a type to which anyone who occupies himself much with series is naturally led.

The source of all the questions which we propose to discuss resides in the inadequacy of the criteria. Those which are necessary *and* sufficient for convergence (the main criterion **81**) are of so general a nature, that in particular cases the convergence can only rarely be ascertained by their means. All our remaining tests (comparison tests or transformations of comparison tests) were *sufficient* criteria only, and they only enabled us to recognise as convergent series which converge at least as rapidly as the comparison series employed. The question at once arises:

1. *Does a series exist which converges less rapidly than any other?* **178.**
This question is already answered, in the negative, by the theorem **175**, 4. In fact, when Σc_n converges, so does $\Sigma c_n' = \sum \frac{c_n}{r_{n-1}^{\frac{1}{2}}}$, though, obviously, less rapidly than Σc_n, as $c_n : c_n' = r_{n-1}^{\frac{1}{2}} \to 0$.

The question is answered almost more simply by *J. Hadamard*[83], who takes the series $\Sigma c_n' \equiv \Sigma(\sqrt{r_{n-1}} - \sqrt{r_n})$. Since $c_n = r_{n-1} - r_n$, the ratio $c_n : c_n' = \sqrt{r_{n-1}} + \sqrt{r_n} \to 0$. The accented series converges less rapidly than the unaccented series.

The next question is equally easy to solve:

2. *Does a series exist which diverges less rapidly than any other?* Here again, the theorem of *Abel-Dini* **173** shows us that when Σd_n diverges, so does $\Sigma d_n' = \sum \frac{d_n}{D_n}$, and hence the answer has to be in the negative. In fact as $d_n : d_n' = D_n \to +\infty$, the theorem provides, for each given divergent series, another whose divergence is not so rapid.

These circumstances, together with our preliminary remarks, show that

3. *No comparison test can be effective with all series.*

Closely connected with this, we have the following question, raised and also answered, by *Abel*[84]:

---

[83] Acta mathematica, Vol. 18, p. 319. 1894.
[84] J. f. d. reine u. angew. Math., Vol. 3, p. 80. 1828.

4. *Can we find positive numbers* p_n, *such that, simultaneously,*

$$\left.\begin{array}{l} \text{a) } p_n a_n \to 0 \\ \text{b) } p_n a_n \geqq \alpha > 0 \end{array}\right\} \text{ are sufficient conditions for } \left\{\begin{array}{l} \text{convergence} \\ \text{divergence} \end{array}\right\}$$

of every possible series of positive terms?

It again follows from the theorem of *Abel-Dini* that this is not the case. In fact, if we put $a_n = \frac{\alpha}{p_n}$, $\alpha > 0$, the series Σa_n necessarily diverges, and hence so does $\Sigma a_n' \equiv \Sigma \frac{a_n}{s_n}$, where $s_n = a_1 + \cdots + a_n$. But, for the latter, $p_n a_n' = \frac{\alpha}{s_n} \to 0$.

The object of the comparison tests was, to some extent, the construction of the *widest* possible conditions *sufficient* for the determination of the convergence or divergence of a series. Conversely, it might be required to construct the *narrowest* possible conditions *necessary* for the convergence or divergence of a series. The only information we have so far gathered on this subject is that $a_n \to 0$ is necessary for convergence. It will at once occur to us to ask:

5. *Must the terms* a_n *of a convergent series tend to zero with any particular rapidity?* It was shown by *Pringsheim*[35] that this is not the case. However slowly the numbers p_n may tend to $+\infty$, we can invariably construct convergent series Σc_n for which

$$\overline{\lim}\, p_n c_n = +\infty.$$

Indeed *every* convergent series $\Sigma c_n'$, by a suitable rearrangement, will produce a series Σc_n to support this statement[36].

Proof. We assume given the numbers p_n, increasing to $+\infty$, and the convergent series $\Sigma c_n'$. Let us choose the indices $n_1, n_2, \ldots, n_\nu, \ldots$ *odd* and such that

$$\frac{1}{p_{n_\nu}} < \frac{c'_{2\nu-1}}{\nu} \qquad (\nu = 1, 2, \ldots)$$

and let us write $c_{n_\nu} = c'_{2\nu-1}$, filling in the remaining c_n's with the terms $c_3', c_4', \ldots$ in their original order. The series Σc_n is obviously a rearrangement of $\Sigma c_n'$. But

$$p_n c_n > \nu$$

whenever n becomes equal to one of the indices n_ν. Accordingly, as asserted,

$$\overline{\lim}\, p_n c_n = +\infty.$$

The underlying fact in this connection is simply that the behaviour or a *sequence* of the form $(p_n c_n)$ bears no essential relation to that of

[35] Math. Annalen, Vol. 35, p. 344. 1890.

[36] Cf. Theorem **82**, 3, which takes into account a sort of decrease on the average of the terms a_n.

the *series* Σc_n — i. e. with the sequence of *partial sums* of this series, — since the latter, though not the former, may be fundamentally altered by a rearrangement of its terms.

6. Similarly, *no condition of the form* $\overline{\lim}\, p_n d_n > 0$ *is necessary for the divergence of* Σd_n, *however rapidly* the positive numbers p_n may increase to $+\infty$[37]. On the contrary, *every* divergent series $\Sigma d_n'$, provided its terms tend to 0, becomes, on being suitably rearranged, a series Σd_n (still divergent, of course) for which $\underline{\lim}\, p_n d_n = 0$. — The proof is easily deduced on the same lines as the preceding.

The following question goes somewhat further:

7. *Does a scale of comparison tests exist which is sufficient for all cases?* More precisely: Given a number of convergent series

$$\Sigma c_n^{(1)}, \quad \Sigma c_n^{(2)}, \ldots, \Sigma c_n^{(k)}, \ldots$$

each of which converges less rapidly than the preceding, with e. g.

$$\frac{c_n^{(k+1)}}{c_n^{(k)}} \to +\infty, \qquad \text{for fixed } k.$$

(The logarithmic scale affords an example of such series.) *Is it possible to construct a series converging less rapidly than* **any** *of the given series?* The answer is in the affirmative[38]. The actual construction of such a series is indeed not difficult. With a suitable choice of the indices $n_1, n_2, \ldots, n_k, \ldots$, the series

$$c_n \equiv c_1^{(1)} + c_2^{(1)} + \cdots + c_{n_1}^{(1)} + c_{n_1+1}^{(2)} + \cdots + c_{n_2}^{(2)} + c_{n_2+1}^{(3)} + \cdots$$
$$+ c_{n_3}^{(3)} + c_{n_3+1}^{(4)} + \cdots$$

is itself of the kind required. We need only choose these indices so large that if we denote by $r_n^{(k)}$ the remainder, after the nth term, of the series $\Sigma c_n^{(k)}$,

$$\begin{array}{llll}
\text{for every } n \geqq n_1, & \text{we have } r_n^{(2)} < \frac{1}{2} & \text{with } c_n^{(2)} > 2\, c_n^{(1)} \\
\text{,, ,, } n \geqq n_2 > n_1, & \text{,, ,, } r_n^{(3)} < \frac{1}{2^2} & \text{,, } c_n^{(3)} > 2\, c_n^{(2)} \\
\cdots\cdots & \cdots\cdots & \cdots\cdots \\
\text{,, ,, } n \geqq n_k > n_{k-1} & \text{,, ,, } r_n^{(k+1)} < \frac{1}{2^k} & \text{,, } c_n^{(k+1)} > 2\, c_n^{(k)} \\
\cdots\cdots & \cdots\cdots & \cdots\cdots
\end{array}$$

The series Σc_n is certainly convergent, for each successive portion of it belonging to one of the series $\Sigma c_n^{(k)}$ is certainly less than the

[37] *Pringsheim*, loc. cit. p. 357.

[38] For the logarithmic scale, this was shewn by *P. du Bois-Reymond* (J. f. d. reine u. angew. Math., Vol. 76, p. 88. 1873). The above extended solution is due to *J. Hadamard* (Acta math., Vol. 18, p. 325. 1894).

remainder of this series, starting with the same initial term, i. e. $< \frac{1}{2^k}$ $(k = 2, 3, \ldots)$. On the other hand, for *every fixed k*,

$$\frac{c_n}{c_n^{(k)}} \to +\infty;$$

in fact for $n > n_q$ $(q > k)$ we have obviously $\frac{c_n}{c_n^{(k)}} > 2^{q-k}$. This proves all that was required. — In particular, there are series converging more slowly than all the series of our logarithmic scale[39].

8. We may show, quite as simply, that, given a number of divergent series $\Sigma d_n^{(k)}$, $k = 1, 2, \ldots$, each diverging less rapidly than the preceding, with, specifically, $d_n^{(k+1)} \div d_n^{(k)} \to 0$, say, there are always divergent series Σd_n diverging less rapidly than *every one* of the series $\Sigma d_n^{(k)}$.

All the above remarks bring us near to the question whether and to what extent the terms of convergent series are fundamentally distinguishable from those of divergent series. In consequence of 7. and 8., we shall no longer be surprised at the observation of *Stieltjes*:

9. *Denoting by* $(\varepsilon_1, \varepsilon_2, \ldots)$ *an arbitrary monotone descending sequence with limit* 0, *a convergent series* Σc_n *and a divergent series* Σd_n *can always be specified, such that* $c_n = \varepsilon_n d_n$. — In fact, if $\varepsilon_n \to 0$ monotonely, $p_n = \frac{1}{\varepsilon_n} \to +\infty$ monotonely. The series

$$p_1 + (p_2 - p_1) + \cdots + (p_n - p_{n-1}) + \cdots,$$

whose partial sums are the numbers p_n, is therefore divergent. By the theorem of *Abel-Dini*, the series

$$\sum_{n=1}^{\infty} d_n \equiv \sum_{n=1}^{\infty} \frac{p_{n+1} - p_n}{p_{n+1}}$$

is also divergent. But the series $\Sigma c_n \equiv \Sigma \varepsilon_n d_n \equiv \Sigma \left(\frac{1}{p_n} - \frac{1}{p_{n+1}}\right)$ is convergent by **131**. —

The following remark is only a re-statement in other words of the above:

10. *However slowly* $p_n \to +\infty$, *there is a convergent series* Σc_n *and a divergent series* Σd_n *for which* $d_n = p_n c_n$.

In this respect, the two remarks due to *Pringsheim*, given in 5. and 6., may be formulated even more forcibly as follows:

[39] The missing initial terms of these series may be assumed to be each replaced by unity.

11. *However rapidly Σc_n may converge, there are always divergent series,* — ***indeed divergent series with monotonely diminishing terms of limit 0,*** — *for which*

$$\underline{\lim}\, \frac{d_n}{c_n} = 0.$$

Thus Σd_n must have an infinite number of terms essentially *smaller* than the corresponding terms of Σc_n. Conversely:

However rapidly Σd_n may diverge, provided only $d_n \to 0$, there are always convergent series Σc_n for which $\overline{\lim}\, \frac{c_n}{d_n} = +\infty$.

We have only to prove the former statement. Here a series Σd_n of the form

$$\sum_{n=0}^{\infty} d_n \equiv c_1 + c_1 + \cdots + c_1 + \frac{1}{2} c_{n_1} + \frac{1}{2} c_{n_1} + \cdots + \frac{1}{2} c_{n_1}$$
$$+ \frac{1}{3} c_{n_2} + \frac{1}{3} c_{n_2} + \cdots + \frac{1}{3} c_{n_2} + \frac{1}{4} c_{n_3} + \cdots$$

is of the required kind, if the increasing sequence of indices $n_1, n_2, \ldots$ be chosen suitably and the successive groups of equal terms contain respectively $n_1, (n_2 - n_1), (n_3 - n_2), \ldots$ terms. In fact, in order that this series may diverge, it is sufficient to choose the number of terms in each group so large that their sum > 1, and in order that the sequence of terms in the series be monotone, it is sufficient to choose $n_k > n_{k-1}$ so large that $c_{n_k} < c_{n_{k-1}}$ $(k = 1, 2, \ldots; n_0 = 1)$ as is always possible, since $c_n \to 0$. As the ratio $\frac{d_n}{c_n}$ has the value $\frac{1}{k+1}$ for $n = n_k$, it follows that $\underline{\lim}\, \frac{d_n}{c_n} = 0$, as required.

In the preceding remarks we have considered only convergence or divergence *per se*. It might be hoped that with narrower requirements, e. g. that the terms of the series should diminish monotonely, a correspondingly greater amount of information could be obtained. Thus, as we have seen, for a convergent series Σc_n whose terms diminish monotonely, we have $n c_n \to 0$. *Can more than this be asserted?* The answer is in the negative (cf. Rem. 5):

12. *However slowly the positive numbers p_n may increase to $+\infty$, there are always convergent series of monotonely diminishing terms for which*

$$n\, p_n\, c_n$$

not only does not tend to 0, but has $+\infty$ for upper limit[40].

---

[40] *Pringsheim,* loc. cit. In particular it was much discussed whether for convergent series of positive terms, diminishing monotonely, the expression $n \log n \cdot c_n$ must $\to 0$; the opinion was held by many, as late as 1860, that $n \log n \cdot c_n \to 0$ was necessary for convergence.

The *proof* is again quite easy. Choose indices $n_1 < n_2 < \cdots$ such that

$$p_{n_\nu} > 4^\nu \qquad (\nu = 1, 2, \ldots)$$

and write

$$c_1 = c_2 = \cdots = c_{n_1} = \frac{1}{n_1 \sqrt{p_{n_1}}},$$

$$c_{n_1+1} = c_{n_1+2} = \cdots = c_{n_2} = \frac{1}{n_2 \sqrt{p_{n_2}}},$$

$$\cdots\cdots\cdots\cdots$$

$$c_{n_{\nu-1}+1} = \cdots\cdots = c_{n_\nu} = \frac{1}{n_\nu \sqrt{p_{n_\nu}}},$$

$$\cdots\cdots\cdots\cdots$$

The groups of terms here indicated contribute successively less than $\frac{1}{2}, \frac{1}{2^2}, \ldots, \frac{1}{2^\nu}, \ldots$ to the sum of the series Σc_n, so that this series will converge. On the other hand, for each $n = n_\nu$ we have

$$n\, p_n\, c_n = \sqrt{p_n},$$

so that, as was required,

$$\overline{\lim}\, n \cdot p_n \cdot c_n = +\infty.$$

13. These remarks may easily be multiplied and extended in all possible directions. They make it clear that it is *quite useless* to attempt to introduce anything of the nature of a *boundary between convergent and divergent series*, as was suggested by *P. du Bois-Reymond*. The notion involved is of course vague at the outset. But in whatever manner we may choose to render it precise, it will never correspond to the actual circumstances. We may illustrate this on the following lines, which obviously suggest themselves[41].

a) As long as the terms of the series Σc_n and Σd_n are subjected to no restriction (excepting that of being > 0), the ratio $\frac{c_n}{d_n}$ is capable of assuming all possible values, as besides the inevitable relation

$$\underline{\lim}\, \frac{c_n}{d_n} = 0 \qquad \text{we may also have} \qquad \overline{\lim}\, \frac{c_n}{d_n} = +\infty.$$

The polygonal graphs by which the two sequences (c_n) and (d_n) may be represented, in accordance with 7, 6, can therefore intersect at an indefinite number of points (which may grow more and more numerous, to an arbitrary extent).

[41] A detailed and careful discussion of all the questions belonging to the subject will be found in *Pringsheim*'s work mentioned on p. 2, and also in his writings in the Math. Ann. Vol. 35 and in the Münch. Ber. Vol. 26 (1896) and 27 (1897), to which we have repeatedly referred.

b) By our remark 11, this remains true when the two sequences (c_n) and (d_n) are *both monotone*, in which case the graphs above referred to are *both monotone descending* polygonal lines. It is therefore certainly not possible to draw a line stretching to the right, with the property that *every* sequence of type (c_n) has a graph, *no part of which lies above* the line in question, and every sequence of type (d_n) a graph, *no part of which lies below* this line, — even if the two graphs are monotone and are considered only from some point situated at a sufficiently great distance to the right.

14. Notes 11 and 12 suggest the question whether the statements there made remain unaltered if the terms of the constructed series Σc_n and Σd_n are not merely *simply monotone* as above, but *fully monotone* in the sense of p. 263. This question has been answered in the affirmative by *H. Hahn* [42].

§ 42. Systematization of the general theory of convergence.

The element of chance inherent in the theory of convergence as developed so far gave rise to various attempts to systematize the criteria from more general points of view. The first extensive attempts of this kind were made by *P. du Bois-Reymond* [43], but were by no means brought to a conclusion by him. *A. Pringsheim* [44] has been the first to accomplish this, in a manner satisfactory both from a theoretical and a practical standpoint. We propose to give a short account of the leading features of the developments due to him [45].

All the criteria set forth in these chapters have been comparison tests, and their common source is to be found in the two comparison tests of the first and second kinds, **157** and **158**. The former, namely

$$\text{(I)} \qquad a_n \leqq c_n \quad : \quad \mathfrak{C}, \qquad a_n \geqq d_n \quad : \quad \mathfrak{D},$$

is undoubtedly the simplest and most natural test imaginable; not so that of the second kind, given originally in the form

$$\text{(II)} \qquad \frac{a_{n+1}}{a_n} \leqq \frac{c_{n+1}}{c_n} \quad : \quad \mathfrak{C}, \qquad \frac{a_{n+1}}{a_n} \geqq \frac{d_{n+1}}{d_n} \quad : \quad \mathfrak{D}.$$

---

[42] *H. Hahn*, Über Reihen mit monoton abnehmenden Gliedern, Monatsheft f. Math. u. Physik, Vol. 33, pp. 121—134, 1923.

[43] J. f. d. reine u. angew. Math. Vol. 76, p. 61. 1873.

[44] Math. Ann. Vol. 35, pp. 297—394. 1890.

[45] We have all the more reason for dispensing with details in this connexion, seeing *Pringsheim*'s researches have been developed by the author himself in a very complete, detailed, and readily accessible form.

In considering the ratio of two successive terms of a series we are already going beyond what is directly provided by the series itself. We might therefore in the first instance endeavour to construct further types of tests by means of other combinations of two or more terms of the series. This procedure has, however, not yielded any criterion of interest in the study of general types of series.

If we restrict our consideration to the ratio of two terms, it is still possible to assign a number of other forms to the criterion of the second kind; e. g. the inequalities may be multiplied by the positive factors a_n or c_n without altering their significance. We shall return to this point later. Except for these relatively unimportant transformations, however, we must regard (I) and (II) as the fundamental forms of all criteria of convergence and divergence[46]. All conceivable special comparison tests will be obtained by introducing in (I) and (II) all conceivable convergent and divergent series, and, if necessary, carrying out transformations of the kind just indicated.

The task of systematizing the general theory of convergence will accordingly involve above all that of providing a general survey of all conceivable convergent and divergent series.

This problem of course cannot be solved in a literal sense, since the behaviour of every series would be determined thereby. We can only endeavour to reduce it to factors in themselves easier to survey and therefore not appearing so urgently to require further treatment. *Pringsheim* shows — and this is essentially the starting point of his investigations — that *a systematization of the general theory of convergence can be fully carried out when we assume as given the totality of all monotone sequences of (positive) numbers increasing to* $+\infty$.

Such a sequence will be denoted by (p_n); thus

$$0 < p_0 \leqq p_1 \leqq p_2 \leqq \cdots \quad \text{and} \quad p_n \to +\infty.$$

In principle, the problem is solved by the two following simple remarks:

a) ***Every divergent series*** Σd_n *is expressible in the form*

$$\sum_{n=0}^{\infty} d_n \equiv p_0 + (p_1 - p_0) + \cdots + (p_n - p_{n-1}) + \cdots$$

(each in one and only one way) in terms of a suitable sequence of type (p_n). *Also, every series of this form is divergent.*

---

[46] Thus — since (as seen in **160**, 1, 2) (II) is a consequence of (I) — it is ultimately from (I) that all the rest follows.

b) ***Every convergent series***[47] Σc_n *is expressible in the form*

$$\sum_{n=0}^{\infty} c_n \equiv \left(\frac{1}{p_0} - \frac{1}{p_1}\right) + \left(\frac{1}{p_1} - \frac{1}{p_2}\right) + \cdots + \left(\frac{1}{p_n} - \frac{1}{p_{n+1}}\right) + \cdots$$

(each in one and only one way) in terms of a suitable sequence of type (p_n). *Also, every series of this form is convergent*[48].

In fact, when these statements have been established, we have only to substitute, in the two comparison tests (I) and (II),

$$\frac{p_{n+1} - p_n}{p_n \cdot p_{n+1}} \quad \text{and} \quad (p_n - p_{n+1})$$

respectively for c_n and d_n, to obtain *in principle* all conceivable tests of the first and second kinds: All particular criteria must necessarily follow by more or less obvious transformation from the tests so obtained; for this very reason, the former can never present anything *fundamentally* new. They become of considerable importance, however, in that they give deeper insight into the connexion between the various criteria and state the latter in a coherent form, and also apply them in practice. Herein lies the chief value of the whole method. It would accordingly be well worth our while to describe the details of the construction of special criteria exactly; but for the reasons given, we shall abide by our plan of giving only a brief account.

1. The typical forms a) and b) must be regarded as undoubtedly **180.** the simplest imaginable forms for convergent and divergent series. But we can obviously replace them by many other forms, thereby altering the outward form of the criteria in various ways. For instance, by the theorem of *Abel-Dini* **173**,

$$\sum \frac{p_n - p_{n-1}}{p_n} \quad \text{and} \quad \sum \frac{p_n - p_{n-1}}{p_{n-1}}$$

diverge with $\Sigma(p_n - p_{n-1})$, while at the same time, by *Pringsheim*'s theorem **174**,

$$\sum \frac{p_n - p_{n-1}}{p_n \cdot p_{n-1}^{\varrho}} \quad \text{and} \quad \sum \frac{p_n - p_{n-1}}{p_n^{1+\varrho}}$$

converge for $\varrho > 0$. With a few restrictions of little importance, all divergent and convergent series are also expressible in one of these new forms.

2. Since the only condition to be satisfied by the numbers p_n, in the typical forms of divergent and convergent series which we are

---

[47] Unless the terms are all 0 from some stage on.

[48] The proofs of these two statements are so easy that we need not go into them further.

considering, is that they are to increase monotonely to $+\infty$, we may of course write $\log p_n$, $\log_2 p_n$, ... or generally $F(p_n)$ instead of p_n, where $F(x)$ denotes any function defined for $x > 0$ and increasing monotonely (in the strict sense) to $+\infty$ with x. This again leads to criteria which, though not *essentially* new, are *formally* so when the p_n's are specially chosen. It is easy to verify that the first named types of series diverge or converge more and more slowly, as $p_n \to +\infty$ more and more slowly; by replacing p_n successively e. g. by $\log p_n$, $\log_2 p_n$, ..., we therefore obtain a means of constructing *scales of criteria* [49]. The case $p_n = n$ naturally calls for consideration on account of its peculiar simplicity; the development of the ideas indicated above for this particular case forms the main contents of §§ 37 and 38.

3. A further advantage of this method is due to the fact that *one and the same* sequence (p_n) will serve to represent both a divergent and a convergent series. The criteria therefore naturally occur in pairs. E. g. every comparison test of the first kind may be deduced from the *pair of tests*:

$$a_n \begin{cases} \leqq \dfrac{p_n - p_{n-1}}{p_n \cdot p_{n-1}} & : \quad \mathfrak{C} \\[2ex] \geqq \dfrac{p_n - p_{n-1}}{p_{n-1}} & : \quad \mathfrak{D} \end{cases}$$

and similarly for other typical forms of series.

4. The right hand sides can be combined to form a single *disjunctive criterion*, if we introduce a modification, arbitrary in character in so far as it is not necessarily suggested by the general trend of ideas, but otherwise of a simple nature. We see at once, for instance, that the series

$$\Sigma \frac{p_n - p_{n-1}}{p_n^\alpha} \quad \text{and} \quad \Sigma \frac{p_n - p_{n-1}}{\alpha^{p_n}}$$

converge when $\alpha > 1$ and diverge when $\alpha \leqq 1$. For the first of these series the proof has just been given; and the second has all its terms less than the first if $\alpha > 1$, while if $\alpha = 1$, and hence for all $\alpha \geqq 1$, it is immediately seen to be divergent. The pair of criteria set up in 3. may accordingly be replaced by the following disjunctive criterion:

$$a_n \begin{Bmatrix} \leqq \\ \geqq \end{Bmatrix} \frac{p_n - p_{n-1}}{p_n^\alpha} \quad \text{with} \quad \begin{cases} \alpha > 1 & : \quad \mathfrak{C} \\ \alpha \leqq 1 & : \quad \mathfrak{D} \end{cases}$$

[49] The usual passage from p_n direct to $\log p_n$, $\log_2 p_n$, ..., is again quite an arbitrary step, of course. Theorems **77** and **175**, 2 render the step natural, however. Between e. g. p_n and $\log p_n$, we could easily introduce intermediary stages, for instance $e^{\sqrt{\log p_n}}$, which increases less rapidly than p_n, — in fact less rapidly than any fixed positive power of p_n, however small its exponent, — yet more rapidly than every fixed positive power of $\log p_n$, however large its exponent.

and, in all essentials[50], also by:

$$a_n \left\{ \begin{matrix} \leqq \\ \geqq \end{matrix} \right\} \frac{p_n - p_{n-1}}{\alpha^{p_n}} \quad \text{with} \quad \left\{ \begin{matrix} \alpha > 1 & : & \mathfrak{C} \\ \alpha \leqq 1 & : & \mathfrak{D} \end{matrix} \right.$$

It is remarkable that in the criteria of *convergence* arising through these transformations, the assumption $p_n \to +\infty$ is no longer necessary at all. It is sufficient that (p_n) should be *monotone*. In fact, if (p_n) is *bounded*, the *convergence* of $\sum (p_n - p_{n-1})$, and hence that of $\sum \frac{p_n - p_{n-1}}{p_n^\alpha}$ and $\sum \frac{p_n - p_{n-1}}{\alpha^{p_n}}$ for arbitrary $\alpha > 0$, follows from that of (p_n), as $(p_n^{-\alpha})$ and (α^{-p_n}) are also bounded sequences. These convergence tests[51] thus possess a special degree of generality, similar to that of *Kummer*'s[52] criterion of the second kind, mentioned below in 7.

6. From this disjunctive criterion — *as indeed in general from any criterion* — others may again be deduced by various transformations, though the criteria so obtained can be new only *in form*. For these transformations we can of course lay down no general rule; new ways may always be found by skill and intuition. This is the reason for the great number of criteria which ultimately remain outside the scope of any given systematization.

It is obvious that every inequality may be multiplied by arbitrary positive factors without altering its meaning; similarly we may form the same function $F(x)$ of either member, provided $F(x)$ be monotone increasing (in the stricter sense), — in particular we may take logarithms, roots, etc. of either side. E. g. the last disjunctive criterion may therefore be put into the form

$$\frac{\log (p_n - p_{n-1}) - \log a_n}{p_n} \left\{ \begin{matrix} \geqq \beta > 0 & : & \mathfrak{C} \\ \leqq 0 & : & \mathfrak{D} \end{matrix} \right.$$

or

$$\sqrt[p_n]{\frac{a_n}{p_n - p_{n-1}}} \left\{ \begin{matrix} \leqq \vartheta < 1 & : & \mathfrak{C} \\ \geqq 1 & : & \mathfrak{D}. \end{matrix} \right.$$

We see at a glance that by this means we obtain a general framework for the criteria of the preceding sections which were set up by assuming $p_n \equiv n$ or $\equiv \log_p n$.

---

[50] The equivalence is not complete, i. e. with the same sequence (p_n) as basis, the new criterion is not so effective as the old one; in fact, the divergence of $\sum \frac{p_n - p_{n-1}}{p_n}$, for instance, may be inferred from the old criterion, but not from the new one.

[51] *Pringsheim*: Math. Ann., Vol. 35, p. 342. 1890.

[52] Journ. f. d. reine u. angew. Math., Vol. 13, p. 78. 1835.

6. Substantially the same remarks remain valid, when we substitute $\frac{p_n - p_{n-1}}{p_n \cdot p_{n-1}}$ for c_n and $p_n - p_{n-1}$ for d_n in the fundamental criterion of the second kind (II), or perform any of the other typical substitutions for c_n and d_n there. In this way we obtain the *most general form of the criteria of the second kind.*

7. We may observe (cf. Rem. 4.) that here again, after carrying out a simple transformation, we may so frame the convergence test that it combines with the divergence test to form a single disjunctive criterion. The convergence test requires in the first instance that, for every sufficiently large n,

$$\frac{c_{n+1}}{c_n} - \frac{a_{n+1}}{a_n} \geqq 0 \qquad \text{or} \qquad \frac{1}{c_n} - \frac{a_{n+1}}{a_n} \frac{1}{c_{n+1}} \geqq 0.$$

If here we replace c_n by $\frac{p_n - p_{n-1}}{p_n \cdot p_{n-1}}$, the former inequality reduces to

$$\frac{p_{n+1} - p_n}{p_{n+1} \cdot p_n} \cdot \frac{p_n \cdot p_{n-1}}{p_n - p_{n-1}} - \frac{a_{n+1}}{a_n} \geqq 0;$$

as p_n cancels out, the typical terms of a *divergent* series automatically appear, so that the convergence test reduces to

$$d_{n+1}\left(\frac{1}{d_n} - 1\right) - \frac{a_{n+1}}{a_n} \geqq 0 \qquad : \qquad \mathfrak{C}$$

or

$$\frac{d_{n+1}}{d_n} - \frac{a_{n+1}}{a_n} \geqq d_{n+1} \qquad : \qquad \mathfrak{C}.$$

Finally, if we take into account the fact that $\Sigma \varrho d_n$ $(\varrho > 0)$ diverges with Σd_n, the criterion takes the form:

$$\frac{1}{d_n} - \frac{a_{n+1}}{a_n} \cdot \frac{1}{d_{n+1}} \geqq \varrho > 0 \qquad : \qquad \mathfrak{C}.$$

Now the original criterion is certainly satisfied by the assumption

$$\frac{1}{c_n} - \frac{a_{n+1}}{a_n} \cdot \frac{1}{c_{n+1}} \geqq \varrho > 0.$$

It thus appears that in this form — slightly less general than the original form — of the convergence test, it is absolutely indifferent whether a *convergent* series or a *divergent* series is introduced as comparison series. Hence, still more generally, the c_n's and d_n's in the above forms of the criterion may be replaced by *any* (positive) numbers b_n; thus we may write:

$$b_n - \frac{a_{n+1}}{a_n} b_{n+1} \geqq \varrho > 0 \qquad : \qquad \mathfrak{C}.$$

This extremely general criterion is due to *E. Kummer*[53].

On the other hand,

$$\frac{1}{d_n} - \frac{a_{n+1}}{a_n} \cdot \frac{1}{d_{n+1}} \begin{cases} \geqq \varrho > 0 & : \quad \mathfrak{C} \\ \leqq 0 & : \quad \mathfrak{D} \end{cases}$$ **181.**

represents a *disjunctive criterion of the second kind* which immediately follows, as the part relative to divergence is merely a slight transformation of (II).

All further details will be found in the papers and treatise by *A. Pringsheim.* The sequences of ideas sketched above can of course lead only to criteria having the nature of comparison tests of the first or second kinds, though all criteria of this character may be developed thereby. The integral test **176** and *Ermakoff*'s test **177** of course could not occur in the considerations of this section, as they do not possess the character in question.

Exercises on Chapter IX.

133. Prove in the case of each of the following series that the given indications of convergence or divergence are correct:

a) $\sum \frac{1 \cdot 3 \ldots (2n-1)}{2 \cdot 4 \ldots (2n)} \frac{1}{2n+1} \quad : \quad \mathfrak{C};$

b) $\sum \left(\frac{1 \cdot 3 \ldots (2n-1)}{2 \cdot 4 \ldots 2n}\right)^{\alpha} \quad$ for $\quad \begin{cases} \alpha > 2 & : \mathfrak{C}, \\ \alpha \leqq 2 & : \mathfrak{D}; \end{cases}$

c) $\sum \left(\frac{1}{n} - \log \frac{n+1}{n}\right) \quad : \quad \mathfrak{C};$

d) $\sum \left(\frac{1}{n \log n \log_2 n} - \log \frac{\log_2 (n+1)}{\log_2 n}\right) \quad : \quad \mathfrak{C};$

e) $\sum \frac{(x+1)(2x+1)\ldots(nx+1)}{(y+1)(2y+1)\ldots(ny+1)} \quad$ for $\quad \begin{cases} y > x > 0 & : \mathfrak{C}, \\ x \geqq y > 0 & : \mathfrak{D}. \end{cases}$

---

[53] It was given by *Kummer* as early as 1835 (Journ. f. d. reine u. angew. Math., Vol. 13, p. 172) though with a restrictive condition which was first recognized as superfluous by *U. Dini* in 1867. Later it was rediscovered several times and gave rise, as late as 1888, to violent contentions on questions of priority. *O. Stolz* (Vorlesungen über allgem. Arithmetik, Vol. 1, p. 259) was the first to give the following extremely simple proof, by means of which the criterion was first rendered fully intelligible:

Direct proof: The criterion is that from some stage on

$$a_n b_n - a_{n+1} \cdot b_{n+1} \geqq \varrho\, a_n.$$

It follows in particular that the products $a_n b_n$ diminish monotonely and therefore tend to a definite limit $\gamma \geqq 0$. By **131**, $\sum \frac{1}{\varrho} (a_n b_n - a_{n+1} b_{n+1})$ is thus a convergent series of positive terms. And as its terms are not less than the corresponding terms of Σa_n, this series is also convergent.

134. For every fixed p, the expression

$$\left[\sum^{n} \frac{1}{\nu \log \nu \ldots \log_p \nu} - \log_{p+1} n\right]$$

has a definite limit C_p when $n \to +\infty$, if the summation commences with the first integer for which $\log_p n > 1$.

135. For every fixed ϱ in $0 < \varrho < 1$, the expression

$$\sum_{\nu=1}^{n} \left[\frac{1}{\nu^{1-\varrho}} - \frac{n^{\varrho}}{\varrho}\right]$$

has a definite limit γ_ϱ when $n \to +\infty$.

136. If $x_n \to \xi$, it follows that

$$\left[\frac{x_{pn+q}}{pn+q} + \frac{x_{p(n+1)+q}}{p(n+1)+q} + \cdots + \frac{x_{pp'n+q}}{pp'n+q}\right] \to \frac{1}{p} \xi \log p',$$

where p, p', and q denote given natural numbers.

137. If Σd_n is divergent, with $d_n \to 0$, and if the D_n's are its partial sums we have

$$\sum_{\nu=1}^{n} d_\nu D_\nu \cong \frac{1}{2} D_n.$$

138. If Σa_n has monotonely diminishing terms, it is certainly divergent when $p \cdot a_{pn} - a_n \geqq 0$ for a fixed p and every sufficiently large n.

139. If $0 < d_n < 1$ for every n, the two series

$$\Sigma d_{n+1} [(1-d_0)(1-d_1)\ldots(1-d_n)]^{\varrho},$$

$$\Sigma \frac{d_{n+1}}{[1+d_0)(1+d_1)\ldots(1+d_n)]^{\varrho}},$$

are convergent, for every $\varrho > 0$.

140. Give a direct proof, without the use of *Ermakoff*'s test and without the help of the integral calculus, of the criterion

$$\underline{\lim} \frac{2^n a_{2^n}}{a_n} \begin{cases} <1 & : \mathfrak{C} \\ >2 & : \mathfrak{D} \end{cases}$$

for series of monotonely diminishing terms.

141. If the convergence of a series Σa_n follows from one of the criteria of the logarithmic scale **164**, II, then, as $n \to +\infty$,

$$[n \log n \log_2 n \ldots \log_k n] \cdot a_n \to 0$$

and diminishes monotonely from a certain stage on, whatever the value of the positive integer k may be.

Chapter X.

Series of arbitrary terms.

§ 43. Tests of convergence for series of arbitrary terms.

With series of positive terms, the study of convergence and divergence was capable of systematization to some extent; in the case of series of arbitrary terms, all attempts of this kind have to be abandoned. The reason lies not so much in insufficient development of the theory, as in the essence of the matter itself. A series of arbitrary terms may converge, without converging absolutely[1]. Indeed this is practically the only case which will interest us here, as the question of absolute convergence reduces, by **85**, to the study of a series of positive terms. We therefore need only consider the case in which either the series is actually not absolutely convergent or its absolute convergence cannot be demonstrated by any of the previously acquired means. If a series is conditionally convergent, however, this convergence is dependent on the mode of succession of the terms as well as on their individual values; any comparison test which we might set up would therefore have to concern the series as a whole, and not merely its terms individually, as before. This ultimately means that each series has to be examined by itself and we cannot obtain a general method of approach valid for them all.

Accordingly we have to be content to establish criteria with a more restricted field of validity. The chief instrument for the purpose is the formula known as

°***Abel's* partial summation**[2]. *If* $a_0, a_1, \ldots$ *and* $b_0, b_1, \ldots$ *denote arbitrary numbers, and we write* **182.**

$$a_0 + a_1 + \cdots + a_n = A_n \qquad (n \geqq 0)$$

then for every $n \geqq 0$ *and every* $k \geqq 1$,

$$\sum_{\nu=n+1}^{n+k} a_\nu b_\nu = \sum_{\nu=n+1}^{n+k} A_\nu (b_\nu - b_{\nu+1}) - A_n \cdot b_{n+1} + A_{n+k} \cdot b_{n+k+1}.$$

Proof. We have

$$a_\nu b_\nu = (A_\nu - A_{\nu-1}) b_\nu = A_\nu (b_\nu - b_{\nu+1}) - A_{\nu-1} b_\nu + A_\nu b_{\nu+1};$$

by summation from $\nu = n+1$ to $\nu = n+k$, the statement at once follows[3].

°**Supplements.** 1. *The formula continues to hold when* $n = -1$, *if we put* $A_{-1} = 0$. **183.**

---

[1] The case in which the series may be transformed into one with positive terms only, by means of a "finite number of alterations" (v. **82**, 4) or by a change of sign of all its terms, of course requires no special treatment.

[2] Journ. f. d. reine u. angew. Math. Vol. 1, p. 314. 1826.

[3] It is sometimes more convenient to write the formula in the form

$$\sum_{\nu=n+1}^{n+k} a_\nu b_\nu = \sum_{\nu=n+1}^{n+k-1} A_\nu (b_\nu - b_{\nu+1}) - A_n b_{n+1} + A_{n+k} b_{n+k}.$$

2. *If c denotes an arbitrary constant, and* $A_\nu' = A_\nu + c$, *we have also*:

$$\sum_{\nu=n+1}^{n+k} a_\nu b_\nu = \sum_{\nu=n+1}^{n+k} A_\nu' (b_\nu - b_{\nu+1}) - A_n' b_{n+1} + A'_{n+k} \cdot b_{n+k+1}$$

— for $a_\nu = A_\nu - A_{\nu-1} = A_\nu' - A'_{\nu-1}$.

Accordingly, in *Abel*'s partial summation we "may" increase or diminish all the A_ν's by any constant amount. This is equivalent to altering a_0.

Abel's partial summation enables us to deduce a number of tests of convergence for series of the form $\Sigma a_\nu b_\nu$ almost immediately[4]. In the first place, it provides the following general

184. ° **Theorem.** *The series* $\Sigma a_\nu b_\nu$ *certainly converges, if*

1) *the series* $\Sigma A_\nu (b_\nu - b_{\nu+1})$ *converges, and*

2) $\lim\limits_{p\to+\infty} A_p \cdot b_{p+1}$ *exists.*

Proof. *Abel*'s partial summation gives for $n = -1$:

$$\sum_{\nu=0}^{k} a_\nu b_\nu = \sum_{\nu=0}^{k} A_\nu (b_\nu - b_{\nu+1}) + A_k b_{k+1},$$

for every $k \geqq 0$; making $k \to +\infty$, the statement follows, in view of the two hypotheses. — The relation just written down shows further that

$$s = s' + l$$

where $\quad \Sigma a_\nu b_\nu = s, \quad \Sigma A_\nu (b_\nu - b_{\nu+1}) = s', \quad \lim A_p b_{p+1} = l.$

In particular, $s = s'$ if, and only if, $l = 0$.

The theorem does not solve the question as to the convergence of the series $\Sigma a_\nu b_\nu$, since it merely reduces it to two new questions; but these are in many cases simpler to treat. The result is in any case a far-reaching one, and it enables us immediately to deduce the following more special criteria, which are comparatively easy to apply.

° **1. *Abel's* test**[5]. $\Sigma a_\nu b_\nu$ *is convergent if* Σa_ν *converges and* (b_n) *is monotone*[6] *and bounded*[7].

---

[4] We can of course reduce *any* series to this form, as any number can be expressed as the product of two other numbers. Success in applying the above theorem will depend on the skill with which the terms are so split up.

[5] loc. cit. — *Abel*'s test provides a sufficient condition to be satisfied by (b_n), in order that the convergence of Σa_n may involve that of $\Sigma a_n b_n$. *J. Hadamard* (Acta math., Vol. 27, p. 177. 1903) gives *necessary and sufficient* conditions; cf. *E. B. Elliot* (Quarterly Journ., Vol. 37, p. 222. 1906), who gives various refinements.

[6] In anticipation of the extension to complex numbers (v. p. 397) it may be emphasized already that a sequence of numbers assumed to be monotone is necessarily real.

[7] In other words: *A convergent series "may" be multiplied, term by term, by factors forming a bounded and monotone sequence.* — Theorem **184** and the criteria deduced from it all deal with the question: By what factors *may* the terms of a convergent series be multplied so that a convergent series results? And by what factors *must* the terms of a divergent series be multiplied, so that the resulting series may be convergent?

Proof. By hypothesis (A_n) and (b_n), (v. **46**), and hence also $(A_n b_{n+1})$, are convergent. On the other hand, by **131**, the series $\Sigma (b_\nu - b_{\nu+1})$ is convergent, and indeed absolutely convergent, as its terms all have the same sign, in consequence of the monotony of (b_n). It follows, by **87**, 2, that the series $\Sigma A_\nu (b_\nu - b_{\nu+1})$ is also convergent, since a convergent sequence is certainly bounded. The two conditions of theorem **184** are accordingly fulfilled and $\Sigma a_\nu b_\nu$ is convergent.

°**2.** ***Dirichlet's*** **test**[8]. *$\Sigma a_\nu b_\nu$ is convergent if Σa_ν has bounded partial sums and (b_n) is a monotone null sequence.*

Proof. By the same reasoning as above, $\Sigma A_\nu (b_\nu - b_{\nu+1})$ is convergent. Further, as (A_n) is bounded, $(A_n b_{n+1})$ is a null sequence if (b_n) is, i. e. it is certainly convergent. The two conditions of **184** are again fulfilled.

°**3. Tests of *du Bois-Reymond*[9] and *Dedekind*[10]**

a) *$\Sigma a_\nu b_\nu$ is convergent if $\Sigma (b_\nu - b_{\nu+1})$ converges absolutely and Σa_ν converges, at least conditionally.*

Proof. By **87**, 2, $\Sigma A_\nu \cdot (b_\nu - b_{\nu+1})$ also converges, as (A_n) is certainly bounded. Since further

$$(b_0 - b_1) + (b_1 - b_2) + \ldots + (b_{n-1} - b_n) = b_0 - b_n$$

tends to a limit when $n \to +\infty$, so does b_n itself; $\lim A_n$ exists by hypothesis, and the existence of $\lim A_n b_{n+1}$ follows.

b) *$\Sigma a_\nu b_\nu$ is convergent if $\Sigma (b_\nu - b_{\nu+1})$ converges absolutely and Σa_ν has bounded partial sums, provided $b_n \to 0$.*

Proof. $\Sigma A_\nu (b_\nu - b_{\nu+1})$ is again convergent and $A_n b_{n+1} \to 0$.

Examples and Applications. **185**

1. The convergence of Σa_n involves, by *Abel*'s test, that of $\Sigma \frac{a_n}{n}$,

$$\Sigma \frac{a_n}{\log n}, \quad \Sigma \frac{a_n}{\log_p n}, \quad \Sigma \frac{n+1}{n} a_n, \quad \Sigma \sqrt[n]{n} \cdot a_n, \quad \Sigma \left(1 + \frac{1}{n}\right)^n \cdot a_n.$$

2. $\Sigma (-1)^n$ has bounded partial sums. Hence if (b_n) is a monotone null sequence,

$$\Sigma (-1)^n b_n$$

---

[8] Vorlesungen über Zahlentheorie, 1st edition, Brunswick 1863, § 101.

[9] Antrittsprogramm d. Univ. Freiburg, 1871. — The designation above adopted for the three tests is rather a conventional one, as all three are substantially due to *Abel*. For the history of these criteria cf. *A. Pringsheim*, Math. Ann., Vol. 25, p. 423. 1885.

[10] § 143 of the work referred to in footnote 8.

converges by *Dirichlet*'s test. This is a fresh proof of *Leibniz*'s criterion for series with alternately positive and negative terms (**82**,5).

3. Given positive integers $k_0, k_1, k_2, \ldots$ such that $\Sigma(-1)^{k_n}$ has bounded partial sums — for this the excess of the number of even integers over that of odd integers among the n first exponents $k_1, k_2, \ldots, k_n$ has to remain bounded as $n \to +\infty$ — the series

$$\Sigma(-1)^{k_n} b_n$$

converges, if (b_n) denotes any null sequence.

4. If Σa_n is convergent, the power series $\Sigma a_n x^n$ is convergent for $0 \leqq x \leqq +1$, since the factors x^n form a monotone and bounded sequence. If Σa_n merely has bounded partial sums, the power series at any rate converges for every x such that $0 \leqq x < 1$, since x^n then tends to 0 monotonely.

5. The series $\Sigma \sin n x$ and $\Sigma \cos n x$ have bounded partial sums, the first for *every* (fixed) real x and the second for every (fixed) real x not a multiple of 2π. This follows from the following elementary but important formula, valid [11] for every $x \neq 2 k \pi$:

$$\sin(\alpha + x) + \sin(\alpha + 2x) + \cdots + \sin(\alpha + n x) = \frac{\sin n \frac{x}{2} \sin\left(\alpha + (n+1)\frac{x}{2}\right)}{\sin \frac{x}{2}}.$$

The proof of the formula is given in **201**. For $\alpha = 0$, we get

$$\sin x + \sin 2x + \cdots + \sin n x = \frac{\sin n \frac{x}{2} \cdot \sin (n+1)\frac{x}{2}}{\sin \frac{x}{2}}, \qquad (x \neq 2 k \pi)$$

and for $\alpha = \frac{\pi}{2}$,

$$\cos x + \cos 2x + \cdots + \cos n x = \frac{\sin n \frac{x}{2} \cdot \cos (n+1)\frac{x}{2}}{\sin \frac{x}{2}}, \quad (x \neq 2 k \pi).$$

From this the boundedness of the partial sums can be inferred at once.

Thus if $\Sigma(b_n - b_{n+1})$ converges absolutely and $b_n \to 0$, we conclude from the criterion 3b that

$$\Sigma b_n \sin n x \quad \text{converges } \textit{for every } x,$$
$$\Sigma b_n \cos n x \quad \text{converges } \textit{for every } x \neq 2 k \pi.$$

In particular [12], this is the case when b_n diminishes monotonely to 0.

6. If the b_n's are positive, and if we may write

$$\frac{b_{n+1}}{b_n} = 1 - \frac{\alpha}{n} - \frac{\beta_n}{n^{1+\delta}},$$

where $\delta > 0$ and (β_n) is *bounded*, then $\Sigma(-1)^n b_n$ *converges if, and only if*, $\alpha > 0$. In fact, *if* $\alpha > 0$, it follows from these hypotheses that $\frac{b_{n+1}}{b_n} < 1$ from some stage on, i. e. (b_n) decreases monotonely, and the convergence of the series in question is therefore secured by 2., if we can show that $b_n \to 0$. The proof of this is similar to that of the parallel fact in **170**, 1: If $0 < \alpha' < \alpha$, we have for every sufficiently large ν, say $\nu \geqq m$,

$$\frac{b_{\nu+1}}{b_\nu} < 1 - \frac{\alpha'}{\nu}.$$

---

[11] For $x = 2 k \pi$, the sum has obviously the value $n \sin \alpha$, for *all* n's.

[12] *Malmstén, C. J.*: Nova acta Upsaliensis (2), Vol. 12, p. 255. 1844.

Writing down this inequality for $\nu = m, m+1, \ldots, n-1$ and multiplying together, we obtain

$$b_n < b_m \cdot \prod_{\nu=m}^{n-1} \left(1 - \frac{\alpha'}{\nu}\right).$$

From the divergence of the harmonic series, it follows as in **170**, 1 that $b_n \to 0$.

In the case $\alpha < 0$, b_n must for similar reasons increase monotonely from some stage on, so that $\sum (-1)^n b_n$ certainly cannot converge. Finally, when $\alpha = 0$, we deduce in precisely the same way as on p. 289, that b_n *cannot* tend to 0 and the series therefore cannot converge.

7. If a series of the form $\sum \frac{a_n}{n^x}$ — such series are known as *Dirichlet* series; we shall investigate them in more detail later on (§ 58, A) — is convergent for a particular value of x, say $x = x_0$, it also converges for every $x > x_0$, for $\left(\frac{1}{n^{x-x_0}}\right)$ is a monotone null sequence. This simple application of *Abel*'s test, by reasoning quite similar to that employed for power series (**93**), leads to the theorem: *Every series of the form* $\sum \frac{a_n}{n^x}$ *possesses a definite abscissa of convergence* λ *with the property that the series converges whenever* $x > \lambda$ *and diverges whenever* $x < \lambda$. (For further details, v. § 58, A.)

General Remarks. **186.**

1. We have already mentioned the fact that the magnitude of the *individual term* in an arbitrary series is not conclusive with regard to convergence. In particular, two series $\sum a_n$ and $\sum b_n$, whose terms are asymptotically equal, i. e. such that $\frac{a_n}{b_n} \to 1$, need not exhibit the same behaviour as regards convergence (cf. **70**, 4).

Thus e. g. for

$$a_n = \frac{(-1)^n}{n} + \frac{1}{n \log n} \quad \text{and} \quad b_n = \frac{(-1)^n}{n} \qquad (n = 2, 3, \ldots)$$

we have

$$\frac{a_n}{b_n} = 1 + \frac{(-1)^n}{\log n} \to 1.$$

But $\sum b_n$ is convergent and $\sum a_n$ divergent, since $\sum (a_n - b_n)$ diverges by **79**, 2.

2. *If the series* $\sum a_n$ *is non-absolutely convergent,* (cf. p. 136, footnote 9), *its positive and its negative terms, taken separately, form two divergent series.* More precisely, let $p_n = a_n$ when $a_n > 0$, and $= 0$ when $a_n \leqq 0$, and similarly let $q_n = -a_n$ when $a_n < 0$, and $= 0$ when $a_n \geqq 0$.[13] The two series $\sum p_n$ and $\sum q_n$ are series of positive terms, the first containing only the positive terms of $\sum a_n$ and the second only the absolute values of the negative terms of $\sum a_n$, in either case with the places unchanged, while their other terms are all 0. *Both these series are divergent.* In fact, as every partial sum of $\sum a_n$ is the difference of two suitable partial sums of $\sum p_n$ and $\sum q_n$, it follows at once that if $\sum p_n$ and $\sum q_n$ were both convergent, so would $\sum |a_n|$ be (by **70**), contrary to hypothes s; and if the one were convergent, the other divergent, the partial sums of $\sum a_n$

---

[13] Thus $p_n = \frac{|a_n| + a_n}{2}$, $q_n = \frac{|a_n| - a_n}{2}$.

would tend to $-\infty$ or $+\infty$ (according as Σp_n or Σq_n is assumed convergent), which is again contrary to hypothesis.

3. By the preceding remark, a conditionally convergent series, or rather the sequence formed by its partial sums, is exhibited as the difference of two monotone increasing sequences of numbers tending to infinity[14]. As regards the rapidity with which these increase, we may easily establish the following

Theorem. *The partial sums of Σp_n and Σq_n are asymptotically equal.*

In fact, we have

$$\frac{p_1 + p_2 + \cdots + p_n}{q_1 + q_2 + \cdots + q_n} - 1 = \frac{a_1 + a_2 + \cdots + a_n}{q_1 + q_2 + \cdots + q_n};$$

since the numerator in the latter ratio remains bounded, while the denominator increases to $+\infty$ with n, this ratio tends to 0, which proves the result.

4. The relative frequency of positive and negative terms in a *conditionally* convergent series Σa_n for which $|a_n|$ *diminishes monotonely* is subject to the following elegant theorem, due to *E. Cesàro*[15]: The limit, if it exists, of the ratio $\frac{P_n}{Q_n}$ of P_n, the *number of positive terms*, to Q_n, the *number of negative terms* a_ν, for $\nu \leqq n$, is necessarily 1.

§ 44. Rearrangement of conditionally convergent series.

The fundamental distinction between absolutely and non-absolutely convergent series has already been made clear in **89**, 2. This is, that the behaviour of non-absolutely convergent series depends essentially on the *order* of the terms in the series, so that for these series the commutative law of addition no longer holds. The proof consisted in showing that a non-absolutely convergent series could, by a mere *rearrangement* in the order of its terms, be transformed into a divergent series. This result may now be considerably elaborated. In fact it may be shewn that by a suitable rearrangement any *prescribed* behaviour, as regards convergence or divergence, may be induced. The theorem which we obtain is

187. Riemann's rearrangement theorem. *If Σa_n is a conditionally convergent series, we may, by a suitable rearrangement* (v. **27**, 3), *deduce a series $\Sigma a_n'$ with any one of the following properties:*

---

[14] It is best to avoid, as being far too superficial in character, the mode of expression which may be found in some writings: "the sum of a conditionally convergent series is given in the form $\infty - \infty$."

[15] Rom. Acc. Lincei Rend. (4), Vol. 4, p. 133. 1888. — Cf. a Note by *G. H. Hardy*, Messenger of Math. (2), Vol. 41, p. 17. 1911, and one by *H. Rademacher*, Math. Zeitschr., Vol. 11, pp. 276—288. 1921.

a) *to converge to an arbitrary*[16] *prescribed sum* s';

b) *to diverge to* $+\infty$ *or to* $-\infty$;

c) *to exhibit as upper and lower limits of its partial sums two arbitrary numbers* μ *and* $\varkappa$, *with* $\mu \geqq \varkappa$.

Proof. It suffices to prove c), since a) and b) are particular cases of c), the former for $\varkappa = \mu = s'$ and the latter for $\varkappa = \mu = +\infty$ or $= -\infty$.

To prove c), let $(\varkappa_n)$ be any sequence tending to $\varkappa$ and (μ_n) any sequence tending to μ, with $\mu_n > \varkappa_n$ and[17] $\mu_1 > 0$.

Let us denote by $p_1, p_2, \ldots$ the terms in $\Sigma a_n \equiv a_1 + a_2 + \cdots$ which are $\geqq 0$, in the order in which they occur, and by $q_1, q_2, \ldots$ the absolute values of those which are < 0, again in their proper order, thus slightly modifying the definition in **186**, 2. The series Σp_n and Σq_n only differ from those in **186**, 2 by the absence of a number of zero terms, and are accordingly both divergent, with positive terms which tend to 0. We proceed to show that a series of the type

$$p_1 + p_2 \cdots + p_{m_1} - q_1 - q_2 - \cdots - q_{k_1} + p_{m_1+1} + \cdots + p_{m_2}$$
$$- q_{k_1+1} - \cdots - q_{k_2} + p_{m_2+1} + \cdots$$

will satisfy all the requirements. Such a series is clearly a rearrangement of the given series, and is indeed one which leaves unaltered the order of the positive terms relatively *to one another* and that of the negative terms relatively *to one another*.

Let us choose the indices $m_1 < m_2 < \ldots$, $k_1 < k_2 < \ldots$, in the above series, so that:

1) the partial sum whose last term is p_{m_1} has a value $> \mu_1$, while that ending one term earlier is $\leqq \mu_1$;

2) the partial sum whose last term is $-q_{k_1}$ has a value $< \varkappa_1$, while that ending one term earlier is $\geqq \varkappa_1$;

3) the partial sum whose last term is p_{m_2} has a value $> \mu_2$, while that ending one term earlier is $\leqq \mu_2$;

---

[16] *Riemann, B.*: Abh. d. Ges. d. Wiss. z. Göttingen, Vol. 13, p. 97. 1866—68. The statements b) and c) are obvious supplementary propositions.

[17] This is clearly possible in any number of ways. In fact, if $\varkappa = \mu$ with a finite value s', say, take $\varkappa_n = s' - \frac{1}{n}$ and $\mu_n = s' + \frac{1}{n}$, — taking μ_1 even larger, if necessary. If $\varkappa = \mu = +\infty\ (-\infty)$, take $\varkappa_n = n\ (-n)$ and $\mu_n = \varkappa_n + 2$. If, finally, $\varkappa < \mu$, take any $(\varkappa_n)$ and (μ_n) tending to $\varkappa$ and μ; from some stage on, $\varkappa_n < \mu_n$, and by a finite number of alterations, we can arrange that this may be the case from the beginning, and also that $\mu_1 > 0$.

4) the partial sum whose last term is $-q_{k_2}$ has a value $<\varkappa_2$, while that ending one term earlier is $\geqq \varkappa_2$;

and so on.

This can always be arranged; for by taking a sufficient number of positive terms, the partial sum may be made as large as we please, and by allowing a sufficient number of negative ones to follow, the partial sum may again be depressed below any assigned value. On the other hand, at least *one* term must be taken at each stage, since $\varkappa_n < \mu_n$; so every term of the original series really does occur in the new series.

Let $\Sigma a_n'$ denote the definite rearrangement of Σa_n so obtained; the partial sums of $\Sigma a_n'$ have the prescribed upper and lower limits. In fact, if for brevity we denote by $\tau_1, \tau_2, \ldots$, the partial sums whose last terms are $p_{m_1}, p_{m_2}, \ldots$ and by $\sigma_1, \sigma_2, \ldots$, those whose last terms are $-q_{k_1}, -q_{k_2}, \ldots$, we have

$$|\sigma_\nu - \varkappa_\nu| < q_{k_\nu} \quad \text{and} \quad |\tau_\nu - \mu_\nu| < p_{m_\nu}.$$

Since $p_n \to 0$ and $q_n \to 0$, it follows that $\sigma_\nu \to \varkappa$ and $\tau_\nu \to \mu$, so that $\varkappa$ and μ certainly represent values of accumulation of the partial sums of Σa_n. Now a partial sum s_n' of $\Sigma a_n'$, which is neither a σ_ν nor a τ_ν, has necessarily a value *between* those of two successive partial sums of this special type; hence s_n' can have no value of accumulation outside the interval $\varkappa \ldots \mu$, (or different from the common value of $\varkappa$ and μ if these coincide). In other words, μ and $\varkappa$ are themselves the upper and the lower limit of the partial sums, q. e. d.

Various researches of an analogous nature were started in different directions as a consequence of this theorem. *M. Ohm* [18] and *O. Schlömilch* [19] investigated the effect of rearrangement on the special series $1 - \frac{1}{2} + \frac{1}{3} - \frac{1}{4} + - \ldots$, in particular the case in which p positive terms are followed by q negative terms throughout (cf. Exercise 148). *A. Pringsheim* [20] was the first, however, to aim at general results for the case in which the *relative* frequency of the positive and negative terms in a conditionally convergent series is modified according to definite prescribed rules. *E. Borel* [21] investigated the opposite problem, as to what rearrangements in a conditionally convergent series do *not* alter its sum. Later, *W. Sierpiński* [22] showed that if $\Sigma a_n = s$ converges conditionally and $s' < s$, the series can be made to have the sum s' by rearranging only the *positive* terms in the series, leaving all the negative terms with unaltered *place* and *order*, while similarly it can be made to have any sum $s'' > s$ by rearranging only the negative terms. (The proof is not so simple.)

§ 45. Multiplication of conditionally convergent series.

We showed in the preceding section, thus completing the considerations of **89**, 2, that the commutative law of addition no longer holds for series which converge only conditionally. We have also seen

---

[18] Antrittsprogramm, Berlin, 1839. [19] Zeitschr. f. Math. u. Phys., Vol. 18, p. 520. 1873. [20] Math. Ann., Vol. 22, p. 455. 1883. [21] Bulletin des sciences mathém. (2), Vol. 14, p. 97. 1890. [22] Bull. internat. Ac. Sciences Cracovie, p. 149. 1911.

already (end of § 17), in an example due to *Cauchy*, that the distributive law does not in general subsist, so that the product of two such series Σa_n and Σb_n may no longer be formed according to the elementary rules. The question remained unsolved, however, whether the product series Σc_n (with $c_n = a_0 b_n + a_1 b_{n-1} + \cdots + a_n b_0$) might not continue to converge under less stringent conditions for $\Sigma a_n = A$ and $\Sigma b_n = B$, and to have the sum $A \cdot B$. In § 17, it was required that both Σa_n and Σb_n should converge absolutely.

In this connection, we have first the

° Theorem of ***Mertens***[23]. *If at least **one** of the two convergent series $\Sigma a_n = A$ and $\Sigma b_n = B$ converges absolutely, Σc_n converges and $= A \cdot B$.* **188.**

Proof. We have only to show that, with increasing n, the partial sums

$$C_n = c_0 + c_1 + \ldots + c_n$$
$$= a_0 b_0 + (a_0 b_1 + a_1 b_0) + \ldots + (a_0 b_n + a_1 b_{n-1} + \ldots + a_n b_0)$$

tend to $A \cdot B$ as limit. We may assume that Σa_n is, of the two series, the one that converges absolutely. If we denote by A_n the partial sums of Σa_n, by B_n those of Σb_n, we have

$$C_n = a_0 \cdot B_n + a_1 B_{n-1} + \cdots + a_n B_0,$$

or, if we put $B_n = B + \beta_n$,

$$= A_n \cdot B + (a_0 \cdot \beta_n + a_1 \beta_{n-1} + \cdots + a_n \beta_0).$$

Since $A_n \cdot B \to A \cdot B$, it only remains to show that when Σa_n is absolutely convergent and $\beta_n \to 0$, the expressions

$$\omega_n = a_n \cdot \beta_0 + a_{n-1} \beta_1 + \cdots + a_0 \beta_n$$

form a null sequence. But this is an immediate consequence of **44**, 9b; we have only to put $x_n = \beta_n$ and $y_n = a_n$ there. Thus the theorem is proved.

Finally, we shall answer the question *whether the product series Σc_n, **if** convergent, necessarily has the sum $A \cdot B$.*

The answer is in the affirmative, as the following theorem shows:

° Theorem of ***Abel***[24]. *If the three series Σa_n, Σb_n and $\Sigma c_n \equiv \Sigma(a_0 b_n + \cdots + a_n b_0)$ are convergent, and A, B, and C are their sums, we have $A \cdot B = C$.* **189.**

1. Proof. The theorem follows immediately from *Abel*'s limit theorem (**100**) and was first proved by *Abel* in this way. If we write

$$\Sigma a_n x^n = f_1(x), \quad \Sigma b_n x^n = f_2(x), \quad \Sigma c_n x^n = f_3(x),$$

[23] J. f. d. reine u. angew. Math., Vol. 79, p. 182. 1875. — An extension was given by *T. J. Stieltjes* (Nouv. Annales (3), Vol. 6, p. 210. 1887).

[24] J. f. d. reine u. angew. Math., Vol. 1, p. 318. 1826.

these three power series (cf. **185**, 4) certainly converge absolutely for $0 \leqq x < 1$, and for these values of x, the relation

(a) $$f_1(x) \cdot f_2(x) = f_3(x)$$

holds. The assumed convergence of Σa_n, Σb_n and Σc_n implies, by *Abel's* limit theorem **100**, that each of the three functions tends to a limit when $x \to +1$ from the left; and

$$f_1(x) \to A = \Sigma a_n, \quad f_2(x) \to B = \Sigma b_n, \quad f_3(x) \to C = \Sigma c_n.$$

Since the relation (a) holds for all the values of x concerned, it follows (by § 19, Theorem 1) that it must hold in the limit:

$$A \cdot B = C.$$

— We may also dispense with the use of functions and adopt the following

2. Proof due to *Cesàro*[25]. It was shown above that

$$C_\nu = a_0 B_\nu + a_1 B_{\nu-1} + \cdots + a_\nu B_0.$$

From this it follows that

$$C_0 + C_1 + \cdots + C_n = A_0 B_n + A_1 B_{n-1} + \cdots + A_n B_0.$$

Dividing both sides of this equality by $n+1$ and letting $n \to +\infty$, we obtain C as limit on the left hand side (by **43**, 2) and $A \cdot B$ as limit on the right (by **44**, 9a). Hence $A \cdot B = C$, q. e. d.

In consequence of this interesting theorem, with which we shall again be concerned later on, any further elaboration of the question of multiplication of series has only to deal with the problem *whether* the series Σc_n converges. Into these investigations we do not, however, propose to enter[26].

Examples and Applications.

1. It follows from $\frac{\pi}{4} = \sum_{n=0}^{\infty} \frac{(-1)^n}{2n+1} \equiv 1 - \frac{1}{3} + \frac{1}{5} - \frac{1}{7} + \cdots$, by the preceding theorem, that

$$\frac{\pi^2}{16} = \sum_{n=0}^{\infty} (-1)^n \left(\frac{1}{1 \cdot (2n+1)} + \frac{1}{3 \cdot (2n-1)} + \cdots + \frac{1}{(2n+1) \cdot 1} \right),$$

provided the series thus obtained converges.

---

[25] Bull. des sciences math. (2), Vol. 14, p. 114. 1890.

[26] Theorems of the kind in question have been proved by *A. Pringsheim* (Math. Ann., Vol. 21, p. 340. 1883), and in connection with the latter's work, by *A. Voss* (ibid. Vol. 24, p. 42. 1884) and *F. Cajori* (Bull. of the Americ. Math. Soc., Vol. 8, p. 231. 1901-2 and Vol. 9, p. 188. 1902-3). — Cf. also § 66 of *A. Pringsheim's* treatise, Vorlesungen über Zahlen- und Funktionenlehre (Leipzig 1916), to which we have already referred more than once. *G. H. Hardy* (Proc. London Math. Soc. (2), vol. 6, p. 410, 1908) has proved a particularly elegant example of a related group of much more fundamental theorems.

Now

$$\frac{1}{(2p+1)(2n+1-2p)} = \frac{1}{2(n+1)}\left(\frac{1}{2p+1} + \frac{1}{2n-2p+1}\right),$$

so that the generic term of the new series has the value

$$\frac{(-1)^n}{n+1}\left(1+\frac{1}{3}+\cdots+\frac{1}{2n+1}\right).$$

Since $\frac{1}{2n+1}$ tends *monotonely to zero*, so does its arithmetic mean

$$\frac{1}{n+1}\left(1+\frac{1}{3}+\cdots+\frac{1}{2n+1}\right),$$

and the new series therefore *does* converge by *Leibniz*'s test **82**, 5. We thus have

(a) $$\sum_{n=0}^{\infty}\frac{(-1)^n}{n+1}\left(1+\frac{1}{3}+\cdots+\frac{1}{2n+1}\right) = \frac{\pi^2}{16}.$$

2. In a precisely similar manner, we deduce (v. **120**), by squaring the series $\log 2 = 1 - \frac{1}{2} + \frac{1}{3} - + \ldots$,

(b) $$\sum_{k=1}^{\infty}\frac{(-1)^{k-1}}{k+1}\left(1+\frac{1}{2}+\cdots+\frac{1}{k}\right) = (\log 2)^2.$$

3. The result obtained in 1. provides a fresh mode of approach to the equation $\sum_{k=1}^{\infty}\frac{1}{k^2} = \frac{\pi^2}{6}$, which has occupied us repeatedly before now (v. **136** and **156**). To see this, we first prove the following

Theorem. *Let* $(a_0, a_1, a_2, \ldots)$ *be a monotone sequence of positive numbers, for which* Σa_n^2 *is convergent. Then the series*

$$1.\ \sum_{n=0}^{\infty}(-1)^n a_n = s; \qquad 2.\ \sum_{n=0}^{\infty} a_n a_{n+p} = \delta_p, \qquad p = 1, 2, \ldots,$$

and

$$3.\ \sum_{p=1}^{\infty}(-1)^p \delta_p = \Delta,$$

also converge, with

(c) $$\sum_{n=0}^{\infty} a_n^2 = s^2 - 2\Delta.$$

Proof. Since Σa_n^2 converges, $a_n \to 0$; accordingly the series 1 converges by *Leibniz*'s test. As $a_n a_{n+p} \leqq a_n^2$ for every $p \geqq 1$, and Σa_n^2 converges, the series 2 are also convergent for $p \geqq 1$. Further, as $a_{n+p+1} \leqq a_{n+p}$, we have $\delta_{p+1} \leqq \delta_p$. The series 3 will accordingly converge if $\delta_p \to 0$. Now given $\varepsilon > 0$, we can choose m so that $a_{m+1}^2 + a_{m+2}^2 + \cdots < \frac{\varepsilon}{2}$; for every sufficiently large p, we shall then have

$$\delta_p < a_0 a_p + a_1 a_{p+1} + \cdots + a_m a_{p+m} + \frac{\varepsilon}{2} < a_p(a_0 + a_1 + \cdots + a_m) + \frac{\varepsilon}{2} < \varepsilon.$$

Hence $\delta_p \to 0$ and the series 3 also converges. Let us now form the array

$$\begin{array}{l}
a_0^2 - a_0 a_1 + a_0 a_2 - a_0 a_3 + - \cdots \\
- a_1 a_0 + a_1^2 - a_1 a_2 + a_1 a_3 - + \cdots \\
+ a_2 a_0 - a_2 a_1 + a_2^2 - a_2 a_3 + - \cdots \\
- a_3 a_0 + a_3 a_1 - a_3 a_2 + a_3^2 - + \cdots \\
\cdots\cdots\cdots\cdots\cdots\cdots
\end{array}$$

and let S_n denote the sum of the products $\pm a_\lambda a_\mu$ for which λ *and* μ *are* $\leqq n$. These obviously fill up a square in the upper left hand corner of the array, and

$$S_n = (a_0 - a_1 + - \cdots + (-1)^n a_n)^2 \to s^2.$$

On the other hand, the sum of all the (primary) diagonals which contain at least one product $a_\lambda a_\mu$ belonging to that square, is clearly

$$T_n = \sum_{\nu=0}^{\infty} a_\nu^2 + 2(-\delta_1 + \delta_2 - + \cdots + (-1)^n \delta_n).$$

Hence, to obtain (c), it now suffices to prove that $T_n - S_n \to 0$. By writing out the above array in a more detailed fashion, we see, moreover, that

$$\begin{aligned}(-1)^n (T_n - S_n) = {} & 2[a_3 a_{n+1} + a_2 a_{n+2} + \cdots] - 2[a_2 a_{n+1} + a_3 a_{n+2} + \cdots] \\ & + 2[a_3 a_{n+1} + a_4 a_{n+2} + \cdots] - + \cdots \\ & + (-1)^{n-1} \cdot 2[a_n a_{n+1} + a_{n+1} a_{n+2} + \cdots] \\ & + (-1)^n \left[a_{n+1}^2 + a_{n+2}^2 + \cdots\right].\end{aligned}$$

This we write for brevity

$$= \alpha_1 - \alpha_2 + \alpha_3 - + \cdots + (-1)^{n-1} \alpha_n + (-1)^n \beta_n,$$

and as $\alpha_1 \geqq \alpha_2 \geqq \cdots \geqq \alpha_n > 0$, we have (cf. **81** c, 3)

$$|T_n - S_n| \leqq \alpha_1 + \beta_n \leqq \delta_n + \beta_n;$$

thus, as was asserted, $T_n - S_n \to 0$ and therefore $\Sigma a_n^2 = s^2 - 2\Delta$.

4. If, in 3., we now take $a_n = \dfrac{1}{2n+1}$, the hypotheses are obviously all fulfilled, and we have

$$\sum_{n=0}^{\infty} \frac{1}{(2n+1)^2} = \frac{\pi^2}{16} + 2(\delta_1 - \delta_2 + \delta_3 - + \cdots).$$

But in this case, we have, by **133**, 1,

$$\delta_p = \sum_{n=0}^{\infty} \frac{1}{(2n+1)(2n+2p+1)} = \frac{1}{2p}\left(1 + \frac{1}{3} + \cdots + \frac{1}{2p-1}\right)$$

for every $p \geqq 1$, and hence

$$\sum_{n=0}^{\infty} \frac{1}{(2n+1)^2} = \frac{\pi^2}{16} + \sum_{n=0}^{\infty} \frac{(-1)^n}{n+1}\left(1 + \frac{1}{3} + \cdots + \frac{1}{2n+1}\right).$$

By the equality (a) proved in 1., the right hand side $= \dfrac{\pi^2}{8}$. By the method used to deduce **137** from **136**, the equality $\displaystyle\sum_{k=1}^{\infty} \frac{1}{k^2} = \frac{\pi^2}{6}$ follows at once

The fresh proof thus obtained for this relation may be regarded as the most elementary of all known proofs, since it borrows nothing from the theory of functions except the *Leibniz* series **122**. The main idea of the proof goes back to *Nicolaus Bernoulli*[27].

Exercises on Chapter X.

142. Determine the behaviour of the following series:

a) $\displaystyle\sum_{n=1}^{\infty} \frac{(-1)^{[\sqrt{n}]}}{n}$, b) $\displaystyle\sum_{n=1}^{\infty} \frac{(-1)^{[\sqrt{n}]}}{n^x}$,

c) $\displaystyle\Sigma x \left(\frac{\sin n x}{n x}\right)^2$, d) $\displaystyle\Sigma \sin \frac{x}{n}$,

[27] Comment. Ac. Imp. scient. Petropolitanae, Vol. X, p. 19. 1738.

e) $\Sigma(-1)^n \sin\frac{x}{n}$, f) $\Sigma \sin^2\frac{x}{n}$,

g) $\Sigma \sin(n^2 x)$, h) $\Sigma \sin(n!\,\pi x)$,

i) $\Sigma\frac{(-1)^n}{x+\log n}$, k) $\Sigma\frac{\sin^2 n x}{n}$,

l) $\Sigma\left(1+\frac{1}{2}+\cdots+\frac{1}{n}\right)\frac{\sin n x}{n}$, m) $\Sigma\, \alpha_n \sin n x \cos^2 n x$.

In the last series, (α_n) is a monotone null sequence. The series g) does not converge *unless* $x = k\pi$; the series h) converges for all rational values of x, also e. g. for $x = e$, $= (2k+1)e$, $= \frac{2k}{e}$, $= \sin 1$, $= \cos 1$, and for

$$x = \frac{1}{2}\frac{1}{4!} - \frac{1}{5!} + \frac{1}{2}\frac{1}{6!} - \frac{1}{7!} + \frac{1}{2}\frac{1}{8!} - + \cdots$$

and many other special values of x. Indicate values of x for which it certainly diverges.

143.
$$\sum_{n=1}^{\infty}\left[\frac{1}{x+2n-1}+\frac{1}{x+2n}-\frac{1}{x+n}\right] = \log 2,$$
for *every* $x > 0$.

144. If $(n a_n)$ and $\Sigma n(a_n - a_{n+1})$ converge, the series Σa_n also converges.

145. a) If Σa_n and $\Sigma|b_n - b_{n+1}|$ both converge, or b), if Σa_n has bounded partial sums, $\Sigma|b_n - b_{n+1}|$ converges and $b_n \to 0$, then for every integer $p \geqq 1$ the series $\Sigma a_n b_n^p$ is convergent.

146. The conditions of the test **184**, 3 are in a certain sense necessary, as well as sufficient, for the convergence of $\Sigma a_n b_n$: If it be required that for a given (b_n), $\Sigma a_n b_n$ *always* converges with Σa_n, the *necessary and* sufficient condition is that $\Sigma|b_n - b_{n+1}|$ should converge. — Show also that it makes little difference in this connection whether we require that $\Sigma|b_n - b_{n+1}|$ converges or merely that (b_n) is monotone.

147. If Σa_n converges, and if p_n increases monotonely to $+\infty$ in such a way that Σp_n^{-1} is divergent, we have

$$\overline{\underline{\lim}}\ \frac{p_1 a_1 + p_2 a_2 + \cdots + p_n a_n}{n}\ \begin{cases}\geqq 0,\\ \leqq 0.\end{cases}$$

148. Let a_n tend to 0 monotonely, and assume that $\lim n a_n$ exists. If we write $\sum_{n=0}^{\infty}(-1)^n a_n = s$, and now rearrange this series (cf. Ex. 51) so as to have alternately p positive and q negative terms:

$$a_0 + a_2 + \cdots + a_{2p-2} - a_1 - a_3 - \cdots - a_{2q-1} + a_{2p} + \cdots,$$

the sum s' of the new series satisfies the relation

$$s' = s + \frac{1}{2}\lim(n a_n)\cdot\log\frac{p}{q}.$$

149. A necessary and sufficient condition for the convergence of the product series

$$\Sigma\, c_n \equiv \Sigma\,(a_0 b_n + a_1 b_{n-1} + \cdots + a_n b_0)$$

of two convergent series Σa_n, Σb_n, is that the numbers

$$\varrho_n = \sum_{\nu=1}^{n} a_\nu\,(b_n + b_{n-1} + \cdots + b_{n-\nu+1})$$

should form a null sequence.

150. If (a_n) and (b_n) are monotone sequences with limit 0, the *Cauchy*'s product series of $\Sigma(-1)^n a_n$ and $\Sigma(-1)^n b_n$ is convergent if, and only if, the numbers $\sigma_n = a_n(b_0 + b_1 + \cdots + b_n)$ and $\tau_n = b_n(a_0 + a_1 + \cdots + a_n)$ also form a null sequence.

151. The two series $\sum \frac{(-1)^n}{n^\alpha}$ and $\sum \frac{(-1)^n}{n^\beta}$, $\alpha > 0$, $\beta > 0$, may be multiplied together by *Cauchy*'s rule if, and only if, $\alpha + \beta > 1$.

152. If (a_n) and (b_n) are monotone null sequences, *Cauchy*'s product of the series $\Sigma(-1)^n a_n$ and $\Sigma(-1)^n b_n$ certainly converges if $\Sigma a_n b_n$ converges. A necessary and sufficient condition for the convergence of the product series is that $\Sigma(a_n b_n)^{1+\varrho}$ should converge for every $\varrho > 0$.

153. If, for every sufficiently large n, we can write

$$a_n = n^{\alpha_0} \cdot (\log n)^{\alpha_1} \cdot (\log_2 n)^{\alpha_2} \cdots (\log_r n)^{\alpha_r},$$
$$b_n = n^{\beta_0} \cdot (\log n)^{\beta_1} \cdot (\log_2 n)^{\beta_2} \cdots (\log_s n)^{\beta_s},$$

and if Σb_n converges, we have, provided a_n is not equal to b_n for every n,

$$(a_0 b_n + a_1 b_{n-1} + \cdots + a_n b_0) \cong a_n \cdot \left(\sum_{\nu=0}^{\infty} b_\nu\right).$$

Chapter XI.

Series of variable terms (Sequences of functions).

§ 46. Uniform convergence.

Thus far, we have almost exclusively taken into consideration series whose terms were given (constant) numbers. It was only in particularly simple cases that the value of the terms depended on the choice of a definite quantity, or variable. Such was the case e. g. when we were considering the geometric series Σa^n or the harmonic series $\Sigma \frac{1}{n^\alpha}$; their behaviour was dependent on the choice of a or of α. A more general example is that of the power series $\Sigma a_n x^n$, where the number x had to be given, before we could attack the problem of its convergence or divergence. This type of case will now be generalized in the following obvious way: we shall consider series whose terms depend *in any manner* on a variable x, i. e. are *functions* of this variable. We accordingly denote these terms by $f_n(x)$ and consider series of the form $\Sigma f_n(x)$.

A function of x, in the general case, is defined only for certain values of x (v. § 19, Def. 1); for our purposes, it will be sufficient to assume that the functions $f_n(x)$ are defined in one or more (open or closed) intervals. For the given series to have a meaning for any value of x at all, we have to require that at least *one point* x belongs to the intervals of definition of all the functions $f_n(x)$. We shall, however, at once lay down the condition that there exists at least *one interval*, in which *all* the functions $f_n(x)$ are simultaneously defined. For every particular x in this interval, the terms of the series $\Sigma f_n(x)$ are in any case all determinate numbers, and the question of its convergence can be raised. We shall now assume further that an interval J (possibly smaller than the former) exists, for every point of which the series $\sum f_n(x)$ is found to converge.

○ **Definition**[1]. *An interval J will be called an* ***interval of convergence*** *of the series $\Sigma f_n(x)$ if, at every one of its points (including one, both, or neither of its endpoints), all the functions $f_n(x)$ are defined and the series converges.* **190.**

Examples and Illustrations.

1. For the geometric series Σx^n, the interval $-1 < x < +1$ is an interval of convergence, and no larger interval of convergence exists outside it.

2. A power series $\Sigma a_n (x - x_0)^n$, — provided it converges at one point at least, other than x_0, — always possesses an interval of convergence of the form $(x_0 - r) \ldots (x_0 + r)$, inclusive or exclusive of one or both endpoints. When r is properly chosen, no further interval of convergence exists outside that one.

3. The harmonic series $\sum \frac{1}{n^x}$ has as interval of convergence the semi-axis $x > 1$, with no further interval of convergence outside it.

4. As a series is no more than a symbolic expression for a certain sequence of numbers, so the series $\Sigma f_n(x)$ represents no more than a different symbolic form for a ***sequence of functions***, namely that of its partial sums

$$s_n(x) = f_0(x) + f_1(x) + \cdots + f_n(x).$$

In principle, it is therefore immaterial whether the terms of the series or its partial sums are assigned, as each set determines the other uniquely. Thus, in principle, it also does not matter whether we speak of infinite series of variable terms or of ***sequences of functions***. *We shall accordingly state our definitions and theorems only for the case of series and leave it to the student to formulate them for the case of sequences of functions*[2].

5. For the series

$$\sum_{n=1}^{\infty} f_n(x) \equiv \frac{x^2}{1+x^2} + \left(\frac{x^4}{1+x^4} - \frac{x^2}{1+x^2}\right) + \left(\frac{x^6}{1+x^6} - \frac{x^4}{1+x^4}\right) + \cdots$$

---

[1] For the case of complex numbers and functions, we have here to substitute throughout the word *region* for the word *interval* and *boundary points of the region* for *endpoints of the interval*. With this modification, the sign ○ has the same significance in this chapter as previously.

[2] Occasionally, however, the definitions and theorems will also be applied to *sequences of functions*.

we have

$$s_n(x) = \frac{(x^2)^n}{1+(x^2)^n}.$$

The series converges for every real x. Clearly, indeed, we have

a) $s_n(x) \to 0$, if $|x| < 1$,

b) $s_n(x) \to 1$, if $|x| > 1$, and

c) $s_n(x) \to \frac{1}{2}$, if $|x| = 1$.

6. On the other hand,

$$s_n(x) = (2 \sin x)^n$$

defines a series with an infinity of separate intervals of convergence; for $\lim s_n(x)$ obviously exists if, and only if, $-\frac{1}{2} < \sin x \leqq \frac{1}{2}$, i. e. if

$$-\frac{\pi}{6} < x \leqq \frac{\pi}{6} \quad \text{or} \quad \frac{5\pi}{6} \leqq x < \frac{7\pi}{6}$$

or if x lies in an interval deduced from these by a displacement through an integral multiple of 2π. The sum of the series $= 0$ throughout the interior of the interval and $= 1$ at the included endpoint.

7. The series $\sin x + \frac{\sin 2x}{2} + \frac{\sin 3x}{3} + \cdots$ converges, by **185**, 5, for every real x; the series $\cos x + \frac{\cos 2x}{2} + \frac{\cos 3x}{3} + \cdots$ converges for every real $x \neq 2k\pi$.

If a given series of the form $\Sigma f_n(x)$ is convergent in a determinate interval J, there corresponds to every point of J a perfectly definite value of the sum of the series. This sum accordingly (§ 19, Def. 1) is itself a function of x, which is *defined or represented by the series*. When the latter function is the chief centre of interest, it is also said to be *expanded in the series in question*. In this sense, we write

$$F(x) = \sum_{n=0}^{\infty} f_n(x).$$

In the case of power series and of the functions they represent (v. Chapters V and VI), these ideas are already familiar to us.

The most important question to be solved, when a series of variable terms is given, will usually be whether, and to what extent, properties belonging to all the functions $f_n(x)$, i. e. to the terms of the given series, are transferred to its sum.

Even the simple examples given above show that this need not be the case for any of the properties which are of particular interest in the case of functions. The geometric series shows that all the functions $f_n(x)$ may be bounded, without $F(x)$ being so; the power series for $\sin x$, $x > 0$, shows that every $f_n(x)$ may be monotone, without $F(x)$ being so; example 5 shows that every $f_n(x)$ may be continuous, without $F(x)$ being so, and the same example illustrates the corresponding fact for differentiability. It is easy to construct an example showing that the property of integrability may also disappear.

For instance, let

$$s_n(x) \begin{cases} = 1 & \text{for every rational } x \text{ expressible as a fraction with denominator (positive and)} \leqq n, \\ = 0 & \text{for every other } x. \end{cases}$$

Then $s_n(x)$, for each n, — and consequently $f_n(x)$, for each n, — is integrable over any bounded interval, as it has only a finite number of discontinuities in such an interval (cf. § 19, theorem 13). Also $\lim s_n(x) = F(x)$ exists for every x. In fact, if x is rational, say $= \frac{p}{q}$ ($q > 0$, p and q prime to one another), we have, for every $n \geqq q$, $s_n(x) = 1$ and hence $F(x) = 1$. If, on the other hand, x is irrational, $s_n(x) = 0$ for every n and so $F(x) = 0$. Thus $\Sigma f_n(x) = \lim s_n(x)$ defines the function

$$F(x) \begin{cases} = 1 & \text{for a rational } x, \\ = 0 & \text{for an irrational } x. \end{cases}$$

This function is not integrable, for it is discontinuous[3] for every x.

Even by these few examples, we are led to see that a quite new category of problems arises with the consideration of series of variable terms. We have to investigate *under what supplementary conditions this or the other property of the terms $f_n(x)$ is transferred to the sum $F(x)$*. It is clear from the examples cited that the mere *fact* of convergence does not secure this, — the cause must reside in the *mode* of convergence. A concept of the greatest importance in this respect is that known as ***uniform convergence*** of a series $\Sigma f_n(x)$ in one of its intervals of convergence or in part of such an interval.

This idea is easy to explain, but its underlying nature is not so readily grasped. We shall therefore first illustrate the matter somewhat intuitively, before proceeding to the abstract formulation:

Let $\sum\limits_{n=0}^{\infty} f_n(x)$ converge, and have for sum $F(x)$, in an interval J, $a \leqq x \leqq b$; we shall speak of the graph of the function $y = s_n(x) = f_0(x) + \cdots + f_n(x)$ as being *the nth curve of approximation* and of the graph of the function $y = F(x)$

[3] We may modify this definition a little by taking $s_n(x) = 1$ for all rational x's whose denominators are factors of $n!$, and $= 0$ elsewhere; the rational x's in question comprise, for each n, a definite number of other values besides the integers $\leqq n$ used above. We then obtain as $\lim s_n(x)$ the *same* function $F(x)$ as above. In this case, however, both $s_n(x)$ and $F(x)$ may be represented in terms of a closed expression, by the usual means; in fact, we have $s_n(x) = \lim\limits_{k\to\infty} (\cos^2 n!\,\pi x)^k$, and therefore

$$F(x) = \lim_{n\to\infty} [\lim_{k\to\infty} (\cos^2 n!\,\pi x)^k].$$

This curious example of a function, *discontinuous everywhere*, yet obtainable by a repeated passage to the limit from continuous functions, is due to *Dirichlet*.

as *the limiting curve.* The fact of the convergence of $\sum_{n=0}^{\infty} f_n(x)$ to $F(x)$ in J then appears to imply that for increasing n, the curves of approximation lie closer and closer to the limiting curve. This, however, is only a very imperfect description of what actually occurs. In fact, the convergence in J implies only, in the first instance, that *at each individual point* there is convergence; all we can say, to begin with, is therefore that when any definite abscissa x is singled out (and kept fixed) the corresponding ordinates of the curves of approximation approach, as n increases, the ordinate of the limiting curve for the same abscissa. There is no reason why the curve $y = s_n(x)$, *as a whole*, should lie closer and closer to the limiting curve. This statement sounds rather paradoxical, but an example will immediately make it clear.

The series whose partial sums for $n = 1, 2, \ldots$ have the values

$$s_n(x) = \frac{n x}{1 + n^2 x^2},$$

certainly converges in the interval $1 \leqq x \leqq 2$. In fact, in that interval,

$$0 < s_n(x) < \frac{n x}{n^2 x^2} \leqq \frac{1}{n}.$$

The limiting curve is therefore the stretch $1 \leqq x \leqq 2$ on the axis of x. The nth curve of approximation lies above this stretch and, by the above inequality, at a distance of less than $\frac{1}{n}$ from the limiting curve, *throughout the whole of the interval* $1 \leqq x \leqq 2$. For large n's, the distance *all along the curve* is therefore very small.

In this case, therefore, matters are much as we should expect; the position is entirely altered if we consider the same series in the interval $0 \leqq x \leqq 1$. We still have $\lim s_n(x) = 0$ at every point of this interval[4], so that the limiting curve is the corresponding portion of the x-axis. But in this case the nth approximation curve no longer lies close to the limiting curve throughout the interval, for *any* n (however large). For $x = \frac{1}{n}$, we have always $s_n(x) = \frac{1}{2}$, so that, for *every* n, the approximation curve in the interval from 0 to 1 has a hump of height $\frac{1}{2}$!! The graph of the curve $y = s_4(x)$ has the following appearance:

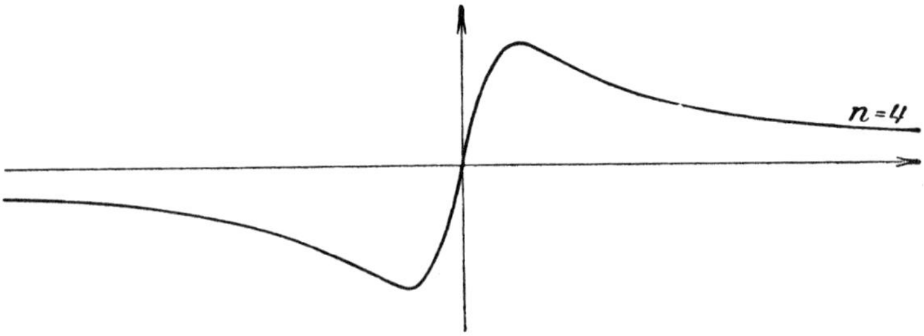

Fig. 4.

[4] In fact, for $x > 0$ we have $0 < s_n(x) < \frac{1}{n x}$ as before, i. e. $< \varepsilon$ for every $n > \frac{1}{\varepsilon x}$; for $x = 0$, $s_n(x) = 0$ even *permanently*.

The curve $y = s_{40}(x)$, however, corresponds more nearly to the following graph:

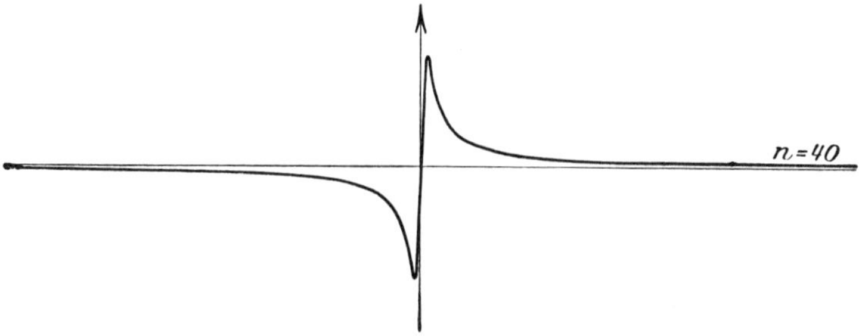

Fig. 5.

For larger n's, the hump in question — without diminishing in height — becomes compressed nearer and nearer to the ordinate-axis. The approximation curve springs more and more steeply upwards[5] from the origin to the height $\frac{1}{2}$, which it attains for $x = \frac{1}{n}$, only to drop down again almost as rapidly to within a very small distance of the x-axis.

The beginner, to whom this phenomenon will appear very odd, should take care to get it quite clear in his mind that the ordinates of the approximation curves do nevertheless, for every *fixed* x, ultimately shrink up to the point on the x-axis, so that we do have, for every *fixed* x, $\lim s_n(x) = 0$. If x is given a fixed value (however small), the disturbing hump of the curve $y = s_n(x)$ will ultimately, i. e. for sufficiently large n's, be situated entirely *to the left* of the ordinate through x (though still *to the right* of the y-axis) and on this ordinate the curve will again have already dropped very close to the x-axis[6].

Therefore the convergence of our series will be called *uniform* in the interval $1 \leqq x \leqq 2$, but not in the interval $0 \leqq x \leqq 1$.

We now proceed to the abstract formulation: Suppose $\Sigma f_n(x)$ possesses an interval of convergence J; it is convergent for every individual point of J, for instance at $x = x_0$; this means that if we write $F(x) = s_n(x) + r_n(x)$ and assume $\varepsilon > 0$ arbitrarily given, there is a number n_0 such that, for every $n > n_0$,

$$|r_n(x_0)| < \varepsilon.$$

Of course the number n_0, as was already emphasized (v. **10**, rem. 3), depends on the choice of ε. But n_0 now *depends on the choice of x_0 also.* In fact for some points of J the series will in general converge more

[5] At the origin, its slope is $s_n'(0) = n$.

[6] If we take, say, $x = \frac{1}{1000}$ and $n = 1000000$, the abscissa of the highest point of the hump is $\frac{1}{1000000}$, and at our point x the curve has already dropped to a height $< \frac{1}{1000}$.

rapidly than for others[7]. By analogy with **10**, 3, we shall therefore write $n_0 = n_0(\varepsilon, x_0)$; or more simply, dispensing with the index 0 and with the special emphasis on the dependence on ε, we shall say: Given $\varepsilon > 0$ and given x in the interval J, a number $n(x)$ can always be assigned, such that for every $n > n(x)$,

$$|r_n(x)| < \varepsilon.$$

If we now assume $n(x)$ — still for the definite given ε — chosen, say as an integer, as small as possible, its value is then uniquely defined by the value of x; as such it represents a function of x. In a certain sense, its value may be considered as *a measure of the rapidity of convergence* of the series at the point x. We now define as follows:

191. °**Definition of uniform convergence (1st form).** *The series $\Sigma f_n(x)$ convergent in the interval J, is said to be* ***uniformly*** *convergent in the sub-interval J' of J, if the function $n(x)$ defined above is bounded in J', for each value*[8] *of ε.* — Supposing we then have $n(x) < N$ in J — this N will of course depend on the choice of ε, like the numbers $n(x)$ themselves — we may also say:

°**2nd (principal) form of the definition.** *A series $\Sigma f_n(x)$, convergent in the interval J, is said to be uniformly convergent in a sub-interval J' of J, if, given ε,* ***a single*** *number $N = N(\varepsilon)$ can be assigned independently of x, such that*

$$|r_n(x)| < \varepsilon,$$

not only (as formerly) for every $n > N$, but also for every x in J'. We also say that the remainders $r_n(x)$ tend uniformly to 0 in J'.

Illustrations and Examples.

1. Uniformity of convergence invariably concerns a whole *interval*, never an isolated point[9].

2. A series $\Sigma f_n(x)$ convergent in an interval J does not necessarily converge *uniformly* in any sub-interval of J.

3. If the power series $\Sigma a_n(x - x_0)^n$ has the positive radius r and if $0 < \varrho < r$, the series is uniformly convergent in the closed sub-interval J' of

---

[7] The student should compare, for instance, the rapidity of convergence of the geometric series Σx^n (i. e. the rapidity with which the remainder diminishes as n increases) for the values $x = \frac{1}{100}$ and $x = \frac{99}{100}$.

[8] If, that is to say, the above-mentioned *measure* of the rapidity of convergence evinces no unduly great irregularities in the interval J'. — In particular cases J' may of course consist of the *complete* interval J.

[9] More generally, it may have reference to *sets* of points *more than finite* in number.

its interval of convergence, defined by $-\varrho \leqq x - x_0 \leqq +\varrho$. In fact, as the point $x = x_0 + \varrho$ lies in the interior of the interval of convergence of the power series, the latter is absolutely convergent at that point. But if $\Sigma a_n \varrho^n$ converges absolutely, we can, given $\varepsilon > 0$, choose $N = N(\varepsilon)$ so that for every $n > N$

$$|a_{n+1}| \cdot \varrho^{n+1} + |a_{n+2}| \cdot \varrho^{n+2} + \cdots < \varepsilon.$$

Also, since $|x - x_0| \leqq \varrho$ for every x in J', we have

$$|r_n(x)| \leqq |a_{n+1}| \cdot \varrho^{n+1} + |a_{n+2}| \cdot \varrho^{n+2} + \cdots.$$

Thus for $n > N$, we certainly have $|r_n(x)| < \varepsilon$, whatever the position of x in J' may be.

The result we have obtained is as follows:

○ **Theorem.** *A power series $\Sigma a_n (x - x_0)^n$ of positive radius r converges uniformly in every sub-interval of the form $|x - x_0| \leqq \varrho < r$ of its interval of convergence.*

4. The above example enables us to make ourselves understood, if we formulate the definition of uniform convergence a little more loosely, as follows: *$\Sigma f_n(x)$ is said to be uniformly convergent in J', if it is possible to make a statement about the value of the remainder, in the form "$|r_n(x)| < \varepsilon$", valid for all positions of x simultaneously.*

5. The series $\sum_{n=1}^{\infty} \frac{\sin n x}{n^2}$ is uniformly convergent for every value of x; for, whatever the position of x may be,

$$|r_n(x)| \leqq \frac{1}{(n+1)^2} + \frac{1}{(n+2)^2} + \cdots,$$

whence the rest may be inferred by 4.

6. The geometric series is *not* uniformly convergent in the *whole* interval of convergence $-1 < x < +1$. For

$$r_n(x) = x^{n+1} + x^{n+2} + \ldots = \frac{x^{n+1}}{1-x};$$

however large N may be chosen, we can always find an $r_n(x)$ with $n > N$ and $0 < x < 1$, for which e. g. $r_n(x) > 1$.

If, for instance, we choose any fixed $n > N$, then as $x \to 1 - 0$ we have

$$r_n(x) = \frac{x^{n+1}}{1-x} \to +\infty.$$

Hence $r_n(x) > 1$ for all x in a definite interval of the form $x_0 < x < 1$.

7. The above clears up the meaning of the statement: $\Sigma f_n(x)$ is *not* uniformly convergent in a portion J' of its interval of convergence. A special value of ε, say the value $\varepsilon_0 > 0$, exists, such that an index n greater than any assigned N may be found, so that the inequality $|r_n(x)| < \varepsilon_0$ is *not* satisfied for some suitably chosen x in J'.

8. With reference to the curves of approximation $y = s_n(x)$, our definition clearly implies that, with increasing n, the curve should lie arbitrarily close to the limiting curve *throughout the portion* which lies above J'. If, for any given $\varepsilon > 0$, we draw the two curves $y = F(x) \pm \varepsilon$, the approximation curves $y = s_n(x)$ will ultimately, for every sufficiently large n, come to lie entirely within the strip bounded by the two curves.

9. The distinction between uniform and non-uniform convergence, and the great significance of the former in the theory of infinite series, were first recognized (almost simultaneously) by *Ph. L. v. Seidel* (Abh. d. Münch. Akad., p. 383, 1848) and by *G. G. Stokes* (Transactions of the Cambridge Phil. Soc., Vol. 8, p. 533. 1848). It appears, however, from a paper by *K. Weierstrass*, unpublished till 1894 (Werke, Vol. 1, p. 67), that the latter must have drawn the distinction as early as 1841. The concept of uniform convergence did not become common property till much later, chiefly through the lectures of *Weierstrass*.

Other forms of the definition of uniform convergence.

°**3rd form.** *$\Sigma f_n(x)$ is said to be uniformly convergent in J' if, in whatever way we may choose the sequence*[10] *(x_n) in the interval J', the corresponding remainders*

$$r_n(x_n)$$

invariably** form a **null sequence[11]**.**

We can verify as follows that this definition is equivalent to the preceding:

a) Suppose that the conditions of the 2nd form of the definition are fulfilled. Then, given ε, we can always determine N so that $|r_n(x)| < \varepsilon$ for every $n > N$ and *every* x in J'; in particular

$$|r_n(x_n)| < \varepsilon \quad \text{for every} \quad n > N;$$

hence $r_n(x_n) \to 0$.

b) Suppose, conversely, that the conditions of the 3rd form are fulfilled. Thus for every (x_n) belonging to J', $r_n(x_n) \to 0$. The conditions of the 2nd form must then be satisfied also. In fact, if this were not the case, — if a number $N = N(\varepsilon)$ with the properties formulated there did *not* exist for *every* $\varepsilon > 0$, — this would imply that for some *special* ε, say $\varepsilon = \varepsilon_0$, *no* number N had these properties; above any number N, however large, there would be at least one other index n such that, for some suitable point $x = x_n$ in J', $|r_n(x_n)| \geqq \varepsilon_0$. Let n_1 be an index such that $|r_{n_1}(x_{n_1})| \geqq \varepsilon_0$. Above n_1 there would be another index n_2, such that $|r_{n_2}(x_{n_2})| \geqq \varepsilon_0$ for a suitable corresponding point x_{n_2}, and so on. We can choose (x_n) in J' so that the points $x_{n_1}, x_{n_2}, \ldots$ belong to (x_n), in which case

$$r_n(x_n)$$

---

[10] The sequence need not converge, but may occupy any position in J'.

[11] Should each of the functions $|r_n(x)|$ attain a maximum in J', we may choose x_n in particular so that $|r_n(x_n)| = \text{Max}\,|r_n(x)|$; our definition thus takes the special form: *$\Sigma f_n(x)$ is said to be uniformly convergent in J' if the maxima $\text{Max}\,|r_n(x)|$ in J' form a null sequence.*

If the function $|r_n(x)|$ does not attain a maximum in J', it has, however, a definite upper bound μ_n. We may also formulate the definition in the general form:

° Form 3a. *$\Sigma f_n(x)$ is said to be uniformly convergent in J' if $\mu_n \to 0$.* (Proof?)

will certainly *not* form a null sequence, contrary to hypothesis. Our assumption that the conditions of the 2nd form could not be fulfilled is inadmissible; the 3rd form of the definition is completely equivalent to the 2nd.

In the previous forms of the definition, it was always the *remainder of the series* which we estimated, the series being already assumed to converge. By using *portions* of the series instead of infinite remainders (v. **81**) the definition of uniform convergence may be stated so as to include that of convergence. We obtain the following definition:

° **4th form.** *A series $\Sigma f_n(x)$ is said to be uniformly convergent in the interval J' if, given $\varepsilon > 0$, we can assign a number $N = N(\varepsilon)$ depending only on ε, and independent of x, such that*

$$|f_{n+1}(x) + f_{n+2}(x) + \dots + f_{n+k}(x)| < \epsilon$$

for every $n > N$, every $k \geqq 1$ and every x in J'. For if the conditions of this definition are satisfied, then it follows firstly (by **81**) that $\Sigma f_n(x)$ converges for each fixed x in J'. In the inequality, we may make k tend to ∞, and we find that $|r_n(x)| \leqq \varepsilon$ for each x in J'. Conversely, if $|r_n(x)| \leqq \varepsilon$ for all $n > N$ and all x in J', then for all these n, all $k \geqq 1$, and all x in J', we have

$$|f_{n+1}(x) + \dots + f_{n+k}(x)| = |r_n(x) - r_{n+k}(x)| \leqq 2\epsilon.$$

This shows, however, that if the series $\Sigma f_n(x)$ satisfies the conditions of the 4th form, it also satisfies those of the 2nd form, and conversely. — We may finally express this definition in the following form (cf. **81** a):

° **5th form.** *A series $\Sigma f_n(x)$ is said to be uniformly convergent in the interval J' if, when positive integers $k_1, k_2, k_3, \dots$ and points $x_1, x_2, x_3, \dots$ of J' are chosen arbitrarily, the quantities*

$$[f_{n+1}(x_n) + f_{n+2}(x_n) + \dots + f_{n+k_n}(x_n)]$$

invariably form a null sequence[12].

Further Examples and Illustrations.

1. The student should examine afresh the behaviour of the series $\Sigma f_n(x)$, with **192.**

$$s_n(x) = \frac{n\,x}{1 + n^2 x^2}$$

a) in the interval $1 \leqq x \leqq 2$,
b) in the interval $0 \leqq x \leqq 1$ (cf. the considerations on pp. 330—1).

2. For the series

$$1 + (x - 1) + (x^2 - x) + \dots + (x^n - x^{n-1}) + .$$

[12] By **51**, we might even write $[f_{\nu_{n+1}}(x_n) + \dots + f_{\nu_{n+k_n}}(x_n)]$ for the above, where the ν_n's are any integers tending to $+\infty$. Exactly as in **81**, we may speak of a *sequence of portions*, except that here we may substitute a different value of x in each portion. The statement we then obtain is: *A series $\Sigma f_n(x)$ is said to be **uniformly** convergent in J' if **every** sequence of portions of the series forms a null sequence.* Similarly: *A **sequence of functions** $s_n(x)$ is said to be uniformly convergent in J' if every difference-sequence is a null sequence.*

we have obviously $s_n(x) = x^n$. The series accordingly converges in the interval J: $-1 < x \leqq +1$, in particular in the sub-interval J': $0 \leqq x \leqq 1$. Here

$$F(x) \begin{cases} = 0 & \text{for} \quad 0 \leqq x < 1. \\ = 1 & \text{for} \quad x = 1. \end{cases}$$

The convergence in this interval is **not** uniform. It is not so even in J'': $0 \leqq x < 1$; for here $r_n(x) = F(x) - s_n(x) = -x^n$. We have only to choose in J'' (hence in J') the sequence of points

$$x_n = 1 - \frac{1}{n} \qquad (n = 1, 2, \ldots)$$

to have $r_n(x_n) = -\left(1 - \frac{1}{n}\right)^n \to -\frac{1}{e}$, so that the series cannot converge uniformly[13]. — This may be made clear geometrically by examining the position of successive curves of approximation, as illustrated by the accompanying figure:

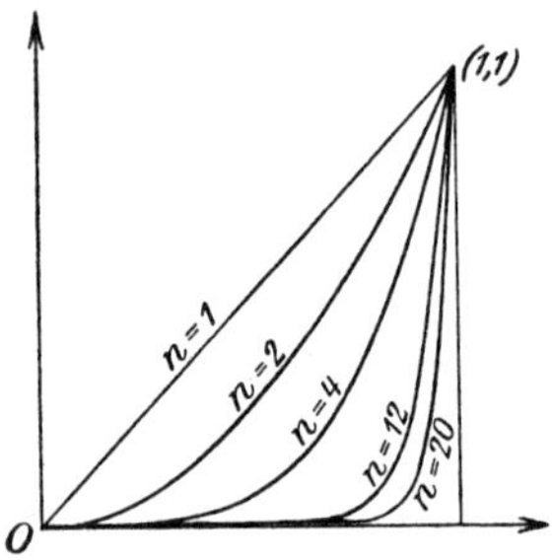

Fig. 6.

For large values of n, the curve $y = s_n(x)$ remains, *almost* throughout the whole interval, quite close to the x-axis, which represents the limiting curve. Just before the ordinate $x = +1$, it rises abruptly until it reaches its terminal point (1, 1). However large a value may be assumed for n, the curve $y = s_n(x)$ will never remain close to the limiting curve throughout the *entire*[14] interval J'' (or J').

3. In the preceding example, we could almost expect *a priori* that the convergence would not be uniform, as $F(x)$ itself has a "jump" of height 1 at the endpoint of the interval. The case was different with the example treated on p. 330. An example similar to the latter, but even more striking, is the following: Consider the series for which

$$s_n(x) = n\,x\,e^{-\frac{1}{2}n x^2} \qquad (n = 1, 2, \ldots).$$

For $x = 0$, we have $s_n(0) = 0$, for every n; for $x \neq 0$, the number $e^{-\frac{1}{2}x^2}$ is positive and less than 1, so that (by **38**, 1) $s_n(x) \to 0$. Our series is therefore convergent for every x and its sum is $F(x) = 0$, i. e. the limiting curve coincides with the x-axis. The convergence is not in the least uniform, however, if we consider an interval containing the origin. Thus, for $x_n = \frac{1}{\sqrt{n}}$,

$$r_n(x_n) = F(x_n) - s_n(x_n) = -n\,x_n \cdot e^{-\frac{1}{2}n x_n^2} = -\sqrt{n} \cdot e^{-\frac{1}{2}} = -\sqrt{\frac{n}{e}},$$

which certainly does *not* $\to 0$. The approximation curves have a similar

---

[13] For $x_n = \left(1 - \frac{1}{n^2}\right)$, we even have $r_n(x_n) \to 1$.

[14] In spite of this, it is easy to see that for every *fixed* x (in $0 < x < 1$) the values $s_n(x)$ diminish to 0 as n increases, so that the abrupt rise to the height 1 occurs to the *right* of x, however near x may be taken to $+1$, provided only that n is chosen sufficiently large.

appearance to those in Figs. 4 and 5, with this modification, that the height of the hump now increases indefinitely with n; this is because[15]

$$s_n\left(\frac{1}{\sqrt{n}}\right)=\sqrt{\frac{n}{e}}\to+\infty.$$

4. We must emphasize particularly that uniform convergence does not require each of the functions $f_n(x)$ to be individually bounded. The series $\frac{1}{x}+1+x+x^2+\cdots$, for instance, is *uniformly* convergent in $0<x\leqq\frac{1}{2}$, with the sum $\frac{1}{x(1-x)}$, since the remainders have the value

$$|r_n(x)|=\left|\frac{x^n}{1-x}\right|\leqq\frac{1}{2^{n-1}}.$$

The first term of this series (as also the limiting function) is *not* bounded in the interval in question. (Cf., however, theorem 4. below.)

With a view to *calculation with uniformly convergent series*, it is convenient to formulate the following theorems specially, although the proofs are so simple that we may leave them to the reader:

○ **Theorem 1.** *If the p series $\Sigma f_{n1}(x)$, $\Sigma f_{n2}(x)$, ..., $\Sigma f_{np}(x)$ are, simultaneously, uniformly convergent in the same interval J, (p is a definite whole number), the series $\Sigma f_n(x)$ for which*

$$f_n(x)=c_1 f_{n1}(x)+c_2 f_{n2}(x)+\cdots+c_p f_{np}(x)$$

is also uniformly convergent in that interval, if $c_1, c_2, \ldots, c_p$ denote any constants. (I. e.: Uniformly convergent series may be multiplied by constant factors and then added term by term.)

○ **Theorem 2.** *If $\Sigma f_n(x)$ is uniformly convergent in J, so is the series $\Sigma g(x) f_n(x)$, where $g(x)$ denotes any function defined and bounded in the interval J. (I. e.: A uniformly convergent series may be multiplied term by term by a bounded function.)*

○ **Theorem 3.** *If not merely $\Sigma f_n(x)$, but $\Sigma|f_n(x)|$ is uniformly convergent in J, then so is the series $\Sigma g_n(x) f_n(x)$, provided that when m is suitably chosen, the functions $g_{m+1}(x)$, $g_{m+2}(x)$, ..., are uniformly bounded in J, — i. e. provided we can find an integer $m>0$ and a number $G>0$ such that $|g_n(x)|<G$ for every x in J and every $n>m$. (I. e.: A series which still converges uniformly when its terms are taken in absolute value may be multiplied term by term by any functions all but a finite number of which, at most, are uniformly bounded in J.)*

---

[15] The point for which $x=\frac{1}{\sqrt{n}}$ is actually the maximum point of the curve $y=s_n(x)$, as may be inferred from $s_n'=(n-n^2x^2)e^{-\frac{1}{2}nx^2}=0$.

° Theorem 4. *If $\Sigma f_n(x)$ converges uniformly in J, then for a suitable m the functions $f_{m+1}(x), f_{m+2}(x), \ldots$ are uniformly bounded in J and converge uniformly to 0.*

° Theorem 5. *If the functions $g_n(x)$ converge uniformly to 0 in J, so do the functions $\gamma_n(x)\, g_n(x)$, where the functions $\gamma_n(x)$ are any functions defined in J and — with the possible exception of a finite number of them — uniformly bounded in J.*

We may give as a model the proofs of Theorems 3 and 4:

Proof of Theorem 3. By hypothesis, given $\varepsilon > 0$, we can determine $n_0 > m$ so that for every $n > n_0$ and every x in J,

$$|f_{n+1}(x)| + |f_{n+2}(x)| + \ldots < \frac{\varepsilon}{G}.$$

For the same n's and x's we then have

$$|g_{n+1} f_{n+1} + \ldots| \leqq |g_{n+1}| \cdot |f_{n+1}| + \ldots < G(|f_{n+1}| + \ldots) < \varepsilon.$$

This proves all that was required.

Proof of Theorem 4. By hypothesis, there exists an m such that, for every $n \geqq m$ and every x in J, $|r_n(x)| < \frac{1}{2}$. Hence for $n > m$ and every x in J,

$$|f_n(x)| = |r_{n-1}(x) - r_n(x)| \leqq |r_{n-1}| + |r_n| < 1,$$

which proves the first part of the theorem. If we now choose $n_0 > m$ so that for every $n \geqq n_0$ and every x in J, $|r_n(x)| < \frac{1}{2}\varepsilon$, ($\varepsilon$ being previously assigned) the second part follows in quite a similar way.

§ 47. Passage to the limit term by term.

Whereas we saw on pp. 328—9 that the fundamental properties of the functions $f_n(x)$ do not in general hold for the function $F(x)$ represented by $\Sigma f_n(x)$, we shall now show that, roughly speaking, this *will* be the case when the series is uniformly convergent [16]

We first give the following simple theorem, which becomes particularly important in applications:

193. ° Theorem 1. *If the series $\Sigma f_n(x)$ is uniformly convergent in an interval and if its terms $f_n(x)$ are continuous at a point x_0 of this interval, the function $F(x)$ represented by the series is also continuous at this point* [17].

---

[16] We may, however, mention at once that uniform convergence still only represents a *sufficient* condition in the following theorems and is not in general *necessary*.

[17] If x_0 is an endpoint of the interval J, only *one-sided* continuity can of course be asserted at x_0 for $F(x)$, but of course only the corresponding one-sided continuity need be assumed at x_0 for $f_n(x)$.

Proof. Given $\varepsilon > 0$, we have (in accordance with § 19, Def. 6b) to show that a number $\delta = \delta(\varepsilon) > 0$ exists such that

$$|F(x) - F(x_0)| < \varepsilon \quad \text{for every } x \text{ with} \quad |x - x_0| < \delta$$

in the interval. Now we may write

$$F(x) - F(x_0) = s_n(x) - s_n(x_0) + r_n(x) - r_n(x_0).$$

By the assumed fact of uniform convergence, we can choose $n = m$ so large that, for *every* x in the interval, $|r_m(x)| < \frac{\varepsilon}{3}$. Then

$$|F(x) - F(x_0)| \leqq |s_m(x) - s_m(x_0)| + \tfrac{2}{3}\varepsilon.$$

The integer m being thus determined, $s_m(x)$ is the sum of a fixed number of functions continuous at x_0, and is therefore (by § 19, Theorem 3) itself continuous at x_0. We can accordingly choose δ so small that for every x in the interval for which $|x - x_0| < \delta$, we have

$$|s_m(x) - s_m(x_0)| < \frac{\varepsilon}{3}.$$

For the same x's we then have

$$|F(x) - F(x_0)| < \varepsilon,$$

which establishes the continuity of $F(x)$ at x_0.

○ **Corollary.** *If $\Sigma f_n(x) = F(x)$ is uniformly convergent in an interval, and if the functions $f_n(x)$ are all continuous throughout the interval, then so is $F(x)$.*

In connection with example 3 of **191**, 2, we have in the above a fresh proof of the continuity of the function represented by a power series in its interval of convergence.

If we use the lim-definition of continuity (v. § 19, Def. 6) instead of the ε-definition, the statement of the theorem may be put into the form:

$$\lim_{x \to x_0} \left(\sum_{n=0}^{\infty} f_n(x)\right) = \sum_{n=0}^{\infty} \left(\lim_{x \to x_0} f_n(x)\right).$$

In this form it appears as a special case of the following much more elaborate theorem:

○ **Theorem 2.** *We assume that the series $F(x) = \sum_{n=0}^{\infty} f_n(x)$ is uniformly convergent in the open interval*[18] *$x_0 \ldots x_1$ and that the limit, when x approaches x_0 from the interior of the interval*[19], **194.**

$$\lim_{x \to x_0} f_n(x) = a_n$$

[18] x_0 may be $>$ or $< x_1$. Whether the series remains convergent at x_0, and indeed whether the functions $f_n(x)$ are defined there at all, is immaterial for the present theorem.

[19] We are therefore concerned here, as also in the two subsequent statements, with a one-sided limit.

exists. The series $\sum_{n=0}^{\infty} a_n$ *then converges and* $\lim F(x)$, *when* $x \to x_0$ *in the above manner, exists. Moreover, if we write* $\Sigma a_n = A$, *we have*

$$\lim_{x\to x_0} F(x) = A,$$

or, otherwise,

$$\lim_{x\to x_0} \left(\sum_{n=0}^{\infty} f_n(x)\right) = \sum_{n=0}^{\infty} \left(\lim_{x\to x_0} f_n(x)\right).$$

(The latter form is expressed shortly by saying: *In the case of uniform convergence, we may proceed to the limit term by term.*)

Proof. Given $\varepsilon > 0$, first choose n_1, (v. 4th form of the definition **191**) so that for every $n > n_1$, every $k \geqq 1$ and every x in our interval,

$$|f_{n+1}(x) + \cdots + f_{n+k}(x)| < \varepsilon.$$

Let us for the moment keep n and k fixed, and make $x \to x_0$. By § 19, Theorem 1a, it follows that

$$|a_{n+1} + a_{n+2} + \cdots + a_{n+k}| \leqq \varepsilon.$$

And this is true for every $n > n_1$ and every $k \geqq 1$. *Hence* Σa_n *is convergent.* Let us denote the partial sums of this series by A_n and its sum by A. It is easy to see now that $F(x) \to A$. If, for a given ε, n_0 is determined so that, for every $n > n_0$, we not only have

$$|r_n(x)| < \frac{\varepsilon}{3}, \quad \text{but also} \quad |A - A_n| < \frac{\varepsilon}{3},$$

then, for a (fixed) $m > n_0$,

$$|F(x) - A|$$

$$= |(s_m(x) - A_m) - (A - A_m) + r_m(x)| \leqq |s_m(x) - A_m| + \frac{\varepsilon}{3} + \frac{\varepsilon}{3}.$$

As $x \to x_0$ involves $s_m(x) \to A_m$, we can determine δ so that

$$|s_m(x) - A_m| < \frac{\varepsilon}{3}$$

for every x belonging to the interval, such that $0 < |x - x_0| < \delta$. For these x's, we then also have

$$|F(x) - A| < \varepsilon,$$

which proves all that we required.

If (x_n) is chosen arbitrarily in the interval of uniform convergence, it follows from

$$F(x_n) = s_n(x_n) + r_n(x_n)$$

and $r_n(x_n) \to 0$ (v. **191**, 3rd form) that the sequences $F(x_n)$ and $s_n(x_n)$ will invariably exhibit the *same* behaviour as regards convergence or divergence, and that if they converge, the limits will coincide. We may contrast this with the case of the series, already seen to be non-uniformly convergent, whose partial sums are $s_n(x) = \frac{n x}{1 + n^2 x^2}$. If here we take $x_n = \frac{1}{n}$, we have $F(x_n) = 0$, i. e. it is convergent with the limit 0, whereas $s_n(x_n) = \frac{1}{2}$, i. e. it also converges, but with the limit $\frac{1}{2}$. The two sequences do *not* have the same behaviour.

Theorem 3. *The series $F(x) = \Sigma f_n(x)$ is assumed* ***uniformly*** *convergent in the interval J, and all the functions $f_n(x)$ are supposed* ***integrable*** *over the closed sub-interval J': $a \leqq x \leqq b$, so that $F(x)$ is also continuous in that sub-interval. Then $F(x)$ is also integrable over J' and the integral of $F(x)$ over the interval J' may then be obtained by term-by-term integration, i. e.* **195.**

$$\int_a^b F(x)\,dx \quad or \quad \int_a^b \left[\sum_{n=0}^{\infty} f_n(x)\right] dx = \sum_{n=0}^{\infty}\left[\int_a^b f_n(x)\,dx\right].$$

(More precisely: The series on the right hand side is also convergent and has for its sum the required integral of $F(x)$.

Proof. Given $\varepsilon > 0$, we determine m so large that for every $n > m$ and every x in $a \ldots b$,

$$|r_n(x)| < \frac{\varepsilon}{4(b-a)}.$$

Since $s_m(x)$ is the sum of a finite number of integrable functions, it is itself integrable over J'. By § 19, theorem 11, we can therefore divide the interval J' into p parts $i_1, i_2, \ldots, i_p$ such that, if σ_ν denotes the oscillation of $s_m(x)$ in i_ν, we have

$$\sum_{\nu=1}^{p} i_\nu \sigma_\nu < \frac{\varepsilon}{2}.$$

Now the oscillation of $r_m(x)$ is certainly $< \frac{\varepsilon}{2(b-a)}$, by the manner in which m was determined. Also the oscillation of the sum of two functions is never greater than the sum of the oscillations of the two functions. So for the same subdivision $i_1, i_2, \ldots, i_p$ of the interval $a \ldots b$, we have

$$\sum_{\nu=1}^{p} i_\nu \bar{\sigma}_\nu < \varepsilon,$$

where $\bar{\sigma}_\nu$ denotes the oscillation of $F(x)$ in i_ν. Thus (again by § 19, theorem 11) $F(x)$ also is integrable over J'. Furthermore, as $F = s_n + r_n$, we have, for every $n \geqq m$,

$$\left|\int_a^b F(x)\,dx - \int_a^b s_n(x)\,dx\right| = \left|\int_a^b r_n(x)\,dx\right| < \frac{\varepsilon}{4} < \varepsilon,$$

— the latter by § 19, theorem 21. Now $s_n(x)$ is the sum of a finite number of functions; applying § 19, theorem 22, we therefore at once obtain

$$\left|\int_a^b F(x)\,dx - \sum_{\nu=0}^{n}\int_a^b f_\nu(x)\,dx\right| < \varepsilon.$$

This, however, implies the convergence of $\Sigma \int\limits_a^b f_\nu(x)\,dx$ and the identity of its sum with the corresponding integral of $F(x)$.

Matters are not so simple in the case of term-by-term differentiation.

In **190**, 7, we saw, for instance, that the series

$$\sum_{n=1}^{\infty} \frac{\sin n x}{n}$$

converges for every x, and so represents a function $F(x)$ defined for every real x. The terms of this series are, without exception, continuous and differentiable functions. If we differentiate term by term, we obtain the series

$$\sum_{n=1}^{\infty} \cos n x,$$

which is divergent[20] for every x. — Even if a series converges uniformly for every x, as for instance the series

$$\sum_{n=1}^{\infty} \frac{\sin n x}{n^2}$$

(cf. Example 5, **191**, 2), the position is no better, since on differentiating term by term we obtain

$$\sum_{n=1}^{\infty} \frac{\cos n x}{n},$$

a series which diverges e. g. for $x = 0$.

The theorem on term-by-term differentiation must accordingly be of a different stamp. It runs as follows:

196. **Theorem 4.** *Given*[21] *a series* $\sum\limits_{n=0}^{\infty} f_n(x)$ *whose terms are differentiable in the interval* $J \equiv a \ldots b$, $(a < b)$; *if the series*

$$\sum_{n=0}^{\infty} f_n'(x),$$

deduced from it by differentiating term by term, converges uniformly in J, *then so does the given series, provided it converges at least at one point of* J. *Further, if* $F(x)$ *and* $\varphi(x)$ *are the functions represented by the two series,* $F(x)$ *is differentiable, and we have*

$$F'(x) = \varphi(x).$$

In other words, with the given hypotheses, the series may be differentiated term by term.

---

[20] The formulae established on p. 357 give, for every $x \neq 2k\pi$,

$$\frac{1}{2} + \cos x + \cos 2x + \cdots + \cos n x = \frac{\sin\left(n + \frac{1}{2}\right) x}{2 \sin \frac{x}{2}}.$$

[21] As regards the convergence of the series, no assumption is made in the first instance.

Proof. a) Let c denote a point of J (existent by hypothesis) for which $\Sigma f_n(c)$ converges. By the first mean value theorem of the differential calculus (§ 19, theorem 8)

$$\sum_{\nu=n+1}^{n+k}(f_\nu(x)-f_\nu(c))=(x-c)\cdot\sum_{\nu=n+1}^{n+k}f_\nu'(\xi),$$

where ξ denotes a suitable point between x and c. Given $\varepsilon>0$, we can, by hypothesis, choose n_0 so that for every $n>n_0$, every $k\geqq 1$, and *every* x in J,

$$\left|\sum_{\nu=n+1}^{n+k}f_\nu'(x)\right|<\frac{\varepsilon}{b-a}.$$

Under the same conditions, we therefore have

$$\left|\sum_{\nu=n+1}^{n+k}(f_\nu(x)-f_\nu(c))\right|<\varepsilon.$$

This shows that $\Sigma(f_n(x)-f_n(c))$, and hence $\Sigma f_n(x)$ itself, is uniformly convergent in the whole interval J and accordingly represents a definite function $F(x)$ in that interval.

b) Now let x_0 be a special point of J and write

$$\frac{f_\nu(x_0+h)-f_\nu(x_0)}{h}=g_\nu(h), \qquad (\nu=0,1,2,\ldots).$$

These functions are defined for every $h\gtrless 0$ for which x_0+h belongs to J. As above, we may write

$$\sum_{\nu=n+1}^{n+k}g_\nu(h)=\sum_{\nu=n+1}^{n+k}f_\nu'(x_0+\vartheta h) \qquad (0<\vartheta<1)$$

and we find, as in a), that

$$\sum_{n=0}^{\infty}g_n(h)$$

converges uniformly for all these values of h. This series represents the function

$$\frac{F(x_0+h)-F(x_0)}{h}.$$

By theorem 2, we may let $h\to 0$ term by term, and we conclude that $F'(x_0)$ exists, with

$$F'(x_0)=\sum_{n=0}^{\infty}(\lim_{h\to 0}g_n(h))=\sum_{n=0}^{\infty}f_n'(x_0).$$

This signifies that $F'(x_0)=\varphi(x_0)$, as asserted.

Examples and Remarks.

1. If $\Sigma a_n(x-x_0)^n$ has the radius $r>0$ and if $0<\varrho<r$, the series $\Sigma n\,a_n(x-x_0)^{n-1}$ converges uniformly for every $|x-x_0|\leqq\varrho$. By theorem 4, the given power series accordingly represents a function which is differentiable for every $|x-x_0|\leqq\varrho$. For any particular x, with $|x-x_0|<r$, which we may choose to consider, we can determine $\varrho<r$ so that $|x-x_0|<\varrho<r$. The

function represented by $\Sigma a_n (x - x_0)^n$ therefore remains differentiable at every point of the open interval $|x - x_0| < r$.

2. The function represented by $\sum_{n=1}^{\infty} \frac{\sin n x}{n^3}$ is differentiable for every x and its derived function is $\sum_{n=1}^{\infty} \frac{\cos n x}{n^2}$. (Cf. Example 5, **191**, 2.)

3. The condition of uniform convergence is certainly sufficient in all four theorems. But it remains questionable whether it is also necessary.

a) In the case of the continuity-theorem 1 or its corollary, this is certainly not so. The series considered in **192**, 2 and 4 have everywhere-continuous terms and represent everywhere-continuous functions themselves. Yet their convergence was not uniform. The framing of necessary and sufficient conditions is not exactly easy. *S. Arzelà* (Rendiconti Accad. Bologna, (1), Vol. 19, p. 85. 1883) was the first to do so in a satisfactory manner. A simplified proof of the main theorem enunciated by him will be found in *G. Vivanti* (Rendiconti del circ. matem. di Palermo, Vol. 30, p. 83. 1910). In the case in which the functions $f_n(x)$ are *positive*, it has been shown by *U. Dini* that uniform convergence is also *necessary* for the continuity of $F(x)$. Cf. Ex. 158.

b) The fact that in theorem **195** on term-by-term integration uniform convergence is again not a *necessary* condition may also be verified by various examples. Taking the series $\sum_{n=1}^{\infty} f_n(x)$ discussed on pp. 330-1, whose partial sums are

$$s_n(x) = \frac{n x}{1 + n^2 x^2},$$

and whose sum is $F(x) = 0$, we see at once that

$$\sum_{\nu=1}^{n} \int_0^1 f_\nu(x)\, dx = \int_0^1 s_n(x)\, dx \to 0 = \int_0^1 F(x)\, dx.$$

Thus term-by-term integration leads to the correct result. In the case of the series **192**, 3, however, in which we also have $\int_0^1 F(x)\, dx = 0$, term-by-term integration gives, on the contrary,

$$\sum_{\nu=0}^{n} \int_0^1 f_\nu(x)\, dx = \int_0^1 s_n(x)\, dx = 1 - e^{-\frac{1}{2} n} \to 1.$$

In this case, therefore, term-by-term integration is not allowed.

§ 48. Tests of uniform convergence.

Now that we are acquainted with the meaning of the concept of uniform convergence, we shall naturally inquire how we can determine whether a given series does or does not converge uniformly in the whole or a part of its interval of convergence. However difficult it may be — and we know it often is so — to determine the mere convergence of a given series, the difficulties will of course be considerably enhanced when the question of uniform convergence is approached. The test which is the most important for applications, because it is the easiest to handle, is the following:

○ ***Weierstrass'* test.** *If each of the functions $f_n(x)$ is defined and* **197.**
bounded in the interval J, — say

$$|f_n(x)| \leqq \gamma_n$$

throughout J — and if the series $\Sigma \gamma_n$ (of positive terms) converges, the series $\Sigma f_n(x)$ converges uniformly in J.

Proof. If the sequence (x_n) is chosen arbitrarily in J, we have $|f_{n+1}(x_n) + f_{n+2}(x_n) + \cdots + f_{n+k_n}(x_n)| \leqq \gamma_{n+1} + \gamma_{n+2} + \cdots + \gamma_{n+k_n}$.

By **81**, 2, the right hand side $\to 0$ when $n \to \infty$; hence so does the left. By **191**, 5th form, $\Sigma f_n(x)$ is therefore uniformly convergent in J.

Examples.

1. In the example **191**, 3 we have already made use of the substance of *Weierstrass'* test.

2. The harmonic series $\sum \frac{1}{n^x}$, which converges for $x > 1$, is uniformly convergent on the semi-axis $x \geqq 1 + \delta$, where δ is any positive number. In fact, for such x's,

$$\left|\frac{1}{n^x}\right| \leqq \frac{1}{n^{1+\delta}} = \gamma_n,$$

where $\Sigma \gamma_n$ converges. This proves the statement.

The function represented by the harmonic series — known as *Riemann*'s ζ-function and denoted by $\zeta(x)$ — is therefore certainly continuous for every[22] $x > 1$.

3. Differentiating the harmonic series term by term, we deduce the series

(*) $$-\sum_{n=1}^{\infty} \frac{\log n}{n^x}.$$

This again is uniformly convergent in $x \geqq 1 + \delta > 1$. In fact, for every sufficiently large n, $\frac{\log n}{n^{\delta/2}} < 1$ (by **38**, 4); for these n's and for every $x \geqq 1 + \delta$, we then have

$$\left|\frac{\log n}{n^x}\right| \leqq \frac{1}{n^{1+\delta/2}} \cdot \frac{\log n}{n^{\delta/2}} < \frac{1}{n^{1+\delta/2}} = \gamma_n.$$

Riemann's ζ-function is accordingly differentiable for every $x > 1$, and its derivative is represented by the series (*).

4. If Σa_n converges absolutely, the series

$$\Sigma a_n \cos n x \quad \text{and} \quad \Sigma a_n \sin n x$$

are uniformly convergent for *every* x, since e. g. $|a_n \cos n x| \leqq a_n = \gamma_n$. These series accordingly define functions *continuous* everywhere.

In spite of its great practical importance, *Weierstrass*' test will necessarily be applicable only to a restricted class of series, since it

---

[22] In fact, if we consider a *special* $x > 1$, we can always assume $\delta > 0$ chosen so that $x > 1 + \delta$.

requires in particular that the series investigated should converge *absolutely*. When this is not the case, we have to make use of more delicate tests, which we construct by analogy with those of § 43. The most powerful means for the purpose is again *Abel*'s partial summation formula. On lines quite similar to those already followed, we first obtain from it the

198. °**Theorem.** *A series of the form* $\sum_{n=0}^{\infty} a_n(x) \cdot b_n(x)$ *certainly converges uniformly in the interval* J, *if, in* J,

1) $\sum_{\nu=0}^{\infty} A_\nu \cdot (b_\nu - b_{\nu+1})$ *is uniformly convergent (as a series) and*

2) $(A_n \cdot b_{n+1})$ *is uniformly convergent (as a sequence)*[23].

Here the functions $A_n = A_n(x)$ *denote the partial sums of* $\sum a_n(x)$.

Proof. As formerly — we have merely to interpret the quantities a_ν, b_ν and A_ν as no longer numbers, but functions of x — we first have

$$\sum_{\nu=n+1}^{n+k} a_\nu \cdot b_\nu = \sum_{\nu=n+1}^{n+k} A_\nu \cdot (b_\nu - b_{\nu+1}) + (A_{n+k} \cdot b_{n+k+1} - A_n \cdot b_{n+1}).$$

Letting x and k vary in any manner with n, we have on the left a *sequence of portions*

$$\sum_{\nu=n+1}^{n+k_n} a_\nu(x_n) \cdot b_\nu(x_n)$$

of the series $\sum a_\nu b_\nu$, and on the right the corresponding one relative to the series $\sum A_\nu (b_\nu - b_{\nu+1})$, and a *difference-sequence* of the sequence $(A_n \cdot b_{n+1})$. Since by hypothesis the latter sequences *always* tend to 0 (v. **191**, 5th form), it follows that so does the sequence on the left. This (again by **191**, 5), proves the statement.

Exactly as in § 43, the above theorem, which is still very general in character, leads to the following more special, but more easily manageable tests[24]:

°**1. *Abel*'s test.** $\sum a_\nu(x) \cdot b_\nu(x)$ *is uniformly convergent in* J, *if* $\sum a_\nu(x)$ *converges uniformly in* J, *if further, for every fixed value of* x, *the numbers* $b_n(x)$ *form a real monotone sequence*[25] *and if, for*

---

[23] a_n, b_n, A_n are now always *functions* of x defined in the interval J; only for brevity we often leave the variable x unmentioned. — For the notion of the uniform convergence of a *sequence of functions* cf. **190**, 4.

[24] For simplicity's sake, we name these criteria after the corresponding ones for constant terms. — Cf. p. 315, footnote 8.

[25] Cf. footnote to **184**, 1.

every n and every x in J, the functions $b_n(x)$ are less in absolute value than one and the same number[26] K.

Proof. Let us denote by $\alpha_n(x)$ the remainder corresponding to the partial sum $A_n(x)$; i. e. $\sum_{n=0}^{\infty} a_\nu(x) = A_n(x) + \alpha_n(x)$. In the formula of *Abel*'s partial summation, we may (by the supplement **183**) substitute $-\alpha_\nu$ for A_ν, and we obtain

$$\sum_{\nu=n+1}^{n+k} a_\nu \cdot b_\nu = -\sum_{\nu=n+1}^{n+k} \alpha_\nu \cdot (b_\nu - b_{\nu+1}) - (\alpha_{n+k} \cdot b_{n+k+1} - \alpha_n \cdot b_{n+1});$$

it therefore again suffices to show that both $\Sigma \alpha_\nu (b_\nu - b_{\nu+1})$ and $(\alpha_n \cdot b_{n+1})$ converge uniformly in J. However, the $\alpha_n(x)$'s, as remainders of a uniformly convergent series, tend *uniformly* to 0 and the $b_\nu(x)$'s remain $< K$ in absolute value for every x in J; it follows that $(\alpha_n \cdot b_{n+1})$ also converges uniformly to 0 in J. On the other hand, if we consider the portions

$$T_n = \sum_{\nu=n+1}^{n+k} \alpha_\nu(x) \cdot (b_\nu(x) - b_{\nu+1}(x)),$$

we can easily show that these tend *uniformly* to 0 in J, — thereby completing the proof of the uniform convergence in J of the series under discussion. In fact, if $\bar{\alpha}_\nu$ denotes the upper bound of $\alpha_\nu(x)$ in J, $\bar{\alpha}_\nu \to 0$ (v. form 3a). Thus if ε_n is the largest of the numbers $\bar{\alpha}_{n+1}$, $\bar{\alpha}_{n+2}, \ldots$, this ε_n also $\to 0$ and

$$|T_n| < \varepsilon_n \cdot \sum_{\nu=n+1}^{n+k} |b_\nu - b_{\nu+1}| \leqq \varepsilon_n \cdot |b_{n+1} - b_{n+k+1}| \leqq 2K \cdot \varepsilon_n$$

involves the fact that $T_n \to 0$ uniformly in J.

°**2.** ***Dirichlet*'s test.** $\sum_{n=0}^{\infty} a_n(x) \cdot b_n(x)$ *is uniformly convergent in J, if the partial sums of the series $\Sigma a_n(x)$ are uniformly bounded*[26] *in J and if the functions $b_n(x)$ converge uniformly to 0 in J, the convergence being monotone for every fixed x.*

Proof. The hypotheses and **192**, 5 immediately involve the uniform convergence (again to 0) of $(A_n \cdot b_{n+1})$. If, further, K' denotes

---

[26] The $b_n(x)$'s form, for a fixed x, a sequence of *numbers* $b_0(x), b_1(x), \ldots$; for a fixed n, however, $b_n(x)$ is *a function of* x, defined in J. The above assumption may, then, be expressed as follows: All the sequences, for the various values of x, shall be *uniformly bounded* with regard to all these values of x; in other words, each one is bounded and there is a number K which is *simultaneously* a bound above *for them all*. Or again: All the functions defined in J for the various values of n shall be uniformly bounded with regard to all these values of n; i. e. each function is bounded, and a number K exists which *simultaneously* exceeds *them all* in absolute value.

a number greater than all the $|A_n(x)|$'s for every x, we have

$$\left|\sum_{\nu=n+1}^{n+k} A_\nu\cdot(b_\nu - b_{\nu+1})\right| \leqq K'\cdot \sum_{\nu=n+1}^{n+k} |b_\nu - b_{\nu+1}| \leqq K'\cdot|b_{n+k+1} - b_{n+1}|.$$

In whatever way x and k may depend on n, the right hand side will tend to 0 by the hypotheses, hence also the left. This proves the uniform convergence in J of the series under consideration.

The monotony of the convergence of $b_n(x)$ for fixed x has only been used in each of these tests to enable us to obtain convenient upper estimations of the portions $\Sigma|b_\nu - b_{\nu+1}|$. By slightly modifying the hypotheses with the same end in view, we obtain

°3. **Two tests of *du Bois-Reymond* and *Dedekind*.**

a) *The series $\Sigma\, a_\nu(x)\cdot b_\nu(x)$ is uniformly convergent in J, if both $\Sigma\, a_\nu$ and $\Sigma\,|b_\nu - b_{\nu+1}|$ converge uniformly in J and if, at the same time, the functions $b_n(x)$ are uniformly bounded in J.*

Proof. We use the transformation

$$\sum_{\nu=n+1}^{n+k} a_\nu\cdot b_\nu = -\sum_{\nu=n+1}^{n+k} \alpha_\nu\cdot(b_\nu - b_{\nu+1}) - (\alpha_{n+k}\cdot b_{n+k+1} - \alpha_n\cdot b_{n+1}).$$

As the remainders $\alpha_\nu(x)$ now converge uniformly to 0, we have, for every $\nu > m$, say, and every x in J, $|\alpha_\nu(x)| < 1$. Hence for every $n \geqq m$,

$$\left|\sum_{\nu=n+1}^{n+k} \alpha_\nu\cdot(b_\nu - b_{\nu+1})\right| \leqq \sum_{\nu=n+1}^{n+k} |b_\nu - b_{\nu+1}|;$$

the expression on the right — even if x and k are made to depend on n, in any manner — now tends to 0 as n increases, hence so does the expression on the left. That $\alpha_n\cdot b_{n+1}$ tends uniformly to 0 in J follows, by **192**, 5, from the fact that $\alpha_n(x)$ does and that the $b_n(x)$'s are uniformly bounded in J.

b) *The series $\Sigma\, a_\nu(x)\cdot b_\nu(x)$ is uniformly convergent in J if the series $\Sigma\,|b_\nu - b_{\nu+1}|$ converges uniformly in J, and the series $\Sigma\, a_\nu$ has uniformly bounded partial sums, provided the functions $b_n(x)\to 0$ uniformly in J.*

Proof. From the hypotheses, it again follows at once that $A_n b_{n+1}$ converges uniformly (to 0) in J. Further, if K' once more denotes a number greater than all the $|A_n(x)|$'s for every x,

$$\left|\sum_{\nu=n+1}^{n+k} A_\nu\cdot(b_\nu - b_{\nu+1})\right| \leqq K'\cdot \sum_{\nu=n+1}^{n+k} |b_\nu - b_{\nu+1}|,$$

whence, on account of our present hypotheses, the uniform convergence in J of the series $\Sigma\, A_\nu(b_\nu - b_{\nu+1})$ may at once be inferred.

Examples and Illustrations.

1. In applications, one or other of the two functions $a_n(x)$ and $b_n(x)$ will often reduce to a constant, for every n; it will usually be the former. Now a series of constant terms Σa_ν must, if it converges, of course be regarded as *uniformly convergent in every interval;* for, its terms being independent of x, so are its portions, and any upper estimation valid for the latter is valid *ipso facto* for *every* x. Similarly the partial sums of a series of constant terms Σa_ν, if bounded, must be accounted *uniformly* bounded in every interval. **199.**

2. Let (a_n) be a sequence of *numbers* with Σa_n convergent, and let $b_n(x) = x^n$. The series $\Sigma a_n x^n$ is *uniformly* convergent in $0 \leqq x \leqq 1$, for the conditions of *Abel*'s test are fulfilled in this interval. In fact, Σa_n, as remarked in 1., is uniformly convergent; further, for every *fixed* x in the interval, (x^n) is monotone and $|x^n| \leqq 1$. — By the theorem **194** on term-by-term passage to the limit, we may therefore conclude that

$$\lim_{x\to 1-0} (\Sigma a_n x^n) = \Sigma \left(\lim_{x\to 1-0} a_n x^n\right), \quad \text{i. e.}$$
$$= \Sigma a_n.$$

This gives a fresh proof of *Abel*'s limit theorem **100**.

3. The functions $b_n(x) = \dfrac{1}{n^x}$ also form a sequence bounded uniformly in J (namely, again $\leqq 1$), and monotone for every *fixed* x. Hence, as above, we deduce that

$$\lim_{x\to +0} \Sigma \frac{a_n}{n^x} = \Sigma a_n,$$

if Σa_n denotes a convergent series of constant terms. (*Abel*'s limit theorem or *Dirichlet* series.)

4. Let $a_n(x) = \cos nx$ or $= \sin nx$, and $b_n(x) = \dfrac{1}{n^\alpha}$, $\alpha > 0$. The series

$$\sum_{n=1}^{\infty} a_n(x) \cdot b_n(x) = \sum_{n=1}^{\infty} \frac{\cos nx}{n^\alpha} \quad \text{or} \quad \sum_{n=1}^{\infty} \frac{\sin nx}{n^\alpha}, \quad (\alpha > 0),$$

then satisfy the conditions of *Dirichlet*'s test in every interval of the form[27] $\delta \leqq x \leqq 2\pi - \delta$, where δ denotes a positive number $< \pi$.

In fact, by **185**, 5, the partial sums of $\Sigma a_n(x)$ are uniformly bounded in the interval $\left(\text{we may take } K = \dfrac{1}{\sin \frac{1}{2}\delta}\right)$ and $b_n(x)$ tends monotonely to 0, — *uniformly*, because b_n does not depend on x. If (b_n) denotes *any monotone null sequence*, it follows for the same reason that

$$\Sigma b_n \cos nx \quad \text{and} \quad \Sigma b_n \sin nx$$

are *uniformly* convergent in the same intervals (cf. **185**, 5). — All these series accordingly represent functions which are defined and continuous[28] for every

[27] Or in intervals obtained from the above by displacement through an integral multiple of 2π.

[28] Every *fixed* $x \neq 2k\pi$ may indeed be regarded as belonging to an interval of the above form, if δ is suitably chosen (cf. p. 343, example 1, and p. 345, footnote).

$x \neq 2k\pi$. Whether the continuity subsists at the excluded points $x = 2k\pi$ we cannot at once determine, — not even in the case of the series $\Sigma b_n \sin n x$, although it certainly converges at these points (cf. **216**, 4).

§ 49. *Fourier* series.

A. *Euler's* formulae.

Among the fields to which we may apply the considerations developed in the preceding sections, one of the most important, and also one of the most interesting in itself, is provided by *the theory of Fourier series*, and more generally by that of *trigonometrical series*, into which we now propose to enter [29].

By a trigonometrical series is meant any series of the form

$$\frac{1}{2} a_0 + \sum_{n=1}^{\infty} (a_n \cos n x + b_n \sin n x),$$

with *constant* [30] a_n and b_n. If such a series converges in an interval of the form $c \leqq x < c + 2\pi$, it converges, in consequence of the periodicity of the trigonometrical functions, for *every* real x, and accordingly represents *a function defined for all values of* x and periodic with the period 2π. We have already come across trigonometrical series convergent everywhere, for instance, the series, occurring a few lines back,

$$\sum_{n=1}^{\infty} \frac{\sin n x}{n^\alpha}, \quad \alpha > 0; \quad \sum_{n=1}^{\infty} \frac{\cos n x}{n^\alpha}, \quad \alpha > 1; \quad \text{etc.}$$

We have never been in a position, so far, to determine the sum of any of these series for all values of x. It will appear very soon, however, that trigonometrical series are capable of representing the most curious types of functions — such as one would not have ventured to call functions at all in *Euler*'s time, as they may exhibit discontinuities and irregularities of the most complicated description, so that they seem rather to represent a patchwork of several functions than to form *one* individual function.

---

[29] More or less detailed and extensive accounts of the theory are to be found in most of the larger text books on the differential calculus (in particular, that referred to on p. 2, by *H. v. Mangoldt* and *K. Knopp*, Vol. 3, 8th ed., Part 8, 1944). For separate accounts, we may refer to *H. Lebesgue*, Leçons sur les séries trigonométriques, Paris 1906, and to the particularly elementary Introduction to the theory of Fourier's series, by *M. Bôcher*, Annals of Math. (2), Vol. 7, pp. 81—152. 1906. A particularly detailed account of the theory is given by *E. W. Hobson*, The theory of functions of a real variable and the theory of Fourier series, Cambridge, 2nd ed., Vol. 1, 1921, and Vol. 2, 1926. The comprehensive works of *L. Tonelli*, Serie trigonometriche, Bologna 1928, and *A. Zygmund*, Trigonometrical series, Warsaw 1935, are quite modern treatments; the little volume by *W. Rogosinski*, Fouriersche Reihen, Sammlung Göschen 1930, is particularly attractive and contains a wealth of matter.

[30] It is only for reasons of convenience that $\frac{1}{2} a_0$ is written instead of a_0.

Thus we shall see later on (v. **210**a) that e. g.

$$\sum_{n=1}^{\infty} \frac{\sin n x}{n} \begin{cases} = 0 \text{ for } x = k\pi, \ (k = 0, \pm 1, \pm 2, \ldots), \text{ but} \\ = \dfrac{(2k+1)\pi - x}{2} \text{ for } 2k\pi < x < 2(k+1)\pi; \end{cases}$$

the function represented by this series thus has a graph of the following type:

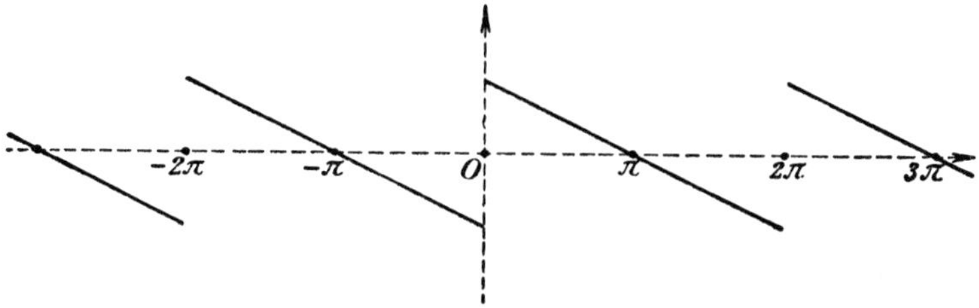

Fig. 7.

Similarly, we shall see (v. 209) that

$$\sum_{n=0}^{\infty} \frac{\sin(2n+1)x}{2n+1} \begin{cases} = 0 \quad \text{for } n = k\pi, \quad \text{but} \\ = \dfrac{\pi}{4} \quad \text{for } 2k\pi < x < (2k+1)\pi, \quad \text{and} \\ = -\dfrac{\pi}{4} \quad \text{for } (2k+1)\pi < x < 2(k+1)\pi; \end{cases}$$

thus the function represented by the series has a graph of the type:

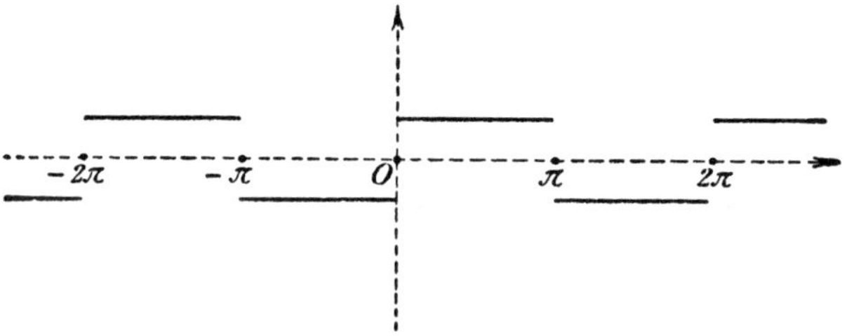

Fig. 8.

In either case, the graph of the function consists of separated stretches (unclosed at either end) and of isolated points.

However, the circumstance that simple trigonometrical series such as the above are capable of representing functions which are themselves altogether discontinuous and "pieced together", is precisely what was chiefly responsible for the thorough revision to which the concept of function, and thence the whole foundation of analysis, came to be subjected at the beginning of the 19th century. We shall see that trigonometrical series are capable of representing most of the so-called "arbitrary functions"[31]; in this respect, they constitute a far more powerful instrument in higher analysis than power series.

---

[31] Of course the concept of an "arbitrary function" is not sharply defined. The term usually denotes a function which cannot be assigned by means of a single *closed* formula (i. e. one avoiding the use of limiting processes) in terms

We will mention only incidentally that the range of this instrument is by no means restricted to pure mathematics. Quite the contrary: such series were first obtained in theoretical physics, in the course of investigations on periodic motion, i. e. chiefly in acoustics, optics, electrodynamics, and the theory of heat; *Fourier*, in his *Théorie de la chaleur* (1822) instituted the first more thorough study of certain trigonometrical series, — although he did not discover any of the fundamental results of their theory.

What functions can be represented by trigonometrical series and by what means can we obtain the representation of a given function, supposing this to be feasible?

In order to lead up to a solution of this question, let us first assume that we have been able to represent a particular function $f(x)$ by a trigonometrical series convergent everywhere:

$$f(x) = \frac{1}{2} a_0 + \sum_{n=1}^{\infty} (a_n \cos n x + b_n \sin n x).$$

On account of the periodicity of the sine and cosine functions, $f(x)$ is then necessarily periodic with the period 2π, and it is sufficient, therefore, to consider any interval of length 2π. *We choose this interval, for all that follows, to be* $0 \leqq x \leqq 2\pi$, — where one of the end-points may, moreover, be omitted.

The function $f(x)$ is then represented in this interval by a convergent series of continuous functions. We know that $f(x)$ may none the less be discontinuous, although it also will be continuous if the series in question converges *uniformly* in the interval. For the moment, we will assume this to be the case.

With these hypotheses, we obtain a relationship between $f(x)$ and the coefficients a_n and b_n which was conjectured by *Euler*:

---

of the so-called elementary functions alone, — i. e. in particular, it denotes a function which is apparently built up from separate portions of simple functions of this type, like the functions given as examples in the text, or the following, defined for every real x:

$$\left.\begin{aligned} f(x) &= k \\ f(x) &= k + (x-k)^2 \\ f(x) &= x - k \end{aligned}\right\} \begin{aligned} &\text{in } k \leqq x < k+1 \\ &(k = 0, \pm 1, \pm 2, \ldots) \end{aligned}$$

$$f(x) = \begin{cases} 0 & \text{for irrational } x \\ x & \text{for rational } x, \end{cases}$$

etc. Cf., however, the "arbitrary" function expressed by means of limiting processes on p. 329, footnote. Not until it was found that even a perfectly "arbitrary" function such as these could be represented by *a single* (relatively simple) expression, as for instance by our trigonometrical series or by other limiting processes, — did any necessity arise for regarding it as being actually *one* function, instead of a mere patchwork of several functions.

Theorem 1. *The series* **200.**

$$\frac{1}{2}a_0 + \sum_{n=1}^{\infty}(a_n \cos n x + b_n \sin n x)$$

is assumed ***uniformly*** *convergent*[32] *in the interval* $0 \leqq x \leqq 2\pi$, *with the sum* $f(x)$. *Then for* $n = 0, 1, 2, \ldots$, *we have*

$$a_n = \frac{1}{\pi}\int_0^{2\pi} f(x)\cos n x\,dx, \quad b_n = \frac{1}{\pi}\int_0^{2\pi} f(x)\sin n x\,dx.$$

(***Euler*** or ***Euler-Fourier formulae***)[33].

Proof. As is known by elementary considerations, the following formulae[34] hold for every integral p and q ($\geqq 0$):

$$\text{a)}\quad \int_0^{2\pi} \cos p x \cdot \cos q x\,dx \begin{cases} = 0 & \text{for } p \neq q \\ = \pi & \text{for } p = q > 0 \\ = 2\pi & \text{for } p = q = 0 \end{cases}$$

$$\text{b)}\quad \int_0^{2\pi} \cos p x \cdot \sin q x\,dx = 0$$

$$\text{c)}\quad \int_0^{2\pi} \sin p x \cdot \sin q x\,dx \begin{cases} = 0 & \text{for } p \neq q \text{ and } p = q = 0 \\ = \pi & \text{for } p = q > 0. \end{cases}$$

Let us multiply the series for $f(x)$, which is *uniformly* convergent in $0 \leqq x \leqq 2\pi$, by $\cos p x$; by **192**, 2 the uniformity of the conver gence is not destroyed, and after performing the multiplication we may accordingly (v. **195**) integrate term by term from 0 to 2π. We immediately obtain:

$$\int_0^{2\pi} f(x)\cos p x\,dx \begin{cases} = \frac{1}{2}a_0 \int_0^{2\pi} \cos p x\,dx & \text{for } p = 0 \\ = a_p \int_0^{2\pi} \cos p x \cdot \cos p x\,dx & \text{for } p > 0, \end{cases}$$

---

[32] In consequence of the periodicity of $\cos x$ and $\sin x$, it is then, *ipso facto, uniformly* convergent *for every* x.

[33] This designation is a purely conventional one; historical remarks are given by *H. Lebesgue,* loc. cit., p. 23; *A. Sachse,* Versuch einer Geschichte der trigonometrischen Reihen, Inaug.-Diss., Göttingen 1879; *P. du Bois-Reymond,* in his answer to the last-named paper; as well as very extensively by *H. Burkhardt,* Trigonometrische Reihen und Integrale bis etwa 1850 (Enzyklop. d. math. Wiss., Vol. II, 1, Parts 7 and 8, 1914—15).

[34] We have only to transform the product of the two functions in the integrand into a sum in accordance with the known addition theorems, $\left(\text{e. g. } \cos p x \cdot \cos q x = \frac{1}{2}[\cos(p-q)x + \cos(p+q)x]\right)$, in order to be able to integrate straight away.

i. e. in either case

$$a_p = \frac{1}{\pi}\int_0^{2\pi} f(x)\cos p x\, dx;$$

for the remaining terms give, on integration, the value 0. In the same way, multiplying the assumed expansion of $f(x)$ by $\sin px$ and then integrating, we at once deduce the second of *Euler*'s formulae

$$b_p = \frac{1}{\pi}\int_0^{2\pi} f(x)\sin p x\, dx.$$

The value of this theorem is diminished by the number of assumptions required to carry out the proof. Also, it gives no indication how to determine *whether a given function can be expanded in a trigonometrical series at all, or, if it can, what the values of the coefficients will be.*

However, the theorem suggests the following mode of procedure: Let $f(x)$ be an arbitrary function defined in the interval $0 \leqq x \leqq 2\pi$, and *integrable in Riemann's sense* in the interval. In that case the integrals in *Euler*'s formulae certainly have a meaning, by § 19, theorem 22, and give definite values for a_n and b_n. We therefore note that these numbers, exist, on the single hypothesis that $f(x)$ is integrable. The numbers $\frac{1}{2}a_0, a_1, a_2, \ldots$ and $b_1, b_2, \ldots$ thus defined by Euler's formulae will be called the ***Fourier constants*** or ***Fourier coefficients*** of the function $f(x)$. The series

$$\frac{1}{2}a_0 + \sum_{n=1}^{\infty}(a_n \cos n x + b_n \sin n x)$$

may now be written down, although this implies nothing as regards its possible convergence. This series will be called (without reference to its behaviour or to the value of its sum, if existent) ***the Fourier series generated by, or belonging to, $f(x)$***, and this is expressed symbolically by

$$f(x) \sim \frac{1}{2}a_0 + \sum_{n=1}^{\infty}(a_n \cos n x + b_n \sin n x).$$

This formula accordingly implies no more than that certain constants a_n, b_n, have been deduced from $f(x)$ (assumed only to be integrable) by means of *Euler*'s formulae, and that then the above series has been written down[35].

---

[35] The symbol "$\sim$" has of course no connection here with the symbol introduced in **40**, Definition 5, for "asymptotically proportional". There is no fear of confusion.

From theorem **1.** and the manner in which this series was derived, we have, it is true, some justification for the hope that the series may converge and have $f(x)$ for its sum.

Unfortunately, this is not the case in general. (Examples will be met with very shortly.) On the contrary, the series *may not* converge in the whole interval, nor even at any single point; and *if* it does so, the sum is not necessarily $f(x)$. It is impossible to say off-hand when the one or the other case may occur; it is this circumstance which prevents the theory of *Fourier* series from being entirely a simple subject, but which, on the other hand, renders it extraordinarily fascinating; for here entirely new problems arise, and we are faced with a fundamental property of functions which appears to be essentially new in character: the property of producing a *Fourier* series whose sum is equal to the function itself. The next task is then to elucidate the connection between this new property and the old ones, — viz. continuity, monotony, differentiability, integrability, and so on. More concretely stated, the problems which arise are therefore as follows:

1. *Is the Fourier series of a given (integrable) function $f(x)$ convergent for some or all values of x in $0 \leqq x \leqq 2\pi$?*

2. *If it converges, does the Fourier series of $f(x)$ have for its sum the value of the generating function?*

3. *If the Fourier series converges at all points of the interval $\alpha \leqq x \leqq \beta$, is the convergence uniform in this interval?*

As it is conceivable that a trigonometrical expansion of $f(x)$ might be obtained by other means than that of *Euler*'s formulae, we may also raise the further question at once:

4. *Is it possible for a function which is capable of expansion in a trigonometrical series to possess several such expansions, — in particular, can it possess another trigonometrical expansion besides the possible Fourier expansion provided by Euler's formulae?*

It is not very easy to find answers to all these questions; in fact no *complete* answer to any of them is known at the present day. It would take us too far to treat all four questions in accordance with modern knowledge. We shall turn our attention chiefly to the first two; the third we shall touch on only incidentally, and we shall leave the last almost entirely out of account [36].

---

[36] It should be noted, however, that the fourth question is answered under extremely general hypotheses by the fact that two trigonometrical series which converge in $0 \leqq x \leqq 2\pi$ cannot represent the same function in that interval without being entirely identical. And if $f(x)$, the function represented, is integrable over $0 \ldots 2\pi$, its *Fourier* coefficients are equal to the coefficients of the trigonometrical expansion; cf. *G. Cantor* (J. f. d. reine u. angew. Math., Vol. 72, p. 139. 1870) and *P. du Bois-Reymond* (Münch. Abh., Vol. 12, Section I, p. 117. 1876).

With the designations introduced above, the content of Theorem 1 may be expressed as follows:

Theorem 1a. *If a trigonometrical series converges uniformly in* $0 \leqq x \leqq 2\pi$ *(i. e. for all x), it is the Fourier series of the function represented by it, and this function*[37] *admits of no other representation by a trigonometrical series converging uniformly in* $0 \leqq x \leqq 2\pi$.

The fact that the *Fourier* series of an integrable function does not necessarily converge will be seen further on; that even when it does converge, it need not have $f(x)$ for its sum, is obvious from the fact that two *different* functions $f_1(x)$ and $f_2(x)$ may very well have identically the same *Fourier* constants; in fact two integrable functions have the same integral (and therefore the same *Fourier* constants; i. e. the same *Fourier* series), if they coincide, for instance, for all rational values of x, without coinciding everywhere (v. § 19, theorem 18). The fact that in an interval of convergence the series need not converge uniformly is shown by the example already used above; for the series $\sum \frac{\sin n x}{n}$ converges everywhere (v. **185**, 5), and if the convergence were uniform, say in the interval $-\delta \leqq x \leqq \delta$, $\delta > 0$, it would have to represent a continuous function in that interval, by **193**. This is not the case, however, as we mentioned before on p. 351 and will prove later on p. 375.

These few remarks suffice to show that the questions formulated above are not of a simple nature. In answering them, we shall follow the line adopted by *G. Lejeune-Dirichlet*, who took the first notable step towards a solution of the above questions, in his paper *Sur la convergence des séries trigonométriques*[38].

B. *Dirichlet's* integral.

We proceed to attack the first of the proposed problems, namely, the question of convergence:

If the *Fourier* series $\frac{1}{2} a_0 + \Sigma (a_n \cos n x + b_n \sin n x)$ generated by a given integrable function $f(x)$, — i. e. with coefficients given in terms of $f(x)$ by *Euler*'s formulae, — is to converge at the point $x = x_0$, its partial sums

$$s_n(x_0) = \frac{1}{2} a_0 + \sum_{\nu=1}^{n} (a_\nu \cos \nu x_0 + b_\nu \sin \nu x_0)$$

must tend to a limit when $n \to +\infty$. It is often possible to determine whether or no this is the case, by expressing $s_n(x_0)$ in the form of a definite integral as follows:

[37] This function is then (by **193**, Corollary) everywhere continuous.

[38] Journ. f. d. reine u. angew. Math., Vol. 4, p. 157. 1829.

For $\nu \geqq 1$, the function $a_\nu \cos \nu x_0 + b_\nu \sin \nu x_0$ is represented by[39]

$$= \left[\frac{1}{\pi}\int_0^{2\pi} f(t)\cos\nu t\,dt\right]\cos\nu x_0 + \left[\frac{1}{\pi}\int_0^{2\pi} f(t)\sin\nu t\,dt\right]\sin\nu x_0$$

$$= \frac{1}{\pi}\int_0^{2\pi} f(t)\cdot\cos\nu(t-x_0)\,dt.$$

Thus

$$s_n(x_0) = \frac{1}{2\pi}\int_0^{2\pi} f(t)\,dt + \frac{1}{\pi}\int_0^{2\pi} f(t)\cos(t-x_0)\,dt + \cdots$$

$$+ \frac{1}{\pi}\int_0^{2\pi} f(t)\cdot\cos n(t-x_0)\,dt$$

$$= \frac{1}{\pi}\int_0^{2\pi} f(t)\cdot\left[\frac{1}{2} + \cos(t-x_0) + \cos 2(t-x_0) + \cdots + \cos n(t-x_0)\right] dt.$$

We now take the important step of replacing the sum of the $(n+1)$ terms in brackets by a single closed expression. We have indeed[40] for every $z \neq 2k\pi$, for every α and all positive integral m's,

$$\cos(\alpha+z) + \cos(\alpha+2z) + \cdots + \cos(\alpha+mz) \qquad \textbf{201.}$$

$$= \frac{\sin\left(\alpha + \overline{2m+1}\,\frac{z}{2}\right) - \sin\left(\alpha+\frac{z}{2}\right)}{2\sin\frac{z}{2}}$$

$$= \frac{\sin m\frac{z}{2}\cdot\cos\left(\alpha+\overline{m+1}\,\frac{z}{2}\right)}{\sin\frac{z}{2}},$$

---

[39] In order to distinguish the parameter of integration from the fixed point x_0, we henceforth denote the former by t.

[40] Proof. If the expression on the left is denoted by C_m, we have

$$2\sin\frac{z}{2}\cdot C_m = \sum_{\nu=1}^{m} 2\sin\frac{z}{2}\cos(\alpha+\nu z)$$

$$= \sum_{\nu=1}^{m}\left[-\sin\left(\alpha+\overline{2\nu-1}\,\frac{z}{2}\right) + \sin\left(\alpha+\overline{2\nu+1}\,\frac{z}{2}\right)\right]$$

$$= -\sin\left(\alpha+\frac{z}{2}\right) + \sin\left(\alpha+\overline{2m+1}\,\frac{z}{2}\right)$$

$$= 2\sin m\frac{z}{2}\cdot\cos\left(\alpha+\overline{m+1}\,\frac{z}{2}\right).$$

Moreover the above formula continues to hold for $z = 2k\pi$, provided we attribute to the ratio on the right hand side the limiting value for $z \to 2k\pi$, i. e. the value $m\cos\alpha$.

from which many analogous formulae may be deduced as particular cases[41]. Taking $\alpha = 0$, $z = t - x_0$, $m = n$, we obtain

$$\frac{1}{2} + \cos(t - x_0) + \cdots + \cos n(t - x_0)$$

$$= \frac{1}{2} + \frac{\sin(2n+1)\frac{t-x_0}{2} - \sin\frac{t-x_0}{2}}{2\sin\frac{t-x_0}{2}} = \frac{\sin(2n+1)\frac{t-x_0}{2}}{2\sin\frac{t-x_0}{2}}.$$

Accordingly[42],

$$\text{(a)} \qquad s_n(x_0) = \frac{1}{2\pi}\int_0^{2\pi} f(t)\cdot\frac{\sin(2n+1)\frac{t-x_0}{2}}{\sin\frac{t-x_0}{2}}\,dt.$$

Finally, we may transform this expression somewhat. The function $f(x)$ need only be defined in the interval $0 \leqq x \leqq 2\pi$ and integrable over this interval. The latter property remains unaltered if we merely modify the value of $f(2\pi)$ (cf. § 19, theorem 17). We will equate it to $f(0)$ and define $f(x)$ further, for every x such that

$$2k\pi \leqq x \leqq 2(k+1)\pi, \quad (k = \pm 1, \pm 2, \ldots),$$

by:

$$f(x) = f(x - 2k\pi).$$

[41] For subsequent use, we mention the following:

1. $\frac{\pi}{2} + \alpha$ substituted for α gives:

$$\sin(\alpha + z) + \sin(\alpha + 2z) + \cdots + \sin(\alpha + mz) = \frac{\sin m\frac{z}{2}\cdot\sin\left(\alpha + \overline{m+1}\,\frac{z}{2}\right)}{\sin\frac{z}{2}};$$

2. $\alpha = 0$ gives: $\cos z + \cos 2z + \cdots + \cos mz = \dfrac{\sin m\frac{z}{2}\cdot\cos\overline{m+1}\,\frac{z}{2}}{\sin\frac{z}{2}}$;

3. $\alpha = \frac{\pi}{2}$ gives: $\sin z + \sin 2z + \cdots + \sin mz = \dfrac{\sin m\frac{z}{2}\cdot\sin\overline{m+1}\,\frac{z}{2}}{\sin\frac{z}{2}}$;

4. $z = 2x$, $\alpha = \gamma - x$, give:
$$\cos(\gamma + x) + \cos(\gamma + 3x) + \cdots + \cos(\gamma + \overline{2m-1}\cdot x) = \frac{\sin mx\cdot\cos(\gamma + mx)}{\sin x};$$

5. $z = 2x$, $\alpha = \frac{\pi}{2} + \gamma - x$, give:
$$\sin(\gamma + x) + \sin(\gamma + 3x) + \cdots + \sin(\gamma + \overline{2m-1}\cdot x) = \frac{\sin mx\cdot\sin(\gamma + mx)}{\sin x}.$$

[42] For $t = x_0$, as we observed once before, we should attribute to the sine-ratio the limiting value for $t \to x_0$, here $(2n+1)$.

Our function $f(x)$ is now defined for all real values of x and we have arranged for it to be periodic with period 2π. Now for any function $\varphi(x)$ periodic with period 2π, we have (by § 19, theorem 19), whatever the values of c and c' may be,

$$\int_0^{2\pi} \varphi(t)\,dt = \int_c^{c+2\pi} \varphi(t)\,dt = \int_0^{2\pi} \varphi(c'+t)\,dt \quad \text{and} \quad \int_\alpha^\beta \varphi(t)\,dt = \int_{\alpha+2\pi}^{\beta+2\pi} \varphi(t)\,dt.$$

As the integrand in (a) is now a function of this type, we have

$$s_n(x_0) = \frac{1}{2\pi}\int_0^{2\pi} f(x_0+t)\cdot\frac{\sin(2n+1)\frac{t}{2}}{\sin\frac{t}{2}}\,dt.$$

If we split up this integral into the parts relative to the intervals 0 to π and π to 2π, substituting $-t$ for t in the second, the latter becomes

$$-\frac{1}{2\pi}\int_{-\pi}^{-2\pi} f(x_0-t)\cdot\frac{\sin(2n+1)\frac{t}{2}}{\sin\frac{t}{2}}\,dt,$$

i. e. by the above remark with regard to $\int_\alpha^\beta \varphi(t)\,dt$

$$\frac{1}{2\pi}\int_0^{+\pi} f(x_0-t)\cdot\frac{\sin(2n+1)\frac{t}{2}}{\sin\frac{t}{2}}\,dt,$$

and we accordingly obtain

$$s_n(x_0) = \frac{1}{\pi}\int_0^{\pi} \frac{f(x_0+t)+f(x_0-t)}{2}\cdot\frac{\sin(2n+1)\frac{t}{2}}{\sin\frac{t}{2}}\,dt.$$

Substituting $2t$ for t, we are ultimately led to the formula

$$s_n(x_0) = \frac{2}{\pi}\int_0^{\frac{\pi}{2}} \frac{f(x_0+2t)+f(x_0-2t)}{2}\cdot\frac{\sin(2n+1)t}{\sin t}\,dt. \qquad \textbf{202.}$$

This is *Dirichlet's integral*[43], by which the partial sums of the *Fourier* series generated by $f(x)$ may be expressed. We may therefore state, as our first important result, the theorem:

Theorem 2. *In order that the Fourier series generated by a function $f(x)$, integrable (hence bounded) and periodic with period 2π, may*

---

[43] We designate as *Dirichlet*'s integrals all integrals of either of the two forms

$$\int_0^a \varphi(t)\frac{\sin kt}{\sin t}\,dt \qquad \text{or} \qquad \int_0^a \varphi(t)\frac{\sin kt}{t}\,dt.$$

converge at a point x_0, it is necessary and sufficient that Dirichlet's integral

$$\frac{2}{\pi}\int_0^{\frac{\pi}{2}} \frac{f(x_0+2t)+f(x_0-2t)}{2}\cdot\frac{\sin(2n+1)t}{\sin t}\,dt$$

should tend to a (finite) limit as $n\to+\infty$. This limit is then the sum of the Fourier series at the point x_0.

Let us denote this sum by $s(x_0)$. The second question (p. 355), concerning the sum of the *Fourier* series, when convergent, may be included in our present considerations and our result may be put in a form still more advantageous in the sequel, by expressing the quantity $s(x_0)$ in the form of a *Dirichlet* integral also. As

$$\frac{1}{2}+\cos t+\cos 2t+\cdots+\cos nt=\frac{\sin(2n+1)\frac{t}{2}}{2\sin\frac{t}{2}},$$

we have

$$\int_0^{2\pi}\frac{\sin(2n+1)\frac{t}{2}}{2\sin\frac{t}{2}}\,dt=\pi,$$

or, effecting the same transformations as before with the general integral,

(b) $$\frac{2}{\pi}\int_0^{\frac{\pi}{2}}\frac{\sin(2n+1)t}{\sin t}\,dt=1.$$

Multiplying this equation[44] by $s(x_0)$, we finally obtain, by subtraction from 202,

$$s_n(x_0)-s(x_0)=\frac{2}{\pi}\int_0^{\frac{\pi}{2}}\left[\frac{f(x_0+2t)+f(x_0-2t)}{2}-s(x_0)\right]\frac{\sin(2n+1)t}{\sin t}\,dt.$$

Our preceding theorem may now be expressed as follows:

203. Theorem 2a. *In order that the Fourier series generated by a function $f(x)$, integrable and periodic with period 2π, should converge to the sum $s(x_0)$ at the point x_0, it is necessary and sufficient that, as $n\to+\infty$, Dirichlet's integral*

$$\frac{2}{\pi}\int_0^{\frac{\pi}{2}}\varphi(t;x_0)\,\frac{\sin(2n+1)t}{\sin t}\,dt$$

---

[44] This equation may also be obtained from **202**, by substituting $f(x)\equiv 1$; this gives $a_0=2$ and, for every $n\geqq 1$, $a_n=b_n=0$, i. e. $s_n(x_0)=1$ for every n and every x_0.

should tend to 0, *where for brevity we have put*

$$\left[\frac{f(x+2t)+f(x-2t)}{2}-s(x)\right]=\varphi(t;x).$$

Although this theorem by no means solves questions 1 and 2 in such a manner that the answer in given concrete cases lies ready to hand, yet it furnishes an entirely new method of attack for their solution. Indeed the same may be said with regard to the third of the questions proposed on p. 355, for theorem 2a may at once be modified to the following:

Theorem 3. *On the assumption that the partial sums* $s_n(x)$ *converge to* $s(x)$ *at every point of the interval* $\alpha \leqq x \leqq \beta$, *they will converge uniformly to this limit in the interval, if, and only if, the integral, depending on* x,

$$\frac{2}{\pi}\int_0^{\frac{\pi}{2}} \varphi(t,x)\cdot\frac{\sin(2n+1)t}{\sin t}\,dt$$

tends uniformly to 0 *as* $n \to +\infty$ *in* $\alpha \leqq x \leqq \beta$, *that is to say if, given* $\varepsilon > 0$, *we can assign* $N = N(\varepsilon)$ *so that this integral is less than* ε *in absolute value for every* $n > N$ *and every* x *in* $\alpha \leqq x \leqq \beta$.

Before we make use of theorem 2 to construct immediate tests of convergence for *Fourier* series, we proceed first to transform and simplify this theorem in various ways. For this purpose, we begin by proving the following theorems, which apparently lead us rather off the track, but also claim considerable interest in themselves.

Theorem 4. *If* $f(x)$ *is integrable over* $0 \ldots 2\pi$, *and if* (a_n) *and* (b_n) *are its Fourier constants, then* $\sum_{n=1}^{\infty}(a_n^2+b_n^2)$ *converges.*

Proof. The integral

$$\int_0^{2\pi}\left[f(t)-\sum_{\nu=1}^{n}(a_\nu\cos\nu t+b_\nu\sin\nu t\right]^2dt$$

is $\geqq 0$, as its integrand is never negative. On the other hand, it is

$$\int_0^{2\pi}[f(t)]^2\,dt-2\,\Sigma\left[a_\nu\int_0^{2\pi}f(t)\cos\nu t\,dt\right]-2\,\Sigma\left[b_\nu\int_0^{2\pi}f(t)\sin\nu t\,dt\right]$$
$$+\int_0^{2\pi}[\Sigma(a_\nu\cos\nu t+b_\nu\sin\nu t)]^2\,dt$$

$$=\int_0^{2\pi}[f(t)]^2\,dt-2\pi\,\Sigma\,a_\nu^2-2\pi\,\Sigma\,b_\nu^2+\pi\,\Sigma\,a_\nu^2+\pi\,\Sigma\,b_\nu^2$$

$$=\int_0^{2\pi}[f(t)]^2\,dt-\pi\,\Sigma(a_\nu^2+b_\nu^2),$$

where each summation is extended from $\nu = 1$ to $\nu = n$. Since this expression is non-negative, we have

$$\sum_{\nu=1}^{n}(a_\nu^2+b_\nu^2)\leqq\frac{1}{\pi}\int_0^{2\pi}[f(t)]^2\,dt.$$

Thus the partial sums of the series (of positive terms) in question are bounded and the series is convergent, as asserted.

The above contains in particular

Theorem 5. *The Fourier constants* (a_n) *and* (b_n) *of an integrable function form a null sequence.*

From this, we may deduce quite simply the further

Theorem 6. *If* $\psi(t)$ *is integrable in the interval* $a \leqq t \leqq b$, *then*

$$A_n = \int_a^b \psi(t) \cos n t \, dt \to 0,$$

$$B_n = \int_a^b \psi(t) \sin n t \, dt \to 0.$$

Proof. If a and b both belong to one and the same interval of the form $2k\pi \leqq t \leqq 2(k+1)\pi$, we define $f(t) = \psi(t)$ in $a \leqq t \leqq b$ and $f(t) = 0$ at the remaining points of the first-named interval; for every other real t, $f(t)$ is defined so as to be periodic with the period 2π. Then

$$A_n = \int_a^b \psi(t) \cos n t \, dt = \int_0^{2\pi} f(t) \cos n t \, dt = \pi a_n$$

and similarly $B_n = \pi b_n$, where a_n and b_n denote the *Fourier* constants of the function $f(t)$. By theorem 5, A_n and B_n therefore $\to 0$. If a and b do not fulfil the above condition, we can split up the interval $a \leqq t \leqq b$ into a finite number of portions, each of which satisfies the condition. A_n and B_n then appear as the sum of a (fixed) finite number of terms, each of which tends to 0 as $n \to \infty$. Hence A_n and B_n do the same[45].

This important theorem will enable us to simplify the problem of the convergence of *Dirichlet*'s integral[46].

Supposing δ chosen arbitrarily with $0 < \delta < \frac{\pi}{2}$, the function

$$\psi(t) = \frac{\varphi(t; x_0)}{\sin t} = \frac{\frac{1}{2}[f(x_0 + 2t) + f(x_0 - 2t)] - s(x_0)}{\sin t}$$

[45] This important theorem appears intuitively plausible if we imagine the curve $y = \psi(t) \cos n t$ to be drawn for large values of n: We isolate a small interval $\alpha \ldots \beta$ in which $\psi(t)$ has an almost negligible oscillation (is practically constant) and proceed to choose n so large that the number of oscillations of $\cos n t$ is fairly large in the interval; in that case, the arc of the curve $y = \psi(t) \cos n t$ corresponding to $\alpha \ldots \beta$ will enclose positive and negative areas in approximately equal numbers and of approximately the same size, so that the integral is almost 0.

[46] Of course theorem 6 may be proved quite directly, without first proving theorem 4. The latter is, however, an equally important theorem in the theory, even though, as it happens, we shall not need it again in the sequel.

is integrable in $\delta \leqq t \leqq \frac{\pi}{2}$. Hence, for fixed δ,

c) $$\int_{\delta}^{\frac{\pi}{2}} \psi(t) \sin(2n+1)t \cdot dt \to 0.$$

The *Dirichlet* integral of theorem 2a will therefore tend to 0 as limit as $n \to \infty$, if, and only if — for a fixed, but in itself arbitrary, value of $\delta > 0$ — the new integral

$$\frac{2}{\pi} \int_0^{\delta} \varphi(t; x_0) \frac{\sin(2n+1)t}{\sin t} dt$$

tends to 0 as n increases. Now the latter integral only involves the values of $f(x_0 \pm 2t)$ in $0 \leqq t \leqq \delta$, i. e. of $f(x)$ in $x_0 - 2\delta \leqq x \leqq x_0 + 2\delta$. Since $\delta > 0$ may be assumed arbitrarily small, this remarkable result contains at the same time the following

Theorem 7. (***Riemann***'s **theorem.**[47]) *The behaviour of the Fourier series of $f(x)$ at the point x_0 depends only on the values of $f(x)$ in the neighbourhood of x_0. This neighbourhood may be assumed as small as we please.* **204.**

In order to illustrate this peculiar theorem, we may mention the following consequence of it: Consider all possible functions $f(x)$ (integrable in $0 \ldots 2\pi$) which coincide at a point x_0 of the interval $0 \ldots 2\pi$ and in some neighbourhood of this point, however small, possibly varying with the particular function. Then the *Fourier* series *of all these functions* — however much they may differ outside the neighbourhood in question — must, at x_0 itself, either *all* converge or *all* diverge, and in the former case they have the same sum $s(x_0)$ (which may or may not be equal to $f(x_0)$).

After inserting these remarks, we proceed to re-formulate the criterion obtained above, which we may henceforth substitute for theorem 2:

Theorem 8. *The necessary and sufficient condition for the Fourier series of $f(x)$ to converge at x_0 to the sum $s(x_0)$, is that for an arbitrarily chosen positive $\delta < \frac{\pi}{2}$, Dirichlet's integral*

$$\frac{2}{\pi} \int_0^{\delta} \varphi(t; x_0) \frac{\sin(2n+1)t}{\sin t} dt$$

should tend to 0 as n increases[48].

---

[47] Über die Darstellbarkeit einer Funktion durch eine trigonometrische Reihe, Hab.-Schrift, Göttingen 1854 (Werke, 2nd ed. p. 227).

[48] As regards *uniformity* of convergence, we can assert nothing straight away, since we are ignorant as to whether the integral (c) above considered, which tends to 0 as n increases, for every *fixed* x_0, will do so *uniformly* for every x of a specified interval on the x-axis. Actually this is the case, but we do not propose to enter into the question further.

There is no difficulty in showing that the denominator $\sin t$ in the last integrand may be replaced by t. In fact the difference between the original integral and the one so obtained, i. e. the integral

$$\frac{2}{\pi}\int_0^\delta \varphi(t; x_0)\left[\frac{1}{\sin t}-\frac{1}{t}\right]\cdot \sin(2n+1)t\cdot dt,$$

automatically tends to 0 as n increases, by theorem 6, — because $\frac{1}{\sin t}-\frac{1}{t}$ is continuous and bounded[49], and hence integrable, in $0 < t \leqq \delta$.

Thus we may finally state:

205. **Theorem 9.** *The necessary and sufficient condition for the Fourier series of a function $f(x)$, periodic with the period 2π and integrable over $0 \ldots 2\pi$, to converge to $s(x_0)$ at the point x_0, is that for an arbitrarily chosen positive $\delta\left(<\frac{\pi}{2}\right)$, the sequence of the values of the integral*

$$\frac{2}{\pi}\int_0^\delta \varphi(t; x_0)\frac{\sin(2n+1)t}{t}\,dt$$

forms a null sequence. Here $\varphi(t; x_0)$ has the same meaning as in theorem 2a. In another form, the condition is that, given $\varepsilon > 0$, we can assign $\delta < \frac{\pi}{2}$ and $N > 0$, so that[50] *for every $n > N$,*

$$\left|\frac{2}{\pi}\int_0^\delta \varphi(t; x_0)\frac{\sin(2n+1)t}{t}\,dt\right| < \varepsilon.$$

C. Conditions of convergence.

Our preliminary investigations have prospered so far that the first two questions of p. 355 may now be attacked directly. By the above, these are completely reduced to the following problem:

Given a function $\varphi(t)$, integrable in $0 \leqq t \leqq \delta$, what further conditions must this function fulfil in order that the integrals[51]

$$J_k = \frac{2}{\pi}\int_0^\delta \varphi(t)\cdot\frac{\sin kt}{t}\,dt$$

---

[49] In fact, $\frac{1}{\sin t}-\frac{1}{t} = \frac{t-\sin t}{t\cdot\sin t} = \frac{\frac{1}{6}t - + \cdots}{1 - + \cdots}$ in the interval, and thus itself tends to 0 as $t \to 0$.

[50] The student should make it quite clear to himself that the second formulation is actually equivalent to the first, although δ need only be determined *after* the value of ε has been chosen.

[51] For $t = 0$, we attribute to $\frac{\sin kt}{t}$ in the integrand the value k.

should tend to a limit as k increases, and what, in that case, is the value of this limit?[52]

Since in this integral, δ has a fixed but arbitrarily small value, the answer to this question depends *only* — cf. *Riemann*'s theorem 7 — on the behaviour of $\varphi(t)$ *immediately to the right of* 0, say in an interval of the form $0 < t < \delta_1 (\leqq \delta)$. We may accordingly inquire also: *What properties must $\varphi(t)$ possess immediately to the right of* 0, *in order that the limit in question may exist?*

A large number of sufficient conditions for this have been found, of which we shall only explain two, the great generality of which renders them sufficient for most purposes. The first of these was established by *Dirichlet* in the above-named paper (v. p. 356) and was the first exact condition of convergence in the theory of *Fourier* series, in which *Dirichlet*'s work is altogether fundamental. The second is due to *U. Dini* and was discovered in 1880.

1. ***Dirichlet***'s rule. *If $\varphi(t)$ is monotone to the right of* 0, *— i. e. in an interval of the form $0 < t < \delta_1 (\leqq \delta)$ — then the limit in question exists, and we have* **206.**

$$\lim_{k \to +\infty} J_k = \lim_{k \to +\infty} \frac{2}{\pi} \int_0^\delta \varphi(t) \cdot \frac{\sin k t}{t} \, dt = \varphi_0,$$

where φ_0 denotes the (right hand) limiting value $\lim_{t \to +0} \varphi(t)$, which certainly exists with the assumptions made[53].

Proof. 1) In the first place,

$$\lim_{x \to +\infty} \int_0^x \frac{\sin t}{t} \, dt = \int_0^\infty \frac{\sin t}{t} \, dt = \frac{\pi}{2}.$$

The existence of this limit, i. e. the convergence of the improper integral, follows simply from the fact that, given $\varepsilon > 0$, and any two values x' and x'' both $> \frac{3}{\varepsilon}$, we have (by § 19, theorem 26)

$$\int_{x'}^{x''} \frac{\sin t}{t} \, dt = \left[-\frac{\cos t}{t} \right]_{x'}^{x''} - \int_{x'}^{x''} \frac{\cos t}{t^2} \, dt,$$

hence

$$\left| \int_{x'}^{x''} \frac{\sin t}{t} \, dt \right| \leqq \frac{1}{x'} + \frac{1}{x''} + \int_{x'}^{x''} \frac{dt}{t^2} < 3 \cdot \frac{\varepsilon}{3} = \varepsilon.$$

---

[52] There is no simplification in observing that it would suffice for k to tend to $+\infty$ through *odd* integral values.

[53] In fact, as $\varphi(t)$ is integrable, it is certainly bounded, and by hypothesis it is monotone in $0 < t < \delta_1$. — Furthermore φ_0 need *not* $= \varphi(0)$.

Now, as we saw on p. 360, equation (b), the integrals

$$i_n = \int_0^{\frac{\pi}{2}} \frac{\sin(2n+1)t}{\sin t}\,dt$$

for $n = 0, 1, 2, \ldots$, are all $= \frac{\pi}{2}$. Therefore we also have $i_n \to \frac{\pi}{2}$. On the other hand, the numbers

$$i_n' = \int_0^{\frac{\pi}{2}} \left(\frac{1}{\sin t} - \frac{1}{t}\right) \sin(2n+1)t \cdot dt$$

(cf. the developments on p. 364) form a null sequence, by theorem 6. Accordingly we also have

$$i_n'' = i_n - i_n' = \int_0^{\frac{\pi}{2}} \frac{\sin(2n+1)t}{t}\,dt \to \frac{\pi}{2}.$$

Since, however (v. § 19, theorem 25),

$$i_n'' = \int_0^{(2n+1)\frac{\pi}{2}} \frac{\sin t}{t}\,dt,$$

this implies that the above-named limit has the value $\frac{\pi}{2}$.

2) By 1), a constant K' exists such that

$$\left|\int_0^x \frac{\sin t}{t}\,dt\right| \leqq K'$$

for every $x \geqq 0$, and therefore a constant $K (= 2K')$ exists such that

$$\left|\int_a^b \frac{\sin t}{t}\,dt\right| \leqq K$$

for every a, b such that $0 \leqq a \leqq b$.

3) Suppose ε given > 0 and choose a positive $\delta' \leqq \delta_1$, so that

$$|\varphi(\delta') - \varphi_0| < \frac{\varepsilon}{3K}.$$

Writing

$$\frac{2}{\pi}\int_0^{\delta'} \varphi(t)\frac{\sin kt}{t}\,dt = J_k',$$

we then have $J_k - J_k'$ tending to 0 as $k \to +\infty$, by theorem 6, and we can accordingly choose k' so large that $|J_k - J_k'| < \frac{\varepsilon}{3}$ for every $k > k'$. Further,

$$\text{(d)}\quad J_k' = \frac{2}{\pi}\int_0^{\delta'} [\varphi(t) - \varphi_0]\cdot\frac{\sin k t}{t}\,dt + \frac{2}{\pi}\varphi_0\cdot\int_0^{\delta'} \frac{\sin k t}{t}\,dt = J_k'' + J_k'''.$$

For the second of these two quantities, we have

$$J_k''' = \frac{2}{\pi}\varphi_0\cdot\int_0^{k\delta'}\frac{\sin t}{t} \to \frac{2}{\pi}\varphi_0\cdot\int_0^{\infty}\frac{\sin t}{t}\,dt = \varphi_0$$

and we may accordingly choose $k_0 > k'$ so large that

$$|J_k''' - \varphi_0| < \frac{\varepsilon}{3}$$

for every $k > k_0$. For J_k'', the first of the two quantities on the right of (d), we use the second mean value theorem of the integral calculus § 19, theorem 27), which gives, for a suitable non-negative $\delta'' \leqq \delta'$,

$$J_k'' = \frac{2}{\pi}\int_0^{\delta'} [\varphi(t) - \varphi_0]\cdot\frac{\sin k t}{t}\,dt = \frac{2}{\pi}[\varphi(\delta') - \varphi_0]\cdot\int_{\delta''}^{\delta'}\frac{\sin k t}{t}\,dt.$$

The latter integral $= \int_{k\delta''}^{k\delta'}\frac{\sin t}{t}\,dt$ and therefore remains $< K$ in absolute value, by 2). Accordingly

$$|J_k''| \leqq \frac{2}{\pi}\cdot\frac{\varepsilon}{3K}\cdot K < \frac{\varepsilon}{3}.$$

Combining the three results of this paragraph, by means of

$$J_k = (J_k - J_k') + J_k'' + J_k''',$$

we see that, given $\varepsilon > 0$, we can choose k_0 so that, for every $k > k_0$,

$$|J_k - \varphi_0| \leqq |J_k - J_k'| + |J_k''| + |J_k''' - \varphi_0| \leqq 3\cdot\frac{\varepsilon}{3} = \varepsilon.$$

Thereby the statement is completely established.

2. *Dini*'s rule. *If* $\lim\limits_{t\to+0}\varphi(t) = \varphi_0$ *exists, and if for every positive* $\tau < \delta$, *the integrals*

$$\int_\tau^\delta \frac{|\varphi(t) - \varphi_0|}{t}\,dt$$

remain less than a fixed positive number[54], *then* $\lim\limits_{k\to+\infty} J_k$ *exists and* $= \varphi_0$.

---

[54] More shortly: If the integral $\int_0^\delta \frac{|\varphi(t) - \varphi_0|}{t}\,dt$, which is improper at 0, has a meaning.

Proof. When τ decreases to 0, the above integral increases monotonely but remains bounded; it therefore tends to a definite limit as $\tau \to 0$, which we denote for brevity by

$$\int_0^\delta \frac{|\varphi(t) - \varphi_0|}{t} \, dt.$$

Given $\varepsilon > 0$, we may choose a *positive* $\delta' < \delta$ so small that

$$\int_0^{\delta'} \frac{|\varphi(t) - \varphi_0|}{t} \, dt < \frac{\varepsilon}{3}.$$

Writing, as in the previous proof,

$$J_k' = \frac{2}{\pi} \int_0^{\delta'} \varphi(t) \frac{\sin k t}{t} \, dt \quad \text{and} \quad J_k' = J_k'' + J_k''',$$

the difference $(J_k - J_k')$ tends to 0, by theorem 6, and we may choose k' so large that $|J_k - J_k'| < \frac{\varepsilon}{3}$ for every $k > k'$. Further, as we saw before, with a suitable choice of $k_0 > k'$ we also have

$$|J_k''' - \varphi_0| = \left| \frac{2}{\pi} \varphi_0 \cdot \int_0^{\delta'} \frac{\sin k t}{t} \, dt - \varphi_0 \right| < \frac{\varepsilon}{3}$$

for every $k > k_0$. Finally,

$$|J_k''| = \left| \frac{2}{\pi} \int_0^{\delta'} [\varphi(t) - \varphi_0] \cdot \frac{\sin k t}{t} \, dt \right| < \int_0^{\delta'} \frac{|\varphi(t) - \varphi_0|}{t} \, dt,$$

i. e. when δ' is suitably chosen, $|J_k''|$ also remains $< \frac{\varepsilon}{3}$. Thus, precisely as before, we conclude that, for every $k > k_0$,

$$|J_k - \varphi_0| < \varepsilon,$$

which proves the validity of *Dini*'s rule.

We may easily deduce from it the two following conditions.

3. *Lipschitz*'s rule. *If two positive numbers A and α exist, such that*[55]

$$|\varphi(t) - \varphi_0| < A \cdot t^\alpha$$

for every t in $0 < t \leqq \delta$, then $J_k \to \varphi_0$.

Proof.

$$\int_\tau^\delta \frac{|\varphi(t) - \varphi_0|}{t} \, dt < A \int_\tau^\delta t^{\alpha - 1} \, dt < A \cdot \frac{\delta^\alpha}{\alpha},$$

---

[55] The "*Lipschitz*-condition", $|\varphi(t) - \varphi_0| < A \cdot t^\alpha$ as $t \to 0$, itself implies that $\lim_{t \to +0} \varphi(t) = \varphi_0$ exists.

so that for every positive $\tau < \delta$ the former integral remains less than a fixed number and in consequence of *Dini*'s rule $J_k \to \varphi_0$, as required.

4th rule. *If* $\varphi'(0)$ *exists*[56] *and therefore* $\lim\limits_{t\to+0} \varphi(t) = \varphi_0 = \varphi(0)$ *exists, then* $J_k \to \varphi_0$.

Proof. The existence of

$$\lim_{t\to+0} \frac{\varphi(t)-\varphi(0)}{t}$$

implies the boundedness of this ratio in an interval of the form $0 < t < \delta_1$, i. e. the fulfilment of a *Lipschitz*-condition with $\alpha = 1$. Hence $J_k \to \varphi_0$, as asserted.

The following corollary to these conditions is immediately obtained:

Corollary. *If* $\varphi(t)$ *can be split up into the sum of two or more functions, each of which satisfies the conditions of one of the four rules above, then* $\lim\limits_{t\to+0} \varphi(t) = \varphi_0$ *again exists, and the Dirichlet integrals* J_k *of the function* $\varphi(t)$ *tend to* φ_0.

The above rules may at once be transferred to the *Fourier* series of an integrable function $f(x)$, which we assume from the first to be given in $0 \leqq x < 2\pi$ and to be extended to all other real values of x by the equation

$$f(x \pm 2\pi) = f(x).$$

In order that the *Fourier* series generated by $f(x)$ should converge to a sum $s(x_0)$ at the point x_0, the integrals

$$J_n = \frac{2}{\pi}\int_0^\delta \varphi(t; x_0) \frac{\sin(2n+1)t}{t} dt$$

must, by theorem 9 (**205**), form a null sequence, where, as before,

$$\varphi(t; x_0) = \frac{1}{2}[f(x_0 + 2t) + f(x_0 - 2t)] - s(x_0).$$

This form of the criterion shows, over and above *Riemann*'s theorem **204**, that *neither* the behaviour of $f(x)$ immediately to the right of x_0, *nor* that immediately to the left of x_0, have *in themselves* any influence whatever on the behaviour of the *Fourier* series of $f(x)$ at x_0. *What is important is that the behaviour of* $f(x)$ *to the right of* x_0 *should stand in a certain relation to that on the left of* x_0, *namely, such that the function*

$$\varphi(t) = \varphi(t; x_0) = \frac{1}{2}[f(x_0 + 2t) + f(x_0 - 2t)] - s(x_0)$$

---

[56] It suffices that $\varphi'(0)$ should exist as the *right hand* differential coefficient (v. § 19, Def. 10), as in fact the possible values of $\varphi(t)$ for $t \leqq 0$ do not come into account.

should possess the necessary and sufficient properties[57] *for the existence of the limit of Dirichlet's integrals* J_k (**206**) *relative to* $\varphi(t)$.

It is not known what these properties are. The four conditions given above for the convergence of *Dirichlet*'s *integrals* furnish us, however, with the same number of *sufficient* conditions for the convergence, at a special point x_0, of the *Fourier series* of a function $f(x)$. Each of these conditions requires, in the first instance, that the function

$$\varphi(t) = \varphi(t; x_0) = \frac{1}{2}[f(x_0 + 2t) + f(x_0 - 2t)] - s(x_0)$$

should tend to a limit φ_0. A *common assumption* for all the rules which we are about to set up is accordingly the following: *The limit*

$$\text{(g)} \qquad \lim_{t \to +0} \frac{1}{2}[f(x_0 + 2t) - f(x_0 - 2t)]$$

must exist. The value of this limit, by theorem 2, will then also be the sum of the *Fourier* series of $f(x)$ at x_0, if the latter converges. This convergence is ensured if the function

$$\varphi(t) = \varphi(t; x_0) = \frac{1}{2}[f(x_0 + 2t) + f(x_0 - 2t)] - s(x_0),$$

considered as a function of t, fulfils one of the four conditions given above. At the same time, the value φ_0 in those conditions must, by theorem 2a, be 0. We accordingly assume that the two following conditions are satisfied:

207. **1st assumption.** *The function $f(x)$ is defined and integrable (hence bounded) in the interval $0 \leqq x < 2\pi$ and its definition is extended to all real values of x by means of the relation*

$$f(x) = f(x + 2k\pi), \qquad k = \pm 1, \pm 2, \ldots$$

2nd assumption. *The limit*

$$\lim_{t \to +0} \frac{1}{2}[f(x_0 + 2t) + f(x_0 - 2t)],$$

where x_0 denotes an arbitrary real number, but is kept fixed throughout, exists[58], *and its value is denoted by $s(x_0)$, so that the function*

$$\varphi(t) = \varphi(t; x_0) = \frac{1}{2}[f(x_0 + 2t) + f(x_0 - 2t)] - s(x_0)$$

has a right hand limit $\lim_{t \to +0} \varphi(t) = 0$.

With these joint assumptions, we have the following four criteria for the convergence of the *Fourier* series of $f(x)$ at the point x_0:

---

[57] Define e. g. $f(x)$ as *entirely arbitrary* to the right of x_0 (but integrable in an interval of the form $x_0 < x < x_0 + \delta$) and, in $x_0 - \delta < x < x_0$, let $f(x) = 1 - f(2x_0 - x)$ say. The *Fourier* series of $f(x)$ at x_0 is convergent with the sum $\frac{1}{2}$. (Proof, for instance, by means of *Dirichlet*'s rule **208**, 1 below.)

[58] The *two-sided* limit then necessarily also exists.

1. *Dirichlet*'s rule. *If $\varphi(t)$ is monotone in an interval of the form $0 < t < \delta_1$, the Fourier series of $f(x)$ converges at x_0 and its sum*[59] *is equal to $s(x_0)$.* **208.**

2. *Dini*'s rule. *If for a fixed (otherwise arbitrary) positive number δ the integrals*

$$\int_{\tau}^{\delta} \frac{|\varphi(t)|}{t}\,dt$$

remain less than a fixed number for every τ such that $0 < \tau < \delta$, the Fourier series of $f(x)$ converges at x_0 and its sum is $s(x_0)$.

3. *Lipschitz*'s rule. *The same is true, if instead of requiring that the integrals should be bounded, we stipulate that two positive numbers A and α should exist, such that, for every t such that $0 < t < \delta$,*

$$|\varphi(t)| < A \cdot t^{\alpha}.$$

4th rule. *The same is true, if instead of the Lipschitz-condition we require that $\varphi(t)$ should possess a right hand differential coefficient at 0.*

The application of these rules is made considerably easier by the following corollaries:

Corollary 1. The function $f(x)$ also fulfils the assumptions 1 and 2 and its *Fourier* series converges at x_0 to the sum $s(x_0)$, if $f(x)$ can be split up into the sum of two or any fixed number of functions, each of which satisfies these two joint assumptions (for a suitable s) and in some neighbourhood of x_0 fulfils the conditions of one of the above rules.

Corollary 2. Similarly, it suffices to stipulate in place of assumption 2 that each of the two (one-sided) limits

$$\lim_{t\to+0} f(x_0 + 2t) = f(x_0 + 0) \quad \text{and} \quad \lim_{t\to+0} f(x_0 - 2t) = f(x_0 - 0)$$

should exist, and that the two functions

$$\varphi_1(t) = f(x_0 + 2t) - f(x_0 + 0) \quad \text{and} \quad \varphi_2(t) = f(x_0 - 2t) - f(x_0 - 0)$$

should each, individually, satisfy the conditions of one of the four rules. The *Fourier* series of $f(x)$ is then convergent at x_0 and has the sum $s(x_0) = \frac{1}{2}[f(x_0 + 0) + f(x_0 - 0)]$.

One or two special cases, which, however, are of particular importance in applications, may be mentioned in the following further corollaries:

---

[59] In case it converges at x_0, the *Fourier* series of a function $f(x)$ satisfying the assumptions **207** accordingly has the sum $f(x_0)$ if, and only if, the limit $s(x_0)$, whose existence is stipulated in the second assumption, $= f(x_0)$. Similarly in the case of the following rules.

Corollary 3. If $f(x)$ satisfies the first assumption and is monotone both to the right and to the left of x_0, the limits mentioned in the preceding corollary exist, and the *Fourier* series of $f(x)$ converges at x_0 to the sum $s(x_0) = \frac{1}{2}[f(x_0+0) + f(x_0-0)]$. — Hence, still more particularly:

Corollary 4. The *Fourier* series of a function $f(x)$ which satisfies the first assumption will converge at the point x_0 and its sum will be the value $f(x)$ of the function at that point, if $f(x)$ is *continuous* at x_0 and monotone on either side of x_0.

Corollary 5. If $f(x)$ satisfies the first assumption, and the two limits $f(x_0 \pm 0)$ exist; if, further, both the (one-sided) limits

$$\lim_{h \to +0} \frac{f(x_0+h) - f(x_0+0)}{h} \quad \text{and} \quad \lim_{h \to +0} \frac{f(x_0-h) - f(x_0-0)}{h}$$

exist; then the *Fourier* series of $f(x)$ will converge at x_0 and will have the sum $s(x_0) = \frac{1}{2}[f(x_0+0) + f(x_0-0)]$.

Corollary 6. The *Fourier* series of a function $f(x)$ which satisfies the first assumption will converge, and will have as its sum the value of the function, at any point x_0 at which $f(x)$ is differentiable.

§ 50. Applications of the theory of *Fourier* series.

As we see from the rules of convergence developed above, extremely general classes of functions are represented by their *Fourier* series. This we propose to illustrate by a number of examples.

The function $f(x)$ to be expanded must always be given in the interval $0 \leqq x < 2\pi$ and must possess the period 2π: $f(x \pm 2\pi) = f(x)$. The corresponding *Fourier* series is then, in general, obtained in the form

$$\frac{1}{2} a_0 + \sum_{n=1}^{\infty} (a_n \cos nx + b_n \sin nx).$$

In particular cases, the sine- or cosine-terms may be absent. In fact, if $f(x)$ is an *even* function,

$$f(-x) = f(2\pi - x) = f(x),$$

(the graph of $f(x)$ is symmetrical with respect to *the straight lines* $x = k\pi$, $(k = 0, \pm 1, \pm 2, \ldots)$, and therefore

$$\pi \cdot b_n = \int_0^{2\pi} f(x) \sin nx \, dx = \int_0^{\pi} + \int_{\pi}^{2\pi} = 0,$$

as is evident if we replace x by $2\pi - x$ in the second of these two partial integrals. The *Fourier* series of $f(x)$ thus reduces to a pure cosine-series. If, on the other hand, $f(x)$ is an *odd* function,

$$f(-x) = f(2\pi - x) = -f(x),$$

(the graph of $f(x)$ is symmetrical with respect to *the points* $x = k\pi$, $k = 0, \pm 1, \pm 2, \ldots$), and therefore

$$\pi \cdot a_n = \int_0^{2\pi} f(x) \cos n x \, dx = 0,$$

as is equally evident. Thus here the *Fourier* series of $f(x)$ reduces to a pure sine-series.

There are accordingly three different ways in which an arbitrary given function $F(x)$, which is defined and integrable in $a \leqq x \leqq b$, may be prepared for the generation of a *Fourier* series.

1st method. If $b - a \geqq 2\pi$, a portion of length 2π is cut out of the interval (a, b), say $\alpha \leqq x < \alpha + 2\pi$, and the origin is transferred to the point α; we thus obtain a function $f(x)$ defined in $0 \leqq x < 2\pi$. It is then defined for the whole x-axis [60] by means of the condition of periodicity $f(x \pm 2\pi) = f(x)$. If $b - a < 2\pi$, define $f(x)$ to be constant $= F(b)$ in $b \leqq x < a + 2\pi$ and proceed as before [61].

2nd method. Precisely as above, define a function $f(x)$ in $0 \leqq x \leqq \pi$ (not 2π) by means of $F(x)$, put $f(x) = f(2\pi - x)$ in $\pi \leqq x \leqq 2\pi$, and then define $f(x)$ for all further x's by the condition of periodicity.

3rd method. Define $f(x)$ as above for $0 < x < \pi$, put $f(0) = f(\pi) = 0$, but put $f(x) = -f(2\pi - x)$ in $\pi < x < 2\pi$; then again define $f(x)$ for all further x's by the condition of periodicity.

The three functions which are obtained by these methods from a given function $F(x)$, and which are now suitable for the generation of a *Fourier* series, we shall distinguish as $f_1(x)$, $f_2(x)$, $f_3(x)$. Whereas $f_2(x)$ will certainly give a pure cosine-series and $f_3(x)$ a pure sine-series, $f_1(x)$ will lead, as a rule, to a *Fourier* series of the general form (unless, in fact, $f_1(x)$ is itself already an odd or an even function).

Since our rules of convergence enable us to recognize the convergence only at points x_0 for which

$$\lim_{t \to +0} \frac{1}{2} [f(x_0 + 2t) + f(x_0 - 2t)]$$

exists, it will be advisable to modify our functions further at the

---

[60] If $b - a > 2\pi$, a portion of the curve $y = F(x)$ is left out of the representation altogether. If we wish to avoid this, we need only alter the unit of measurement on the x-axis so that the interval of definition of $F(x)$ has the length 2π; i. e. we substitute $a + \dfrac{b-a}{2\pi} x$ for x.

[61] Or else give the interval of definition of $F(x)$ the exact length 2π by modifying the unit of measurement on the x-axis.

junctions $2k\pi$ by writing

$$f(0) = f(2k\pi) = \lim_{x \to +0} \frac{1}{2}[f(x) + f(2\pi - x)]$$

whenever this limit exists. (This is certainly the case for $f_3(x)$, and provides the condition $f_3(0) = f_3(2k\pi) = 0$.) If this limit does not exist, the functional value $f(2k\pi)$ does not come into account, as with our resources we cannot discover whether the *Fourier* series converges there or not. — For corresponding reasons we have already put $f_3(\pi) = 0$ above.

We now go on to concrete examples.

209. 1. Example. $F(x) \equiv a \neq 0$. Here

$f_1(x) \equiv f_2(x) \equiv a$, while we have to put

$$f_3(x) = \begin{cases} 0 & \text{for } x = 0 \text{ and } x = \pi, \\ a & \text{,, } 0 < x < \pi, \\ -a & \text{,, } \pi < x < 2\pi. \end{cases}$$

Dirichlet's conditions are evidently fulfilled at *every* point (inclusive of the junctions), for each of the three functions. The expansions obtained must accordingly converge everywhere and must represent the functions themselves. For $f_1(x)$ and $f_2(x)$, however, they are trivial, as they reduce to the constant term $\frac{1}{2}a_0 = a$. For $f_3(x)$, however, we obtain:

$$b_n = \frac{1}{\pi}\int_0^{2\pi} f_3(x)\sin nx\,dx = \frac{a}{\pi}\int_0^{\pi}\sin nx\,dx - \frac{a}{\pi}\int_{\pi}^{2\pi}\sin nx\,dx = \frac{2a}{\pi}\int_0^{\pi}\sin nx\,dx,$$

i. e.

$$b_n = \begin{cases} 0 & \text{for even values of } n, \\ \dfrac{4a}{n\pi} & \text{for odd values of } n. \end{cases}$$

The expansion accordingly is

$$f_3(x) = \frac{4a}{\pi}\left[\sin x + \frac{\sin 3x}{3} + \frac{\sin 5x}{5} + \cdots\right]$$

or

$$\sin x + \frac{\sin 3x}{3} + \frac{\sin 5x}{5} + \cdots = \begin{cases} +\frac{\pi}{4} & \text{in } 0 < x < \pi, \\ 0 & \text{at } 0 \text{ and at } \pi, \\ -\frac{\pi}{4} & \text{in } \pi < x < 2\pi. \end{cases}$$

This establishes the second of the examples given on p. 351, and provides the sum of this curious series, of whose convergence we were already aware (v. **185**, 5)[62]. For $x = \frac{\pi}{2}, \frac{\pi}{6}, \frac{\pi}{3}$, we obtain special series, with the first of which we are familiar in an entirely different connection (v. **122**):

$$1 - \frac{1}{3} + \frac{1}{5} - \frac{1}{7} + - \cdots = \frac{\pi}{4},$$

$$1 + \frac{1}{5} - \frac{1}{7} - \frac{1}{11} + \frac{1}{13} + \frac{1}{17} - - + + \cdots = \frac{\pi}{3},$$

$$1 - \frac{1}{5} + \frac{1}{7} - \frac{1}{11} + \frac{1}{13} - + \cdots = \frac{\pi}{2\sqrt{3}}.$$

2. Example. $F(x) = a\,x$, $(a \neq 0)$. Here

$$f_1(x) = \begin{cases} a\,x & \text{in} \quad 0 < x < 2\pi, \\ a\,\pi & \text{at} \quad 0 \text{ and at } 2\pi, \end{cases}$$

$$f_2(x) = \begin{cases} a\,x & \text{in} \quad 0 \leqq x \leqq \pi, \\ a(2\pi - x) & \text{in} \quad \pi \leqq x \leqq 2\pi, \end{cases}$$

$$f_3(x) = \begin{cases} a\,x & \text{in} \quad 0 < x < \pi, \\ 0 & \text{at} \quad 0 \text{ and at } \pi, \\ -a(2\pi - x) & \text{in} \quad \pi < x < 2\pi. \end{cases}$$

After an easy calculation, the expansion of $f_1(x)$ gives:

$$\text{(a)} \quad \sin x + \frac{\sin 2x}{2} + \frac{\sin 3x}{3} + \frac{\sin 4x}{4} + \cdots \qquad \textbf{210.}$$

$$= \begin{cases} \dfrac{\pi - x}{2} & \text{in } 0 < x < 2\pi, \\ 0 & \text{at } 0 \text{ and at } 2\pi, \end{cases}$$

which establishes the first of the examples of p. 351. Similarly, the expansion of $f_3(x)$ gives

$$\text{(b)} \quad \sin x - \frac{\sin 2x}{2} + \frac{\sin 3x}{3} - \frac{\sin 4x}{4} + - \cdots$$

$$= \begin{cases} \dfrac{1}{2}x & \text{in } 0 \leqq x < \pi, \\ 0 & \text{at } \pi, \\ \dfrac{1}{2}x - \pi & \text{in } \pi < x \leqq 2\pi \end{cases}$$ [60].

Or, more shortly,
$$= \begin{cases} \dfrac{x}{2} & \text{in } -\pi < x < \pi, \\ 0 & \text{at } \pm\pi. \end{cases}$$

---

[62] This and the following examples are already found, for the most part, in *Euler*'s writings. Many others have been given by *Fourier*, *Legendre*, *Cauchy*, *Frullani*, *Dirichlet* and others. They are collected together, in a convenient form for reference, in *H. Burkhardt*, Trigonometrische Reihen und Integrale bis etwa 1850, Enzyklopädie d. math. Wiss., Vol. II A, pp. 902—920.

The function $f_2(x)$, however, provides the expansion:

(c) $$\frac{\cos x}{1^2}+\frac{\cos 3x}{3^2}+\frac{\cos 5x}{5^2}+\cdots=\begin{cases}\dfrac{\pi^2}{8}-\dfrac{\pi x}{4} & \text{in } 0\leqq x\leqq \pi,\\[2mm] \dfrac{\pi x}{4}-\dfrac{3\pi^2}{8} & \text{in } \pi\leqq x\leqq 2\pi.\end{cases}$$

The first of these expansions gives for $x=\frac{\pi}{2}$ the known series for $\frac{\pi}{4}$; the third, for $x=0$, gives the series, also previously known to us (**137**),

$$1+\frac{1}{3^2}+\frac{1}{5^2}+\frac{1}{7^2}+\cdots=\frac{\pi^2}{8},$$

from which we may immediately deduce the relation

$$1+\frac{1}{2^2}+\frac{1}{3^2}+\frac{1}{4^2}+\cdots=\frac{\pi^2}{6},$$

previously established (**136**, **156** and **189**) in an entirely different way[63]. — On comparing the two results, we obtain the remarkable fact that in $0<x\leqq\pi$ the function x is capable of the *two Fourier* expansions

211. $$x=\begin{cases}\pi-2\left[\dfrac{\sin x}{1}+\dfrac{\sin 2x}{2}+\dfrac{\sin 3x}{3}+\cdots\right] & \text{and}\\[2mm] \dfrac{\pi}{2}-\dfrac{4}{\pi}\left[\dfrac{\cos x}{1^2}+\dfrac{\cos 3x}{3^2}+\dfrac{\cos 5x}{5^2}+\cdots\right].\end{cases}$$

With a view to penetrating still further into the significance of these results, it is well to sketch the graphs of the function $f(x)$ and a few of the corresponding curves of approximation. This we must leave to the reader, and we shall only draw attention to the following phenomenon:

The convergence of the series **210** c is uniform for all x's; not so that of the series **210** a and b, since their sums are discontinuous, the

[63] A fifth proof, quite different again, is as follows: The expansion **123** is uniformly convergent in $0\leqq x\leqq 1$, by the stipulations made in **123**, together with **199**, 2. Putting $x=\sin t$, we see that the expansion

$$t=\sum_{n=0}^{\infty}(-1)^n\binom{-\frac12}{n}\frac{\sin^{2n+1}t}{2n+1}$$

is uniformly convergent in $0\leqq t\leqq\frac{\pi}{2}$ and may therefore be integrated term-by-term over that interval. Now

$$\int_0^{\pi/2}\sin^{2n+1}t\,dt=\frac{2\cdot4\ldots(2n)}{3\cdot5\ldots(2n+1)}=\frac{1}{(-1)^n(2n+1)\binom{-\frac12}{n}};$$

this is shewn by a recurrence process, or by writing $\cos t=z$ and using Example 117 b. Hence at once

$$\frac{\pi^2}{8}=\sum_{n=0}^{\infty}\frac{1}{(2n+1)^2}.$$

This method was essentially given by *Euler*. (Cf. the note referred to in the footnote 38 to **156**.)

first at 0, the second at π. In the former case, the approximation curves lie close to the zigzag line representing the limiting curve along the whole of its length, whilst in (a) and (b) the corresponding state of affairs does not and cannot occur (cf. **216**, 4).

3. Example. $F(x) = \cos \alpha x$ (α arbitrary[64], but $\neq 0, \pm 1, \pm 2, \ldots$).

a) We first form the function $f_2(x)$, and accordingly define

$$f_2(x) = \begin{cases} \cos \alpha x & \text{in} \quad 0 \leqq x \leqq \pi \\ \cos \alpha (2\pi - x) & \text{in} \quad \pi \leqq x \leqq 2\pi; \end{cases}$$

thus $f_2(x)$ is a function continuous everywhere, which by *Dirichlet*'s rule will also generate a *Fourier* series continuous everywhere, which represents the function, and is necessarily a pure cosine-series. Here we have

$$\pi a_n = 2 \int_0^\pi \cos \alpha x \cos n x \, d x = \int_0^\pi [\cos(\alpha + n) x + \cos(\alpha - n) x] \, d x;$$

hence, as α was assumed not to be an integer,

$$\pi a_n = (-1)^n \frac{2 \alpha \sin \alpha \pi}{\alpha^2 - n^2}.$$

Therefore the function $f_2(x)$ in $0 \leqq x \leqq 2\pi$, or in other words *the function* $\cos \alpha x$ *in* $-\pi \leqq x \leqq +\pi$, is represented by the series:

$$\cos \alpha x = \frac{\sin \alpha \pi}{\pi} \left[\frac{1}{\alpha} - \frac{2\alpha}{\alpha^2 - 1^2} \cos x + \frac{2\alpha}{\alpha^2 - 2^2} \cos 2x - + \ldots \right]. \qquad \mathbf{212.}$$

For $x = \pi$, we obtain from this the expansion **117**, previously deduced from entirely different sources:

$$\pi \frac{\cos \alpha \pi}{\sin \alpha \pi} = \pi \cot \alpha \pi = \frac{1}{\alpha} + \frac{2\alpha}{\alpha^2 - 1^2} + \frac{2\alpha}{\alpha^2 - 2^2} + \cdots.$$

We thus enter the sphere of the developments of § 24. Of course the other series expansions there deduced may also be obtained directly from our new source. Thus **212** gives for $x = 0$

$$\frac{\pi}{\sin \alpha \pi} = \frac{1}{\alpha} - \frac{2\alpha}{\alpha^2 - 1^2} + \frac{2\alpha}{\alpha^2 - 2^2} - \frac{2\alpha}{\alpha^2 - 3^2} + - \cdots.$$

Subtracting the cotangent expansion obtained just before, we further obtain

$$\pi \frac{1 - \cos \alpha \pi}{\sin \alpha \pi} = \pi \tan \frac{\alpha \pi}{2} = - \frac{4\alpha}{\alpha^2 - 1^2} - \frac{4\alpha}{\alpha^2 - 3^2} - \frac{4\alpha}{\alpha^2 - 5^2} - \cdots;$$

and so on.

b) If we now similarly construct an odd function $f_3(x)$ from $F(x) = \cos \alpha x$, we have

$$f_3(x) = \begin{cases} \cos \alpha x & \text{in} \quad 0 < x < \pi, \\ 0 & \text{at} \quad 0 \text{ and at } \pi, \\ -\cos \alpha (2\pi - x) & \text{in} \quad \pi < x < 2\pi. \end{cases}$$

---

[64] Because otherwise the cosine-expansion would become trivial.

Here α may also assume integral values without reducing the result to a trivial one. The coefficients b_n are obtained from integrals whose value is easily worked out, and they lead to the following expansions, valid in $0 < x < \pi$:

213. α) for $\alpha \neq 0, \pm 1, \pm 2, \ldots$

$$\cos \alpha x = \frac{1+\cos \alpha \pi}{\pi}\left[\frac{2}{1^2-\alpha^2}\sin x + \frac{6}{3^2-\alpha^2}\sin 3x + \frac{10}{5^2-\alpha^2}\sin 5x + \cdots\right]$$
$$+ \frac{1-\cos \alpha \pi}{\pi}\left[\frac{4}{2^2-\alpha^2}\sin 2x + \frac{8}{4^2-\alpha^2}\sin 4x + \cdots\right];$$

β) for $\alpha = \pm p =$ integer

$$\cos p x = \begin{cases} \frac{4}{\pi}\left[\frac{1}{1^2-p^2}\sin x + \frac{3}{3^2-p^2}\sin 3x + \cdots\right], & \text{if } p \text{ is } \textit{even}, \\ \frac{4}{\pi}\left[\frac{2}{2^2-p^2}\sin 2x + \frac{4}{4^2-p^2}\sin 4x + \cdots\right], & \text{if } p \text{ is } \textit{odd}. \end{cases}$$

From all the above series, innumerable numerical series may be deduced by taking particular values of x and α.

4. The treatment of $F(x) = \sin \alpha x$ leads to quite similar expansions.

5. If the function $F(x) = -\log\left(2 \sin \frac{x}{2}\right)$ is arranged for the generation of a pure cosine series, we obtain the expansion, valid in $0 < x < \pi$,

214. $$\cos x + \frac{\cos 2x}{2} + \frac{\cos 3x}{3} + \cdots = -\log\left(2 \sin \frac{x}{2}\right).$$

It has, however, to be shewn by a special investigation that the result holds in spite of the fact that the function is unbounded in the neighbourhood of the points 0 and 2π, and therefore is not (properly) integrable. (Cf. § 55, V below, where this will follow quite simply in another way.)

6. Example. $F(x) = e^{\alpha x} + e^{-\alpha x}$, $\alpha \neq 0$, is to be expanded in a cosine series. We have therefore to take

$$f_2(x) = \begin{cases} F(x) & \text{in } 0 \leqq x \leqq \pi \\ F(2\pi - x) & \text{in } \pi \leqq x \leqq 2\pi. \end{cases}$$

After working out the extremely easy integrals giving the coefficients a_n, we obtain

215. $$\frac{\pi}{2}\frac{e^{\alpha x}+e^{-\alpha x}}{e^{\alpha\pi}-e^{-\alpha\pi}} = \frac{1}{2\alpha} - \frac{\alpha}{\alpha^2+1^2}\cos x + \frac{\alpha}{\alpha^2+2^2}\cos 2x - + \cdots,$$

which is valid in $-\pi \leqq x \leqq +\pi$. If we substitute e. g. $x = \pi$ and write t for $2\alpha\pi$ for simplicity, we are led, after a few simple transformations, to the relation, — valid for every $t \neq 0$ —

$$\frac{1}{t}\left[\frac{1}{1-e^{-t}} - \frac{1}{t} - \frac{1}{2}\right] = 2\sum_{n=1}^{\infty}\frac{1}{(2n\pi)^2+t^2},$$

i. e. to an "expansion in partial fractions" of this remarkable function; its expansion in power series we can at once deduce from § 24, 4, where the function $\frac{t}{2}\,\frac{e^t+1}{e^t-1}$ was considered, — for our function reduces to the latter by multiplying by t^2 and adding 1.

Various remarks.

The very fact that trigonometrical series are capable of representing extremely general types of functions renders the question as to the limits of this capacity doubly interesting. As was already remarked, *necessary and sufficient* conditions for a function to be representable by its *Fourier* series are not known. On the contrary, we find ourselves obliged to consider this as a fundamental property of functions, new of its kind, for all attempts to build it up directly by means of the other fundamental properties (continuity, differentiability, integrability, etc.) have so far failed. We must deny ourselves the satisfaction of supporting this statement in all details by working out relevant examples, but we should nevertheless like to put forward a few of the facts in this connection.

1. One of the conjectures which will naturally be made at first sight is that **216.**
all continuous functions are representable by their *Fourier* series. *This is not the case,* as *du Bois-Reymond* was the first to show by an example (Gött. Nachr. 1873, p. 571) [65].

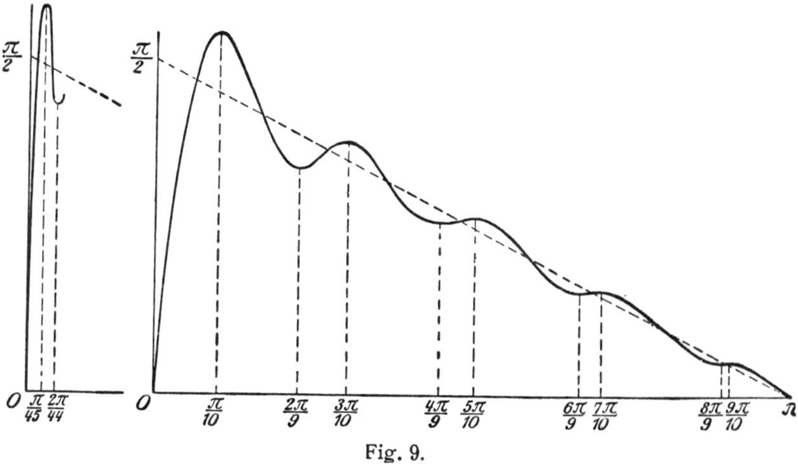

Fig. 9.

2. On the other hand, to assume the function differentiable as well as continuous is more than is necessary, as is shown by *Weierstrass'* [66] example of a uniformly convergent trigonometrical series, viz.

$$\sum_{n=1}^{\infty} a^n \cos(b^n \pi x) \qquad \left(0 < a < 1,\ b \text{ a positive integer, } ab > 1 + \frac{3}{2}\pi\right),$$

which accordingly is the *Fourier* series of its sum (v. **200**, 1 a), but which represents a function that is continuous but nowhere differentiable.

---

[65] We now have simpler examples than that mentioned above. E. g. *L. Fejér* has given a very clear and beautiful example (J. f. d. reine u. angew. Math., Vol. 137, p. 1. 1909).

[66] Abhandlungen zur Funktionenlehre, Werke, Vol. 2, p. 223. (First published 1875.)

3. Whether continuous functions exist whose *Fourier* series are *everywhere* divergent is not at present known.

4. A specially remarkable phenomenon is that known as *Gibbs'* phenomenon [67], which was first discovered (by *J. W. Gibbs*) in connection with the series **210** a: The curves of approximation $y = s_n(x)$ overshoot the mark, so to speak, in the neighbourhood of $x = 0$. More precisely, let us denote by ξ_n the abscissa of the greatest maximum [68] of $y = s_n(x)$ between 0 and π and let η_n be the corresponding ordinate. Then $\xi_n \to 0$; but η_n does *not* $\to \frac{\pi}{2}$, as we should expect, but tends to a value g equal to $\frac{\pi}{2}(1{\cdot}17898\ldots)$. Thus it appears that the limiting configuration to which the curves $y = s_n(x)$ approximate contains, besides the graph of the function **210** a (p. 351, fig. 7), a stretch of the y-axis, between the ordinates $\pm g$, whose length exceeds the "jump" of the function by nearly $\frac{2}{11}$. In fig. 9, the nth approximation curve is drawn for $n = 9$ in the interval $0 \ldots \pi$, and for $n = 44$ the initial portion is given.

§ 51. Products with variable terms.

Given a product of the form

$$\prod_{n=1}^{\infty}(1 + f_n(x)),$$

whose terms are functions of x, we shall define (in complete analogy with the theory of series) as an *interval of convergence of the product*, an interval J at every point of which *all the functions $f_n(x)$ are defined* and *the product itself is convergent.*

Thus e. g. the products

$$\prod_{n=1}^{\infty}\left(1 - \frac{x^2}{n^2}\right),\quad \prod_{n=1}^{\infty}\left(1 + \frac{x^2}{n^2}\right),\quad \prod_{n=1}^{\infty}\left(1 + (-1)^n \frac{x}{n}\right),\quad \prod_{n=2}^{\infty}\left(1 + \frac{x}{n \log^2 n}\right),\ldots$$

are convergent for every real x, and the same is true of any product of the form $\prod(1 + a_n x)$, if Σa_n is either absolutely convergent (v. **127**, theorem 7) or a conditionally convergent series for which Σa_n^2 converges absolutely (**127**, theorem 9).

For every x in J, the product then has a specific value and therefore defines a determinate function $F(x)$ in J. We again say: *the product represents the function $F(x)$ in J*, or: *$F(x)$ is expanded in the given product in J.* The main question is as before: *how far do the fundamental properties (of continuity, differentiability, etc.) belonging to the terms $f_n(x)$ still hold for the function $F(x)$ represented by the product?* Here again the

---

[67] *J. W. Gibbs*, Nature, Vol. 59 (London 1898-99), p. 606. — Cf. also *T. H. Gronwall*, Über die *Gibbs*sche Erscheinung, Math. Annalen, Vol. 72, p. 228, 1912.

[68] The maxima in the interval occur at $x = \frac{\pi}{n+1}, \frac{3\pi}{n+1}, \frac{5\pi}{n+1}, \ldots$, the first being the greatest maximum. The minima occur at $x = \frac{2\pi}{n}, \frac{4\pi}{n}, \ldots$.

answer will be that this is the case in the widest measure, as long as the products considered are *uniformly* convergent.

What the definition of uniform convergence of a product is to be is almost obvious if we refer to the corresponding definition for series, since in either case we are essentially concerned with *sequences of functions* (cf. **190**, 4). However, we shall set down the definition corresponding to the 4th form (**191**, 4) for series:

° Definition [69]. *The product $\Pi(1+f_n(x))$ is said to be uniformly convergent in an interval J, if, given $\varepsilon > 0$,* ***a single*** *number $N = N(\varepsilon)$* ***depending only on*** ε, ***not on*** x, *can be chosen so that* **217.**

$$|(1+f_{n+1}(x))(1+f_{n+2}(x))\ldots(1+f_{n+k}(x))-1| < \varepsilon$$

for every $n > N$, every $k \geqq 1$ and ***every*** [70] *x in J.*

It is not difficult to show that with this definition as basis the theorems of § 47 hold substantially for infinite products [71]. We will, however, leave the details to the student, while we prove a few theorems which are less far-reaching, but which will amply suffice for all our applications, and which have the advantage of providing us at the same time with *criteria* for the uniformity of the convergence of a product. We first have

° Theorem 1. *The product $\Pi(1+f_n(x))$ converges uniformly in J and represents a continuous function in that interval, if the functions $f_n(x)$ are all continuous in J and the series $\Sigma|f_n(x)|$ converges uniformly in J.* **218.**

Proof. If $\Sigma|f_n(x)|$ converges in J, so does the product $\Pi(1+f_n(x))$, by **127**, theorem 7; indeed, it converges absolutely. Let $F(x)$ denote the function it represents. Let us choose m so large that

$$|f_{m+1}(x)|+|f_{m+2}(x)|+\ldots+|f_{m+k}(x)| < 1$$

for ***every*** *x in J and* ***every*** *$k \geqq 1$*; this is possible, by hypothesis. Consider the product

$$\prod_{n=m+1}^{\infty}(1+f_n(x)),$$

[69] The symbol ∘ in this section again holds only with the same restrictions as in §§ 46—48; cf. p. 327, footnote 1.

[70] This definition includes that of convergence. If the latter be assumed, we may speak of the "remainder" $r_n(x) = (1+f_{n+1}(x))(1+f_{n+2}(x))\ldots$ and define uniform convergence as follows: $\Pi(1+f_n(x))$ is said to converge uniformly in J, if for every (x_n) in J, however chosen, $r_n(x)\to 1$.

[71] Writing $\prod_{\nu=1}^{m}(1+f_\nu(x)) = P_m(x)$ and $\prod_{\nu=m+1}^{\infty}(1+f_\nu(x)) = F_m(x)$, we may quite easily deduce e. g. the continuity of $F(x)$ at x_0 from that of the functions $f_\nu(x)$ there, by means of the relation

$$\begin{aligned}F(x)-F(x_0) &= P_m(x)\cdot F_m(x) - P_m(x_0)\cdot F_m(x_0)\\ &= [P_m(x)-P_m(x_0)]\,F_m(x) + [F_m(x)-F_m(x_0)]\,P_m(x_0).\end{aligned}$$

and denote its partial products by $p_n(x)$, $n > m$. Let $F_m(x)$ be the function represented by this product. We have (cf. **190**, 4)

$$F_m = p_{m+1} + (p_{m+2} - p_{m+1}) + \ldots + (p_n - p_{n-1}) + \ldots$$
$$= p_{m+1} + p_{m+1} \cdot f_{m+2} + p_{m+2} \cdot f_{m+3} + \ldots + p_{n-1} \cdot f_n + \ldots,$$

i. e. $F_m(x)$ is also expressible by an infinite series — as is indeed evident from § 30. Now (by **192**, theorem 3) this series *converges uniformly in J*. In fact, for every $n > m$, we have

$$|p_n| \leqq (1+|f_{m+1}|)\cdot(1+|f_{m+2}|)\ldots(1+|f_n|) < e^{|f_{m+1}|+|f_{m+2}|+\cdots} < e < 3,$$

and

$$\sum_{n=m+1}^{\infty} |f_n(x)|$$

is uniformly convergent in J, by hypothesis. Accordingly the sequence of its partial sums, i. e. the *sequence of functions $p_n(x)$*, tends uniformly to F_m in J, so that the product $\prod_{n=m+1}^{\infty}(1+f_n(x))$ is seen to converge uniformly in J, and this property is not affected when we prefix the first m factors.

By **193**, $F_m(x)$ is necessarily continuous in J, since the terms of the series which represents it are all continuous in that interval. The same is then true of the function

$$F(x) = (1 + f_1(x)) \ldots (1 + f_m(x))\, F_m(x), \qquad \text{q. e. d.}$$

A similar proof holds for

° **Theorem 2.** *If the functions $f_n(x)$ are all differentiable in J and if not only $\Sigma|f_n(x)|$, but $\Sigma|f_n'(x)|$ converges uniformly in J, then $F(x)$ is also differentiable in J. Moreover its differential coefficient at every point of J where $F(x) \neq 0$ is given by* [72]

$$\frac{F'(x)}{F(x)} = \sum_{n=1}^{\infty} \frac{f_n'(x)}{1 + f_n(x)}.$$

Proof. The proof may be put in a form analogous to that of the previous theorem; however, in order to make other methods of attack familiar, we will conduct the proof by means of the logarithmic function, as follows. Let us choose m so large that

$$|f_{m+1}(x)| + |f_{m+2}(x)| + \ldots < \frac{1}{2}$$

---

[72] If $g(x)$ is differentiable at a special point x and $g(x) \neq 0$ there, the ratio $\frac{g'(x)}{g(x)}$ is called the *logarithmic differential coefficient* of $g(x)$, because it is $\frac{d}{dx} \log |g(x)|$. For $g(x) = g_1(x) \cdot g_2(x) \ldots g_k(x)$, we have, as is well known,

$$\frac{g'(x)}{g(x)} = \frac{g_1'(x)}{g_1(x)} + \frac{g_2'(x)}{g_2(x)} + \ldots + \frac{g_k'(x)}{g_k(x)},$$

provided that the functions $g_\lambda(x)$ are all differentiable at the point x in question.

for every x in J, so that, in particular, for every $n > m$,

$$|f_n(x)| < \frac{1}{2}.$$

By **127**, theorem 8, the series

$$\sum_{n=m+1}^{\infty} \log(1 + f_n(x))$$

is then absolutely convergent in J. The series obtained from it by differentiating term by term,

$$\sum_{n=m+1}^{\infty} \frac{f_n'(x)}{1 + f_n(x)},$$

is indeed also uniformly (and absolutely) convergent in J. For since $|f_n(x)| < \frac{1}{2}$ for every $n > m$, $|1 + f_n(x)| > \frac{1}{2}$ and therefore $\left|\frac{1}{1 + f_n(x)}\right| < 2$, so that the uniform convergence of the last series follows from that of $\Sigma |f_n'(x)|$. Accordingly (by **196**)

$$\frac{F_m'(x)}{F_m(x)} = \sum_{n=m+1}^{\infty} \frac{f_n'(x)}{1 + f_n(x)},$$

if, as before, we put

$$\prod_{n=m+1}^{\infty} (1 + f_n(x)) = F_m,$$

i. e.

$$\sum_{n=m+1}^{\infty} \log(1 + f_n(x)) = \log F_m(x).$$

Since finally

$$F(x) = (1 + f_1(x)) \ldots (1 + f_m(x)) \cdot F_m(x),$$

and the last factor on the right has been seen to be differentiable in J, $F(x)$ itself is differentiable in J. If, further, $F(x) \neq 0$, the last relation leads at once to the required result, by the rule of differentiation mentioned in the preceding footnote.

Applications.

°1. The product **219.**

$$F_m(x) = \prod_{n=m+1}^{\infty} \left(1 - \frac{x^2}{n^2}\right) \qquad (m > 0)$$

is uniformly convergent in every bounded interval, since, with $f_n(x) = -\frac{x^2}{n^2}$,

$$\Sigma |f_n(x)| = \Sigma \left|\frac{x^2}{n^2}\right| = |x|^2 \cdot \Sigma \frac{1}{n^2}$$

is evidently a uniformly convergent series in that interval. The product accordingly defines a function $F_m(x)$ continuous everywhere, which, in particular, is never zero in $|x| < m + 1$. This function is also differentiable, for $\Sigma |f_n'(x)| = 2|x| \Sigma \frac{1}{n^2}$

is uniformly convergent in every bounded interval. Hence for $|x| < m + 1$

$$\frac{F_m'(x)}{F_m(x)} = \frac{1}{x} + \sum_{n=m+1}^{\infty} \frac{2x}{x^2 - n^2}.$$

By **117**, this however implies

$$\frac{F_m'(x)}{F_m(x)} = \pi \cot \pi x - \frac{1}{x} - \sum_{n=1}^{m} \frac{2x}{x^2 - n^2} = \frac{G_m'(x)}{G_m(x)},$$

where $G_m(x)$ denotes the function

$$\frac{\sin \pi x}{x \prod_{n=1}^{m}\left(1 - \frac{x^2}{n^2}\right)},$$

interpreting this expression as equal, for $x = 0, \pm 1, \ldots, \pm m$, to its limit (obviously existent and $\neq 0$) as x tends to these values. (The corresponding convention is made for the middle term in the relation immediately preceding.) If however two functions $F(x)$ and $G(x)$ have their logarithmic derivatives equal in an interval, in which the two functions never vanish, it follows that they can only differ by a constant factor ($\neq 0$). Hence, in $|x| < m + 1$,

$$\sin \pi x = c \cdot x \cdot \prod_{n=1}^{\infty}\left(1 - \frac{x^2}{n^2}\right),$$

where c is a suitable constant. To determine its value, we need only divide the last relation by x and let $x \to 0$. The left hand side then $\to \pi$, while the right hand side $\to c$, because the product is continuous at $x = 0$. Accordingly $c = \pi$ and we have, first for $|x| < m + 1$, but hence, as m was arbitrary, for all x,

$$\boldsymbol{\sin \pi x = \pi x \cdot \prod_{n=1}^{\infty}\left(1 - \frac{x^2}{n^2}\right)}.$$

This product, and those discussed below in 2 and 4, as well as the remarkable product **257**, 9, and many other fundamental expansions in products, are due to *Euler*.

2. For $\cos \pi x$ we now find, without further calculation,

$$\boldsymbol{\cos \pi x} = \frac{\sin 2\pi x}{2 \sin \pi x} = \frac{2\pi x \cdot \prod\left(1 - \frac{4x^2}{n^2}\right)}{2\pi x \cdot \prod\left(1 - \frac{x^2}{n^2}\right)} = \boldsymbol{\prod_{k=1}^{\infty}\left(1 - \frac{4x^2}{(2k-1)^2}\right)}.$$

3. The sine-product for special values of x leads to important numerical product expansions. E. g. for $x = \frac{1}{2}$,

$$1 = \frac{\pi}{2} \cdot \prod\left(1 - \frac{1}{4n^2}\right) = \frac{\pi}{2} \cdot \prod_{n=1}^{\infty}\left(\frac{(2n-1)(2n+1)}{2n \cdot 2n}\right).$$

As $\frac{2n+1}{2n} \to 1$, we may clearly omit the brackets, and we accordingly write

$$\frac{\pi}{2} = \frac{2 \cdot 2 \cdot 4 \cdot 4 \cdot 6 \cdot 6 \cdot 8 \cdot 8 \ldots}{1 \cdot 3 \cdot 3 \cdot 5 \cdot 5 \cdot 7 \cdot 7 \cdot 9 \ldots}$$

(***Wallis' Product***)[73].

Since it follows from this that

$$\lim_{k \to +\infty} \left(\frac{2}{1}\right)^2 \cdot \left(\frac{4}{3}\right)^2 \cdot \ldots \cdot \left(\frac{2k}{2k-1}\right)^2 \cdot \frac{1}{2k+1} = \frac{\pi}{2}$$

or

$$\frac{2}{1} \cdot \frac{4}{3} \cdot \frac{6}{5} \cdot \ldots \cdot \frac{2k}{2k-1} \cdot \frac{1}{\sqrt{k}} \to \sqrt{\pi}$$

[73] Arithmetica infinitorum, Oxford 1656. (Cf. pp. 218—9, footnote 1.)

we obtain at the same time the remarkable asymptotic relation

$$\frac{1.3.5\ldots(2n-1)}{2.4.6\ldots2n}=\frac{1}{2^{2n}}\binom{2n}{n}=(-1)^n\binom{-\frac{1}{2}}{n}\cong\frac{1}{\sqrt{\pi n}}$$

for the ratio of the middle coefficient in the binomial expansion of the $(2n)^{\text{th}}$ power to the sum 2^{2n} of *all* the coefficients of this expansion. or for the coefficient of x^n in the expansion of $\frac{1}{\sqrt{1-x}}$.

○4. The sequence of functions

$$g_n(x)=\frac{x(x+1)(x+2)\ldots(x+n)}{n!\,n^x}=\frac{1}{n^x}\cdot x\left(1+\frac{x}{1}\right)\left(1+\frac{x}{2}\right)\ldots\left(1+\frac{x}{n}\right),$$

(v. **128** 4) cannot be immediately replaced by a product of the form $\Pi(1+f_n(x))$, as $\Pi\left(1+\frac{x}{n}\right)$ diverges for $x\neq0$. However, this divergence is of such a kind that

$$\left(1+\frac{x}{1}\right)\left(1+\frac{x}{2}\right)\ldots\left(1+\frac{x}{n}\right)\sim e^{x\left(1+\frac{1}{2}+\cdots+\frac{1}{n}\right)}.$$

By **128**, 2 and **42**, 3, this implies that

$$\frac{x\cdot\left(1+\frac{x}{1}\right)\left(1+\frac{x}{2}\right)\ldots\left(1+\frac{x}{n}\right)}{n^x}=g_n(x)$$

tends, as $n\to\infty$, to a specific limit, finite and $\neq0$; — the latter, of course, only if $x\neq0,-1,-2,\ldots$. Accordingly

$$\lim\frac{1}{g_n(x)}=\Gamma(x)$$

is a definite number for every $x\neq0,-1,-2,\ldots$. The function of x so defined is called the ***Gamma-function*** **(Γ-*function*)**. It was introduced into analysis by *Euler* (see above) and, next to the elementary functions, is one of the most important in analysis. Further investigation of its properties lies outside the scope of this book. (Cf., however, pp. 439—440 and p. 530.)

Exercises on Chapter XI.

I. Arbitrary series of variable terms.

154. Let (nx) denote the difference between nx and the integer *nearest* to x, or the value $+\frac{1}{2}$, if nx lies exactly in the middle of the interval between two consecutive integers. The series $\sum\frac{(nx)}{n^2}$ is uniformly convergent for all x's. The function represented by it, however, is discontinuous for $x=\frac{2p+1}{2q}$, (p, q integers), while it is continuous for all other rational values of x and for all irrational values of x.

155. If $a_n\to0$,

$$\sum a_n\cdot x\left(\frac{\sin nx}{nx}\right)^2$$

converges uniformly for all x's. Does this remain true for $a_n\equiv1$?

156. The products

a) $\prod\left(1+(-1)^n\frac{x}{n}\right)$, b) $\prod\cos\frac{x}{n}$,

c) $\prod\left(1+\sin^2\frac{x}{n}\right)$, d) $\prod\left(1+(-1)^n\sin\frac{x}{n}\right)$,

converge uniformly in every bounded interval.

157. The series whose partial sums have the values $s_n(x)=\frac{x^{2n}}{1+x^{2n}}$ converges for every x. Is this convergence uniform in every interval? Draw the curves of approximation.

158. A series $\sum f_n(x)$ of continuous *positive* functions certainly converges uniformly if it represents a continuous function $F(x)$. (Cf. p. 344, Rem. 3.)

159. Does $\sum\frac{x}{n(1+nx^2)}$ converge uniformly in every interval? Is the function it represents continuous?

160. In the proof of **111**, a situation of the following kind occurred: An expression of the form

$$F(n)=a_0(n)+a_1(n)+\cdots+a_k(n)+\cdots+a_{p_n}(n)$$

is considered, in which, for every *fixed* k, the term $a_k(n)$ tends to a limit α_k as n increases. At the same time, the number of terms increases, $p_n\to\infty$. May we infer that

$$\lim_{n\to\infty}F(n)=\sum_{k=0}^{\infty}\alpha_k,$$

provided the series on the right converges? Show that this is certainly permissible if, for every k and every n,

$$|a_k(n)| \text{ remains } <\gamma_k \quad\text{and}\quad \sum\gamma_k$$

converges. — Formulate the corresponding theorem for infinite products. — (Cf. Exercise 15, where such term-by-term passages to the limit were not allowed.)

161. The two series

$$x+\frac{x^3}{3}-\frac{x^4}{2}+\frac{x^5}{5}+\frac{x^7}{7}-\frac{x^8}{4}++-++-\cdots$$

$$x+\frac{x^3}{3}-\frac{x^2}{2}+\frac{x^5}{5}+\frac{x^7}{7}-\frac{x^4}{4}++-++-\cdots$$

are both convergent for $0\leqq x\leqq 1$ and have the same sum $\frac{3}{2}\log 2$ for $x=+1$. What is their behaviour when $x\to 1-0$? — Examine the two series, convergent for $x\geqq 1$,

$$1+\frac{1}{3^x}-\frac{2}{4^x}+\frac{1}{5^x}+\frac{1}{7^x}-\frac{2}{8^x}++-\cdots$$

$$1+\frac{1}{3^x}-\frac{1}{2^x}+\frac{1}{5^x}+\frac{1}{7^x}-\frac{1}{4^x}++-\cdots$$

for $x\to+1+0$.

162. The series $\sum_{n=1}^{\infty}\left[\frac{x^n}{n}-\frac{x^{2n-1}}{2n-1}-\frac{x^{2n}}{2n}\right]$ converges in $0\leqq x\leqq 1$. What is its sum? Is its convergence uniform?

163. Show that, for $x \to 1+0$,

a) $\lim (x-1) \sum_{n=1}^{\infty} \frac{1}{n^x} = 1$,

b) $\lim \left[\sum_{n=1}^{\infty} \frac{1}{n^x} - \frac{1}{(x-1)} \right] = C$ (v. **176**, 1).

164. Show that, for $x \to 1-0$,

a) $\sum_{n=1}^{\infty} \frac{(-1)^{n-1}}{n} \cdot \frac{x^n}{1+x^n} \to \frac{1}{2} \log 2$,

b) $(1-x) \cdot \sum_{n=1}^{\infty} (-1)^{n-1} \frac{x^n}{1-x^{2n}} \to \frac{1}{2} \log 2$,

c) $(1-x) \cdot \sum_{n=1}^{\infty} (-1)^{n-1} \frac{n x^n}{1-x^{2n}} \to -\frac{1}{4}$.

165. The series whose partial sums have the values $s_n(x) = \frac{n x}{1+n^2 x^4}$ may *not* be integrated term by term over an interval with endpoint 0. Draw the curves of approximation.

II. *Fourier* series.

166. May we deduce from the series **210** a, by integration term by term:

a) $\sum_{n=1}^{\infty} \frac{\cos 2\pi n x}{n^2} = \left(x^2 - x + \frac{1}{6}\right) \cdot \pi^2$,

b) $\sum_{n=1}^{\infty} \frac{\sin 2\pi n x}{n^3} = \left(\frac{2}{3} x^3 - x^2 + \frac{1}{3} x\right) \cdot \pi^3$,

c) $\sum_{n=1}^{\infty} \frac{\cos 2\pi n x}{n^4} = \left(-\frac{1}{3} x^4 + \frac{2}{3} x^3 - \frac{1}{3} x^2 + \frac{1}{90}\right) \cdot \pi^4$, etc.?

In which intervals are these relations valid? (Cf. **297**.)

167. In the same way, deduce from **210** c the relations

a) $\sum_{n=1}^{\infty} \frac{\sin (2n-1) x}{(2n-1)^3} = \frac{\pi x}{8} (\pi - x)$,

b) $\sum_{n=1}^{\infty} \frac{\cos (2n-1) x}{(2n-1)^4} = \frac{\pi}{48} \left(\frac{\pi}{2} - x\right) (\pi^2 + 2\pi x - 2x^2)$.

What would be the results of further integrations? In which intervals are these expansions valid?

168. From **209**, **210**, and the relations in the two preceding exercises, deduce the following further expansions and determine their exact intervals of validity:

a) $\cos x - \frac{\cos 3x}{3} + \frac{\cos 5x}{5} - + \cdots = \pm \frac{\pi}{4}$,

b) $\cos x - \frac{\cos 3x}{3^3} + \frac{\cos 5x}{5^3} - + \cdots = \frac{\pi}{8} \left(\frac{\pi^2}{4} - x^2\right)$,

c) $\sin x - \frac{\sin 3x}{3^4} + \frac{\sin 5x}{5^4} - + \cdots = \frac{\pi x}{8} \left(\frac{\pi^2}{4} - \frac{x^2}{3}\right)$, etc.

169. From **215**, deduce further expansions by substituting $\pi - x$ for x or by differentiating term by term. Is the latter operation allowed? What are the new series so obtained?

170. What are the sine-series and the cosine-series for e^{ax}? What is the complete *Fourier* expansion of $e^{\sin x}$? Show that the latter is of the form

$$\frac{1}{2}a_0 + b_1 \sin x - a_2 \cos 2x - b_3 \sin 3x + a_4 \cos 4x + b_5 \sin 5x - - + + \cdots$$

where a_ν and b_ν are positive.

171. If x and y are positive and $< \pi$,

$$\sum_{n=1}^{\infty} \frac{\sin nx \cos ny}{n} = \begin{cases} \frac{\pi}{2} - \frac{x}{2}, & \text{if } x > y, \\ \frac{\pi}{4} - \frac{x}{2}, & \text{if } x = y, \\ -\frac{x}{2}, & \text{if } x < y. \end{cases}$$

172. Determine the values of the integrals

$$\int_0^{\frac{\pi}{2}} \frac{\sin x}{x}\, dx \quad \text{and} \quad \int_0^{\pi} \frac{\sin x}{x}\, dx.$$

(The former $= 1{\cdot}37498\ldots$, the latter $= 1{\cdot}8519\ldots$.)

173. For every x and every n,

$$\left| \sin x + \frac{\sin 2x}{2} + \cdots + \frac{\sin nx}{n} \right| \leqq \int_0^{\pi} \frac{\sin x}{x}\, dx,$$

where the bound on the right hand side cannot be diminished (cf. the preceding exercise).

(Further exercises on special *Fourier* series will be given in the next chapter.)

Chapter XII.

Series of complex terms.

§ 52. Complex numbers and sequences.

After we have discussed in detail, as in Chapter I, the modes of formation of all the concepts essential for building up the system of real numbers, no new difficulties are raised by the introduction of further types of numbers. Since the (ordinary) complex numbers and their algebra are known to the reader, we may accordingly be content with briefly mentioning one or two main points here.

220. 1. It was shown in § 4 that the system of real numbers is incapable of any further extension, and is, moreover, the only system of symbols satisfying the conditions which we laid down for a number system. Yet the system of complex numbers is a system of symbols to which the name of number system is applied. This apparent contradiction is easily removed. For our definition of the number concept was in a certain sense an arbitrary one, as we emphasized on p. 12, footnote 16: A series of properties which appeared to us essential in the case of rational numbers was raised *to the rank of characteristic properties of numbers in general,* and the result justified our doing this, in so far as we were able actually to construct a system — in all essentials, a single one, — which possessed all these properties.

If we desire to attribute to other systems the character of a system of numbers, we must therefore of necessity diminish the list of characteristic properties which we set up in **4**, 1—4. The question arises which of these properties may be dispensed with first of all; i. e. which of them may be missing from a system of symbols without its becoming impossible to regard the latter as a number system.

2. Among the properties **4** of a system of symbols, the first with which we may dispense, without fear of the system losing the character of a number system entirely, are the laws of order and monotony. These are based, by **4**, 1, on the fact that of two different numbers of the system, the one can always be called less than the other, and the latter greater than the former. If we drop this distinction and in **4** replace *both* the symbols $<$ and $>$ by $\neq$, it appears that the modified conditions **4** are satisfied by another more general system of symbols, namely *the system of ordinary complex numbers,* but that no other system substantially different from the latter can satisfy them.

3. Accordingly, the system of (ordinary) complex numbers is a system of symbols — which, as is known, may be assumed to be of the form $x + yi$, where x and y are real numbers, and i is a symbol whose manipulation is regulated by the *single* condition $i^2 = -1$, — for which the fundamental laws of arithmetic **2** remain valid without exception, provided the symbols $<$ and $>$ are suitably replaced throughout by $\neq$. In short: Except for the last-named restriction, we may work formally with complex numbers exactly as with real numbers.

4. In a known manner (cf. p. 8), complex numbers may be brought into (1, 1) correspondence with the points of a plane and may thus be represented by these: with the complex number $x + yi$ we associate the point (x, y) of an xy-plane. Every calculation may then be interpreted geometrically. Instead of representing the number $x + yi$ by the point (x, y), it is often more convenient to represent it by a directed line (vector) coincident in magnitude and direction with the line from $(0, 0)$ to (x, y).

5. Complex numbers will be denoted in the sequel by a single letter: $z, \zeta, a, b, \ldots$; and unless the contrary is expressly mentioned or follows without ambiguity from the context, such letters will *invariably denote complex numbers.*

6. By the absolute value (or modulus) $|z|$ of the complex number $x+yi$, is meant the non-negative real value $\sqrt{x^2+y^2}$; by its amplitude ($\operatorname{am} z$, $z \neq 0$), we mean the angle φ for which both $\cos\varphi = \frac{x}{|z|}$ and $\sin\varphi = \frac{y}{|z|}$. When we calculate with absolute values, the rules **3**, II, 1—4 hold unchanged, while 5. loses all meaning.

Since we may accordingly operate, broadly speaking, in precisely the same ways with complex as with real numbers, by far the greater part of our previous investigations may be carried out in an entirely analogous manner in the realm of complex numbers, or transferred to the latter, as the case may be. The only considerations which will have to be omitted or suitably modified are those in which the numbers themselves (not merely their absolute values) are connected by the symbol $<$ or $>$.

In order to avoid repetitions, which this parallel course would otherwise involve, *we have prefixed the sign* ° *to all definitions and theorems, from Chapter II onwards, which remain valid word for word when arbitrary real numbers are replaced by complex numbers, (this validity extending equally to the proofs, with a few small alterations which will be explained immediately)*. We need only glance rapidly over the whole of our preceding developments and indicate at each place what modification is required when we transfer them to the realm of complex numbers. A few words will also be said on the subject of the somewhat different geometrical *representation*.

Definition **23** remains unaltered. A sequence of numbers will now be represented by a *sequence of points* (each counted once or more than once) *in the plane*. If it is bounded (**24**, 1), none of its points lie outside a circle of (suitably chosen) radius K with origin at 0.

Definition **25**, that of a null sequence, and the theorems **26**, **27**, and **28** relating to such null sequences remain entirely unaltered.

The sequences (z_n) with

$$z_n = \left(\frac{1+i}{2}\right)^n, \quad = \left(\frac{1+i\sqrt{3}}{3}\right)^n, \quad = \frac{i}{n}, \quad = \frac{(-i)^n}{n}, \quad (n = 1, 2, 3, \ldots)$$

are examples of null sequences whose terms are not all real. The student should form an exact idea of the position of the corresponding sets of points and prove that the sequences are actually null sequences.

The definitions in § 7 of roots, of powers in the general sense, and of logarithms were essentially based on the laws of order for real numbers. They cannot, therefore, be transferred to the realm of complex numbers in that form (cf. § 55 below).

The fundamental notions of the convergence and divergence of a sequence of numbers (**39** and **40**, 1) still remain unaltered, although the *representation* of $z_n \to \zeta$ now becomes the following[1]: If a circle of arbitrary (positive) radius ε is described about the *point* ζ as centre, we can always assign a (positive) number n_0 such that all terms of the sequence (z_n) with index $n > n_0$ lie within the given circle. The remark **39**, 6 (1st half) therefore holds word for word, *provided we interpret the ε-neighbourhood of a complex number ζ as being the circle mentioned above.*

In setting up the definitions **40**, 2, 3, the symbols $<$ and $>$ played an essential part; they cannot, therefore, be retained unaltered. And although it would not be difficult to transfer their main content to the complex realm, we will drop them entirely, and accordingly in the complex realm we shall call every non-convergent sequence *divergent*[2].

Theorems **41**, 1 to 12, and the important group of theorems **43**, with the exception of theorem 3, remain word for word the same, together with all the proofs.

The most important of these theorems were the *Cauchy-Toeplitz* limit-theorems **43**, 4 and 5, and since we have in the meantime gained complete familiarity with infinite series, we shall formulate them once more in this place, with the extension indicated in **44**, 10, and for complex numbers.

Theorem 1. *The coefficients of the matrix* **221.**

$$\text{(A)} \qquad \begin{pmatrix} a_{00}, & a_{01}, & a_{02}, & \ldots, & a_{0n}, & \ldots \\ a_{10}, & a_{11}, & a_{12}, & \ldots, & a_{1n}, & \ldots \\ a_{20}, & a_{21}, & a_{22}, & \ldots, & a_{2n}, & \ldots \\ \cdot & \cdot & \cdot & \cdot & \cdot & \cdot \\ a_{k0}, & a_{k1}, & a_{k2}, & \ldots, & a_{kn}, & \ldots \\ \cdot & \cdot & \cdot & \cdot & \cdot & \cdot \end{pmatrix}$$

are assumed to satisfy the two conditions:

(a) *the terms in each* column *form a null sequence, i. e. for every fixed $n \geqq 0$,*

$$a_{kn} \to 0 \quad as \quad k \to \infty.$$

---

[1] For complex numbers and sequences, we preferably use in the sequel the letters $z, \zeta, Z, \ldots$.

[2] We might say, in the case $|z_n| \to +\infty$, that (z_n) is definitely divergent with the limit ∞, or tends or diverges (or even converges) to ∞. That would be quite a consistent definition, such as is indeed constantly made in the theory of functions. However, it evidently involves a small inconsistency relative to the use of the terms in the real domain, that e. g. the sequence of numbers $(-1)^n n$ should be called definitely or indefinitely convergent, according as it is considered in the complex or in the real domain. And even though, with a little attention, this may not give us any trouble, we prefer to avoid the definition here.

(b) *there exists a constant K such that the sum of the absolute values of any number of terms in any one row remains less than K, i. e., for every fixed* $k \geqq 0$, *and any* n:

$$|a_{k0}| + |a_{k1}| + \cdots + |a_{kn}| < K.$$

Under these conditions, when $(z_0, z_1, \ldots)$ *is any null sequence, the numbers*

$$z_k' = a_{k0} z_0 + a_{k1} z_1 + \cdots \equiv \sum_{n=0}^{\infty} a_{kn} z_n$$

also form a null sequence[3].

Theorem 2. *The coefficients* a_{kn} *of the matrix* (A), *besides satisfying the two conditions* (a) *and* (b), *are assumed to satisfy the further condition*[3]

(c) $$\sum_{n=0}^{\infty} a_{kn} = A_k \to 1 \quad \text{as} \quad k \to \infty.$$

In this case, if $z_n \to \zeta$, *we have also*

$$z_k' = a_{k0} z_0 + a_{k1} z_1 + \cdots \equiv \sum_{n=0}^{\infty} a_{kn} z_n \to \zeta.$$

(For applications of this theorem, see more especially **233**, as well as §§ 60, 62 and 63.)

Unfortunately, we lose the first of the two main criteria of § 9, which was the more useful of the two. Moreover, the proof of the second main criterion cannot be transferred to the case of complex numbers, as it makes use of theorems of order throughout. In spite of this, we shall at once see *that the second main criterion itself — in all its forms — remains valid for complex numbers.* The proof may be conducted in two different ways: either we reduce the new (complex) theorem to the old (real) one, or we construct fresh foundations for the proof of the new theorem, by extending the developments of § 10 to complex numbers. Both ways are equally simple and may be indicated briefly:

1. *The reduction of complex sequences to real sequences* is most easily accomplished by splitting up the terms into their real and imaginary parts. If we write $z_n = x_n + i y_n$ and $\zeta = \xi + i\eta$, we have the following theorem, which completely reduces the question of the convergence or divergence of complex sequences to the corresponding real problem:

222. **Theorem 1.** *The sequence* $(z_n) \equiv (x_n + i y_n)$ *converges to* $\zeta = \xi + i\eta$ *if, and only if, the real parts* x_n *converge to* ξ ***and*** *the imaginary parts* y_n *converge to* η.

---

[3] In consequence of (b), $A_k = \sum_n a_{kn}$ is absolutely convergent and therefore, as the z_n's are bounded, by **41**, Theorem 2, the series $\sum_n a_{kn} z_n = z_k'$ is also absolutely convergent.

Proof. a) If $x_n \to \xi$ and $y_n \to \eta$, $(x_n - \xi)$ and $(y_n - \eta)$ are null sequences. By **26**, 1, the same is true of $i(y_n - \eta)$ and, by **28**, 1, of

$$(x_n - \xi) + i(y_n - \eta), \quad \text{i. e. of } (z_n - \zeta).$$

b) If $z_n \to \zeta$, $|z_n - \zeta|$ is a null sequence; since[4]

$$|x_n - \xi| \leqq |z_n - \zeta| \quad \text{and} \quad |y_n - \eta| \leqq |z_n - \zeta|,$$

$(x_n - \xi)$ and $(y_n - \eta)$ are also null sequences, by **26**, 2, i. e. we have both

$$x_n \to \xi \quad \text{and} \quad y_n \to \eta.$$

The theorem is established.

The theorem at which we are aiming follows immediately:

Theorem 2. *For the convergence of a complex sequence (z_n), the conditions of the second main criterion* **47** *are again necessary and sufficient, — namely, that, for every choice of $\varepsilon > 0$, we should be able to assign n_0 so that*

$$|z_{n'} - z_n| < \varepsilon,$$

for every $n > n_0$ and every $n' > n_0$.

Proof. a) If (z_n) converges, so do (x_n) and (y_n) by the preceding theorem. As these are real sequences, we may apply **47**, and, given $\varepsilon > 0$, we may choose n_1 and n_2 so that

$$|x_{n'} - x_n| < \frac{\varepsilon}{2} \text{ for every } n > n_1 \text{ and every } n' > n_1,$$

and

$$|y_{n'} - y_n| < \frac{\varepsilon}{2} \text{ for every } n > n_2 \text{ and every } n' > n_2.$$

Taking n_0 greater than n_1 and n_2, we have accordingly, for every $n > n_0$ and every $n' > n_0$,

$$|z_{n'} - z_n| = |(x_{n'} - x_n) + i(y_{n'} - y_n)| \leqq |x_{n'} - x_n| + |y_{n'} - y_n| < \frac{\varepsilon}{2} + \frac{\varepsilon}{2} = \varepsilon.$$

The conditions of our theorem are therefore *necessary.*

b) If, conversely, (z_n) fulfils the conditions of the theorem, — i. e. given $\varepsilon > 0$, we can determine n_0 so that $|z_{n'} - z_n| < \varepsilon$, provided only that n and n' are both $> n_0$, — we have also, for the same n and n' (by our last footnote)

$$|x_{n'} - x_n| < \varepsilon \quad \text{and} \quad |y_{n'} - y_n| < \varepsilon.$$

---

[4] We have in general

$$|\Re(z)| \leqq |z| \quad \text{and} \quad |\Im(z)| \leqq |z|$$

since

$$\left.\begin{matrix} x^2 \\ y^2 \end{matrix}\right\} \leqq x^2 + y^2, \quad \text{i. e.} \quad \left.\begin{matrix} |x| \\ |y| \end{matrix}\right\} \leqq \sqrt{x^2 + y^2} = |z|.$$

By **47**, this implies that (x_n) and (y_n) are convergent, so that (z_n) must also converge, by the preceding theorem; the conditions of our theorem are therefore also *sufficient*.

2. *Direct treatment of complex sequences.* In the treatment of real sequences, *nests of intervals* constituted our most frequent resource. In the complex domain, *nests of squares* will render us the same services:

223. **Definition.** *Let $Q_0, Q_1, Q_2, \ldots$ denote squares, whose sides will for simplicity be assumed parallel to the coordinate-axes. If each square is entirely contained in the preceding and if the lengths $l_0, l_1, \ldots$ of the sides form a null sequence, we shall say that the squares form **a nest**.*

For nests of squares, we have the

Theorem. *There exists one and only one point belonging to all the squares of a given nest of squares. (Principle of the innermost point.)*

Proof. Let the left hand bottom corner of Q_n be denoted by $a_n + i a_n'$ and the right hand upper corner by $b_n + i b_n'$. A point $z = x + i y$ belongs to the square Q_n if, and only if[5],

$$a_n \leqq x \leqq b_n \quad \text{and} \quad a_n' \leqq y \leqq b_n' .$$

Now, in consequence of our hypotheses, the intervals $J_n = a_n \ldots b_n$ on the x-axis, and similarly the intervals $J_n' = a_n' i \ldots b_n' i$ on the y-axis, form a *nest of intervals*. There is therefore exactly one point ξ on the x-axis and exactly one point $i\eta$ on the y-axis belonging to all the intervals of the corresponding nest. But this means that there is also *exactly one* point $\zeta = \xi + i\eta$, belonging to all the squares Q_n.

We are now in a position to transfer definition **52** and theorem **54** to the complex domain:

224. **Definition.** *If (z_n) is an arbitrary sequence, ζ is said to be a **limiting point or point of accumulation** of the sequence if, given an arbitrary $\varepsilon > 0$, the relation*

$$|z_n - \zeta| < \varepsilon$$

is satisfied for an infinity of values of n (in particular, for at least one $n >$ any given n_0).

225. **Theorem.** *Every bounded sequence possesses at least one limiting point.* ***(Bolzano-Weierstrass Theorem.)***

Proof. Suppose $|z_n| < K$ and draw the square Q_0 whose sides lie on the parallels to the axes through $\pm K$ and $\pm iK$. *All* the z_n's

[5] This statement at the same time expresses, in pure arithmetical language, the relations of magnitude framed in geometrical form in the theorem and definition **223**.

are contained in it, i. e. certainly an infinity of z_n's. Q_0 is divided by the coordinate-axes into four equal squares. One at least of the four must contain an infinity of z_n's. (In fact, if there were only a finite number in each, there would also be only a finite number in Q_0, which is not the case.) Let Q_1 denote the first quarter[6] which has this property. This we again proceed to divide into four equal squares, denoting by Q_2 the first quarter which contains an infinity of points z_n, and so on. The sequence $Q_0, Q_1, Q_2, \ldots$ forms a *nest of squares,* since each Q_n lies within the preceding and the lengths of the sides form a null sequence, namely $\left(2K \cdot \frac{1}{2^n}\right)$. Let ζ denote the innermost point of this nest[7]; ζ is a point of accumulation of (z_n). For if ε is given > 0 and m is chosen so that the side of Q_m is less than $\frac{\varepsilon}{2}$, the whole of the square Q_m lies within the ε-neighbourhood of ζ, and, with it, an infinite number of points z_n also lie in this neighbourhood. Therefore ζ is a point of accumulation of (z_n), and the existence of such a point is established.

The validity of the second main criterion for the complex domain, — i. e. of the theorem **222**, 2, formulated above — may now be established once more, but without any appeal to the "real" theorems, on the same lines as in **47**.

Proof. a) If $z_n \to \zeta$, i. e. $(z_n - \zeta)$ is a null sequence, we can determine n_0 so that

$$|z_n - \zeta| < \frac{\varepsilon}{2} \quad \text{and} \quad |z_{n'} - \zeta| < \frac{\varepsilon}{2}$$

provided only that n and n' are simultaneously $> n_0$ [see part a) of the proof of **47**]. For these n's and n''s, we therefore also have

$$|z_n - z_{n'}| \leqq |z_{n'} - \zeta| + |z_n - \zeta| < \varepsilon.$$

The condition is accordingly *necessary.*

b) If, conversely, the ε-condition is fulfilled, (z_n) is certainly bounded. In fact, if $m > n_0$ and $n > m$,

$$|z_n - z_m| < \varepsilon,$$

i. e. every z_n with $n > m$ lies in the circle of radius ε round z_m. Taking K to be larger than all the m numbers $|z_1|, |z_2|, \ldots, |z_{m-1}|, |z_m| + \varepsilon$, we have $|z_n| < K$ for *every* n.

[6] We regard the four quarters as numbered in the order in which the four quadrants of the xy-plane are habitually taken.

[7] The process of obtaining this point corresponds exactly to the *method of successive bisection* so often applied in the real domain.

By our preceding theorem, it follows that (z_n) has at least one limiting point ζ. Supposing there exists a second limiting point $\zeta' \neq \zeta$, choose

$$\varepsilon = \frac{1}{3}|\zeta' - \zeta|,$$

which is positive. By **224**, the definition of limiting point, we can choose n_0 as large as we please and yet have an $n > n_0$ for which $|z_n - \zeta| < \varepsilon$ and also an $n' > n_0$ for which $|z_{n'} - \zeta'| < \varepsilon$. Thus above any number n_0, however large, there exist a pair of indices n and n' for which[8]

$$|z_{n'} - z_n| > \varepsilon.$$

This contradicts our hypothesis. Accordingly ζ must be the *unique* limiting point, and outside the circle of radius ε round ζ there is only a finite number of points z_n. If n_0 is suitably chosen, we therefore have $|z_n - \zeta| < \varepsilon$ for *every* $n > n_0$, and consequently $z_n \to \zeta$. The condition of the theorem is therefore sufficient also[9].

§ 53. Series of complex terms.

As a series Σa_n of complex terms must obviously be interpreted as the sequence of its partial sums, the basis for the extension of our theory of infinite series has already been provided by the above.

Corresponding to **222**, 1, we have first the

226. Theorem. *A series Σa_n of complex terms is convergent if, and only if, the series $\Sigma \Re(a_n)$ of the real parts of its terms* **and** *the series $\Sigma \Im(a_n)$ of their imaginary parts converge separately. Further, if these two series have the sums s' and s'' respectively, the sum of Σa_n is $s = s' + i s''$.*

In accordance with **222**, 2 the second principal criterion (**81**) for the convergence of infinite series remains unaltered in all its forms, and, at the same time, the theorems **83** deduced from it, on the algebra of convergent series, also retain their full validity.

Since, in the same way, theorem **85** also remains unchanged, we shall, as before, distinguish between *absolute* and *non-absolute* convergence of series of complex terms (Def. **86**).

---

[8] $$z_{n'} - z_n = (\zeta' - \zeta) + (z_{n'} - \zeta') + (\zeta - z_n),$$
hence
$$|z_{n'} - z_n| \geqq |\zeta' - \zeta| - |z_{n'} - \zeta'| - |z_n - \zeta| > 3\varepsilon - \varepsilon - \varepsilon = \varepsilon.$$

[9] Hence we may also say: (z_n) converges if, and only if, it is bounded and possesses only one point of accumulation. This is then at the same time the limit of the sequence.

Here again we have the

Theorem. *The series Σa_n of complex terms is* ***absolutely*** *con-* **227.**
vergent if, and only if, ***both*** *the series $\Sigma \Re(a_n)$ and $\Sigma \Im(a_n)$ are absolutely convergent.*

The **proof** results simply from the fact that every complex number $z = x + iy$ satisfies the inequalities (cf. p. 393, footnote 4)

$$\left.\begin{array}{l}|x| \\ |y|\end{array}\right\} \leqq |z| \leqq |x| + |y|.$$

In consequence of this simple theorem, it is at once clear that, with series of complex terms as with real series, the order of the terms is immaterial if the series converges absolutely (Theorem **88**, 1).

If, however, Σa_n is not absolutely convergent, either $\Sigma \Re(a_n)$ or $\Sigma \Im(a_n)$ must be conditionally convergent. By a suitable rearrangement of the terms, the convergence of the series Σa_n may therefore be destroyed in any case, as in the proof of theorem **89**, 2, that is: *In the case of series of complex terms also, the convergence, when it is not absolute, depends essentially on the order of succession of the terms.* (Regarding the extension to series of complex terms of *Riemann*'s rearrangement theorem § 44, cf. the remarks on the following page.)

The next theorems, **89**, 3 and 4, as also the main rearrangement theorem **90**, which relate to absolutely convergent series, still remain valid, without modification or addition, for series of complex terms.

Since the determination of the absolute convergence of a series is a question relating to series of positive terms, *the whole theory of series of positive terms* is again enlisted for the study of series of *complex* terms: Everything that was proved for absolutely convergent series of real terms may be utilized for absolutely convergent series of complex terms.

If we omit power series from consideration for the present, we observe, on looking over the later sections of Part II (§§ 18—27), that the developments of Chapter X are the first for which there is any question of transference to series of complex terms.

Abel's partial summation **182**, being of a purely formal nature, and its *corollary* **183**, of course hold also for complex numbers, and so does the convergence-test **184** which was based directly on them. The special forms of this test may also all be retained, provided we keep to the convention agreed on in **220**, 5, in accordance with which all sequences assumed to be monotone are real. In the case of *du Bois-Reymond*'s and *Dedekind*'s tests, even this precaution becomes unnecessary: they hold word for word and without any restriction for arbitrary series of the form $\Sigma a_n b_n$, with complex a_n and b_n.

Riemann's rearrangement theorem (§ 44) is, on the contrary, essentially a "real" theorem. In fact, if a series Σa_n of complex terms is not absolutely convergent, so is one at least of the two series $\Sigma \Re(a_n)$ and $\Sigma \Im(a_n)$, by **227**. By a suitable rearrangement, we can therefore, in accordance with *Riemann*'s theorem, produce in *one* of these two series a prescribed type of convergence or divergence. But the other one of the two series will be rearranged in precisely the same manner, and there is no immediate means of foreseeing what the effect of the rearrangement on this series or on Σa_n itself will be. — It has recently been shown, however, that if Σa_n is not absolutely convergent, it may be transformed by a suitable rearrangement into a series, again convergent, whose sum may be prescribed to have either any value in the whole complex plane or any value on a particular straight line in this plane, according to the circumstances of the case[10].

The theorems **188** and **189** of *Mertens* and *Abel* on multiplication of series (§ 45) again remain valid word for word, together with the proofs. For the second of these theorems we must, it is true, rely on the second proof (*Cesàro*'s) alone, as we have provisionally skipped the consideration of power series (cf. later **232**).

At this point we are in possession of the whole machinery required for the mastery of series of complex terms and we can at once proceed to the most important of its applications.

Before doing so, however, we shall first deduce the following extremely far-reaching criterion.

228. **Weierstrass' criterion**[11]. *A series* $\sum_{n=0}^{\infty} a_n$ *of complex terms, for which*

$$\frac{a_{n+1}}{a_n} = 1 - \frac{\alpha}{n} - \frac{A_n}{n^\lambda}$$

with A_n *bounded,* — where α *is complex and arbitrary, and*[12] $\lambda > 1$, —

[10] We thus have the following very elegant theorem, which in a certain sense completes the solution of the rearrangement problem: The "range of summation" of a series Σa_n of complex terms — i. e. the set of values which may be obtained as sums of convergent rearrangements of Σa_n — is either a definite point, or a definite straight line, or the entire plane. Other cases cannot occur. A proof is given by *P. Lévy* (Nouv. Annales (4), Vol. 5, p. 506, 1905), but an unexceptionable statement of the proof is not found earlier than in *E. Steinitz* (Bedingt konvergente Reihen und konvexe Systeme, J. f. d. reine u. angew. Math., Vol. 143, 1913; Vol. 144, 1914; Vol. 146, 1915).

For the (more restricted) result that every conditionally convergent series $\Sigma a_n = s$ can be rearranged to give another convergent series $\Sigma a_n' = s'$ with $s' \neq s$, *W. Threlfall* has given a fairly short proof (Bedingt konvergente Reihen, Math. Zschr., Vol. 24, p. 212, 1926).

[11] J. f. d. reine u. angew. Math., Vol. 51, p. 29, 1856; Werke I, p. 185.

[12] An equality of this kind may of course always be assumed; we need only write $A_n = n^\lambda\left(1 - \frac{\alpha}{n} - \frac{a_{n+1}}{a_n}\right)$ as a definition. What is essential in the condition is here, as previously (cf. footnote to **166**), that when α and λ are suitably chosen the A_n's should be *bounded*. — It is substantially the same thing to assume that $a_n/a_{n+1} = 1 + \alpha/n + B_n/n^\lambda$ with $\lambda > 1$ and B_n bounded.

is absolutely convergent if, and only if, $\Re(\alpha) > 1$. *For* $\Re(\alpha) \leqq 0$ *the series is invariably divergent. If* $0 < \Re(\alpha) \leqq 1$, *both the series*

$$\sum_{n=0}^{\infty} |(a_n - a_{n+1})| \quad \textit{and} \quad \sum_{n=0}^{\infty} (-1)^n a_n$$

are convergent[13].

Proof. 1. Let $\alpha = \beta + i\gamma$ and let us first assume $\beta = \Re(\alpha) > 1$. In that case, if $|A_n| < K$, say, we write, as is permissible,

$$\left|\frac{a_{n+1}}{a_n}\right| \leqq \left|1 - \frac{\beta + i\gamma}{n}\right| + \frac{K}{n^\lambda};$$

and it follows at once that, if β' is any number such that $1 < \beta' < \beta$,

$$\left|\frac{a_{n+1}}{a_n}\right| \leqq 1 - \frac{\beta'}{n}$$

for every sufficiently large n. By *Raabe*'s test, the series $\Sigma |a_n|$ is therefore convergent.

2. Now suppose $\Re(\alpha) = \beta \leqq 1$. In that case, since

$$\left|\frac{a_{n+1}}{a_n}\right| \geqq 1 - \frac{\beta}{n} - \frac{K}{n^\lambda}$$

for sufficiently large values of n, it follows from *Gauss*'s test **172** that $\Sigma |a_n|$ is divergent.

3 a. If, on the other hand, $\Re(\alpha) = \beta < 0$, our last inequality shows that then

$$\left|\frac{a_{n+1}}{a_n}\right| > 1.$$

Therefore Σa_n must now diverge.

3b. If $\Re(\alpha) = \beta = 0$, i. e.

$$\left|\frac{a_{n+1}}{a_n}\right| = 1 - \frac{i\gamma}{n} - \frac{A_n}{n^\lambda}$$

it is easy to verify that we then have

$$\left|\frac{a_{n+1}}{a_n}\right| = 1 - \frac{A_n'}{n^{\lambda'}}$$

where $\lambda' > 1$ and is the smaller of the two numbers 2 and λ, and the A_n''s are again *bounded*. Accordingly, if c denotes a suitable constant,

$$\left|\frac{a_{n+1}}{a_n}\right| \geqq 1 - \frac{c}{n^{\lambda'}} > 0$$

---

[13] As regards the series Σa_n itself, it was shown by *Weierstrass*, l.c., that this is also divergent whenever $\Re(\alpha) \leqq 1$. The proof is somewhat troublesome. — A further more exact investigation of the series Σa_n itself in the case $0 \leqq \Re(\alpha) \leqq 1$ is given by *A. Pringsheim* (Archiv d. Math. und Phys. (3), Vol. 4, pp. 1—19, in particular pp. 13—17. 1902), *J. A. Gmeiner*, Monatshefte f. Math. u. Phys., Vol. 19, pp. 149—163. 1908.

for every $n \geqq m$, say. It follows by multiplication that

$$\left|\frac{a_n}{a_m}\right| = \left|\frac{a_{m+1}}{a_m}\right| \cdots \left|\frac{a_n}{a_{n-1}}\right| > \prod_{\nu=m}^{n-1}\left(1-\frac{c}{\nu^{\lambda'}}\right) > \prod_{\nu=m}^{\infty}\left(1-\frac{c}{\nu^{\lambda'}}\right) = C_m > 0.$$

Hence $|a_n| > C_m \cdot |a_m|$, for every $n > m$, and a_n cannot tend to 0, so that Σa_n again diverges (cf. **170**, 1).

4. If, finally, $\Re(\alpha) = \beta > 0$, we have to show that both the series

$$\Sigma |a_n - a_{n+1}| \quad \text{and} \quad \Sigma(-1)^n a_n$$

are convergent. Now as in **1.** we have, for every sufficiently large n,

$$\left|\frac{a_{n+1}}{a_n}\right| < 1 - \frac{\beta'}{n}, \quad \text{with} \quad 0 < \beta' < \beta,$$

so that $|a_n|$ diminishes monotonely from some stage on, and therefore tends to a definite limit $\geqq 0$. Accordingly,

a) the series $\Sigma(|a_n| - |a_{n+1}|)$ is convergent, by **131**, and has, moreover, all its terms positive for sufficiently large n's. Now

$$\frac{|a_n - a_{n+1}|}{|a_n| - |a_{n+1}|} = \frac{\left|1 - \frac{a_{n+1}}{a_n}\right|}{1 - \left|\frac{a_{n+1}}{a_n}\right|} \leqq \frac{\left|\frac{\alpha}{n} + \frac{A_n}{n^\lambda}\right|}{\frac{\beta'}{n}};$$

since the fraction on the right hand side tends to the positive limit $\frac{|\alpha|}{\beta'}$ when $n \to +\infty$, that on the left is, for every sufficiently large n, less than a suitable constant A. By **70**, 2, this means that $\Sigma|a_n - a_{n+1}|$ converges with $\Sigma(|a_n| - |a_{n+1}|)$. — We can show more precisely, however, that

b) $a_n \to 0$. For it again follows, by multiplication, from

$$\left|\frac{a_{n+1}}{a_n}\right| < 1 - \frac{\beta'}{n} \qquad (n \geqq m)$$

that

$$\left|\frac{a_n}{a_m}\right| < \left(1 - \frac{\beta'}{m}\right)\left(1 - \frac{\beta'}{m+1}\right) \cdots \left(1 - \frac{\beta'}{n-1}\right).$$

The right hand side (by **126**, 2) tends to 0 as $n \to +\infty$, hence (cf. **170**, 1) we must have $a_n \to 0$. Now the series

$$(a_0 - a_1) + (a_2 - a_3) + (a_4 - a_5) + \cdots \equiv \sum_{k=0}^{\infty}(a_{2k} - a_{2k+1})$$

is a sub-series of $\Sigma(a_n - a_{n+1})$ and therefore converges *absolutely*, by a); also, since $|a_n| + |a_{n+1}| \to 0$ with a_n, we may omit the brackets, by **83**, supplement to theorem 2. This proves the convergence of $\Sigma(-1)^n a_n$.

This theorem enables us to deduce easily the following further theorem, which will be of use to us shortly:

Theorem. *If, as in the preceding theorem,* **229.**

$$\frac{a_{n+1}}{a_n} = 1 - \frac{\alpha}{n} - \frac{A_n}{n^\lambda} \qquad \begin{cases} \alpha \text{ arbitrary, } \lambda > 1, \\ (A_n) \text{ bounded,} \end{cases}$$

the series $\sum a_n z^n$ *is absolutely convergent for* $|z| < 1$, *divergent for every* $|z| > 1$, *and for the points of the circumference* $|z| = 1$, *the series will*

a) *converge absolutely, if* $\Re(\alpha) > 1$,

b) *converge conditionally, if* $0 < \Re(\alpha) \leqq 1$, *except possibly*[14] *for the single point* $z = +1$,

c) *diverge, if* $\Re(\alpha) \leqq 0$.

Proof. Since

$$\left|\frac{a_{n+1} z^{n+1}}{a_n z^n}\right| \to |z|,$$

the statements relative to $|z| \gtrless 1$ are immediately verified. For $|z| = 1$, the statement a) is an immediate consequence of the convergence of $\sum |a_n|$ ensured by the preceding theorem. Similarly c) is an immediate consequence of the fact established above, that in this case $|a_n|$ remains greater than a certain positive number for every sufficiently large n.

Finally, if $0 < \Re(\alpha) \leqq 1$ and $z \neq +1$, the convergence of $\sum a_n z^n$ follows from *Dedekind*'s test **184**, 3. For we proved in the preceding theorem that $\sum |a_n - a_{n+1}|$ converges and $a_n \to 0$; that the partial sums of $\sum z^n$ are bounded, for every (fixed) $z \neq +1$ on the circumference $|z| = 1$, follows simply from the fact that for every n

$$|1 + z + z^2 + \cdots + z^n| = \left|\frac{1 - z^{n+1}}{1 - z}\right| \leqq \frac{2}{|1 - z|}.$$

§ 54. Power series. Analytic functions.

The term "power series" is again used here to denote a series of the form $\sum a_n z^n$, or, more generally, of the form $\sum a_n (z - z_0)^n$, where now both the coefficients a_n and the quantities z and z_0 may be complex.

The theory of these series developed in §§ 18 to 21 remains valid without any essential modification. In transferring the considerations of those sections, we may therefore be quite brief.

Since the theorems **93**, 1 and 2 remain entirely unaltered in the new domain, the same is true of the fundamental theorem **93** itself, on the behaviour of power series in the real domain. Only the geometrical *interpretation* is somewhat different: The power series $\sum a_n z^n$

---

[14] If we take into account *Pringsheim*'s result mentioned in the preceding footnote, we may state here, more definitely: *except for* $z = +1$.

converges — indeed absolutely — for every z interior to the circle of radius r round the origin 0, while it diverges for all points outside that circle. This circle is called the ***circle of convergence*** of the power series — and the name *radius* applied to the number r thus becomes, for the first time, completely intelligible. Its magnitude is given as before by the *Cauchy-Hadamard* theorem **94**.

Regarding convergence on the circumference of the *circle* of convergence, we can no more give a general verdict than we could regarding the behaviour at the endpoints of the *interval* of convergence in the case of real power series. (The examples which follow immediately will show that this behaviour may be of the most diverse nature.)

The remaining theorems of § 18 also retain their validity unaltered.

230. Examples.

1. $\sum z^n$; $r=1$. In the interior of the *unit circle*, the series is convergent, with the sum $\frac{1}{1-z}$. On the boundary, i. e. for $|z|=1$, it is *everywhere divergent*, as z^n does not $\to 0$ there.

2. $\sum \frac{z^n}{n^2}$; $r=1$. This series[15] remains (absolutely) convergent at *all* the boundary points $|z|=1$.

3. $\sum \frac{z^n}{n}$; $r=1$. The series is certainly not convergent for *all* the boundary points, for $z=1$ gives the divergent series $\sum \frac{1}{n}$. However, it is also not divergent for *all* these points, since $z=-1$ gives a convergent series. In fact, theorem **229** of the preceding section shows, more precisely, that the series must *converge conditionally* at all points of the circumference $|z|=1$ different from $+1$; for we have here

$$\frac{a_n}{a_{n-1}}=\frac{n-1}{n}=1-\frac{1}{n}.$$

The same result may also be deduced directly from *Dirichlet*'s test **184**, 2, since $\sum z^n$ has bounded partial sums for $z \neq +1$ and $|z|=1$ (cf. the last formula of the preceding section) and $\frac{1}{n}$ tends monotonely to 0. As $\sum\left|\frac{z^n}{n}\right|=\sum\frac{1}{n}$, the convergence can, however, only be conditional[16].

4. $\sum \frac{z^{4n}}{4n}$; $r=1$. This series diverges at the four boundary points ± 1 and $\pm i$, and converges conditionally at every other point of the boundary.

---

[15] If $\sum a_n z^n$ has ***real coefficients*** (as in most of the subsequent examples) this power series of course has the same radius as the real power series $\sum a_n x^n$.

[16] These facts regarding convergence may also be deduced from **185**, 5, by splitting up the series into its real and imaginary parts. Conversely, however, the above mode of reasoning provides a new proof of the convergence of these two real series.

5. For $\sum \frac{z^n}{n!}$, $r = +\infty$. For $\sum n!\, z^n$, $r = 0$; thus this series converges nowhere but at $z = 0$.

6. The series $\sum_{k=0}^{\infty} (-1)^k \frac{z^{2k}}{(2k)!}$ and $\sum_{k=0}^{\infty} (-1)^k \frac{z^{2k+1}}{(2k+1)!}$ are everywhere convergent.

7. A power series of the general form $\sum a_n (z - z_0)^n$ converges absolutely at all interior points of the circle of radius r round z_0, and diverges outside this circle, where r denotes the radius of $\sum a_n z^n$.

Before proceeding to examine the properties of power series in more detail, we may insert one or two remarks on

Functions of a complex variable.

If to every point z within a circle $\mathfrak{K}$ (or more generally, a domain[17] $\mathfrak{G}$) a value w is made to correspond in any particular manner, we say that *a function $w = f(z)$ of the complex variable z is given in this circle (or domain)*. The correspondence may be brought about in a great number of ways (cf. the corresponding remark on the concept of a real function, § 19, Def. 1); in all that follows, however, the functional value will almost always be capable of expression by an explicit formula in terms of z, or else will be the sum of a convergent series whose terms are explicitly given. Numerous examples will occur very shortly; for the moment we may think of the value w, for instance, which at each point z within the circle of convergence of a given power series represents the sum of the series at that point.

The concepts of the limit, the continuity, and the differentiability of a function are those which chiefly interest us in this connection, and their definitions, in substance, follow precisely the same lines as in the real domain:

1. Definition of limit. If the function $w = f(z)$ is defined[18] for **231.**
every z in a neighbourhood of the fixed point ζ, we say that

$$\lim_{z \to \zeta} f(z) = \omega$$

or

$$f(z) \to \omega \quad \text{for} \quad z \to \zeta,$$

---

[17] A strict definition of the word "domain" is not needed here. In the sequel, we shall always be concerned with the interior of plane areas bounded by a finite number of straight lines or arcs of circles, in particular with circles and half-planes.

[18] $f(z)$ need not be defined at the point ζ itself, but only for all z's which satisfy the condition $0 < |z - \zeta| < \varrho$. The δ of the above definition must then of course be assumed $< \varrho$.

if, given an arbitrary $\varepsilon > 0$, we can assign $\delta = \delta(\varepsilon) > 0$ so that

$$|f(z) - \omega| < \varepsilon$$

for every z satisfying the condition $0 < |z - \zeta| < \delta$; or — which comes to exactly the same thing[19] — if for *every* sequence (z_n) converging to ζ, whose terms lie in the given neighbourhood of ζ and do not coincide with ζ, the corresponding functional values $w_n = f(z_n)$ converge to ω.

If we consider the values of $f(z)$, not at *all* the points of a neighbourhood of ζ, but only at those which lie, for instance, on a particular arc of a curve ending at ζ, or in an angle with its vertex at ζ, or, more generally, which belong to a set of points M, for which ζ is a point of accumulation, — we say that $\lim f(z) = \omega$ or $f(z) \to \omega$ *as* $z \to \zeta$ *along that arc, or within that angle, or in that set* M, if the above conditions are fulfilled, at least for all points z of the set M which come into consideration in the process.

2. Definition of continuity. If the function $w = f(z)$ is defined in a neighbourhood of ζ *and at* ζ *itself,* we say that $f(z)$ *is continuous at the point* ζ, if

$$\lim_{z \to \zeta} f(z)$$

exists and is equal to the value of the function at ζ, i. e. if $f(z) \to f(\zeta)$. We may also define the continuity of $f(z)$ at ζ when z is restricted to an arc of a curve containing the point ζ, or an angle with its vertex at ζ, or any other set of points M that contains ζ and of which ζ is a limiting point; the definitions are obvious from 1.

3. Definition of differentiability. If the function $w = f(z)$ is defined in a neighbourhood of ζ and at ζ itself, $f(z)$ is said to be *differentiable* at ζ, if the limit

$$\lim_{z \to \zeta} \frac{f(z) - f(\zeta)}{z - \zeta}$$

exists in accordance with 1. Its value is called the differential coefficient of $f(z)$ at ζ and is denoted by $f'(\zeta)$. (Here again the mode of variation of z may be subjected to restrictions.)

We must be content with these few definitions concerning the general functions of a complex variable. The study of these functions in detail constitutes the object of the so-called theory of *functions,* one of the most extensive domains of modern mathematics, into which we of course cannot enter further in this place[20].

---

[19] Same proof as in the real domain.

[20] A rapid view of the most important fundamental facts of the theory of functions may be obtained from two short tracts by the author: Funktionen-

The above explanations are abundantly sufficient to enable us to transfer the most important of the developments of §§ 20 and 21 to power series with complex terms.

In fact, those developments remain valid without exception for our present case, if we suitably change the words "interval of convergence" to "circle of convergence" throughout. Theorem 5 (**99**) is the only one to which we can form no analogue, since the concept of integral has not been introduced for functions of a complex argument. All this is so simple that the reader will have no trouble, on looking through these two sections again, *to interpret them as if they had been intended from the first to relate to power series with complex terms.*

At the most, a few remarks may be necessary in connection with *Abel's* limit theorem **100** and theorem **107** on the reversion of a power series. In the case of the latter, the convergence of the series $y + \beta_2 y^2 + \cdots$, and hence of the series $y + b_2 y^2 + \cdots$, which satisfied the conditions of the theorem, were only proved for *real* values of y. This is clearly sufficient, however, as we have thereby proved that this power series has a positive radius of convergence, which is all that is required.

As regards *Abel's* limit theorem, we may even — corresponding to the greater degree of freedom of the variable point z — prove more than before, and for this reason we will go into the matter once more:

Let us suppose $\Sigma a_n z^n$ to be a given power series, not everywhere convergent, but with a positive radius of convergence. We first observe that, exactly as before, we may assume this radius $= 1$ without introducing any substantial restriction. On the circumference of the circle of convergence, $|z| = 1$, we assume that at least one point z_0 exists at which the series continues to converge. Here again we may assume that z_0 is the special point $+1$. In fact, if $z_0 \neq +1$, we need only put

$$a_n z_0^n = a_n';$$

the series $\Sigma a_n' z^n$ also has the radius 1 and converges at the point $+1$.

The proof originally given, where everything may now be interpreted as "complex", then establishes the

Theorem. *If the power series $\Sigma a_n z^n$ has the radius* 1 *and remains convergent at the point $+1$ of the unit circle, and if $\Sigma a_n = s$, then we also have* **232.**

$$\lim_{z \to +1} (\Sigma a_n z^n) = s$$

if z approaches the point $+1$ along the positive real axis from the origin[21] 0.

---

theorie, I. Teil, Grundlagen der allgemeinen Theorie, 4th ed., Leipzig 1930; II. Teil, Anwendungen und Weiterführung der allgemeinen Theorie, 4th ed., Leipzig 1931 (Sammlung Göschen, Nos. 668 and 703).

[21] We are therefore dealing with a limit of the kind mentioned above in **231**, 1.

233. We can now easily prove more than this:

Extension of *Abel's* theorem. *With the conditions of the preceding theorem, the relation*

$$\lim_{z \to +1} (\Sigma\, a_n z^n) = s$$

remains true if the mode of approach of z to $+1$ is restricted only by the condition that z should remain within the unit circle and in the angle between two arbitrary (fixed) rays which penetrate into the interior of the unit circle, starting from the point $+1$ (see Fig. 10).

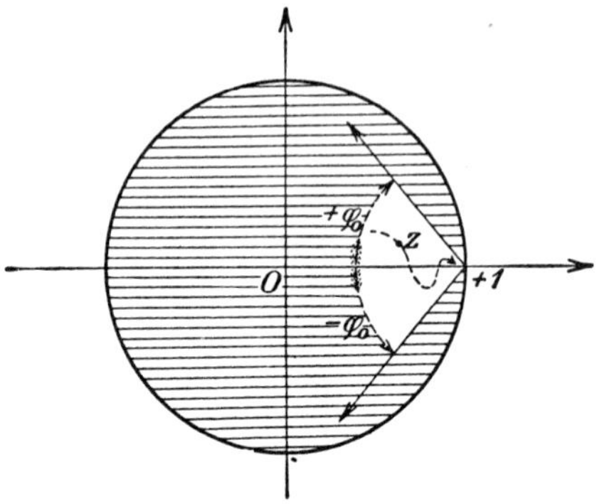

Fig. 10.

The **proof** will be conducted quite independently of previous considerations, so that we shall thus obtain a third proof of *Abel*'s theorem.

Let $z_0, z_1, \ldots, z_k, \ldots$ be any sequence of points of limit $+1$ in the described portion of the unit circle We have to show that

$$f(z_k) \to s$$

if, as before, we write $\Sigma\, a_n z^n = f(z)$. In *Toeplitz*' theorem **221**, 2, choose for a_{kn} the value

$$a_{kn} = (1 - z_k) \cdot z_k^{\,n}$$

and apply the theorem to the sequence of partial sums

$$s_n = a_0 + a_1 + \cdots + a_n,$$

which, by hypothesis, converges to s. It follows immediately that

$$\sum_{n=0}^{\infty} (1 - z_k) \cdot z_k^{\,n} \cdot s_n = (1 - z_k) \cdot \sum_{n=0}^{\infty} s_n z_k^{\,n} = \sum_{n=0}^{\infty} a_n z_k^{\,n} = f(z_k)$$

also tends to s as k increases. This proves the statement, provided we can show that the chosen numbers a_{kn} satisfy the conditions (a), (b) and (c) of **221**. Now (a) is clearly fulfilled, as $z_k \to 1$, and the sum of the kth row is now $A_k = (1 - z_k) \sum_{n=0}^{\infty} z_k^{\,n} = 1$, so that (c) is fulfilled. Finally (b) requires the existence of a constant K such that

$$|1 - z| \cdot \sum_{n=0}^{\infty} |z^n| = \frac{|1 - z|}{1 - |z|} < K$$

for all points $z = z_k \neq +1$ in the angle (or any sector-shaped portion of it with its vertex at $+1$). It only remains, therefore, to establish the existence of such a constant. This reduces (v. Fig. 10) to proving the following statement: *If* $z = 1 - \varrho(\cos\varphi + i\sin\varphi)$ *with* $|\varphi| \leqq \varphi_0 < \frac{\pi}{2}$ *and* $0 < \varrho \leqq \varrho_0 < 2\cos\varphi_0$, *a constant* $A = A(\varphi_0, \varrho_0)$ *exists, depending only on* φ_0 *and* ϱ_0, *such that*

$$\frac{|1-z|}{1-|z|} \leqq A$$

for every z of the type described. In the proof of this statement, it is sufficient to assume $\varrho_0 = \cos\varphi_0$, and in that case we may at once show that $A = \frac{2}{\cos\varphi_0}$ is a constant of the desired kind. In fact, the statement then runs:

$$\frac{\varrho}{1-\sqrt{1-2\varrho\cos\varphi+\varrho^2}} \leqq \frac{2}{\cos\varphi_0}$$

or

$$-2\varrho\cos\varphi + \varrho^2 \leqq -\varrho\cos\varphi_0 + \frac{1}{4}\varrho^2\cos^2\varphi_0,$$

for $0 < \varrho \leqq \cos\varphi_0$ and $|\varphi| \leqq \varphi_0$. By replacing φ by φ_0 and ϱ^2 by $\varrho\cos\varphi_0$ on the left hand side, the latter is increased; therefore it certainly suffices to show that

$$-\varrho\cos\varphi_0 \leqq -\varrho\cos\varphi_0 + \frac{1}{4}\varrho^2\cos^2\varphi_0,$$

— which is obviously true. — This extension of *Abel*'s theorem to "complex modes of approach" or "approach within an angle" is due to *O. Stolz* [22].

This completes the extension to the case of complex numbers of all the theorems of §§ 20 and 21 — with the single exception of the theorem on integration, which we have not defined in the present connection. In particular, it is thereby established that a power series in the interior of its circle of convergence defines a function of a complex variable, which is continuous and differentiable — the latter "term by term" and as often as we please — in that domain, and accordingly possesses the two properties which above all others are required, in the case of a function, for all purposes of practical application. For this reason, and on account of their great importance in further developments of the theory, a special name has been reserved for functions representable in the neighbourhood of a point

[22] Zeitschrift f. Math. u. Phys., Vol. 20, p. 369, 1875. In recent years the question of the converse of *Abel*'s theorem has been the object of numerous investigations, — i. e. the question, under what (minimum of) assumptions relating to the coefficients a_n, the existence of the limit of $f(z)$ as $z \to 1$ (within the angle) entails the convergence of Σa_n. An exhaustive survey of the present state of research in this respect is given in papers by *G. H. Hardy* and *J. E. Littlewood*, Abel's theorem and its converse, Proc. Lond. Math. Soc. (2), I. Vol. 18, pp. 205—235, 1920; II. Vol. 22, pp. 254—269, 1923; III. Vol. 25, pp. 219—236, 1926. — Cf. also theorems **278** and **287**.

z_0 by a power series $\Sigma a_n (z - z_0)^n$. They are said to be *analytic* or *regular at* z_0. By **99**, such a function is then analytic at every other interior point of the circle of convergence; it is therefore said simply to be *analytic or regular in this circle*[23]. In particular, a series everywhere convergent represents a function regular in the *whole plane*, which is therefore shortly called an *integral function.*

All the theorems which we have proved about functions expressed by power series are theorems about analytic functions. Only the two following, which are of special importance in the sequel, need be expressly formulated again.

234. 1. *If two functions are analytic in one and the same circle, then so are (by § 21) their sum, their difference, and their product.*

For the *quotient* the corresponding statement is primarily true (by **105**, 4) only if the function in the denominator is not zero at the centre of the circle, and provided, if necessary, that this circle is replaced by a smaller one.

2. *If two functions, analytic in one and the same circle, coincide in a neighbourhood, however small, of its centre (or indeed at all points of a set having this centre as point of accumulation), the two functions are completely identical in the circle* (Identity theorem for power series **97**).

Besides stating these two theorems, which are new only in form, we shall prove the following important theorem, which gives us some information on the connection between the moduli of the coefficients of a power series and the modulus of the function it represents:

235. Theorem. *If* $f(z) = \sum_{n=0}^{\infty} a_n (z - z_0)^n$ *converges for* $|z - z_0| < r$, *then*

$$|a_p| \leqq \frac{M}{\varrho^p} \qquad (p = 0, 1, 2, \ldots),$$

if $0 < \varrho < r$ *and* $M = M(\varrho)$ *is a number which* $|f(z)|$ *never exceeds along the circumference* $|z - z_0| = \varrho$. *(Cauchy's inequality.)*

Proof[24]. We first choose a complex number η, of modulus $= 1$, for which however $\eta^q \neq 1$ for any integral[25] exponent $q \gtrless 0$. Now we consider the function

$$g(z) = a \cdot (z - z_0)^k$$

---

[23] A function is accordingly said to be "analytic" or "regular" in a circle $\mathfrak{K}$ when it can be represented by a power series which converges in this circle.

[24] The following very elegant proof is due to *Weierstrass* (Werke II, p. 224) and dates as far back as 1841. *Cauchy* (Mémoire lithogr., Turin 1831) proved the formula indirectly by means of his expression for $f(z)$ in the form of an integral. The existence of a constant M that $|f(z)|$ never exceeds on $|z - z_0| = \varrho$ is practically obvious, of course, since $M = \Sigma |a_n| \varrho^n$ clearly has this property. *This* M is obviously also such that $|a_n| \varrho^n \leqq M$. But the above theorem states that *every* M that $|f(z)|$ never exceeds has the property that $|a_n| \varrho^n$ is always $\leqq M$.

[25] Such numbers η of course exist, for if $\eta = \cos(\alpha\pi) + i \sin(\alpha\pi)$, then $\eta^q = \cos(q\alpha\pi) + i\sin(q\alpha\pi)$; this is never 1 if α is chosen irrational.

for a specific integral value of the exponent $k \lessgtr 0$ and an arbitrary constant coefficient a. If we denote by $g_0, g_1, g_2, \ldots$ the values of this function for $z = z_0 + \varrho \cdot \eta^\nu$, $\nu = 0, 1, 2, \ldots$, we have for $n \geqq 1$

$$g_0 + g_1 + \cdots + g_{n-1} = a \cdot \varrho^k \cdot \frac{1 - \eta^{kn}}{1 - \eta^k},$$

hence

$$\left| \frac{g_0 + g_1 + \cdots + g_{n-1}}{n} \right| \leqq \frac{2}{n} \cdot \varrho^k \cdot \left| \frac{a}{1 - \eta^k} \right|.$$

The expression on the right hand side contains only constants, besides the denominator n; it therefore follows that the arithmetic mean

$$\frac{g_0 + g_1 + \cdots + g_{n-1}}{n} \to 0$$

as n increases. In the case $k = 0$, we should be concerned with the identically constant function $g(z) \equiv a$, for which

$$\frac{g_0 + g_1 + \cdots + g_{n-1}}{n} \to a,$$

since the ratio is equal to a for every n, in this case. If we consider the rather more general function

$$g(z) = \frac{b_{-l}}{(z - z_0)^l} + \frac{b_{-l+1}}{(z - z_0)^{l-1}} + \cdots + \frac{b_{-1}}{z - z_0} + b_0 + b_1 (z - z_0) + \cdots$$
$$\cdots + b_m (z - z_0)^m,$$

where l and m are fixed integers $\geqq 0$, and now form the arithmetic mean

$$\frac{g_0 + g_1 + \cdots + g_{n-1}}{n},$$

(where, as before, $g_\nu = g(z_0 + \varrho\, \eta^\nu)$, $\nu = 0, 1, \ldots$), this clearly $\to b_0$, by the two cases just treated. If, further, it is known that the function $g(z)$, for *every* z of the circumference $|z - z_0| = \varrho$, is never greater than a certain constant K, we have also

$$\left| \frac{g_0 + g_1 + \cdots + g_{n-1}}{n} \right| \leqq \frac{n\,K}{n} = K,$$

and therefore also

$$|b_0| \leqq K.$$

With these preliminary remarks, the proof of the theorem is now quite simple: Let p be a specific integer $\geqq 0$. As $\sum |a_n| \varrho^n$ converges, given $\varepsilon > 0$, we can determine $q > p$ so that

$$|a_{q+1}| \varrho^{q+1} + |a_{q+2}| \varrho^{q+2} + \cdots < \varepsilon.$$

A fortiori, we then have for all values of z such that $|z - z_0| = \varrho$,

$$\Big|\sum_{n=q+1}^{\infty} a_n (z - z_0)^n\Big| < \varepsilon$$

and therefore, for the same values of z,

$$\Big|\sum_{n=0}^{q} a_n (z - z_0)^n\Big| < M + \varepsilon,$$

if M has the meaning given in the text. Accordingly, on the circumference $|z - z_0| = \varrho$,

$$\left|\frac{a_0}{(z-z_0)^p} + \cdots + \frac{a_{p-1}}{z-z_0} + a_p + a_{p+1}(z - z_0) + \cdots + a_q (z - z_0)^{q-p}\right| \leqq \frac{M+\varepsilon}{\varrho^p}.$$

The function between the modulus signs is of the kind just considered. The inequality $|b_0| \leqq K$ there obtained now becomes

$$|a_p| \leqq \frac{M+\varepsilon}{\varrho^p},$$

and, as ε was arbitrary and > 0, we have, in fact, (cf. footnote to **41**, 1)

$$|a_p| \leqq \frac{M}{\varrho^p},$$

q. e. d.

§ 55. The elementary analytic functions.

I. Rational functions.

1. The rational function $w = \frac{1}{1-z}$ is expressible as a power series for every centre $z_0 \neq +1$:

$$\frac{1}{1-z} = \frac{1}{1 - z_0 - (z - z_0)} = \frac{1}{1-z_0} \cdot \frac{1}{1 - \frac{z - z_0}{1 - z_0}} = \sum_{n=0}^{\infty} \frac{1}{(1-z_0)^{n+1}} \cdot (z - z_0)^n;$$

and this series converges for $|z - z_0| < |1 - z_0|$ i. e. for every z nearer to z_0 than $+1$; in other words, the circle of convergence of the series is the circle with centre z_0 passing through the point $+1$. The function $\frac{1}{1-z}$ is thus analytic at every point different from $+1$.

With reference to this example, we may briefly draw attention to the following phenomenon, which becomes of fundamental importance in the theory of functions: If the geometric series Σz^n, whose circle of convergence is the unit circle, is expanded by *Taylor*'s theorem about a new centre z_1 within the unit circle, we could assert with certainty, by that theorem, that the new series converges at least in the circle of centre z_1 which touches the unit circle on the inside. We now see that the circle of convergence of the new series may very possibly extend beyond the boundary of the old. This will always be the case, in fact, when z_1 is not real and positive. If z_1 is real and negative, the new circle will indeed include the old one entirely. (Cf. footnote to **99**, p. 176.)

2. Since a rational integral function

$$a_0 + a_1 z + a_2 z^2 + \cdots + a_m z^m$$

may be regarded as a power series, convergent everywhere, such functions are *analytic in the whole plane.* Hence the rational functions of general type

$$\frac{a_0 + a_1 z + \cdots + a_m z^m}{b_0 + b_1 z + \cdots + b_k z^k}$$

are analytic at all points of the plane at which the denominator is not 0, — i. e. everywhere, with the exception of a finite number of points. Their expansion in power series at a point z_0, at which the denominator is $\neq 0$, is obtained as follows: If z is replaced by $z_0 + (z - z_0)$ both in the numerator and denominator of such a function, these being then rearranged in powers of $(z - z_0)$, the function takes the form

$$\frac{a_0' + a_1'(z - z_0) + \cdots + a_m'(z - z_0)^m}{b_0' + b_1'(z - z_0) + \cdots + b_k'(z - z_0)^k},$$

where, on account of our assumption, $b_0' \neq 0$. We may now carry out the division in accordance with **105**, 4 and expand the quotient in the required power series[26] of the form $\Sigma c_n (z - z_0)^n$.

II. The exponential function.

The series

$$1 + \frac{z}{1!} + \frac{z^2}{2!} + \cdots + \frac{z^n}{n!} + \cdots$$

is a power series converging everywhere, and therefore defines a function regular in the whole plane, i. e. an *integral function.* To every point z of the complex plane there corresponds a definite number w, the sum of the above series.

This function, which for real values of z has the value e^z as defined in **33**, may be used to define powers of the base e (and then further those of any *positive* base) for all complex exponents:

---

[26] An alternative method consists in first splitting up the function into partial fractions. Leaving out of account any part which represents a rational integral function, we are then concerned with the sum of a finite number of fractions of the form

$$\frac{A}{(z-a)^q} = \frac{A}{(-a)^q} \cdot \left(\frac{1}{1 - \frac{z}{a}}\right)^q,$$

each of which we may, by 1, expand separately in a power series of the form $\Sigma c_n (z - z_0)^n$, provided $z_0 \neq a$. — This method enables us to see, moreover, that the radius of the resulting expansion will be equal to the distance of z_0 from the nearest point at which the denominator of the given function vanishes.

236. Definition. *For all real or complex exponents, the meaning to be attributed to the power e^z is defined, without ambiguity, by the relation*

$$e^z = 1 + \frac{z}{1!} + \frac{z^2}{2!} + \cdots + \frac{z^n}{n!} + \cdots.$$

And if p is any positive number, p^z shall denote the value determined, without ambiguity, by the formula

$$p^z = e^{z \log p},$$

where $\log p$ *is the (real) natural logarithm of p as defined*[27] *in* **36.** (For a non-positive base b, the power b^z can no longer be *uniquely* defined; cf., however, **244.**)

As there was no meaning attached *per se* to the idea of powers with complex exponents, we may interpret them in any manner we please. Reasons of suitability and convenience can alone determine the choice of a particular interpretation. That the definition just given is a thoroughly suitable one, results from formula **91**, example 3 (leaving out of account the obvious requirement that the new definition must coincide with the old one for real values of the exponent[28]); this formula was proved by means of a multiplication of series, the validity of which holds equally for real and complex variables, and the formula must accordingly also hold for any complex exponent; it is

237. $$e^{z_1} \cdot e^{z_2} = e^{z_1 + z_2}$$

whence also

$$p^{z_1} \cdot p^{z_2} = p^{z_1 + z_2}.$$

This important fundamental law for the algebra of powers therefore certainly remains true. At the same time it provides us with the key to the further study of the function e^z.

238. 1. Calculation of e^z. For real y's, we have

$$e^{iy} = \sum_{n=0}^{\infty} \frac{(iy)^n}{n!} = \sum_{k=0}^{\infty} (-1)^k \frac{y^{2k}}{(2k)!} + i \sum_{k=0}^{\infty} (-1)^k \frac{y^{2k+1}}{(2k+1)!}$$
$$= \cos y + i \sin y.$$

---

[27] It may be noted how far removed this definition is from the elementary definition "x^k is the product of k factors all equal to x". — At first sight, there is no knowing *what value* belongs e. g. to 2^i; yet this value is in any case uniquely determined by the above definition.

[28] By **234**, 2, there can exist *no other* function than the function e^z just defined which is regular in the neighbourhood of the origin and coincides on the real axis $z = x$ with the function e^x defined by **33**. For this reason we may indeed say that every definition of e^z differing from the above would necessarily be unsuitable.

Hence it follows that, for $z = x + iy$,

$$e^z = e^{x+iy} = e^x \cdot e^{iy} = e^x (\cos y + i \sin y).$$

By means of this formula[29] the value of e^z may easily be determined for all complex z's.

This formula enables us, besides, to obtain in a convenient and complete manner an idea of the values which the function e^z assumes at the various points of the complex plane (in short, of its *stock of values*). We note the following facts.

2. *We have* $|e^z| = e^{\Re(z)} = e^x$. In fact

$$|e^{iy}| = |\cos y + i \sin y| = \sqrt{\cos^2 y + \sin^2 y} = 1,$$

hence $|e^z| = |e^x| \cdot |e^{iy}| = e^x$, because $e^x > 0$ and the second factor $= 1$. Similarly,

$$\operatorname{am} e^z = \Im(z) = y,$$

also from the formula **238**, 1 just used.

3. *e^z has the periods $2k\pi i$,* that is to say, for all values of z,

$$e^z = e^{z+2\pi i} = e^{z+2k\pi i}, \qquad (k \gtreqless 0, \text{ integral}).$$

For if we increase z by $2\pi i$ its imaginary part y increases by 2π, while its real part remains unaltered, and by 1. and § 24, 2, this leaves the value of the function unchanged. Every value which e^z is able to assume accordingly occurs in the strip $-\pi < \Im(z) = y \leq \pi$, or in any strip which may be obtained from it by a parallel translation. Every such strip is called a *period-strip*; Fig. 11 represents the first-named of these strips.

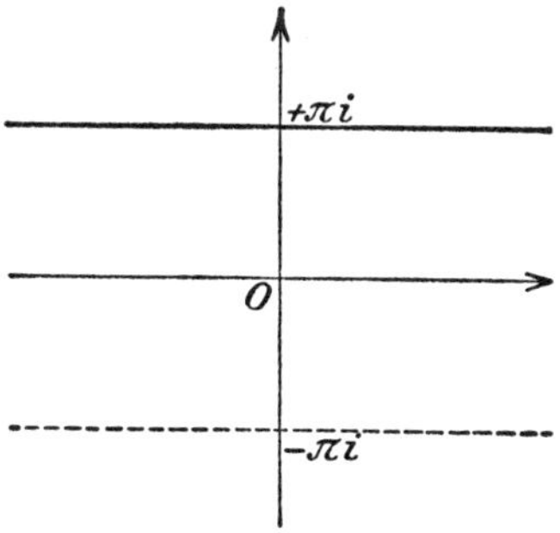

Fig. 11.

4. *e^z has no other period,* — indeed, more precisely: if between two *special* numbers z_1 and z_2 we have the relation

$$e^{z_1} = e^{z_2},$$

this necessarily implies that

$$z_2 = z_1 + 2k\pi i.$$

For we first infer that $e^{z_2 - z_1} = 1$; then we note that if

$$e^z = e^{x+iy} = e^x (\cos y + i \sin y) = 1,$$

[29] *Euler:* Intr. in Analysin inf. Vol. I, § 138, 1748.

we must by 2. have $e^x = 1$, hence $x = 0$. Further, we also have

$$\cos y + i \sin y = 1$$

i. e.

$$\cos y = 1, \quad \sin y = 0,$$

hence $y = 2 k \pi$. Thus, as asserted,

$$z = z_2 - z_1 = 2 k \pi i.$$

5. *e^z assumes every value $w \neq 0$ once and only once in the period strip*; or: the equation $e^z = w_1$, for given $w_1 \neq 0$, has one and only one solution in that strip.

If $w_1 = R_1 (\cos \Phi_1 + i \sin \Phi_1)$ with $R_1 > 0$, the number

$$z_1 = \log R_1 + i \Phi_1$$

is certainly *a* solution of $e^z = w_1$, as

$$e^{z_1} = e^{\log R_1} e^{i \Phi_1} = R_1 (\cos \Phi_1 + i \sin \Phi_1) = w_1.$$

By 3., the numbers

$$z_1 + 2 k \pi i \qquad (k = 0, \pm 1, \pm 2, \ldots)$$

are also solutions of the same equation, and by 4. no other solutions can exist. Now k may always be chosen, in one and only one way, so that

$$-\pi < \mathfrak{J}(z_1 + 2 k \pi i) \leqq + \pi, \qquad \text{q. e. d.}$$

6. *The value 0 is never assumed by e^z*; for, by **237**,

$$e^z \cdot e^{-z} = 1,$$

so that e^z can never be 0.

7. The derivative $(e^z)'$ of e^z is again $= e^z$, as follows at once by differentiating term-by-term the power series that defines e^z.

8. From **238**, 1, we also deduce the special values

$$e^{2\pi i} = 1, \quad e^{\pi i} = -1, \quad e^{\frac{\pi i}{2}} = i, \quad e^{\frac{-\pi i}{2}} = -i.$$

III. The functions cos z and sin z.

In the case of the trigonometrical functions, we can again use the expansions in power series convergent everywhere to define the functions

239. for complex values of the variable.

Definition. *The sum of the power series, convergent everywhere,*

$$1 - \frac{z^2}{2!} + \frac{z^4}{4!} - + \ldots + (-1)^k \frac{z^{2k}}{(2k)!} + \cdots,$$

is denoted by **cos** z, *that of the power series, also convergent everywhere,*

$$\frac{z}{1!} - \frac{z^3}{3!} + \frac{z^5}{5!} - + \ldots + (-1)^k \frac{z^{2k+1}}{(2k+1)!} + \cdots,$$

by **sin** z, *— for every complex z.*

For real $z = x$, this certainly gives us the former functions $\cos x$ and $\sin x$. We have only to verify, as before, whether these definitions are suitable ones, in the sense that the functions defined, — *which are analytic in the whole plane, i. e. integral functions,* — possess the same fundamental properties as the real functions[30] $\cos x$ and $\sin x$. That this is again the case, to the fullest extent, is shown by the following statement of their main properties:

1. *For every complex z, we have the formulae* **240.**

$$\cos z + i \sin z = e^{iz}, \qquad \cos z - i \sin z = e^{-iz},$$

whence further

$$\cos z = \frac{e^{iz} + e^{-iz}}{2}, \qquad \sin z = \frac{e^{iz} - e^{-iz}}{2i}.$$

(***Euler's formulae***).

The proof follows immediately by replacing the functions on both sides by the power series which define them.

2. *The addition theorems remain valid for complex values of z:*

$$\cos(z_1 + z_2) = \cos z_1 \cos z_2 - \sin z_1 \sin z_2,$$
$$\sin(z_1 + z_2) = \cos z_1 \sin z_2 + \sin z_1 \cos z_2,$$

This follows from 1., since by **237**

$$e^{i(z_1+z_2)} = e^{iz_1} \cdot e^{iz_2},$$

and the latter involves

$$\begin{aligned}\cos(z_1 + z_2) &+ i \sin(z_1 + z_2)\\ &= (\cos z_1 + i \sin z_1)(\cos z_2 + i \sin z_2)\\ &= (\cos z_1 \cos z_2 - \sin z_1 \sin z_2) + i(\cos z_1 \sin z_2 + \sin z_1 \cos z_2).\end{aligned}$$

Substituting $-z_1$ and $-z_2$ for z_1 and z_2, and taking into account the fact that $\cos z$ is an even, $\sin z$ an odd function, we obtain a similar formula, which differs from the last only in that i appears to be changed to $-i$ on either side. Addition and subtraction of the two relations give us the required addition formulae.

3. The fact that the addition theorems for our two integral functions are formally the same as those for the functions $\cos x$ and $\sin x$ of the real variable x, not only sufficiently justifies our designating these functions by $\cos z$ and $\sin z$, but shows, at the same time, that *the entire formal machinery of the so-called goniometry,* since it is evolved from the addition theorems, *remains unaltered.* In particular,

---

[30] Here again a remark analogous to that on p. 412, footnote 28, may be made.

we have the formulae

$$\cos^2 z + \sin^2 z = 1, \qquad \cos 2z = \cos^2 z - \sin^2 z,$$
$$\sin 2z = 2 \sin z \cos z, \qquad \text{etc.}$$

valid without change *for every complex z*.

4. *The period-properties of the functions are also retained in the complex domain.* For it follows from the addition theorems that

$$\cos(z + 2\pi) = \cos z \cdot \cos 2\pi - \sin z \cdot \sin 2\pi = \cos z,$$
$$\sin(z + 2\pi) = \cos z \cdot \sin 2\pi + \sin z \cdot \cos 2\pi = \sin z.$$

5. *The functions* $\cos z$ *and* $\sin z$ *possess no other zeros in the complex domain besides those already known in the real domain*[31]. In fact, $\cos z = 0$ necessarily involves, by 1., $e^{iz} = -e^{-iz}$ or

$$e^{2iz} = -1 = e^{\pi i}$$

i. e.

$$e^{2iz - \pi i} = 1.$$

By **238**, 4, this can only occur when

$$2iz - \pi i = 2k\pi i \qquad \text{or} \qquad z = (2k+1)\frac{\pi}{2}.$$

Similarly, $\sin z = 0$ implies $e^{iz} = e^{-iz}$, or $e^{2iz} = 1$, i. e. $2iz = 2k\pi i$, or $z = k\pi$, q. e. d.

6. *The relation* $\cos z_1 = \cos z_2$ *is satisfied if, and only if,* $z_2 = \pm z_1 + 2k\pi$, — i. e. under the same condition as in the real domain. *Similarly* $\sin z_1 = \sin z_2$ *if, and only if,* $z_2 = z_1 + 2k\pi$ *or* $z_2 = \pi - z_1 + 2k\pi$. It follows in fact from

$$\cos z_1 - \cos z_2 = -2 \sin\frac{z_1 + z_2}{2} \sin\frac{z_1 - z_2}{2} = 0,$$

by 5., that either $\frac{z_1 + z_2}{2}$ or $\frac{z_1 - z_2}{2}$ must $= k\pi$; similarly it follows from

$$\sin z_1 - \sin z_2 = 2 \cos\frac{z_1 + z_2}{2} \sin\frac{z_1 - z_2}{2} = 0,$$

by 5., that either $\frac{z_1 - z_2}{2} = k\pi$ or $\frac{z_1 + z_2}{2} = (2k+1)\frac{\pi}{2}$.

7. *The functions* $\cos z$ *and* $\sin z$ *assume every complex value* w *in the period-strip, i. e. in the strip* $-\pi < \Re(z) \leq +\pi$; the equations $\cos z = w$ and $\sin z = w$ have indeed exactly *two* solutions in that strip, if $w \neq \pm 1$, but only *one*, if $w = \pm 1$.

---

[31] Or in other words: The sum of the power series $1 - \frac{z^2}{2!} + - \cdots = 0$ if, and only if, z has one of the values $(2k+1)\frac{\pi}{2}$, $k = 0, \pm 1, \pm 2, \ldots$; and similarly for the sine series.

Proof. In order to have $\cos z = w$, we must have $e^{iz} + e^{-iz} = 2w$ or $e^{iz} = w + \sqrt{w^2 - 1}$. (Here $\sqrt{r(\cos\varphi + i\sin\varphi)}$ is defined as one of the two numbers, for instance $r^{\frac{1}{2}}\left(\cos\frac{\varphi}{2} + i\sin\frac{\varphi}{2}\right)$, whose square is the quantity under the radical sign.) Since in any case[32] $w + \sqrt{w^2 - 1} \neq 0$, there certainly exists a complex number z' such that $-\pi < \mathfrak{I}(z') \leqq +\pi$, for which $e^{z'} = w + \sqrt{w^2 - 1}$, — by **238**, 5. Writing $-iz' = z$, we have $-\pi < \mathfrak{R}(z) \leqq +\pi$ and $e^{iz} = w + \sqrt{w^2 - 1}$, or $\cos z = w$. This equation therefore certainly has at least one solution in the period-strip. By 6., however, a second solution, different from it, (viz. $-z$), exists in the period-strip if, and only if, $z \neq 0$ and $\neq \pi$, i. e. $w \neq \pm 1$.

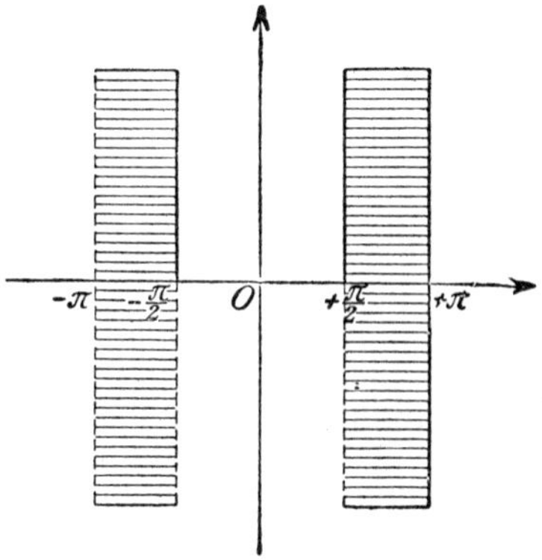

Fig. 12.

We reason in precisely the same manner with regard to the equation $\sin z = w$. In this case, we can also easily convince ourselves that there is always one and only one solution of the equation in the portion of the period-strip left *unshaded* in Fig. 12, if we include the parts of the rim indicated in black, but omit the parts represented by the dotted lines (see VI below).

8. For the derivatives, we have as in the real case,

$$(\cos z)' = -\sin z, \quad (\sin z)' = \cos z.$$

IV. The functions cot z and tan z.

1. Since $\cos z$ and $\sin z$ are analytic in the whole plane, the functions

$$\cot z = \frac{\cos z}{\sin z} \quad \text{and} \quad \tan z = \frac{\sin z}{\cos z}$$

will also be regular in the whole plane, with the exception of the points $k\pi$ for the former and $(2k+1)\frac{\pi}{2}$ for the latter, which are the zeros of $\sin z$ and $\cos z$ respectively. Their expansions in power series may be obtained by carrying out the division of the cosine and sine series. Since this operation is of a purely formal nature, the result must be the same as it was in the real domain. Accordingly, by § 24, 4, where the result of this division was obtained by a special artifice, we have

$$z\cot z = \sum_{k=0}^{\infty} (-1)^k \frac{2^{2k} B_{2k}}{(2k)!} z^{2k}, \qquad \textbf{241.}$$

$$\tan z = \sum_{k=1}^{\infty} (-1)^{k-1} \frac{2^{2k}(2^{2k}-1) B_{2k}}{(2k)!} z^{2k-1}.$$

---

[32] In fact, since $w^2 - 1 \neq w^2$, $\sqrt{w^2 - 1} \neq \pm w$.

On account of **94** and **136**, we are now also in a position to determine the exact radius of convergence of these series. The absolute value of the coefficient of z^{2k} in the first series, by **136**, is

$$(-1)^{k-1}\frac{2^{2k}\cdot B_{2k}}{(2k)!}=\frac{2\cdot 2^{2k}}{(2\pi)^{2k}}\cdot\sum_{n=1}^{\infty}\frac{1}{n^{2k}}.$$

Its $(2k)^{\text{th}}$ root is

$$\frac{1}{\pi}\cdot\sqrt[2k]{2\cdot s_{2k}},$$

if s_{2k} denotes the sum $\sum\frac{1}{n^{2k}}$. The latter lies between 1 and 2 for every $k=1, 2, \ldots$ (for it is $\frac{\pi^2}{6}$ when $k=1$, and is less than this for every other k, but >1); therefore

$$\sqrt[2k]{(-1)^{k-1}\frac{2^{2k}B_{2k}}{(2k)!}}\to\frac{1}{\pi},$$

and the radius of the cot-series $=\pi$, by **94**. Similarly that of the tan-series is found to be $\frac{\pi}{2}$.

2. $\cot z$ *and* $\tan z$ *have the period* π. For $\cos z$ and $\sin z$ *both* change in sign alone when z is increased by π. Here again we may show, more precisely, that

$$\cot z_1=\cot z_2 \quad\text{and}\quad \tan z_1=\tan z_2$$

involve

$$z_2=z_1+k\pi \qquad (k=0, \pm 1, \ldots).$$

In fact, it follows from

$$\cot z_1-\cot z_2=\frac{\cos z_1}{\sin z_1}-\frac{\cos z_2}{\sin z_2}=\frac{\sin(z_2-z_1)}{\sin z_1\cdot\sin z_2}$$

that in the case of the first equation $\sin(z_2-z_1)=0$, i. e. $z_2-z_1=k\pi$. Similarly in the case of the second.

3. *In the "period-strip", i. e. in the strip* $-\frac{\pi}{2}<\Re(z)\leqq+\frac{\pi}{2}$, $\cot z$ *and* $\tan z$ *assume every complex value* $w\neq\pm i$ *just once; the values* $w=\pm i$ *are never assumed.* To see this, write $e^{2iz}=\zeta$. The equation $\cot z=w$ then becomes

$$i\frac{\zeta+1}{\zeta-1}=w \quad\text{or}\quad \zeta=\frac{w+i}{w-i}.$$

For each $w\neq\pm i$, ζ is a definite complex number $\neq 0$ and (by II, 5) there accordingly exists a z' such that $-\pi<\Im(z')\leqq\pi$, for which $e^{z'}=\zeta$. For $z=-i\frac{z'}{2}$, we then have

$$-\frac{\pi}{2}<\Re(z)\leqq\frac{\pi}{2} \quad\text{and}\quad \cot z=w,$$

i. e. z is a solution of the latter equation in the prescribed strip. By 2. there can be no other solution in this strip. The impossibility of a solution for $\cot z = \pm i$ results from the fact that these equations both involve

$$\cot^2 z + 1 = 0$$

which cannot be satisfied by any value of z, as $\cos^2 z + \sin^2 z = 1$. — For $\tan z$ the procedure is quite similar.

4. *The expansion in partial fractions deduced in* § 24, 5 *for the cotangent in the real domain remains valid in the same form for every complex z different from* $0, \pm 1, \pm 2, \ldots$ (and similarly for the expansions of $\tan z$, $\frac{1}{\sin z}$ etc.). Indeed the complete reasoning given there may be interpreted in the "complex" sense, without altering a single word[33]. In particular, for every z satisfying the above condition,

$$\pi z \cot \pi z = 1 + z \sum_{n=1}^{\infty} \left[\frac{1}{z-n} + \frac{1}{z+n}\right] = 1 + 2 z^2 \sum_{n=1}^{\infty} \frac{1}{z^2 - n^2}.$$

Now

$$\pi z \cot \pi z = i \pi z \frac{e^{i\pi z} + e^{-i\pi z}}{e^{i\pi z} - e^{-i\pi z}} = i \pi z \frac{e^{2i\pi z} + 1}{e^{2i\pi z} - 1};$$

it follows, if we substitute z for $2 i \pi z$, that

$$\frac{z}{2} \frac{e^z + 1}{e^z - 1} = 1 + \sum_{n=1}^{\infty} \frac{2 z^2}{z^2 + 4 \pi^2 n^2};$$

hence we obtain the expansion

$$\frac{1}{z}\left[\frac{1}{1 - e^{-z}} - \frac{1}{z} - \frac{1}{2}\right] = \sum_{n=1}^{\infty} \frac{2}{z^2 + 4 \pi^2 n^2},$$

valid for every complex $z \neq 2 k \pi i$ ($k \gtrless 0$, integer). This is the extension to the complex variable z of the remarkable expansion in partial fractions obtained on p. 378, and it exhibits the true connection between this expansion and that of $\cot z$, which previously seemed rather fortuitous.

V. The logarithmic series.

In § 25, we saw that the series

$$y = \sum_{n=1}^{\infty} \frac{(-1)^{n-1}}{n} x^n$$

represents for every $|x| < 1$ the inverse function of the exponential function $e^y - 1$; i. e. substituting for y in

$$y + \frac{y^2}{2!} + \frac{y^3}{3!} + \cdots$$

[33] It was precisely for this purpose that at the time we framed some of our estimates in a form somewhat different from that required for the real domain (e. g. those on pp. 206—207 to which footnote 26 refers).

the above series and rearranging (as is certainly allowed) in powers of x, we reduce the new series simply to x. This fact — because it is purely formal in character — necessarily remains when complex quantities are considered. Hence, for every $|z| < 1$,

$$e^w - 1 = z \quad \text{or} \quad e^w = 1 + z,$$

if w denotes the sum of the series

$$(L) \qquad w = \sum_{n=1}^{\infty} \frac{(-1)^{n-1}}{n} z^n.$$

We now adopt for the complex domain the

242. Definition. *A number a is said to be a natural logarithm of c, in symbols,*

$$a = \log c,$$

if $e^a = c$.

In accordance with II, 5, we may then assert that every complex number $c \neq 0$ possesses one, and only one, logarithm whose imaginary part lies between $-\pi$ exclusive and $+\pi$ inclusive (to the number 0, however, by II, 6, no logarithm can be assigned at all). This uniquely defined value will be more especially referred to as the *principal value* of the natural logarithm of c. Besides this value, there is an infinity of other logarithms of c, since with $e^a = c$ we have also $e^{a+2k\pi i} = c$; thus if a is the principal value of the logarithm of c, the numbers

$$a + 2k\pi i \qquad (k \gtreqless 0, \text{ integer})$$

must also be called logarithms of c. These values of the logarithm (for $k \neq 0$) are called its subsidiary values [34]. By **238**, 4 there can be no further logarithms of c. We have, for each of its values,

$$\Re(\log c) = \log|c|, \quad \Im(\log c) = \text{am } c,$$

if in the first of these relations $\log|c|$ denotes the (single-valued) real logarithm of the positive number $|c|$, and the second is interpreted as meaning that, taken as a whole, all the values of the one side are equal to all the values of the other.

With these definitions, we may assert in any case that the above series (L) provides *a* logarithm of $(1 + z)$. But we may at once prove more, namely the

243. Theorem. *The logarithmic series (L) gives, at each point of the unit circle (including its rim, with the exception of the point -1), the principal value of* $\log(1 + z)$.

---

[34] If c is real and positive, the principal value of $\log c$ coincides with the (real) natural logarithm as formerly defined (**36**, Def.).

Proof. That the series converges for each $z \neq -1$ for which $|z| \leqq 1$ was shown in **230**, 3. (We have only to put $-z$ for z there.) For this z, $\operatorname{am}(1+z)$ has precisely that value ψ for which

$$-\frac{\pi}{2} < \psi < +\frac{\pi}{2}.$$

Hence we have, for the imaginary part of the sum w of the series (L),

$$(\mathfrak{I}) \qquad \mathfrak{I}(w) = \mathfrak{I}(\log(1+z)) = \psi + 2ki,$$

with integral k. Now w is a continuous function of z in $|z| < 1$, and assumes the value 1 for $z = 0$. Hence $\mathfrak{I}(w)$ too is a continuous function in $|z| < 1$. Therefore, in the equation $(\mathfrak{I})$, k must have the same value for all these z. But for $z = 0$ we have clearly to take $k = 0$; hence this is its value in the whole of $|z| < 1$. Finally we learn from the application of *Abel*'s limit theorem that the sum of our series is still equal to the principal value of $\log(1+z)$ at the points $z \neq -1$ for which $|z| = 1$.

VI. The inverse sine series.

We saw in III, 7 that the equation $\sin w = z$, for a given complex $z \neq \pm 1$, has exactly *two* solutions, — for $z = \pm 1$ exactly *one*, — in the strip $-\pi < \mathfrak{R}(\pi) \leqq +\pi$. The two solutions (by III, 6) are symmetrical, either with respect to $+\frac{\pi}{2}$ or $-\frac{\pi}{2}$; accordingly, we may assert more precisely that the equation $\sin w = z$, for an arbitrary given z (inclusive of ± 1), has *one and only one* solution in the strip

$$-\frac{\pi}{2} \leqq \mathfrak{R}(w) \leqq +\frac{\pi}{2},$$

if the lower portions of its rim, from the real axis downwards, are omitted (cf. Fig. 12, where the parts of the rim not counted with the strip are drawn in dotted lines, and the others are marked by a continuous black line). This value of the solution of the equation $\sin w = z$, which is thus uniquely defined for *every* complex z, is called the *principal value* of the function

$$w = \sin^{-1} z.$$

All the remaining values are contained, by III, 6, in the two formulae

$$\sin^{-1} z + 2k\pi,$$

$$\pi - \sin^{-1} z + 2k\pi,$$

and may be called *subsidiary values* of the function.

For real values of x such that $|x| \leqq 1$, the series **123**,

$$y = x + \frac{1}{2}\frac{x^3}{3} + \frac{1\cdot 3}{2\cdot 4}\frac{x^5}{5} + \cdots$$

represents the inverse series of the sine power series

$$y - \frac{y^3}{3!} + \frac{y^5}{5!} - + \cdots.$$

Exactly the same considerations as in V. for the case of the logarithmic series now show that, for complex values of z such that $|z| \leqq 1$, the series

$$w = z + \frac{1}{2}\frac{z^3}{3} + \frac{1\cdot 3}{2\cdot 4}\frac{z^5}{5} + \cdots$$

is the inverse series of the sine power series $w - \frac{w^3}{3!} + - \cdots$. It therefore gives at any rate *one* of the values of $\sin^{-1} z$. That this actually is the *principal value*, may be seen from the fact that, for $|z| \leqq 1$, $z \neq \pm 1$,

$$|\Re(\sin^{-1} z)| \leqq |\sin^{-1} z| \leqq |z| + \frac{1}{2}\frac{|z|^3}{3} + \frac{1\cdot 3}{2\cdot 4}\cdot\frac{|z|^5}{5} + \cdots$$
$$= \sin^{-1}|z| \leqq \sin^{-1} 1 = \frac{\pi}{2},$$

— a condition which the principal value alone fulfils.

VII. The inverse tangent series.

The equation $\tan w = z$, as we know from IV, 3, has for every given $z \neq \pm i$ one and only one solution in the strip $-\frac{\pi}{2} < \Re(w) \leqq +\frac{\pi}{2}$. This is called the *principal value* of the function

$$w = \tan^{-1} z$$

the other values of which (by IV, 2) are then obtained from the formula $\tan^{-1} z + k\pi$. The equations $\tan z = \pm i$ have no solutions whatever.

Almost word for word the same considerations as above again show that, for $|z| < 1$, the series

$$\text{(A)} \qquad w = z - \frac{z^3}{3} + \frac{z^5}{5} - + \cdots$$

gives *one* of the solutions of $\tan w = z$. To show that this is actually the principal value of $\tan^{-1} z$, we have to show that the real part of the sum of the series lies between $-\frac{\pi}{2}$ (exclusive) and $+\frac{\pi}{2}$ (inclusive). This remains true for every $z \neq \pm i$ on $|z| = 1$, as well as for $|z| < 1$, and is proved as follows:

The sum w of the series (A), as may be seen by substituting the log-series, is

$$w = \frac{1}{2i}\log(1 + iz) - \frac{1}{2i}\log(1 - iz)$$

for every $|z| \leqq 1$, $z \neq \pm i$, where principal values are taken for both logarithms. Accordingly,

$$\Re(w) = \frac{1}{2} \Im \log(1 + iz) - \frac{1}{2} \Im \log(1 - iz);$$

by **243**, both terms of the difference lie between $-\frac{\pi}{4}$ and $+\frac{\pi}{4}$, hence $\Re(w)$ lies between $-\frac{\pi}{2}$ and $+\frac{\pi}{2}$, the two extreme values being excluded in either case. Thus the series (A) certainly represents the principal value of $\tan^{-1} z$, provided $|z| \leqq 1$ and $z \neq \pm i$, q. e. d.

VIII. The binomial series.

To complete our present treatment of the special power series investigated in the real domain, we have only to consider the binomial series

$$(1 + x)^{\alpha} = \sum_{n=0}^{\infty} \binom{\alpha}{n} x^n$$

in the case where the quantities occurring there — i. e. the *exponent* α as well as the *variable* x — assume complex values. We start with the

Definition. *The name of **principal value of the power** b^a, where a and b denote any complex numbers, with $b \neq 0$ as the only condition, is given to the number uniquely defined by the formula* **244.**

$$b^a = e^{a \log b}$$

when $\log b$ *is given its principal value.* — By choosing other values of $\log b$, we obtain further values of the power, which may be called its *subsidiary values.* All these values are contained in the formula

$$b^a = e^{a[\log b + 2k\pi i]},$$

each value being represented exactly once, if $\log b$ is given its principal value and k takes all integral values $\gtreqless 0$.

Remarks and Examples.

1. A power b^a accordingly has an infinite number of values in general, but possesses one and only one principal value.

2. The symbol i^i, for instance, denotes the infinity of numbers (all *real* numbers, moreover)

$$e^{i(\log i + 2k\pi i)} = e^{i\left(\frac{\pi i}{2} + 2k\pi i\right)} = e^{-\frac{\pi}{2} - 2k\pi}, \quad (k = 0, \pm 1, \pm 2, \ldots)$$

of which $e^{-\frac{\pi}{2}}$ is the principal value of the power i^i.

3. The only case in which a power b^a will *not* have an infinite number of values is that in which

$$e^{a \cdot 2k\pi i} \qquad (k = 0, \pm 1, \pm 2, \ldots)$$

gives only a *finite* number of values; this will occur if, and only if, $k \cdot a$ assumes, for $k = 0, \pm 1, \pm 2, \ldots$, only a finite number of *essentially different* values. Here two numbers are described (just for the moment) as *essentially different* if, and only if, they do *not* differ merely by a (real) integer. Now this is the case if, and only if, a is a real rational number, as may be seen at once; and the number of "essentially different" values which may in this case be assumed by $k \cdot a$ is given by the smallest positive denominator with which a may be written in fractional form.

4. It follows that $b^{\frac{1}{m}} = \sqrt[m]{b}$, where m is a positive integer, has exactly m different values, one of which is quite definitely distinguished as the principal value.

5. The number of different values of b^a will reduce to one, by 3. and 4., if, and only if, a is a rational number of denominator 1, i. e. a real integer. For all real integral exponents (but for these alone), the power thus remains now as before a single-valued symbol.

6. If b is positive and a real, the value formerly defined (v. **33**) as the power b^a is now the *principal value* of this power.

7. Similarly, the values defined in **236** for e^z and p^z, $(p > 0)$, are now, more precisely, the *principal values* of these powers. In themselves, these symbols would represent, for complex values of z, an infinity of values, in accordance with our last definition. *Nevertheless, we shall keep in future to the convention that e^z, and generally p^z for any positive p, shall represent the value defined by* **236**, *i. e. the principal value only.*

8. The following theorems will show that it is consistent to define b^a also for $b = 0$ when $\Re(a) > 0$. The value attributed to the power in that case is 0 (uniquely).

After making these preliminary preparations, we proceed to prove the following far-reaching

245. Theorem[35]. *For any complex exponent α and any complex z in $|z| < 1$, the binomial series*

$$\sum_{n=0}^{\infty} \binom{\alpha}{n} z^n \equiv 1 + \binom{\alpha}{1} z + \binom{\alpha}{2} z^2 + \cdots + \binom{\alpha}{n} z^n + \cdots$$

converges and has for sum the principal value of the power $(1+z)^\alpha$.

Proof. The convergence follows word for word as in the case of real z's and α's (v. pp. 209—210), so that we have only to prove the statement as to the sum of the series. Now for real x's such that $|x| < 1$, and real α's, we may substitute

$$\alpha \sum_{n=1}^{\infty} \frac{(-1)^{n-1}}{n} x^n = \alpha \log(1 + x)$$

[35] *Abel*: J. f. d. reine u. angew. Math., V. 1, p. 311. 1826.

for y in the exponential series $e^y = 1 + y + \frac{y^2}{2!} + \cdots$ and so obtain, after rearranging in powers of x (allowed by **104**), the power series for $e^{\alpha \log(1+x)} = (1+x)^\alpha$, i. e. the binomial series $\sum \binom{\alpha}{n} x^n$. Let us proceed in this manner, purely formally in the first instance, assuming α complex and writing z for x; i. e. we substitute

$$w = \alpha \cdot \sum_{n=1}^{\infty} \frac{(-1)^{n-1}}{n} z^n \quad \text{in} \quad e^w = \sum_{n=1}^{\infty} \frac{w^n}{n!}$$

and rearrange in powers of z. We necessarily obtain — without reference as yet to any question of convergence — the series

$$\sum_{n=0}^{\infty} \binom{\alpha}{n} z^n,$$

whose sum would therefore be proved to be $e^{\alpha \log(1+z)} = (1+z)^\alpha$ (where the principal value is taken for the logarithm and hence for the power also), if we could show that the rearrangement carried out was permissible. Now by **104**, this is certainly so; in fact the exponential series converges everywhere and the series $\alpha \sum \frac{(-1)^{n-1}}{n} z^n$ remains convergent for $|z| < 1$ when α and all the terms of the series are replaced by their absolute values. This proves the theorem in its full extent.

If we split up $(1+z)^\alpha$ into its real and imaginary parts, we obtain a formula due to *Abel*, which is complicated in appearance, but which for that very reason shows how far-reaching a result is contained in the preceding theorem, and from which we also obtain a means for *evaluating* the power $(1+z)^\alpha$. Writing $z = r(\cos\varphi + i \sin\varphi)$ and $\alpha = \beta + i\gamma$, $0 < r < 1$, φ, β, γ all real, and writing

$$1 + z = R(\cos\Phi + i \sin\Phi),$$

we have

$$R = \sqrt{1 + 2r\cos\varphi + r^2}, \quad \Phi = \text{principal value}^{36} \text{ of } \tan^{-1} \frac{r \sin\varphi}{1 + r\cos\varphi}.$$

With these values of R and Φ, we thus obtain

$$\begin{aligned}(1+z)^\alpha &= e^{(\beta+\gamma i)[\log R + i\Phi]} \\ &= R^\beta \cdot e^{-\gamma\Phi} \cdot [\cos(\beta\Phi + \gamma \log R) + i \sin(\beta\Phi + \gamma \log R)].\end{aligned}$$

For the case $|z| < 1$, theorem **245** and the remark just made completely answer the question as to the sum of the binomial series. We have now only to consider the points of the circumference $|z| = 1$. From *Abel*'s theorem, together with the continuity of the principal value of $\log(1+z)$ for every $z \neq -1$ in $|z| \leq 1$ and the continuity of the exponential function, we at once deduce the

[36] Φ has accordingly to be chosen between $+\frac{\pi}{2}$ and $-\frac{\pi}{2}$.

246. **Theorem.** *At every point of the rim $|z| = 1$ of the unit circle, at which the binomial series continues to converge, except possibly for $z = -1$, its sum remains now as previously the principal value of $(1+z)^\alpha$.*

The determination whether, and for what values of α and z, the binomial series continues to converge *on the rim* of the unit circle presents no difficulties after the preparations made in this respect (and chiefly for this purpose) in § 53. The theorem we have is the following, which sums up the entire question once more:

247. **Theorem.** *The binomial series $\sum_{n=0}^{\infty}\binom{\alpha}{n} z^n$ reduces, for real integral values of $\alpha \geqq 0$, to a finite sum, and has then the (ipso facto unique) value $(1+z)^\alpha$; in particular for $\alpha = 0$ it has the value 1 (also when $z = -1$). If α does not have one of these values, the series converges absolutely for $|z| < 1$ and diverges for $|z| > 1$, while it exhibits the following behaviour on the circumference $|z| = 1$:*

a) *if $\Re(\alpha) > 0$, it converges absolutely at all points on the circumference;*

b) *if $\Re(\alpha) \leqq -1$, it diverges at all these points;*

c) *if $-1 < \Re(\alpha) \leqq 0$, it diverges at $z = -1$ and converges conditionally at every other point of the circumference.*

The sum of the series when it converges is invariably the principal value of $(1+z)^\alpha$; in particular, its value is 0 in the case $z = -1$.

Proof. Writing $(-1)^n\binom{\alpha}{n} = a_{n+1}$, we have

$$\frac{a_{n+1}}{a_n} = -\frac{\binom{\alpha}{n}}{\binom{\alpha}{n-1}} = \frac{n-(\alpha+1)}{n} = 1 - \frac{\alpha+1}{n};$$

hence theorem **229** may be applied, and the validity of a), b) and c) follows immediately. Only the case of the point $z = -1$, i. e. the convergence of the series

$$\sum_{n=0}^{\infty}(-1)^n\binom{\alpha}{n}$$

requires special investigation. Now

$$1 - \binom{\alpha}{1} + \binom{\alpha}{2} = 1 - \alpha + \frac{\alpha(\alpha-1)}{1\cdot 2} = (1-\alpha)\left(1-\frac{\alpha}{2}\right),$$

$$1 - \binom{\alpha}{1} + \binom{\alpha}{2} - \binom{\alpha}{3} = (1-\alpha)\left(1-\frac{\alpha}{2}\right) - \frac{\alpha}{3}\,\frac{(\alpha-1)(\alpha-2)}{1\cdot 2}$$
$$= (1-\alpha)\left(1-\frac{\alpha}{2}\right)\left(1-\frac{\alpha}{3}\right),$$

and in general, as may at once be verified by induction:

$$1-\binom{\alpha}{1}+\binom{\alpha}{2}-+\cdots+(-1)^n\binom{\alpha}{n}=(1-\alpha)\left(1-\frac{\alpha}{2}\right)\cdots\left(1-\frac{\alpha}{n}\right);$$

the partial *sums* of our series are equal to the partial *products,* with the same index n, of the product $\prod_{n=1}^{\infty}\left(1-\frac{\alpha}{n}\right)$. The behaviour of this product is immediately evident. In fact:

1. If $\Re(\alpha)=\beta>0$, choose β' such that $0<\beta'<\beta$; for every sufficiently large n, say $n\geqq m$,

$$\left|1-\frac{\alpha}{n}\right|<1-\frac{\beta'}{n},$$

hence

$$\left|\left(1-\frac{\alpha}{m}\right)\left(1-\frac{\alpha}{m+1}\right)\cdots\left(1-\frac{\alpha}{n}\right)\right|<\left(1-\frac{\beta'}{m}\right)\left(1-\frac{\beta'}{m+1}\right)\cdots\left(1-\frac{\beta'}{n}\right).$$

By **126**, 2, it follows at once that the partial products, and hence the partial sums of our series, tend to 0. *The series therefore converges*[37] *to the sum* 0.

2. If, however, $R(\alpha)=-\beta<0$, we have

$$\left|1-\frac{\alpha}{n}\right|>1+\frac{\beta}{n},$$

whence it again follows by multiplication that

$$\left|\left(1-\frac{\alpha}{1}\right)\left(1-\frac{\alpha}{2}\right)\cdots\left(1-\frac{\alpha}{n}\right)\right|>\left(1+\frac{\beta}{1}\right)\left(1+\frac{\beta}{2}\right)\cdots\left(1+\frac{\beta}{n}\right),$$

and hence that the left hand side tends to ∞. *The series therefore diverges in this case.*

3. If, finally, $\Re(\alpha)=0$, $\alpha=-i\gamma$, say, with $\gamma\lessgtr 0$, the nth partial sum of our series is

$$(1+i\gamma)\left(1+\frac{i\gamma}{2}\right)\cdots\left(1+\frac{i\gamma}{n}\right).$$

The fact that this value tends to no limit as $n\to+\infty$ may be proved most speedily in the present connection as follows: On account of the absolute convergence of the series $\sum\left(\frac{i\gamma}{n}\right)^2$, we have, by § 29, theorem 10:

$$\left(1+\frac{i\gamma}{1}\right)\left(1+\frac{i\gamma}{2}\right)\cdots\left(1+\frac{i\gamma}{n}\right)\sim e^{i\gamma\left(1+\frac{1}{2}+\cdots+\frac{1}{n}\right)}.$$

Letting $n\to+\infty$, the right hand side evidently tends to no limit; on the contrary, the points which it represents for successive values of n circulate incessantly round the circumference of the unit circle in a constant sense, the interval between successive points becoming smaller

---

[37] The mere *convergence* of $\sum(-1)^n\binom{\alpha}{n}$ follows already from **228** and we see that the convergence is absolute when $\Re(\alpha)>0$. It is the fact of the sum being 0 which requires the artifice employed above for its detection.

and smaller at each turn. In view of the asymptotic relationship, the same is therefore true of the left hand side. *Hence our series* $\sum (-1)^n \binom{\alpha}{n}$ *also diverges when* $\Re(\alpha) = 0$. Thus theorem **247** is established in all its parts, the behaviour of the binomial series is determined for every value of z and of α, and its sum for all points of convergence is given by means of a "closed expression".

§ 56. Series of variable terms. Uniform convergence. *Weierstrass'* theorem on double series.

The fundamental remarks on series of variable terms

$$\sum_{n=0}^{\infty} f_n(z)$$

are substantially the same for the complex as for the real domain (v. § 46); but instead of the common *interval* of definition we must now assume a common *region* of definition, which for simplicity — this is also quite sufficient for most purposes — we shall suppose to be a *circle* (cf. p. 403, footnote 17). We accordingly assume that

1. *A circle* $|z - z_0| < r$ *exists, in which the functions* $f_n(z)$ *are all defined.*

2. *For every individual* z *in the circle* $|z - z_0| < r$, *the series*

$$\sum_{n=0}^{\infty} f_n(z)$$

is convergent.

The series $\sum f_n(z)$ then has, for every z in the circle, a definite sum, whose value therefore defines a function of z (in the sense of the definition on p. 403). We accordingly write

$$\sum_{n=0}^{\infty} f_n(z) = F(z).$$

The same problems as those discussed in §§ 46 and 47 for the case of real variables arise in connection with the functions represented by complex series of variable terms. In the real domain, however, it is of the greatest importance, both for the theory and its applications, to make use of the concept of function in its most general form, while in the complex domain this has not been found profitable. The usual restriction, which is sufficiently wide for all ordinary purposes, is to consider *analytic* functions only. We therefore assume further that

3. *The functions* $f_n(z)$ *are all analytic in the circle* $|z - z_0| < r$, *i. e. expressible by power series with* z_0 *as centre and radius not less than a fixed number* r.

We then speak for brevity of ***series of analytic functions***[38]; the chief problem concerning such a series is the following: *Is the function $F(z)$ which it represents analytic in the circle $|z - z_0| < r$, or not?* Precisely as in the real domain, it may be shown by examples that without further assumptions this need not be the case. On the other hand, the desired behaviour of $F(z)$ may be ensured by stipulating (cf. § 47, first paragraph) that the series converges *uniformly*. The definition for this is almost word for word a repetition of **191**:

Definition (2nd form[39]). *A series $\Sigma f_n(z)$, all of whose terms are defined in the circle $|z - z_0| < r$ or in the circle $|z - z_0| \leqq r$, and which converges in this circle, is said to converge uniformly in this circle if, for every $\varepsilon > 0$, it is possible to choose a* ***single*** *number $N > 0$ (independent, therefore, of z) such that* **248.**

$$|f_{n+1}(z) + f_{n+2}(z) + \cdots| \equiv |r_n(z)| < \varepsilon$$

for every $n > N$ and every z in the circle considered.

Remarks.

1. Uniformity of convergence is here considered relative to all the points of an open or closed circle[40]. Of course other types of region or indeed arcs of curves or any other set $\mathfrak{M}$ of points, *not merely finite in number*, may be taken as a basis for the definition. The definition remains the same in substance. — In applications, we shall usually be concerned with the case in which the terms $f_n(z)$ are defined, and the series $\Sigma f_n(z)$ converges, at every point interior to a circle $|z - z_0| < r$ (or a domain $\mathfrak{G}$), but the convergence is uniform only in a *smaller circle* $|z - z_0| < \varrho$, where $\varrho < r$, (or in a *smaller subdomain* $\mathfrak{G}_1$, which, *together with its boundary*, belongs to the interior of $\mathfrak{G}$).

2. If the power series $\Sigma a_n (z - z_0)^n$ has the radius r, and $0 < \varrho < r$, the series is uniformly convergent in the (closed) circle $|z - z_0| \leqq \varrho$. Proof word for word as on p. 333.

3. If r is the exact radius of convergence of $\Sigma a_n (z - z_0)^n$, the convergence is *not* necessarily uniform in the circle $|z - z_0| < r$. Example: the geometric series; proof on p. 333.

4. Exactly as before, we may verify that our definition is completely equivalent to the following:

---

[38] Here again we may remark (cf. **190**, 4) that there is no substantial difference between the treatment of *series of variable terms* and that of *sequences of functions*. A *series* $\Sigma f_n(z)$ is equivalent to the sequence of its partial sums $s_0(z), s_1(z), \ldots,$ — and a *sequence* of functions $s_n(z)$ is equivalent to the series $s_0(z) + (s_1(z) - s_0(z)) + \cdots$. For simplicity, we shall hereafter formulate all definitions and theorems for *series* alone; the student will easily be able to enunciate them for *sequences*.

[39] This definition corresponds to the former 2nd *form*. The 1st form **191** may here be omitted, as it did not appear essential for the *application* of the concept of uniform convergence, but only for its *introduction*.

[40] The set of points of a circle (or, for short, the circle itself) is said to be *closed* or *open* according as the points of the circumference are regarded as included in the set or not.

3rd form. $\Sigma f_n(z)$ *is said to be* ***uniformly*** *convergent in* $|z - z_0| \leqq \varrho$ *(or in the set* $\mathfrak{M}$*), if, for every choice of points* z_n *belonging to this circle (or set), the corresponding remainders* $r_n(z_n)$ ***always*** *form a null sequence.*

The 4th and 5th forms of the definition (p. 335) also remain entirely unaltered and we may dispense with a special statement of them here.

On the other hand, it is impossible to give as impressive a geometrical representation of uniform and non-uniform convergence of a series as in the real domain.

We are now in a position to formulate and prove the theorem announced.

249. ***Weierstrass'*** **theorem on double series**[41]. *We suppose given a series*

$$\sum_{k=0}^{\infty} f_k(z)$$

each of whose terms $f_k(z)$ *is analytic at least for* $|z - z_0| < r$*, so that the expansions*[42]

$$\begin{array}{l} f_0(z) = a_0^{(0)} + a_1^{(0)}(z - z_0) + \cdots + a_n^{(0)}(z - z_0)^n + \cdots \\ f_1(z) = a_0^{(1)} + a_1^{(1)}(z - z_0) + \cdots + a_n^{(1)}(z - z_0)^n + \cdots \\ \cdots\cdots\cdots\cdots\cdots\cdots\cdots\cdots\cdots\cdots \\ f_k(z) = a_0^{(k)} + a_1^{(k)}(z - z_0) + \cdots + a_n^{(k)}(z - z_0)^n + \cdots \\ \cdots\cdots\cdots\cdots\cdots\cdots\cdots\cdots\cdots\cdots \end{array}$$

all exist and converge at least for $|z - z_0| < r$*. Further, we assume that the series* $\Sigma f_k(z)$ *converges* ***uniformly*** *in the circle* $|z - z_0| \leqq \varrho$*, for* ***every*** $\varrho < r$*, so that the series converges, in particular, everywhere within the circle* $|z - z_0| < r$*, and represents a definite function* $F(z)$ *there. It may then be shown that:*

1. *The coefficients in a vertical column form a convergent series:*

$$\sum_{k=0}^{\infty} a_n^{(k)} = A_n \qquad (\textit{fixed } n, = 0, 1, 2, \ldots).$$

2. $\sum_{n=0}^{\infty} A_n (z - z_0)^n$ *converges for* $|z - z_0| < r$.

3. *For* $|z - z_0| < r$*, the function*

$$F(z) = \sum_{k=0}^{\infty} f_k(z)$$

is again ***analytic****, with*

$$F(z) = \sum_{n=0}^{\infty} A_n (z - z_0)^n.$$

---

[41] Werke, Vol. 1, p. 70. The proof dates from the year 1841.

[42] The upper index, in the coefficient $a_n^{(k)}$, indicates the place occupied in the given series by the corresponding function, while the lower index relates to the position, in the expansion of this function, of the term to which the coefficient belongs.

4. *For* $|z - z_0| < r$ *and for every (fixed)* $\nu = 1, 2, \ldots,$

$$F^{(\nu)}(z) = \sum_{k=0}^{\infty} f_k^{(\nu)}(z)$$

i. e. the successive derived functions of $F(z)$ *may be obtained by term-by-term differentiation of the given series, and each of the new series converges uniformly in every circle* $|z - z_0| \leqq \varrho$, with $\varrho < r$.

Remarks.

1. If we direct our attention primarily to expansions in power series, the theorem simply states that *with the assumptions detailed above, an infinite number of power series "may" be added term by term.* If on the other hand we look rather at the analytic character of the various functions, we have the following

Theorem. *If each of the functions* $f_k(z)$ *is regular for* $|z - z_0| < r$ *and the series* $\Sigma f_k(z)$ *converges uniformly in* $|z - z_0| \leqq \varrho$, *for every* $\varrho < r$, *then this series represents an analytic function* $F(z)$, *regular in the circle* $|z - z_0| < r$. *The successive derived functions* $F^{(\nu)}(z)$ *of* $F(z)$, *for every* $\nu \geqq 1$, *are represented, in that circle, by the series* $\Sigma f_k^{(\nu)}(z)$, *obtained from* $\Sigma f_k(z)$ *by differentiating term by term,* ν *times in succession. Each of these series converges uniformly in every circle* $|z - z_0| \leqq \varrho$, *with* $\varrho < r$.

2. The assumption that $\Sigma f_k(z)$ converges in $|z - z_0| \leqq \varrho$ for every $\varrho < r$ is satisfied, for instance, by every power series $\Sigma c_k (z - z_0)^k$ with radius of convergence r. It is also satisfied e. g. by the series $\sum \frac{z^k}{1 - z^k}$ for $r = 1$; cf. § 58, C.

3. The first of our four statements shows that the present theorem cannot be proved simply as an application of *Markoff*'s transformation of series; for the latter *assumes* the convergence of the columns, — here this is *deduced* from the other hypotheses.

Proof. 1. Let an index m, a positive $\varrho < r$ and an $\varepsilon > 0$ be chosen to be kept *fixed* throughout. By hypothesis, we can determine a k_0 such that, throughout $|z - z_0| \leqq \varrho$,

$$|s_{k'} - s_k| < \varepsilon' = \varepsilon \cdot \varrho^m,$$

for every k such that $k' > k > k_0$, if we write

$$s_k = s_k(z) = f_0(z) + \ldots + f_k(z).$$

Now the function $s_{k'}(z) - s_k(z)$ is a definite power series, whose m^{th} coefficient is

$$a_m^{(k+1)} + a_m^{(k+2)} + \cdots + a_m^{(k')}.$$

By *Cauchy*'s inequality **235**, we therefore have

$$|a_m^{(k+1)} + a_m^{(k+2)} + \cdots + a_m^{(k')}| \leqq \frac{\varepsilon'}{\varrho^m} = \varepsilon.$$

Hence the series

(a) $$a_m^{(0)} + a_m^{(1)} + \cdots + a_m^{(k)} + \cdots \equiv \sum_{k=0}^{\infty} a_m^{(k)}$$

is convergent, by **81**. Let A_m be its sum. As m could be chosen arbitrarily, the first of our statements is thus established.

2. Now let M' be the maximum[43] of $|s_{k_0+1}(z)|$ along the circumference $|z - z_0| = \varrho$. We have then for every $k > k_0$ on the same circumference

$$|s_k(z)| \leqq |s_{k_0+1}(z)| + |s_k(z) - s_{k_0+1}(z)| \leqq M' + \varepsilon' = M.$$

Again, using *Cauchy*'s inequality, we obtain, for every $n = 0, 1, 2, \ldots$,

$$|a_n^{(0)} + a_n^{(1)} + \cdots + a_r^{(k)}| \leqq \frac{M}{\varrho^n},$$

whatever the value of k. Hence

$$|A_n| \leqq \frac{M}{\varrho^n},$$

and $\sum\limits_{n=0}^{\infty} A_n (z - z_0)^n$ therefore converges for $|z - z_0| < \varrho$. Since the only restriction on ϱ was that it should be $< r$, the series must even converge for $|z - z_0| < r$. (In fact, if z is any *determinate* point satisfying the inequality $|z - z_0| < r$, it is always possible to assume ϱ to be chosen so that $|z - z_0| < \varrho < r$.) Let us for the moment denote by $F_1(z)$ the function represented by the series $\Sigma A_n (z - z_0)^n$; it is thus, by its definition, an analytic function in $|z - z_0| < r$.

3. We have now to show that $F_1(z) \equiv F(z)$, so that $F(z)$ is itself an analytic function regular in $|z - z_0| < r$. For this purpose, we choose, as in the first part of our proof, a positive $\varrho' < r$, a positive ϱ in $\varrho' < \varrho < r$, and an $\varepsilon > 0$, fixed. We can determine k_0 so that, for all z in $|z - z_0| \leqq \varrho$,

$$|s_{k'} - s_k| < \varepsilon' = \varepsilon \cdot \frac{\varrho - \varrho'}{\varrho}$$

for every k such that $k' > k > k_0$. By *Cauchy*'s inequality, it follows as before that, for $k' > k > k_0$ and *for every* $n \geqq 0$,

$$|a_n^{(k+1)} + a_n^{(k+2)} + \cdots + a_n^{(k')}| < \frac{\varepsilon'}{\varrho^n}.$$

Making $k' \to +\infty$, we infer that, for every $k > k_0$ and every $n \geqq 0$,

$$|A_n - (a_n^{(0)} + a_n^{(1)} + \cdots + a_n^{(k)})| \leqq \frac{\varepsilon'}{\varrho^n}.$$

Now the expression between the modulus signs is the nth coefficient in the expansion of $F_1(z) - \sum\limits_{\nu=0}^{k} f_\nu(z)$ in powers of $(z - z_0)$. Hence we have, for $|z - z_0| < \varrho$:

$$\left|F_1(z) - \sum_{\nu=0}^{k} f_\nu(z)\right| \leqq \varepsilon' \cdot \left[1 + \frac{|z - z_0|}{\varrho} + \frac{|z - z_0|^2}{\varrho^2} + \cdots\right].$$

The right hand side is, for $|z - z_0| \leqq \varrho'$,

$$\leqq \varepsilon' \cdot \left[1 + \frac{\varrho'}{\varrho} + \left(\frac{\varrho'}{\varrho}\right)^2 + \cdots\right] = \varepsilon' \cdot \frac{\varrho}{\varrho - \varrho'} = \varepsilon.$$

---

[43] $|s_{k_0+1}(z)|$ is a continuous function of am $z = \varphi$ along the circumference in question and (φ being real) attains a definite maximum on this circumference.

Thus, when $\varepsilon > 0$ and $\varrho' < \varrho < r$ have been chosen arbitrarily, we can determine k_0 so that

$$\left| F_1(z) - \sum_{\nu=0}^{k} f_\nu(z) \right| < \varepsilon$$

for every $k > k_0$ and every $|z - z_0| \leqq \varrho'$. This implies, however, that for these values of z

$$F_1(z) = \sum_{\nu=0}^{\infty} f_\nu(z), \quad \text{i. e.} \quad = F(z).$$

The numbers ϱ' and ϱ were subjected to no restriction other than $0 < \varrho' < \varrho < r$; hence (as above) it follows that the equation holds for every z interior to the circle $|z - z_0| < r$.

4. We write

$$\begin{aligned}
f_0'(z) &= a_1^{(0)} + 2\,a_2^{(0)}(z - z_0) + 3\,a_3^{(0)}(z - z_0)^2 + \cdots \\
f_1'(z) &= a_1^{(1)} + 2\,a_2^{(1)}(z - z_0) + 3\,a_3^{(1)}(z - z_0)^2 + \cdots \\
&\cdots\cdots\cdots\cdots\cdots\cdots\cdots\cdots\cdots\cdots \\
\hline
&\quad A_1 + 2\,A_2\;(z - z_0) + 3\,A_3\;(z - z_0)^2 + \cdots,
\end{aligned}$$

where the sum of the coefficients in any one column converges to the value written immediately below them. Just as in 3. (we have only to begin our evaluations with $\varepsilon' = \left(\frac{\varrho - \varrho'}{\varrho}\right)^2 \cdot \varepsilon$) we deduce that for $|z - z_0| \leqq \varrho' < \varrho < r$ *and every* $k > k_0$,

$$\left| F'(z) - \sum_{\nu=0}^{k} f_\nu'(z) \right| \leqq \varepsilon'\left[1 + 2\frac{\varrho'}{\varrho} + 3\frac{\varrho'^2}{\varrho^2} + \cdots\right] = \varepsilon' \cdot \frac{\varrho^2}{(\varrho - \varrho')^2} = \varepsilon.$$

Hence for those values of z, $F'(z) = \sum_{k=0}^{\infty} f_k'(z)$. Indeed, by the same reasoning as before, this series converges uniformly in $|z - z_0| < r$, for *every* $\varrho < r$. If we write down the corresponding system of series for the νth derived functions, we obtain, in the same manner:

$$F^{(\nu)}(z) = \sum_{k=0}^{\infty} f_k^{(\nu)}(z) \qquad (\nu = 1, 2, \ldots, \text{fixed})$$

for every $|z - z_0| < r$; i. e. the series $\Sigma f_k^{(\nu)}(z)$ obtained by differentiating term by term, ν times in succession, converges in the whole circle $|z - z_0| < r$ (and converges uniformly in every circle $|z - z_0| \leqq \varrho < r$) and gives the νth derived function of $F(z)$ there.

Remarks.

1. A few examples of particular importance will be discussed in detail in the next section but one.

2. The fact of assuming the convergence uniform in a circular domain is immaterial for the most essential part of the theorem: If G is a domain of arbitrary shape [44] and if every point z_0 of the domain is the centre of a circle $|z - z_0| \leqq \varrho$ (for some ϱ) which belongs entirely to the domain, is such that each term of the series $\Sigma f_k(z)$ is analytic there, and is a circle of uniform convergence of the given series, then this series also represents a function $F(z)$ analytic in the domain in question, whose derived functions may be obtained by differentiation term by term. — Examples of this will also be given in § 58.

[44] Cf. p. 403, footnote 17.

§ 57. Products with complex terms.

The developments of Chapter VII were conducted in such a way that all definitions and theorems relating to products with "arbitrary" terms hold without alteration when we admit *complex* values for the factors. In particular the definition of convergence **125** and the theorems 1, 2 and 5 connected with it, as well as the proofs of the latter, remain entirely unchanged. There is also nothing to modify in **127**, the definition of absolute convergence, and the related theorems 6 and 7. On the other hand, some doubt might arise as to the literal transference of theorem 8 to the complex domain. Here again, however, everything may be interpreted as "complex", provided we agree to take $\log(1+a_n)$ to mean the *principal value* of the logarithm, for every sufficiently large n. The reasoning requires care, and we shall therefore carry out the proof in full:

250. Theorem. *The product $\Pi(1+a_n)$ converges if, and only if, the series, starting with a suitable index m,*

$$\sum_{n=m+1}^{\infty} \log(1+a_n),$$

whose terms are the principal values of $\log(1+a_n)$, *converges. If L_m is the sum of this series, we have, moreover,*

$$\prod_{n=1}^{\infty}(1+a_n) = (1+a_1)(1+a_2)\ldots(1+a_m)\cdot e^{L_m}.$$

Proof. a) *The conditions are sufficient.* For if the series $\sum_{n=m+1}^{\infty} \log(1+a_n)$, with the principal values of the logarithms, is convergent, its partial sums s_n, $(n>m)$, tend to a definite limit L, and consequently, since the exponential function is continous at every point,

$$e^{s_n} = (1+a_{m+1})(1+a_{m+2})\cdots(1+a_n) \to e^L$$

i. e. it certainly tends to a value $\neq 0$. Hence the product is convergent in accordance with the definition **125** and has the value stated.

b) *The conditions are necessary.* For, if the product converges, given a positive ε, which we may assume <1, we can determine n_0 so that

$$\text{(a)} \qquad |(1+a_{n+1})(1+a_{n+2})\cdots(1+a_{n+k})-1| < \frac{\varepsilon}{2}$$

for every $n \geqq n_0$ and every $k \geqq 1$. We then have, in particular, $|a_n| < \frac{\varepsilon}{2} < \frac{1}{2}$ for every $n > n_0$, and the inequality $|a_n| < \frac{1}{2}$ is thus certainly fulfilled for every n greater than a certain index m. We may now show further that for the same values of n and k (using the principal values of the logarithms)[45]

(b) $$\left| \sum_{\nu=n+1}^{n+k} \log(1+a_\nu) \right| < \varepsilon$$

and therefore the series $\sum_{n=m+1}^{\infty} \log(1+a_n)$ is convergent. In fact, as $|a_\nu| < \frac{\varepsilon}{2}$ for every $\nu > n_0$, we also have[46], for these values of ν,

(c) $$|\log(1+a_\nu)| < \varepsilon,$$

and likewise, by (a),

$$|\log[(1+a_{n+1})\ldots(1+a_{n+k})]| < \varepsilon$$

for every $n \geqq n_0$ and every $k \geqq 1$. Accordingly, for some suitable integer[47] q, we certainly have

$$|\log(1+a_{n+1}) + \log(1+a_{n+2}) + \ldots + \log(1+a_{n+k}) + 2q\pi i| < \varepsilon,$$

and it only remains to show that q may in every case be taken $= 0$. Now if we take any particular $n \geqq n_0$, this is certainly true for $k = 1$, by (c). It follows that it is true for $k = 2$. For in the expression

$$\log(1+a_{n+1}) + \log(1+a_{n+2}) + 2q\pi i$$

the modulus of either of the two first terms $< \varepsilon$, by (c), and by (d) the modulus of the whole expression has to be $< \varepsilon$; as $\varepsilon < 1$, q cannot, therefore, be an integer different from 0. For corresponding reasons, it also follows that for $k = 3$ the integer q must be 0, and this is then easily seen by induction to be true for every k. This establishes the theorem.

The part of theorem **127**, 8 relating to absolute convergence may also be immediately transferred to the complex domain, — viz.

the series $\sum_{n=m+1}^{\infty} \log(1+a_n)$ *and the product* $\prod_{n=m+1}^{\infty} (1+a_n)$

are simultaneously absolutely or non-absolutely convergent, in every case. Similarly the theorems 9—11 of §§ 29 and 30 remain valid. In fact, it remains true for complex a_n's of modulus $< \frac{1}{2}$ that in

$$\log(1+a_n) = a_n + \vartheta_n a_n^2$$

---

[45] The logarithms are always taken to have their *principal values* in what follows.

[46] In fact, for $|z| < \frac{1}{2}$,

$$|\log(1+z)| \leqq |z| + \frac{|z|^2}{2} + \ldots \leqq |z| + |z|^2 + \ldots = \frac{|z|}{1-|z|} \leqq 2|z|.$$

[47] For the principal value of the logarithm of a product is not necessarily the sum of the principal values of the logarithms of the factors, but may differ from this sum by a multiple of $2\pi i$. Thus e. g. $\log i = \frac{\pi i}{2}$, but

$$\log(i \cdot i \cdot i \cdot i) = \log 1 = 0,$$

if we take principal values throughout.

the quantities ϑ_n are bounded, — since when $|z| < \frac{1}{2}$

$$\log(1+z) = z + \left[-\frac{1}{2} + \frac{z}{3} - \frac{z^2}{4} + - \cdots\right] \cdot z^2,$$

while the expression in square brackets clearly has its modulus < 1 for those z's.

Finally, the remarks on the general connection between series and products also hold without alteration, since they were purely formal in character.

251. Examples.

1. $\prod\left(1+\frac{i}{n}\right)$ is divergent. For $\sum |a_n|^2 \equiv \sum \frac{1}{n^2}$ is convergent, so that by § 29, theorem 10, the partial products

$$p_n = \left(1+\frac{i}{1}\right)\left(1+\frac{i}{2}\right)\cdots\left(1+\frac{i}{n}\right) \sim e^{i\left(1+\frac{1}{2}+\cdots+\frac{1}{n}\right)};$$

the right hand expression represents, for successive values of n, points on the circumference of the unit circle, which circulate incessantly round this circumference at shorter and shorter intervals. p_n therefore tends to no limiting value. (Cf. pp. 427—8.)

2. $\prod_{n=0}^{\infty} \frac{n(n+1)+(1+i)}{n(n+1)+(1-i)} = -1$. In fact, the nth partial product is at once seen to be $\frac{1+(n+1)i}{1-(n+1)i}$, which $\to -1$.

3. For $|z| < 1$, $\prod_{n=0}^{\infty}\left(1+z^{2^n}\right) = \frac{1}{1-z}$. In fact the absolute) convergence of this product is obvious by **127**, 7 and its nth partial product multiplied by $(1-z)$ is

$$(1-z)(1+z)(1+z^2)(1+z^4)\ldots(1+z^{2^n}) = 1 - z^{2^{n+1}},$$

which tends to 1.

The consideration of products whose terms are functions of a complex variable,

$$\prod_{n=1}^{\infty}(1+f_n(z)),$$

— like that of series of variable terms in the preceding section, — will be restricted to the simplest, but also the most important case, in which the functions $f_n(z)$ are all analytic in one and the same circle $|z-z_0| < r$ (i. e. possess an expansion in power series convergent in that circle) and in which the product also converges everywhere in the circle. The product then represents a definite function $F(z)$ in the circle, which is said, conversely, to be *expanded in the given product.*

We next enquire under what convenient conditions the function $F(z)$ represented by the product is also analytic in the circle $|z-z_0| < r$. For the great majority of applications, the following theorem is sufficient:

Theorem. *If the functions* $f_1(z), f_2(z), \ldots, f_n(z), \ldots$ *are all analytic at least in the (fixed) circle* $|z - z_0| < r$; *if, further, the series* **252.**

$$\sum_{n=1}^{\infty} |f_n(z)|$$

converges uniformly in the smaller circle $|z - z_0| \leqq \varrho$, *for every positive* $\varrho < r$; *then the product* $\Pi(1 + f_n(z))$ *converges everywhere in* $|z - z_0| < r$ *and represents a function* $F(z)$ *which is itself analytic in that circle.*

The proof follows the same line of argument as that of the continuity theorem **218**, 1 almost word for word. To establish the convergence and analytic character of the product at a particular point z_1 in the circle $|z - z_0| < r$, we choose a $\varrho < r$ and prove the two facts first for every z of the circle $|z - z_0| < \varrho$. The series $\Sigma |f_n(z)|$ converges uniformly in the whole of $|z - z_0| \leqq \varrho$, so that the product $\Pi(1 + f_n(z))$ certainly converges there (indeed absolutely). Choose m so large that

$$|f_{m+1}(z)| + |f_{m+2}(z)| + \cdots + |f_n(z)| < 1$$

for every $n > m$ and every $|z - z_0| \leqq \varrho$; then for all these n's and z's,

$$|p_n(z)| = |(1 + f_{m+1}(z)) \ldots (1 + f_n(z))| \leqq e^{|f_{m+1}(z)| + \cdots + |f_n(z)|} < 3.$$

It follows precisely as on p. 382 that the series

$$p_{m+1} + (p_{m+2} - p_{m+1}) + \cdots + (p_n - p_{n-1}) + \cdots$$

converges *uniformly* in $|z - z_0| \leqq \varrho$. As all the terms of this series are analytic in $|z - z_0| < r$, the series itself, by **249**, therefore represents a function $F_m(z)$ analytic in $|z - z_0| < \varrho$. Hence

$$F(z) = \prod_{n=1}^{\infty} (1 + f_n(z)) = (1 + f_1(z)) \ldots (1 + f_m(z)) \cdot F_m(z)$$

is also an analytic function, regular in that circle.

From the above considerations, we may deduce two further theorems, which provide an analogue to *Weierstrass*' theorem on double series:

Theorem 1. *With the assumptions of the preceding theorem, the expansion in power series of* $F(z)$ *may be obtained by expanding the product term by term.* More precisely, *we know that the (finite) product* **253.**

$$P_k(z) = \prod_{\nu=1}^{k} (1 + f_\nu(z))$$

may be expanded in a power series of centre z_0 *which converges for* $|z - z_0| < r$, *since this is the case with each of the functions* $f_1, f_2, \ldots$.

Let the expansion be

$P_k(z) = A_0^{(k)} + A_1^{(k)}(z - z_0) + A_2^{(k)}(z - z_0)^2 + \cdots + A_n^{(k)}(z - z_0)^n + \cdots.$

Then for each (fixed) $n = 0, 1, 2, \ldots,$ *the limit*

$$\lim_{k\to+\infty} A_n^{(k)} = A_n$$

exists, and

$$F(z) = \prod_{k=1}^{\infty}(1 + f_k(z)) = \sum_{n=0}^{\infty} A_n (z - z_0)^n.$$

Proof. By § 46, theorem 2, the uniform convergence, in $|z - z_0| \leqq \varrho$, of the series

$$p_{m+1} + (p_{m+2} - p_{m+1}) + \cdots,$$

used in the preceding proof, implies the uniform convergence in the same circle of the series[48]

$$P_1(z) + [P_2(z) - P_1(z)] + \cdots + [P_k(z) - P_{k-1}(z)] + \cdots.$$

Applying *Weierstrass*' theorem on double series to this series, we obtain precisely the theorem stated.

Finally we prove a theorem about the derived function of $F(z)$, quite similar to **218**, 2:

Theorem 2. *For every z in $|z - z_0| < r$ for which $F(z) \neq 0$, we have*

$$\frac{F'(z)}{F(z)} = \sum_{n=1}^{\infty} \frac{f_n'(z)}{1 + f_n(z)},$$

i. e. the series on the right hand side converges for all these values of z and gives the ratio on the left hand side, the logarithmic differential coefficient of $F(z)$.

Proof. We saw that the expansion

$$F(z) = P_1(z) + (P_2(z) - P_1(z)) + \cdots$$

was uniformly convergent in $|z - z_0| \leqq \varrho < r$. By **249**,

$$F'(z) = P_1'(z) + (P_2'(z) - P_1'(z)) + \cdots,$$

which implies that

$$P_n'(z) \to F'(z)$$

at every point in the circle. If at a particular point $F(z) \neq 0$, we have $P_n(z) \neq 0$ for each n, and hence by **41**, 11,

$$\frac{P_n'(z)}{P_n(z)} \to \frac{F'(z)}{F(z)}.$$

---

[48] For the remainders of the latter series only differ from those of the former in that they contain the common factor $P_m(z)$, which is a continuous function for every z in the circle $|z - z_0| \leqq \varrho$, and hence is bounded in this closed circle.

Since, however, $$\frac{P_n'(z)}{P_n(z)} = \sum_{\nu=1}^{n} \frac{f_\nu'(z)}{1+f_\nu(z)},$$

this is precisely what our theorem asserts.

Examples.

1. If Σa_n is any *absolutely* convergent series of constant terms, the product **254.**

$$\prod_{n=1}^{\infty} (1 + a_n z)$$

represents a function regular in the whole plane, by **252**. By **253**, its expansion, in power series, which is convergent everywhere, is

$$1 + A_1 z + A_2 z^2 + A_3 z^3 + \cdots + A_k z^k + \cdots$$

with

$$A_1 = \sum_{\lambda=1}^{\infty} a_\lambda, \qquad A_2 = \sum_{\lambda_1 < \lambda_2}^{\infty} a_{\lambda_1} \cdot a_{\lambda_2}, \qquad A_3 = \sum_{\lambda_1 < \lambda_2 < \lambda_3}^{\infty} a_{\lambda_1} \cdot a_{\lambda_2} \cdot a_{\lambda_3},$$

$$\ldots, A_k = \sum_{\lambda_1 < \cdots < \lambda_k}^{\infty} a_{\lambda_1} \cdot a_{\lambda_2} \ldots a_{\lambda_k}, \quad \ldots .$$

Here the indices $\lambda_1, \lambda_2, \ldots, \lambda_k$ independently take for their values all the natural numbers, subject only to the condition $\lambda_1 < \lambda_2 < \cdots < \lambda_k$. The existence of the sums $A_1, A_2, \ldots$ is secured by theorem **253** itself; it is also easy to verify that they are independent of the order of the terms. — It was by applying this theorem that *Euler*[49] and later *C. G. J. Jacobi*[50] were led to an abundance of most remarkable formulae.

2. We have

$$\sin \pi z = \pi z \cdot \prod_{n=1}^{\infty} \left(1 - \frac{z^2}{n^2}\right),$$

where the product on the right hand side converges in the whole plane. The proof is word for word the same as that given in **219**, 1 for a real variable.

3. Taking $z = i$ in the above sine product, we obtain

$$\pi i \prod_{n=1}^{\infty} \left(1 + \frac{1}{n^2}\right) = \sin \pi i = \frac{e^{-\pi} - e^{\pi}}{2i}$$

or

$$\prod_{n=1}^{\infty} \left(1 + \frac{1}{n^2}\right) = \frac{e^{\pi} - e^{-\pi}}{2\pi}.$$

(Cf. however the extremely easy evaluation of $\prod \left(1 - \frac{1}{n^2}\right)$ in **128**, 6).

4. The sequence of functions

$$g_n(z) = \frac{z(z+1)(z+2)\cdots(z+n)}{n!\, n^z}, \qquad n = 1, 2, \ldots,$$

converges for every z in the whole plane. — In fact

$$g_n(z) = z\left(1 + \frac{z}{1}\right)\left(1 + \frac{z}{2}\right)\cdots\left(1 + \frac{z}{n}\right) \cdot n^{-z};$$

by **127**, theorem 10,

$$\left(1 + \frac{z}{1}\right)\left(1 + \frac{z}{2}\right)\cdots\left(1 + \frac{z}{n}\right) \sim e^{z\left(1 + \frac{1}{2} + \cdots + \frac{1}{n}\right)};$$

[49] Introductio in analysin inf. Vol. 1, Chap. 15. 1748.
[50] Fundamenta nova, Königsberg 1829.

also, by **128**, 2, the numbers $\gamma_n = \left(1 + \frac{1}{2} + \cdots + \frac{1}{n}\right) - \log n$ tend, as $n \to +\infty$, to *Euler*'s constant C, so that the right hand expression, — which is

$$e^{z(\log n + \gamma_n)} = n^z e^{\gamma_n z}, \text{ —}$$

when divided by n^z, tends to a definite limit as $n \to +\infty$. This proves the statement. — Further, the limit, $K(z)$ say, becomes 0 only for $z = 0, -1, -2, \ldots$. Excluding these values, we have, for all other values of z,

$$\lim_{n \to +\infty} \frac{1}{g_n(z)} = \lim_{n \to +\infty} \frac{n!\, n^z}{z(z+1)(z+2)\cdots(z+n)} = \frac{1}{K(z)} = \Gamma(z).$$

This function of a complex variable z (restricted only to be $\neq 0, -1, -2, \ldots$) is the so-called ***Gamma-function*** $\boldsymbol{\Gamma(z)}$ which we have already defined on p. 385 for real values of the argument.

We proceed to show that $K(z)$ is analytic in the whole plane (i. e. an *integral* function). For this, it suffices to show that the series

$$K(z) = g_1(z) + (g_2(z) - g_1(z)) + \cdots + (g_n(z) - g_{n-1}(z)) + \cdots$$

converges uniformly in every circle $|z| \leqq \varrho$. Now

$$g_n(z) - g_{n-1}(z) = g_{n-1}(z)\left[\left(1 + \frac{z}{n}\right)\left(1 - \frac{1}{n}\right)^z - 1\right];$$

also a constant A exists [51] such that $|g_\nu(z)| \leqq A$ for every $\nu = 1, 2, 3, \ldots$ and every $|z| \leqq \varrho$, and further, we may write (see p. 283 and p. 442, footnote 54)

$$\left(1 - \frac{1}{n}\right)^z = 1 - \frac{z}{n} + \frac{\vartheta_n(z)}{n^2}$$

where $|\vartheta_n(z)|$ remains less than some constant B for every $n = 2, 3, \ldots$ and

---

[51] Let $|z| \leqq \varrho$ and $n > m \geqq 2\varrho$. Then

$$g_n(z) = z\left(1 + \frac{z}{1}\right)\cdots\left(1 + \frac{z}{m}\right)\cdot\left(1 + \frac{z}{m+1}\right)\cdots\left(1 + \frac{z}{n}\right)\cdot n^{-z}$$

$$= z\left(1 + \frac{z}{1}\right)\cdots\left(1 + \frac{z}{m}\right)\cdot e^{z\left(\frac{1}{m+1} + \cdots + \frac{1}{n} - \log n\right)} \cdot e^{\frac{\eta_{m+1}}{(m+1)^2} + \cdots + \frac{\eta_n}{n^2}},$$

where $\log\left(1 + \frac{z}{\nu}\right) = \frac{z}{\nu} + \frac{\eta_\nu}{\nu^2}$. As $\left|\frac{z}{\nu}\right| < \frac{1}{2}$ (cf. p. 435) we have $|\eta_\nu| \leqq |z|^2 \leqq \varrho^2$, and the last factor in the preceding expression therefore remains $< e^{\varrho^2 \cdot \frac{\pi^2}{6}} = A_3$, for every $|z| \leqq \varrho$ and every $n > m$. Similarly the last factor but one (see p. 295), also remains less than a fixed number A_2. As the remaining factor is also always less than a fixed number A_1 for every $|z| \leqq \varrho$, it follows that $|g_n(z)| \leqq A_1 \cdot A_2 \cdot A_3$ for all these values of z and every $n > m$. On the other hand, the first m functions $|g_1(z)|, |g_2(z)|, \ldots, |g_m(z)|$ also remain bounded for every $|z| \leqq \varrho$; the existence of the number A as asserted in the text is thus established.

If z is restricted to lie in a circle $\mathfrak{K}$, in the interior and on the boundary of which $z \neq 0, -1, -2, \ldots$ and $|z| \leqq \varrho$, then for every $n > m$

$$\frac{1}{g_n(z)} = \frac{1}{z\left(1 + \frac{z}{1}\right)\cdots\left(1 + \frac{z}{m}\right)} \cdot e^{-z\left(\frac{1}{m+1} + \cdots + \frac{1}{n} - \log n\right)} \cdot e^{-\frac{\eta_{m+1}}{(m+1)^2} - \cdots - \frac{\eta_n}{n^2}}$$

From this we infer in exactly the same way that a constant A' exists such that $\left|\frac{1}{g_n(z)}\right| < A'$ in $\mathfrak{K}$, for every $n = 1, 2, \ldots$.

every $|z| \leqq \varrho$. Thus for all these z's and n's,

$$|g_n(z) - g_{n-1}(z)| \leqq A \cdot \left| -\frac{z^2}{n^2} + \frac{\vartheta_n(z)}{n^2} + \frac{z \cdot \vartheta_n(z)}{n^3} \right| \leqq \frac{C}{n^2},$$

where C is a suitable constant. By **197**, it follows that the series for $K(z)$ converges *uniformly* in the circle $|z| \leqq \varrho$, — indeed the series of absolute values $\Sigma |g_n(z) - g_{n-1}(z)|$ does so, — and, by **249**, $K(z)$ is analytic in the whole plane.

§ 58. Special classes of series of analytic functions.

A. *Dirichlet*'s series.

A *Dirichlet series* is a series of the form[52]

$$\sum_{n=1}^{\infty} \frac{a_n}{n^z}.$$

Here the terms — as exponential functions — are analytic in the whole plane. The chief question will therefore be to determine whether and where the series converges and, in particular, whether and where it converges uniformly. We have

Theorem 1. *To every Dirichlet series there corresponds a real number λ — known as the* **abscissa of convergence** *of the series — such that the series converges when $\Re(z) > \lambda$ and diverges when $\Re(z) < \lambda$. The number λ may also be $-\infty$ or $+\infty$; in the former case the series converges everywhere, in the latter nowhere. Further, if $\lambda \neq +\infty$ and $\lambda' > \lambda$, the series is uniformly convergent in every circle of the half-plane $\Re(z) \geqq \lambda'$ and accordingly the series, by Weierstrass' theorem* **249**, *represents a function analytic and regular in every such circle and hence in the half-plane*[53] $\Re(z) > \lambda$. **255.**

The proof follows a line of argument similar to that used in the case of power series (cf. **93**). We first show that if the series converges at a point z_0, it converges at every other point z for which $\Re(z) > \Re(z_0)$. As however

$$\sum \frac{a_n}{n^z} = \sum \frac{a_n}{n^{z_0}} \cdot \frac{1}{n^{z-z_0}},$$

it suffices, by **184**, 3a, to show that the series

$$\sum_{n=1}^{\infty} \left| \frac{1}{n^{z-z_0}} - \frac{1}{(n+1)^{z-z_0}} \right| \equiv \sum_{n=1}^{\infty} \frac{1}{(n+1)^{\Re(z-z_0)}} \cdot \left| \left(1 + \frac{1}{n}\right)^{z-z_0} - 1 \right|$$

---

[52] More generally, a series is called a *Dirichlet* series when it is of the form $\sum \frac{a_m}{p_n^z}$ or of the form $\sum a_n e^{-\lambda_n z}$, where the p_n's are *positive numbers and the λ_n's any real numbers increasing monotonely to* $+\infty$.

[53] The existence of the *half-plane of convergence* was proved by *J. L. W. V. Jensen* (Tidskrift for Mathematik (5), Vol. 2, p. 63. 1884); the uniformity of the convergence and thereby the analytic character of the function represented were pointed out by *E. Cahen* (Annales Éc. Norm. sup. (3), Vol. 11, p. 75. 1894).

is convergent. Writing (for a *fixed* exponent $(z - z_0)$)

$$\left(1+\frac{1}{n}\right)^{z-z_0} = 1 + \frac{\vartheta_n}{n},$$

the numbers $\vartheta_n \to (z - z_0)$, as is at once seen[54]; they are therefore certainly *bounded*, $|\vartheta_n| < A$, say. The nth term of the above series is therefore

$$< \frac{A}{n^{1+\Re(z-z_0)}},$$

and the series is accordingly convergent when $\Re(z - z_0) > 0$.

As a corollary, we have the statement: If a *Dirichlet* series is divergent at a point $z = z_1$, it is divergent at every other point whose real part is less than that of z_1. Supposing that a given *Dirichlet* series does not converge everywhere or nowhere, the existence of the limiting abscissa λ is inferred (as in **93**) as follows: Let z' be a point of divergence and z'' a point of convergence of the series, and choose $x_0 < \Re(z')$ and $y_0 > \Re(z'')$, — both real. For $z = x_0$ the series will diverge, for $z = y_0$ it will converge. Now apply the method of successive bisection, word for word as in **93**, to the interval $J_0 = x_0 \ldots y_0$ on the real axis. The value λ so obtained will be the required abscissa.

Now suppose $\lambda' > \lambda$ (for $\lambda = -\infty$, λ' may therefore be any real number); if z is restricted to lie in a domain G in which $\Re(z) \geqq \lambda'$ and $|z| \leqq R$, — so that in general G will take the shape of a segment of a circle, — our series is *uniformly* convergent in that domain. To show this, let us choose a point z_0 for which $\lambda < \Re(z_0) < \lambda'$; as before, we write

$$\sum \frac{a_n}{n^z} = \sum \frac{a_n}{n^{z_0}} \cdot \frac{1}{n^{z-z_0}}.$$

---

[54] More generally, we may at once observe that if $|z| \leqq \frac{1}{2}$ and $|w| \leqq R$, and if we write, taking the principal value,

$$(1+z)^w = 1 + zw + \vartheta \cdot z^2,$$

the factor ϑ, which depends on z and w, remains less than a *fixed* constant for all the values allowed for z and w. — Proof:

$$(1+z)^w = e^{w \log(1+z)} = e^{w(z+\eta z^2)}, \quad \text{with } \eta = -\frac{1}{2} + \frac{z}{3} - \frac{z^2}{4} + \cdots.$$

For every $|z| \leqq \frac{1}{2}$, we therefore have $|\eta| \leqq 1$; hence in

$$e^{w(z+\eta z^2)} = 1 + wz(1+\eta z) + \frac{w^2 z^2 (1+\eta z)^2}{2!} + \cdots$$

$$= 1 + wz + \left[w\eta + \frac{w^2(1+\eta z)^2}{2!} + \frac{z w^3 (1+\eta z)^3}{3!} + \cdots\right] \cdot z^2$$

the expression in square brackets, which was denoted by ϑ, satisfies the inequality

$$|\vartheta| \leqq e^{2R}.$$

This is at once obvious if we replace all the quantities in the brackets by their absolute values and then replace $|\eta|$ and $|z|$ by 1 and finally $|w|$ by R.

$\sum \frac{a_n}{n^{z_0}}$ is a convergent series of constant terms; by **198**, 3a it therefore suffices to show that

$$\sum_{n=1}^{\infty} \left| \frac{1}{n^{z-z_0}} - \frac{1}{(n+1)^{z-z_0}} \right|$$

converges uniformly in the domain in question and that the factors $\frac{1}{n^{z-z_0}}$ are uniformly bounded in G. Now, writing $\lambda' - \Re(z_0) = \delta\,(>0)$,

$$\left| \frac{1}{n^{z-z_0}} - \frac{1}{(n+1)^{z-z_0}} \right| \leqq \frac{1}{n^{\delta}} \cdot \left| \left(1+\frac{1}{n}\right)^{z-z_0} - 1 \right|.$$

Using the evaluation given in the preceding footnote (or else directly, by expanding $\left(1+\frac{1}{n}\right)^{z-z_0} = e^{(z-z_0)\log\left(1+\frac{1}{n}\right)}$ in powers of $(z-z_0)$) we now see that a constant A certainly exists such that the difference within the modulus signs on the right hand side of the above inequality is in absolute value

$$< \frac{A}{n}$$

for every z in our domain and every $n = 1, 2, 3, \ldots$. The whole expression on the right is thus

$$< \frac{A}{n^{1+\delta}}.$$

On the other hand, since $\left| \frac{1}{n^{z-z_0}} \right| \leqq \frac{1}{n^{\delta}}$, the factors $\frac{1}{n^{z-z_0}}$ are uniformly bounded in G. By **198**, 3a, this proves that the *Dirichlet* series is uniformly convergent in the domain stated, and hence, in particular, that every *Dirichlet* series represents a function which is analytic in the interior of the region of convergence of the series (the half-plane $\Re(z) > \lambda$).

From

$$\sum \left| \frac{a_n}{n^{z}} \right| = \sum \left| \frac{a_n}{n^{z_0}} \right| \cdot \left| \frac{1}{n^{z-z_0}} \right|$$

it follows at once that if a *Dirichlet* series converges *absolutely* at a point z_0, it does so at any point z for which $\Re(z) > \Re(z_0)$, and if it does *not* converge absolutely at z_0, then it cannot do so at any point z for which $\Re(z) < \Re(z_0)$. Just as before we obtain

Theorem 2. *There exists a definite real number l (which may also be $+\infty$ or $-\infty$) such that the Dirichlet series converges absolutely for $\Re(z) > l$, but not for $\Re(z) < l$.*

Of course we have $\lambda \leqq l$; over and above this, the relative positions of the two straight lines $\Re(z) = \lambda$ and $\Re(z) = l$ is subject to the following

Theorem 3. *We have in every case* $l - \lambda \leqq 1$.

Proof. If $\sum \frac{a_n}{n^{z_0}}$ is convergent and $\Re(z) > \Re(z_0) + 1$, then $\sum \frac{a_n}{n^z}$ is absolutely convergent, for $\left|\frac{a_n}{n^z}\right| = \left|\frac{a_n}{n^{z_0}}\right| \frac{1}{n^{\Re(z-z_0)}}$ with $\Re(z - z_0) > 1$. This proves the statement at once.

Remarks and Examples.

256. 1. If a *Dirichlet* series is not merely everywhere or nowhere convergent the situation will in general be as follows: the half-plane $\Re(z) < \lambda$ of divergence of the series is followed by *a strip* $\lambda < \Re(z) < l$ *of conditional convergence of the series*; the breadth of this strip is in any case at most 1, and in the remaining half-plane $\Re(z) > l$, the series converges absolutely.

2. It may be shown by easy examples that the difference $l - \lambda$ may assume any value between 0 and 1 (both inclusive), and that the behaviour on the bounding lines $\Re(z) = \lambda$ and $\Re(z) = l$ may vary in different cases.

3. The two series $\sum \frac{1}{2^n \cdot n^z}$ and $\sum \frac{2^n}{n^z}$ provide simple examples of *Dirichlet* series which converge everywhere and nowhere.

4. $\sum \frac{1}{n^z}$ has the abscissa of convergence $\lambda = 1$; thus it represents an analytic function, regular in the half-plane $\Re(z) > 1$. It is known as *Riemann*'s ζ-function (v. **197**, 2, 3) and is used in the analytical theory of numbers, on account of its connection with the distribution of prime numbers (see below, Rem. 9)[55].

5. Just as the radius of a power series can be deduced directly from its coefficients (theorem **94**), so we may infer from the coefficients of a given *Dirichlet* series what positions the two limiting straight lines occupy. We have the following

Theorem. *The abscissa of convergence* λ *of the Dirichlet series* $\sum \frac{a_n}{n^z}$ *is invariably given by the formula*

$$\lambda = \overline{\lim_{x \to +\infty}} \frac{1}{x} \log \left| a_{u+1} + a_{u+2} + \cdots + a_v \right|$$

where x *increases continuously and*

$$[e^{[x]}] = u, \quad [e^x] = v.$$

Substituting a_n *for* a_n *in this formula, we obtain* l, *the limiting abscissa of absolute convergence*[56].

6. A concise account of the most important results in the theory of *Dirichlet*'s series may be found in *G. H. Hardy* and *M. Riesz*, Theory of *Dirichlet*'s series, Cambridge 1915.

---

[55] A detailed investigation of this remarkable function (as well as of arbitrary *Dirichlet* series) is given by *E. Landau*, Handbuch der Lehre von der Verteilung der Primzahlen, Leipzig 1909, 2 Vols., in *E. Landau*, Vorlesungen über Zahlentheorie, Leipzig 1927, 3 Vols., and in *E. C. Titchmarsh*, The Zeta-Function of Riemann, Cambridge 1930.

[56] As regards the proof, we must refer to a note by the author: "Über die Abszisse der Grenzgeraden einer Dirichletschen Reihe" in the Sitzungsberichte der Berliner Mathematischen Gesellschaft (Vol. X, p. 2, 1910).

7. By repeated term-by-term differentiation of a *Dirichlet* series $F(z) = \sum \frac{a_n}{n^z}$, we obtain the *Dirichlet* series

$$(-1)^\nu \sum_{n=1}^{\infty} \frac{a_n (\log n)^\nu}{n^z} \qquad \text{(fixed } \nu).$$

As an immediate consequence of *Weierstrass*' theorem on double series, these necessarily cannot have a larger abscissa of convergence than the original series, and, owing to the additional factors $\log^\nu n$, they can obviously not have a smaller one either. They represent, in the interior of the half-plane of convergence, the derived functions $F^{(\nu)}(z)$.

8. By **255**, the function represented by a *Dirichlet* series can be expanded in a power series about any point interior to the half-plane of convergence as centre. The expansion itself is provided by *Weierstrass*' theorem on double series. If, for instance, it is required to expand the function $\zeta(z) = \sum_{k=1}^{\infty} \frac{1}{k^z}$ about $z_0 = +2$ as centre, we have for $k = 2, 3, \ldots$

$$\frac{1}{k^z} = \frac{1}{k^2} \cdot \frac{1}{k^{z-2}} = \frac{1}{k^2} \cdot e^{-(z-2)\log k} = \frac{1}{k^2} \sum_{n=0}^{\infty} (-1)^n \frac{(\log k)^n}{n!} (z-2)^n \qquad (k \text{ fixed}),$$

and this continues to hold for $k = 1$ provided we interpret $(\log 1)^0$ as having the value 1. Hence for $n \geqq 0$

$$A_n = \frac{(-1)^n}{n!} \sum_{k=1}^{\infty} \frac{(\log k)^n}{k^2} \qquad (n \text{ fixed}),$$

which gives the desired expansion

$$\zeta(z) = \sum_{n=0}^{\infty} \frac{(-1)^n}{n!} \left[\sum_{k=1}^{\infty} \frac{(\log k)^n}{k^2} (z-2)^n \right] = \frac{\pi^2}{6} - \left(\sum_{k=1}^{\infty} \frac{\log k}{k^2} \right) (z-2) + - \cdots.$$

9. *For* $\Re(z) > 1$, **257.**

the series $\sum_{n=1}^{\infty} \frac{1}{n^z}$ *and the product* $\prod \frac{1}{1-p^{-z}}$

(where p takes for its values all the prime numbers 2, 3, 5, 7, ... in succession) have everywhere the same value, and accordingly both represent the Riemann ζ-function $\zeta(z)$. (*Euler*, 1737; v. Introd. in analysin, p. 225.)

Proof. Let z be a definite point such that $\Re(z) = 1 + \delta > 1$. By our remark 4 and **127**, 7, the series and product certainly converge absolutely at this point. We have only to prove that they have the same value. Now

$$\frac{1}{1-p^{-z}} = 1 + \frac{1}{p^z} + \frac{1}{p^{2z}} + \frac{1}{p^{3z}} + \cdots;$$

multiplying these expansions together, for all prime numbers $p \leqq N$, — where N denotes an integer kept fixed for the moment, — the (finite) product so obtained is

$$\prod_{p \leqq N} \frac{1}{1-p^{-z}} = \sum_{n=1}^{N} \frac{1}{n^z} + \sum_{n=N+1}^{\infty}{}' \frac{1}{n^z},$$

where the accent on the Σ indicates that only some, and not all, of the terms of the series written down are taken. Here we have made use of the elementary proposition that every natural number $\geqq 2$ can be expressed in one and only one way as a product of powers of distinct primes (provided only positive integral exponents are allowed and the order of succession of the factors is left out of account). Accordingly

$$\left| \prod_{p \leqq N} \frac{1}{1-p^{-z}} - \sum_{n=1}^{N} \frac{1}{n^z} \right| \leqq \sum_{n=N+1}^{\infty} \frac{1}{n^{1+\delta}}.$$

On the right hand side we have the remainder of a convergent series, which tends to 0 when $N \to +\infty$. This proves the equality of the values of the *infinite* product and of the *infinite* series, as was required.

10. By **257**, we have for $\Re(z) > 1$

$$\frac{1}{\zeta(z)} = \prod_p (1-p^{-z}) = \prod_p \left(1 - \frac{1}{p^z}\right) = \sum_{n=1}^{\infty} \frac{\mu(n)}{n^z}$$

where

$\mu(1) = 1$, $\mu(2) = -1$, $\mu(3) = -1$, $\mu(4) = 0$, $\mu(5) = -1$, $\mu(6) = +1, \ldots$ and generally $\mu(n) = 0$, $+1$, or -1 according as n is divisible by the square of a prime number, or is a product of an even number of primes, all *different*, or of an odd number of primes, all *different*. The product-expansion of the ζ-function also shows that for $\Re(z) > 1$, we always have $\zeta(z) \neq 0$. The curious coefficients $\mu(n)$ are known as *Möbius*' coefficients. There is no superficial regularity in the mode of succession of the values 0, $+1$, -1 among the numbers $\mu(n)$.

11. Since $\zeta(z) = \sum \frac{1}{n^z}$ converges absolutely for $\Re(z) > 1$, we may form the square $(\zeta(z))^2$ by multiplying the series by itself term by term and re-arranging in order of increasing denominators (as is allowed by **91**). We thus obtain

$$\zeta^2(z) = \sum_{n=1}^{\infty} \frac{\tau_n}{n^z},$$

where τ_n denotes the *number of divisors of* n. — These examples may suffice to explain the importance of the ζ-function in problems in the theory of numbers.

B. Faculty series.

A *faculty series* (*of the first kind*) is a series of the form

$$\text{(F)} \qquad \sum_{n=1}^{\infty} \frac{n!\, a_n}{z(z+1)\ldots(z+n)},$$

which of course has a meaning only if $z \neq 0, -1, -2, \ldots$. The questions of convergence, elucidated in the first instance by *Jensen*, are completely solved by the following

258. Theorem of ***Landau***[57]. *The faculty series* (F) *converges — with the exclusion of the points* $0, -1, -2, \ldots$ — *wherever the "associated" Dirichlet series*

$$\sum_{n=1}^{\infty} \frac{a_n}{n^z}$$

converges, and conversely the latter converges wherever the series (F) *converges. The convergence is uniform in a circle for either series, when it is so for the other, provided the circle contains none of the points* $0, -1, -2, \ldots$ *either in its interior or on its boundary.*

[57] Über die Grundlagen der Theorie der Fakultätenreihen. Münch. Ber. Vol. 36, pp. 151—218. 1906.

Proof. 1. We first show that the convergence of the *Dirichlet* series at any particular point $\neq 0, -1, -2, \ldots$ involves that of the faculty series at the same point. As

$$\frac{n!\, a_n}{z(z+1)\ldots(z+n)} = \frac{a_n}{n^z} \cdot \frac{1}{g_n(z)},$$

if $g_n(z)$ has the same significance as in **254**, example 4, it is sufficient, by **184**, 3a, to show that the series

$$\sum_{n=1}^{\infty} \left| \frac{1}{g_n(z)} - \frac{1}{g_{n+1}(z)} \right| = \sum_{n=1}^{\infty} \frac{|g_{n+1}(z) - g_n(z)|}{|g_n(z) \cdot g_{n+1}(z)|}$$

is convergent. Now $\dfrac{1}{g_n(z)}$ tends to a finite limit as n increases, namely to the value $\Gamma(z)$; hence, in particular, this factor remains *bounded* for all values of n (z being fixed). Hence it suffices to establish the convergence of the series

$$\sum_{n=1}^{\infty} |g_n(z) - g_{n+1}(z)|.$$

But this has been done already in **254**, example 4.

2. The fact that the convergence of the faculty series at any point involves that of the *Dirichlet* series follows in precisely the same manner, as again, by **184**, 3a, everything turns on the convergence of

$$\sum |g_n(z) - g_{n+1}(z)|.$$

3. Now let $\mathfrak{K}$ be a circle in which the *Dirichlet* series converges absolutely and which contains none of the points $0, -1, -2, \ldots$, either as interior or boundary points. We have to show that the faculty series also converges uniformly in that circle. By **198**, 3a, this again reduces to proving that

$$\sum_{n=1}^{\infty} \left| \frac{g_{n+1}(z) - g_n(z)}{g_n(z) \cdot g_{n+1}(z)} \right|$$

is uniformly convergent in $\mathfrak{K}$ and that the functions $1/g_n(z)$ remain uniformly bounded in $\mathfrak{K}$. The uniform convergence of

$$\sum_{n=1}^{\infty} |g_{n+1}(z) - g_n(z)|$$

was already established in **254**, 4. Also it was shown on p. 440, footnote 51, that there exists a constant A' such that

$$\left| \frac{1}{g_n(z)} \right| < A'$$

for every z in $\mathfrak{K}$ and every n. This is all that is required. (Cf. § 46, theorem 3.)

4. The converse, that the *Dirichlet* series converges uniformly in every circle in which the faculty series does so, follows at once by **198**, 3a from the uniform convergence of the series $\sum |g_{n+1}(z) - g_n(z)|$ and the uniform boundedness of the functions $g_n(z)$ in the circle, both of which were established in **254**, 4.

Examples.

1. The faculty series

$$\sum_{n=1}^{\infty} \frac{1}{2^{n+1}} \frac{n!}{z(z+1)\ldots(z+n)}$$

converges at *every* point of the plane $\neq 0, -1, \ldots$. For the *Dirichlet* series

$$\sum_{n=1}^{\infty} \frac{1}{2^{n+1} \cdot n^z}$$

is evidently convergent everywhere.

As

$$\Delta \frac{1}{x} = \frac{1}{x} - \frac{1}{x+1} = \frac{1}{x(x+1)},$$

$$\Delta^2 \frac{1}{x} = \frac{2!}{x(x+1)(x+2)}, \quad \ldots, \quad \Delta^k \frac{1}{x} = \frac{k!}{x(x+1)\ldots(x+k)},$$

the given faculty series results simply, by *Euler*'s transformation **144**, from the series

$$\sum_{n=0}^{\infty} \frac{(-1)^n}{z+n} \equiv \frac{1}{z} - \frac{1}{z+1} + \frac{1}{z+2} - + \cdots.$$

2. It is also easily seen (cf. pp. 265—6) that for $\Re(z) > 0$

$$\frac{1}{z^2} = \frac{0!}{z(z+1)} + \frac{1!}{z(z+1)(z+2)} + \cdots + \frac{(n-1)!}{z(z+1)\cdots(z+n)} + \cdots,$$

i. e.

$$\frac{1}{z^2} = \sum_{n=1}^{\infty} \frac{1}{n} \cdot \frac{n!}{z(z+1)\cdots(z+n)}.$$

To show this, we have only to subtract the terms of the right hand side *successively* from the left hand side. After the nth subtraction we have

$$\frac{n!}{z^2(z+1)(z+2)\ldots(z+n)} = \frac{1}{z \cdot n^z} \cdot \frac{n!\, n^z}{z(z+1)\ldots(z+n)},$$

and this, by **254**, example 4, tends to 0 when $n \to \infty$, provided $\Re(z) > 0$. (*Stirling*: Methodus differentialis, London 1730, p. 6 seqq.)

C. *Lambert*'s series.

A *Lambert* series is a series of the form [58]

$$\sum_{n=1}^{\infty} a_n \frac{z^n}{1-z^n}.$$

If we again inquire what is the precise region of convergence of the series, it must first be noted that for every z for which $z^n - 1$ can be equal to zero, an infinite number of the terms of the series become meaningless. For this reason, the circumference of the unit circle will be entirely excluded from consideration [59] while we discuss the

[58] A more extensive treatment of this type of series is to be found in a paper by the author: Über *Lambert*sche Reihen. Journ. f. d. reine u. angew. Mathem., Vol. 142, pp. 283—315. 1913.

[59] This does not imply that this series may not converge at some points z_1 of this circumference, for which $z_1^n \neq +1$ for every $n \geqq 1$. This may actually happen; but we will not consider the case here.

question of convergence of these series, and the points inside and outside the circle will be examined separately. We have the following theorem, which completely solves the question of convergence in this respect:

Theorem. *If Σa_n converges, the Lambert series converges for every z whose modulus is $\neq 1$. If Σa_n is not convergent, the Lambert series converges at precisely the same points as the "associated" power series $\Sigma a_n z^n$ — provided $|z| \neq 1$ as before.* **259.**

Further, the convergence is uniform in every circle $\mathfrak{K}$ which lies completely (circumference included) within one of the regions of convergence of the series and contains no point of modulus 1.

Proof. 1. Suppose Σa_n *divergent.* The radius r of $\Sigma a_n z^n$ is in that case necessarily $\leqq 1$ and we have to show first that the *Lambert* series and the associated power series converge and diverge together for every $|z| < 1$, and that the *Lambert* series diverges for $|z| > 1$.

Now
$$\Sigma a_n z^n \equiv \Sigma a_n \frac{z^n}{1-z^n} \cdot (1 - z^n)$$
and
$$\Sigma a_n \frac{z^n}{1-z^n} \equiv \Sigma a_n z^n \cdot \frac{1}{1-z^n}.$$

Accordingly, it suffices, by **184**, 3a, to establish the convergence of the two series
$$\Sigma |(1 - z^{n+1}) - (1 - z^n)| = \Sigma |z^n - z^{n+1}| = |1 - z| \cdot \Sigma |z^n|$$
and
$$\Sigma \left| \frac{1}{1-z^{n+1}} - \frac{1}{1-z^n} \right| = |1 - z| \cdot \Sigma \frac{|z^n|}{|(1-z^n)(1-z^{n+1})|}$$
for $|z| < 1$. The first of these facts is obvious, however, while the second follows from the remark that for $|z| < 1$, we have $|1 - z^n| > \frac{1}{2}$ for all sufficiently large n's.

On the other hand, if the *Lambert* series converged at a point z_0, where $|z_0| > 1$, the power series
$$\Sigma \frac{a_n}{1 - z_0^n} z^n$$
would converge for $z = z_0$, and by **93**, theorem **1**, would have also to converge for $z = +1$. Hence the series
$$\Sigma \frac{a_n}{1-z_0^n} - \Sigma a_n \frac{z_0^n}{1-z_0^n} = \Sigma a_n$$
would also have to converge, which is contrary to hypothesis.

Finally, the fact that the *Lambert* series converges uniformly in $|z| \leqq \varrho < r$ may at once be inferred from the corresponding fact in the case of the power series $\Sigma |a_n z^n|$, by § 46, 2, — in virtue of the inequality
$$\left| a_n \frac{z^n}{1-z^n} \right| \leqq \frac{1}{1-\varrho} |a_n z^n|.$$

The case where Σa_n diverges is thus completely dealt with.

2. Now suppose Σa_n convergent, so that $\Sigma a_n z^n$ has a radius $r \geqq 1$. The *Lambert* series is certainly convergent for every $|z| < 1$ and indeed uniformly so for all values of z such that $|z| \leqq \varrho < 1$.

For $|z| \geqq \varrho' > 1$, we have

$$\Sigma a_n \frac{z^n}{1-z^n} = -\Sigma a_n - \Sigma a_n \frac{\left(\frac{1}{z}\right)^n}{1-\left(\frac{1}{z}\right)^n};$$

and as $\left|\frac{1}{z}\right| \leqq \frac{1}{\varrho'} < 1$, this reduces the later assertions to the preceding ones, and the theorem is therefore established in all its parts.

By the above, a very simple connection exists, in the case where Σa_n is convergent, between the sum of the series at a point z outside the unit circle and the same sum at the point $\frac{1}{z}$ inside it. Accordingly it will suffice if we consider only that region of convergence of the series which lies inside the unit circle. This is either the circle $|z| < r$ or the unit circle $|z| < 1$ itself, according as the radius r of the series $\Sigma a_n z^n$ is < 1 or $\geqq 1$. Let r_1 denote the radius of this perfectly definite region of convergence.

The terms of a *Lambert* series are analytic functions regular in $|z| < r_1$, and for every positive $\varrho < r_1$, the series is uniformly convergent in $|z| \leqq \varrho$: hence we may apply *Weierstrass'* theorem on double series to obtain the expansion in power series of the function represented by a *Lambert* series in $|z| < r_1$. We have

$$\begin{array}{llllllll}
a_1 \frac{z}{1-z} = a_1 z & + a_1 z^2 & + a_1 z^3 & + a_1 z^4 & + a_1 z^5 & + a_1 z^6 & + a_1 z^7 & + \cdots \\
a_2 \frac{z^2}{1-z^2} = & a_2 z^2 & & + a_2 z^4 & & + a_2 z^6 & & + \cdots \\
a_3 \frac{z^3}{1-z^3} = & & a_3 z^3 & & & + a_3 z^6 & & + \cdots \\
a_4 \frac{z^4}{1-z^4} = & & & a_4 z^4 & & & & + \cdots \\
\dots & \dots & \dots & \dots & \dots & \dots & \dots & \dots
\end{array}$$

and we may add all these series together term by term. In the kth row, a given power z^n will occur if, and only if, n is a multiple of k, or k a divisor of n. Therefore A_n, the coefficient of z^n in the resulting series, will be equal to the sum of those coefficients a_ν whose suffix ν is a divisor of n (including 1 or n). This we write symbolically[60]

$$A_n = \sum_{d/n} a_d,$$

and we then have, for $|z| < r_1$,

$$\sum_{n=1}^{\infty} a_n \frac{z^n}{1-z^n} = \sum_{n=1}^{\infty} A_n z^n.$$

---

[60] In words: the sum of all a_d's for which d is a divisor of n.

Examples. **260.**

1. $a_n = 1$. Here A_n is equal to the *number* of divisors of n, which (as in **257**, example 11) we denote by τ_n; then

$$\sum_{n=1}^{\infty} \frac{z^n}{1-z^n} \equiv \sum_{n=1}^{\infty} \tau_n z^n \qquad (|z|<1)$$
$$= z + 2z^2 + 2z^3 + 3z^4 + 2z^5 + 4z^6 + 2z^7 + 4z^8 + \dots.$$

In this curious power series, the terms z^n whose exponents are *prime numbers* are distinguished by the coefficient 2. It was due to the misleadingly close connection between this special *Lambert* series and the problem of primes that this series (as a rule called simply *the Lambert series*) [61] played a considerable part in the earlier attempts to deal with this problem. But nothing of importance was obtained in this manner for some time. Only quite recently *N. Wiener* [62] succeeded by this means in proving the famous prime number theorem.

2. $a_n \equiv n$. Here A_n is equal to the *sum* of all the divisors of n, which we will denote by τ_n'. Thus for $|z| < 1$

$$\sum_{n=1}^{\infty} n \frac{z^n}{1-z^n} = \sum_{n=1}^{\infty} \tau_n' z^n = z + 3z^2 + 4z^3 + 7z^4 + 6z^5 + 12z^6 + \dots.$$

3. The relation $A_n = \sum_{d/n} a_d$ is uniquely reversible, i. e. for given A_n's, the coefficients a_n can be determined in one and only one way so as to satisfy the relation. We then have in fact

$$a_n = \sum_{d/n} \mu\left(\frac{n}{d}\right) \cdot A_d,$$

where $\mu(k)$ denotes the *Möbius* coefficients defined in **257**, example 10, whose values are 0, $+1$ and -1. In consequence of this fact, not only can a *Lambert* series always be expanded in a power series, but conversely every power series may be expressed as a *Lambert* series, provided it vanishes for $z = 0$, i. e. $A_0 = 0$. But it should be observed that a relation of the form

$$\sum A_n z^n = \sum a_n \frac{z^n}{1-z^n}$$

need not remain true for $|z| > 1$, even when both series converge there.

4. For instance, if $A_1 = 1$ and every other $A_n = 0$,

$$a_n = \mu(n),$$

and we have the curious identity

$$z = \sum_{n=1}^{\infty} \mu(n) \frac{z^n}{1-z^n} \qquad (|z|<1).$$

5. Similarly, we find the representation, valid for $|z| < 1$,

$$\frac{z}{(1-z)^2} = \sum_{n=1}^{\infty} \varphi(n) \frac{z^n}{1-z^n},$$

where $\varphi(n)$ denotes the number of integers less than n and prime to n, — a number introduced by *Euler*.

6. Writing $\sum_{n=1}^{\infty} a_n \frac{z^n}{1-z^n} = f(z)$ and $\sum_{n=1}^{\infty} a_n z^n = g(z)$, and grouping the terms by *diagonals* in the double expansion of the *Lambert* series on p. 450 (which is allowed), we obtain

$$f(z) = g(z) + g(z^2) + \dots = \sum_{m=1}^{\infty} g(z^m).$$

---

[61] *Lambert, J.H.* Anlage zur Architektonik, Vol. 2, p. 507. Riga 1771.

[62] *Wiener, N.*, a new method in Tauberian theorems, J. Math. Massachusetts, Vol. 7, pp. 161—184, 1928, and Tauberian theorems, Ann. of Math. (2), Vol. 33, pp. 1—100, 1932.

7. For $a_n = (-1)^{n-1}$, $= n$, $= (-1)^{n-1} n$, $= \frac{1}{n}$, $= \frac{(-1)^{n-1}}{n}$, $= \alpha^n, \ldots$, we obtain in this way, successively, the following remarkable identities, valid for $|z| < 1$, in which the summations are taken from $n = 1$ to ∞:

$$\text{a)}\quad \Sigma(-1)^{n-1}\frac{z^n}{1-z^n} = \Sigma\frac{z^n}{1+z^n},$$

$$\text{b)}\quad \Sigma n\frac{z^n}{1-z^n} = \Sigma\frac{z^n}{(1-z^n)^2},$$

$$\text{c)}\quad \Sigma(-1)^{n-1}\cdot n\frac{z^n}{1-z^n} = \Sigma\frac{z^n}{(1+z^n)^2},$$

$$\text{d)}\quad \Sigma\frac{1}{n}\frac{z^n}{1-z^n} = \Sigma\log\frac{1}{1-z^n},$$

$$\text{e)}\quad \Sigma\frac{(-1)^{n-1}}{n}\cdot\frac{z^n}{1-z^n} = \Sigma\log(1+z^n),$$

$$\text{f)}\quad \Sigma\alpha^n\frac{z^n}{1-z^n} = \Sigma\frac{\alpha z^n}{1-\alpha z^n} \qquad (|\alpha| < 1),$$

etc.

8. In the two identities d) and e) we have on the right hand side a series of logarithms (for which of course we take the principal values); thus simple connections can be established between certain *Lambert* series and infinite products. E. g. from the two identities in question:

$$\Pi(1-z^n) = e^w, \quad \text{with} \quad w = -\Sigma\frac{1}{n}\frac{z^n}{1-z^n},$$

$$\Pi(1+z^n) = e^w, \quad \text{with} \quad w = \Sigma\frac{(-1)^{n-1}}{n}\frac{z^n}{1-z^n}.$$

9. As an interesting numerical example we may mention the following: Taking $u_0 = 0$, $u_1 = 1$, and for every $n > 1$, $u_n = u_{n-1} + u_{n-2}$, we obtain *Fibonacci*'s sequence (cf. 6, 7)

$$0,\ 1,\ 1,\ 2,\ 3,\ 5,\ 8,\ 13,\ 21,\ 34,\ 55,\ \ldots.$$

We then have

$$\sum_{k=1}^{\infty}\frac{1}{u_{2k}} = 1 + \frac{1}{3} + \frac{1}{8} + \frac{1}{21} + \frac{1}{55} + \cdots = \sqrt{5}\left[L\left(\frac{3-\sqrt{5}}{2}\right) - L\left(\frac{7-3\sqrt{5}}{2}\right)\right],$$

where $L(x)$ denotes the sum of the *Lambert* series[63] $\Sigma\frac{x^n}{1-x^n}$. The proof is based on the fact, which is easily established, that

$$u_\nu = \frac{\alpha^\nu - \beta^\nu}{\alpha - \beta} \qquad (\nu = 0, 1, 2, \ldots),$$

where α and β are the roots of the quadratic equation $x^2 - x - 1 = 0$. (Cf. Ex. 114.)

Exercises on Chapter XII[64].

174. Suppose $z_n \to \zeta$ and $b_n \to b \neq 0$. Under what conditions may we infer that $b_n^{z_n} \to b^\zeta$?

175. Suppose $z_n \to \infty$ (i. e. $|z_n| \to +\infty$). Under what conditions may we then infer that

$$\text{a)}\quad \left(1 + \frac{z}{z_n}\right)^{z_n} \to e^z,$$

$$\text{b)}\quad z_n\cdot(z^{1/z_n} - 1) \to \log z?$$

[63] *Landau, E.*: Bull. de la Soc. math. de France, Vol. 27, p. 298. 1899.

[64] In these exercises, wherever the contrary does not follow clearly from the context, all numbers are to be regarded as *complex*.

176. The principal value of z^i remains, for all values of z, less in absolute value than some fixed bound.

177. If
$$z_n = \sum_{\nu=0}^{n} (-1)^\nu \binom{z}{\nu},$$
either $z_n \to 0$ or $\frac{1}{z_n} \to 0$, according as $\Re(z) > 0$ or < 0. What is the behaviour of (z_n) when $\Re(z) = 0$?

178. Let a, b, c, d be four constants for which $a\,d - b\,c \neq 0$ and let z_0 be arbitrary. Investigate the sequence of numbers $(z_0, z_1, z_2, \ldots)$ given by the recurrence formula
$$z_{n+1} = \frac{a\,z_n + b}{c\,z_n + d} \qquad (n = 0, 1, 2, \ldots).$$
What are the necessary and sufficient conditions that (z_n) or $\left(\frac{1}{z_n}\right)$ should converge? And if neither of the two converges, under what conditions can z_p become $= z_0$ again for some index p? When are all the z_n's identically equal?

179. Let a be given $\neq 0$ and z_0 chosen arbitrarily, and write for each $n \geqq 0$
$$z_{n+1} = \frac{1}{2}\left(z_n + \frac{a}{z_n}\right).$$
(z_n) converges if, and only if, z_0 does not lie on the perpendicular to the straight line joining the two values of $\sqrt{a}$ through its middle point. If this condition is fulfilled, (z_n) converges to the value of $\sqrt{a}$ *nearest* to z_0. What is the behaviour of (z_n) when z_0 lies *on* the perpendicular in question?

180. The series $\sum \frac{1}{n^{1+i\gamma}}$ does *not* converge for *any* real γ; the series $\sum \frac{1}{n^{1+i\gamma} \log n}$, on the other hand, *does* converge for *every* real $\gamma \neq 0$.

180 a. The refinement of *Weierstrass*'s theorem **228** that was mentioned in footnote 13, p. 399, may be proved as follows in connection with the foregoing example: From the assumptions, it follows, firstly, that we may write
$$\frac{(n+1)^\alpha a_{n+1}}{n^\alpha a_n} = 1 + \frac{B_n}{n^{\lambda'}} \qquad (\lambda' = \text{Min}(\lambda, 2) > 1),$$
where the B_n's are bounded; hence, secondly, that we may write
$$a_n = \frac{c}{n^\alpha}\left(1 + \frac{C_n}{n^{\lambda'-1}}\right),$$
with c constant and the C_n's bounded. The factors $b_n = \left(1 + \frac{C_n}{n^{\lambda'-1}}\right)^{-1}$ satisfy the assumptions of the test **184**, 3. If $\sum a_n$ were to converge, then $\sum a_n b_n = \sum \frac{c}{n^\alpha}$ would also have to converge, contrary to the preceding example and theorem **255**.

181. For a fixed value of z and a suitable determination of the logarithm, does
$$\left[\frac{1}{z+1} + \ldots + \frac{1}{z+n} - \log(z+n)\right]$$
tend to a limit as $n \to +\infty$?

182. For every fixed z with $0 < \Re(z) < 1$,
$$\lim_{n\to\infty}\left[1 + \frac{1}{2^z} + \frac{1}{3^z} + \ldots + \frac{1}{n^z} - \frac{n^{1-z}}{1-z}\right]$$
exists (cf. Ex. 135).

183. The function $(1-z)\cdot\sin\left(\log\frac{1}{1-z}\right)$ may be expanded in a power series $\sum a_n z^n$ for $|z| < 1$, if we take the principal value for the logarithm. Show that this series still converges absolutely for $|z| = 1$.

184. If z tends to $+1$ from within the unit circle, and "within the angle", we have

a) $1 - z + z^4 - z^9 + z^{16} - + \cdots \to \frac{1}{2}$;

b) $(1 - z)[1 + z + z^4 + z^9 + \cdots]^2 \to \frac{\pi}{4}$;

c) $\frac{1}{\log \frac{1}{1-z}}[z + z^p + z^{p^2} + z^{p^3} + \cdots] \to \frac{1}{\log p}$;

d) $(1 - z)^{p+1}[z + 2^p z^2 + 3^p z^3 + \cdots] \to \Gamma(p + 1)$;

e) $\frac{\Sigma a_n z^n}{\Sigma b_n z^n} \to \lim \frac{a_n}{b_n}$,

provided the right hand limit exists, b_n is positive for each n, and Σb_n is divergent.

185. Investigate the behaviour of the following power series on the circumference of the unit circle:

a) $\sum \frac{(-1)^{[\sqrt{n}]}}{n} z^n$; b) $\sum \frac{z^n}{n^{\alpha + i\beta}}$, $\alpha > 0$;

c) $\sum \frac{z^n}{(n + a)^{\alpha + i\beta}}$; d) $\sum \frac{z^n}{n \log n}$;

e) $\sum \frac{\varepsilon_n}{n \log n} z^n$, where ε_n has the same meaning as in Ex. 47.

186. If $\Sigma\, a_n z^n$ converges for $|z| < 1$ and its sum is numerically $\leqq 1$ for all such values of z, then $\Sigma |a_n|^2$ converges and its sum is $\leqq 1$.

187. The power series

a) $\sum \frac{z^n}{n}$, b) $\sum \frac{z^{2k-1}}{2k-1}$,

c) $\sum (-1)^{k-1} \frac{z^{2k-1}}{2k-1}$, d) $\sum (-1)^{n-1} \frac{z^n}{n}$,

e) $\sum (-1)^n \frac{z^n}{(n+1)(n+2)}$, f) $\sum (-1)^n \binom{-\frac{1}{2}}{n} z^n$,

g) $\sum \frac{z^n}{(n-1)\cdot(n+1)}$, h) $\sum \frac{z^{2n}}{(2n-1)\cdot 2n}$

all have the unit circle as circle of convergence. On the circumference, they also converge in general, i. e. with the possible exception of isolated points. Try to express their sums by means of closed expressions involving elementary functions; separate the real and imaginary parts by writing $z = r(\cos x + i \sin x)$, and write down the trigonometrical expansions so obtained for $r < 1$ and for $r = 1$ separately. For which values of x do they converge? What are their sums? Are they the *Fourier* series of their sums?

188. What are the sums of the following series:

a) $\sum \frac{\cos nx \cos ny}{n}$; b) $\sum \frac{\cos nx \sin ny}{n}$;

c) $\sum \frac{\sin nx \sin ny}{n}$,

and of the three further series obtained by giving the terms of the above series the sign $(-1)^n$?

189. Proceeding with the geometric series Σz^n as in Ex. 187, but leaving $r < 1$, we obtain the expressions

a) $$\sum_{n=0}^{\infty} r^n \cos n x = \frac{1 - r\cos x}{1 - 2r\cos x + r^2};$$

b) $$\sum_{n=1}^{\infty} r^n \sin n x = \frac{r \sin x}{1 - 2r\cos x + r^2}.$$

Deduce from them the further expansions

c) $$\sum_{n=1}^{\infty} \frac{\cos n x}{(2\cos x)^n} = \cos 2x,$$

d) $$\sum_{n=1}^{\infty} \frac{\sin n x}{(2\cos x)^n} = \sin 2x$$

and indicate the exact intervals of validity.

190. In Exercise 187a the following expansion will have been obtained, among others:

$$\sum_{n=1}^{\infty} \frac{r^n}{n} \sin n x = \tan^{-1}\left(\frac{r\sin x}{1 - r\cos x}\right).$$

Deduce from it the expansions

a) $$\sum_{n=1}^{\infty} (-1)^{n-1} r^n \sin^n x \cdot \sin n x = \tan^{-1}(r + \cot x) - \left(\frac{\pi}{2} - x\right),$$

b) $$\sum_{n=1}^{\infty} \frac{\cos^n x \sin n x}{n} = \frac{\pi}{2} - x$$

and determine the exact intervals of validity.

191. Determine the exact regions of convergence of the following series:

a) $\displaystyle\sum \frac{(-1)^n}{z+n}$, b) $\displaystyle\sum \frac{1}{1-z^n}$,

c) $\displaystyle\sum \frac{z^n}{p_n{}^n}$, d) $\displaystyle\sum \frac{1}{p_n{}^z}$,

e) $\displaystyle\sum \left[\frac{1}{z - p_n} + \frac{1}{p_n} + \frac{z}{p_n{}^2} + \cdots + \frac{z^{n-1}}{p_n{}^n}\right]$,

f) $\displaystyle\sum \left(\frac{z}{p_n}\right)^{[\log n]}$, g) $\displaystyle\sum \left(\frac{z}{p_n}\right)^{[\log\log n]}$,

where (p_n) is real and increases monotonely to $+\infty$.

192. Establish the relations

a) $$\sum \frac{z^n}{1-z^n} = \sum z^{n^2} \cdot \frac{1+z^n}{1-z^n};$$

b) $$\sum (-1)^{n-1} \frac{z^{2n-1}}{1-z^{2n-1}} = \sum \frac{z^n}{1+z^{2n}},$$

where the summation begins with $n = 1$.

193. Corresponding to *Landau*'s theorem (258) we have the following: The *Dirichlet* series $\sum (-1)^{n-1}\frac{a_n}{n^z}$ and the so-called binomial coefficient series $\sum a_n \binom{z-1}{n}$ are convergent and divergent together, the points $z = 1, 2, 3, \ldots$ being disregarded.

194. For which values of z does the equation

$$\sum_{n=0}^{\infty} (-1)^n \binom{z}{n} = 0$$

hold good?

195. Determine the exact regions of convergence of the following infinite products:

a) $\prod\left(1-\frac{1}{n^z}\right)$, b) $\prod\left(1-\frac{(-1)^n}{n^z}\right)$,

c) $\prod(1+z^{2n+1})$, d) $\prod(1+n^2 z^n)$;

e) $\prod\left(1-\frac{z^n}{1-z^n}\right)$, f) $\prod\left[\left(1+\frac{z}{n}\right)\left(1-\frac{1}{n}\right)^z\right]$,

g) $\prod\left[\left(1-\frac{z}{z_n}\right)e^{\frac{z}{z_n}+\frac{1}{2}\left(\frac{z}{z_n}\right)^2+\cdots+\frac{1}{n}\left(\frac{z}{z_n}\right)^n}\right]$, if $|z_n| \to +\infty$,

h) $\prod\left(1-\frac{z}{n}\right)$, i) $\prod\left(1-\frac{(-1)^n}{n}z\right)$,

k) $\prod\left(1-\frac{(-1)^n}{2n+1}z\right)$.

196. Determine, by means of the sine product, the values of the products

a) $\prod\left(1+\frac{x^2}{n^2}\right)$, b) $\prod\left(1+\frac{x^4}{n^4}\right)$, c) $\prod\left(1+\frac{x^6}{n^6}\right)$,

for real values of x. The second of these has the value

$$\frac{1}{2\pi^2 x^2}\left[\cosh(\pi x\sqrt{2}) - \cos(\pi x\sqrt{2})\right].$$

Does this continue to hold for complex values of x?

197. The values of the products 195, i) and k), can be determined in the form of a closed expression by means of the Γ-function.

198. For $|z|<1$,

$$\frac{1}{(1-z)(1-z^3)(1-z^5)\ldots} = (1+z)(1+z^2)(1+z^3)\ldots,$$

199. By means of the sine product and the expansion of the cotangent in partial fractions, the following series and product may be evaluated in the form of closed expressions; x and y are real, and the symbol $\sum_{n=-\infty}^{+\infty} f(n)$ indicates the sum of the two series $\sum_{n=0}^{+\infty} f(n)$ and $\sum_{k=1}^{+\infty} f(-k)$, and similarly for the product:

a) $\sum_{n=-\infty}^{+\infty} \frac{1}{(n+x)^2+y^2}$, b) $\sum_{n=-\infty}^{+\infty} \frac{1}{n^4+x^4}$,

c) $\sum_{n=-\infty}^{+\infty} \frac{1}{(x-n)^3}$, d) $\prod_{n=-\infty}^{+\infty}\left(1-\frac{4x^2}{(n+x)^2}\right)$.

Chapter XIII.

Divergent series.

§ 59. General remarks on divergent sequences and the processes of limitation.

The conception of the nature of infinite sequences which we have set forth in all the preceding pages, and especially in §§ 8—11, is of comparatively recent date; for a strict and irreproachable construction of the theory could not be attempted until the concept of the real number had been made clear. But even if this concept and any one general convergence test for sequences of numbers, say our second main criterion, were recognized without proof as practically axiomatic, it nevertheless remains true that the theory of the convergence of infinite sequences, and of infinite series in particular, is far more recent than the extensive use of these sequences and series, and the discovery of the most elegant results of the subject, e. g. by *Euler* and his contemporaries, or even earlier, by *Leibniz*, *Newton* and their contemporaries. To these mathematicians, infinite series appeared in a very natural way as the result of calculation, and forced themselves into notice, so to speak: e. g. the geometric series $1+x+x^2+\cdots$ occurred as the non-terminating result of the division $1/(1-x)$; *Taylor*'s series, and with it almost all the series of Chapter VI, resulted from the principle of equating coefficients or from geometrical considerations. It was in a similar manner that infinite products, continued fractions and all other approximation processes occurred. In our exposition, *the symbol for infinite sequences was created* and then worked with; it was not so originally, these sequences *were there*, and the question was, what could be done with them.

On this account, problems of convergence in the modern sense were at first remote from the minds of these mathematicians[1]. Thus it is not to be wondered at that *Euler*, for instance, uses the geometric series

$$1+x+x^2+\cdots=\frac{1}{1-x}$$

even for $x=-1$ or $x=-2$, so that he unhesitatingly writes[2]

$$1-1+1-1+-\cdots=\frac{1}{2}$$

[1] Cf. the remarks at the beginning of § 41.

[2] This relation is used by *James Bernoulli* (Posit. arithm., Part 3, Basle 1696) and is referred to by him as a "paradoxon non inelegans". For details of the violent dispute which arose in this connection, see the work of *R. Reiff* mentioned in **69**, 8.

or
$$1 - 2 + 2^2 - 2^3 + - \ldots = \frac{1}{3};$$
similarly from $\left(\frac{1}{1-x}\right)^2 = 1 + 2x + 3x^2 + \ldots$ he deduces the relation
$$1 - 2 + 3 - 4 + - \ldots = \frac{1}{4};$$
and a great deal more. It is true that most mathematicians of those times held themselves aloof from such results in instinctive mistrust, and recognized only those which are true in the present-day sense [3]. But they had no clear insight into the reasons why one type of result should be admitted, and not the other.

Here we have no space to enter into the very instructive discussions on this point among the mathematicians of the 17th and 18th centuries [4]. We must be content with stating, e. g. as regards infinite series, that *Euler* always let these stand when they occurred naturally by expanding an analytical expression which itself possessed a definite value [5]. This value was then in every case regarded as the sum of the series.

It is clear that this convention has no precise basis. Even though, for instance, the series $1 - 1 + 1 - 1 + - \ldots$ results in a very simple manner from the division $1/(1-x)$ for $x = -1$ (see above), and therefore should be equated to $\frac{1}{2}$, there is no reason why the *same* series should not result from quite different analytical expressions and why, in view of these other methods of deducing it, it should not be given a different value. The above series may actually be obtained, for $x = 0$, from the function $f(x)$ represented for every $x > 0$ by the *Dirichlet* series
$$f(x) = \sum_{n=1}^{\infty} \frac{(-1)^{n-1}}{n^x} = 1 - \frac{1}{2^x} + \frac{1}{3^x} - \frac{1}{4^x} + \ldots,$$
or from $\frac{1+x}{1+x+x^2} = \frac{1-x^2}{1-x^3} = 1 - x^2 + x^3 - x^5 + x^6 - x^8 + - \ldots$

putting $x = 1$. In view of this latter method of deduction, we should have to take $1 - 1 + \ldots = \frac{2}{3}$, and in the case of the former there is no immediate evidence what value $f(0)$ may have; it *need not* at any rate be $+\frac{1}{2}$.

---

[3] Thus *d'Alembert* says (Opusc. Mathem., Vol. 5, 1768, 35; Mémoire, p. 183): "Pour moi, j'avoue que tous les raisonnements et les calculs fondés sur des séries qui ne sont pas convergentes ou qu'on peut supposer ne pas l'être, me paraîtront toujours très suspects".

[4] For details, see *R. Reiff*, loc. cit.

[5] In a letter to *Goldbach* (7. VIII. 1745) he definitely says: ". . . so habe ich diese neue Definition der Summe einer jeglichen seriei gegeben: Summa cujusque seriei est valor expressionis illius finitae, ex cujus evolutione illa series oritur".

Euler's principle is therefore insecure in any case, and it was only *Euler*'s unusual instinct for what is mathematically correct which in general saved him from false conclusions in spite of the copious use which he made of divergent series of this type[6]. *Cauchy* and *Abel* were the first to make the concept of convergence clear, and to renounce the use of any non-convergent series; *Cauchy* in his *Analyse algébrique* (1821), and *Abel* in his paper on the binomial series (1826), which is expressly based on *Cauchy*'s treatise. At first both hesitated to take this decisive step[7], but finally resolved to do so, as it seemed unavoidable if their reasoning were to be made strict and free from gaps.

We are now in a position to survey the problem from above, as it were; and the matter at once becomes clear when we remember that the symbol for an infinite sequence of numbers — in whatever form it is given, sequence, series, product or otherwise — has, and can have, no meaning whatever *in itself*, but that a meaning was only assigned to it by us, by an arbitrary convention. This convention consisted *firstly* in allowing only *convergent* sequences, i. e. sequences whose terms approached a definite and unique number in an absolutely definite sense; *secondly*, it consisted in associating this number with the infinite sequence, as its *value*, or in regarding the sequence as no more than another symbol (cf. **41**, 1) for the number. However obvious and natural this definition may be, and however closely it may be connected with the way in which sequences occur (e. g. as successive approximations to a result which cannot be obtained directly), a definition of this kind must nevertheless in all circumstances be considered *as an arbitrary one*, and it might even be replaced by quite different definitions. Suitability and success are the only factors which can determine whether one or the other definition is to be preferred; in the nature of the thing itself, that is to say, in the symbol (s_n) of an infinite sequence[8], there is nothing which *necessitates* any preference.

We are therefore quite justified in asking whether the complication which our theory exhibits (in parts at least) may not be due

[6] Cf. on the other hand p. 133, footnote 6.

[7] So far as *Cauchy* is concerned, cf. the preface to his *Analyse algébrique*, in which, among other things, he says: "Je me suis vu forcé d'admettre plusieurs propositions qui paraîtront peut-être un peu dures, par exemple qu'une série divergente n'a pas de somme". As regards *Abel*, cf. his letter to *Holmboe* (16. I. 1826), in which he says: "Les séries divergentes sont, en général, quelque chose de bien fatal, et c'est une honte qu'on ose y fonder aucune démonstration". — As already mentioned (p. 458, footnote 3), *J. d'Alembert* had expressed himself in a similar sense as early as 1768.

[8] (s_n) may be assumed to be any given sequence of numbers, in particular, therefore, the partial sums of an infinite series Σa_n or the partial products of an infinite product. We use the letter s, with its reminder of the word "sum", because infinite series are by far the most important means of defining sequences.

to our interpretation of the symbol (s_n), as the *limit* of the sequence, assumed *convergent*, being an unfavourable one, — however obvious and ready-to-hand it may appear. Other conventions might be drawn up in all sorts of ways, among which more suitable ones might per-
261. haps be found. From this point of view, the general *problem* which presents itself is as follows: *A particular sequence* (s_n) *is defined in some way, either by direct indication of the terms, or by a series or product, or otherwise. Is it possible to associate a "value"* s *with it, in a reasonable way?*

"In a reasonable way" might perhaps be taken to mean that the number s is obtained by a process *closely connected with the previous concept of convergence,* that is to say, with the formation of $\lim s_n = s$. This has been found so extraordinarily efficacious in all the preceding that we will not depart from it to any considerable extent without good reasons.

"In a reasonable way" might also, on the other hand, be interpreted as meaning that the sequence (s_n) is to have such a value s associated with it that *wherever this sequence may occur as the final result of a calculation, this final result shall always, or at least usually, be put equal to* s.

Let us first illustrate these general statements by an example. The series

262.
$$\Sigma(-1)^n \equiv 1 - 1 + 1 - 1 + - \cdots,$$

i. e. the geometric series Σx^n for $x = -1$, or the sequence

$$(s_n) \equiv 1,\ 0,\ 1,\ 0,\ 1,\ 0,\ \ldots,$$

has so far been rejected as divergent, because its terms s_n do not approach a single definite number. On the contrary, they oscillate unceasingly between 1 and 0. This very fact, however, suggests the idea of forming the arithmetic means

$$s_n' = \frac{s_0 + s_1 + \cdots + s_n}{n+1} \qquad (n = 0, 1, 2, \ldots).$$

Since $s_n = \frac{1}{2}[1 + (-1)^n]$, we find that

$$s_n' = \frac{(n+1) + \frac{1}{2}[1 + (-1)^n]}{2(n+1)} = \frac{1}{2} + \frac{1 + (-1)^n}{4(n+1)},$$

so that s_n' (in the former sense) approaches the value $\frac{1}{2}$:

$$\lim s_n' = \frac{1}{2}.$$

By this very obvious process of taking the arithmetic mean, we have accordingly managed, in a perfectly accurate way, to give a meaning to *Euler*'s paradoxical equation $1 - 1 + 1 - + \cdots = \frac{1}{2}$, to associate with the series on the left hand side the number $\frac{1}{2}$ as its "value", or to obtain this number from the series. Whether we can always equate the final result of a calculation to $\frac{1}{2}$ whenever it appears in the form $\Sigma(-1)^n$, cannot of course be determined off-hand. In the case of the expansion $\frac{1}{1-x}=\Sigma x^n$ for $x=-1$, it is certainly so; in the case of $\Sigma\frac{(-1)^{n-1}}{n^x}$ for $x=0$, it is equally true, as may be shown by fairly simple means (cf. Exercise 200); — and a great deal more evidence can be adduced to show that the association of the sequence $1, 0, 1, 0, 1, \ldots$ with the value $\frac{1}{2}$ obtained in the manner described above is "reasonable"[9].

We might therefore, as an experiment, make the following definition. If, and only if, the numbers

$$s_n' = \frac{s_0+s_1+\ldots+s_n}{n+1} \qquad (n=0, 1, 2, \ldots)$$

tend to a limit s in the previous sense, the sequence (s_n), or series Σa_n, will be said to "converge" to the "limit", or "sum", s.

The suitability of this new definition has already been demonstrated in connection with the series $\Sigma(-1)^n$, which now becomes convergent "in the new sense", with the sum $\frac{1}{2}$, — which seems thoroughly reasonable. Two further remarks will illustrate the advantages of this new definition:

1. Every sequence (s_n), convergent in the former sense and of limit s, is so constituted, in virtue of *Cauchy*'s theorem **43**, 2, that it would also have to be called convergent "in the new sense", with the same limit s. The new definition would therefore enable us to accomplish at least all that we could do with the former, while the example of the series $\Sigma(-1)^n$ shows that the new definition is more far-reaching than the old one.

2. If two series, convergent in the old sense, $\Sigma a_n = A$ and $\Sigma b_n = B$, are multiplied together by *Cauchy*'s rule, giving the series $\Sigma c_n \equiv \Sigma(a_0 b_n + a_1 b_{n-1} + \cdots + a_n b_0)$, we know that this series is not necessarily convergent (in the old sense). And the question *when* Σc_n *does* converge presents very considerable difficulties and has not been satisfactorily cleared up so far. The second proof of theorem **189**,

---

[9] From the series (see above) for $\frac{1+x}{1+x+x^2}$ also we can accordingly deduce the value $\frac{2}{3}$ for $x=1$. We have only to observe that the series, written somewhat more carefully, is $1+0\cdot x - x^2 + x^3 + 0\cdot x^4 - x^5 + + - \cdots$, and is therefore $1+0-1+1+0-1++-\cdots$ for $x=1$.

however, shows that in every case

$$\frac{C_0 + C_1 + \cdots + C_n}{n+1} \to AB$$

if C_n denotes the nth partial sum of Σc_n. The meaning of this is that Σc_n *always* converges *in the new sense,* with the sum AB. Here the advantage of the new convention is obvious: A situation which, owing to the insuperable difficulties involved, it was impossible to clear up as long as we kept to the old concept of convergence, may be dealt with exhaustively in a very simple way, by introducing a slightly more general concept of convergence.

We shall very soon become acquainted with other investigations of this kind (see § 61 in particular); first of all, however, we shall make some definitions relating to several fundamental matters:

Besides the formation of the arithmetic mean, we shall become acquainted with quite a number of other processes, which may with success be substituted for the former concept of convergence, for the purpose of associating a number s with a sequence of numbers (s_n). These processes have to be distinguished from one another by suitable designations. In so doing it is advisable to proceed as follows: The former concept of convergence was so natural, and has stood the test so well, that it ought to have a special name reserved for it. Accordingly, the expression: "*convergence* of an infinite sequence (series, product, ...)" shall continue to mean exactly what it did before. If by means of new rules, as, for instance, by the formation of the arithmetic mean described above, a number s is associated with a sequence (s_n), we shall say that the sequence (s_n) is *limitable** by that process, and that the corresponding series Σa_n is *summable* by the process, and we shall call s the *value* of either (or in the case of the series, its *sum* also).

When, however, as will occur directly, we are making use of *several* processes of this kind, we distinguish these by attached initials $A, B, \ldots, V, \ldots$, and speak for instance of a V-process[10]. We shall say that the sequence (s_n) is *limitable* V, and that the series Σa_n is *summable* V; and the number s will be referred to as the V-limit of the sequence or V-sum of the series; symbolically

$$V\text{-}\lim s_n = s, \qquad V\text{-}\Sigma a_n = s.$$

When there is no fear of misunderstanding, we may also express the

* German: *limitierbar.*

[10] In the case of the concept of *integrability* the situation is somewhat similar and it was probably in this connection that the above type of notation was first introduced. Thus we say a function is *integrable* R or *integrable* L according as we are referring to integrability in *Riemann*'s or in *Lebesgue*'s sense.

former of the two statements by the symbolism

$$V(s_n) \to s,$$

which more precisely implies that the *new sequence* deduced from (s_n) by the V-process converges to s.

When, as will usually be the case in what follows, the process admits of a k-fold iteration, or can be graded into different orders, we attach a suffix and speak of a *V_k-limitation process*, a *V_p-summation process*, etc.

In the construction and choice of such processes we shall of course **263.** not proceed quite arbitrarily, but we shall rather let ourselves be guided by questions of suitability. We must give the first place to the fundamental stipulation to be made in this connection, namely that the new definition must not contradict the old one. We accordingly stipulate that any V-process which may be introduced must satisfy the following *permanence condition*:

I. *Every sequence (s_n) convergent in the former sense, with the limit s, must be limitable V with the value s.* Or in other words, $\lim s_n = s$ *must in every case imply*[11] $V\text{-}\lim s_n = s$.

In order that the introduction of a process of this kind may not be superfluous, we further stipulate that the following *extension condition* is to hold:

II. *At least* **one** *sequence (s_n), which diverges in the former sense, must be limitable by the new process.*

Let us call the totality of sequences which are limitable by a particular process the ***range of action*** of this process. The condition II implies that only those processes will be allowed which possess a wider range of action than the ordinary process of convergence. It is precisely the limitation of formerly divergent sequences and the summation of formerly divergent series which will naturally claim the greater part of our attention now.

Finally, if several processes are employed *together,* say a V-process and a W-process *simultaneously,* we should be in danger of hopeless confusion if we did not also stipulate that the following *compatibility condition* should be fulfilled:

III. *If one and the same sequence (s_n) is limitable by two different processes, simultaneously applied, then it must have the* **same** *value by both processes.* In other words, *we must in every case have* $V\text{-}\lim s_n = W\text{-}\lim s_n$, *if both these values exist.*

---

[11] We might also be satisfied if some convergent sequences at least are limitable with unaltered value by the process considered. This is the case e. g. with the E_p-process discussed further on, provided the suffix p is complex.

We shall only consider processes which satisfy these three conditions. Besides these, however, we require some indication whether the association of a value s with the sequence (s_n) effected by a particular V-process is a *reasonable* one in the sense explained above (p. 460). Here widely-varying conditions may be laid down, and the processes which are in current use are of very varied degrees of efficiency in this respect. In the first instance we should no doubt require that the elementary rules of the algebra of convergent sequences (v. § 8) should as far as possible be maintained, i. e. the rules for term-by-term addition and subtraction of two sequences, term-by-term addition of a constant, and term-by-term multiplication by a constant, and the effect of a finite number of alterations (**27**, 4), etc. Next we might perhaps require that if, say, a divergent series Σa_n has associated with it the value s, and if this series is deduced, e. g. from a power series $f(x) = \Sigma c_n x^n$ by substituting a special value x_1 for x, then the number s should bear an appropriate relation to $f(x_1)$ or to $\lim f(x)$ for $x \to x_1$; and similarly for other types of series (*Dirichlet* series, *Fourier* series etc.). *In short, we should require that wherever this series appears as the final result of a calculation, the result should be* s. The greater the number of conditions similar to the above which are satisfied by a
264. particular process — let us call them the *conditions F*, without taking pains to formulate them with absolute precision — and at the same time, the greater the range of action of the process, the greater will be its usefulness and value from our point of view.

We proceed to indicate a few of these processes of limitation which have proved their worth in some way or another.

265. 1. The C_1-, H_1-, or M-process [12]. As described above, **262**, we form the arithmetic means of the terms of a sequence (s_n):

$$\frac{s_0 + s_1 + \ldots + s_n}{n+1} \qquad (n = 0, 1, 2, \ldots)$$

which we will denote by c_n', h_n', or m_n. If these tend to a limit s in the older sense, when $n \to \infty$, we say that (s_n) is ***limitable C_1*** or ***limitable H_1*** or ***limitable M*** with the value s and we write

$$M\text{-}\lim s_n = s \qquad \text{or} \qquad M(s_n) \to s,$$

or use the letters C_1 or H_1 instead of M. The series Σa_n with the partial sums s_n will be called ***summable C_1*** or ***summable H_1*** or ***summable M***, *and s will be called its C_1-, H_1-, or M-sum.*

The sequence of units 1, 1, 1, ... may be considered to be the simplest convergent sequence we can conceive. The process described above consists in comparing, *on the average*, the terms s_n of the sequence

[12] The choice of the letters C and H is explained in the two next sub-sections.

under consideration with those of the sequence of units:

$$c_n' \equiv h_n' \equiv m_n = \frac{s_0 + s_1 + \dots + s_n}{1 + 1 + \dots + 1}.$$

This "averaged" comparison of (s_n) with the unit sequence will be met with again in the case of the following processes.

The usefulness of this process has already been illustrated above by several examples. We have also seen that it satisfies the two conditions **263**, I and II, and III does not come under consideration at the moment. In §§ 60 and 61 it will further be seen that the conditions F (**264**) are also in wide measure fulfilled.

2. ***Hölder*'s process, or the H_p-process**[13]. If with a given sequence (s_n), we proceed from the arithmetic means h_n' just formed to *their* mean

$$h_n'' = \frac{h_0' + h_1' + \dots + h_n'}{n+1} \qquad (n = 0, 1, 2, \dots)$$

and if the sequence (h_n'') has a limit in the ordinary sense, $\lim h_n'' = s$, *we say that*[14] *the sequence s_n is* ***limitable H_2*** *with the value s.*

By **43**, 2, every sequence which is limitable H_1 (and therefore also every convergent sequence), is also limitable H_2, with the same value. The new process therefore satisfies the conditions **263**, I, II and III; moreover, its range is wider than that of the H_1-process, for the series

$$\sum_{n=0}^{\infty} (-1)^n (n+1) \equiv 1 - 2 + 3 - 4 + - \cdots,$$

for instance, is summable H_2 with the sum $\frac{1}{4}$, but not summable H_1 nor convergent. In fact, we have here

$$(s_n) \equiv 1, -1, 2, -2, 3, -3, \dots$$

and

$$(h_n') \equiv 1, 0, \frac{2}{3}, 0, \frac{3}{5}, 0, \dots.$$

These sequences are not convergent. On the other hand, the numbers $h_n'' \to \frac{1}{4}$, as is easily calculated. This is precisely the value which one would expect from

$$\left(\frac{1}{1-x}\right)^2 = \sum_{n=0}^{\infty} (n+1) x^n$$

for $x = -1$.

---

[13] *Hölder, O.*: Grenzwerte von Reihen an der Konvergenzgrenze. Math. Ann., Vol. 20, pp. 535—549. 1882. Here arithmetic means of the kind described are for the first time introduced for a special purpose.

[14] The rest of the notation is formed in the same way, H_2-lim $s_n = s$, H_2-$\Sigma a_n = s$, $H_2(s_n) \to s$, etc. but hereafter we shall not mention it specially.

If the numbers h_n'' do not tend to a unique limit, we proceed to take *their* mean

$$h_n''' = \frac{h_0'' + h_1'' + \ldots h_n''}{n+1} \qquad (n = 0, 1, 2, \ldots)$$

or, in general, for[15] $p \geqq 2$, the mean

$$h_n^{(p)} = \frac{h_0^{(p-1)} + h_1^{(p-1)} + \ldots + h_n^{(p-1)}}{n+1} \qquad (n = 0, 1, 2, \ldots)$$

between the numbers $h_n^{(p-1)}$ obtained at the previous stage; if these new numbers $h_n^{(p)} \to s$, for some definite p, we say that *the sequence* (s_n) *is* ***limitable*** $\boldsymbol{H_p}$ *with the value s.*

It is easy to form sequences which are limitable H_p for any particular given p, but for no smaller value of p than this[16]. This, together with **43**, 2, shows that the H_p-processes not only satisfy the conditions **263**, I—III, but that their range of action is wider for each fixed $p \geqq 2$ than for all smaller values of p. As regards the conditions F, we must again refer to §§ 60 and 61.

3. ***Cesàro's*** **process, or the** $\boldsymbol{C_k}$**-process**[17]. We first write $s_n \equiv S_n^{(0)}$, and also, for each $k \geqq 1$,

$$S_0^{(k-1)} + S_1^{(k-1)} + \ldots + S_n^{(k-1)} = S_n^{(k)}, \quad (n = 0, 1, 2, \ldots)$$

and we now examine the sequence of numbers[18]

$$c_n^{(k)} = \frac{S_n^{(k)}}{\binom{n+k}{k}},$$

for each fixed k. If, for some value of k, $c_n^{(k)} \to s$, we say that *the sequence* (s_n) *is* ***limitable*** $\boldsymbol{C_k}$ *with the value s.*

In the case of the H-process, we cannot obtain simple formulae giving $h_n^{(p)}$ directly in terms of s_n, for larger values of p. In the case of the C-process, this is easily done, for we have

$$S_n^{(k)} = \binom{n+k-1}{k-1} s_0 + \binom{n+k-2}{k-1} s_1 + \ldots + \binom{k-1}{k-1} s_n,$$

---

[15] Or indeed for $p \geqq 1$, provided we agree to put $h_n^{(0)} \equiv s_n$ and take the H_0-process to be ordinary convergence, as we shall do here and in all analogous cases in future.

[16] Write, for instance, $\left(h_n^{(p-1)}\right) \equiv 1, 0, 1, 0, 1, \ldots$ and work backwards to the values of s_n. Other examples will be found in the following sections.

[17] *Cesàro, E.*: Sur la multiplication des séries. Bull. des sciences math. (2), Vol. 14, pp. 114—120. 1890.

[18] The denominators of the right hand side are exactly the values of $S_n^{(k)}$ obtained by starting with the sequence $(s_n) \equiv 1, 1, 1, \ldots$, i. e. they indicate how many of the partial sums s_ν are comprised in $S_n^{(k)}$. Thus the C_k-process again involves an "averaged" comparison between a given sequence (s_n) and the unit sequence.

or if we wish to go back to the series Σa_n, with the partial sums s_n,

$$S_n^{(k)} = \binom{n+k}{k} a_0 + \binom{n+k-1}{k} a_1 + \cdots + \binom{k}{k} a_n .$$

This may be proved quite easily by induction, or by noticing that, by **102**,

$$\sum_{n=0}^{\infty} S_n^{(k-1)} x^n = (1-x) \sum_{n=0}^{\infty} S_n^{(k)} x^n,$$

so that for every integral $k \geqq 0$

$$\sum_{n=0}^{\infty} S_n^{(k)} x^n = \frac{1}{(1-x)^k} \sum_{n=0}^{\infty} s_n x^n = \frac{1}{(1-x)^{k+1}} \sum_{n=0}^{\infty} a_n x^n,$$

whence, by **108**, the truth of the statement follows[19].

In the following sections we shall enter in detail into this process also, which becomes identical with the preceding one ($h_n' \equiv c_n'$) for $p = 1$.

4. ***Abel*'s process, or the *A*-process.** Given a series Σa_n with the partial sums s_n, we consider the power series

$$f(x) = \Sigma a_n x^n = (1-x) \Sigma s_n x^n.$$

If its radius is $\geqq 1$, and if (for real values of x) the limit

$$\lim_{x \to 1-0} \Sigma a_n x^n = \lim_{x \to 1-0} (1-x) \Sigma s_n x^n = s$$

exists, *we say that the series Σa_n is*[20] ***summable A**, and that the sequence (s_n) is **limitable A**, with the value s; in symbols*:

$$A\text{-}\Sigma a_n = s, \quad A\text{-}\lim s_n = s.$$

In consequence of *Abel*'s theorem **100**, this process also fulfils the permanence condition I, and simple examples show that it fulfils the "extension condition" II; for instance, in the case of the series $\Sigma(-1)^n$ already used, the limit for $x \to 1-0$

$$\lim (\Sigma (-1)^n x^n) = \lim \frac{1}{1+x} = \frac{1}{2}$$

exists. Thus *Euler*'s paradoxical equation (p. 457) is again justified

[19] In view of these last formulae, it is fairly natural to allow *non-integral* values > -1 for the suffix k also. Such limitation processes of non-integral order were first consistently introduced and investigated by the author (Grenzwerte von Reihen bei der Annäherung an die Konvergenzgrenze, Inaug.-Diss., Berlin 1907). We shall however not enter into this question, either here in the case of the C-process, or later in that of the other processes considered.

[20] If the product $(1-x)\,\Sigma s_n x^n$ is written in the form

$$\frac{\Sigma s_n x^n}{\Sigma x^n},$$

we see that it is again an "averaged" comparison of the given sequence with the unit sequence which is involved, though in a somewhat different manner.

by this process. If we now use the more precise form

$$A\text{-}\Sigma(-1)^n = \frac{1}{2} \quad\text{or}\quad C_1\text{-}\Sigma(-1)^n = \frac{1}{2},$$

we thus indicate two perfectly definite processes by which the value $\frac{1}{2}$ may be obtained from the series $\Sigma(-1)^n$.

5. *Euler's* process, or the *E*-process. We saw in **144** that if the *first* of the two series

$$\sum_{n=0}^{\infty}(-1)^n a_n \quad\text{and}\quad \sum_{k=0}^{\infty}\frac{\Delta^k a_0}{2^{k+1}}$$

converges, then so does the *second*, and to the same sum. Simple examples show, however, that the second series may quite well converge without the first one doing so:

1. If $a_n \equiv 1$, then $a_0 = 1$ and $\Delta^k a_0 = 0$ for $k \geqq 1$. Accordingly, the two series are

$$1 - 1 + 1 - 1 + - \cdots \quad\text{and}\quad \frac{1}{2} + 0 + 0 + 0 + \cdots$$

the second of which converges to the sum $\frac{1}{2}$.

2. If, for $n = 0, 1, 2, \ldots,$

$$a_n = \quad 1, \quad 2, \quad 3, \quad 4, \ldots,$$

then

$$\Delta a_n = -1, \ -1, \ -1, \ -1, \ldots,$$

and for $k \geqq 2$

$$\Delta^k a_n = \quad 0, \quad 0, \quad 0, \quad 0, \ldots.$$

Accordingly, the two series are

$$1 - 2 + 3 - 4 + - \cdots \quad\text{and}\quad \frac{1}{2} - \frac{1}{4} + 0 + 0 + \cdots,$$

the second of which converges to the sum $\frac{1}{4}$.

3. Similarly for $a_n = (n+1)^3$ we find $\Delta a_0 = -7$, $\Delta^2 a_0 = 12$, $\Delta^3 a_0 = -6$, and, for $k > 3$, $\Delta^k a_0 = 0$. The two series are thus

$$1 - 8 + 27 - 64 + - \cdots \quad\text{and}\quad \frac{1}{2} - \frac{7}{4} + \frac{12}{8} - \frac{6}{16} + 0 + 0 + \cdots,$$

the second of which converges to the sum $-\frac{1}{8}$.

4. For $a_n = 2^n$, $\Delta^k a_0 = (-1)^k$. Thus the two series are:

$$1 - 2 + 4 - 8 + - \cdots \quad\text{and}\quad \frac{1}{2} - \frac{1}{4} + \frac{1}{8} - \frac{1}{16} + - \cdots,$$

the second of which converges to the sum $\frac{1}{3}$, i. e. the sum which we should expect for $x = -2$ from $\frac{1}{1-x} = \Sigma x^n$.

5. For $a_n = (-1)^n z^n$, $\Delta^k a_0 = (1+z)^k$. The two series are therefore

$$\sum_{n=0}^{\infty} z^n \quad\text{and}\quad \sum_{k=0}^{\infty}\frac{(1+z)^k}{2^{k+1}},$$

the second of which converges to the sum $\frac{1}{1-z}$, provided $|z+1| < 2$.

If we start with any series Σa_n, *without* alternately + and − signs, the series

$$\sum_{n=0}^{\infty} a_n', \quad \text{with} \quad a_n' = \frac{1}{2^{n+1}}\left[\binom{n}{0}a_0 + \binom{n}{1}a_1 + \cdots + \binom{n}{n}a_n\right]$$

will be an *Euler*'s transformation of the given series, which we may also obtain as follows: The series Σa_n results from the power series $\Sigma a_n x^{n+1}$ for $x = 1$, hence from

$$\sum_{n=0}^{\infty} a_n \left(\frac{y}{1-y}\right)^{n+1}$$

for $y = \frac{1}{2}$. Expanding the latter in powers of y, before substituting $y = \frac{1}{2}$, we obtain *Euler*'s transformation. In fact

$$\sum_{k=0}^{\infty} a_k x^{k+1} = \sum_{k=0}^{\infty} a_k \left(\frac{y}{1-y}\right)^{k+1} = \sum_{k=0}^{\infty} a_k \sum_{\lambda=0}^{\infty} \binom{k+\lambda}{k} y^{k+\lambda+1}$$
$$= \sum_{n=0}^{\infty}\left\{\sum_{\nu=0}^{n}\binom{n}{\nu}a_\nu\right\} y^{n+1} = \sum_{n=0}^{\infty} a_n'(2y)^{n+1}.$$

In order to adapt this process for use with any sequence (s_n) we write, deviating somewhat from the usual notation,

$$a_0 + a_1 + \cdots + a_{n-1} = s_n \quad \text{for} \quad n \geqq 1, \quad \text{and} \quad s_0 = 0,$$

and also

$$a_0' + a_1' + \cdots + a_{n-1}' = s_n' \quad \text{for} \quad n \geqq 1, \quad \text{and} \quad s_0' = 0.$$

It is now easy to verify that[21] for every $n \geqq 0$

$$s_n' = \frac{1}{2^n}\left[\binom{n}{0}s_0 + \binom{n}{1}s_1 + \cdots + \binom{n}{n}s_n\right].$$

We accordingly make the following definition: *A sequence (s_n) is said to be **limitable E_1** with the value s*, if the sequence (s_n') just defined tends[22] to s. If, without testing the convergence of (s_n'), we write

---

[21] From $\Sigma a_n x^{n+1} = \Sigma a_n'(2y)^{n+1}$ it follows, by multiplication by $\frac{1}{1-x} = \frac{1-y}{1-2y}$, that

$$\sum_{k=0}^{\infty} s_k x^k = (1-y)\sum_{n=0}^{\infty} s_n'(2y)^n.$$

Hence

$$\sum_{n=0}^{\infty} s_n'(2y)^n = \frac{1}{1-y}\sum_{k=0}^{\infty} s_k\left(\frac{y}{1-y}\right)^k = \sum_{k=0}^{\infty}\sum_{\lambda=0}^{\infty}\binom{k+\lambda}{k} s_k y^{k+\lambda}$$
$$= \sum_{n=0}^{\infty}\frac{1}{2^n}\left[\binom{n}{0}s_0 + \binom{n}{1}s_1 + \cdots + \binom{n}{n}s_n\right](2y)^n,$$

whence the relation may at once be inferred.

[22] Here also the denominator 2^n is obtained from the numerator

$$\left[\binom{n}{0}s_0 + \cdots + \binom{n}{n}s_n\right]$$

by replacing each of the s_n's by 1. Thus we are again concerned with an "averaged" comparison, of a definite kind, between the sequence (s_n) and the unit sequence.

$$s_n'' = \frac{1}{2^n}\left[\binom{n}{0} s_0' + \binom{n}{1} s_1' + \cdots + \binom{n}{n} s_n'\right],$$

and in general, for $r \geqq 1$,

$$s_n^{(r)} = \frac{1}{2^n}\left[\binom{n}{0} s_0^{(r-1)} + \binom{n}{1} s_1^{(r-1)} + \cdots + \binom{n}{n} s_n^{(r-1)}\right], \quad (n = 0, 1, 2, \ldots),$$

we shall similarly say that the sequence (s_n) is *limitable* E_r and regard s as its E_r-limit, if, for a particular r, $s_n^{(r)} \to s$.

Our former theorem **144** (see also **44**, 8) then shows in any case that this E-process satisfies the permanence condition I, and the examples given there show that the condition II is also satisfied. This process will be examined further in § 63.

6. *Riesz*'s process, or the $R_{\lambda k}$-process [23]. For making the principle of averaged comparison of the sequence (s_n) with the unit sequence more powerful, — a principle which, as we saw, lies at the basis of all the former limitation processes, — a fairly obvious procedure consists in attributing *arbitrary* weights to the various terms s_n. If $\mu_0, \mu_1, \mu_2, \ldots$ denote any sequence of positive numbers, then

$$s_n' = \frac{\mu_0 s_0 + \mu_1 s_1 + \ldots + \mu_n s_n}{\mu_0 + \mu_1 + \ldots + \mu_n}$$

is a generalized mean of this kind. In the special case of $\mu_n = \frac{1}{n+1}$, we speak of a *logarithmic* mean.

As with the H-, C-, or E-processes, this generalized method of forming the mean may of course be repeated, writing, for instance, as in the C-process,

$$\sigma_n^{(0)} = s_n \quad \text{and} \quad \lambda_n^{(0)} = 1,$$

and then, for $k \geqq 1$,

$$\sigma_n^{(k)} = \mu_0 \sigma_0^{(k-1)} + \mu_1 \sigma_1^{(k-1)} + \ldots + \mu_n \sigma_n^{(k-1)}$$

and

$$\lambda_n^{(k)} = \mu_0 \lambda_0^{(k-1)} + \mu_1 \lambda_1^{(k-1)} + \ldots + \mu_n \lambda_n^{(k-1)},$$

and then proceeding to investigate, for fixed $k \geqq 1$, the ratio

$$\varrho_n^{(k)} = \frac{\sigma_n^{(k)}}{\lambda_n^{(k)}}$$

for $n \to +\infty$. If these tend to a limit s, we might say that (s_n) was limitable [24] $R_{\mu k}$ with the value s. This definition, however, is not in use. The process in question has reached its great importance only by being transformed into a form more readily amenable to analysis, as

[23] *Riesz, M.*: Sur les séries de Dirichlet et les séries entières. Comptes rendus Vol. 149, pp. 909—912. 1909.

[24] Here we add a suffix μ to R_k, the notation of the process, as a reference to the sequence (μ_n) used in the formation of the mean. For $\mu_n \equiv 1$, this process reduces exactly to the C_k-process.

follows: A (complex) function $s(t)$ of the real variable $t \geqq 0$ is defined by

$$s(t) = s_\nu \quad \text{in} \quad \lambda_{\nu-1}^{(1)} < t \leqq \lambda_\nu^{(1)} \qquad (\nu = 0, 1, 2, \ldots;\ \lambda_{-1}^{(1)} = 0)$$

with $s(0) = 0$; then

$$\mu_0 s_0 + \mu_1 s_1 + \ldots + \mu_n s_n = \int_0^{\lambda_n^{(1)}} s(t)\, dt$$

and it is natural to substitute repeated integration for the repeated summation used in the formation of the numbers $\sigma_n^{(k)}$ and $\lambda_n^{(k)}$. A k-ple integration [25] gives

$$\int_0^\omega dt_{k-1} \int_0^{t_{k-1}} \cdots \int_0^{t_1} s(t)\, dt = \frac{1}{(k-1)!} \int_0^\omega s(t) \cdot (\omega - t)^{k-1}\, dt$$

instead of $\sigma_n^{(k)}$. Similarly, instead of the numbers $\lambda_n^{(k)}$, we have to take the values which we obtain by putting $s_n \equiv 1$ in the integrals just written down, i. e.

$$\frac{1}{(k-1)!} \int_0^\omega (\omega - t)^{k-1}\, dt = \frac{\omega^k}{k!}.$$

We should then have to deal with the limit (for fixed k)

$$\lim_{\omega \to +\infty} \frac{k}{\omega^k} \int_0^\omega s(t)\, (\omega - t)^{k-1}\, dt.$$

If this limit exists and $= s$, *the sequence* (s_n) *will be called* ***limitable*** $\boldsymbol{R_{\lambda k}}$ *with the value* s.

Here we cannot enter into a more detailed examination of the question whether the two definitions given for the $R_{\lambda k}$-process are really exactly equivalent, or into the elegant and far-reaching applications of the process in the theory of *Dirichlet*'s series. (For references to the literature, see **266.**)

7. ***Borel*'s process, or the *B*-process.** We have just seen how *Riesz*' process tends to increase the efficiency of the H- or C-processes, by substituting for the method of averaged comparison between the sequence (s_n) and the unit sequence a more general form of this procedure. The range of *Abel*'s process may be enlarged in a similar way by making use of other series instead of the geometric series there used for purposes of comparison. Taking the exponential series as a particular case, and accordingly considering the quotient of the two series

$$\sum_{n=0}^{\infty} s_n \frac{x^n}{n!} \quad \text{and} \quad \sum_{n=0}^{\infty} \frac{x^n}{n!},$$

[25] The equality of the two sides is easily proved by induction, using integration by parts.

that is to say, the product

$$F(x) = e^{-x} \cdot \sum_{n=0}^{\infty} s_n \frac{x^n}{n!}$$

for $x \to +\infty$, we obtain the process introduced by *E. Borel*[26]. In accordance with it we make the following definition: A sequence (s_n) such that the power series $\sum s_n \frac{x^n}{n!}$ converges everywhere and the function $F(x)$ just defined tends to a unique limit s as $x \to +\infty$, *will be called* ***limitable B*** *with the value* s.

In order to illustrate the process to some extent, let us first take $\sum a_n \equiv \sum (-1)^n$ once more; then $s_n = 1$ or 0, according as n is even or odd. Accordingly

$$\sum s_n \frac{x^n}{n!} = 1 + \frac{x^2}{2!} + \frac{x^4}{4!} + \cdots = \frac{e^x + e^{-x}}{2}$$

and we have to deal with the limit

$$\lim_{x \to +\infty} e^{-x} \cdot \frac{e^x + e^{-x}}{2},$$

which is evidently $\frac{1}{2}$. Thus $\sum (-1)^n$ is summable B with the sum $\frac{1}{2}$. More generally, taking $\sum a_n \equiv \sum z^n$, we have, provided only that $z \neq +1$,

$$s_n = \frac{1 - z^{n+1}}{1 - z}$$

and

$$F(x) = e^{-x} \cdot \sum s_n \frac{x^n}{n!} = \frac{1}{1-z} - \frac{z}{1-z} e^{(z-1)x}$$

which $\to \frac{1}{1-z}$ when $x \to +\infty$, provided $\Re(z) < 1$. *Thus the geometric series* $\sum z^n$ *is summable B with the sum* $\frac{1}{1-z}$ *throughout the half-plane*[27] $\Re(z) < 1$.

This process also satisfies the *permanence condition*; for we have

$$\left(e^{-x} \cdot \sum s_n \frac{x^n}{n!}\right) - s = e^{-x} \cdot \sum (s_n - s) \frac{x^n}{n!}.$$

If $s_n \to s$ in the ordinary sense, we can for any given ε choose m so

---

[26] Sur la sommation des séries divergentes, Comptes rendus, Vol. 121, p. 1125. 1895, — and in many Notes in connection with it. A connected account is given in his Leçons sur les séries divergentes, 2nd ed., Paris 1928.

[27] By the C-processes, as shewn in **268**, 8, the geometric series is summable, beyond $|z| < 1$, only for the *boundary points* of the unit circle, $+1$ excepted; by *Euler*'s process it is summable throughout the circle $|z + 1| < 2$, which encloses the unit circle, with a wide margin; by *Borel*'s process it is summable in the whole half-plane $\Re(z) < 1$, — the value in this and the preceding cases being everywhere $\frac{1}{1-z}$.

large that $|s_n - s| < \frac{1}{2}\varepsilon$ for every $n > m$. The expression on the right hand side is then in absolute value

$$\leq e^{-x} \cdot \sum_{n=0}^{\infty} |s_n - s| \cdot \frac{x^n}{n!} \leq e^{-x} \cdot \sum_{n=0}^{m} |s_n - s| \cdot \frac{x^n}{n!} + \frac{\varepsilon}{2},$$

for positive x's. Now the product of e^{-x} and a polynomial of the mth degree tends to 0 when $x \to +\infty$; we can therefore choose ξ so large that this product is $< \frac{1}{2}\varepsilon$ for every $x > \xi$. For these x's the whole expression is then $< \varepsilon$ in absolute value, and our statement is established.

8. The B_r-process. The range of the process just described is, in a certain sense, extended by substituting other series for $\Sigma \frac{x^n}{n!}$, in the first instance $\Sigma \frac{x^{rn}}{(r\,n)!}$, say, where r is some fixed integer > 1. *We accordingly say that a sequence* (s_n) *is* ***limitable B_r*** *with the value* s if the quotient of the two functions

$$\sum_{n=0}^{\infty} s_n \frac{x^{rn}}{(r\,n)!} \quad \text{and} \quad \sum_{n=0}^{\infty} \frac{x^{rn}}{(r\,n)!}, \quad \text{i. e. the product} \quad r\,e^{-x} \sum_{n=0}^{\infty} s_n \frac{x^{rn}}{(r\,n)!},$$

tends to the limit s when $x \to +\infty$. (We must, of course, assume again here that the first-named series is everywhere convergent.) Thus the B-process, for instance, is quite useless for the sequence $s_n = (-1)^n\, n!$, since here $\Sigma\, s_n \frac{x^n}{n!} = \Sigma\,(-1)^n\, x^n$ does not converge for every x; whereas the series $\Sigma\, s_n \frac{x^{rn}}{(r\,n)!}$ already converges everywhere[28] when we take $r = 2$.

9. *Le Roy*'s process. We have usually interpreted the limitation processes by saying that by means of them we carry out an "averaged" comparison between the given sequence (s_n) and the unit sequence 1, 1, 1, . . . We may look at the matter in a slightly different way. If the numbers s_n are the partial sums of the series $\Sigma\, a_n$, we have to examine, for instance in the C_1-process, the limit of

$$\frac{s_0 + s_1 + \ldots + s_n}{n+1}$$
$$= a_0 + \left(1 - \frac{1}{n+1}\right) a_1 + \left(1 - \frac{2}{n+1}\right) a_2 + \ldots + \left(1 - \frac{n}{n+1}\right) a^n.$$

Here the terms of the series appear multiplied by *variable factors* which reduce the given series to a finite sum, or at any rate to a series convergent in the old sense. By means of these factors, the influence of distant terms is destroyed or diminished; yet as n increases all the factors tend to 1 and thus ultimately involve all the terms to their full extent. The situation is similar in the case of *Abel*'s process, where we were concerned with the limit of $\Sigma\, a_n\, x^n$ for $x \to 1 - 0$; here the effect described above is

[28] This does not mean that the B_r-process ($r > 1$) is more favourable than the B-process for *every* sequence (s_n). On the contrary, there are sequences that are limitable B but not limitable B_2.

brought about by the factors x^n, which, however, increase to 1 as $x \to 1-0$. This principle appears most clearly as the basis of the following process [29]: The series

$$\sum_{n=0}^{\infty} \frac{\Gamma(n x+1)}{n!} a_n$$

is assumed convergent for $0 \leqq x < 1$. If the function which it defines in that interval tends to a limit s as $x \to 1-0$, the series Σa_n may be called summable R to the value s.

This method is not so easily dealt with analytically, and for this reason it is of smaller importance.

10. The most general form of the limitation processes. It will have been noticed that all the processes so far described belong essentially to two types:

1. In the case of the first type, from a sequence (s_n), with the help of a *matrix* (cf. *Toeplitz*' theorem **221**)

$$T = (a_{kn})$$

a *new sequence* of numbers

$$s_k' = a_{k0} s_0 + a_{k1} s_1 + \cdots + a_{kn} s_n + \cdots, \quad (k = 0, 1, 2, \ldots)$$

is formed by combination of the sequence $s_0, s_1, \ldots, s_n, \ldots$ with the successive rows $a_{k0}, a_{k1}, \ldots, a_{kn}, \ldots,$ — the assumption being, of course, that the series on the right hand side represents a definite value, i. e. is convergent (in the old sense) [30]. The sequence $s_0', s_1', \ldots, s_k', \ldots$ will be called for short the *T-transformation* [31] of the sequence (s_n) and its nth term, when there is no fear of ambiguity, will be denoted by $T(s_n)$. If the accented sequence (s_k') is convergent with the limit s, *the given sequence is said to be* ***limitable*** ***T*** *with the value* s. In symbols:

$$T\text{-}\lim s_n = s \qquad \text{or} \qquad T(s_n) \to s.$$

---

[29] *Le Roy:* Sur les séries divergentes, Annales de la Fac. des sciences de Toulouse (2), Vol. 2, p. 317. 1900.

[30] If each row of the matrix T contains only a finite number of terms, this condition is automatically fulfilled. This is the case with the processes 1, 2, 3 and 5.

[31] The series $\Sigma a_k'$, of which the s_k''s are the partial sums, may similarly be called the T-transformation of the series Σa_n with the s_n's as its partial sums. Thus e. g. the series

$$a_0 + \frac{a_1}{1 \cdot 2} + \cdots + \frac{a_1 + 2 a_2 + \cdots + n a_n}{n(n+1)} + \cdots$$

is the C_1-transformation of the series Σa_n. In this sense, all T-processes give more or less remarkable transformations of series, which may very often be of use in *numerical calculations*. (This is particularly the case with the E-process). The transformation of the series may equally, of course, be regarded as the *primary* process and the transformation of the sequence of partial sums may be deduced from it. Indeed it was in this way that we were led to the E-process.

It is at once clear that the processes 1, 2, 3, 5, and the first one described in 6 belong to this type. They differ only in the choice of the matrix T. Theorem **221**, 2 also immediately tells us with what matrices we are certain to obtain limitation processes satisfying the permanence condition [32].

2. In the case of the second type, we deduce from a sequence (s_n), by combining it with a *sequence of functions*

$$(\varphi_n) \equiv \varphi_0(x), \quad \varphi_1(x), \quad \ldots, \quad \varphi_n(x), \ldots,$$

the *function*

$$F(x) = \varphi_0(x)\, s_0 + \varphi_1(x)\, s_1 + \ldots + \varphi_n(x)\, s_n + \ldots,$$

where we assume, say, that each of the functions $\varphi_n(x)$ is defined for every $x > x_0$ and that the series $\Sigma \varphi_n(x)\, s_n$ converges for each of these values of x. In that case $F(x)$ is also defined for every $x > x_0$, and we may investigate the existence of the limit $\lim\limits_{x\to+\infty} F(x)$. If the limit exists and $= s$, *the sequence* (s_n) *will be called* [33] ***limitable*** $\boldsymbol{\varphi}$ *with the value* s.

By analogy with **221**, 2, we shall at once be able to assign conditions under which a process of this type will satisfy the permanence condition. This will certainly be the case if a) for every fixed n,

$$\lim_{x\to+\infty} \varphi_n(x) = 0,$$

if b) a constant K exists such that

$$|\varphi_0(x)| + |\varphi_1(x)| + \ldots + |\varphi_n(x)| < K$$

for every $x > x_0$ and all n's, and if c) for $x \to +\infty$

$$\lim \{\Sigma \varphi_n(x)\} = 1.$$

It will be noticed that these conditions correspond exactly to the assumptions [34] a), b) and c) of theorem **221**, 2. The proof, which is quite analogous to that of this theorem, may therefore be left to the reader.

Borel's process evidently belongs to this type, with $\varphi_n(x) = e^{-x} \cdot \dfrac{x^n}{n!}$. The same may be said of *Abel*'s process, if the interval $0 \ldots +\infty$

---

[32] The importance of theorem **221**, 2 lies chiefly in the fact that the conditions a), b) and c) of the theorem are not merely *sufficient*, but actually *necessary* for its general validity. We cannot enter into the question (v. p. 74, footnote 19), but we may observe that in consequence of this fact, the T-processes whose matrix satisfies the conditions mentioned are the *only* ones which fulfil the permanence condition.

[33] In all essentials this is the scheme by means of which *O. Perron* (Beiträge zur Theorie der divergenten Reihen, Math. Zschr. Vol. 6, pp. 286—310. 1920) classifies all the summation processes.

[34] Like these they are not only *sufficient*, but also *necessary* for the general validity of the theorem. Further details in *H. Raff*, Lineare Transformationen beschränkter integrierbarer Funktionen. Math. Zeitschr. Vol. 41, pp. 605—629. 1936.

is projected into the interval $0 \dots 1$ which is used in the latter, that is, if the series $(1-x)\sum s_n x^n$ is replaced by the series

$$F(x) = \frac{1}{1+x}\sum_{n=0}^{\infty} s_n \left(\frac{x}{1+x}\right)^n$$

and the latter is examined for $x \to +\infty$. — In an equally simple manner, it may be seen that *Le Roy*'s process belongs to this type.

The second type of limitation process contains the first as a particular case, obtained when x assumes integral values $\geqq 0$ only $(\varphi_n(k) = a_{kn})$. We merely use a continuous parameter in the one case, and a discontinuous one in the other. Conversely, in view of § 19, def. 4a, the continuous passage to the limit may be replaced by a discontinuous one, and hence the φ-processes may be exhibited as a sub-class of the T-processes. These remarks, however, are of little use: in further methods of investigation the two types of process nevertheless remain essentially different.

It is not our intention to investigate all the processes which come under these two headings from the general points of view indicated above. Let us make only the following remarks: We have already pointed out what conditions the matrix T or sequence of functions (φ_n) must fulfil, in order that the limitation process based on it may satisfy the permanence condition **263**, I. Whether the conditions **263**, II and III are also fulfilled, will depend on further hypotheses regarding the matrix T or sequence (φ_n); this question is accordingly best left to a separate investigation in each case. The question as to the extent to which the conditions F (**264**) are fulfilled, cannot be attacked in a general way either, but must be specially examined for each process. One important property alone is common to all the T- and φ-processes, namely their *linear character*: If two sequences (s_n) and (t_n) are limitable in accordance with one and the same process, the first with the value s, and the second with the value t, then the sequence $(a\,s_n + b\,t_n)$, whatever the constants a and b may be, is also limitable by the same process, with the value $a\,s + b\,t$. The proof follows immediately from the way in which the process is constructed. Owing to this theorem, all the simplest rules of the algebra of convergent sequences (term-by-term addition of a constant, term-by-term multiplication by a constant, term-by term addition or subtraction of two sequences) remain formally unaltered. On the other hand, we must expressly emphasize the fact that the theorem on the influence of a finite number of alterations (**42**, 7) does *not* necessarily remain valid[35].

---

[35] For this, the following simple example relating to the B-process was first given by *G. H. Hardy*: Let s_n be defined by the expansion

$$\sin(e^x) = \sum_{n=0}^{\infty} s_n \frac{x^n}{n!}.$$

Since $e^{-x} \cdot \sin(e^x) \to 0$ as $x \to +\infty$, the sequences $s_0, s_1, s_2, \dots$ is limitable B

If we wished to give a general and fairly complete survey of the present state of the theory of divergent series, we should now be obliged to enter into a more detailed investigation of the processes which we have described. To begin with, we should have to deal with the questions whether, and to what extent, the individual processes do actually satisfy the stipulations **263**, II, III and **264**; we should have to obtain necessary and sufficient conditions for a series to be summable by a particular process; we should have to find the relations between the ways in which the various processes act, and go further into the questions indicated in No. 10, etc. Owing to lack of space it is of course out of the question to investigate all this in detail. We must be content with examining a few of the processes more particulary; — we choose the H-, C-, A-, and E-processes. At the same time we will so arrange the choice of subjects that as far as possible all questions and all methods of proof which play a part in the complete theory may at least be indicated.

For the rest we must refer to the original papers, of which we may mention the following, in addition to those mentioned in the footnotes of this section and of the following sections: **266.**

1. The following give a general survey of the group of problems:

Borel, E.: Leçons sur les séries divergentes, 2nd ed., Paris 1928.

Bromwich, T. J. I'A.: An introduction to the theory of infinite series. London 1908: 2nd ed. 1926.

Hardy, G. H., and *S. Chapman*: A general view of the theory of summable series. Quarterly Journal Vol. 42, p. 181. 1911.

Chapman, S.: On the general theory of summability, with applications to Fourier's and other series. Ibid., Vol. 43, p. 1. 1911.

Carmichael, R. D.: General aspects of the theory of summable series. Bull. of the American Math. Soc. Vol. 25, pp. 97—131. 1919.

Knopp, K.: Neuere Untersuchungen in der Theorie der divergenten Reihen. Jahresber. d. Deutschen Math.-Ver. Vol. 32, pp. 43—67. 1923.

2. A more detailed account of the $R_{\lambda, k}$-process, which is *not* specially considered in the following sections, is given by

Hardy, G. H., and *M. Riesz*: The general theory of Dirichlet's series. Cambridge 1915.

The B-process is dealt with in the books by *Borel* and *Bromwich* mentioned under 1., and also in more detail by

Hardy, G. H.: The application to Dirichlet's series of Borel's exponential method of summation. Proceedings of the Lond. Math. Soc. (2) Vol. 8, pp. 301 to 320. 1909.

---

with the value 0. By differentiation of the relation above, we obtain

$$\cos(e^x) = e^{-x} \cdot \sum_{n=0}^{\infty} s_{n+1} \frac{x^n}{n!};$$

this shows, since $\cos(e^x)$ tends to no limit when $x \to +\infty$, that the sequence $s_1, s_2, s_3, \ldots$ is *not* limitable B at all!

Hardy, G. H., and *J. E. Littlewood*: The relations between Borel's and Cesàro's methods of summation. Ibid., (2) Vol. **11**, pp. 1—16. 1913.

Hardy, G. H., and *J. E. Littlewood*: Contributions to the arithmetic theory of series. Ibid., (2) Vol. 11, pp. 411—478. 1913.

Hardy, G. H., and *J. E. Littlewood*: Theorems concerning the summability of series by Borel's exponential method. Rend. del Circolo Mat. di Palermo Vol. 41, pp. 36—53. 1916.

Doetsch, G.: Eine neue Verallgemeinerung der Borelschen Summabilitätstheorie. Inaug.-Diss., Göttingen 1920.

3. Apart from the books mentioned under 1., a full account of the theory of divergent series is to be found in

Bieberbach, L.: Neuere Untersuchungen über Funktionen von komplexen Variablen. Enzyklop. d. math. Wissensch. Vol. II, Part C, No. 4. 1921.

4. Finally, the general question of the classification of limitation processes is dealt with in the following papers:

Perron, O.: Beitrag zur Theorie der divergenten Reihen. Math. Zeitschr. Vol. 6, pp. 286—310. 1920.

Hausdorff, F.: Summationsmethoden und Momentenfolgen I und II. Math. Zeitschr. Vol. 9, p. 74 seqq. and p. 280 seqq. 1920.

Knopp, K.: Zur Theorie der Limitierungsverfahren. Math. Zeitschr. Vol. 31; 1st communication pp. 97—127, 2nd communication pp. 276—305. 1929.

§ 60 The *C*- and *H*-processes.

Of all the summation processes briefly sketched in the preceding section, the *C*- and *H*-processes — and especially the process of limitation by arithmetic means of the first order, which is the same in both — are distinguished by their great simplicity; they have, moreover, proved of great importance in the most diverse applications. We shall accordingly first examine these processes in somewhat greater detail.

267. In the case of the *H*-process, *Cauchy*'s theorem **43**, 2 shows that, for $p \geqq 1$, $h_n^{(p-1)} \to s$ implies[36] $h_n^{(p)} \to s$, so that the range of the H_p-process contains that of the H_{p-1}-process. The corresponding fact holds in the case of the C-process:

Theorem 1. *If a sequence is limitable C_{k-1} with the value s, $(k \geqq 1)$, it is also limitable C_k with the same value.* In symbols: *From $c_n^{k-1} \to s$, it follows that $c_n^{(k)} \to s$.* (*Permanence theorem for the C-process.*)

---

[36] Cf. p. 466, footnote 15. By the 0th degree of a transformation, higher degrees of which are introduced, we mean the original sequence.

Proof. By definition (v. **265**, 3)

$$c_n^{(k)} \equiv \frac{S_n^{(k)}}{\binom{n+k}{k}} \equiv \frac{S_0^{(k-1)}+\cdots+S_n^{(k-1)}}{\binom{k-1}{k-1}+\cdots+\binom{n+k-1}{k-1}}$$
$$= \frac{\binom{k-1}{k-1} c_0^{(k-1)}+\cdots+\binom{n+k-1}{k-1} c_n^{(k-1)}}{\binom{k-1}{k-1}+\cdots+\binom{n+k-1}{k-1}},$$

whence by **44**, 2 the statement immediately follows.

Accordingly, to every sequence which is limitable C_p, for some suitable suffix p, there corresponds a *definite* integer k such that the sequence is limitable C_k but is not limitable C_{k-1}. (If the sequence is convergent from the first, we of course take $k=0$.) We then say that the sequence is *exactly* limitable C_k.

Examples of the C_k-limitation Process[37]. **268.**

1. $\sum_{n=0}^{\infty}(-1)^n$ is summable C_1 with the value $\frac{1}{2}$. Proof above, **262**.

2. $\sum_{n=0}^{\infty}(-1)^n\binom{n+k}{k}$ is exactly summable C_{k+1}, to the value $s=\frac{1}{2^{k+1}}$.

In fact, for $a_n \equiv (-1)^n\binom{n+k}{k}$, we have by **265**, 3

$$\sum S_n^{(k)} x^n = \left(\frac{1}{1-x}\right)^{k+1} \cdot \sum (-1)^n \binom{n+k}{k} x^n = \left(\frac{1}{1-x}\right)^{k+1}\left(\frac{1}{1+x}\right)^{k+1}$$
$$= \left(\frac{1}{1-x^2}\right)^{k+1} = \sum \binom{\nu+k}{k} x^{2\nu}.$$

Accordingly

$$S_n^{(k)} = \binom{\nu+k}{k} \text{ or } =0, \text{ according as } n=2\nu \text{ or } =2\nu+1.$$

Hence both for $n=2\nu$ *and* for $n=2\nu+1$,

$$S_n^{(k+1)} = \binom{k}{k}+\binom{1+k}{k}+\cdots+\binom{\nu+k}{k} = \binom{\nu+k+1}{k+1},$$

whence the statement follows immediately.

3. The series $\Sigma(-1)^n(n+1)^k \equiv 1-2^k+3^k-4^k+-\cdots$, summable C_1 to the value $\frac{1}{2}$ for $k=0$, by Example 1., is for each $k \geqq 1$ exactly summable C_{k+1} to the sum $s=\frac{2^{k+1}-1}{k+1}B_{k+1}$, if B_ν denotes the νth of *Bernoulli*'s numbers. The *fact* of the summability indeed follows directly from Example 2. For the moment denoting the series there summed by Σ_k, we at once see, from the linear character of our process (v. p. 476), that the series, obtained from Σ_k

[37] As a result of the equivalence theorem established immediately below these examples hold unaltered for the H_k-limitation processes. On account of the explicit formulae for $S_n^{(k)}$ and $c_n^{(k)}$, given in **265**, 3, to which there is no analogue in the H-process, the C-process is usually preferred.

by term-by-term addition, of the form

$$c_0 \Sigma_0 + c_1 \Sigma_1 + \ldots + c_k \Sigma_k$$

is exactly summable C_{k+1} if $c_0, c_1, \ldots, c_k$ denote any constants, with $c_k \neq 0$. Now the c_ν may obviously be chosen so that we obtain precisely the series $\Sigma (-1)^n (n+1)^k$. The *value* s is most easily obtained by A-summation; see **288**, 1.

4. The series $\frac{1}{2} + \cos x + \cos 2x + \ldots + \cos nx + \ldots$ is summable C_1 to the sum 0, provided $x \neq 2k\pi$.

Proof. By **201**,

$$s_n = \frac{1}{2} + \cos x + \cos 2x + \ldots + \cos nx = \frac{\sin\left(n+\frac{1}{2}\right)x}{2 \sin \frac{x}{2}},$$

for each $n = 0, 1, 2, \ldots$; hence

$$s_0 + s_1 + \ldots + s_n = \frac{1}{2\sin\frac{x}{2}}\left(\sin\frac{x}{2} + \sin 3\frac{x}{2} + \ldots + \sin(2n+1)\frac{x}{2}\right) = \frac{\sin^2 (n+1)\frac{x}{2}}{2\sin^2\frac{x}{2}}$$

and consequently

$$\left| \frac{s_0 + s_1 + \ldots + s_n}{n+1} \right| \leqq \frac{1}{n+1} \cdot \frac{1}{2\sin^2\frac{x}{2}}.$$

For a fixed $x \neq 2k\pi$, the expression on the right tends to 0 as n increases, which proves what was stated. — This is our first example of a summable series with variable terms. The function represented by its "sum" $\equiv 0$ in every interval not containing any of the points $2k\pi$. At the excluded points, the series is *definitely divergent* to $+\infty$!

5. The series $\sin x + \sin 2x + \sin 3x + \ldots$ is obviously convergent with the sum 0, for $x = k\pi$. For $x \neq k\pi$ it is no longer convergent, but it is summable C_1, and it then[38] has the "sum" $\frac{1}{2}\cot\frac{x}{2}$.

Proof. From the relation

$$s_n = \sin x + \ldots + \sin nx = \frac{1}{2}\cot\frac{x}{2} - \frac{\cos(2n+1)\frac{x}{2}}{2\sin\frac{x}{2}},$$

the statement follows as in 4.

6. $\cos x + \cos 3x + \cos 5x + \ldots$ is summable C_1 to the sum 0, for $x \neq k\pi$.

7. $\sin x + \sin 3x + \sin 5x + \ldots$ is also summable C_1 to the sum $\frac{1}{2\sin x}$, for $x \neq k\pi$.

8. $1 + z + z^2 + \ldots$ is summable C_1 on the circumference $|z| = 1$, excepting only for $z = +1$, and the sum is $\frac{1}{1-z}$. (Examples 4 and 5 result from this by separating real and imaginary parts.) Here, in fact,

$$s_n = \frac{1}{1-z} - \frac{z^{n+1}}{1-z}, \text{ so that } \frac{s_0 + s_1 + \ldots + s_n}{n+1} = \frac{1}{1-z} - \frac{1}{n+1}\,\frac{z(1-z^{n+1})}{(1-z)^2},$$

whence the statement can be inferred at a glance.

---

[38] The graph of this function thus exhibits "infinitely great jumps" at the points $2k\pi$.

9. The series $\frac{1}{(1-z)^k} = \sum_{n=0}^{\infty} \binom{n+k-1}{k-1} z^n$ remains summable C_k to the sum $\frac{1}{(1-z)^k}$ on the circumference $|z| = 1$, provided only $z \neq +1$. For the corresponding quantities $S_n^{(k)}$ are, by **265**, 3, the coefficients of x^n in the expansion of

$$\frac{1}{(1-x)^{k+1}} \cdot \frac{1}{(1-x\,z)^k} = \frac{a}{(1-x)^{k+1}} + \ldots,$$

(the right hand side being the expansion in partial fractions of the left hand side). All the partial fractions after the one written down contain in the denominator the kth power of $(1-x)$ or $(1-x\,z)$ at most. Hence, multiplying by $(1-x)^{k+1}$ and letting $x \to 1$, we at once obtain $a = \frac{1}{(1-z)^k}$. Accordingly

$$\sum_{n=0}^{\infty} S_n^{(k)} x^n = \sum_{n=0}^{\infty} \left[\frac{1}{(1-z)^k} \binom{n+k}{k} + \ldots\right] x^n,$$

where it is sufficient to know that the supplementary terms within the square bracket involve binomial coefficients of the order n^{k-1} with respect to n at most. Therefore, as $n \to +\infty$,

$$\frac{S_n^{(k)}}{\binom{n+k}{k}} \to \frac{1}{(1-z)^k}, \qquad \text{q. e. d.}$$

Since the H-process outwardly seems to bear a certain relationship to the C-process, it is natural to ask whether their effects are distinguishable or not. We shall see that the two ranges of action coincide completely. Indeed we have the following theorem, due to the author [39] and to *W. Schnee* [40]:

Theorem 2. *If a sequence* (s_n) *is, for some particular k, limitable* [41] H_k *to the value s, it is also summable* C_k *to the same value s and conversely.* In **269.**
symbols:

$$h_n^{(k)} \to s \quad \textit{always involves} \quad c_n^{(k)} \to s,$$

and conversely. (*Equivalence theorem for the C- and H-processes.*)

Many proofs have been given for this theorem [42], among which that of *Schur* [43] is probably the clearest and best adapted to the nature of the

---

[39] Cf. the paper cited on p. 467, footnote 19.

[40] *Schnee, W.*: Die Identität des *Cesàro*schen und *Hölder*schen Grenzwertes. Math. Ann. Vol. 67, pp. 110—125. 1909.

[41] Since for $k = 1$ the theorem is trivial, we may assume $k \geqq 2$ in the sequel.

[42] A detailed bibliography, for this theorem and its numerous proofs, may be found in the author's papers: I. Zur Theorie der C- und H-Summierbarkeit. Math. Zeitschr. Vol. 19, pp. 97—113. 1923; II. Über eine klasse konvergenzerhaltender Integraltransformationen und den Äquivalenzsatz der C- und H-Verfahren, ibid. Vol. 47, pp. 229—264. 1941; III. Über eine Erweiterung des Äquivalenzsatzes der C- und H-Verfahren und eine Klasse regulär wachsender Funktionen, ibid. Vol. 49, pp. 219—255. 1943.

[43] *Schur, I.*: Über die Äquivalenz der *Cesàro*schen und *Hölder*schen Mittelwerte. Math. Ann. Vol. 74, pp. 447—458. 1913. Also: Einige Bemerkungen zur Theorie der unendlichen Reihen, Sitzber. d. Berl. Math. Ges., Vol. 29, pp. 3—13. 1929.

problem. Combined with a skilful artifice of *A. F. Andersen* [44], the proof becomes particularly simple.

We next show that the equivalence theorem is contained in the following theorem, simpler in appearance:

270. Theorem 2a. *If (z_n), for $k \geq 1$, is limitable C_k with the value ζ, the sequence* **270** *of the arithmetic means* $z_n' = \dfrac{z_0 + z_1 + \ldots + z_n}{n+1}$ *is limitable C_{k-1} with the value ζ, and conversely.*

By this theorem, each of the k relations

$$\begin{aligned} c_n^{(k)} \equiv C_k(s_n) \quad &\to s \\ C_{k-1}(h_n') \quad &\to s \\ \cdots\cdots\cdots \\ C_2(h_n^{(k-2)}) &\to s \\ C_1(h_n^{(k-1)}) \equiv h_n^{(k)} &\to s \end{aligned}$$

is in fact a consequence of any of the others; in particular, the first is a consequence of the last. But that is what the equivalence theorem states.

It suffices, therefore, to prove Theorem 2 a. But this follows immediately from the two relations connecting the C_k- and C_{k-1}-transformations of the sequence (z_n) with those of the sequence (z_n'), viz.

$$\text{(I)}\quad C_k(z_n) = k\,C_{k-1}(z_n') - (k-1)\,C_k(z_n'),$$

$$\text{(II)}\quad C_{k-1}(z_n') = \frac{1}{k}\,C_k(z_n) + \left(1 - \frac{1}{k}\right)\frac{C_k(z_0) + C_k(z_1) + \ldots + C_k(z_n)}{n+1}.$$

For if, in the *first* place, we have $C_{k-1}(z_n') \to \zeta$, then, by Theorem 1, we have also $C_k(z_n') \to \zeta$. Hence by (I),

$$C_k(z_n) \to k\,\zeta - (k-1)\,\zeta = \zeta.$$

If, in the *second* place, $C_k(z_n) \to \zeta$, then, by **43**, 2, so do the arithmetic means

$$\frac{C_k(z_0) + C_k(z_1) + \ldots + C_k(z_n)}{n+1} \to \zeta,$$

and, with equal ease, (II) provides that [45]

$$C_{k-1}(z_n') \to \frac{1}{k}\zeta + \left(1 - \frac{1}{k}\right)\zeta = \zeta.$$

Accordingly all reduces to verifying the two relations (I) and (II), and this may be done for instance as follows:

---

[44] *Andersen, A. F.*: Bemerkung zum Beweis des Herrn *Knopp* für die Äquivalenz der *Cesàro*- und *Hölder*-summabilität. Math. Zeitschr. Vol. 28, pp. 356—359. 1928.

[45] If M denotes the operation of taking the arithmetic mean of a sequence, the above relations (I) and (II) may be written in the short and comprehensive form

$$\text{(I)}\qquad C_k = k\,C_{k-1}\,M - (k-1)\,C_k\,M,$$
$$\text{(II)}\qquad C_k = k\,C_{k-1}\,M - (k-1)\,M\,C_k.$$

Each of these follows from the other if it is known that the C_k-transformation and the process of taking the arithmetic mean are two *commutable* operations.

In **265**, 3, the iterated sums $S_n^{(k)}$ were formed, to define the C_k-transformation of a sequence (s_n). Let us denote these sums more precisely by $S_n^{(k)}(s)$, and use the corresponding symbols when starting with other sequences. The identity

$$\frac{1}{(1-y)^k}\sum_{n=0}^{\infty} z_n y^n = \frac{1}{(1-y)^{k-1}}\sum_{n=0}^{\infty}(z_0+z_1+\dots+z_n)\,y^n$$
$$= \frac{1}{(1-y)^{k-1}}\sum_{n=0}^{\infty}(n+1)\,z_n'\,y^n$$

then implies

$$S_n^{(k)}(z) = \binom{n+k-2}{k-2}1\,.\,z_0' + \dots + \binom{n+k-2-\nu}{k-2}(\nu+1)\,z_\nu' + \dots$$
$$+\binom{k-2}{k-2}(n+1)\,z_n'.$$

Here write $\qquad \nu+1 = (n+k)-(n+k-1-\nu),$

and observe that

$$\binom{n+k-2-\nu}{k-2}(n+k-1-\nu) = (k-1)\binom{n+k-1-\nu}{k-1}.$$

It then follows further that

$$(*)\qquad S_n^{(k)}(z) = (n+k)\,S_n^{(k-1)}(z') - (k-1)\,S_n^{(k)}(z').$$

Dividing by $\binom{n+k}{k}$, we deduce at once the relation (I).

On the other hand, by the definition of the quantities $S_n^{(k)}$, we have

$$S_n^{(k-1)} = S_n^{(k)} - S_{n-1}^{(k)} \qquad (n=0,1,\dots;\ S_{-1}^{(k)}=0).$$

Substituting in (*), we get

$$(**)\qquad S_n^{(k)}(z) = (n+1)\,S_n^{(k)}(z') - (n+k)\,S_{n-1}^{(k)}(z'),$$

and hence, dividing by $\binom{n+k}{k}$,

$$C_k(z_n) = (n+1)\,C_k(z_n') - n\,C_k(z'_{n-1}).$$

Substituting in turn $0, 1, \dots, n$ for n in this relation, and adding, we obtain finally

$$\frac{C_k(z_0)+C_k(z_1)+\dots+C_k(z_n)}{n+1} = C_k(z_n').$$

Put into words, this relation signifies that the arithmetic mean of the C_k-transformations of a sequence is equal to the C_k-transformation of its arithmetic means, or, as we say for short, the C_k-transformation and the process of forming the arithmetic mean are two *commutable* operations [46].

[46] Cf. preceding footnote 45.

Now if we substitute for $C_k(z_n')$ in (I) the expression just found, we obtain (II) at once. This completes the proof of the Equivalence Theorem.

After thus establishing the equivalence of the C-process and the H-process, we need only consider one of them. As the C-process is easier to work with analytically, on account of the explicit formulae **265**, 3 for the $S_n^{(k)}$'s, it is usual to give the preference to it.

We next inquire how far its *range of action* extends, i. e. what are the necessary conditions to be satisfied by a sequence in order that it may be limitable C_k. Using the notation, which was introduced by *Landau* and has been generally adopted, $x_n = O(n^\alpha)$, α real, to indicate that the sequence $\left(\frac{x_n}{n^\alpha}\right)$ is *bounded*, and $x_n = o(n^\alpha)$ to indicate that $\left(\frac{x_n}{n^\alpha}\right)$ is a *null sequence*[47], we have the following theorem, which may be interpreted by saying that sequences whose terms increase too rapidly are excluded from C_k-limitation altogether:

271. Theorem 3. *If Σa_n, with partial sums s_n, is summable C_k, then*

$$a_n = o(n^k) \quad \textit{and} \quad s_n = o(n^k).$$

Proof. For $k = 0$, the statement is a consequence of Theorem **82**, 1, which we are generalizing. For $k \geqq 1$, with the notation of **265**, 3, the sequence of numbers

$$\frac{S_n^{(k)}}{\binom{n+k}{k}} = \frac{S_0^{(k-1)} + \ldots + S_n^{(k-1)}}{\binom{n+k}{k}}$$

is convergent. Since $\binom{n+k-1}{k} \cong \binom{n+k}{k}$, the sequence

$$\frac{S_0^{(k-1)} + \ldots + S_{n-1}^{(k-1)}}{\binom{n+k}{k}}$$

is convergent, with the same limit. The difference of the two quotients, viz. $S_n^{k-1} \Big/ \binom{n+k}{k}$, therefore forms a null sequence. As $\binom{n+k}{k} \sim n^k$ this implies that $S_n^{(k-1)} = o(n^k)$. It follows that

$$S_n^{(k-2)} = S_n^{(k-1)} - S_{n-1}^{(k-1)} = o(n^k) + o(n^k) = o(n^k),$$

and similarly[48]

$$S_n^{(k-3)} = o(n^k), \quad \ldots, \quad S_n' = o(n^k), \quad s_n = o(n^k), \quad a_n = o(n^k).$$

---

[47] The first statement thus implies that the quantities $|x_n|$ are of *at most* the same order as const. $\cdot\, n^\alpha$, the second that they are of *smaller* order than n^α, in the way in which they increase to $+\infty$.

[48] The reader will be able to work out quite easily for himself the very simple rules for calculations with the order symbols O and o which are used here and in the sequel.

The intermediary result $S_n^{(k-1)} = o(n^k)$ just obtained in the proof may be interpreted as an even more significant generalization of the theorem in question. In fact, it means that

$$\frac{\binom{n+k-1}{k-1} a_0 + \binom{n+k-2}{k-1} a_1 + \ldots + \binom{k-1}{k-1} a_n}{\binom{n+k}{k}} \to 0.$$

We accordingly have the following elegant analogue of **82**, 1:

Theorem 4. *In a series Σa_n, summable C_k, we necessarily have* **272.**

$$C_k\text{-}\lim a_n = 0.$$

Moreover, even *Kronecker*'s theorem **82**, 3 has its exact analogue, though we shall confine ourselves to the case $p_n = n$:

Theorem 5. *In a series Σa_n, summable C_k, we necessarily have* **273.**

$$C_k\text{-}\lim \left(\frac{a_1 + 2a_2 + \ldots + n a_n}{n+1}\right) = 0.$$

In fact, it follows from the corollary to **270** that $C_k(s_n) \to s$ involves $C_{k-1}\left(\frac{s_0 + s_1 + \ldots + s_n}{n+1}\right) \to s$, and therefore by the permanence theorem $C_k\left(\frac{s_0 + s_1 + \ldots + s_n}{n+1}\right) \to s$. Subtracting this from $C_k(s_n) \to s$, we at once obtain the statement

$$C_k\text{-}\lim \left(s_n - \frac{s_0 + s_1 + \ldots + s_n}{n+1}\right) = C_k\text{-}\lim \left(\frac{a_1 + 2a_2 + \ldots + n a_n}{n+1}\right) = 0.$$

By means of these simple theorems, the range of action of the C_k-process is staked off *on the outside*, as we might say, for the theorems inform us how far *at most* the range may extend into the domain of divergent series. Where this range properly *begins* is a much more delicate question. By this we mean the following: Every series convergent in the usual sense to the value s is also summable C_k (for every $k \geqq 0$) to the same value s. Where is the boundary line, in the aggregate of all series which are summable C_k, between convergent and divergent series? On this point we have the following simple theorem, relating solely to the C_1-process:

Theorem 6. *If the series Σa_n is C_1-summable to sum s, and if* **274.** $\delta_n = \frac{a_1 + 2a_2 + \ldots + n a_n}{n+1} \to 0$, *then Σa_n is in fact convergent with sum s.* For (v. supra)

$$s_n - \frac{s_0 + s_1 + \ldots + s_n}{n+1} = \frac{a_1 + 2a_2 + \ldots + n a_n}{n+1} = \delta_n,$$

whence the proof of the statement is immediate [49]. The last expression

[49] With reference to **262**, 1 (or **43**, Theorem 2), and to **82**, Theorem 3, we may express the theorem as follows: A series Σa_n converges if, and only if, it is C_1-summable with $\delta_n \to 0$.

tends, in particular, to 0 if $a_n = o\left(\frac{1}{n}\right)$. A much deeper result is the fact that $a_n = O\left(\frac{1}{n}\right)$ suffices, i. e.

Theorem 6a. *If a series $\Sigma\, a_n$ is summable C_k and if its terms a_n satisfy the condition*

$$a_n = O\left(\frac{1}{n}\right),$$

then $\Sigma\, a_n$ is convergent. (O-$C_k \to$ K-theorem) [50].

A proof of this theorem may be dispensed with here, since it will follow as a simple corollary of *Littlewood*'s theorem **287**. The direct proof would not be essentially easier than the proof of that theorem.

Application. The series $\sum_{n=1}^{\infty} a_n \equiv \sum_{n=1}^{\infty} \frac{1}{n^{1+\alpha i}}$, $\alpha \gtrless 0$, is *not* convergent, as it is easy to verify, by an argument modelled on the proof on p. 442, footnote 54, that for $n = 1, 2, \ldots$,

$$\frac{1}{n^{1+\alpha i}} = \frac{1}{i\,\alpha}\left[\frac{1}{n^{\alpha i}} - \frac{1}{(n+1)^{\alpha i}}\right] - \frac{\vartheta_n}{n^2}$$

with (ϑ_n) *bounded*. Further, for this series $(n\, a_n)$ is bounded, hence the series cannot be summable C_k to any order.

Closely connected with the preceding, we have the following theorem, where for simplicity we shall confine ourselves to summation of the *first* order.

275. **Theorem 7.** *A necessary and sufficient condition for a series $\Sigma\, a_n$, with partial sums s_n, to be summable C_1 to the sum s, is that the series*

$$\text{(A)} \qquad \sum_{\nu=0}^{\infty} \frac{a_\nu}{\nu+1}$$

---

[50] *Hardy, G. H.*: Theorems relating to the convergence and summability of slowly oscillating series. Proc. Lond. Math. Soc. (2) Vol. 8, pp. 301—320. 1909. Cf. also the author's work I. quoted on p. 481, footnote 42. — The theorem deduces convergence (K) from C-summability. We accordingly call it a $C \to K$ theorem for short, and more precisely an O-$C \to K$ theorem, since an O (that is, the boundedness of a certain sequence) is employed in the determining hypothesis. A theorem of this kind was first proved by *A. Tauber*, — in his case, for the A-process (v. **286**); for this reason, *Hardy* gives the name of "Tauberian theorems" to all theorems in which ordinary convergence is deduced from some type of summability. We shall call them *converse theorems* or, more precisely, limitizing converse or averaging converse theorems.

should be convergent and that for its remainder

$$\varrho_n = \frac{a_{n+1}}{n+2} + \frac{a_{n+2}}{n+3} + \cdots \qquad (n = 0, 1, 2, \ldots)$$

the relation

(B) $$s_n + (n+1)\,\varrho_n \to s$$

holds [51].

If σ_n denotes the partial sums of the series (A), and σ its sum, then (B) asserts that

(B′) $$s - s_n - (n+1)(\sigma - \sigma_n) \to 0,$$

i. e. that the error $(s - s_n)$ is n times as large as the error $(\sigma - \sigma_n)$, except for a difference that decreases to 0 with n.

Proof. I. *If Σa_n is summable C_1*, we have by **183**, since $a_\nu = s_\nu - s_{\nu-1}$,

$$\sum_{\nu=n+1}^{n+p} \frac{a_\nu}{\nu+1} = -\frac{s_n}{n+2} + \sum_{\nu=n+1}^{n+p} \frac{s_\nu}{(\nu+1)(\nu+2)} + \frac{s_{n+p}}{n+p+2},$$

and, since $s_\nu = S_\nu' - S_{\nu-1}'$, on again applying *Abel*'s partial summation this becomes

$$-\frac{s_n}{n+2} - \frac{S_n'}{(n+2)(n+3)}$$

$$+ 2\sum_{\nu=n+1}^{n+p} \frac{S_\nu'}{(\nu+1)(\nu+2)(\nu+3)} + \frac{s_{n+p}}{n+p+2} + \frac{S_{n+p}'}{(n+p+2)(n+p+3)}.$$

As $n \to +\infty$, all five terms of the right hand side tend to 0, whatever the value of p, for by the assumed C_1-summability and theorem 3, $s_n = o(n)$ and $S_n' = O(n)$. Hence (A) holds. At the same time, keeping n fixed and letting $p \to +\infty$, we obtain

$$s_n + (n+2)\,\varrho_n = -\frac{S_n'}{n+3} + 2(n+2)\sum_{\nu=n+1}^{\infty} \frac{S_\nu'}{(\nu+1)(\nu+2)(\nu+3)}.$$

This tends to s, by **221**, because $\dfrac{S_n'}{n+1} \to s$. Hence (B) also holds, since $\varrho_n \to 0$. Thus (A) and (B) are necessary.

II. *Suppose conversely the conditions* (A) *and* (B) *hold good.* Then, if we write τ_n for the expressions on the left in (B), we have

$$\begin{aligned} \tau_{n+1} - \tau_n &= a_{n+1} + (n+2)\,\varrho_{n+1} - (n+1)\,\varrho_n \\ &= \varrho_n + a_{n+1} + (n+2)(\varrho_{n+1} - \varrho_n) \\ &= \varrho_n, \end{aligned}$$

and hence

$$\tau_n = s_n + (n+1)(\tau_{n+1} - \tau_n).$$

---

[51] *Knopp, K.*: Über die Oszillationen einfach unbestimmter Reihen, Sitzungsber. Berl. Math. Ges., Vol. XVI, pp. 45—50. 1917.

Hardy, G. H.: A theorem concerning summable series. Proc. Cambridge Phil. Soc. Vol. 20, pp. 304—307. 1921.

Another proof is given in the author's work I. quoted in footnote 42, and another again in *G. Lyra.* Über einen Satz zur Theorie der C-summierbaren Reihen. Math. Zeitschr. Vol. 45, pp. 559—572. 1939. This latter work has furnished the above proof of the *sufficiency* of (A) and (B) for the C_1-summability of Σa_n.

Consequently

$$s_n = 2\,\tau_n - [(n+1)\,\tau_{n+1} - n\,\tau_n]$$

and therefore

$$\frac{s_0 + s_1 + \dots + s_n}{n+1} = 2\,\frac{\tau_0 + \tau_1 + \dots + \tau_n}{n+1} - \tau_{n+1}.$$

But owing to $\tau_n \to s$, it follows from this that the sequence (s_n) is limitable C_1 to the value s, as required [52].

We shall content ourselves with these general theorems on C-summability [53] and we shall now proceed to a few applications.

Among the introductory remarks (pp. 461—462), it was pointed out that the problem of multiplication of infinite series, which remained very difficult and obscure as long as the old concept of convergence was scrupulously adhered to, may be completely solved in an extremely simple manner when the concept of summability is admitted. For the second proof of *Abel*'s theorem (p. 322) provides the

276. Theorem 8. *Cauchy's product* $\Sigma\, c_n \equiv \Sigma\,(a_0\,b_n + a_1\,b_{n-1} + \dots + a_n\,b_0)$ *of two convergent series* $\Sigma\, a_n = A$ *and* $\Sigma\, b_n = B$ *is always summable* C_1 *to the value* $C = A \cdot B$.

Over and above this, we now have the following more general

277. Theorem 9. *If* $\Sigma\, a_n$ *is summable* C_α *to the value* A *and* $\Sigma\, b_n$ *is summable* C_β *to the value* B, *then their Cauchy product*

$$\Sigma\, c_n \equiv \Sigma\,(a_0\,b_n + a_1\,b_{n-1} + \dots + a_n\,b_0)$$

is certainly summable C_γ *to the value* $C = A \cdot B$, *where* $\gamma = \alpha + \beta + 1$.

Proof. Let us denote by $A_n^{(\alpha)}$, $B_n^{(\beta)}$, $C_n^{(\gamma)}$ the quantities which in the case of our three series correspond to the $S_n^{(k)}$'s of the general C-process as described in **265**, 3. For $|x| < 1$, since

$$\Sigma\, a_n\, x^n \cdot \Sigma\, b_n\, x^n = \Sigma\, c_n\, x^n,$$

we have [54]

$$\frac{1}{(1-x)^{\alpha+1}}\,\Sigma\, a_n\, x^n \cdot \frac{1}{(1-x)^{\beta+1}}\,\Sigma\, b_n\, x^n = \frac{1}{(1-x)^{\gamma+1}}\,\Sigma\, c_n\, x^n.$$

Hence, by **265**, 3,

$$C_n^{(\gamma)} = A_0^{(\alpha)}\,B_n^{(\beta)} + A_1^{(\alpha)}\,B_{n-1}^{(\beta)} + \dots + A_n^{(\alpha)}\,B_0^{(\beta)}.$$

---

[52] The theorem may be established similarly for summability C_k; cf. the paper quoted in footnote 42, p. 481.

[53] A very complete account of the theory is given by *Andersen, A. F.*: Studier over Cesàro's Summabilitetsmetode, Kopenhagen 1921, and *E. Kogbetliantz*, Sommation des séries et intégrales divergentes par des moyennes arithmétiques et typiques, Mémorial des Sciences math., Fasc. 51, Paris 1931.

[54] Since $a_n = o\,(n^\alpha)$, $b_n = o\,(n^\beta)$, the power series employed are absolutely convergent for $|x| < 1$.

But from this the statement required follows immediately, by Theorem **43, 6**. We need only write

$$x_n = \frac{A_n^{(\alpha)}}{\binom{n+\alpha}{n}}, \qquad y_n = \frac{B_n^{(\beta)}}{\binom{n+\beta}{n}}, \qquad a_{n\nu} = \frac{\binom{\nu+\alpha}{\nu}\binom{n-\nu+\beta}{n-\nu}}{\binom{n+\gamma}{n}},$$

in that theorem, so that

$$\frac{C_n^{(\gamma)}}{\binom{n+\gamma}{n}} = \sum_{\nu=0}^{n} a_{n\nu}\, x_\nu\, y_{n-\nu}.$$

By the hypotheses made, we have $x_n \to A$, $y_n \to B$, and the $a_{n\nu}$ clearly satisfy the four requirements of the theorem. Hence the last expression tends to $A\,B$ as $n \to +\infty$.

Examples and Remarks.

1. If the series $\Sigma(-1)^n$ is multiplied by itself $(k-1)$ times in succession, we obtain the series

$$(k-1)!\sum_{n=0}^{\infty}(-1)^n\binom{n+k-1}{k-1} \qquad (k = 1, 2, \ldots).$$

The original series $(k = 1)$ being summable C_1 by **262**, its square is (certainly) summable C_3, its cube summable C_5, etc. However, by **268**, 2, we know that the kth of these series is (exactly) summable C_k.

2. These examples show that the order of summability of the product-series given by theorem 9 is not necessarily the *exact* order, and that in special cases it may actually be too high. This is not surprising, inasmuch as we already know that the product of two convergent series $(k = 0)$ may still be convergent. The determination of the exact order of summability of the product series requires a special investigation in each case.

In conclusion, we will investigate one more theorem which may be materially extended by introducing summability in place of convergence, — namely *Abel*'s limit theorem **100** and its generalization **233**:

Theorem **10.** *If the power series* $f(x) = \Sigma\, a_n z^n$ *is of radius* 1 *and is summable* C_k *to the value* s *at the point* $+1$ *of the circumference of the unit circle, then* **278.**

$$f(z) = (\Sigma\, a_n z^n) \to s,$$

for every mode of approach of z *to* $+1$, *in which* z *remains within an angle of vertex* $+1$, *bounded by two fixed chords of the unit circle* (v. Fig. 10, p. 406).

P r o o f. As in the proof of **233**, we choose any particular sequence of points $(z_0, z_1, \ldots, z_\lambda, \ldots)$ within the unit circle and the angle, and tending to $+1$ as limit. We have to show that $f(z_\lambda) \to s$. Apply *Toeplitz*' theorem **221** to the sequence [55] $\sigma_n = S_n^{(k)} / \binom{n+k}{n}$, which by hypothesis converges to s, using for the matrix $(a_{\lambda n})$

$$a_{\lambda n} = \binom{n+k}{k} (1 - z_\lambda)^{k+1} \cdot z_\lambda^n.$$

We deduce at once that the transformed sequence also tends to s:

$$\sigma'_\lambda = \sum_{n=0}^{\infty} a_{\lambda n} \sigma_n = (1 - z_\lambda)^{k+1} \cdot \sum_{n=0}^{\infty} S_n^{(k)} \cdot z_\lambda^n \to s.$$

Since $\Sigma S_n^{(k)} z^n = \frac{1}{(1-z)^{k+1}} f(z)$, this is exactly what our statement implied. For this proof to be correct, we have, however, still to verify that the chosen matrix $(a_{\lambda n})$ satisfies the conditions (a), (b) and (c) of the theorems **221**. Since $z_\lambda \to 1$, this is obvious for (a); and, since

$$A_\lambda = \sum_{n=0}^{\infty} a_{\lambda n} = (1 - z_\lambda)^{k+1} \sum_{n=0}^{\infty} \binom{n+k}{k} z_\lambda{}^n = (1 - z_\lambda)^{k+1} \cdot \frac{1}{(1-z_\lambda)^{k+1}} = 1,$$

(c) is also fulfilled. The condition (b) requires the existence of a constant K' such that

$$\sum_n |a_{\lambda n}| = \left(\frac{|1 - z_\lambda|}{1 - |z_\lambda|}\right)^{k+1} < K'$$

for every λ. By the considerations on p. 406, this is obviously the case with $K' = K^{k+1}$, if K has the meaning there laid down.

For $k = 0$, this is exactly the proof of *Abel*'s theorem as carried out on pp. 406—407, in the generalized form of *Stolz*. For $k = 1$, we obtain an extension of this theorem, first indicated by *G. Frobenius* [56], and for $k = 2, 3, \ldots$ we obtain further degrees of generalization, due in substance to *O. Hölder* [57] — taking H_k- instead of C_k-summability and approaching along the radius instead of within the angle only — and first expressed in the form proved above (though with entirely different proofs) by *E. Lasker* [58] and *A. Pringsheim* [59].

By this theorem 10, we have, in particular, $\lim (\Sigma a_n x^n) = s$, for *real* x's increasing to $+1$, and accordingly we can express the essential content of the theorem in the following short form, which is more in keeping with the context:

279. Theorem 11. *The C_k-summability of a series Σa_n to a value s always involves its A-summability to the same value.*

---

[55] k is now the fixed order of the assumed summability.
[56] Journ. f. d. reine u. angew. Math. Vol. 89, p. 262. 1880.
[57] Cf. the paper cited on p. 465, footnote 13.
[58] Phil. Trans. Roy. Soc., Series (A), Vol. 196, p. 431, London 1901.
[59] Acta mathematica Vol. 28, p. 1. 1904.

With the exception of the C_1-summation of *Fourier* series, which will be considered more fully in the following section, further applications of the C_k-process of summation mostly penetrate too deeply into the theory of functions to permit us to discuss them in any detail. We should, however, like to give some account, without detailed proofs, of an application which has led to specially elegant results. This is the application of C_k-summation to the theory of *Dirichlet* series.

The *Dirichlet* series

$$f(z) = \sum_{n=1}^{\infty} \frac{(-1)^{n-1}}{n^z}$$

is [60] convergent for every z for which $\Re(z) > 0$, divergent for every other z. At the point 0, however, where it reduces to the series $\sum_{n=1}^{\infty}(-1)^{n-1}$, it is summable C_1 to the sum $\frac{1}{2}$; at the point -1, where it reduces to $\sum_{n=1}^{\infty}(-1)^{n-1} n$, it is (cf. p. 465) summable C_2 to the sum $\frac{1}{4}$; and the indications given in **268**, 3 show that for $z = -(k-1)$ the series is summable C_k to the sum $\frac{2^k - 1}{k} B_k$, for every integral value of $k \geqq 2$.

This property of being summable C_k, for a suitable k, outside its region of convergence $\Re(z) > 0$, is not restricted to the points mentioned; it can be shown by relatively simple means that our series is summable C_k for every z with $\Re(z) > -k$. Moreover the order of summability is *exactly* k throughout the strip

$$-k < \Re(z) \leqq -(k-1).$$

Thus in addition to the boundary of *convergence*, we have boundaries of *summability* of successive orders, the domain in which the series is *certainly* summable to order k being, in fact, the half-plane

$$\Re(z) > -k \qquad (k = 0, 1, 2, \ldots).$$

Whereas formerly it was only with each point of the right hand half-plane $\Re(z) > 0$ that we could associate a "sum" of the series $\sum \frac{(-1)^{n-1}}{n^z}$, we now associate such a sum with *every* point of the entire plane, thus defining a function of z *in the whole plane*. In a way quite analogous to that used for points within the domain of *convergence* of *Dirichlet*'s series, further investigations now show that these functional values also repre-

---

[60] $f(z) = \left(1 - \frac{2}{2^z}\right) \cdot \zeta(z)$, where $\zeta(z) = \sum_{n=1}^{\infty} \frac{1}{n^z}$ is *Riemann*'s ζ-function. (Cf. **256**, 4, 9, 10 and 11.)

sent an analytic function in the domain of summability — i. e. in the whole plane. *Our series therefore defines an integral function* [61].

Quite analogous properties of summability belong in general to every *Dirichlet* series [62]

$$\sum_{n=1}^{\infty} \frac{a_n}{n^z}.$$

Besides the boundary of *convergence* $\Re(z) = \lambda$, — or λ_0, as we shall now prefer to write, since convergence coincides with C_0-summability, — we have the boundaries $\Re(z) = \lambda_k$ for C_k-summability, $k = 1, 2, \ldots$. They are defined by the condition that the series is certainly summable to the kth order for $\Re(z) > \lambda_k$, but no longer so for $\Re(z) < \lambda_k$. We of course have $\lambda_0 \geqq \lambda_1 \geqq \lambda_2 \geqq \ldots$, and the numbers λ_k therefore tend either to $-\infty$ or to a definite finite limit. Denoting this in either case by Λ, the given *Dirichlet* series is summable C_k for every z with $\Re(z) > \Lambda$, where k is suitably chosen, and its sum defines an analytic function which is regular in this domain. If Λ is *finite*, the straight line $\Re(z) = \Lambda$ is called the *boundary of summability* of the series.

For the investigation of the more general *Dirichlet* series, (v. p. 441, footnote 52)

$$\Sigma\, a_n\, e^{-\lambda_n s},$$

it has been found more convenient to use *Riesz*' $R_{\mu k}$-summation. Cf. the tract by *Hardy* and *Riesz* mentioned in **266, 2.**

§ 61. Application of C_1-summation to the theory of *Fourier* series.

The processes described above possess the obvious advantage of all summation processes, namely, that many infinite series which previously had to be rejected as meaningless are henceforth given a useful meaning, with the result that the field of application of the theory of infinite series is considerably enlarged. Apart from this, the extremely satisfactory nature of these processes from a theoretical point of view lies in the fact that many obscure and confusing situations suddenly become very simple when these processes are introduced. The first example of this was afforded by the problem of the multiplication of infinite series (see p. 461–2, also p. 488). But the application of C_1-summation which

---

[61] From this it follows fairly simply that for *Riemann*'s ζ-function the difference $\zeta(z) - \dfrac{1}{z-1}$ is an *integral* function, — an important result.

[62] *Bohr, H.*: Über die Summabilität *Dirichlet*scher Reihen, Gött. Nachr. 1909, p. 247, and: Bidrag til de Dirichletske Räkkers Theori, Dissert., Kopenhagen 1910.

is, perhaps, the most elegant in this respect, as well as the most important in practice, is the application to the theory of *Fourier* series, due to *L. Fejér* [63]. As we have seen (pp. 369—370), the question of the necessary and sufficient conditions under which the Fourier series of an integrable function converges and represents the given function is one which presents very great difficulties. In particular, it is not known e. g. what type of necessary and sufficient conditions a function *continuous* at a point x_0 must satisfy at that point in order that its *Fourier* series may converge there and represent the functional value in question. In § 49, C, we became acquainted with various criteria for this; but all of these were sufficient conditions only. It was for a long time supposed that *every* function $f(x)$ which is continuous at x_0 possesses a *Fourier* series which converges at that point and has the sum $f(x_0)$ there. An example given by *du Bois-Reymond* (see **216**, 1) was the first to discredit this supposition. *The Fourier series of a function which is continuous at x_0 may actually diverge at that point.*

The question becomes still more difficult, if we require only — as the minimum of hypotheses regarding $f(x)$ — that the (integrable) function $f(x)$ should be such that the limit

$$\lim_{t \to +0} \tfrac{1}{2} [f(x_0 + 2t) + f(x_0 - 2t)] = s(x_0)$$

exists. *What are the necessary and sufficient conditions which must be fulfilled by $f(x)$ in order that its Fourier series may converge at x_0 and have the sum $s(x_0)$?*

As was pointed out, this question is not yet solved by any means. Nevertheless, this obscure and confusing situation is cleared up very satisfactorily when the consideration of the summability of *Fourier* series — C_1-summability is quite sufficient — is substituted for that of their convergence. In fact we have the following elegant

Theorem of Fejér. *If a function $f(x)$, which is integrable in* **280.**
$0 \leqq x \leqq 2\pi$ *and periodic with period 2π, is such that the limit*

$$\lim_{t \to 0} \tfrac{1}{2} [f(x_0 + 2t) + f(x_0 - 2t)] = s(x_0)$$

exists, then its Fourier series is always summable C_1 at this point, to the value $s(x_0)$.

Proof. Let

$$\frac{1}{2} a_0 + \sum_{n=1}^{\infty} (a_n \cos n x_0 + b_n \sin n x_0)$$

---

[63] *Fejér, L.*: Untersuchungen über die *Fourier*schen Reihen. Math. Ann. Vol. 58, p. 51. 1904.

be the *Fourier* series of $f(x)$ at the point x_0; we know, from pp. 356—359, that the nth partial sum may be expressed by

$$s_n = s_n(x_0) = \frac{2}{\pi}\int_0^{\frac{\pi}{2}} \frac{1}{2}[f(x_0+2t)+f(x_0-2t)]\,\frac{\sin(2n+1)t}{\sin t}\,dt \qquad (n = 0, 1, \ldots).$$

Consequently, for $n = 1, 2, \ldots$,

$$s_0 + s_1 + \ldots + s_{n-1}$$
$$= \frac{2}{\pi}\int_0^{\frac{\pi}{2}} \frac{1}{2}[f(x_0+2t)+f(x_0-2t)]\,\frac{\sin t + \sin 3t + \ldots + \sin(2n-1)t}{\sin t}\,dt.$$

Now by **201**, 5, we have, for $t \neq k\pi$,

$$\sin t + \sin 3t + \ldots + \sin(2n-1)t = \frac{\sin^2 nt}{\sin t},$$

and this continues to hold for $t = k\pi$, if we take the right hand side to be in this case the limit of the ratio for $t \to k\pi$, which is evidently 0. Hence [64]

$$\sigma_{n-1} = \frac{s_0 + s_1 + \ldots + s_{n-1}}{n}$$
$$= \frac{2}{n\pi}\int_0^{\frac{\pi}{2}} \frac{1}{2}[f(x_0+2t)+f(x_0-2t)]\left(\frac{\sin nt}{\sin t}\right)^2 dt.$$

As contrasted with *Dirichlet*'s integral, the critical factor $\frac{\sin nt}{\sin t}$ occurs to the second power in the above integral — which is called *Fejér*'s integral for short — and therefore the latter can never change sign; to this and to the fact that the whole is multiplied by $\frac{1}{n}$ the success of the subsequent part of the proof is due. If then the limit

$$\lim_{t\to+0} \frac{1}{2}[f(x_0+2t)+f(x_0-2t)] = s(x_0) = s$$

exists, *Fejér*'s theorem simply states that $\sigma_n \to s$.

We observe that

$$\int_0^{\frac{\pi}{2}} \left(\frac{\sin nt}{\sin t}\right)^2 dt = n\frac{\pi}{2},$$

since [65] the integrand is

$$\sum_{\nu=1}^{n} \frac{\sin(2\nu-1)t}{\sin t}$$

[64] Here the arithmetic mean of the numbers s_n is denoted by $\sigma_n = \sigma_n(x)$ instead of by $s_n' = s_n'(x)$, to avoid confusion with the notation for differentiation.

[65] The value of the integral may also be inferred directly from *Fejér*'s integral itself, for $f(x) = 1$, for which $a_0 = 2$ and the remaining *Fourier* constants $= 0$.

and each term of this, when integrated from 0 to $\frac{\pi}{2}$, contributes the value $\frac{\pi}{2}$, — since

$$\frac{\sin(2\nu-1)t}{\sin t} = 1 + 2\cos 2t + 2\cos 4t + \cdots + 2\cos 2(\nu-1)t.$$

Hence we may write

$$s = \frac{2}{n\pi}\int_0^{\frac{\pi}{2}} s \cdot \left(\frac{\sin nt}{\sin t}\right)^2 dt$$

and therefore

$$\sigma_{n-1} - s = \frac{2}{n\pi}\int_0^{\frac{\pi}{2}} \left[\frac{f(x_0+2t)+f(x_0-2t)}{2} - s\right] \cdot \left(\frac{\sin nt}{\sin t}\right)^2 dt.$$

By hypothesis, the expression in square brackets tends to 0 when $t \to +0$. In order to prove that σ_{n-1} or $\sigma_n \to s$, it is therefore sufficient to show that

If $\varphi(t)$ is integrable in $0 \ldots \frac{\pi}{2}$ and **281.**

$$\lim_{t\to+0} \varphi(t) = 0,$$

then

$$\frac{2}{n\pi}\int_0^{\frac{\pi}{2}} \varphi(t) \cdot \left(\frac{\sin nt}{\sin t}\right)^2 dt \to 0$$

as n increases.

Now this follows from a very simple train of inequalities. As $\varphi(t) \to 0$, we can determine $\delta < \frac{\pi}{2}$, for a given $\varepsilon > 0$, so that $|\varphi(t)| < \frac{\varepsilon}{2}$ for every t such that $0 < t \leqq \delta$. Then

$$\left|\frac{2}{n\pi}\int_0^{\delta} \varphi(t) \cdot \left(\frac{\sin nt}{\sin t}\right)^2 dt\right| \leqq \frac{\varepsilon}{2} \cdot \frac{2}{n\pi}\int_0^{\delta} \left(\frac{\sin nt}{\sin t}\right)^2 dt < \frac{\varepsilon}{2},$$

since the last integral has a positive integrand, and therefore remains less than the integral of the same function over the whole range 0 to $\frac{\pi}{2}$. On the other hand, a constant M exists such that $|\varphi(t)|$ remains $< M$ throughout $0 < t < \frac{1}{2}\pi$. Consequently

$$\left|\frac{2}{n\pi}\int_{\delta}^{\frac{\pi}{2}} \varphi(t) \cdot \left(\frac{\sin nt}{\sin t}\right)^2 dt\right| \leqq \frac{2M}{n\pi} \cdot \frac{\pi}{2} \cdot \frac{1}{\sin^2\delta}.$$

On the right hand side, everything but n is fixed, and we can therefore choose n_0 so large that this expression becomes $< \frac{1}{2}\varepsilon$ for every $n > n_0$. We then have

$$|\sigma_{n-1} - s| < \varepsilon$$

for these n's; hence $\sigma_n \to s$. Thus *Fejér*'s theorem is completely established[66].

282. **Corollary 1.** If $f(x)$ is continuous in the interval $0 \leqq x \leqq 2\pi$, and if further $f(0) = f(2\pi)$, *then the Fourier series of $f(x)$ is summable C_1 to the sum $f(x)$, for every x.* For the hypotheses of *Fejér*'s theorem are now certainly fulfilled for every x, and $s(x) = f(x)$ everywhere. We assume, as usual, that the function $f(x)$ is defined in the intervals $2k\pi \leqq x \leqq 2(k+1)\pi$, for $k = \pm 1, \pm 2, \ldots$, by means of the periodicity condition, $f(x) = f(x - 2k\pi)$.

We now further state:

Corollary 2. *With the conditions of the preceding corollary, the C_1-summability, which has been established for all x's, is, moreover, uniform for all x's, i. e. the sequence of functions $\sigma_n(x)$ tends uniformly to $f(x)$ for all x's.* In other words: Given $\varepsilon > 0$, we can determine *one* number N such that for every $n > N$, irrespective of the position of x, we have[67]

$$|\sigma_n(x) - f(x)| < \varepsilon.$$

P r o o f. We have only to show that the inequalities in the proof of the theorem can be arranged so as to hold for *every* x. Now

$$\varphi(t) = \varphi(t, x) = \frac{1}{2}[f(x + 2t) - f(x)] + \frac{1}{2}[f(x - 2t) - f(x)];$$

since $f(x)$ is periodic and is continuous everywhere, it is uniformly continuous for all x's (cf. § 19, theorem 5), and, given ε, we can choose *one* $\delta > 0$ such that

$$|f(x \pm 2t) - f(x)| < \frac{\varepsilon}{2}$$

for every $|t| < \delta$, and *every* x. This implies that for all these t's

$$|\varphi(t)| = |\varphi(t, x)| < \frac{\varepsilon}{2}$$

irrespective of x; hence, as before,

$$\left| \frac{2}{n\pi} \int_0^\delta \varphi(t) \left(\frac{\sin nt}{\sin t}\right)^2 dt \right| < \frac{\varepsilon}{2}.$$

Further, since $f(x)$ is periodic and is continuous everywhere, it is bounded, say $|f(x)| < K$ for every x. It follows at once that for *all* t's and *all* x's,

$$|\varphi(t)| = |\varphi(t, x)| < 2K$$

---

[66] Note in passing that the curves of approximation $y = \sigma_n(x)$ do *not* exhibit *Gibbs'* phenomenon (v. **216**, 4). (*Fejér, L.*: Math. Annalen, Vol. 64, p. 273. 1907.)

[67] The corresponding statement holds, moreover, in the case of the general theorem of *Fejér* for every closed interval entirely contained, together with its end-point, in an open interval in which $f(x)$ is continuous.

and hence, as before,

$$\left|\frac{2}{n\pi}\int_0^{\frac{\pi}{2}} \varphi(t)\cdot\left(\frac{\sin n t}{\sin t}\right)^2 d t\right| \leqq \frac{1}{n}\cdot\frac{2K}{\sin^2\delta}.$$

Now we can actually determine *one* number N such that the last expression remains $< \frac{1}{2}\varepsilon$ for every $n \geqq N$. For these n's we therefore have $|\sigma_{n-1} - s| < \varepsilon$, so that, as asserted, we can associate with every given ε *one* number N such that

$$|\sigma_n(x) - f(x)| < \varepsilon$$

for every $n > N$, irrespective of the position of x.

As an easy application, the following important theorem results from the above theorems:

Weierstrass's Approximation. If $F(x)$ is a function continuous in the closed interval $a \leqq x \leqq b$, and if $\varepsilon > 0$ is arbitrarily assigned, then there is always a polynomial $P(x)$ with the property that, in $a \leqq x \leqq b$, **282a.**

$$|F(x) - P(x)| < \varepsilon.$$

Proof. Put $F\left(a + \frac{b-a}{\pi}x\right) = f(x)$. Then $f(x)$ is defined and continuous in $0 \leqq x \leqq \pi$. In $\pi \leqq x \leqq 2\pi$, write as in § 50, 2nd method, $f(x) = f(2\pi - x)$. Define $f(x)$ for all other x by the periodicity condition $f(x + 2\pi) = f(x)$. Then $f(x)$ is everywhere continuous. Now, for this $f(x)$, let $\sigma_n(x)$ have the meaning laid down in the statement of the preceding theorem. An index m may then be found such that

$$|f(x) - \sigma_m(x)| < \frac{\varepsilon}{2}$$

for all x. This $\sigma_m(x)$ is the sum of a finite number of expressions of the form $a\cos p x + b\sin q x$; hence it can be expanded in a power series convergent everywhere, by means of the power series of § 24. Let

$$c_0 + c_1 x + \ldots + c_n x^n + \ldots$$

denote this expansion. Since it converges uniformly in $0 \leqq x \leqq \pi$, we can determine a finite k so that the polynomial

$$c_0 + c_1 x + \ldots + c_k x^k = p(x)$$

satisfies the inequality $\quad |\sigma_m(x) - p(x)| < \frac{\varepsilon}{2}$

throughout $0 \leqq x \leqq \pi$. Hence it satisfies

$$|f(x) - p(x)| < \varepsilon.$$

Putting finally $\quad p\left(\frac{x-a}{b-a}\pi\right) = P(x),$

we see that $P(x)$ is a polynomial of the required kind, since, throughout $a \leqq x \leqq b$,

$$|F(x) - P(x)| < \varepsilon.$$

§ 62. The A-process.

The last theorem of § 60 has already shown that the range of action of the A-process embraces that of *all* the C_k-processes. In this respect it is superior to the C- and H-processes. Also, it is not difficult to give examples of series which are summable A but not summable C_k to any order k, however large. We need only consider $\Sigma a_n x^n$, the expansion in power series of

$$f(x) = e^{\frac{1}{1-x}}$$

at the point $x = -1$. Since obviously $\lim f(x)$ exists for $x \to -1+0$ and $= \sqrt{e}$, the series $\Sigma (-1)^n a_n$ is summable A to the value $\sqrt{e}$. If, however, it were summable C_k, for some specific k, by **271** we should require to have $a_n = o(n^k)$. Now a particular coefficient a_n is obtained by adding together the coefficients of x^n in the expansions of the individual terms of the series, which is uniformly convergent for $|x| \leqq \varrho < 1$:

$$e^{\frac{1}{1-x}} = 1 + \frac{1}{1-x} + \frac{1}{2!}\frac{1}{(1-x)^2} + \cdots + \frac{1}{\nu!}\frac{1}{(1-x)^\nu} + \cdots$$

(v. **249**). As all the coefficients in these expansions are positive, a_n is certainly greater than the coefficient of x^n in the expansion of a single term. Picking out the $(k+2)^{\text{th}}$ term, we see that

$$a_n > \frac{1}{(k+2)!}\binom{n+k+1}{k+1} > \frac{n^{k+1}}{(k+2)!\,(k+1)!}.$$

For a fixed k, a_n/n^k therefore *cannot* tend to 0; on the contrary, it tends to $+\infty$.

Although the A-process is thus more powerful than all the C_k-processes taken together, it is, nevertheless, restricted by the very simple stipulation that in order that it may be applicable to a series Σa_n, the series $\Sigma a_n x^n$ and $\Sigma s_n x^n$ must converge for $|x| < 1$:

283. Theorem 1. *If the series Σa_n, with partial sums s_n, is summable A, we necessarily have*

$$\overline{\lim} \sqrt[n]{|a_n|} \leqq 1 \quad \text{and} \quad \overline{\lim} \sqrt[n]{|s_n|} \leqq 1$$

or, what comes to exactly the same thing,

$$a_n = O((1+\varepsilon)^n) \quad \text{and} \quad s_n = O((1+\varepsilon)^n),$$

for every $\varepsilon > 0$, however small.

In this we have a companion to theorem 3 of § 60; but theorems 4 and 5 of that section also have literal analogues in this connection:

284. Theorem 2. *In a series Σa_n, which is summable A, we necessarily have*

$$A\text{-}\lim a_n = 0 \quad \textit{and indeed} \quad A\text{-}\lim \left(\frac{a_1 + 2a_2 + \cdots + n a_n}{n+1}\right) = 0.$$

Proof. The first of these two relations indicates that $(1-x)\,\Sigma\, a_n x^n$ must tend to 0 as $x \to 1-0$. This is almost obvious, since by hypothesis $\Sigma\, a_n x^n \to s$. The truth of the second statement follows, on the same lines as the proof of **273**, from the two relations

$$(*)\quad (1-x)\cdot\Sigma\, s_n x^n \to s \quad\text{and}\quad (1-x)\cdot\Sigma\,\frac{s_0+s_1+\ldots+s_n}{n+1}\,x^n \to s$$

by subtraction; the first of these is nothing more than an explicit form of the hypothesis that $\Sigma\, a_n$ is summable A, while the second is quite easily deduced from it. In fact, from $(1-x)\,\Sigma\, s_n x^n \to s$, we first infer that

$$(1-x)^2\,\Sigma\,(s_0+s_1+\ldots+s_n)\,x^n \to s,$$

by **102**. That the second of the relations (*) follows from this, is a special case of the following simple theorem:

Auxiliary theorem. *If, for $x \to 1-0$, a function $f(x)$, which is integrable in $0 \leqq x \leqq 1$, satisfies the limiting relation* **285.**

$$(1-x)^2 f(x) \to s,$$

then, for $x \to 1-0$, we also have

$$(1-x)\int_0^x f(t)\,dt = (1-x)\,F(x) \to s.$$

The proof follows immediately from the rule known as l'Hospital's, by which

$$\lim_{x\to 1-0}\frac{F(x)}{G(x)} = \lim_{x\to 1-0}\frac{F'(x)}{G'(x)},$$

provided the right hand side exists and $G(x)$ is positive and tends to $+\infty$ as $x \to 1-0$. The direct proof is as follows [68]:

Put $(1-x)^2 f(x) = s + \varrho(x)$. The function $\varrho(x)$ tends to 0 as $x \to +1-0$, and so for any given $\varepsilon > 0$, we can assign an x_1 in $0 < x_1 < 1$, such that $|\varrho(x)|$ remains $< \frac{\varepsilon}{2}$ for $x_1 < x < 1$. We then have, for these values of x,

$$|(1-x)\,F(x) - s| \leqq (1-x)\cdot|s| + (1-x)\cdot\left|\int_0^{x_1}\frac{\varrho(x)}{(1-x)^2}\,dx\right| + \frac{\varepsilon}{2}.$$

From this the statement follows in the usual way.

By these theorems 1 and 2 we have to some extent fixed *outer* limits to the range of action of the A-process. As before (cf. the developments

[68] The proof is on quite similar lines to that in **43**, 1 and 2. — The meaning of the assertion under consideration, that the first of the relations (*) implies the second, may also be stated thus:

$$A\text{-}\lim s_n = s \quad\text{implies}\quad A\,C_1\text{-}\lim s_n = s.$$

For in the case of the second relation we are concerned with the successive application first of the C_1-process, and then of the A-process, for the limitation of (s_n).

on p. 485—6), the question as to the point, beyond the region of series which actually converge, at which its action *begins* is a much more delicate one. In this connection we have the following theorem due to *A. Tauber*[69]:

286. Theorem 3. *A series Σa_n, which is summable A, and for which $n a_n \to 0$, i. e. for which*

$$a_n = o\left(\frac{1}{n}\right),$$

is convergent in the usual sense. (*o-A → K-theorem.*)

Proof. If we are given $\varepsilon > 0$, we can choose $n_0 > 0$ so that for every $n > n_0$

$$\text{a)} \quad |n a_n| < \frac{\varepsilon}{3}, \qquad \text{b)} \quad \frac{|a_1| + 2|a_2| + \ldots + n|a_n|}{n} < \frac{\varepsilon}{3},$$

$$\text{c)} \quad \left|f\left(1 - \frac{1}{n}\right) - s\right| < \frac{\varepsilon}{3} \qquad (f(x) = \Sigma a_n x^n).$$

(Here a) and c) can be satisfied by hypothesis, and b) by referring to **43**, 2.) For these n's and for every positive $x < 1$, we then have

$$s_n - s = f(x) - s + \sum_{\nu=1}^{n} a_\nu (1 - x^\nu) - \sum_{\nu=n+1}^{\infty} a_\nu x^\nu.$$

If we now observe that in the first of the sums

$$(1 - x^\nu) = (1 - x)(1 + x + \ldots + x^{\nu-1}) \leqq \nu (1 - x),$$

and in the second $|a_\nu| = \frac{|\nu a_\nu|}{\nu} < \frac{\varepsilon}{3n}$, it follows that

$$|s_n - s| \leqq |f(x) - s| + (1 - x) \sum_{\nu=1}^{n} |\nu a_\nu| + \frac{\varepsilon}{3n(1-x)},$$

for every positive $x < 1$. Choosing, in particular, $x = 1 - \frac{1}{n}$, we obtain, by a), b) and c),

$$|s_n - s| < \frac{\varepsilon}{3} + \frac{\varepsilon}{3} + \frac{\varepsilon}{3} = \varepsilon$$

for every $n > n_0$. Hence $s_n \to s$, q. e. d.

In this proof, if we interpret ε as being, not an *arbitrary prescribed* positive number, but a *suitably chosen* (sufficiently large) one, then we may infer the following corollary:

Corollary. *A series Σa_n, summable A, with $(n a_n)$ bounded, i. e. one for which*

$$a_n = O\left(\frac{1}{n}\right),$$

has bounded partial sums.

On account of the great similarity between this theorem and theorem 6 of § 60, it appears likely that an *O-A → K*-theorem also holds, i. e. one

[69] *Tauber, A.*: Ein Satz aus der Theorie der unendlichen Reihen, Monatshefte f. Math. u. Phys., Vol. 8, pp. 273—277. 1897. Cf. p. 486, footnote 50.

which deduces the convergence of Σa_n from its A-summability, by assuming, as regards the a_n's, merely the fact that they are $O\left(\frac{1}{n}\right)$. This theorem is actually true. It goes very much deeper, however, and was proved for the first time in 1910, by *J. E. Littlewood*[70]:

Theorem 4. *A series Σa_n, which is summable A, and whose terms satisfy the relation* **287.**

$$a_n = O\left(\frac{1}{n}\right),$$

— i. e. for which $(n\, a_n)$ is bounded, — is convergent in the ordinary sense. (O-$A \to K$-theorem.)

Before going on to the proof, we may mention that this theorem contains, as a corollary, Theorem 6 of § 60, as already stated there. For if a series is summable C_k, then by Theorem 11, § 60, it is also summable A. Every series, therefore, that satisfies the assumptions of Theorem 6, § 60, also satisfies those of Littlewood's theorem just stated, and is therefore convergent.

Previously known proofs of Littlewood's theorem were very complicated, in spite of the number of researches devoted to it[71], till in 1930 *J. Karamata*[72] found a surprisingly simple proof. We shall preface his argument with the following obvious lemma:

Lemma. *Let ϱ and ε be arbitrary real numbers, and let $f(t)$ denote the following function*[73], *defined and integrable (in the Riemann sense) over the interval* $0 \leqq t \leqq 1$ (v. Fig. 13):

$$f(t) = \begin{cases} 0 \text{ in } 0 \leqq t < e^{-(1+\varrho)}, \\ \frac{1}{t} \text{ in } e^{-(1+\varrho)} \leqq t < e^{-1}, \\ 0 \text{ in } e^{-1} \leqq t \leqq 1. \end{cases}$$

Then there exist two polynomials $p(t)$ and $P(t)$ for which

(a) $$p(t) \leqq f(t) \leqq P(t) \quad \textit{in} \quad 0 \leqq t \leqq 1,$$

(b) $$\int_0^1 (P(t) - p(t))\, dt < \varepsilon.$$

---

[70] The converse of *Abel*'s theorem on power series: Proc. Lond. Math. Soc. (2) Vol. 9, pp. 434—448. 1911.

[71] Besides the paper just mentioned, cf. *E. Landau*, Darstellung u. Begründung einiger neuerer Ergebnisse d. Funktionentheorie. 1st ed. pp. 45—46. 1916; 2nd ed. pp. 57—62. 1929.

[72] *Karamata, J.*, Über die Hardy-Littlewoodsche Umkehrung des *Abel*schen Stetigkeitssatzes. Math. Zeitschr. Vol. 32, pp. 319—320. 1930.

[73] The theorem holds unaltered for *every* function integrable in the sense of *Riemann*.

Proof. Let $OAA'B'BE$ (cf. the rough diagram, Fig. 13) be the graph of the function $f(t)$, so that A and A' have the abscissa $e^{-(1+\varrho)}$, while that of B and B' is e^{-1}. Now choose a positive δ less than the abscissa of A, less than half the difference between the abscissae of A and B, and furthermore

$$< \frac{\varepsilon}{4} e^{-(1+\varrho)}.$$

On the graph, mark the points A_1, A_2 with the abscissae $e^{-(1+\varrho)} \pm \delta$, and the points B_1, B_2 with the abscissae $e^{-1} \pm \delta$. Then the lines OAA_2B_1BE

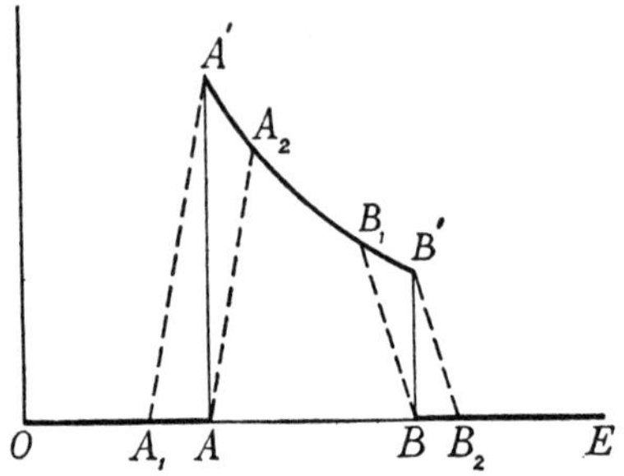

Fig. 13.

and $OA_1A'B'B_2E$ (with A_2B_1 and $A'B'$ taken along the curve $\frac{1}{t}$, the other portions being straight) are the graphs of two continuous functions $g(t)$ and $G(t)$ respectively, for which, obviously,

$$\text{(a')} \qquad g(t) \leqq f(t) \leqq G(t) \quad \text{in} \quad 0 \leqq t \leqq 1,$$

$$\text{(b')} \qquad \int_0^1 (G(t) - g(t))\, dt < \frac{\varepsilon}{2}.$$

By *Weierstrass*'s approximation (**282 a**), there exists a polynomial $p(t)$ that differs by less than $\frac{\varepsilon}{4}$ from the continuous function $g(t) - \frac{\varepsilon}{4}$ in the interval $0 \leqq t \leqq 1$:

$$\left| g(t) - \frac{\varepsilon}{4} - p(t) \right| < \frac{\varepsilon}{4} \quad \text{in} \quad 0 \leqq t \leqq 1.$$

Similarly there exists a polynomial $P(t)$ that differs by less than $\frac{\varepsilon}{4}$ from $G(t) + \frac{\varepsilon}{4}$ there:

$$\left| G(t) + \frac{\varepsilon}{4} - P(t) \right| < \frac{\varepsilon}{4} \quad \text{in} \quad 0 \leqq t \leqq 1.$$

These polynomials clearly satisfy the conditions (a) and (b) of the lemma.

Proof of Littlewood's theorem.

I. By the corollary to theorem 3, the sequence (s_n), under present hypotheses, is certainly *bounded.*

In proceeding with the proof, it will be no restriction to assume the *terms of the series* Σa_n *to be real.* For, once the theorem is proved for real series, it can be inferred immediately for series of complex terms by splitting these up into their real and imaginary parts.

II. *Let* ϱ *be given* > 0, *and put* $[(1+\varrho)\,n] = k\,(n) = k$. *Let* s_n *denote as usual the partial sums of* Σa_n, *and, for* $n > \frac{1}{\varrho}$, *write* [74]

$$\operatorname*{Max}_{n<\nu\leqq k} |\, s_\nu - s_n \,| = \mu_n(\varrho),$$

and

$$\overline{\lim_{n\to+\infty}}\, \mu_n(\varrho) = \mu(\varrho).$$

Then $\mu(\varrho) \to 0$ *as* $\varrho \to 0$.

Indeed, for $n < \nu \leqq k$, we have

$$\begin{aligned} |\, s_\nu - s_n \,| &= |\, a_{n+1} + a_{n+2} + \dots + a_\nu \,| \\ &\leqq (k-n)\,\mathrm{Max}\,(|\, a_{n+1}\,|,\ |\, a_{n+2}\,|,\ \dots,\ |\, a_k \,|). \end{aligned}$$

If $|\, a_r \,|$ be this maximum, it follows further that

$$|\, s_\nu - s_n \,| \leqq \frac{k-n}{r}\cdot r\,|\, a_r \,| \leqq \varrho \cdot r\,|\, a_r \,|.$$

Now, $(n\,a_n)$ being assumed bounded, there exists a constant K such that $n\,|\, a_n \,| < K$ for all n, and so

$$\mu_n(\varrho) = \operatorname*{Max}_{n<\nu\leqq k} |\, s_\nu - s_n \,| \leqq \varrho\, K.$$

Thus

$$\mu(\varrho) \leqq \varrho\, K,$$

whence the statement follows.

III. *Suppose the sequence* (s_n) *is*

(i) *bounded on one side, say* $s_n \geqq -M$, $(M \geqq 0)$;

(ii) *limitable* A, *say* $(1-x)\sum_{\nu=0}^{\infty} s_\nu x^\nu \to s$, *for* $x \to 1-0$.

---

[74] The symbol Max $(t_1, t_2, \dots, t_p)$, or Max t_ν $(1 \leqq \nu \leqq p)$, denotes the largest of the numbers $t_1, t_2, \dots, t_p$ (assumed real).

Then[75] *if* $f(t)$ *denotes the function defined in the lemma,*

$$(*) \qquad (1-x)\sum_{\nu=0}^{\infty} s_\nu f(x^\nu)\cdot x^\nu \to s\int_0^1 f(t)\,dt, \quad \text{i. e.} = \varrho\, s.$$

For, by 2, we have, for every integer $k \geqq 0$,

$$(1-x^{k+1})\sum_{\nu=0}^{\infty} s_\nu \cdot (x^{k+1})^\nu \to s,$$

as $x \to 1-0$, so

$$(1-x)\sum_{\nu=0}^{\infty} s_\nu (x^\nu)^k \cdot x^\nu \to \frac{s}{k+1}.$$

Now if $Q(x) = b_0 + b_1 x + \ldots + b_q x^q$ is any polynomial, it follows at once that

$$(**) \quad (1-x)\sum_{\nu=0}^{\infty} s_\nu\, Q(x^\nu)\, x^\nu \to s\left(\frac{b_0}{1}+\frac{b_1}{2}+\ldots+\frac{b_q}{q+1}\right) = s\cdot\int_0^1 Q(t)\,dt.$$

Now let ϵ denote any positive number. Then a pair of polynomials $p(t)$, $P(t)$ can be assigned, by the lemma, so that

$$\text{(a)} \qquad p(t) \leqq f(t) \leqq P(t) \quad \text{in} \quad 0 \leqq t \leqq 1,$$

$$\text{(b)} \qquad \int_0^1 (P(t)-p(t))\,dt < \varepsilon.$$

First assume $M = 0$, *so that* $s_\nu \geqq 0$ *for all* ν; then

$$(1-x)\sum_{\nu=0}^{\infty} s_\nu p(x^\nu)\cdot x^\nu \leqq (1-x)\sum_{\nu=0}^{\infty} s_\nu f(x^\nu)\cdot x^\nu \leqq (1-x)\sum_{\nu=0}^{\infty} s_\nu P(x^\nu)\cdot x^\nu.$$

For $x \to 1-0$, it follows by (**), that

$$s\cdot\int_0^1 p(t)\,dt \leqq \overline{\underline{\lim}}\,(1-x)\sum_{\nu=0}^{\infty} s_\nu f(x^\nu)\cdot x^\nu \leqq s\int_0^1 P(t)\,dt.$$

By (a) and (b), the integrals on the left and on the right differ from each other and from $\int_0^1 f(t)\,dt$ by less than ε. Hence

$$\left|\overline{\underline{\lim}}\,(1-x)\sum_{\nu=0}^{\infty} s_\nu f(x^\nu)\cdot x^\nu - s\int_0^1 f(t)\,dt\right| \leqq s\cdot\varepsilon.$$

Since $\varepsilon > 0$ was arbitrary, it follows that the statement (*) is true for non-negative s_n.

---

[75] This is *J. Karamata*'s Main Theorem. Both theorem and proof apply unaltered to any function integrable in the *Riemann* sense over $0 \leqq t \leqq 1$, — except for the special *value* $\varrho\, s$ of the integral $\int_0^1 f(t)\,dt$ in our case.

If however $M > 0$, apply the theorem as so far proved to the two sequences $t_n \equiv s_n + M$ and $u_n \equiv M$ instead of to s_n. Subtracting the results, we get (*) in its full generality.

IV. In III (*), put $x = e^{-\frac{1}{n}}$ on the left hand side, and refer back to the definition of $f(t)$ in the lemma; writing as before $[(1+\varrho)n] = k(n) = k$, we infer that as $n \to +\infty$,

$$(1 - e^{-\frac{1}{n}}) \sum_{\nu=n+1}^{k} s_\nu \to \varrho\, s.$$

Since
$$n(1 - e^{-\frac{1}{n}}) \to 1 \text{ and } \frac{k-n}{n} \to \varrho,$$

we have then
$$\frac{1}{k-n} \sum_{\nu=n+1}^{k} s_\nu \to s.$$

Writing, therefore,
$$\frac{1}{k-n} \sum_{\nu=n+1}^{k} s_\nu - s = \delta_n,$$

we have $\delta_n \to 0$, and

$$s - s_n = \frac{1}{k-n} \sum_{\nu=n+1}^{k} (s_\nu - s) - \delta_n.$$

Hence

$$|s - s_n| \leqq \operatorname*{Max}_{n<\nu\leqq k} |s_\nu - s_n| + |\delta_n| = \mu_n(\varrho) + |\delta_n|.$$

Making $n \to +\infty$, we deduce

$$\overline{\lim}\, |s - s_n| \leqq \mu(\varrho);$$

and as this holds for every $\varrho > 0$, it follows by II that, for $\varrho \to +0$,

$$\overline{\lim}\, |s - s_n| = 0,$$

i. e.
$$s_n \to s.$$

This completes the proof.

288. Examples and Applications.

1. Every series which is summable C is also summable A, to the same value. This often enables us to determine the values of series which are summable C. Thus in **268**, 3 we saw that the series $\Sigma(-1)^n(n+1)^k$ are summable C_{k+1}; by means of the A-summation process we can now obtain the *values* of these series, which occur conveniently as the kth derivatives of the geometric series $\Sigma(-1)^n x^n$, when the exponential function is inserted by substituting $x = e^{-t}$. In this way we obtain the series

$$e^{-t} - e^{-2t} + e^{-3t} - + \dots,$$

convergent for $t > 0$. The sum of this series is

$$= \frac{e^{-t}}{1+e^{-t}} = \frac{1}{e^t+1} = \frac{e^t-1}{e^{2t}-1} = \frac{e^t+1-2}{e^{2t}-1} = \frac{1}{e^t-1} - \frac{2}{e^{2t}-1}$$

$$= \frac{1}{t}\cdot\frac{t}{e^t-1} - \frac{1}{t}\,\frac{2t}{e^{2t}-1}.$$

For a sufficiently small $t > 0$, these last fractions may be expanded in power series by **105, 5**; the first terms of the two expansions cancel each other and we obtain

$$e^{-t} - e^{-2t} + - \dots + (-1)^n e^{-(n+1)t} + \dots = -\sum_{n=0}^{\infty} \frac{2^{n+1}-1}{(n+1)!} B_{n+1}\, t^n.$$

Differentiating k times in succession with respect to t, we further obtain

$$e^{-t} - 2^k e^{-2t} + - \dots + (-1)^n (n+1)^k e^{-(n+1)t} + \dots$$

$$= (-1)^{k+1} \sum_{n=k}^{\infty} \frac{2^{n+1}-1}{(n+1)!} B_{n+1}\cdot n(n-1)\dots(n-k+1)\cdot t^{n-k}.$$

Now, letting t diminish and $\to +0$, we at once obtain on the right hand side [76]

$$(-1)^{k+1}\frac{2^{k+1}-1}{k+1} B_{k+1}.$$

Putting $e^{-t} = x$ on the left hand side, we see that we are dealing with a power series of radius 1; when t decreases to 0, x increases to $+1$. The value just obtained is therefore by definition the A-sum of the series

$$1 - 2^k + 3^k - + \dots + (-1)^n (n+1)^k + \dots$$

for integral $k \geqq 0$. And as this series was seen to be summable C_{k+1} in **268, 3**, we have thus obtained its C_{k+1}-sum also, by **279**.

2. If the function represented by a power series $\Sigma c_n z^n$ of radius r is *regular* at a point z_1 of the circumference of the unit circle, $\lim f(x z_1)$ for positive increasing $x \to +1$ certainly exists and $= f(z_1)$. At every such point the series $\Sigma a_n \equiv \Sigma c_n z_1^n$ is therefore summable A and its A-sum is the functional value $f(z_1)$.

---

[76] For $k > 0$, the sign $(-1)^{k+1}$ may simply be omitted, by footnote 4, p. 237.

3. Combining the preceding remark with theorem 4, we get the statement: If $f(x) = \Sigma c_n z^n$ converges for $|z| < 1$ and $(n c_n)$ is bounded, then the series continues to converge (in the ordinary sense) at every point z_1, on the circumference of the unit circle, at which $f(z)$ is regular.

4. *Cauchy*'s product $\Sigma c_n \equiv \Sigma (a_0 b_n + \cdots + a_n b_0)$ of two series Σa_n and Σb_n, which are summable A to the values A and B, is also summable A, to the value $C = AB$, as an immediate consequence of the definition of A-summability.

5. With regard to the series $\sum_{n=1}^{\infty} \frac{1}{n^{1+\alpha i}}$, $\alpha \gtrless 0$, we have already seen in **274** that these do *not* converge, and that they are not summable C_k to any order k. By *Littlewood*'s theorem 4, we may now add that they cannot be summable A either.

§ 63. The E-process.[77]

The E_1-process was introduced on the strength of *Euler*'s transformation of series (**144**). Starting from any series Σa_n (*not* having alternately $+$ and $-$ signs), we should have to write

$$\frac{1}{2^{n+1}}\left[\binom{n}{0} a_0 + \binom{n}{1} a_1 + \cdots + \binom{n}{n} a_n\right] = a_n'$$

and we should have to consider $\Sigma a_n'$ as the E_1-transformation of Σa_n. We had agreed to depart from the usual notation so far as to write[78] $s_0 = 0$ and $s_n = a_0 + a_1 + \ldots + a_{n-1}$ for $n > 0$, — and similarly for the accented series. Then (v. **265**, 5)

$$s_n' = \frac{1}{2^n}\left[\binom{n}{0} s_0 + \binom{n}{1} s_1 + \cdots + \binom{n}{n} s_n\right]$$

is the E_1-transformation of the sequence (s_n). Applying this again, we obtain for the E_2-transformation, after an easy calculation,

$$\Sigma a_n'', \quad \text{with} \quad a_n'' = \frac{1}{4^{n+1}}\left[\binom{n}{0} 3^n a_0 + \binom{n}{1} 3^{n-1} a_1 + \cdots + \binom{n}{n} a_n\right]$$

the partial sums of which are now

$$a_0'' + a_1'' + \cdots + a_{n-1}'' = s_n''$$

$$= \frac{1}{4^n}\left[\binom{n}{0} 3^n s_0 + \binom{n}{1} 3^{n-1} s_1 + \cdots + \binom{n}{n} s_n\right], \quad (n > 0).$$

[77] A detailed investigation of this process is to be found in two papers by the author, *Über das Eulersche Summierungsverfahren* (I: Mathemat. Zeitschr. Vol. 15, pp. 226—253. 1922; II: ibid., Vol. 18, pp. 125—156. 1923). Complete proofs of all the theorems mentioned in this section are given there.

[78] It may be verified without much difficulty that in the case of the E_1-process "a finite number of alterations" is allowed, as in the case of convergent series. (A proof, into which we shall not enter here, is given in the first of the two papers mentioned in the preceding footnote.) Consequently the shifting of indices has no effect on the result of the limitation process.

For the E_p-transformation we obtain in the same way the series

$$\Sigma\, a_n^{(p)}$$

with terms

$$a_n^{(p)} = \frac{1}{(2^p)^{n+1}}\left[\binom{n}{0}(2^p-1)^n a_0 + \binom{n}{1}(2^p-1)^{n-1} a_1 + \cdots + \binom{n}{n} a_n\right]$$

and partial sums [79] $(n > 0)$

$$a_0^{(p)} + a_1^{(p)} + \cdots + a_{n-1}^{(p)} = s_n^{(p)}$$

$$= \frac{1}{(2^p)^n}\left[\binom{n}{0}(2^p-1)^n s_0 + \binom{n}{1}(2^p-1)^{n-1} s_1 + \cdots + \binom{n}{n} s_n\right].$$

The examples given in **265**, 5 have already illustrated the action of the E_1-process; the last of them shows that the range of the E_1-process is considerably wider than those of the C- and H-processes. By analogy with that example, we may form the E_p-transformation of the geometric series Σz^n, and we shall obtain

$$\sum_{n=0}^{\infty}\left[\frac{1}{(2^p)^{n+1}}\sum_{\nu=0}^{n}\binom{n}{\nu}(2^p-1)^{n-\nu} z^\nu\right] = \frac{1}{2^p}\sum_{n=0}^{\infty}\left(\frac{2^p-1+z}{2^p}\right)^n.$$

This series converges [80], — to the sum $\frac{1}{1-z}$, — if, and only if, $|z+(2^p-1)| < 2^p$, i. e. if z lies within the circle of radius 2^p round the point $-(2^p-1)$. Evidently *every* point in the half-plane $\Re(z) < 1$ can be made to lie inside such a circle, by taking the exponent p sufficiently large. We may accordingly say: *The geometric series Σz^n is summable E_p to a suitable order p for each point z interior to the half-plane $\Re(z) < 1$.* The sum is in every case $\frac{1}{1-z}$, i. e. it is the analytical extension of the function defined by the series in the unit circle.

The case of any power series is quite similar, but in order to carry out the proofs we require assistance from the more difficult parts of function theory. We shall therefore content ourselves with indicating the most tangible results [81]:

The power series $\Sigma c_n z^n$ is assumed to have a finite positive radius of convergence, and the function which it represents in its circle of convergence is denoted by $f(z)$. This function we suppose analytically extended along every ray am $z = \varphi =$ const. until we reach the first singular point of $f(z)$ on this ray, which we shall denote by ζ_φ. (If there is no singular point at all on the ray, it may be left entirely out of account.) For a particular integer $p > 0$ we now describe the circle

$$\left|\frac{z}{\zeta_\varphi} + 2^p - 1\right| < 2^p,$$

which corresponds to the one occurring in the case of the geometric series, and which we shall denote by K_φ. The points common to all the circles K_φ

---

[79] The formula of this transformation suggests that, for the order p, the restriction to integers $\geqq 0$ might be removed. Here, however, we shall not enter into the question of these non-integral orders. Cf. p. 467, footnote 18.

[80] This series is then, moreover, *absolutely* convergent.

[81] As regards the proof, see p. 507, footnote 77.

will, in the simplest cases (i. e. when there are only a small number of singular points), make up a curvilinear polygon whose boundary consists of arcs of the different circles, and in every case they will form a definite set of points which we denote by $\mathfrak{G}_p$. We then have the

Theorem. *For each fixed p, $\Sigma c_n z^n$ is summable E_p at every interior point of $\mathfrak{G}_p$ and the E_p-transformation of $\Sigma c_n z^n$ is indeed absolutely convergent at that point. The numerical values thus associated with every interior point of $\mathfrak{G}_p$ form the analytical extension of the element $\Sigma c_n z^n$ into the interior of $\mathfrak{G}_p$. Outside $\mathfrak{G}_p$, the E_p-transformation of $\Sigma c_n z^n$ is divergent.* **289.**

After noting these examples, we now return to the general question, within what bounds the range of action of the E-process lies; in this as in further investigations, we shall restrict ourselves to the first order, i. e. to the E_1-process. The cases of E-summation of higher orders are, however, quite analogous.

Since E_1-summability of a series Σa_n means, by definition, the convergence of its E_1-transformation

$$\Sigma \frac{1}{2^{n+1}}\left[\binom{n}{0} a_0 + \cdots + \binom{n}{n} a_n\right],$$

the general term of the latter must necessarily tend to 0:

$$\frac{1}{2^n}\left[\binom{n}{0} a_0 + \binom{n}{1} a_1 + \cdots + \binom{n}{n} a_n\right] \to 0,$$

which we may now write, for short,

$$E_1\text{-}\lim a_n = 0.$$

In this form, we again have an exact analogue to **82**, 1 and **272** or **284**. *Kronecker*'s theorem **82**, 3 also has its analogue here. For if (s_n) is a sequence which is limitable E_1 with the value s, its E_1-transformations $E_1(s_n) \equiv s_n' \to s$. The arithmetic means

$$\frac{s_0' + s_1' + \cdots + s_n'}{n+1}$$

of the latter therefore also tend to s; we may denote them for short by $C_1 E_1(s_n)$, since they are obtained by applying in succession first the E_1-transformation and then the C_1-transformation. Now it is easy to show by direct calculation — we prefer to leave this to the reader —, that we obtain exactly the same result if we apply the C_1-transformation *first*, and then the E_1-transformation, i. e. if we form the sequence $E_1 C_1(s_n) = E_1\left(\frac{s_0 + s_1 + \cdots + s_n}{n+1}\right)$: *We have $C_1 E_1(s_n) \equiv E_1 C_1(s_n)$; the two transformations are completely identical*[82]. Thus we also

[82] By calculation, the identity to be proved is at once reduced to the relation

$$\binom{k+1}{n+1} + \binom{k+1}{n+2} + \cdots + \binom{k+1}{k+1} = \binom{k}{n} + 2\binom{k-1}{n} + \cdots + 2^{k-n}\binom{n}{n},$$

for $0 \leq n \leq k$, which is easily seen to be true. — On account of the property in question, the E_1- and C_1-transformations are said to be ***commutable.*** The corresponding property holds good for E_p- and C_q-transformations of any order; *in every case*, $E_p C_q(s_n) \equiv C_q E_p(s_n)$. Cf. p. 482, footnote 45, and p. 483.

have

$$E_1\left(\frac{s_0+s_1+\cdots+s_n}{n+1}\right) \dashrightarrow s,$$

and subtracting this from

$$E_1(s_n) \to s,$$

we obtain, exactly as on p. 485, the relation stated, namely

$$E_1\text{-}\lim \frac{a_1+2a_2+\cdots+n a_n}{n+1} = 0,$$

as a necessary condition for the E_1-summability of the series Σa_n. To determine further what the condition E_1-lim $a_n = 0$ implies as regards the order of magnitude of the terms a_n, we deduce from

$$a_n' = \frac{1}{2^{n+1}}\left[\binom{n}{0} a_0 + \cdots + \binom{n}{n} a_n\right]$$

the expression for the a_n's in terms of the a_n''s:

$$a_n = (-1)^n \cdot 2\left[\binom{n}{0} a_0' - \binom{n}{1} 2 a_1' + - \cdots + (-1)^n \binom{n}{n} 2^n a_n'\right],$$

whence, as $a_n' \to 0$, it at once follows, by **43**, 5, that

$$\frac{a_n}{3^n} \to 0 \quad \text{or} \quad a_n = o(3^n).$$

If we carry out the corresponding calculation for s_n and s_n', we similarly find that $s_n = o(3^n)$. Summing up, we therefore have

290. **Theorem 1.** *The four conditions*

$$E_1\text{-}\lim a_n = 0, \qquad E_1\text{-}\lim \frac{a_1+2a_2+\cdots+n a_n}{n+1} = 0,$$

$$a_n = o(3^n) \quad \textit{and} \quad s_n = o(3^n)$$

are necessary in order that the series Σa_n, with partial sums s_n, may be summable E_1.

A comparison of this theorem with the theorems **271** and **283** and the examples for **265**, 5 shows that the range of the E_1-process is considerably more extensive than those of the C- and A-processes; the E_1-process is a good deal more powerful than these. The question, however, which in the case of the C- and A-processes led to the theorems **274** and **287**, here reveals what may be described as *a loss of sensitiveness* in the E_1-summation process, as compared with the C- and A-processes. We in fact have

291. **Theorem 2.** *If the series Σa_n, with partial sums s_n, is summable E_1 to the value s, so that the numbers*

(A) $$s_n' = \frac{1}{2^n}\left[\binom{n}{0} s_0 + \binom{n}{1} s_1 + \cdots + \binom{n}{n} s_n\right] \to s,$$

and if, besides this, we have

(B) $$a_n = o\left(\frac{1}{\sqrt{n}}\right),$$

then the series Σa_n is convergent with the sum s. (o-$E_1 \to K$-*theorem*).

Proof. We form the difference

$$s'_{4n} - s_{2n} = \frac{1}{2^{4n}} \sum_{\nu=0}^{4n} \binom{4n}{\nu} (s_\nu - s_{2n}),$$

and we split up the expression on the right hand side into three parts: $T_1 + T_2 + T_3$. T_1 is to denote the part from $\nu = 0$ to $\nu = n$, T_3 the part from $\nu = 3n$ to $\nu = 4n$, and T_2 the remaining part in the middle. In virtue of (B) there certainly exists — roughly estimated — a constant K_1 such that $|s_n| \leqq K_1 \sqrt{n}$. Hence there also exists a constant K such that

$$|s_\nu - s_{2n}| \leqq K \sqrt{n}$$

for every n, provided $0 \leqq \nu \leqq 4n$. Hence $|T_1|$ and $|T_3|$ are both

$$< \frac{1}{2^{4n}} \cdot K \cdot \sqrt{n} \sum_{\nu=0}^{n} \binom{4n}{\nu} < \frac{K\sqrt{n}}{2^{4n}} (n+1) \binom{4n}{n}.$$

Now for every integer $k > 1$, we have [83]

$$e \left(\frac{k}{e}\right)^k < k! < e\,k \left(\frac{k}{e}\right)^k,$$

and accordingly [84]

$$\frac{1}{2^{4n}} \binom{4n}{n} < \frac{1}{2^{4n}} \cdot \frac{4n\, 4^{4n}}{e\, 3^{3n}} < 2n \left(\frac{16}{27}\right)^n.$$

Therefore T_1 and T_3 both tend to 0 as n increases.

In T_2, i. e. for $n < \nu < 3n$, we have, by (B),

$$|s_\nu - s_{2n}| < \frac{\varepsilon_n}{\sqrt{n}} |2n - \nu|,$$

if ε_n denotes the largest of the values $|a_{n+1}|\sqrt{n+1}, |a_{n+2}|\sqrt{n+2}, \ldots$; ε_n must tend to 0 as n increases. Therefore

$$|T_2| \leqq \frac{\varepsilon_n}{\sqrt{n}} \frac{1}{2^{4n}} \sum_{\nu=n+1}^{3n-1} |2n - \nu| \binom{4n}{\nu} < \frac{2\varepsilon_n}{2^{4n}\sqrt{n}} \sum_{\nu=0}^{2n} (2n - \nu) \binom{4n}{\nu}.$$

This last sum is however easily seen to have the value $n \binom{4n}{2n}$; it suffices to separate $2n - \nu$ into $2n$ and $-\nu$. Thus

$$|T_2| \leqq \frac{2\varepsilon_n \sqrt{n}}{2^{4n}} \binom{4n}{2n}.$$

---

[83] This somewhat rough estimation for $k!$, which, however, is often useful, is most simply obtained by multiplying together all the inequalities

$$\left(1 + \frac{1}{\nu}\right)^\nu < e < \left(1 + \frac{1}{\nu}\right)^{\nu+1}$$

(see **46 a**) for $\nu = 1, 2, \ldots, k-1$.

[84] Substitute $\binom{4n}{n} = \frac{(4n)!}{n!\,(3n)!}$ and use in the numerator the upper estimate for $k!$, in the denominator the lower estimate.

Thus, as $\varepsilon_n \to 0$, we have, by **219**, 3,

$$T_2 \to 0.$$

Summing up, we therefore have

$$T_1 + T_2 + T_3 = s'_{4n} - s_{2n} \to 0.$$

Now by hypothesis $s'_{4n} \to s$, hence $s_{2n} \to s$ also; and further, since $a_\nu \to 0$ by (B), it finally follows that

$$s_n \to s,$$

q. e. d.[85].

In conclusion, we shall also consider the question of the E-summability of the product of two series which are summable E_1, as well as that of the relation of the range of action of the E-process to that of the C-process.

As regards the multiplication problem, we have two theorems, which are the exact analogues of *Mertens*' theorem **188** and *Abel*'s theorem **189**. We confine ourselves to the development of the former and we therefore proceed to prove

292. Theorem 3. *Let the two series Σa_n and Σb_n be assumed to be summable E_1, i. e. let their E_1-transformations, which we shall denote by $\Sigma a_n'$ and $\Sigma b_n'$, be convergent. If* one *at least of the two latter series converges absolutely, then Cauchy's product*

$$\Sigma c_n = \Sigma (a_0 b_n + a_1 b_{n-1} + \cdots + a_n b_0)$$

is also summable E_1, and between A, B, and C, the E_1-sums of the three series, we have the relation $A \cdot B = C$.

Proof. By **265**, 5, for $x = \frac{y}{1-y}$ and for all sufficiently small values of x (v. theorem 1), we have

$$f_1(x) = \Sigma a_n x^{n+1} = \Sigma a_n' (2y)^{n+1},$$
$$f_2(x) = \Sigma b_n x^{n+1} = \Sigma b_n' (2y)^{n+1},$$
$$f_3(x) = \Sigma c_n x^{n+1} = \Sigma c_n' (2y)^{n+1}.$$

On the other hand

$$f_1(x) \cdot f_2(x) = x \cdot f_3(x).$$

Thus we have the identity

$$(2y) \cdot \Sigma (a_0' b_n' + a_1' b'_{n-1} + \cdots + a_n' b_0')(2y)^{n+1} = \frac{y}{1-y} \Sigma c_n' (2y)^{n+1},$$

---

[85] The theorems **274** and **287** suggest that an O-$E \to K$-theorem may also hold here, i. e. one which enables us to infer the convergence of Σa_n from its E_1-summability, provided that $a_n = O\left(\frac{1}{\sqrt{n}}\right)$. This is actually the case, but the proof is so much more difficult than the above that we must omit it here. (Cf. the second of the papers referred to on p. 507, footnote 77.)

whence, besides $c_0' = 2\,a_0'\,b_0'$, we obtain the general formula for $n \geqq 1$:
$c_n' = 2\,(a_0'\,b_n' + a_1'\,b_{n-1}' + \cdots + a_n'\,b_0') - (a_0'\,b_{n-1}' + \cdots + a_{n-1}'\,b_0')$.
Since by hypothesis *one* at least of the two series $\Sigma\,a_n'$ and $\Sigma\,b_n'$ is absolutely convergent, *Cauchy*'s product, $\Sigma(a_0'\,b_n' + \cdots + a_n'\,b_0')$, of these two series is convergent and $= A \cdot B$, by **188**. From the last-obtained expression for c_n', the convergence of $\Sigma\,c_n'$ follows at once, and for its sum C we obtain

$$C = 2\,AB - AB = AB,$$

q. e. d.[86]

Finally, we shall examine the question of the relation between the C- and E-processes. It is very easy to show, in the first instance, that the processes fulfil the compatibility condition **263**, III; i. e. that we have

Theorem 4. *If a series is summable C_1 and also summable E_1, the two processes give it the same value.* **293.**

Proof. If (c_n') is the C_1-transformation and (s_n') the E_1-transformation of (s_n), both these sequences are convergent, by hypothesis: say $c_n' \to c'$, $s_n' \to s'$. Since both processes satisfy the permanence condition, the C_1-transformation of (s_n') also converges to s':

$$\frac{s_0' + s_1' + \cdots + s_n'}{n+1} \to s'$$

and the E_1-transformation of (c_n') converges to c':

$$\frac{1}{2^n}\left[\binom{n}{0} c_0' + \cdots + \binom{n}{n} c_n'\right] \to c'.$$

With the abbreviated notation, these two relations are

$$C_1\,E_1\,(s_n) \to s' \quad \text{and} \quad E_1\,C_1\,(s_n) \to c'.$$

But, as was pointed out on p. 509, these two sequences are identical, so that s' must be equal to c', q. e. d.

We have already seen (p. 508), from the example of the geometric series $\Sigma\,z^n$, that the E_1-process is considerably more powerful than the C_1-process. In fact, we cannot sum the geometric series by the latter anywhere outside the unit circle, while the E_1-process enables us to sum it at every point of the circle $|z+1| < 2$. But this must not be interpreted to mean that the range of action of the E_1-process completely includes that of the C_1-process, much less those of *all* C_k-processes. *On the contrary, it is easy to give an instance of a se-*

[86] The form of this proof suggests that in certain cases it will be convenient to introduce the concept of *absolute* summability: A series $\Sigma\,a_n$ will be said to be *absolutely* summable E_1 if $\Sigma\,a_n'$, its E_1-transformation, converges *absolutely.*

quence (s_n) *which is limitable* C_1, *but not limitable* E_1. The sequence

$$(s_n) \equiv 0,\ 1,\ 0,\ 0,\ 2,\ 0,\ 0,\ 0,\ 0,\ 3,\ 0,\ \ldots$$

is of this type, where $s_{\nu^2} = \nu$ and every other $s_n = 0$ (i. e. for every index n which is not a perfect square).

This sequence is limitable C_1 with the value $\frac{1}{2}$. For the largest values of the arithmetic means $\frac{s_0 + s_1 + \cdots + s_n}{n+1}$ are obviously attained for $n = \nu^2$ and the least for $n = \nu^2 - 1$. The latter $= \frac{\nu(\nu-1)}{2\nu^2}$, the former $= \frac{(\nu+1)\nu}{2(\nu^2+1)}$, both of which $\to \frac{1}{2}$; i. e. $C_1(s_n) \to \frac{1}{2}$.

If the same sequence were also limitable E_1 we should therefore require to have $E_1(s_n) \to \frac{1}{2}$: but, for $n = \nu^2$, the $(2n)^{\text{th}}$ term of the E_1-transformation is

$$\frac{1}{2^{2n}}\left[\binom{2n}{0} s_0 + \cdots + \binom{2n}{n} s_n + \cdots + \binom{2n}{2n} s_{2n}\right] \geqq \frac{1}{2^{2n}} \binom{2n}{n} \sqrt{n}.$$

The expression on the right hand side tends to $\frac{1}{\sqrt{\pi}}$ by **219**, 3, so that for all sufficiently large n's the terms remain $> \frac{5}{9} > \frac{1}{2}$ and $E_1(s_n)$ cannot $\to \frac{1}{2}$. The sequence (s_n) is therefore *not* limitable E_1. We may accordingly state:

294. **Theorem 5.** *Of the two ranges, that of the* C_1*-process and that of the* E_1*-process, neither contains the other entirely. There are series which can be summed by the* C_1*-process, but not by the* E_1*-process, and conversely*[87].

This circumstance raises the further question: which series, summable C_1, can be summed by the E_1-process? Little is as yet known on this subject, and we shall content ourselves with mentioning the following theorem:

295. **Theorem 6.** *If* Σa_n *is summable* C_1, *with*

$$\frac{s_0 + s_1 + \cdots + s_n}{n+1} = s + o\left(\frac{1}{\sqrt{n}}\right),$$

then Σa_n *is also summable* E_1. ($o \cdot C_1 \to E_1$*-theorem*[88].)

---

[87] The statement remains the same when higher orders of both processes are considered.

[88] The corresponding O-theorem does *not* hold, as has already been shown by the example on theorem 5, where the arithmetic mean is actually $\frac{1}{2} + O\left(\frac{1}{\sqrt{n}}\right)$.

Proof. Writing $s_n - s = \sigma_n$, the sequence (σ_n) is limitable C_1 with the value 0. Put

$$\frac{\sigma_0 + \sigma_1 + \cdots + \sigma_n}{n+1} = \sigma_n{}';$$

by the hypotheses, we then have not only $\sigma_n{}' \to 0$, but also $\sqrt{n}\,\sigma_n{}' \to 0$. Now we have the following general inequality, due to *Abel*: If $\sigma_0, \sigma_1, \ldots, \sigma_n$ are *any* numbers, $\sigma_\nu{}' = \frac{\sigma_0 + \sigma_1 + \cdots + \sigma_\nu}{\nu+1}$, $(\nu = 0, 1, \ldots, n)$, the corresponding arithmetic means; if, further, τ is a number greater than all the $(n+1)$ quantities $|\sigma_\nu{}'|$, for $\nu = 0, 1, \ldots, n$, and τ_k a number greater than all the quantities $|\sigma_\nu{}'|$, for $k \leqq \nu \leqq n$; then, given any set of $(n+1)$ positive numbers $\alpha_0, \alpha_1, \ldots, \alpha_n$, which increase monotonely to the term α_m and decrease monotonely from that term on, we have the following inequality [89], if $0 < p < m < n$:

$$\left| \frac{\alpha_0 \sigma_0 + \alpha_1 \sigma_1 + \cdots + \alpha_n \sigma_n}{\alpha_0 + \alpha_1 + \cdots + \alpha_n} \right| \leqq \tau_m + \frac{\tau p \alpha_p + (\tau_p + \tau_m) m \alpha_n}{\alpha_0 + \alpha_1 + \cdots + \alpha_n}.$$

Applying this, for a *fixed* $n > 8$, to the numbers σ_ν, $\sigma_\nu{}'$ introduced above, and taking $\alpha_\nu = \binom{n}{\nu}$, $\nu = 0, 1, 2, \ldots, n$, we can choose for p the greatest integer $\leqq \frac{n}{4}$, i. e. $p = \left[\frac{n}{4}\right]$, and we may similarly take $m = \left[\frac{n}{2}\right]$. We then obtain

$$\frac{1}{2^n}\left[\binom{n}{0}\sigma_0 + \binom{n}{1}\sigma_1 + \cdots + \binom{n}{n}\sigma_n\right] \leqq \tau_m + \frac{\tau}{2^n}\, p \binom{n}{p} + \frac{2\tau_p}{2^n}\, m \binom{n}{m}.$$

This inequality will hold *a fortiori* if we assume τ to be greater than *all* the quantities $|\sigma_\nu{}'|$ and τ_k greater than *all* the quantities $|\sigma_\nu{}'$ with $\nu \geqq k$. Now, by **219**, 3, the greatest term $\binom{n}{m}$ of the values $\binom{n}{\nu}$ satisfies the limiting relation $\frac{\sqrt{n}}{2^n}\binom{n}{m} \to \frac{1}{\sqrt{\pi}}$, so that $\frac{\sqrt{n}}{2^n}\binom{n}{m}$ is certainly < 1 from some stage on. Since from some stage on we have also $\sqrt{n} < 3\sqrt{p}$, it follows that, for all sufficiently large n,

$$\frac{2\tau_p}{2^n}\, m \binom{n}{m} < \tau_p \sqrt{n} < 3\,\tau_p \sqrt{p}.$$

---

[89] In fact, $\sigma_\nu = (\nu + 1)\,\sigma'_\nu - \nu\,\sigma'_{\nu-1}$ and therefore, taking $\alpha_{n+1} = 0$,

$$\sum_{\nu=0}^{n} \alpha_\nu \sigma_\nu = \sum_{\nu=0}^{n} (\nu + 1)\,\sigma'_\nu\,(\alpha_\nu - \alpha_{\nu+1}) = \sum_{\nu=0}^{p-1} + \sum_{\nu=p}^{m-1} + \sum_{\nu=m}^{n}.$$

Noting that $(\alpha_\nu - \alpha_{\nu+1})$ is negative in the first and second sums and positive in the third, it follows that

$$\left|\sum_{\nu=0}^{n} \alpha_\nu \sigma_\nu\right| \leqq \tau\, p\, \alpha_p + \tau_p\, m\, \alpha_m + \tau_m\,(m\,\alpha_m + \alpha_m + \alpha_{m+1} + \ldots + \alpha_n),$$

whence the above relation follows directly.

Since $\tau_p \sqrt{p}$ was to tend to 0, since, further, p and m tend to $+\infty$ as n does, and at the same time $\frac{p}{2^n}\binom{n}{p} \to 0$, it follows that when $n \to +\infty$

$$\frac{1}{2^n}\left[\binom{n}{0}\sigma_0 + \binom{n}{1}\sigma_1 + \cdots + \binom{n}{n}\sigma_n\right] \to 0$$

i. e.

$$E_1(s_n) \to s,$$

q. e. d.

Exercises on Chapter XIII.

200. With the help of example 119 (p. 270), prove the fact mentioned on p. 461, namely that

$$\lim_{x\to+0}\sum_{n=1}^{\infty}\frac{(-1)^{n-1}}{n^x} = \frac{1}{2}.$$

What summation process related to the A-process might be deduced from this? Define it and indicate some of its properties.

201. Is the condition

$$C_k\text{-}\lim\left(\frac{a_1 + 2a_2 + \cdots + n a_n}{n+1}\right) = 0,$$

given in **273**, substantially equivalent

a) to $C_{k+1}\text{-}\lim(n a_n) = 0$, b) to $C_k\text{-}\lim\left(\frac{a_1 + 2a_2 + \cdots + n a_n}{n}\right) = 0$?

202. With reference to the relations (*) in the proof of **284**, show that in general $A\cdot\lim s_n = s$ always involves $A\,C_k\text{-}\lim s_n = s$, i. e.

$$(1-x)\,\Sigma s_n x^n \to s \quad \text{always involves} \quad (1-x)\sum_{n=0}^{\infty}\frac{S_n^{(k)}}{\binom{n+k}{k}}x^n \to s,$$

(for $x \to 1-0$).

203. Show similarly that $B\text{-}\lim s_n = s$ always involves $B\,C_k\text{-}\lim s_n = 0$, i. e.

$$e^{-x}\sum_{n=0}^{\infty}\frac{S_n^{(k)}}{\binom{n+k}{k}}\cdot\frac{x^n}{n!} \to s$$

for $x \to +\infty$.

204. Are the conclusions mentioned in **202** and **203** reversible, e. g. does the relation $(1-x)\sum\frac{s_0 + s_1 + \cdots + s_n}{n+1}x^n \to s$ imply, conversely, that $(1-x)\,\Sigma s_n x^n \to s$?

205. The series $\sum_{n=0}^{\infty}\binom{n+k-1}{k-1}z^n$ is summable C_k for $|z| = 1$, $z \neq +1$. Put $z = \cos\varphi + i\sin\varphi$, separate the real and imaginary parts and write down the trigonometrical series summed in this way, as well as their respective values. E. g.,

$$1 + 2\cos x + 3\cos 2x + 4\cos 3x + \cdots = \frac{1}{2} - \frac{1}{4\sin^2\frac{x}{2}}$$

$$\cos x + 2\cos 2x + 3\cos 3x + \cdots = -\frac{1}{4\sin^2\frac{x}{2}}, \quad \text{etc.}$$

206. If (a_n) is a positive monotone null sequence, and if we put

$$a_0 + a_1 + a_2 + \cdots + a_n = b_n,$$

the series

$$b_0 - b_1 + b_2 - b_3 + - \cdots$$

is summable C_1 to the sum $s = \frac{1}{2} \Sigma (-1)^n a_n$.

207. If we write $1 + \frac{1}{2} + \frac{1}{3} + \cdots + \frac{1}{n} = h_n$, it follows from the preceding exercise that the C_1-sum

$$h_1 - h_2 + h_3 - h_4 + - \cdots = \frac{1}{2} \log 2,$$

and similarly that the C_1-sum

$$\log 2 - \log 3 + \log 4 - + \cdots = \frac{1}{2} \log \frac{\pi}{2}.$$

208. If Σa_n is convergent or summable C_1 with the sum s, the following series is always *convergent* with the sum s:

$$a_0 + \sum_{n=1}^{\infty} \frac{a_1 + 2a_2 + \cdots + n a_n}{n(n+1)} = s.$$

209. If Σa_n is known to be summable C_1 and $\Sigma n |a_n|^2$ is convergent, then Σa_n is itself convergent.

210. Prove the following extensions of *Frobenius'* theorem (p. 490): If $\sum_{n=1}^{\infty} a_n$ is summable C_1 to the sum s, then for $z \to +1$ (within the angle)

$$\sum_{n=1}^{\infty} a_n z^{n^2} \to s \quad \text{and} \quad \sum_{n=0}^{\infty} a_n z^{n^3} \to s$$

and in general, for every fixed integer $p \geqq 1$,

$$\sum_{n=0}^{\infty} a_n z^{n^p} \to s.$$

But $\Sigma a_n z^{n!}$ does not necessarily tend to s, as may be shown by the example

$$\sum_{n=0}^{\infty} (-1)^n x^{n!}$$

for real $x \to 1 - 0$. (Hint: The maximum of $t - t^n$, and the value of t for which it is attained, both $\to +1$ from the left as n increases.)

211. For every *real* series Σa_n, for which $\Sigma a_n x^n$ converges in $0 \leqq x < 1$, we have

$$\varliminf_{n \to +\infty} \frac{s_0 + s_1 + \ldots + s_n}{n+1} \leqq \varlimsup_{x \to 1-0} (1 - x) \sum_{n=0}^{\infty} s_n x^n \leqq \varlimsup_{n \to +\infty} \frac{s_0 + s_1 + \ldots s_n}{n+1}$$

(Cf. Theorem **161**.)

212. With reference to *Fejér*'s theorem, show that the arithmetic means $\sigma_n(x)$ considered there do not exhibit *Gibbs'* phenomenon. (Cf. p. 496, footnote 66.)

213. The product of two series which are summable E_1 is *invariably* summable $E_1 C_1$.

214. If Σa_n is summable E_1, then $\Sigma a_n x^n$ is also summable E_1 for $0 \leqq x < 1$ and

$$\lim_{x \to 1-0} E_1\text{-}\Sigma a_n x^n = E_1\text{-}\Sigma a_n.$$

215. Give a general proof of the commutability of the E_p- and C_q-processes.

216. Deduce the o-$E_p \to K$-theorem (what is its statement?) by induction from the o-$E_1 \to K$-theorem.

Chapter XIV.

Euler's summation formula and asymptotic expansions.

§ 64. *Euler's* summation formula.

A. The summation formula.

The range of action of all the summation processes with which we became acquainted in the last chapter was limited. It is only when the terms a_n of Σa_n, the divergent series under consideration, do not increase too rapidly as n increases that we can sum the series. Thus in the case of the B-process, it is necessary that $\Sigma \frac{a_n}{n!} x^n$ should be convergent everywhere, i. e. that $\sqrt[n]{\frac{|a_n|}{n!}}$ or $\frac{1}{n}\sqrt[n]{|a_n|}$ should tend to zero. Hence the B-process cannot be used e. g. for the series

$$\sum_{n=0}^{\infty}(-1)^n n! = 1 - 1! + 2! - 3! + 4! - + \ldots + (-1)^n n! + \ldots .$$

Series like this one, and even more rapidly divergent series, occurred, however, in early investigations of the most varied kind. In order to deal conclusively with them by the methods used hitherto, we should have to introduce still more powerful processes, such as the B_r-process. However, no essential results have been obtained in this way.

At a fairly early stage in the development of the subject other methods were indicated, which in certain cases lead more conveniently to results useful both in theory and in practice. In the case of the numerical evaluation of the sum of an alternating series $\Sigma(-1)^n a_n$, in which the a_n's constitute a positive monotone null sequence, we observed (see pp. 250 and 251) that the remainder r_n always has the same sign as the first term neglected, and, moreover, that it is less than this term in absolute value. Thus in the calculation of the partial sums we need only continue until the terms have decreased down to the required degree of accuracy. A somewhat similar state of affairs exists in the case of the series

$$e^{-x} = 1 - x + \frac{x^2}{2} - + \ldots + (-1)^n \frac{x^n}{n!} + \ldots, \quad x > 0,$$

since the terms $\frac{x^n}{n!}$ likewise decrease monotonely when $n > x$. We can therefore write

$$e^{-x} = 1 - x + \frac{x^2}{2!} - + \ldots + (-1)^n \frac{x^n}{n!} + (-1)^{n+1} \vartheta \frac{x^{n+1}}{(n+1)!},$$

for every $n > x$, where ϑ stands for a value between 0 and 1, depending on x and n, but is otherwise undetermined. It is impossible in practice, however, actually to calculate e^{-x} from this formula when x is large, for e. g. when $x = 1000$, the thousandth term is equal to $\frac{10^{3000}}{1000!}$. As $1000!$ is a number with 2568 digits (for the calculation see below, p. 529), the term under consideration is greater than 10^{431}, so that the evaluation of the sum of the series cannot be carried out in practice. From the theoretical point of view, on the other hand, the series fulfils all requirements, since its terms, which (for large values of x) at first increase very rapidly, nevertheless end by decreasing to zero, and that for *every value of* x. Hence *any degree of accuracy whatever* can be obtained in theory.

The circumstances are exactly the reverse, if we know that the value of a function $f(x)$ is represented by the formula

$$f(x) = 1 - \frac{1!}{x} + \frac{2!}{x^2} - + \ldots + (-1)^n \frac{n!}{x^n} + (-1)^{n+1} \vartheta \frac{(n+1)!}{x^{n+1}},$$

$$0 < \vartheta < 1,$$

for every n. The series $\Sigma(-1)^n \frac{n!}{x^n}$, whose partial sums appear in this formula, *diverges for every* x: but *in contrast to nearly all the divergent series met with in the last chapter*, the terms of the series (for large values of x) at first decrease very rapidly — the series at first behaves like a convergent one — and it is only later on that they increase rapidly and without limit. Hence we can calculate e. g. $f(1000)$ to about ten decimal places with great ease; we have only to find an n for which $\frac{(n+1)!}{1000^{n+1}} < \frac{1}{2} 10^{-10}$. As this is true even for $n = 3$, the value sought is given by

$$1 - \frac{1}{10^3} + \frac{2}{10^6} - \frac{6}{10^9}$$

to the desired degree of accuracy. Thus it happens here that an expansion in powers, which takes the form of an infinite series which is divergent everywhere and very rapidly so, nevertheless yields useful numerical results, *because it appears along with its remainder*. We are not in a position, however, — not even in theory — to obtain *any degree of accuracy what-*

ever in the evaluation of $f(x)$, since $f(x)$ is given by its expansion only with an error of the order of one of the terms of the series. The degree of accuracy therefore cannot be lowered below the value of the *least* term of the series. (A least term certainly exists, seeing that the terms finally increase.) As the example shows, however, in suitable circumstances all practical requirements may be satisfied.

Series of the type described were produced for the first time by *Euler*'s summation formula [1], which we shall now consider more closely.

If the terms $a_0, a_1, \ldots, a_n, \ldots$ of a series [2] are the values of a function $f(x)$ for $x = 0, 1, \ldots, n, \ldots$, we have already proved by the integral test (**176**) that in certain circumstances there is a relation between the partial sums $s_n = a_0 + a_1 + \ldots + a_n$ and the integrals

$$J_n = \int_0^n f(x)\,dx.$$

Euler's summation formula throws further light on this relation. If $f(x)$ possesses a continuous differential coefficient in $0 \leqq x \leqq n$, then, for $\nu = 0, 1, \ldots, n-1$,

$$\int_\nu^{\nu+1} (x - \nu - \tfrac{1}{2}) f'(x)\,dx = \left[(x - \nu - \tfrac{1}{2}) f(x)\right]_\nu^{\nu+1} - \int_\nu^{\nu+1} f(x)\,dx.$$

Now, for each of the values ν, we can put $\nu = [x]$ in the integrand on the left, at least for $\nu \leqq x < \nu + 1$. Since, however, by § 19, theorem 17, the one value $x = \nu + 1$ does not matter, we get

$$\tfrac{1}{2}(f_\nu + f_{\nu+1}) = \int_\nu^{\nu+1} f(x)\,dx + \int_\nu^{\nu+1} (x - [x] - \tfrac{1}{2}) f'(x)\,dx.$$

(To simplify the writing, we denote by f_ν and $f_\nu^{(k)}$ respectively the values of $f(x)$ and of its derivative $f^{(k)}(x)$ for integral values $x = \nu$.) Adding these relations for the relevant values of ν, and adjoining the term $\frac{1}{2}(f_0 + f_n)$, we finally obtain the formula

---

[1] With regard to the summation formula cf. footnote 3, p. 521. — The phenomenon described above was first noticed by *Euler* (Commentarii Acad. sc. Imp. Petropolitanae, Vol. 11 (year 1739), p. 116, 1750); *A. M. Legendre* gave the name of *semi-convergent series* to series which exhibit this phenomenon. This name has survived to the present time, especially in astronomical literature, but nowadays it is being superseded by the term "*asymptotic series*", which was introduced by *H. Poincaré* on account of another property of such series.

[2] In the subsequent remarks all the quantities are to be real.

$$f_0+f_1+\cdots+f_n=\int_0^n f(x)\,dx+\tfrac{1}{2}(f_0+f_n)+\int_0^n \left(x-[x]-\tfrac{1}{2}\right)f'(x)\,dx. \quad \mathbf{296.}$$

This in fact is *Euler*'s summation formula in its simplest form [3]. It gives a closed expression for the difference between the sum $f_0+f_1+\ldots+f_n$ and the corresponding integral $\int_0^n f(x)\,dx$.

We shall denote the function which appears in the last integrand by $P_1(x)$:

$$P_1(x)=x-[x]-\frac{1}{2}.$$

This is essentially the same function as the one which we met with in one of the first examples of Fourier expansions (see pp. 351, 375).

It is periodic, with period 1, and for every non-integral value of x we have

$$P_1(x)=-\sum_{n=1}^{\infty}\frac{\sin 2n\pi x}{n\pi}.$$

A simple example to begin with will illustrate the importance of this formula. If $f(x)=\frac{1}{1+x}$, we obtain, by replacing n by $n-1$,

$$1+\frac{1}{2}+\cdots+\frac{1}{n}=\log n+\frac{1}{2}+\frac{1}{2n}-\int_1^n\frac{P_1(x)}{x^2}\,dx.$$

We may substitute the latter integral for $\int_0^{n-1}\frac{P_1(x)}{(1+x)^2}\,dx$, since $P_1(x+1)=P_1(x)$. As $P_1(x)$ is bounded in $x\geqq 1$, the integral obviously converges when $n\to\infty$, and we find that

$$\lim_{n\to\infty}\left(1+\frac{1}{2}+\cdots+\frac{1}{n}-\log n\right)=C=\frac{1}{2}-\int_1^{\infty}\frac{P_1(x)}{x^2}\,dx.$$

---

[3] The formula, in its general form **298**, originated with *Euler*, who mentioned it in passing in the Commentarii Acad. Petrop., Vol. 6 (years 1732–3, published 1738) and illustrated it by a few examples. In Vol. 8 (year 1736, published 1741) he gives a proof of the formula. *C. Maclaurin* uses the formula in several places in A Treatise of Fluxions (Edinburgh 1742), and seems to have discovered it independently. The formula became well-known, especially through *Euler*'s Institutiones calculi differentialis, in the fifth chapter of which it is proved and illustrated by examples. For long it was known as *Maclaurin*'s formula, or the *Euler-Maclaurin* formula; it is only recently that *Euler*'s undoubted priority has been established.

The remainder — which is most essential — was first added by *S. D. Poisson* (v. Mémoires Acad. scienc. Inst. France, Vol. 6, year 1823, published 1827). The particularly simple proof given in the text is due to *W. Wirtinger* (Acta mathematica, Vol. 26, p. 255, 1902).

An up-to-date, detailed, and expanded treatment is to be found in *N.E. Nörlund*'s Differenzenrechnung, Berlin 1924, especially in chapters II—V.

We already know that this limit exists, from **128**, 2. Now we have a new proof of this fact, and in addition we have an expression in the form of an integral for *Euler*'s constant C, by means of which we can evaluate the constant numerically.

From the formula **296**, i. e.

$$(*)\quad f_0 + f_1 + \dots + f_n = \int_0^n f(x)\,dx + \tfrac{1}{2}(f_0 + f_n) + \int_0^n P_1(x) f'(x)\,dx,$$

integration by parts leads to more advantageous representations. In order to be in a position to carry it out, we must first assume that $f(x)$ has continuous derivatives of all the orders which occur in what follows; then we have to select an indefinite integral of $P_1(x)$, and an integral of the latter, and so on. By suitable choice of the constants of integration the further calculations are greatly simplified. We shall follow *Wirtinger*[4] and set

$$P_2(x) = +\sum_{n=1}^{\infty} \frac{2\cos 2n\pi x}{(2n\pi)^2}.$$

Then $P_2'(x) = P_1(x)$, for every non-integral value of x, and $P_2(0) = \frac{1}{2\pi^2}\sum_{n=1}^{\infty}\frac{1}{n^2} = \frac{1}{12}$. Moreover, $P_2(x)$ is continuous throughout, and has the period 1. We now proceed to set

$$P_3(x) = +\sum_{n=1}^{\infty} \frac{2\sin 2n\pi x}{(2n\pi)^3},$$

whence we have $P_3'(x) = P_2(x)$ for *every* value of x, $P_3(0) = 0$, and in general

297. (a)
$$\begin{cases} P_{2\lambda}(x) = (-1)^{\lambda-1} \displaystyle\sum_{n=1}^{\infty} \frac{2\cos 2n\pi x}{(2n\pi)^{2\lambda}}, \\[2ex] P_{2\lambda+1}(x) = (-1)^{\lambda-1} \displaystyle\sum_{n=1}^{\infty} \frac{2\sin 2n\pi x}{(2n\pi)^{2\lambda+1}}. \end{cases}$$

Then, for $\lambda = 1, 2, \dots$, all these functions are throughout continuous and continuously differentiable, and have the period 1; and we have

(b)
$$\begin{cases} P'_{k+1}(x) = P_k(x), \\[1ex] P_{2\lambda}(0) = (-1)^{\lambda-1} \displaystyle\sum_{n=1}^{\infty} \frac{2}{(2n\pi)^{2\lambda}} = \frac{B_{2\lambda}}{(2\lambda)!}, \qquad P_{2\lambda+1}(0) = 0 \end{cases}$$

for $k, \lambda = 1, 2, \dots$ (cf. **136**). As is immediately obvious from the proof, in the interval $0 \leq x \leq 1$ and for $k \geq 2$, the functions $P_k(x)$ are rational integral functions. Besides the fact that $P_1(x) = x - \frac{1}{2}$ in $0 < x < 1$,

[4] Cf. the last footnote.

we have, in $0 \leqq x \leqq 1$,

$$P_2(x) = \frac{x^2}{2} - \frac{x}{2} + \frac{1}{12} \qquad = \frac{x^2}{2!} + \frac{B_1}{1!}\frac{x}{1!} + \frac{B_2}{2!},$$

$$P_3(x) = \frac{x^3}{6} - \frac{x^2}{4} + \frac{x}{12} \qquad = \frac{x^3}{3!} + \frac{B_1}{1!}\frac{x^2}{2!} + \frac{B_2}{2!}\frac{x}{1!},$$

$$P_4(x) = \frac{x^4}{24} - \frac{x^3}{12} + \frac{x^2}{24} - \frac{1}{720} = \frac{x^4}{4!} + \frac{B_1}{1!}\frac{x^3}{3!} + \frac{B_2}{2!}\frac{x^2}{2!} + \frac{B_4}{4!}.$$

Hence in general, as may immediately be established by induction,

$$P_k(x) = \frac{x^k}{k!} + \frac{B_1}{1!}\frac{x^{k-1}}{(k-1)!} + \frac{B_2}{2!}\frac{x^{k-2}}{(k-2)!} + \cdots + \frac{B_k}{k!}$$

$$= \frac{1}{k!}\left\{\binom{k}{0}x^k + \binom{k}{1}B_1 x^{k-1} + \binom{k}{2}B_2 x^{k-2} + \cdots \right.$$

$$\left. + \binom{k}{k-1}B_{k-1}(x) + \binom{k}{k}B_k\right\}$$

or

(c) $$P_k(x) = \frac{1}{k!}(x+B)^k,$$

if we employ the symbolic notation already used in **105**. These are the so-called *Bernoulli's polynomials*[5], which play an important part in many investigations[6]. We shall meet with some of their important properties directly.

First of all, however, we shall improve the formula (*) by means of these polynomials. Integration by parts gives

$$\int_0^n P_1(x) f'(x)\,dx = [P_2 f']_0^n - \int_0^n P_2 f''\,dx$$

$$= \frac{B_2}{2!}(f_n' - f_0') - [P_3 f'']_0^n + \int_0^n P_3 f'''\,dx$$

$$= \frac{B_2}{2!}(f_n' - f_0') + \int_0^n P_3 f'''\,dx$$

[5] They first occur in *James Bernoulli*, Ars conjectandi, Basle 1713. — There the polynomials appear as the result of the special summation problem which will be dealt with later in B, 1.

[6] Many writers call the polynomial $\varphi_k(x) = (x+B)^k - B^k$ the kth *Bernoulli* polynomial; others, again, give this name to the polynomial

$$\psi_k(x) = \frac{(x+B)^{k+1} - B^{k+1}}{k+1}.$$

These differences are unimportant.

and, generally,

$$\int_0^n P_{2\lambda-1} f^{(2\lambda-1)}\,dx = \frac{B_{2\lambda}}{(2\lambda)!}\left(f_n^{(2\lambda-1)} - f_0^{(2\lambda-1)}\right) + \int_0^n P_{2\lambda+1} f^{(2\lambda+1)}\,dx$$

for $\lambda \geqq 1$. Hence, for every $k \geqq 0$, provided only that the derivatives of $f(x)$ involved exist and are continuous, we can write:

298. $$f_0 + f_1 + \cdots + f_n = \int_0^n f(x)\,dx + \tfrac{1}{2}(f_n + f_0)$$
$$+ \frac{B_2}{2!}(f_n' - f_0') + \frac{B_4}{4!}(f_n''' - f_0''') + \cdots$$
$$+ \frac{B_{2k}}{(2k)!}\left(f_n^{(2k-1)} - f_0^{(2k-1)}\right) + R_k,$$

where we put

$$R_k = \int_0^n P_{2k+1}(x) f^{(2k+1)}(x)\,dx$$

for short. This is *Euler's summation formula.*

Remarks.

1. Since in the last integration by parts, namely

$$-\int_0^n P_{2k} f^{(2k)}\,dx = -\left[P_{2k+1} f^{(2k)}\right]_0^n + \int_0^n P_{2k+1} f^{(2k+1)}\,dx,$$

the integrated part vanishes, on account of the fact that $P_{2k+1}(n) = P_{2k+1}(0) = 0$, we may also write

$$R_k = -\int_0^n P_{2k}(x) f^{(2k)}(x)\,dx$$

for the remainder term in the summation formula.

2. If we put $F(a + xh) = f(x)$, the formula takes the somewhat more general form, in which

$$F(a) + F(a+h) + \ldots + F(a+nh)$$

forms the left hand side. The formula may therefore be used for the summation of any equidistant values of a function.

3. With suitable provisos, it is permissible to let $n \to \infty$ in the summation formula. According as Σf_n converges or diverges, we then obtain an expression for the sum of the series or for the growth of its partial sums. The statement is different (on the right hand side) for every value of k.

4. If we let $k \to \infty$, R_k *may* tend to 0. We should then have an infinite series on the right hand side, into which the sum on the left hand side is transformed. This case actually occurs very seldom, however, since, as we are aware (v. p. 237, footnote), *Bernoulli*'s numbers increase very rapidly. The series

$$\sum_{k=1}^{\infty} \frac{B_{2k}}{(2\,k)!}\left(f_n^{(2k-1)} - f_0^{(2k-1)}\right)$$

will turn out divergent for almost all the functions $f(x)$ which occur in applications, no matter what n may be. Thus the formula suggests *a summation process for a certain type of divergent series.* Cf. however the example B. 3 below.

5. Provided that the differences $\left(f_n^{(2\lambda-1)} - f_0^{(2\lambda-1)}\right)$ have the same sign, the series just discussed is an alternating series, since the signs of the numbers $B_{2\lambda}$ are alternating. We shall see that, in spite of the divergence, the above-mentioned evaluation of the remainder of the alternating series remains valid. (Cf. the introductory remarks to this section.)

6. The formula will be useful only in the cases where, for a suitable value of k, R_k is small enough to give the desired degree of accuracy. At first sight, we have only the inequality

$$|P_k(x)| \leqq \frac{2}{(2\,\pi)^k} \sum_{n=1}^{\infty} \frac{1}{n^k} \leqq \frac{4}{(2\,\pi)^k}$$

at our disposal for the estimation of R_k, for $k \geqq 2$: but, as we see subsequently, the inequality also holds for $k = 1$, and by **136** it can be put in the more precise form $|P_k(x)| \leqq \frac{|B_k|}{k!}$ for even values of k.

B. Applications.

1. It is obvious that the most favourable results are obtained when the higher derivatives of $f(x)$ are very small, and especially when they vanish. We therefore first choose $f(x) = x^p$, where p is an integer $\geqq 1$, and we have **299.**

$$1^p + 2^p + 3^p + \ldots + n^p = \int_0^n x^p\,d\,x + \frac{1}{2} n^p + \frac{B_2}{2!} p\, n^{p-1} + \ldots\,.$$

Here the series on the right hand side is to be broken off at the last positive power of n, for $(f_n^{(k)} - f_0^{(k)})$ vanishes not only when $f^{(k)}(x) \equiv 0$, but also (by **297** b) when $f^{(k)}(x)$ is identically equal to a non-vanishing constant. Thus by transferring n^p to the right hand side we have

$$1^p + 2^p + \ldots + (n-1)^p$$
$$= \frac{1}{p+1}\left\{n^{p+1} + \binom{p+1}{1} B_1\, n^p + \binom{p+1}{2} B_2\, n^{p-1} + \ldots\right\},$$

or — since there is no constant term appearing inside the brackets on the right hand side —

$$1^p + 2^p + \cdots + (n-1)^p = \frac{1}{p+1}\{(n+B)^{p+1} - B^{p+1}\}.$$

2. The sums dealt with above can be obtained in quite a different way. If we imagine that each term of the sum

$$1 + e^t + e^{2t} + \cdots + e^{(n-1)t}$$

is expanded in powers of t, the coefficient of $\frac{t^p}{p!}$ is obviously

$$1^p + 2^p + \cdots + (n-1)^p.$$

On the other hand, if we use symbolic notation (cf. **105**, 5), the first sum is equal to

$$\frac{e^{nt}-1}{e^t-1} = \frac{e^{nt}-1}{t}\, e^{Bt} = \frac{e^{(n+B)t} - e^{Bt}}{t}.$$

Hence we immediately obtain the expression

$$\frac{1}{p+1}\left\{(n+B)^{p+1} - B^{p+1}\right\}$$

for the coefficient of $\frac{t^p}{p!}$.

3. If we put $f(x) = e^{\alpha x}$, $n = 1$, we obtain

$$\frac{1}{2}(e^\alpha + 1) = \frac{e^\alpha - 1}{\alpha} + \sum_{\nu=1}^{k} \frac{B_{2\nu}}{(2\nu)!}\alpha^{2\nu-1}(e^\alpha - 1)$$

$$+ \alpha^{2k+1}\int_0^1 P_{2k+1}(x)\cdot e^{\alpha x}\,dx,$$

or

$$\frac{\alpha}{e^\alpha - 1} = 1 - \frac{\alpha}{2} + \sum_{\nu=1}^{k}\frac{B_{2\nu}}{(2\nu)!}\alpha^{2\nu} + \frac{\alpha^{2k+2}}{e^\alpha - 1}\int_0^1 P_{2k+1}(x)\, e^{\alpha x}\,dx.$$

Since we can immediately prove, by **298**, 6, that the remainder tends to zero in this case, provided only that $|\alpha| < 2\pi$, we have, for these values of α,

$$\frac{\alpha}{e^\alpha - 1} = 1 - \frac{\alpha}{2} + \sum_{\nu=1}^{\infty}\frac{B_{2\nu}}{(2\nu)!}\alpha^{2\nu} = \sum_{\lambda=0}^{\infty}\frac{B_\lambda}{\lambda!}\alpha^\lambda,$$

which is the expansion stated in **105**.

Similarly, by putting $f(x) = \cos\alpha x$, $n = 1$, we obtain the expansion **115** for $\frac{\alpha}{2}\cot\frac{\alpha}{2}$.

4. If we put $f(x) = \frac{1}{1+x}$, we have, by replacing n by $(n-1)$,

$$1 + \frac{1}{2} + \cdots + \frac{1}{n} = \log n + \frac{1}{2} + \frac{1}{2n} + \frac{B_2}{2}\left(1 - \frac{1}{n^2}\right) + \frac{B_4}{4}\left(1 - \frac{1}{n^4}\right)$$

$$+ \cdots + \frac{B_{2k}}{2k}\left(1 - \frac{1}{n^{2k}}\right) - (2k+1)! \int_1^n \frac{P_{2k+1}(x)}{x^{2k+2}}\,dx.$$

Since here we may let $n \to \infty$, just as on p. 521 above, we obtain the following refined expression for *Euler*'s constant:

$$C = \frac{1}{2} + \frac{B_2}{2} + \frac{B_4}{4} + \cdots + \frac{B_{2k}}{2k} - (2k+1)! \int_1^\infty \frac{P_{2k+1}(x)}{x^{2k+2}}\,dx.$$

In this case the remainder certainly does not decrease to 0 as k increases; and the series $\sum \frac{B_{2k}}{2k}$ diverges rapidly, — so rapidly that even the corresponding power series $\sum \frac{B_{2k}}{2k} x^{2k}$ diverges everywhere; for, by **136**,

$$|B_{2k}| = \frac{2\,(2k)!}{(2\pi)^{2k}}\,\eta, \quad \text{where} \quad 1 < \eta < 2.$$

Nevertheless, we can evaluate C very accurately by means of the above expression (cf. Rem. 6). If we take e. g. $k = 3$, we have, in the first instance,

$$\text{(a)} \qquad C = \frac{1}{2} + \frac{1}{12} - \frac{1}{120} + \frac{1}{252} - 7! \int_1^\infty \frac{P_7(x)}{x^8}\,dx.$$

If we take only the part of the integral from $x = 1$ to $x = 4$, the absolute value of the error is

$$\leqq 7! \frac{4}{(2\pi)^7} \int_4^\infty \frac{dx}{x^8} = \frac{4 \cdot 7!}{(2\pi)^7 \cdot 7 \cdot 4^7} < 10^{-6}.$$

Hence

$$C = \frac{1459}{2520} - 7! \int_1^4 \frac{P_7(x)}{x^8}\,dx + \frac{\eta}{10^6}, \quad \text{where} \quad |\eta| < 1.$$

The required evaluation of the integral is also given by the first formula written down, for $n = 4$, namely

$$-7! \int_1^4 \frac{P_7(x)}{x^8}\,dx = 1 + \frac{1}{2} + \frac{1}{3} + \frac{1}{4} - \log 4$$

$$- \frac{1459}{2520} - \frac{1}{2 \cdot 4} + \frac{1}{12 \cdot 4^2} - \frac{1}{120 \cdot 4^4} + \frac{1}{252 \cdot 4^6}.$$

Hence

$$0{\cdot}5772146 < C < 0{\cdot}5772168.$$

In this way we can easily obtain C with much greater accuracy than before, and theoretically *to any degree of accuracy whatever*. The reason for this favourable state of affairs lies solely in the fact that we may regard the logarithms as known.

5. We now put $f(x) = \log(1+x)$ and proceed just as we did in the previous examples on pp. 525—7. If we again substitute $(n-1)$ for n, we first of all obtain, from **298** with $k = 0$,

$$\log 1 + \log 2 + \cdots + \log n = \int_1^n \log x\,dx + \tfrac{1}{2}\log n + \int_1^n \frac{P_1(x)}{x}\,dx$$

or

$$\log n! = \left(n + \tfrac{1}{2}\right)\log n - (n-1) + \int_1^n \frac{P_1(x)}{x}\,dx.$$

Integrating by parts, we have

$$\int_1^n \frac{P_1(x)}{x}\,dx = \left[\frac{P_2(x)}{x}\right]_1^n + \int_1^n \frac{P_2(x)}{x^2}\,dx,$$

which shows that the integral converges as $n \to \infty$. Hence we can put

$$(*)\qquad \log n! = \left(n + \tfrac{1}{2}\right)\log n - n + \gamma_n,$$

and we know that

$$\lim_{n\to\infty} \gamma_n = \gamma$$

exists. Its value is obtained as follows: by (*) we have

$$\begin{aligned} 2\log(2\cdot 4\cdot \ldots 2n) &= 2n\log 2 + 2\log n! \\ &= 2n\log 2 + (2n+1)\log n - 2n + 2\gamma_n \\ &= (2n+1)\log 2n - 2n - \log 2 + 2\gamma_n \end{aligned}$$

and

$$\log(2n+1)! = \left(2n + \tfrac{3}{2}\right)\log(2n+1) - (2n+1) + \gamma_{2n+1}.$$

By subtraction

$$\log \frac{2\cdot 4\cdot 6\cdot \ldots 2n}{1\cdot 3\cdot 5\cdot \ldots 2n-1}\cdot\frac{1}{2n+1} = (2n+1)\log\left(1 - \frac{1}{2n+1}\right) - \tfrac{1}{2}\log(2n+1)$$
$$+ 1 - \log 2 + 2\gamma_n - \gamma_{2n+1}.$$

If we now transfer the term $\frac{1}{2}\log(2n+1)$ to the left hand side and let $n\to\infty$, we know, from *Wallis'* product (**219**, 3), that

$$\log\sqrt{\frac{\pi}{2}} = -1+1-\log 2+2\gamma-\gamma,$$

so that

$$\gamma = \log\sqrt{2\pi}\,.$$

Hence, finally, we have

$$(**)\qquad \log n! = \left(n+\frac{1}{2}\right)\log n - n + \log\sqrt{2\pi} - \int_n^\infty \frac{P_1(x)}{x}\,dx.$$

If we multiply by M, the modulus of the *Briggian* logarithms (pp. 256—7), and denote the latter logarithms by Log, we have

$$\operatorname{Log} n! = \left(n+\frac{1}{2}\right)\operatorname{Log} n - n\,M + \operatorname{Log}\sqrt{2\pi} - M\int_n^\infty \frac{P_1(x)}{x}\,dx.$$

This gives, e. g. for $n = 1000$,

$$\operatorname{Log} 1000! = 3001{\cdot}5 - 434{\cdot}29448\ldots + 0{\cdot}39908\ldots - M\int_{1000}^\infty \frac{P_1(x)}{x}\,dx.$$

Since

$$\left|M\int_{1000}^\infty \frac{P_1(x)}{x}\,dx\right| \leqq M\left[\frac{P_2(x)}{x}\right]_{1000}^\infty + \left|M\int_{1000}^\infty \frac{P_2(x)}{x^2}\,dx\right|$$

$$\leqq \frac{4}{(2\pi)^2}\cdot\frac{M}{1000} + \frac{4}{(2\pi)^2}\cdot\frac{M}{1000} < \frac{1}{10000},$$

it follows that

$$\operatorname{Log} 1000! = 2567{\cdot}6046\ldots$$

with an error $< 10^{-4}$ in absolute value, so that 1000! is a number with 2568 digits, which begins with the figures 402....

Just as in the previous example, we can now improve our result (**) considerably by means of integration by parts. Since

$$\int_n^\infty \frac{P_\lambda(x)}{x^\lambda}\,dx = \frac{P_{\lambda+1}(0)}{n^\lambda} + \lambda\int_n^\infty \frac{P_{\lambda+1}(x)}{x^{\lambda+1}}\,dx, \quad (\lambda \geqq 1),$$

after $2k$ steps we obtain

$$\log n! = \left(n+\frac{1}{2}\right)\log n - n + \log\sqrt{2\pi} + \frac{B_2}{1\cdot 2}\cdot\frac{1}{n} + \frac{B_4}{3\cdot 4}\cdot\frac{1}{n^3} + \cdots$$

$$\cdots + \frac{B_{2k}}{(2k-1)\,2k}\cdot\frac{1}{n^{2k-1}} - (2k)!\int_n^\infty \frac{P_{2k+1}(x)}{x^{2k+1}}\,dx.$$

As here the remainder (for fixed k) is less than a certain constant divided by n^{2k}, we can also write the result in the form

$$n! = \left(\frac{n}{e}\right)^n \sqrt{2\pi n}\, e^{\frac{B_2}{1\cdot 2}\cdot\frac{1}{n} + \cdots + \frac{B_{2k}}{(2k-1)\,2k}\cdot\frac{1}{n^{2k-1}} + \frac{A_n}{n^{2k}}},$$

in which the A_n's always (i. e. for every fixed k) form a bounded sequence. The result in either form is usually known as *Stirling*'s formula [7].

6. If we take the somewhat more general form $f(x) = \log(y + x)$, where $y > 0$, *Euler*'s formula for $k = 0$ gives, to begin with,

$$\log y + \log(y+1) + \ldots + \log(y+n) = (y+n)\log(y+n) - n$$
$$- y \log y + \frac{1}{2}\{\log(y+n) + \log y\} + \int_0^n \frac{P_1(x)}{y+x}\,dx.$$

Hence we can obtain a corresponding expression for the gamma-function (v. p. 385 and pp. 439—40) as follows: subtract this equation from the equation (**) in the last example, add $\log n^y$ to both sides, and we obtain

$$\log \frac{n!\,n^y}{y(y+1)\ldots(y+n)} = \left(y - \frac{1}{2}\right)\log y - \left(y + n + \frac{1}{2}\right)\log\frac{y+n}{n}$$
$$+ \log\sqrt{2\pi} - \int_0^n \frac{P_1(x)}{y+x}\,dx - \int_n^\infty \frac{P_1(x)}{x}\,dx.$$

If $n \to \infty$, this relation becomes

$$\log \Gamma(y) = \left(y - \frac{1}{2}\right)\log y - y + \log\sqrt{2\pi} - \int_0^\infty \frac{P_1(x)}{y+x}\,dx.$$

By integrating this expression by parts $2k$ times (or by at once using *Euler*'s formula for any value of k), we deduce the following generalized *Stirling*'s formula [8]:

$$\mathbf{\log \Gamma(y) = \left(y - \frac{1}{2}\right)\log y - y + \log\sqrt{2\pi}}$$
$$\mathbf{+ \frac{B_2}{1\cdot 2}\cdot\frac{1}{y} + \frac{B_4}{3\cdot 4}\,\frac{1}{y^3} + \ldots + \frac{B_{2k}}{(2k-1)(2k)}\cdot\frac{1}{y^{2k-1}}}$$
$$\mathbf{- (2k)!\int_0^\infty \frac{P_{2k+1}(x)}{(y+x)^{2k+1}}\,dx.}$$

7. We now put $f(x) = \frac{1}{(1+x)^s}$, where $x > 0$ and s is arbitrary. As we have already dealt with the cases $s = 1, -1, -2, \ldots$, and the case $s = 0$ is trivial, we shall consider s as being different from any of these values. If we again replace n by $(n - 1)$, *Euler*'s formula now gives

---

[7] *J. Stirling*, Methodus differentialis, London 1730, p. 135. But the fact that the constant γ is $\log\sqrt{2\pi}$ was not discovered till later.

[8] *Stirling* (loc. cit.) gives the formula for the sum

$$\log x + \log(x+a) + \log(x+2a) + \ldots + \log(x+na).$$

$$1+\frac{1}{2^s}+\frac{1}{3^s}+\cdots+\frac{1}{n^s}=\frac{1}{s-1}\left(1-\frac{1}{n^{s-1}}\right)+\frac{1}{2}\left(\frac{1}{n^s}+1\right)$$
$$+\frac{B_2}{2}\binom{s}{1}\left(1-\frac{1}{n^{s+1}}\right)+\cdots+\frac{B_{2k}}{2k}\binom{s+2k-2}{2k-1}\left(1-\frac{1}{n^{s+2k-1}}\right)$$
$$-(2k+1)!\binom{s+2k}{2k+1}\int_1^n\frac{P_{2k+1}(x)}{x^{s+2k+1}}\,dx.$$

If $s>1$ we can let $n\to\infty$, and we obtain the following remarkable expression for *Riemann*'s ζ-function (cf. pp. 345, 444—6, and 491–2):

$$\zeta(s)=\frac{1}{s-1}+\frac{1}{2}+\frac{B_2}{2}\binom{s}{1}+\cdots+\frac{B_{2k}}{2k}\binom{s+2k-2}{2k-1}$$
$$-(2k+1)!\binom{s+2k}{2k+1}\int_1^\infty\frac{P_{2k+1}(x)}{x^{s+2k+1}}\,dx.$$

Since the right hand side has a meaning for $s>-2k$, $s\neq 1$, and since k can take any positive integral value whatever, we immediately infer from the above — the details of the proof belong to the theory of complex functions — that

$$\zeta(s)-\frac{1}{s-1}$$

is an integral transcendental function (cf. p. 492, footnote 61). Further this expression gives the values

$$\zeta(0)=-\frac{1}{2},$$

and for $s=-p$ (p a positive integer), if we suppose that $2k>p$:

$$\zeta(-p)=-\frac{1}{p+1}-B_1+\frac{B_2}{2}\binom{-p}{1}+\frac{B_4}{4}\binom{-p+2}{3}+\cdots.$$

Here the series terminates of itself, and we can write

$$\zeta(-p)=-\frac{1}{p+1}\left\{1+\binom{p+1}{1}B_1+\binom{p+1}{2}B_2+\binom{p+1}{3}B_3+\cdots\right\}$$
$$=-\frac{1}{p+1}(1+B)^{p+1}=-\frac{B_{p+1}}{p+1},$$

where the last step follows from the fact (v. **106**) that

$$(1+B)^{p+1}-B^{p+1}=0.$$

C. The evaluation of remainders.

The evaluation of the remainder in *Euler*'s formula, which for practical purposes is particularly important, we have avoided hitherto. Now, however, the question becomes imperative whether we cannot formulate some general statement as to the magnitude of the remainder in *Euler*'s summation formula. It may be shown that, very generally, *the remainder is of the same sign as, but smaller in absolute value than, the first term neglected,* — i. e. the term which would appear in the summation formula, if we replaced k by $k+1$. This will, moreover, always be the case *if $f(x)$ has a constant sign for $x > 0$ and if $f(x)$ and all its derivatives tend monotonely to 0 as x tends to $+\infty$.*

In order to prove this, we must examine the graph of the function $y = P_k(x)$, $k \geqq 2$, in the interval $0 \leqq x \leqq 1$, somewhat more closely. We assert that the graph is of the type represented in Fig. 14; 1, 2, 3, 4, according as k leaves the remainder 1, 2, 3, or 0, when divided by 4.

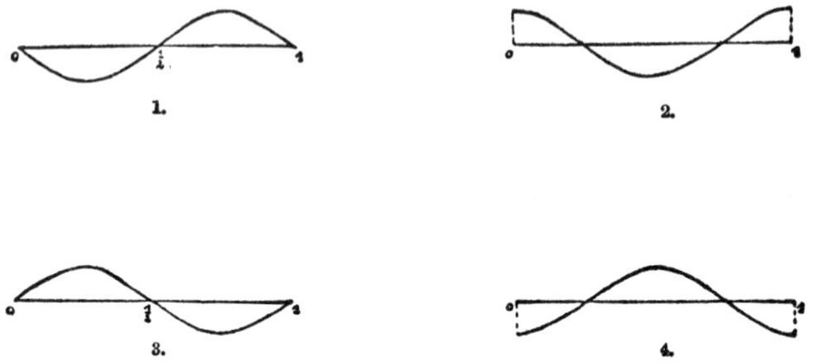

Fig. 14.

More precisely, we assert that the functions with odd suffixes have exactly three zeros of the first order at 0, $\frac{1}{2}$, 1, but those with even suffixes exactly two zeros of the first order *within* the interval, and, moreover, that the functions have the signs shown in the graphs. More shortly: $P_{2\lambda}(x)$ is of the type of the curve $(-1)^{\lambda-1}\cos 2\pi x$ and $P_{2\lambda+1}(x)$ is of the type of the curve $(-1)^{\lambda-1}\sin 2\pi x$.

These statements are proved directly for the suffixes 2, 3, 4, by using the methods which follow, or they can be deduced from the explicit formulae on p. 522. We may therefore assume that the assertions are proved up to $P_{2\lambda}(x)$, $\lambda \geqq 2$, inclusive. It is immediately obvious, by **297**, that $P_{2\lambda+1}(x)$ vanishes for $x = 0, \frac{1}{2}, 1$, and also that

$$P_{2\lambda+1}(1-x) = -P_{2\lambda+1}(x),$$

so that $P_{2\lambda+1}(x)$ is symmetrical with respect to the *point* $x = \frac{1}{2}$, $y = 0$. Thus if $P_{2\lambda+1}(x)$ has another zero, it must have two more at least, i. e. five in all, and $P_{2\lambda}(x)$ must have at least four zeros by *Rolle*'s theorem (§ 19, theorem 8), which is contrary to hypothesis. The sign of $P_{2\lambda+1}(x)$ in $0 < x < \frac{1}{2}$ is the same as that of $P'_{2\lambda+1}(0) = P_{2\lambda}(0)$, that is, the same as that of $B_{2\lambda}$; i. e. the sign is given by $(-1)^{\lambda-1}$.

Since $P'_{2\lambda+2}(x) = P_{2\lambda+1}(x)$, $P_{2\lambda+2}(x)$ has only one stationary value in $0 < x < 1$, namely at $x = \frac{1}{2}$. Its value $P_{2\lambda+2}\left(\frac{1}{2}\right)$ must have the opposite sign to $P_{2\lambda+2}(0)$, for otherwise $P_{2\lambda+2}(x)$ would have a constant sign in $0 \leqq x \leqq 1$, and consequently we should have

$$\int_0^1 P_{2\lambda+2}(x)\,dx = [P_{2\lambda+3}(x)]_0^1 \neq 0,$$

which is certainly not the case, because of the periodicity of our functions. Finally, since $P_{2\lambda+2}(0)$ has the same sign as $B_{2\lambda+2}$, i. e. the sign $(-1)^\lambda$, all our assertions are now established [9]. Since

$$P_{2\lambda}(1-x) = P_{2\lambda}(x),$$

$P_{2\lambda}(x)$ is symmetrical with respect to the *line* $x = \frac{1}{2}$.

Now, if $h(x)$ is a positive and monotone decreasing function for $x \geqq 0$,

$$\int_p^{p+1} P_{2\lambda+1}(x)\,h(x)\,dx \qquad (p \text{ an integer } \geqq 0),$$

obviously has the sign of $P_{2\lambda+1}(x)$ in $0 < x < \frac{1}{2}$, i. e. the sign $(-1)^{\lambda-1}$. For, on account of the symmetry of the graph of $P_{2\lambda+1}(x)$ and the fact that $h(x)$ decreases, we have

$$\left|\int_p^{p+\frac{1}{2}} P_{2\lambda+1}(x)\,h(x)\,dx\right| \geqq \left|\int_{p+\frac{1}{2}}^{p+1} P_{2\lambda+1}(x)\,h(x)\,dx\right|.$$

Hence

$$\int_0^n P_{2\lambda+1}(x)\,h(x)\,dx$$

also has the sign of $(-1)^{\lambda-1}$, so that, in particular, the signs are alternating, if $\lambda = 0, 1, 2, \ldots$. The exact opposite signs occur, of course, when $h(x)$ is always less than 0 and increasing.

Now if we assume that $f(x)$ is defined for $x \geqq 0$, and, together with all its derivatives, tends monotonely to 0 as $x \to \infty$, each of these derivatives is of constant sign [10], and $f^{(2k+1)}(x)$ has the same sign as $f^{(2k+3)}(x)$. The remainder in *Euler*'s summation formula is given by

$$R_k = \int_0^n P_{2k+1}(x)\,f^{(2k+1}(x)\,dx.$$

---

[9] The fact that only zeros of the first order come under consideration follows immediately from the relation

$$P'_{k+1}(x) = P_k(x).$$

[10] The possibility that one of these derivatives is always $= 0$ from some point onwards is to be included here.

Hence R_k and R_{k+1} have opposite signs, and therefore R_k and $(R_k - R_{k+1})$ have the same sign, and we have, moreover,

$$|R_k| \leqq |R_k - R_{k+1}|.$$

Now, by *Euler*'s summation formula,

$$R_k = (f_0 + f_1 + \ldots + f_n) - \int_0^n f(x)\,dx - \ldots - \frac{B_{2k}}{(2k)!}\left(f_n^{(2k-1)} - f_0^{(2k-1)}\right),$$

whence it follows that

$$R_k - R_{k+1} = \frac{B_{2k+2}}{(2k+2)!}\left(f_n^{(2k+1)} - f_0^{(2k+1)}\right).$$

But this is the "first term neglected", so that its sign also is the same as the sign of R_k, whereas its absolute value exceeds the absolute value of R_k, q. e. d.

Thus we have the

300. Theorem: *If $f(x)$ is defined for $x \geqq 0$, and, together with all its derivatives, tends monotonely to 0 as $x \to \infty$, Euler's summation formula may be stated in the simplified form*

$$f_0 + f_1 + \ldots + f_n = \int_0^n f(x)\,dx + \frac{1}{2}(f_n + f_0) + \frac{B_2}{2!}(f_n' - f_0') + \ldots$$

$$\ldots + \frac{B_{2k}}{(2k)!}\left(f_n^{(2k-1)} - f_0^{(2k-1)}\right) + \vartheta\,\frac{B_{2k+2}}{(2k+2)!}\left(f_n^{(2k+1)} - f_0^{(2k+1)}\right),$$

where $0 < \vartheta < 1$.

Thus in this form the series (divergent in general), of which the first few terms appear on the right hand side, effectively possesses the characteristic property of alternating series (mentioned on p. 518) which is particularly convenient for numerical calculations.

Remarks and Examples.

1. As *Cauchy* remarks, the characteristic property of alternating series just mentioned is exhibited by the geometrical series

$$\frac{1}{c+t} = \frac{1}{c} - \frac{t}{c^2} + \frac{t^2}{c^3} - + \ldots, \qquad c > 0,\ t > 0,$$

not only when it converges, but for arbitrary (positive) c and t. For, if we write it with the remainder, i. e. in the form

$$\frac{1}{c+t} = \frac{1}{c} - \frac{t}{c^2} + \ldots + (-1)^n \frac{t^n}{c^{n+1}} + (-1)^{n+1} \frac{t^{n+1}}{c^{n+2}} \cdot \frac{1}{1 + \frac{t}{c}},$$

it is true without exception. For any (positive) c and t, the value of the left hand side is represented by the nth partial sum, except for an error *which has the sign of the first term neglected, but is less than this term in absolute value.*

If we carry out this process with the fractions

$$\frac{2}{4\nu^2\pi^2 + t^2} = 2\left(\frac{1}{(2\nu\pi)^2} - \frac{t^2}{(2\nu\pi)^4} + \frac{t^4}{(2\nu\pi)^6} - + \ldots\right)$$

and add, we see that the value of

$$\sum_{\nu=1}^{\infty} \frac{2}{4\,\nu^2\,\pi^2 + t^2} = \left(\frac{1}{e^t - 1} - \frac{1}{t} + \frac{1}{2}\right)\frac{1}{t}$$

for every $t > 0$ is also equal to the sum

$$\frac{B_2}{2!} + \frac{B_4}{4!}\,t^2 + \dots + \frac{B_{2k}}{(2\,k)!}\,t^{2k-2} + \vartheta\,\frac{B_{2k+2}}{(2\,k+2)!}\,t^{2k},$$

where all that is known about ϑ is that it lies in the interval $0 \dots 1$.

If we now multiply by e^{-xt} and integrate from 0 to $+\infty$, it follows, since

$$\int_0^{\infty} t^{2\lambda}\,e^{-xt}\,d\,t = \frac{1}{x^{2\lambda+1}}\int_0^{\infty} \tau^{2\lambda}\,e^{-\tau}\,d\tau = \frac{(2\,\lambda)!}{x^{2\lambda+1}},$$

that

$$\int_0^{\infty}\left(\frac{1}{e^t - 1} - \frac{1}{t} + \frac{1}{2}\right)\frac{e^{-xt}}{t}\,d\,t = \frac{B_2}{1\cdot 2}\cdot\frac{1}{x} + \frac{B_4}{3\cdot 4}\cdot\frac{1}{x^3} + \dots + \frac{B_{2k}}{(2\,k-1)\,(2\,k)}\cdot\frac{1}{x^{2k-1}}$$

$$+\,\vartheta_1\,\frac{B_{2k+2}}{(2\,k+1)\,(2\,k+2)}\cdot\frac{1}{x^{2k+1}}, \qquad 0 < \vartheta_1 < 1.$$

2. By B. 6 the function

$$\log\Gamma(x) - \left\{\left(x - \frac{1}{2}\right)\log x - x + \log\sqrt{2\,\pi}\right\}$$

also can be equated to the expression found in 1, for the remainder term used in B. 6 may be replaced by the one just written down, by **300**. But we may not conclude from this, without further examination, that

$$\log\Gamma(x) = \left(x - \frac{1}{2}\right)\log x - x + \log\sqrt{2\,\pi} + \int_0^{\infty}\left(\frac{1}{e^t - 1} - \frac{1}{t} + \frac{1}{2}\right)\frac{e^{-xt}}{t}\,dt$$

(cf. **301**, 4). We have indeed proved that both sides agree *very closely* for *large* values of x; but we may not conclude from the previous considerations that they are actually *equal* for *any* value of x. (In fact, however, the equation written above is true.)

3. Just as before, we can also briefly evaluate the remainder in the examples 4, 5, 6, 7 of section B., from the fact that the remainder has the same sign as, but is smaller in absolute value than, the "first term neglected". For it is immediately obvious that the functions $f(x)$ used in these examples satisfy the hypotheses of the theorem **300**.

§ 65. Asymptotic series.

We now return to the introductory remarks of § 64, A. The series which we obtain from *Euler*'s summation formula in examples 4—7, by continuing the expansion to infinity instead of writing down the remainder, are divergent. In the cases when they are power series in $\frac{1}{x}$ or $\frac{1}{n}$, we can say, more precisely, that they are power series which diverge everywhere. In spite of this, they can be employed in practice, since examination of the remainder shows that the error corresponding to a particular partial sum is *smaller in absolute value than, and of the same sign as,* the first term neglected. Now at first these terms decrease, and become even very small for large values of the variable; it is only later on that they increase to a high value. Hence the series can be used for numerical calculations in spite of its divergence; with limited accuracy, to be sure, but with an accuracy which is often close enough to be sufficient for the most refined practical purposes (in Astronomy in particular)[11]. Moreover, the larger the variable is, the more readily does the series yield the results just mentioned. More precisely: if (as in B, 5 and 6) the expansion obtained from *Euler*'s summation formula is of the form

$$f(x) = g(x) + a_0 + \frac{a_1}{x} + \frac{a_2}{x^2} + \cdots,$$

not only do we have

$$f(x) - \left(g(x) + a_0 + \frac{a_1}{x} + \cdots + \frac{a_k}{x^k}\right) \to 0$$

as $x \to \infty$, for every fixed k, but even

$$x^k \left[f(x) - \left(g(x) + a_0 + \frac{a_1}{x} + \cdots + \frac{a_k}{x^k}\right)\right] \to 0.$$

A general investigation of this property of the expansions was made almost simultaneously by *Th. J. Stieltjes*[12] and *H. Poincaré*[13]. Following the older usage, *Stieltjes* calls our series *semi-convergent,* a term which emphasizes the fact that so far as numerical purposes are concerned they behave almost like convergent series. *Poincaré,* on the other hand, speaks of *asymptotic series,* thus putting the last-mentioned property, which can be accurately defined, in the foreground. The older term has not held its ground, although it is often used, especially in astronomical literature. The reason is that it clashes with the terminology which is customary, particularly in France, whereby our *conditionally* convergent series are called *semi-convergent.* We shall therefore adopt *Poincaré*'s term, and we proceed to set up the following exact

---

[11] *Euler,* who makes no mention of remainders whatever, frankly regards the left hand side of **298** as the sum of the divergent series on the right hand side. Thus he writes $C = \frac{1}{2} + \frac{B_2}{2} + \frac{B_4}{4} + \cdots$ without hesitation, on account of **299**, 4. This interpretation is not valid, however, even from the general view-point of § 59, for the investigations of § 64 have provided no process by which the sum in question may be obtained from the partial sums of the series by a *convergent* process, as was always the case in Chap. XIII.

[12] *Stieltjes, Th. J.*: Recherches sur quelques séries semi-convergentes, Annales de l'Ec. Norm. Sup. (3), Vol. 3, pp. 201—258. 1886.

[13] *Poincaré, H.*: Sur les intégrales irregulières des équations linéaires, Acta mathematica, Vol. 8, pp. 295—344. 1886.

Definition. *A series of the form* $a_0 + \frac{a_1}{x} + \frac{a_2}{x^2} + \cdots$ *(which need* **301.**
not converge for any value of x*) is called an asymptotic representation (or expansion) of a function* $F(x)$ *which is defined for every sufficiently large positive value of* x*, if, for every (fixed)* $n = 0, 1, 2, \ldots,$

$$\left[F(x) - \left(a_0 + \frac{a_1}{x} + \frac{a_2}{x^2} + \cdots + \frac{a_n}{x^n}\right)\right] x^n \to 0$$

as $x \to +\infty$*: and we shall write symbolically*

$$F(x) \sim a_0 + \frac{a_1}{x} + \frac{a_2}{x^2} + \cdots.$$

Remarks and Examples.

1. Here the coefficients a_n are not bound to satisfy any conditions, since the series $\sum \frac{a_n}{x^n}$ need not converge. They may be complex, in fact, if $F(x)$ is a complex function of the real variable x. The variable may also be complex, in which case x must approach infinity along a fixed radius am x = constant; for the asymptotic expansion may be different for each radius. In what follows we shall set these generalizations aside and henceforward suppose all the quantities to be real.

On the other hand, it frequently happens that the function $F(x)$ is defined for integral values of the variable only; e. g.

$$1^p + 2^p + \ldots + x^p, \qquad 1 + \frac{1}{2} + \ldots + \frac{1}{x}.$$

In such cases we shall usually denote the variable by $k, \nu, n, \ldots$ Then $F(x)$ simply represents a sequence, the terms of which are asymptotically expressed as functions of the integral variables.

2. If the series $\sum \frac{a_n}{x^n}$ *does* converge for $x > R$, and represents the function $F(x)$, the series is obviously an asymptotic representation of $F(x)$ in this case also. Thus examples of asymptotic representation can be obtained from any convergent power series.

3. The question whether a function $F(x)$ possesses an asymptotic representation, and what the values of the coefficients are, is immediately settled in theory by the fact that the successive limiting values (for $x \to +\infty$)

$$F(x) \to a_0,$$
$$(F(x) - a_0)\, x \to a_1,$$
$$\left(F(x) - a_0 - \frac{a_1}{x}\right) x^2 \to a_2,$$
$$\cdots\cdots\cdots\cdots$$

must exist. In fact, however, the decision can seldom be made in this way; but these simple considerations show that any function can have only one asymptotic expansion.

4. On the other hand, for $f(x) = e^{-x}$, $x > 0$, all the a_n's are zero, since

$$x^k e^{-x} \to 0$$

for every integral $k \geqq 0$, when $x \to \infty$. Thus

$$e^{-x} \sim 0 + \frac{0}{x} + \frac{0}{x^2} + \cdots,$$

a result which shows that different functions may have the same asymptotic expansion. Thus, if $F(x)$ has an asymptotic representation, e. g.

$$F(x) + e^{-x}, \quad F(x) + a\,e^{-bx} \ (b > 0), \ \ldots,$$

have the same asymptotic representation.

It was for this reason that we could not infer that the two functions mentioned in **300**, 2 were identical.

5. Geometrically speaking, the curves

$$y = a_0 + \frac{a_1}{x} + \cdots + \frac{a_n}{x^n} \quad \text{and} \quad y = F(x)$$

have contact of at least the nth order at infinity; and the contact becomes closer as n increases.

6. For applications it is advantageous to use the notation

$$F(x) \sim f(x) + g(x)\left(a_0 + \frac{a_1}{x} + \frac{a_2}{x^2} + \cdots\right),$$

where $f(x)$ and $g(x)$ are any two functions which are defined for sufficiently large values of x, and such that, further, $g(x)$ never vanishes. This notation is intended essentially to express that

$$\frac{F(x) - f(x)}{g(x)} \sim a_0 + \frac{a_1}{x} + \frac{a_2}{x^2} + \cdots.$$

Some of the examples worked out in § 64, B may be regarded as giving the asymptotic expansions, in this sense, of the functions involved, for we may now write

a) $1 + \frac{1}{2} + \cdots + \frac{1}{n} \sim \log n + C + \frac{1}{2n} - \frac{B_2}{2} \cdot \frac{1}{n^2} - \frac{B_4}{4} \cdot \frac{1}{n^4} - \cdots;$

b) $\log n! \sim \left(n + \frac{1}{2}\right) \log n - n + \log \sqrt{2\pi} + \frac{B_2}{1 \cdot 2} \cdot \frac{1}{n} + \frac{B_4}{3 \cdot 4} \cdot \frac{1}{n^3} + \cdots;$

c) $\log(\Gamma(x)) \sim \left(x - \frac{1}{2}\right) \log x - x + \log \sqrt{2\pi} + \frac{B_2}{1 \cdot 2} \cdot \frac{1}{x} + \frac{B_4}{3 \cdot 4} \cdot \frac{1}{x^3} + \cdots;$

d) $1 + \frac{1}{2^s} + \cdots + \frac{1}{n^s} \sim \left(\zeta(s) - \frac{1}{s-1} \cdot \frac{1}{n^{s-1}}\right)$

$$+ \frac{1}{n^s}\left[\frac{1}{2} - \frac{B_2}{2} \cdot \binom{s}{1} \frac{1}{n} - \frac{B_4}{4} \cdot \binom{s+2}{3} \frac{1}{n^3} - \cdots\right];$$

In the last formula we must have $s \neq 1$; for $s = 1$ it becomes the expansion in a).

Calculations with asymptotic series.

In many respects we can make calculations with asymptotic series just as we do with convergent series.

It is immediately obvious that from

$$F(x) \sim a_0 + \frac{a_1}{x} + \frac{a_2}{x^2} + \cdots$$

and

$$G(x) \sim b_0 + \frac{b_1}{x} + \frac{b_2}{x^2} + \cdots,$$

there results the expansion

$$\alpha F(x) + \beta G(x) \sim \alpha a_0 + \beta b_0 + \frac{\alpha a_1 + \beta b_1}{x} + \frac{\alpha a_2 + \beta b_2}{x^2} + \cdots,$$

where α and β are any constants.

It is almost as easy to see that the product of the functions also possesses an asymptotic expansion, and that

$$F(x)\,G(x) \sim c_0 + \frac{c_1}{x} + \frac{c_2}{x^2} + \cdots$$

if, as in the case of convergent series,

$$a_0 b_n + a_1 b_{n-1} + \cdots + a_n b_0$$

is set equal to c_n. For, by hypothesis, we may write (for fixed n),

$$F(x) = a_0 + \frac{a_1}{x} + \frac{a_2}{x^2} + \cdots + \frac{a_{n-1}}{x^{n-1}} + \frac{a_n + \varepsilon}{x^n},$$

$$G(x) = b_0 + \frac{b_1}{x} + \frac{b_2}{x^2} + \cdots + \frac{b_{n-1}}{x^{n-1}} + \frac{b_n + \eta}{x^n},$$

if by $\varepsilon = \varepsilon(x)$ and $\eta = \eta(x)$ we denote functions which tend to 0 as $x \to +\infty$. But in this case we have

$$\left[F(x)\cdot G(x) - \left(c_0 + \frac{c_1}{x} + \frac{c_2}{x^2} + \cdots + \frac{c_n}{x^n}\right)\right] x^n = a_0 \eta + b_0 \varepsilon$$

$$+ \frac{a_1(b_n + \eta) + a_2 b_{n-1} + \cdots + (a_n + \varepsilon) b_1}{x} + \cdots + \frac{(a_n + \varepsilon)(b_n + \eta)}{x^n},$$

and this obviously tends to 0 as $x \to +\infty$.

Repeated application of this simple result gives

Theorem 1. *If each of the functions* $F_1(x), F_2(x), \ldots, F_p(x)$ *possesses an asymptotic representation, and if* $g(z_1, z_2, \ldots, z_p)$ *is a polynomial, or — if we anticipate what immediately follows — any rational function whatever, of the variables* $z_1, z_2, \ldots, z_p$, *then the function* **302.**

$$F(x) = g(F_1(x), F_2(x), \ldots, F_p(x))$$

also possesses an asymptotic representation; and this is calculated exactly as if all the expansions were convergent series, provided only that the denominator of the rational function does not vanish when the constant terms of the asymptotic expansions are substituted for $z_1, z_2, \ldots, z_p$.

Further, the following theorem also holds:

Theorem 2. *If* $g(z) = \alpha_0 + \alpha_1 z + \ldots + \alpha_n z^n + \ldots$ *is a power series with positive radius* r, *if* $F(x)$ *possesses the asymptotic representation*

$$F(x) \sim a_0 + \frac{a_1}{x} + \frac{a_2}{x^2} + \cdots,$$

and if $|a_0| < r$, *the function of a function*

$$\Phi(x) = g(F(x))$$

— *which is obviously defined for every sufficiently large* x, *since* $F(x) \to a_0$ *as* $x \to +\infty$, *and since* $|a_0| < r$ — *also possesses an asymptotic representation, and this is again calculated exactly as if* $\Sigma \frac{a_n}{x^n}$ *were convergent.*

Proof. In order to calculate the coefficients of the expansion of $\Phi(x)$, when $\Sigma \frac{a_n}{x^n}$ converges for $x > R$, say, we have to set $F(x) = a_0 + f$, and — assuming only that $|a_0| < r$ — we obtain, in the first instance,

$$g(F) = g(a_0 + f) = \beta_0 + \beta_1 f + \cdots + \beta_k f^k + \cdots, \qquad (*)$$

where we put

$$\frac{1}{k!} g^{(k)}(a_0) = \beta_k \qquad (k = 0, 1, 2, \ldots)$$

for short. This expansion (*) converges whenever $|f(x)| < r - |a_0|$, which is certainly the case for every sufficiently large value of x, whether $\Sigma \frac{a_n}{x^n}$ converges or not, since in fact $f(x) \to 0$ as $x \to \infty$. In accordance with the part of theorem 1 which has already been proved, from

$$f(x) \sim \frac{a_1}{x} + \frac{a_2}{x^2} + \cdots$$

we deduce the asymptotic expansion

$$(f(x))^k \sim \frac{a_k^{(k)}}{x^k} + \frac{a_{k+1}^{(k)}}{x^{k+1}} + \cdots \qquad (**)$$

for every $k = 1, 2, 3, \ldots$. Here the quantities $a_\nu^{(k)}$ have quite definite values, obtainable by the product rule for asymptotic expansions (i. e. as for convergent series). We must now substitute these expansions (**) in (*) and arrange the result formally (i. e. again just as if the series (**)

were convergent) in powers of $\frac{1}{x}$. Thus we obtain an expansion of the form

$$A_0 + \frac{A_1}{x} + \frac{A_2}{x^2} + \cdots,$$

where the coefficients are given by

$$A_0 = \beta_0, \quad A_1 = \beta_1 a_1, \quad A_2 = \beta_2 a_2 + \beta_2 a_2^{(2)}, \ldots,$$
$$A_n = \beta_1 a_n + \beta_2 a_n^{(2)} + \ldots + \beta_n a_n^{(n)}, \ldots .$$

It remains to show that $\Sigma \frac{A_n}{x^n}$ is an asymptotic expansion of $\Phi(x)$, that is, that the expression

$$\left[\Phi(x) - \left(A_0 + \frac{A_1}{x} + \frac{A_2}{x^2} + \cdots + \frac{A_n}{x^n}\right)\right] \cdot x^n$$

tends to 0 for fixed n as $x \to \infty$.

Now if $\varepsilon_1 = \varepsilon_1(x)$, $\varepsilon_2 = \varepsilon_2(x), \ldots$ denote certain functions which tend to 0 as $x \to \infty$, it follows from (*) and (**) that

$$\Phi(x) = \beta_0 + \beta_1\left(\frac{a_1}{x} + \cdots + \frac{a_n}{x^n} + \frac{\varepsilon_1}{x^n}\right) + \cdots + \beta_n\left(\frac{a_n^{(n)}}{x^n} + \frac{\varepsilon_n}{x^n}\right)$$
$$+ f^{n+1} \cdot [\beta_{n+1} + \beta_{n+2} f + \cdots].$$

Hence, since f^{n+1} may be put equal to $\frac{\varepsilon_{n+1}}{x^n}$,

$$\Phi(x) - \left(A_0 + \frac{A_1}{x} + \cdots + \frac{A_n}{x^n}\right) = \frac{1}{x^n}[\beta_1 \varepsilon_1 + \beta_2 \varepsilon_2 + \cdots + \beta_n \varepsilon_n]$$
$$+ \frac{\varepsilon_{n+1}}{x^n}[\beta_{n+1} + \beta_{n+2} f + \cdots]$$

and our assertion follows at a glance, for the expression in the last square bracket tends to β_{n+1} as $x \to +\infty$, and the (finitely numerous) ε_ν's tend to 0.

Taking $g(z) = \frac{1}{a_0 + z}$ and replacing $F(x)$ by $F(x) - a_0$, it follows as a particular case, provided only that $a_0 \neq 0$, that

$$\frac{1}{F(x)} \sim \frac{1}{a_0} - \frac{a_1}{a_0^2}\frac{1}{x} + \frac{a_1^2 - a_0 a_2}{a_0^3}\frac{1}{x^2} + \cdots .$$

Hence we "may" divide by asymptotic expansions with non-vanishing constant terms; this completes the proof of theorem 1.

Taking $g(z) = e^z$, we obtain, without any restrictions,

$$e^{F(x)} \sim e^{a_0}\left[1 + \frac{a_1}{x} + \frac{\frac{1}{2}a_1^2 + a_2}{x^2} + \ldots\right].$$

In particular, we may write, by **299**, 5,

$$n! \sim \left(\frac{n}{e}\right)^n \sqrt{2\pi n}\left[1 + \frac{1}{12n} + \frac{1}{288n^2} - \frac{139}{51840}\frac{1}{n^3} + \ldots\right].$$

Term-by-term integration and differentiation are also valid with suitable provisos. We have

Theorem 3. *If* $F(x) \sim a_0 + \frac{a_1}{x} + \frac{a_2}{x^2} + \ldots$ *and if* $F(x)$ *is continuous for* $x \geqq x_0$, *then*

$$\Psi(x) = \int_x^\infty \left(F(t) - a_0 - \frac{a_1}{t}\right) dt \sim \frac{a_2}{x} + \frac{a_3}{2x^2} + \ldots + \frac{a_{n+1}}{n x^n} + \ldots .$$

If $F(x)$ *has a continuous derivative, and if* $F'(x)$ *is known to* ***possess an asymptotic expansion***, *then this expansion may be obtained by differentiating term-by-term,* i. e.

$$F'(x) \sim -\frac{a_1}{x^2} - \frac{2a_2}{x^3} - \ldots - \frac{(n-1)a_{n-1}}{x^n} - \ldots .$$

Proof. Since $t^2\left(F(t) - a_0 - \frac{a_1}{t}\right) \to a_2$ as $t \to +\infty$, the integral which defines the function $\Psi(x)$ always exists for $x \geqq x_0$. Further, we may set $F(t) - a_0 - \frac{a_1}{t} - \ldots - \frac{a_{n+1}}{t^{n+1}} = \frac{\varepsilon(t)}{t^{n+1}}$ ($n \geqq 1$, fixed), where $\varepsilon(t) \to 0$ as $t \to +\infty$. Hence

$$\Psi(x) - \frac{a_2}{x} - \ldots - \frac{a_{n+1}}{n x^n} = \int_x^\infty \frac{\varepsilon(t)}{t^{n+1}} dt.$$

Now if $\bar{\varepsilon}(x)$ denotes the maximum value of $|\varepsilon(t)|$ in $x \leqq t < +\infty$, then $\bar{\varepsilon}(x) \to 0$ also as x increases; and since the last integral $\leqq \frac{\bar{\varepsilon}(x)}{n x^n}$, after multiplication by x^n it likewise tends to 0.

Now if the derivative $F'(x)$, which is continuous for $x \geqq x_0$, possesses an asymptotic expansion

$$F'(x) \sim b_0 + \frac{b_1}{x} + \frac{b_2}{x^2} + \cdots,$$

we have

$$F(x) = \int_{x_0}^x F'(t)\,dt + C_1 = \int_{x_0}^x \left(b_0 + \frac{b_1}{t}\right) dt + \int_{x_0}^x \left(F'(t) - b_0 - \frac{b_1}{t}\right) dt + C_1$$

$$= b_0 x + b_1 \log x + C_2 - \int_x^\infty \left(F'(t) - b_0 - \frac{b_1}{t}\right) dt,$$

where C_1, C_2 are constants. By what we have just proved, and because a function defines its asymptotic expansion uniquely, it follows that $b_0 = b_1 = 0$ and that $b_n = -(n-1)a_{n-1}$ for $n \geqq 2$.

The expansion

$$F(x) = e^{-x}\sin(e^x) \sim 0 + \frac{0}{x} + \frac{0}{x^2} + \cdots$$

exemplifies the fact that $F'(x)$ need not possess an asymptotic expansion, even when $F(x)$ does.

Theorems 1—3 lay the foundation for *Poincaré*'s very fruitful applications of asymptotic series to the solution of differential equations[14]. A detailed account lies outside the plan of this book, however, and we must content ourselves by giving an example of this application of asymptotic series in the following section.

§ 66. Special cases of asymptotic expansions.

The use of asymptotic expansions raises two main questions: first there is the question whether the function under consideration possesses an asymptotic expansion at all, and how it is to be found in a given case (the *expansion problem*); on the other hand, there is the question how *the* function, or rather, *a* function, is to be found, which is represented by a given asymptotic expansion (the *summation problem*). In the case of both questions, the answers available in the present state of knowledge are not completely satisfactory as yet, for although they are very numerous and in part of remarkably wide range, they are somewhat isolated and lack methodical and fundamental connections. This section will therefore consist rather of a collection of representative examples than of a satisfactory solution of the two problems.

A. Examples of the expansion problem.

1. From the theoretical point of view, the expansion of given functions was thoroughly dealt with in the note **301**, 3; but it is only seldom that the required determinations of limits can all be carried out. The method also fails if $\lim_{x \to \infty} F(x)$ does not exist, i. e. if only an asymptotic expansion in the more general sense mentioned in **301**, 6 can be considered. It is only when $f(x)$ and $g(x)$, the functions involved, have been found that we can proceed as in **301**, 3. **303.**

2. We have learned that asymptotic series very frequently arise from *Euler*'s summation formula: but there it is not so much a case of expanding given functions as that by special choice of the function $f(x)$ in the summation formula we are often led to valuable expansions.

3. As we have already emphasized, hitherto perhaps the most important application of asymptotic expansions is *Poincaré*'s use of them in the theory of differential equations[14]. The simple fundamental

[14] A very clear account of the contents of *Poincaré*'s paper, including all the essential points, is given by *E. Borel* in his "Leçons sur les séries divergentes" (v. **266**).

idea is this: suppose we know that a function $y = F(x)$ satisfies a differential equation of the nth order

$$\Phi(x, y, y', \ldots, y^{(n)}) = 0,$$

where Φ denotes a rational function of the variables involved. Now if we know that $y = F(x)$ and its first n derivatives all possess asymptotic representations, the expansions for $y', y'', \ldots, y^{(n)}$ follow, by **302**, Theorem 3, from the first one,

$$y \sim a_0 + \frac{a_1}{x} + \frac{a_2}{x^2} + \cdots.$$

If we substitute these expansions in the differential equation, in accordance with **302**, Theorem 1, we must obtain an expansion which stands for 0, all the coefficients of which must therefore vanish. From the equations obtained in this way, together with the initial conditions, the coefficients and hence the expression for $F(x)$ are in general found.

Thus e. g. the function

$$y = F(x) = e^x \int_x^\infty \frac{e^{-t}}{t}\, dt,$$

which is defined for $x > 0$, has for its derivative

$$y' = F'(x) = e^x \int_x^\infty \frac{e^{-t}}{t}\, dt - e^x\, \frac{e^{-x}}{x} = y - \frac{1}{x},$$

that is, it satisfies the differential equation

$$y' - y + \frac{1}{x} = 0$$

for $x > 0$. It may be proved directly — but we cannot give the details here — that this equation has only one solution y such that y and y' exist for $x > x_0 \geqq 0$ and have an asymptotic representation. If we accordingly set

$$y \sim a_0 + \frac{a_1}{x} + \frac{a_2}{x^2} + \cdots, \quad \text{so that } y' \sim -\frac{a_1}{x^2} - \frac{2a_2}{x^3} - \cdots,$$

we have the equations

$$a_0 = 0, \quad a_1 = 1, \quad a_2 = -a_1, \cdots, \quad a_{n+1} = -n\, a_n, \ldots,$$

whence it follows that

$$a_0 = 0, \quad a_1 = 1, \quad a_2 = -1, \ldots, \quad a_{n+1} = (-1)^n n!, \ldots.$$

We therefore find that

$$F(x) \sim \frac{1}{x} - \frac{1}{x^2} + \frac{2!}{x^3} - \frac{3!}{x^4} + - \cdots.$$

4. The function in the previous example can be asymptotically expanded by another method, which is frequently applicable. If we put $t = u + x$, we have

$$F(x) = \int_0^\infty \frac{e^{-u}}{u+x}\,du.$$

Here, by *Cauchy*'s observation (v. **300**, 1), we can put

$$\frac{1}{u+x} = \frac{1}{x} - \frac{u}{x^2} + \frac{u^2}{x^3} - + \cdots + (-1)^n \frac{u^n}{x^{n+1}} + (-1)^{n+1}\vartheta \frac{u^{n+1}}{x^{n+2}},$$
$$0 < \vartheta < 1,$$

for all positive values of x and u. It follows that

$$F(x) = \frac{1}{x} - \frac{1}{x^2} + \frac{2!}{x^3} - + \cdots + (-1)^n \frac{n!}{x^{n+1}} + (-1)^{n+1}\vartheta_1 \frac{(n+1)!}{x^{n+2}},$$
$$0 < \vartheta_1 < 1.$$

Thus we have again found the expansion in the last example [15].

5. If $f(u)$ is a function which is defined for $u \geqq 0$ and is positive there, and if the integrals

$$\int_0^\infty f(u)\, u^{n-1}\, du = (-1)^{n-1} a_n$$

exist for every integral $n \geqq 1$, we similarly obtain the asymptotic expansion

$$F(x) \sim \frac{a_1}{x} + \frac{a_2}{x^2} + \frac{a_3}{x^3} + \cdots$$

for the function

$$F(x) = \int_0^\infty \frac{f(u)}{u+x}\,du.$$

Moreover, the partial sums of this series represent $F(x)$, except for an error which is less in absolute value than the first term neglected and is of the same sign as the latter. Expansions of this kind have been investigated especially by *Th. J. Stieltjes* [16]. (For further particulars, see below, B, p. 549.)

---

[15] The function $e^{-x} F(x) = \int_x^\infty \frac{e^{-t}\,dt}{t}$, which becomes $-\int_0^{e^{-x}} \frac{dv}{\log v}$ with the transformation $e^{-t} = v$, is known as the *Logarithmic-integral* function of $y = e^{-x}$.

[16] *Stieltjes, Th. J.*: Recherches sur les fractions continues, Annales de la Fac. des Sciences de Toulouse, Vols. 8 and 9, 1894 and 1895.

6. Certain methods requiring the more advanced resources of the theory of functions date back to *Laplace,* but have recently been extended by *E. W. Barnes*[17], *H. Burkhardt*[18], *O. Perron*[19], and *G. Faber*[20]. We cannot go into details, but must content ourselves with the following remarks. *Barnes* gives the asymptotic expansions of many integral functions, e. g. $\sum \frac{x^n}{n!\,(n+\vartheta)}$, $(\vartheta \neq 0, -1, -2, \ldots)$, and similar functions. Besides the expansions we have met with, *O. Perron* obtains as examples the asymptotic expansion in terms of n of certain integrals which occur in the theory of *Keplerian* motion, such as

$$A(n) = \int_{-\pi}^{+\pi} \frac{e^{n(t-\varepsilon \sin t)i}}{1-\varepsilon \cos t}\,dt, \qquad (0 < \varepsilon \leqq 1,\ n \text{ an integer}),$$

$$C(n) = \int_{-\pi}^{+\pi} e^{n(t-\sin t)i}\,dt.$$

From our point of view it is noteworthy that in these examples the terms of the expansion do not proceed by integral powers of $\frac{1}{n}$, but by fractional powers. Thus the expansion of $C(n)$ is of the form

$$C(n) \sim \frac{c_1}{n^{1/3}} + \frac{c_2}{n^{5/3}} + \frac{c_3}{n^{7/3}} + \frac{c_4}{n^{11/3}} + \cdots.$$

This suggests another extension of the definition **301**, 6, which, however, we shall not discuss.

Numerous additional examples of asymptotic expansions of this kind, in particular those of trigonometrical integrals occurring in physical and astronomical investigations, are to be found in the article by *H. Burkhardt*: "Über trigonometrische Reihen und Integrale", in the Enzyklopädie der mathematischen Wissenschaften, Vol. II, 1, pp. 815—1354.

7. An expansion, which was first given by *L. Fejér*[21], and was subsequently treated in detail by *O. Perron*[22], is of a more specialized nature; its object is to deduce an asymptotic representation for the coefficients of the expansion in power series of $e^{\alpha\frac{x}{1-x}}$, or, more gener-

---

[17] *Barnes, E. W.*: The Asymptotic Expansions of Integral Functions defined by Taylor's Series, Phil. Trans. Roy. Soc., A, 206, pp. 249—297. 1906.

[18] *Burkhardt, H.*: Über Funktionen großer Zahlen, Sitzungsber. d. Bayr. Akad. d. Wissensch., pp. 1—11. 1914.

[19] *Perron, O.*: Über die näherungsweise Berechnung von Funktionen großer Zahlen, Sitzungsber. d. Bayr. Akad. d. Wissensch., pp. 191—219. 1917.

[20] *Faber, G.*: Abschätzung von Funktionen großer Zahlen, Sitzungsber. d. Bayr. Akad. d. Wissensch., pp. 285—304. 1922.

[21] *Fejér, L.*: in a paper in Hungarian. 1909.

[22] *Perron, O.*: Über das infinitäre Verhalten der Koeffizienten einer gewissen Potenzreihe, Archiv d. Math. u. Phys. (3), Vol. 22, pp. 329—340. 1914.

ally, of $e^{\alpha/(1-x)^\varrho}$, where $\varrho > 0$ and $\alpha > 0$. We at once find that

$$e^{\alpha \frac{x}{1-x}} = \sum_{k=0}^{\infty} \frac{1}{k!}\left(\alpha \frac{x}{1-x}\right)^k = 1 + c_1 x + c_2 x^2 + \cdots$$

where the coefficients c_n have the values

$$c_n = \sum_{\nu=1}^{n} \binom{n-1}{\nu-1} \frac{\alpha^\nu}{\nu!}.$$

For these *Perron* showed in a later work [23] that they have an asymptotic expansion of the form [24]

$$c_n \sim \frac{\sqrt[4]{\alpha}\, e^{2\sqrt{\alpha n}}}{2\sqrt{\pi e^{\alpha}}\;\sqrt[4]{n^3}}\left(1 + \frac{a_1}{\sqrt{n}} + \frac{a_2}{\sqrt{n^2}} + \frac{a_3}{\sqrt{n^3}} + \cdots\right).$$

8. Finally, we draw attention to the fact that the asymptotic representation of certain functions forms the subject of many profound investigations in the analytical theory of numbers. In fact, our examples **301**, 6, a, b, and d, belong to this class, for the functions expanded have a meaning only for integral values of the variable in the first instance. Just to indicate the nature of such expansions, we give a few more examples, without proof:

a) If $\tau(n)$ denotes the *number of divisors* of n,

$$\frac{\tau(1) + \tau(2) + \cdots + \tau(n)}{n} \sim \log n + (2C - 1) + \cdots,$$

where C is *Euler*'s constant [25]. Regarding the next term [26], practically all that is known is that it is lower in degree than $n^{-\frac{2}{3}}$ but not lower than $n^{-\frac{3}{4}}$.

b) If $\sigma(n)$ denotes the sum of the divisors of n,

$$\frac{\sigma(1) + \sigma(2) + \cdots + \sigma(n)}{n} \sim \frac{\pi^2}{12} n + \cdots.$$

---

[23] *Perron, O.*: Über das Verhalten einer ausgearteten hypergeometrischen Reihe bei unbegrenzten Wachstum eines Parameters. J. reine u. angew. Math. Vol. 151, pp. 63—78. 1921.

[24] An elementary proof of the far less complete result

$$\log c_n \sim 2\sqrt{\alpha n}$$

is given by *K. Knopp* and *I. Schur*: Elementarer Beweis einiger asymptotischer Formeln der additiven Zahlentheorie, Math. Zeitschr., Vol. 24, p. 559. 1925.

[25] *Lejeune-Dirichlet, P. G.*: Über die Bestimmung der mittleren Werte in der Zahlentheorie (1849), Werke, Vol. II, pp. 49—66.

[26] *Hardy, G. H.*: On Dirichlet's Divisor Problem, Proc. Lond. Math. Soc. (2), 15, pp. 1—15. 1915.

c) If $\varphi(n)$ denotes the number of numbers less than n and prime to it,

$$\frac{\varphi(1)+\varphi(2)+\cdots+\varphi(n)}{n} \sim \frac{3}{\pi^2} n + \cdots.$$

d) If $\pi(n)$ denotes the number of primes not greater than n,

$$\pi(n) \sim \frac{n}{\log n} + \cdots.$$

In all these and in many similar cases, it is not known whether a complete asymptotic expansion exists. Hence the relation which we have written down only means that the difference of the right and left hand sides is of smaller order, as regards n, than the last term on the right hand side.

e) If $p(n)$ denotes the number of different ways in which n may be partitioned into a sum of (equal or unequal) positive integers [27],

$$p(n) \sim \frac{1}{4n\sqrt{3}} e^{\frac{\pi}{3}\sqrt{6n}} + \cdots.$$

In this particularly difficult case *G. H. Hardy* and *S. Ramanujan* [28] succeeded by means of very profound investigations in continuing the expansion to terms of the order $\frac{1}{\sqrt[4]{n}}$.

B. Examples of the summation problem.

304. Here we have to deal with the converse question, that of finding a function $F(x)$ whose asymptotic expansion

$$a_0 + \frac{a_1}{x} + \frac{a_2}{x^2} + \cdots$$

is an assigned, everywhere divergent series [29]. The answers to this question are still more isolated and lacking in generality than those of the previous division.

When the function $F(x)$ is found, it has some claim to be regarded as the "sum" of the divergent series $\Sigma \frac{a_n}{x^n}$ in the sense of § 59, since it becomes more and more closely related to the partial sums of the series as their index increases. This is the case only to a very limited extent, however, since, as we have already emphasized, the function $F(x)$ is not

---

[27] E. g. $p(4) = 5$, since 4 admits of the five partitions: 4, $3+1$, $2+2$, $2+1+1$, and $1+1+1+1$.

[28] *Hardy*, *G. H.*, and *S. Ramanujan*: Asymptotic Formulae in Combinatory Analysis, Proc. Lond. Math. Soc. (2), Vol. 17, pp. 75—115. 1917. See also *Rademacher*, *H.*: A Convergent Series for the Partition Function. Proc. Nat. Acad. Sci. U.S.A., Vol. 23, pp. 78—84. 1937.

[29] When the series converges, the required function is defined by the series itself.

defined uniquely by the series. Thus the question how far $F(x)$ behaves like the "sum" of the series can only be investigated in each particular case *a posteriori*.

1. The most important advance in this direction was made by *Stieltjes*[30]. We saw above (v. A, 5), that a function given in the form

$$F(x) = \int_0^\infty \frac{f(u)}{x+u}\,du$$

possesses the asymptotic expansion $\Sigma \frac{a_n}{x^n}$, in which

$$(*) \qquad (-1)^{n-1} a_n = \int_0^\infty f(u)\,u^{n-1}\,du, \qquad (n = 1, 2, 3, \ldots).$$

Conversely, if we are given the expansion $\Sigma \frac{a_n}{x^n}$, with coefficients a_1, a_2, $a_3, \ldots$, and if we can discover a positive function $f(u)$ defined in $u > 0$, for which the integral in (*) has the given values $a_1, -a_2, a_3, \ldots$, for $n = 1, 2, 3, \ldots$, then the function

$$F(x) = \int_0^\infty \frac{f(u)}{x+u}\,du$$

will be a solution of the given summation problem, by A. 5. The problem of finding, given a_n, a function $f(u)$ which satisfies the set of equations (*) is now called *Stieltjes' problem of moments*. *Stieltjes* gives the necessary and sufficient conditions for it to be capable of solution, and, in particular, for the existence of just one solution, with very general assumptions. In particular, if $f(u)$, and hence $F(x)$, is uniquely determined by the problem of moments — such a series $\Sigma \frac{a_n}{x^n}$ is called a *Stieltjes series* for short — we are more justified in claiming $F(x)$ as a sum of the divergent series $\Sigma \frac{a_n}{x^n}$, for instance as its *S-sum*.

Lack of space prevents us from entering into closer details of these very comprehensive investigations. An account which includes everything essential is given by *E. Borel* in his "Leçons sur les séries divergentes", which we have repeatedly referred to. As an example, suppose we are given the series

$$(\dagger) \qquad \frac{1}{x} - \frac{1!}{x^2} + \frac{2!}{x^3} - \frac{3!}{x^4} + - \cdots.$$

---

[30] *Loc. cit.* (footnotes 12, 16), and also in his memoir, Sur la réduction en fraction continue d'une série procédant suivant les puissances descendantes d'une variable, Annales de la Fac. Scienc. Toulouse Vol. 3, H. 1—17. 1889.

The statement of the problem of moments is

$$\int_0^\infty f(u)\, u^{n-1}\, d\, u = (n-1)!, \qquad n = 1, 2, \ldots,$$

which obviously possesses the solution $f(u) = e^{-u}$. In this case we can prove without difficulty that the above is the only solution. Hence in

$$F(x) = \int_0^\infty \frac{e^{-u}}{x+u}\, d\, u$$

we have not only found a function whose asymptotic expansion is the given series, but, in the sense of § 59, we can regard $F(x)$ as the S-sum [31] of the (everywhere) divergent series (†).

2. The appeal to the theory of differential equations is just as useful in the summation problem as in the expansion problem (v. A, 3). Frequently we can write down the differential equation which is formally satisfied by a given series and among the solutions there *may* be a function whose asymptotic expansion is the original series. As a rule, however, matters are not as described above, nor as in A. 3, but the differential equation itself is the primary problem. It is only when this equation can be solved formally by means of an asymptotic series, as was indicated in A, 3, and *provided* we succeed in summing the series directly that we can hope to obtain a solution of the differential equation in this way. Otherwise we must try to deduce the properties of the solution from the asymptotic expansion. *Poincaré*'s researches [32], which were extended later, especially by *A. Kneser* and *J. Horn* [33], deal with this problem, which lies outside the scope of this book.

3. In *Stieltjes*' process the coefficients a_n were recovered, so to speak, from the given series $\sum_{n=1}^{\infty} \frac{a_n}{x^n}$, by replacing a_n by

$$\int_0^\infty (-1)^{n-1} f(u)\, u^{n-1}\, d\, u$$

---

[31] Thus for $x = 1$ we obtain the value

$$s = \int_0^\infty e^{-u} \frac{d\, u}{1+u} = 0{\cdot}596347\ldots$$

for the S-sum of the divergent series $\sum_{n=0}^{\infty} (-1)^n\, n!$ This series had already been studied by *Euler* (who obtained the same value for its sum), *Lacroix*, and *Laguerre*. *Laguerre*'s work formed the starting-point of *Stieltjes*' investigations.

[32] *Poincaré*, *loc. cit.* (footnote 13).

[33] A comprehensive account is given by *J. Horn*: Gewöhnliche Differentialgleichungen, 2nd ed., Leipzig 1927.

and hence the series by

$$\int_0^\infty f(u)\left(\frac{1}{x}-\frac{u}{x^2}+\frac{u^2}{x^3}-+\cdots\right)du.$$

In place of the series Σx^n there now appears the very simple geometrical series, multiplied through by the factor $f(u)$. The solution of the problem of moments is necessary in order to determine $f(u)$, and this is usually not easy. We can, however, make the process more elastic by putting

$$\Sigma\frac{a_n}{x^n}=\Sigma\, c_n\left(\frac{a_n}{c_n}\right)\frac{1}{x^n},$$

and choosing the factors c_n firstly so that the problem of moments

$$c_n=\int_0^\infty f(u)\,u^n\,d\,u$$

is soluble, and secondly so that the power series

$$\Sigma\frac{a_n}{c_n}\left(\frac{u}{x}\right)^n$$

represents a known function. Thus we can for instance link up in this way with *Borel*'s summation process, by putting $c_n = n!$. If the function

$$\Sigma\frac{a_n}{c_n}\left(\frac{u}{x}\right)^n=\Sigma\frac{a_n}{n!}\left(\frac{u}{x}\right)^n=\Phi\left(\frac{u}{x}\right)$$

can be regarded as known, then

$$F(x)=\int_0^\infty e^{-u}\,\Phi\left(\frac{u}{x}\right)du$$

is a solution of the given summation problem [34]. Here we cannot discuss the details of the assumptions under which this method leads to the desired

---

[34] The connection with *Borel*'s summation process can be established as follows. The function

$$y=e^{-x}\sum_{n=0}^{\infty}s_n\frac{x^n}{n!},$$

which was introduced in § 59, 7 for the definition of *Borel*'s process, has for its derivative

$$y'=e^{-x}\sum_{n=0}^{\infty}a_{n+1}\frac{x^n}{n!}.$$

Thus if we set $\sum_{n=0}^{\infty}\frac{a^n}{n!}t^n=\Phi(t)$, as in the text above, we have $y'=e^{-x}\Phi'(x)$, so that

$$y=a_0+\int_0^x e^{-t}\,\Phi'(t)\,d\,t.$$

Hence if the B-sum of $\Sigma\, a_n$ exists, it is given by

$$s=a_0+\int_0^\infty e^{-t}\,\Phi'(t)\,d\,t,$$

— an expression which, with suitable assumptions, can be transformed into $\int_0^\infty e^{-t}\Phi(t)\,dt$ by integration by parts. This corresponds precisely to the value of $F(1)$ deduced in the text.

result. We shall conclude with a few examples of this method of summation: in these, of course, the question whether the function found is really represented by the series must remain unsettled, since we have not proved any general theorems. It can, however, easily be verified *a posteriori*.

a) For the series

$$\frac{1}{x} - \frac{1!}{x^2} + \frac{2!}{x^3} - \frac{3!}{x^4} + - \cdots,$$

which we have already discussed in 1, we have

$$a_n = (-1)^{n-1}(n-1)!, \quad \text{so that} \quad \Phi\left(\frac{u}{x}\right) = \log\left(1 + \frac{u}{x}\right),$$

and accordingly

$$F(x) = \int_0^\infty e^{-u} \log\left(1 + \frac{u}{x}\right) d\,u.$$

By integration by parts, it is easily shown that the function is identical with that discussed in 1.

b) If we are given the asymptotic series

$$1 - \frac{1}{2 \cdot x} + \frac{1 \cdot 3}{2^2 \cdot x^2} - + \cdots + (-1)^n \frac{1 \cdot 3 \cdot 5 \ldots (2n-1)}{2^n \cdot x^n} + \cdots,$$

we have $\Phi\left(\frac{u}{x}\right) = \left(1 + \frac{u}{x}\right)^{-\frac{1}{2}}$, so that

$$F(x) = \sqrt{x} \int_0^\infty \frac{e^{-u}}{\sqrt{u+x}}\, d\,u = 2\, e^x \sqrt{x} \int_{\sqrt{x}}^\infty e^{-t^2}\, d\,t.$$

This provides, further, the asymptotic expansion

$$G(z) = \int_z^\infty e^{-t^2}\, d\,t = \frac{1}{2} e^{-z^2} \left(\frac{1}{z} - \frac{1}{2\,z^3} + \cdots\right),$$

for what is known as *Gauss*'s error-function, which is of special importance in the calculus of probabilities.

c) If we are given the somewhat more general series

$$1 - \alpha \cdot \frac{1}{x} + \alpha(\alpha+1)\frac{1}{x^2} - + \cdots + (-1)^n \alpha(\alpha+1)\cdots(\alpha+n-1)\frac{1}{x^n} + \cdots,$$

with $\alpha > 0$, we have $\Phi\left(\frac{u}{x}\right) = \left(1 + \frac{u}{x}\right)^{-\alpha}$, so that

$$F(x) = x^\alpha \int_0^\infty \frac{e^{-u}}{(u+x)^\alpha}\, du = \frac{1}{\alpha} x^\alpha e^x \int_{x^\alpha}^\infty e^{-t^{\frac{1}{\alpha}}}\, t^{\frac{1}{\alpha}-2}\, d\,t.$$

d) For the series

$$\frac{1}{x} - \frac{2!}{x^3} + \frac{4!}{x^5} - + \cdots$$

we have $\Phi\left(\frac{u}{x}\right) = \tan^{-1}\left(\frac{u}{x}\right)$, so that

$$F(x) = \int_0^\infty e^{-u} \tan^{-1}\left(\frac{u}{x}\right) du = x \int_0^\infty \frac{e^{-u}}{x^2 + u^2} du.$$

If this is regarded as the S-sum of the given divergent series, we obtain e. g. the value

$$s = \int_0^\infty \frac{e^{-u}}{1 + u^2} du = 0{\cdot}6214\ldots$$

for the sum of the series

$$1 - 2! + 4! - 6! + - \ldots.$$

Exercises on Chapter XIV.

217. Generalize this result and prove the following statements:

If $$\frac{2}{e^x + 1} = 1 + \frac{C_1}{1!} x + \frac{C_2}{2!} x^2 + \ldots + \frac{C_n}{n!} x^n + \ldots = e^{Cx}$$

symbolically, we have, in the first instance,

$$C_n = \frac{(1 + 2B)^{n+1} - (2B)^{n+1}}{n+1} = -\frac{2(2^{n+1} - 1) B_{n+1}}{n+1}$$

and

$$(C + 1)^n + C^n = 0 \quad \text{for} \quad n \geqq 1,$$

so that

$$C_0 = 1, \quad C_1 = -\frac{1}{2}, \quad C_2 = 0, \quad C_3 = \frac{1}{4}, \quad C_4 = 0, \quad C_5 = -\frac{1}{2}, \quad \ldots.$$

Using these numbers, we have (again symbolically)

$$1^p - 2^p + 3^p - + \ldots + (-1)^n n^p = \frac{1}{2}\{(-1)^{n-1}(C + 1 + n)^p - C^p\}.$$

218. Generalize the result of Exercise 217 and deduce a formula for the sum

$$f(1) - f(2) + f(3) - + \ldots + (-1)^{n-1} f(n),$$

where $f(x)$ denotes a polynomial.

219. Following **296** and **298**, deduce a formula for

$$f_0 - f_1 + f_2 - + \ldots + (-1)^n f_n.$$

220. a) Following **299**, 3, and using *Euler*'s summation formula, derive the power series expansion for $\frac{x}{\sin x}$.

b) In *Euler*'s summation formula, put

$$f(x) = x \log x, \quad x^2 \log x, \quad x^\alpha \log \alpha, \quad x (\log x)^2, \ldots$$

and investigate the relations so obtained. (Cf. Exercise 224.)

221. *Euler*'s summation formula **298** can of course be used equally well for the evaluation of integrals as for the evaluation of sums. Show in this way that

$$\int_0^4 \frac{e^{-t}}{1+t^2}\, dt = 0{\cdot}620\ldots$$

(v. Ex. 223 below).

222. a) The sum $1 + \frac{1}{2} + \frac{1}{3} + \ldots + \frac{1}{n}$ has the value

$$7{\cdot}485\,470\ldots \text{ for } n = 1000,$$

and the value
$$14{\cdot}392\,726\ldots \text{ for } n = 1\,000\,000.$$

Prove this, first assuming that C is known, and then without assuming a knowledge of C.

b) Prove that $n!$ has the value

$$10^{456\,573}\,.\,2{\cdot}8242\ldots$$

for $n = 10^5$, and the value

$$10^{5\,565\,708}\,.\,8{\cdot}2639\ldots$$

for $n = 10^6$.

c) Prove that $\Gamma\left(x + \frac{1}{2}\right)$ has the value

$$10^{2566}\,.\,1{\cdot}2723\ldots$$

for $x = 10^3$, and the value

$$10^{5\,565\,705}\,.\,8{\cdot}2639\ldots$$

for $x = 10^6$.

d) Without assuming a knowledge of the value of π^2, evaluate

$$\frac{1}{10^2} + \frac{1}{11^2} + \ldots + \frac{1}{n^2} \quad \text{for} \quad n = 10^3,$$

and find the limit of this sum as $n \to \infty$. (We obtain $0{\cdot}104\,166\,83\ldots$, $0{\cdot}105\,166\,33\ldots$).

e) Using d), show that

$$\frac{\pi^2}{6} = 1{\cdot}644\,934\,06\ldots\,.$$

f) Show that

$$\sum_{n=1}^{\infty} \frac{1}{n^3} = 1{\cdot}202\,056\,90\ldots,$$

and that

$$\sum_{n=1}^{\infty} \frac{1}{n^{\frac{3}{2}}} = 2{\cdot}612\,37\ldots\,.$$

g) Prove that $1 + \frac{1}{\sqrt{2}} + \frac{1}{\sqrt{3}} + \ldots + \frac{1}{\sqrt{n}}$ has the value

$$1998{\cdot}540\,14\ldots$$

for $n = 10^6$.

223. Taking § 66, B. 3d as a model, find the S-sum of the following series, for fixed $p = 1, 2, 3, \ldots$:

$$\text{a)} \sum_{n=0}^{\infty} (-1)^n (p\,n)!, \quad \text{b)} \sum_{n=0}^{\infty} (-1)^n (p\,n+1)!, \quad \ldots,$$

$$\text{c)} \sum_{n=0}^{\infty} (-1)^n (p\,n+p-1)!, \qquad \text{d)} \sum_{n=0}^{\infty} \binom{-\frac{1}{2}}{n} n!\, x^n.$$

(Cf. Ex. 221.)

224. Prove the following relationships, stated by *Glaisher*:

$$1^1 \cdot 2^2 \cdot 3^3 \cdot \ldots \cdot n^n \cong A \cdot n^{\frac{n^2}{2} + \frac{n}{2} + \frac{1}{12}} e^{-\frac{1}{4} n^2},$$

where A has the following value:

$$A = 2^{\frac{1}{36}} \pi^{\frac{1}{6}} \exp\left[\frac{1}{3}\left(-\frac{1}{4} C + \frac{1}{3} s_2 - \frac{1}{4} s_3 + - \ldots\right)\right] = 1{\cdot}2824271\ldots,$$

where C is *Euler*'s constant and s_k denotes the sum $\sum_{n=0}^{\infty} \frac{1}{(2n+1)^k}$. (Cf. Exercise 220, b.)

225. For the function

$$F(x) = \sum_{n=0}^{\infty} \frac{(-1)^n}{(x+n)}$$

obtain the asymptotic expansion

$$F(x) \sim \sum_{n=1}^{\infty} \frac{a_n}{x^n} \equiv \frac{1}{2x} + \frac{1}{4x^2} - \frac{1}{8x^4} + \ldots$$

and prove that the coefficient a_n has the value $\frac{2n-1}{n} B_n$ for $n \geqq 2$.

Bibliography.

(This includes some fundamental papers, comprehensive accounts, and textbooks.)

1. *Newton, I.*: De analysi per aequationes numero terminorum infinitas. London 1711 (written in 1669).
2. *Wallis, John*: Treatise of algebra both historical and practical, with some additional treatises. London 1685.
3. *Bernoulli, James*: Propositiones arithmeticae de seriebus infinitis earumque summa finita, with four additions. Basle 1689—1704.
4. *Euler, L.*: Introductio in analysin infinitorum. Lausanne 1748.
5. *Euler, L.*: Institutiones calculi differentialis cum ejus usu in analysi infinitorum ac doctrina serierum. Berlin 1755.
6. *Euler, L.*: Institutiones calculi integralis. St. Petersburg 1768—69.
7. *Gauss, K. F.*: Disquisitiones generales circa seriem infinitam

$$1+\frac{\alpha\cdot\beta}{1\cdot\gamma}x+\frac{\alpha(\alpha+1)\cdot\beta(\beta+1)}{1\cdot2\cdot\gamma(\gamma+1)}x^2+\text{etc.}$$

Göttingen 1812.
8. *Cauchy, A. L.*: Cours d'analyse de l'école polytechnique. Part I. Analyse algébrique. Paris 1821.
9. *Abel, N. H.*: Untersuchungen über die Reihe $1+\frac{m}{1}x+\frac{m(m-1)}{1\cdot2}x^2+\cdots$. Journal für die reine und angewandte Mathematik, Vol. 1, pp. 311 to 339. 1826.
10. *du Bois-Reymond, P.*: Eine neue Theorie der Konvergenz und Divergenz von Reihen mit positiven Gliedern. Journal für die reine und angewandte Mathematik, Vol. 76, pp. 61—91. 1873.
11. *Pringsheim, A.*: Allgemeine Theorie der Divergenz und Konvergenz von Reihen mit positiven Gliedern. Mathematische Annalen, Vol. 35, pp. 297—394. 1890.
12. *Pringsheim, A.*: Irrationalzahlen und Konvergenz unendlicher Prozesse. Enzyklopädie der mathematischen Wissenschaften, Vol. I, 1, 3. Leipzig 1899.
13. *Borel, É.*: Leçons sur les séries à termes positifs. Paris 1902.
14. *Runge, C.*: Theorie und Praxis der Reihen. Leipzig 1904.
15. *Stolz, O.*, and *A. Gmeiner*: Einleitung in die Funktionentheorie. Leipzig 1905.
16. *Pringsheim, A.* and *J. Molk*: Algorithmes illimités de nombres réels. Encyclopédie des Sciences Mathématiques, Vol. I, 1, 4. Leipzig 1907.
17. *Bromwich, T. J. I'A.*: An introduction to the theory of infinite series. London 1908: 2nd ed. 1926.
18. *Pringsheim, A.* and *G. Faber*: Algebraische Analysis. Enzyklopädie der mathematischen Wissenschaften, Vol. II, C. 1. Leipzig 1909.
19. *Fabry, É.*: Théorie des séries à termes constants. Paris 1910.
20. *Pringsheim, A.*, *G. Faber*, and *J. Molk*: Analyse algébrique, Encyclopédie des Sciences Mathématiques, Vol. II, 2, 7. Leipzig 1911.
21. *Stolz, O.*, and *A. Gmeiner*: Theoretische Arithmetik, Vol. II. 2nd edition. Leipzig 1915.
22. *Pringsheim, A.*: Vorlesungen über Zahlen- und Funktionenlehre, Vol. I, 2 and 3. Leipzig, 1916 and 1921. 2nd (unaltered) ed. 1923.

Name and Subject Index.

The references are to pages.

A CATALOG OF SELECTED DOVER BOOKS IN SCIENCE AND MATHEMATICS

A CATALOG OF SELECTED

DOVER BOOKS

IN SCIENCE AND MATHEMATICS

THE FOUR-COLOR PROBLEM: Assaults and Conquest, Thomas L. Saaty and Paul G. Kainen. Engrossing, comprehensive account of the century-old combinatorial topological problem, its history and solution. Bibliographies. Index. 110 figures. 228pp. 5⅜ × 8½. 65092-8 Pa. $6.95

CATALYSIS IN CHEMISTRY AND ENZYMOLOGY, William P. Jencks. Exceptionally clear coverage of mechanisms for catalysis, forces in aqueous solution, carbonyl- and acyl-group reactions, practical kinetics, more. 864pp. 5⅜ × 8½. 65460-5 Pa. $19.95

PROBABILITY: An Introduction, Samuel Goldberg. Excellent basic text covers set theory, probability theory for finite sample spaces, binomial theorem, much more. 360 problems. Bibliographies. 322pp. 5⅜ × 8½. 65252-1 Pa. $8.95

LIGHTNING, Martin A. Uman. Revised, updated edition of classic work on the physics of lightning. Phenomena, terminology, measurement, photography, spectroscopy, thunder, more. Reviews recent research. Bibliography. Indices. 320pp. 5⅜ × 8¼. 64575-4 Pa. $8.95

PROBABILITY THEORY: A Concise Course, Y.A. Rozanov. Highly readable, self-contained introduction covers combination of events, dependent events, Bernoulli trials, etc. Translation by Richard Silverman. 148pp. 5⅜ × 8¼. 63544-9 Pa. $5.95

THE CEASELESS WIND: An Introduction to the Theory of Atmospheric Motion, John A. Dutton. Acclaimed text integrates disciplines of mathematics and physics for full understanding of dynamics of atmospheric motion. Over 400 problems. Index. 97 illustrations. 640pp. 6 × 9. 65096-0 Pa. $17.95

STATISTICS MANUAL, Edwin L. Crow, et al. Comprehensive, practical collection of classical and modern methods prepared by U.S. Naval Ordnance Test Station. Stress on use. Basics of statistics assumed. 288pp. 5⅜ × 8½. 60599-X Pa. $6.95

DICTIONARY/OUTLINE OF BASIC STATISTICS, John E. Freund and Frank J. Williams. A clear concise dictionary of over 1,000 statistical terms and an outline of statistical formulas covering probability, nonparametric tests, much more. 208pp. 5⅜ × 8½. 66796-0 Pa. $6.95

STATISTICAL METHOD FROM THE VIEWPOINT OF QUALITY CONTROL, Walter A. Shewhart. Important text explains regulation of variables, uses of statistical control to achieve quality control in industry, agriculture, other areas. 192pp. 5⅜ × 8½. 65232-7 Pa. $6.95

THE INTERPRETATION OF GEOLOGICAL PHASE DIAGRAMS, Ernest G. Ehlers. Clear, concise text emphasizes diagrams of systems under fluid or containing pressure; also coverage of complex binary systems, hydrothermal melting, more. 288pp. 6½ × 9¼. 65389-7 Pa. $10.95

STATISTICAL ADJUSTMENT OF DATA, W. Edwards Deming. Introduction to basic concepts of statistics, curve fitting, least squares solution, conditions without parameter, conditions containing parameters. 26 exercises worked out. 271pp. 5⅜ × 8½. 64685-8 Pa. $7.95